Micro-organisms
function, form and environment

Micro-organisms

function, form and environment

edited by

Lilian E. Hawker
Ph.D., D.Sc.
Emeritus Professor of Mycology, University of Bristol

and

Alan H. Linton
M.Sc., Ph.D., M.R.C.Path.
Senior Lecturer in Veterinary Bacteriology,
University of Bristol

EDWARD ARNOLD (PUBLISHERS) LTD

© Edward Arnold (Publishers) Ltd., 1971

First published 1971
by Edward Arnold (Publishers) Ltd.
25 Hill Street
London W1X 8LL

First published as a Paperback 1972
Reprinted 1972
Reprinted 1974

ISBN: 0 7131 2372 9

Printed in Great Britain by
William Clowes & Sons, Limited, London, Beccles and Colchester

Preface

This book is intended to be a successor to an earlier one *An Introduction to the Biology of Micro-organisms* published in 1960. It is in no way a second edition of the earlier volume. It does, however, contain some short passages taken from the latter, with the kind permission of our two fellow authors and former colleagues, Professor B. F. Folkes and Dr M. J. Carlile, and it has similar aims, i.e. to survey the whole field of microbiology and to describe it in a manner which we hope will be useful both to University students and to research workers.

Owing to the rapid advances in many sections of microbiology during the last decade, our task has been much more difficult than it was ten years ago. We could not have undertaken it without the collaboration of many of our colleagues at Bristol who have contributed chapters or shorter sections. The book owes much not only to stimulating discussions between the contributors, but also to other colleagues at Bristol and elsewhere who have read sections and given advice or information on particular aspects, or have provided illustrations.

While we have tried to make the subject readily understandable by first-year students, some aspects of it are so complex that there are inevitably a few parts, notably Chapters 3 and 4 and shorter sections in other chapters, which such students may find harder to grasp, but which specialists may welcome as a useful summary.

We would like to thank all those authors, editors, and publishers who have provided us with figures or photographs, either unpublished or from published works. These are acknowledged in the figure captions.

We wish also to thank all our co-authors for their co-operation, and in particular Professor Emeritus K. E. Cooper who, as with the earlier volume, has given us much good advice in the preparation of this one, and Professor M. Richmond, who not only undertook his own contribution at short notice but has read much of the rest of the book and made many helpful suggestions; other Bristol colleagues, in particular Professor W. A. Gillespie, Professor M. Ingram, Dr D. B. Peacock, Dr E. S. Meek and Dr M. P. English, who have read particular chapters and/or given us the benefit of their specialist knowledge; Dr F. Last of the Glasshouse Research Station, Littlehampton, for information on fungal viruses; Dr W. P. K. Findlay and Dr B. Kirsop of the Brewing Industry Research Foundation, Redhill, for information on the use of yeasts in bioassay; Dr B. E. B. Moseley of the Department of General Microbiology, Edinburgh, for information on the effects of radiation; Dr I. Thorpe of the N.A.A.S. Bristol for reading the section on plant viruses; and the secretarial staffs of the Departments of Bacteriology and Botany who have typed the various sections of the book.

Bristol 1970

L. E. HAWKER
A. H. LINTON
(EDITORS)

Contents

2   BIOSYNTHESIS IN MICRO-ORGANISMS

by M. H. Richmond

3   BIOENERGETICS IN MICRO-ORGANISMS

by P. B. Garland and O. T. G. Jones

4   GENETICS AND GENE ACTION
by C. J. Grant and T. G. B. Howe

8   STRUCTURE, BIOLOGY, AND CLASSIFICATION OF VIRUSES

by L. W. Greenham, L. E. Hawker, T. J. Hill, K. B. Linton, M. F. Madelin, A. Mayr-Harting, and F. E. Round

9 STRUCTURE, BIOLOGY, AND CLASSIFICATION OF PROKARYOTIC MICRO-ORGANISMS

by A. H. Linton with R. C. W. Berkeley, M. F. Madelin, and F. E. Round

10   EUKARYOTES

by A. Beckett, L. E. Hawker, M. A. Sleigh, M. F. Madelin and F. E. Round

FINE STRUCTURE OF THE EUKARYOTIC CELL

THE FUNGI

PROTOZOA

22   THE INDUSTRIAL USES OF MICRO-ORGANISMS
     by J. G. Carr

Contributors

A. Beckett, B.Sc., Ph.D. *Lecturer in Botany, Department of Botany, University of Bristol.*

F. W. Beech, B.Sc., Ph.D., D.Sc., A.R.I.C., F.I.Biol. *Reader in Microbiology and Head of Cider Section, Long Ashton Research Station, University of Bristol.*

R. C. W. Berkeley, B.Sc., Ph.D. *Lecturer in Bacteriology, Department of Bacteriology, University of Bristol.*

R. J. W. Byrde, B.Sc., Ph.D. *Research Fellow in Plant Pathology, Long Ashton Research Station, University of Bristol.*

R. Campbell, B.Sc., M.S., Ph.D. *Lecturer in Microbiology within the Department of Botany, University of Bristol.*

J. G. Carr, B.Sc., Ph.D., D.Sc., F.I.Biol. *Reader in Microbiology, Long Ashton Research Station, University of Bristol.*

K. E. Cooper, B.Sc., Ph.D., M.R.C.S., L.R.C.P. *Emeritus Professor of Bacteriology, University of Bristol.*

P. B. Garland, M.A., M.B., B.Ch., Ph.D., *Professor of Biochemistry. University of Dundee, formerly Lecturer in Biochemistry, University of Bristol.*

C. J. Grant, M.A., D.Phil., *Lecturer in Botany, Department of Botany, University of Bristol.*

L. W. Greenham, B.Sc., B.V.Sc., Ph.D., M.R.C.V.S. *Lecturer in Veterinary Bacteriology and Virology, Department of Bacteriology, University of Bristol.*

Lilian E. Hawker, Ph.D., D.Sc. *Emeritus Professor of Mycology, Department of Botany, University of Bristol.*

A. J. Hedges, B.Sc., Ph.D. *Senior Lecturer in Bacteriology, Department of Bacteriology, University of Bristol.*

T. J. Hill, B.V.Sc., Ph.D., M.R.C.V.S. *Lecturer in Bacteriology, Department of Bacteriology, University of Bristol.*

T. G. B. Howe, M.A., Ph.D. *Lecturer in Bacteriology, Department of Bacteriology, University of Bristol.*

O. T. G. Jones, B.Sc., Ph.D. *Reader in Biochemistry, Department of Biochemistry, University of Bristol.*

A. H. Linton, M.Sc., Ph.D., M.R.C.Path. *Senior Lecturer in Veterinary Bacteriology, Department of Bacteriology, University of Bristol.*

K. B. Linton, B.Sc., Ph.D. *Senior Lecturer in Medical Bacteriology, Department of Bacteriology, University of Bristol.*

M. F. Madelin, B.Sc., Ph.D. *Reader in Mycology, Department of Botany, University of Bristol.*

Anna Mayr-Harting, M.D., Ph.D. *formerly Senior Lecturer in Bacteriology, Department of Bacteriology, University of Bristol.*

M. H. Richmond, M.A., Ph.D., Sc.D. *Professor of Bacteriology, Department of Bacteriology, University of Bristol.*

F. E. Round, Ph.D., D.Sc. *Reader in Phycology, Department of Botany, University of Bristol.*

M. A. Sleigh, B.Sc., Ph.D. *Reader in Zoology, Department of Zoology, University of Bristol.*

Introduction

Micro-organisms include several distinct groups of organisms, differing widely in form and life-cycle but resembling one another in their small size (Table 1) and relatively simple structure. Microbiology is concerned with the way micro-organisms live and multiply, with their form and classification and the roles they play in the varied habitats they occupy.

Table 1   Sizes of some micro-organisms. (1 micron (μ) or 1 micrometre (μm) = 0·001 mm; 1 mμ or 1 nanometre (nm) = 0·001 μ).

A particle of foot and mouth virus	28 mμ
A particle of tomato bushy stunt virus	30 mμ
Staphylococcus albus, a spherical bacterium	1·0 μ diam.
Bacillus megaterium, a rod-shaped bacterium	ca. 2·8 × 1·2–1·5 μ
Saccharomyces carlsbergensis, a yeast	5–10·5 × 4–8 μ
Chlamydomonas kleinii, a unicellular alga	28–32 × 8–12 μ

THE BIOCHEMISTRY AND PHYSIOLOGY OF MICRO-ORGANISMS

The first seven chapters of this book consider the way micro-organisms function. Owing to the ease with which many of them can be cultured (p. 144) under controlled conditions and to their relatively simple organization, micro-organisms are favourite material for biochemical and genetical studies and have yielded a wealth of fundamental information. Since numerous introductory textbooks of biochemistry are now readily available, no attempt has been made to deal comprehensively with the chemistry of cell constituents. The first chapter deals with the structure of some, but not all, of the macromolecules found in micro-organisms, but does not consider how they fit into the architecture of the cell. The second chapter considers how these substances are synthesized by the cell and the third describes how the energy required for these activities is supplied by the metabolism of the cells. Synthesis and metabolism are initiated and regulated by the development of stored chemical 'information'. Much of our knowledge of these regulatory systems in living cells has been obtained from genetical experiments with micro-organisms and is discussed in Chapter 4.

Macromolecules are built up into the organized structural components of the cell. The latter may increase in size and complexity up to limits characteristic of the species and may then divide to form two similar daughter cells or may reproduce by some more complex process. The growth of cell constituents, whole cells, multicellular organisms, and populations of organisms is obviously a complex process, depending on genetic potential and environmental factors. The processes of growth and the effect of external factors on them are described in Chapters 5 and 6, while factors bringing about senescence and death of micro-organisms are the subject of Chapter 7.

THE FORM OF MICRO-ORGANISMS

The second part of the book (Chap. 8–11) is concerned with the range of form, life-cycles, and classification of micro-organisms.

Modern biologists recognize that the traditional division of all living things into the two 'kingdoms', plants and animals, is no longer tenable. Micro-organisms include some groups which are plant-like (e.g. the green algae), some which are animal-like (e.g. protozoa) and others which have some characteristics of both kingdoms (e.g. fungi). It is now usual to recognize a third kingdom, the *Protista*, which includes those organisms of small size and relatively lowly differentiation which have hitherto been loosely termed 'micro-organisms' and which are the subject matter of Microbiology.

With the exception of the viruses (Chap. 8), some of the true slime moulds (p. 422) and a few other examples, micro-organisms, like plants and animals, normally consist of cells. In very many of them the thallus comprises only one cell, but some are multicellular, the constituent cells being similar to one another in many species, but becoming differentiated into distinct types of cell of varied form and function in the higher algae and higher fungi. Even the most highly differentiated forms, however, lack the true tissues, formed by cell division in more than one plane, characteristic of higher plants and animals.

Micro-organisms or protists have in the past been classified in a number of more or less unsatisfactory ways, based on a variety of morphological, physiological, or chemical characters. The study of cell organization, made possible by the invention of the electron microscope and the ultra-microtome, and by advances in biochemical techniques, has revealed fundamental resemblances and differences between various micro-organisms and offers a sound basis for their division into major groups and, in many instances, for the further subdivision of these.

It is convenient to classify micro-organisms in three major groups, the *viruses*, the *prokaryotes*, and the *eukaryotes*.

Viruses

The viruses (Chap. 8), capable of replication only within a living host cell, are of simple structure below the level of cellular organization. They have some of the characters of living organisms, notably the power of self replication, and for this, together with their importance in animal and plant pathology, they are studied by microbiologists.

With few exceptions, viruses cannot be seen under the light microscope. The majority of those which have been isolated from the host cell consist only of nucleic acid and protein (p. 223). The electron microscope confirms earlier evidence that mature viruses are in the form of discrete particles, the shape and complexity of which are characteristic of the particular virus.

Prokaryotes (Bacteria and blue-green algae)

The early observation by Cohn (1872) that bacteria and blue-green algae show closer similarity to each other than to other micro-organisms has been confirmed by cytological (including electron microscopy) and biochemical studies. These two groups are now classed together as prokaryotes.

A prokaryotic cell is vastly more complex than a virus particle but is of relatively simple organization compared with that of the cells of higher plants and animals or of eukaryotic micro-organisms. The nucleus is represented by a body consisting largely of DNA but not enclosed in a membrane. Although no chromosomes comparable with those of higher organisms have been convincingly demonstrated, a single chromosome-like structure has been described, and genetic studies with bacteria (p. 99) prove that an orderly redistribution of chromatin takes place at cell division. Typical mitochondria are also absent. It is thought that their function is taken over by less complex membrane systems (p. 286).

In addition to these more or less well-defined groups there are a number of organisms which are usually treated as bacteria, but which do not fit clearly into any major group. These include the Pleuropneumoniae (p. 344) and the Rickettsiae (p. 345).

Eukaryotes

The cells of plants, animals, and many micro-organisms, i.e. fungi, algae (with the exception of the Cyanophyceae or blue-green algae), protozoa, and slime moulds are characterized by a membrane-bound nucleus (p. 359) and the presence of other highly complex organelles, such as mitochondria (p. 365), which are lacking in the cells of prokaryotic species. A summary of some characteristics distinguishing prokaryotic and eukaryotic cells is presented in Table 2; explanations of the terms used will be found in Chapters 9 and 10.

Table 2 Some differences between prokaryotic and eukaryotic cells (after R. Y. Stanier, 1970, unpublished).

	Prokaryotes	*Eukaryotes*
Genetic System		
Location	nucleus	nucleus
		mitochondria
		chloroplasts
Structure of nucleus:		
segregation by unit membrane	−	+
number of chromosomes	1*	>1
chromosomes contain histones	−	+
mitotic division	−	+
Sexuality:		
mechanism of zygote formation	conjugation	conjugation
	transformation	
	transduction	
nature of zygote	partial diploid**	diploid
	(merozygote)	
Protein-synthesizing System		
Nature of ribosomes	70 S†	80 S† (cytoplasm)
		70 S† (organelles)
Cytoplasmic Structures		
Mitochondria	−	+
Chloroplasts	−	+ or −
Lysosomes	−	+ (? all)
Golgi structures	−	+
Endoplasmic reticulum	−	+
Microtubular systems	−	+
(sometimes from centrioles)		
True vacuoles	−	+
(bounded by unit membrane)		
Outer Cell Structures		
Cell membranes:		
contain sterols	−**$_*$	+
contain part of respiratory and		
photosynethic machinery	+**$_{**}$	−
Cell wall:		
frequency	almost universal	+ (some groups)
		− (other groups)
if present, contains		
peptidoglycan (murein)	+	−
Locomotor organelles:		
flagella contain 9 + 2 micro-		
tubules, originate from centriole	−	+ (some groups)
flagella not as above	some eubacteria	
pseudopodia	−	+ (some groups)

  * Usually one major chromosome; the number may vary.
 ** Occasionally complete diploids by conjugation only.
 **$_*$ Low concentrations have been demonstrated in blue-green algae.
 **$_{**}$ Not true of green bacteria.
 † S (Svedberg unit) is the sedimentation coefficient of a particle, being the velocity $(10^{-13}$ cm/sec) per unit centrifugal field.

Micro-organisms of Protista may, therefore, be subdivided as in the scheme below:

Classification of micro-organisms

A	Organization subcellular	VIRUSES
AA	Thallus unicellular, multicellular, or plasmodial	
	B Nucleoplasm not bounded by a membrane	PROKARYOTA
	C Chlorophyll absent, or if present of type different from that of plants	Bacteria (including Actinomycetes)
	CC Chlorophyll present, together with characteristic blue-green pigment, pigments not located in discrete plastids.	Cyanophyceae (Blue-green algae)
	BB Cells or plasmodia containing one or more discrete membrane-bounded nuclei	EUKARYOTA
	D Cell(s) of vegetative thallus (with a few exceptions) possessing a cell wall(s).	
	E Chlorophyll present and located in discrete chloroplasts.	Algae (excluding the 'blue-green' algae)
	EE Chlorophyll absent.	Fungi
	DD Cell(s) of vegetative thallus lacking true cell walls.	
	F Thallus unicellular, remaining so.	Protozoa
	FF Thallus unicellular at first, becoming a plasmodium or pseudoplasmodium and eventually forming a fructification	Slime moulds (Myxomycetes or Mycetozoa)

THE ENVIRONMENT OF MICRO-ORGANISMS

The third section of the book (Chaps. 12–22) considers the relation between micro-organisms and their environment, including both natural habitats and man-made ones. Environmental factors are seldom constant and, moreover, micro-organisms are in competition with others of many different species for food, oxygen, and living space. The metabolic products of one organism may either stimulate or inhibit growth of another. Interaction between the components of a mixed population may thus be highly complex.

Vigorously growing saprophytes, able to utilize dead organic matter, rapidly colonize suitable habitats. Less generally favourable sites are occupied by species which may be unable to compete with more vigorous organisms under optimal conditions for growth, but which can survive through their tolerance of sub-optimal or even unfavourable conditions. The occupation of a particular habitat does not necessarily mean that it is the most favourable one for the occupying species. Thus many, but not all, successful parasites actually grow better in pure culture on artificial media than on the normal host, but cannot survive in nature outside the host owing to competition with saprophytic organisms. Some soil fungi normally grow fairly deep in the soil through their ability to tolerate conditions of poor aeration, but are unable to survive the intense competition in the better-aerated surface layers. Some habitats are rendered more suitable for the growth of micro-organisms by the action of pioneer species, which break down complex materials, thus making food available for a wide range of other organisms. Thus a definite succession of micro-organisms occurs on such a substrate and the pioneer species are often overcome and excluded by the later growth of less specialized ones.

Both as saprophytes and as parasites, micro-organisms are of importance to man. Saprophytes attack stored products of various kinds and may cause serious economic loss (Chap. 21). On the other hand their activities in the soil, notably the breakdown of complex organic materials, is essential to soil fertility (Chap. 12) and thus to all life. Some micro-organisms are employed in industry to bring about desirable chemical syntheses or degradations (Chap. 22). Parasites cause many serious diseases of man, domestic animals, and crop plants. Control of these organisms is both difficult and expensive; the study of the morphology, physiology, and ecology of micro-organisms is essential to the improvement of existing control methods and the introduction of new ones.

METHODS USED IN THE STUDY OF MICRO-ORGANISMS

Despite the diverse forms and habitats of micro-organisms, their small size allows the use of standard methods in their study, thus further justifying the inclusion of unrelated groups within the subject matter of a single discipline, microbiology. Techniques of particular importance and applicable to all micro-organisms are artificial culture and microscopy.

Culture methods

Since the natural environment of living micro-organisms is usually complex and seldom constant the study of their physiology and development is usually carried out under simplified and uniform conditions. Pure cultures of a particular organism are grown in media of known composition and under constant environmental conditions. Such media suffice for many organisms but others require the addition of various complex organic substances. Some organisms cannot be grown in pure culture (i.e. in the absence of living organisms of at least one different species) on simple artificial media. Obligate parasites, such as the rust fungi (p. 399), can be grown normally only on the living host. Some organisms, such as many slime moulds (p. 421), have been grown only in two-membered cultures, that is, in the presence of another living organism. In some examples killed cells will suffice. Nevertheless by the technique of pure culture the raw materials necessary for the synthesis of the complex components of the living cell have been determined, and a great deal has been learnt of the mechanisms by which these syntheses are performed. The response of organisms to environmental changes has also been studied by varying a single factor. The methods of culturing micro-organisms are described more fully in Chapter 5, p. 144.

Microscopy

Microscopy includes techniques common to the study of cell structure of all organisms. The unaided eye cannot comfortably distinguish two points less than 0·2 mm. apart, a resolving power inadequate for the study of most micro-organisms. Magnification is therefore essential to render them visible.

Light microscopy

Micro-organisms, because of their small size, are measured in microns (one micron or $\mu = 0\cdot001$ mm) and millimicrons (one millimicron or $m\mu = 0\cdot001\ \mu$ or $0\cdot000001$ mm). Recently adoption of the MKS system—metres, kilograms, and seconds—has introduced alternative symbols of measurement; the micron is termed a micrometre (μm) and the millimicron, a nanometre (nm). The resolving power of the light microscope is limited by the wavelength of visible light used and, even with the best objectives, particles less than about 300 $m\mu$ cannot be resolved. The particulate nature of the majority of viruses cannot therefore be determined with the ordinary light microscope. Useful observations on larger micro-organisms can, however, be made.

Except for some pigments that are able to absorb light at certain wavelengths, the majority of cell components absorb very little light of the visible region. This means that living cells studied with the light microscope exhibit low contrast. The use of dyes that selectively stain different cell components helps to overcome this limitation by providing some of the necessary contrast. Unfortunately most staining techniques cannot be used with living cells. Instead the cells must be fixed, dehydrated, and the larger ones embedded and sectioned prior to staining, and these often prolonged procedures may introduce a variety of changes in both the chemical and morphological make up of the cell.

Observation of the living cell may be aided by the use of vital stains. The recent use of phase contrast and interference microscope techniques has resulted in further advances in the study of living cells.

Phase contrast microscopy

By this method structures within cells of refractive index slightly different from that of the rest of the cell are rendered visible. Also differences in refractive index between the whole cell and the surrounding medium make the cell clearly visible. The rays of light passing through the object are retarded a fraction of a wavelength compared with similar rays passing through the suspending medium. This produces a difference in 'phase' between the two emerging types of rays. The phase contrast microscope converts phase differences into light intensity, producing light and dark contrast in the image. Thus transparent objects show contrast where differences in refractive index or thickness occur. Many structures, such as chromatin bodies in bacteria, not usually visible in unstained organisms, are thereby revealed (Fig. 9.5A).

The interference microscope

The interference microscope not only reveals cell organelles but enables the cell biologist to make quantitative measurements of such things as the lipid, nucleic acid, and protein contents of the cell.

Dark ground illumination

Another method used for observing certain details of cell structure is darkfield microscopy. By means of a condenser which in its simplest form is fitted with a central circular stop, the object is illuminated with a cone of light without permitting any direct rays to enter the objective. Only light reflected or scattered by the object is taken into the objective with the result that the object appears self-luminous against a black background. The resolving power of the lens system is subject to the same limitations as in the ordinary microscope with direct illumination, but particles below the resolving power of the lenses may be seen as dots of light. This method of microscopy is widely used for the examination of bacterial motility (p. 301), and for organisms which are not readily stained and are too small to be resolved by ordinary light microscopy (e.g. spirochaetes, p. 341, and the larger viruses, p. 221).

Polarization microscopy

Some cell components behave characteristically when observed with polarized light in that the light is transmitted through them with varying velocities. Such structures or materials are termed birefringent because they possess two different indices of refraction corresponding to the different velocities of light transmission. Until the electron microscope was devised and built, polarization microscopy was an important device in the indirect analysis of cell ultrastructure since birefringence is dependent on structural properties smaller than the wavelength of light. In this respect it has been particularly useful in the study of fibrillar structures such as the mitotic apparatus.

Electron microscopy

The electron microscope differs from the optical microscope in using a beam of electrons instead of light rays. Glass is opaque to electrons and the focusing of the electron beam is brought about by circular electromagnets (analogous to the glass lenses of the light microscope). The focus varies with the strength of the magnetic field applied. The object, by scattering the electrons, produces an image which is focused onto a fluorescent viewing screen. The wavelength of the electrons used in the electron microscope is approximately 0.05Å (compared with approximately 5000Å for visible light). In practice, at the present stage of technical proficiency, it is possible to resolve particles as small as 1–2Å; 1 Angstrom (Å) $= 0.1\ \text{m}\mu = 0.1\ \text{nm}$.

Since gas molecules scatter the electron beam, it is necessary with the instruments at present available to examine the object *in vacuo*, and hence only killed organisms can at present be studied; these are subjected to fixation and drying in their preparation, so shrinkage and other artifacts must always be considered.

Some of the first techniques for preparing biological specimens for examination in the electron microscope, involved the use of whole, fixed organisms dried down onto grids and shadowed with a heavy-metal alloy such as gold–palladium. Later, improvements were made in fixation techniques such that finer detail could be resolved in ultrathin sections of material embedded in epoxy resin and stained with heavy-metal stains such as uranyl acetate and lead citrate.

The examination of micro-organisms in the electron microscope has also been facilitated by use of the negative staining technique. In this technique the particles to be examined are embedded in an electron-dense material, e.g. phosphotungstic acid (PTA). The PTA not only outlines the general shape of the particle but also delineates details of fine structure on the surface. This simple technique has been used very successfully in revealing structure almost at the molecular level, especially in the study of viruses.

Attempts have been made to overcome the disadvantages of chemical fixation and dehydration, which as already pointed out may cause a certain amount of structural change in the specimen, by the use of *freeze-etching*. This method involves the deep freezing of specimens in a liquid gas and the subsequent formation of carbon-platinum replicas of fractured surfaces of the material (Fig. 10.4C). Since cells frozen in this way remain viable, it is claimed that freeze-etching enables the investigator to observe the ultrastructure as it appears in the living condition.

Recently a great deal of information has been obtained from observations made on the surfaces of whole or fragmented cells with the *scanning electron microscope*. Here the image is formed by collecting the electrons that are scattered or reflected from the specimen surface which is usually coated with a suitable conductor. The instrument has proved particularly useful in the investigation of such things as fungal spores and diatom cell walls (Figs. 10.12, 10.29).

Autoradiography, biochemical analysis, fractional centrifugation, and X-ray diffraction used in conjunction with the electron microscope have also yielded a wealth of information on all aspects of cell ultrastructure.

The recent development of very high voltage electron microscopes (operating at voltages up to 1·5 MV compared with the usual 50–200 kV) may eventually make it practicable to examine living organisms. The use of these high voltages greatly increases the penetrating power of the electron beam and preliminary observations have been made on intact, hydrated bacteria held in a special thin-walled double-chamber.

BOOKS AND ARTICLES FOR REFERENCE AND FURTHER STUDY

CHARLES, H. P. and KNIGHT, B. C. J. G. (eds.) (1970). Organization and Control in Prokaryotic and Eukaryotic Cells, *20th Symp. Soc. gen. Microbiol.*, Cambridge University Press.

CLARK, J. A., SALSBURY, A. J. and ROWLAND, G. G. (1968). The scanning electron microscope II. *Sci. J.*, **4**, No. 8, 54.

COSSLET, V. E. (1968). Megavolt electron microscopes. *Sci. J.*, **4**, No. 12, 38.

CRUICKSHANK, R. (ed.) (1965). *Medical Microbiology*, 11th Edn. Livingstone, Edinburgh, 1067 pp.

GRIMSTONE, A. V. (1968). *The Electron Microscope in Biology*, Inst. Biol. Studies in Biology, No. 9, Edward Arnold, London, 54 pp.

NEEDHAM, G. H. (1958). *The Practical Use of the Microscope*, Thomas, Springfield, Ill., 493 pp.

STOWARD, P. J. (1968). Fluorescence microscopy. *Sci. J.*, **4**, No. 2, 65.

THORNTON, P. R. (1965). The scanning electron microscope. *Sci. J.*, **1**, No. 9, 66.

1

Macromolecules in Micro-organisms

INTRODUCTION

It is customary to think of bacterial macromolecules solely in terms of single and readily identifiable high molecular weight polysaccharides, lipids, nucleic acids, and proteins, but many more complex situations exist in microbial cells. There are intricate molecular arrangements in which two or more macromolecules are held together in specific ways by complementary ionic and hydrogen bonds. Much of the cell seems to be made up of chemical structures that are more than just an intimate mixture of two types of macromolecule joined together by stable covalent linkages. A good example of this situation is the mucopeptide that forms the rigid structural matrix of the cell wall of many bacterial species. In this molecule there is close cross-linking between polysaccharide and polypeptide chains in such a way that it is no longer possible to define the limits of the molecule accurately. Thus, with molecules of this type, such concepts as molecular weight and size cease to have much meaning. As far as mucopeptide is concerned the entire bacterial cell seems to be enclosed by a single sack-like molecule that might not even have a similar repeating structure over the whole of its area.

Although such complex macromolecules do pose problems as far as isolation and identification are concerned, it is possible to consider simple polysaccharides, lipids, proteins, and nucleic acids as single molecular entities that can be defined with some precision. In the following sections, therefore, the structures of some examples of each of these four groups will be considered before attention is turned to bacterial mucopeptide as an example of one of the more complex and less well-defined types of microbial macromolecule.

POLYSACCHARIDES

All simple polysaccharides have a repeating structure made up either of a single, or alternating types, of monosaccharide. One of the simplest is amylose, a polysaccharide which is found in many types of micro-organism and in which the molecule consists of glucose units joined together to form a long chain (Fig. 1.1). The bonds linking the structure join the C_1 atom of one glucose residue to the C_4 atom of the next by an oxygen bridge, often known as a *glycoside bond*. The use of these bonds with each sugar residue leads to a molecule in which all the monosaccharide units have an identical linkage save the first and the last. In the first, the hydroxyl group on C_4 is unsubstituted, while on the last the —OH group on C_1 is in a similar state (Fig. 1.1).

Most monosaccharide sugars can undergo a molecular rearrangement at C_1 to give a free aldehyde group (Fig. 1.2) and thus uncover reducing properties. However, in amylose, and in many other polysaccharides, the C_1 atom is part of the bond joining the residues of the chain together and under these circumstances is incapable of forming an aldehyde residue. Polysaccharides do not, therefore, show nearly such powerful reducing properties as monosaccharides, the total reducing power of most polymers being derived solely from the single free C_1—OH at one end of the chain. On hydrolysis however, polysaccharides yield monosaccharides with their C_1—OH groups free and this process is accompanied by the parallel appearance of reducing ability.

The amylose molecule (Fig. 1.1) is one of the simplest types of polysaccharide, since it is made up solely from a single repeating glucose unit and all the bonds linking the monosaccharides are identical. In many other polysaccharides, however, two different types of monosaccharide are involved in the chain. An

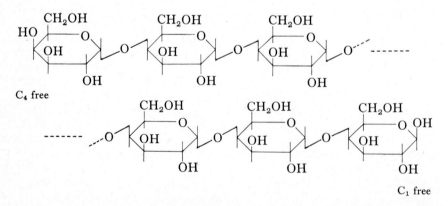

Fig. 1.1 Overall structure of amylose (poly-α-1:4-glucose). The free C_1 end of the molecule is responsible for the reducing properties.

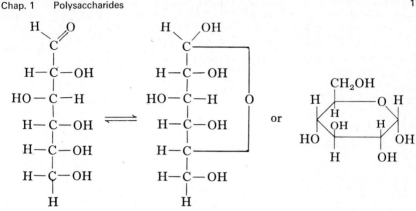

Fig. 1.2 Tautomerism between the straight-chain and the ring form of D-glucose to show how the potential reducing group (—CHO) at C_1 arises.

example of such a molecule is the polysaccharide found in the Type III capsular material from certain strains of pneumococci. In this case the structure consists of chains of alternating glucose and glucuronic acid residues arranged as shown in Fig. 1.3.

A further, and more complex, class of polysaccharides are those that are branched. Branching occurs where a monosaccharide residue in the chain is attached to three other monosaccharide units rather than to two, as in any straight-chain polysaccharide. An example of this type of molecule is the dextran molecule synthesized by *Leuconostoc mesenteroides*. This molecule, as in all dextrans (see Table 1.1), is made up entirely from glucose units which, in this case, are joined by a glycoside bond between C_1 of one glucose residue and C_6 of the next, rather than between C_1 and C_4 as with amylose. The branching points in the *Leuconostoc* dextran are introduced where occasional glycosidic bonds are formed between C_1 and C_4 or between C_1 and C_3 of two glucose

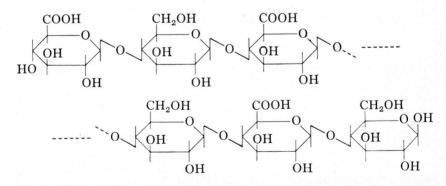

Fig. 1.3 The overall structure of the polysaccharide from the Type III capsular material found in some pneumococci.

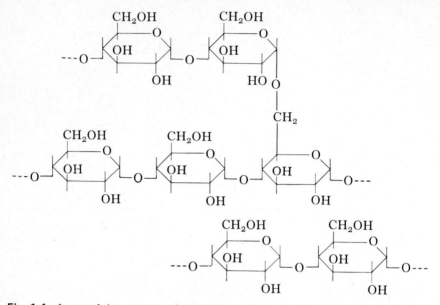

Fig. 1.4 A part of the structure of a branched polysaccharide to show the molecular nature of the branching points.

residues rather than between C_1 and C_6 (see Fig. 1.4). The overall structure is one in which about 5 per cent of the glucose residues in the molecule are triply substituted and consequently the molecule would seem to have a loosely branched structure of the overall pattern shown in Fig. 1.5.

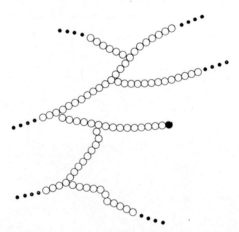

Fig. 1.5 The overall pattern of monosaccharide residues to be found in a hypothetical branched glucan. Open circles: residues with C_1 occupied by glycoside bonds; solid circle: reducing residue by virtue of a free C_1.OH.

MUCOPOLYSACCHARIDES

The term mucopolysaccharide is often used for those polysaccharides containing residues of *N*-acetyl-amino sugars such as *N*-acetyl-glucosamine or *N*-acetyl-galactosamine (Fig. 1.6). One of the simplest mucopolysaccharides, and undoubtedly one of the most important in many moulds since it forms a great part of their cell walls, is *chitin*. This molecule consists of unbroken chains of *N*-acetyl-glucosamine linked from the C_1 of one residue to the C_4 of the next. In point of fact this polymer is usually present in an organism in association with protein but whether there are a small number of covalent bonds between the two, or whether the mixture is stabilized solely by ionic and hydrogen bonds, is unclear at the moment. Strictly speaking therefore it is probably more correct

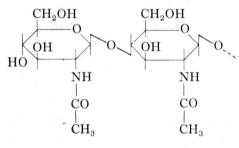

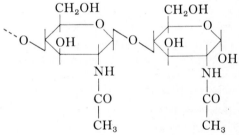

Fig. 1.6   The overall structure of chitin.

Table 1.1   Trivial names and chemical constitution of various microbial polysaccharides.

Trivial name	Constitution	Source
Glucan (general term)	Poly-glucose	Many yeasts and bacteria
Dextran	Poly-1:6-glucose	*Leuconostoc* and many other microbial species
Mannan (general term)	Poly-mannose	Yeasts
Amylose	Poly-α1:4-glucose	Many bacteria
Chitin	Poly-β1:4-*N*-acetyl-glucosamine	Fungi
Cellulose	Poly-β1:4-glucose	*Acetobacter xylinum*

to speak of the polysaccharide moiety of chitin as a poly-1-4-linked *N*-acetyl-glucosamine and reserve the term 'chitin' for the polysaccharide/protein complex.

Not all mucopolysaccharides contain amino-sugars unmixed with non-nitrogenous monosaccharides. As an example, the mucopolysaccharide component of the Type XIV capsular material from pneumococci contains *N*-acetylglucosamine, glucose and galactose in the overall molecular arrangement shown in Fig. 1.7. The various general names used to describe simple polysaccharides usually give some idea of their composition and a list of the most common of these names is given in Table 1.1.

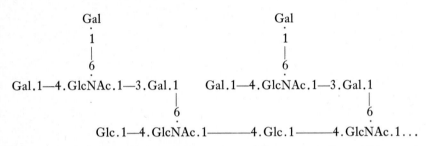

Fig. 1.7 The overall structure of the polysaccharide from the Type XIV capsular material found in some pneumococci.

LIPIDS

All lipids share a common property of being soluble in fat solvents, such as chloroform or ether, and almost insoluble in water. A wide range of compounds with different structures fall within this classification and it is convenient to consider lipids under the following groups:

1. Fatty acids
2. Triglycerides
3. Phospholipids, phosphatidic acids and glycolipids
4. Steroids
5. Carotenoids

Fatty acids

Fatty acids are long chain monocarboxylic acids of the general formula R.COOH. Usually the R-group has a long chain, commonly unbranched, consisting of an even number between 8 and 24 carbon atoms. The fatty acid may be described as *saturated* or *unsaturated* depending on whether the R-group contains any double bonds; acids with one double bond being known as mono-enoic, those with two as di-enoic, and so on. The structure of a number of fatty acids and their trivial names is given in Table 1.2. Little more need be said about their structure at this stage since their greatest importance, apart from the role of the lower examples as metabolic intermediates, is to form part of the higher molecular weight lipids described in the next two sections.

Table 1.2 Trivial names of some of the various fatty acids to be found in microbial cells.

Molecular formula	Common name	Systematic name
Saturated acids:		
$C_4H_8O_2$	Butyric	*n*-Butanoic
$C_6H_{12}O_2$	Caproic	*n*-Hexanoic
$C_{10}H_{20}O_2$	Capric	Decanoic
$C_{12}H_{24}O_2$	Lauric	Dodecanoic
$C_{14}H_{28}O_2$	Myristic	Tetradecanoic
$C_{16}H_{32}O_2$	Palmitic	Hexadecanoic
$C_{18}H_{36}O_2$	Stearic	Octadecanoic
$C_{20}H_{40}O_2$	Arachidic	Eicosadecanoic
$C_{24}H_{48}O_2$	Lignoceric	Tetracosanoic
Unsaturated acids:		
$C_{16}H_{30}O_2$	Palmitoleic	Hexadec-9-enoic
$C_{18}H_{34}O_2$	Oleic	Octadec-9-enoic
Doubly unsaturated acids:		
$C_{18}H_{32}O_2$	Linoleic	Octadeca-9,12-dienoic
$C_{18}H_{30}O_2$	Linolenic	Octadeca-9,12,15-trienoic
$C_{20}H_{32}O_2$	Arachidonic	Eicosa-5,8,11,14-tetraenoic

Triglycerides

Triglycerides consist of molecular complexes of glycerol with fatty acids and the generalized structure of all such molecules is:

$$H_2C.O.OC.R_1$$
$$R_2.CO.O.CH$$
$$H_2C.O.OC.R_3$$

In view of the absence of charged groups in the molecule, these structures are sometimes known as neutral fats in distinction to the phosphatidic acids described below. In practice, some triglycerides have the same R-group in all positions, but in others two or three different R-groups may be represented in a single molecule.

It should be emphasized that fats as isolated from bacterial cells are unlikely to consist of single molecular entities, largely because of the wide variety of different fatty acid residues involved and the consequent difficulty of separating molecules of such close molecular similarity.

Phospholipids

These compounds, like triglycerides, are also substituted glycerols, but all contain phosphorus; many also contain nitrogen. Most of the examples found in bacteria fall into the group known as phosphatidic acids and have the generalized structure:

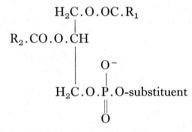

$$H_2C.O.OC.R_1$$
$$R_2.CO.O.CH$$
$$H_2C.O.P.O\text{-substituent}$$

where R_1 and R_2 are fatty acids and the 'substituent' may be choline, ethanolamine, serine, or inositol. Where the substituent is choline, the phosphatidic acids are known as *lecithins*.

Steroids

These are a group of molecules that may be regarded as variants of the basic substance perhydropentanophenanthrene (Fig. 1.8). It must be stressed, however, that they are not synthesized from this molecule but are built up by a

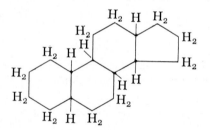

Fig. 1.8 *Cyclo*-perhydropentanophenanthrene—the parent substance to which all steroids are related. It is important to note that this compound is *not* the *biological* precursor of the steroid nucleus.

long series of chemical conversions from acetic acid (for details see a textbook of biochemistry). Steroids are unknown in bacteria but certain derivatives are found in fungi where they may be essential for spore formation. One such compound is ergosterol (Fig. 1.9) and this is of particular industrial importance since chemical modification of this compound in the laboratory allows the synthesis of many medicinally important steroids on a commercial scale (p. 688).

Carotenoids

Carotenoids are a group of lipid substances that have a characteristic orange or red colour. They are compounds responsible for the pigmentation of carrots (hence their name), and also of many bacterial species, particularly the staphylococci and micrococci. All are molecular variants of β-carotene whose structure is as follows:

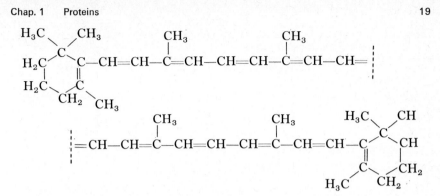

When present in bacteria cells these pigments are usually found in the cell membrane.

In general the investigation of lipids and their role in microbial cells has greatly lagged behind research on other macromolecules, largely because their insolubility in water has raised considerable problems in their investigation by routine chemical and biochemical techniques. It is only since the comparatively recent advent of gas chromotography as an experimental tool that the investigation of these compounds has advanced at an appreciable rate. Another problem connected with investigation of lipids is that the presence of large numbers of molecules of closely related structure and properties makes the problems of isolation, already severe enough because of their insolubility, even more exacting.

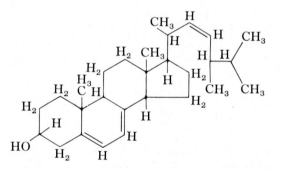

Fig. 1.9   Ergosterol—a steroid with important medicinal uses that is produced by some fungi.

PROTEINS

Proteins may be classified into two types on the basis of their composition. *Simple* proteins consist solely of amino acids while *conjugated* proteins are structures of the same type but with one or more additional molecules attached. This additional molecule is usually known as the *prosthetic group* and is often involved in the physiological function of the protein, particularly if it is an enzyme, and in the immunological specificity of an antigen (p. 562). For example,

Table 1.3 The sidechain substituents (R-groups) of the amino acids commonly found in microbial proteins.

General structure: $H_2N.CHR.COOH$

Amino acid	(abbreviation)	R-group
Glycine	(gly)	H—
Alanine	(ala)	CH_3—
Valine	(val)	$\begin{array}{c} CH_3 \\ \diagdown \\ CH— \\ \diagup \\ CH_3 \end{array}$
Leucine	(leu)	$\begin{array}{c} CH_3 \\ \diagdown \\ CH.CH_2— \\ \diagup \\ CH_3 \end{array}$
Isoleucine	(ileu)	$\begin{array}{c} CH_3.CH_2 \\ \diagdown \\ CH— \\ \diagup \\ CH_3 \end{array}$
Phenylalanine	(phe)	⬡CH_2—
Tyrosine	(tyr)	HO⬡CH_2—
Threonine	(thr)	$CH_3.CH(OH)$—
Serine	(ser)	$HO.CH_2$—
Cysteine	(cys)	$HS.CH_2$—
Methionine	(met)	$CH_3.S.CH_2.CH_2$—

electron transport is mediated by flavine prosthetic groups in the dehydrogenases and pyridoxal phosphate is present as a prosthetic group in many transaminases and amino acid racemases.

In both simple and conjugated proteins, the amino acid part of the structure consists of one or more unbranched polypeptide chains formed from amino acid residues joined together by peptide bonds (Fig. 1.10). Although twenty different types of amino acid may be found in such peptide chains, the common basic structure of all (namely $H_2N.CHR.COOH$, where R is one of 20 different chemical residues—see Table 1.3) ensures that the whole polypeptide chain can be formed by joining the amino acids together with a single type of bond. Examination of a number of proteins has shown that not all amino acids are present in the same molecular proportions in each; indeed their distribution

Table 1.3 continued

General structure: $H_2N.CHR.COOH$		
Amino acid	*(abbreviation)*	*R-group*
Tryptophan	(try)	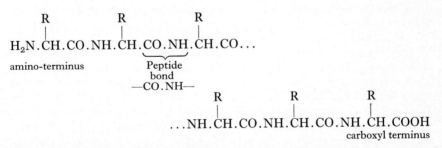
Proline	(pro)	
Aspartic acid	(asp)	$HOOC.CH_2—$
Asparagine	(asn)	$H_2N.OC.CH_2—$
Glutamic acid	(glu)	$HOOC.CH_2.CH_2—$
Glutamine	(gln)	$H_2N.OC.CH_2.CH_2—$
Arginine	(arg)	$H_2N.C.NH.CH_2.CH_2—$
		$\overset{\parallel}{NH}$
Lysine	(lys)	$H_2N.CH_2.CH_2.CH_2.CH_2—$
Histidine	(his)	

Note: The proline molecule, being an imino acid, does not fall into the general structural classification of amino acids; the structure shown here is the whole molecule and not just the R-group.

$$\overset{\overset{R}{|}}{H_2N.CH}.CO.NH.\overset{\overset{R}{|}}{CH}.\underbrace{CO.NH}.\overset{\overset{R}{|}}{CH}.CO...$$

amino-terminus   Peptide bond
                                  $—CO.NH—$

$$...NH.\overset{\overset{R}{|}}{CH}.CO.NH.\overset{\overset{R}{|}}{CH}.CO.NH.\overset{\overset{R}{|}}{CH}.COOH$$

carboxyl terminus

Fig. 1.10   The overall structure of a simple polypeptide chain as found in many proteins. The free $—NH_2$ residue at the left hand end is the *amino terminus*, and the free $—COOH$ at the other end is the *carboxyl terminus*.

Table 1.4 Comparative molecular composition of some bacterial proteins.

	Species		
	Bacillus licheniformis	*Bacillus cereus*	*Staphylococcus aureus*
Approx. Mol. wt.	28,000	31,000	29,000
N-terminal amino acid	Lysine	Aspartic acid	Lysine
Amino acid composition (residues/molecules)			
Lysine	30	23	44
Arginine	19	11	4
Histidine	1	6	3
Aspartic acid	35	25	43
Threonine	18	12	14
Serine	8	5	17
Glutamic acid	28	22	22
Proline	10	10	10
Glycine	14	21	16
Alanine	24	30	21
Valine	13	17	11
Methionine	2	1	2
Isoleucine	13	22	15
Leucine	25	18	22
Tyrosine	5	4	9
Phenylalanine	7	6	7
Cysteine	0	0	0
Tryptophan	0	1	0

All the enzymes are *bacterial penicillinases*, and these proteins have been chosen because they consist of a single polypeptide chain and have a similar molecular weight, thus facilitating comparison between the molecules when the protein composition is quoted in terms of the number of residues/molecule. Despite the fact that all the molecules are penicillinases, the molecular compositions of the three proteins are distinctly different. The differences in composition between *different* types of enzyme in bacterial cells are even more striking.

may vary quite widely (see Table 1.4) and some may even be missing completely. Although it is impossible to generalize, proteins often contain relatively small quantities of tryptophan, methionine, and cysteine but large amounts of the non-polar amino acids valine, leucine and iso-leucine. Some analyses of typical proteins are shown in Table 1.4.

Many proteins consist of only one polypeptide chain (which may vary in length from 20 to about 300 residues in different proteins) but many others contain more than one chain, and some as many as six or eight. In such molecules the individual chains are held together by molecular bridges formed between the thiol groups of two cysteine residues, one in each chain (see Fig. 1.11). Such cross connexions are often referred to as —S.S—(disulphide) bridges and usually occur to the extent of about one to two bridges to every 100 residues, although considerable variation in this value is found.

Although all proteins are similar in that they consist of polypeptide chains that may be cross-linked with —S.S— bridges, this basic structure admits an

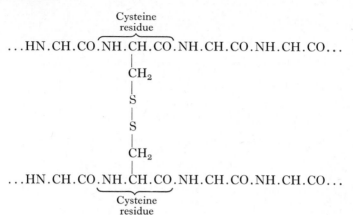

Fig. 1.11 The molecular nature of the —S.S— cross bridges that link different poly-peptide chains (and sometimes two points on the same chain) in proteins.

enormous variety of detailed structure and, consequently, of detailed properties. This variety depends to a certain extent on the number and length of the poly-peptide chains, but chiefly on the type and distribution of the individual amino acid residues along the chains themselves. The chemical nature of the R-groups found in the twenty different types of amino acid (see Table 1.3) show a wide variation in ionic properties, in polarity and in molecular shape (to mention only a few characteristics) and this ensures that each polypeptide chain has a correspondingly wide range of detailed chemical properties depending on the exact distribution of different types of amino acids along its length.

Although the order and nature of the amino acids along the polypeptide chains together with the position of the cross-bridges between the chains (the so-called *primary* and *secondary* structure of the protein) are important factors in the nature of the protein, the crucial element that determines the properties of the molecule is the way in which the structure folds up and this seems to be determined by the amino acid sequence of the polypeptide chains themselves. As an example, Fig. 1.12 shows the primary and secondary structure of egg-white lysozyme, while Fig. 1.13 shows how this structure folds up to form the final, or *tertiary*, structure of the molecule. The nature of this folding of the chains is vitally important since the overall properties of the molecule are determined by the nature of the amino acid chains that are brought into struc-tural juxtaposition with one another and with the external environment in the folded structure and not by the sequence of the amino acids in any one chain.

With many proteins, an examination of the tertiary structure of the protein shows that the amino acids with the polar and ionic side chains tend to be arranged so that these groups project into the surrounding medium from the surface of the protein, while the interior of the structure is largely composed of amino acids that have non-polar side chains. This arrangement means that protein molecules seem to fold into the most condensed shape possible and the hydrophobic nature of their interior suggests that no water molecules can pass through the surface structure of the molecule to the interior.

Amino terminus:

                                        1
Lys.Val.Phe.Gly.Arg.Cys.Glu.Leu.Ala.Ala.Ala...

...Met.Lys.Arg.His.Gly.Leu.Asp.Asn.Tyr.Arg.Gly.Tyr...
                                                  2
...Ser.Leu.Gly.Asn.Try.Val.Cys.Ala.Lys.Phe.Glu.Ser...

...Asn.Phe.Asn.Thr.Gln.Ala.Thr.Asn.Arg.Asn.Thr.Asp...

...Gly.Ser.Thr.Asp.Try.Gly.Ileu.Leu.Gln.Ileu.Asn.Ser...
                          3
...Arg.Try.Try.Cys.Asp.Asn.Gly.Arg.Thr.Pro.Gly.Ser...
              4   3
...Arg.Asn.Leu.Cys.Asn.Ileu.Pro.Cys.Ser.Ala.Leu.Leu...
                                              4
...Ser.Ser.Asp.Ileu.Thr.Ala.Ser.Val.Asn.Cys.Ala.Lys...

...Lys.Ileu.Val.Ser.Asp.Gly.Met.Asn.Ala.Try.Val.Ala...
                                2
...Try.Arg.Asn.Arg.Cys.Lys.Gly.Thr.Asp.Val.Gln.Ala...
              1
...Tyr.Ileu.Arg.Gly.Cys.Arg.Glu.

                                        Carboxyl terminus.

Fig. 1.12 The primary and secondary sequence of egg-white lysozyme. Note that in this molecule there is only one polypeptide chain which is bridged at intervals by —S.S— bridges. There are four such bridges linking the molecule at intervals; they are formed between the pairs of cysteine residues labelled 1, 2, 3, and 4 respectively.

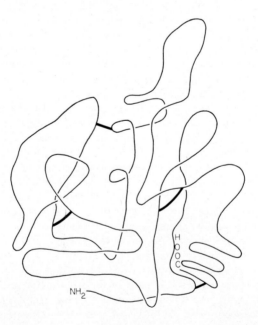

Fig. 1.13 The tertiary folding of the 129 residues of egg-white lysozyme shown in Fig. 1.12. The —S.S— cross bridges are shown as solid bars linking the chain.

Although many protein molecules appear to be approximately spherical when examined by appropriate physico-chemical and crystallographic techniques, this is an over-simplified view of the true situation. Close examination often reveals clefts or pits in the surface of the protein molecule and these can often be shown to be lined with amino acids with a specialized function. Thus, for example, egg-white lysozyme has a cleft on its surface into which the substrate can be shown to fit, and, in this case, the specialized residues lining the cleft seem to be tryptophan molecules. These residues are almost certainly involved in the hydrolytic split of the substrate by the enzyme in the course of its action.

Early theoretical work suggested that amino acids were condensed together into polypeptide chains and formed a clearly defined spiral structure, or helix. Although tests showed that short polypeptide chains, particularly if they contained only a single type of amino acid residue, would form helices of this type, examination of whole proteins has shown that the helical content of the molecules is usually relatively small. Thus in most proteins whose tertiary structure is known in any detail, not more than about 20 per cent of the polypeptide sequence is helical, and most of this structure is buried in the heart of the molecules.

Protein subunits

Although some proteins consist of one or more distinct polypeptide chains joined together by —S.S— bridges, others may be formed by the interaction of two or more subunits. In some cases the subunits have an identical structure but in others two, or even three, different subunits are grouped together to form a molecule. One type of protein that usually consists of subunits is any enzyme that is subject to feed-back inhibition (see p. 44). In this case one subunit is the enzyme *sensu stricto* and is responsible for recognizing and metabolizing the substrate. The other subunit recognizes the inhibitor, which is usually the product of the pathway. The inhibitory action is mediated by interaction between the subunits in such a way that reaction of one subunit with the inhibitor reduces the action of the other against the normal substrate. Such an effect transmitted between subunits is known as an *allosteric* effect.

NUCLEIC ACIDS

Deoxyribonucleic acid (DNA)

DNA is an extremely long chemical thread made up of two strands, but unlike most threads, the strands are not twisted round one another but wound together round an imaginary core to form a double spiral or, more accurately, 'double helix' (Fig. 1.14). Each thread of the molecule consists of a chain of deoxyribose (Fig. 1.15) and phosphate residues arranged alternatively as shown in Fig. 1.16. This arrangement ensures that each individual sugar residue in the chain is joined to two phosphate groups, one substituted on the 3-hydroxyl of the sugar and the other on the 5'-OH group, and that all the phosphate groups in the strand are doubly substituted except the first and the last. Thus one end of a DNA strand carries a single substituted phosphate on C-3 of a deoxyribose

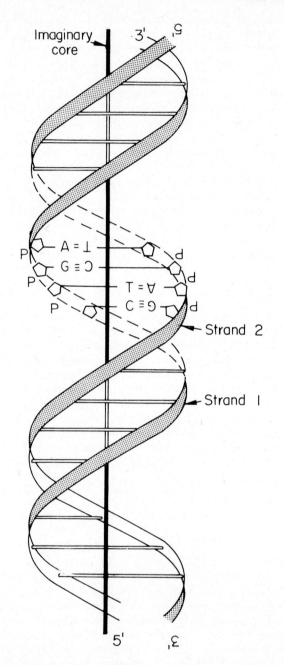

Fig. 1.14 The overall arrangement of residues in the anti-parallel double-strand of a DNA molecule. A = adenine, G = guanine, C = cytosine, T = thymine, and P = phosphate.

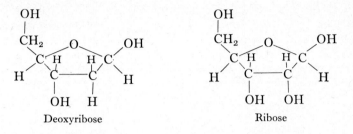

Fig. 1.15 The structure of deoxyribose and ribose sugars.

residue and the other end has a similar phosphate on the C-5 of the sugar. Thus any chemical change, such as an enzyme reaction, that passes along one of the DNA strands can be thought of as moving in a direction that is either $3 \rightarrow 5$ or $5 \rightarrow 3$. This concept of direction in a DNA strand is important when one comes to examine the overall structure of the molecule, for the two strands are arranged

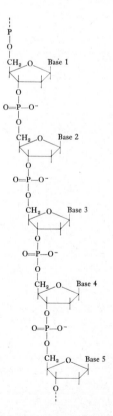

Fig. 1.16 A typical polynucleotide sequence as found in RNA and DNA. In RNA the sugar is *ribose*, while in DNA it is *deoxyribose*.

to lie in the double helical thread so that one lies in the $5 \rightarrow 3$ direction and is always faced by a strand running $3 \rightarrow 5$ (see Fig. 1.14). For this reason, therefore, DNA is said to exist in the form of an *antiparallel double helix*.

In addition to the substitution at the 3 and the 5 positions, each sugar residue in a DNA strand is also substituted at C-1 by one of a number of purine or pyrimidine bases. The bases commonly found in DNA are the purines, adenine (A) and guanine (G), and the pyrimidines, cytosine (C) and thymine (T) (Fig.

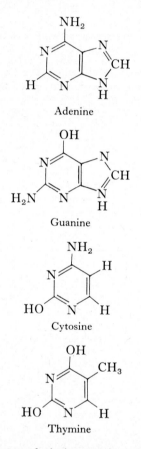

Adenine

Guanine

Cytosine

Thymine

Fig. 1.17 The structure of adenine, guanine, cytosine, and thymine.

1.17). Examination of the overall structure of DNA shows that the double-stranded nature of the molecule is maintained by hydrogen bonding between the purine and pyrimidine bases in such a way that an adenine residue is always faced by a thymine while guanine is faced by cytosine (Fig. 1.18). Thus adenine and guanine are said to be a *complementary base pair*, and share two hydrogen bonds, while guanine and cytosine are a similar complementary pair and share three hydrogen bonds.

The fact that there is always an adenine opposite a thymine and a guanine opposite a cytosine in the molecule means that there is an equal amount of adenine and thymine in DNA and the same is true of guanine and cytosine. However, the ratio of each pair of bases (that is $[A + T]/[G + C]$) varies widely from one bacterial species to another. Analysis of the composition of DNA from a number of bacteria shows that the $[A + T]/[G + C]$ ratio may vary from as much as 70/30 in *Clostridium perfringens* (*welchii*) at one extreme to

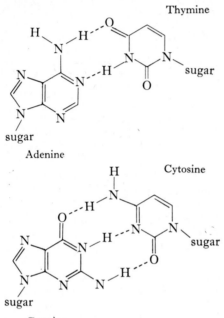

Fig. 1.18 The hydrogen bonding between adenine and thymine and between guanine and cytosine to form the base pairs that constitute part of the DNA double helical structure.

28/72 in *Micrococcus lysodeikticus* and 26/74 in *Actinomyces bovis* (Table 1.5). This wide variation in DNA base composition is confined to bacteria and similar relatively primitive forms of life. All vertebrates have a $[A + T]/[G + C]$ ratio of about 60/40 and this is usually taken to indicate a relatively close evolutionary relationship among the vertebrates when compared with bacteria.

Were it possible to look at the DNA molecule from the side it would appear rather like a screw, but with the difference that the grooves on the molecule (which are analogous to the threads on a screw) are not equally spaced (see Fig. 1.14). The two types of groove on the surface of DNA are normally referred to as the *major* and the *minor* groove, and the second of these two features is important in the action of certain antibacterial substances, notably actinomycin D, that bind to DNA and disrupt its function.

Table 1.5 Percentage of (guanine + cytosine) base pairs in the DNA of some bacteria and rickettsias.

Organisms	%G + C	Organisms	%G + C
Clostridium perfringens		Escherichia coli	52
(welchii)	30	Salmonella typhi	54
Rickettsia prowazekii	32	Aerobacter aerogenes	56
Staphylococcus aureus	34	Brucella abortus	58
Proteus vulgaris	36		60
Diplococcus pneumoniae	38	Pseudomonas aeruginosa	62
	40		64
Bacillus subtilis	42		66
Rickettsia burnetii	44	Mycobacterium tuberculosis	68
	46		70
Corynebacterium acne	48	Micrococcus lysodeikticus	72
Neisseria gonorrhoeae	50	Actinomyces bovis	74

It must be stressed that the structure of DNA as described above has largely been determined by crystallographic techniques on pure DNA, and that there is reasonable evidence that such a rigid, helical structure cannot be accommodated unmodified in a living cell. The molecular weight of the DNA isolated from bacterial cells usually has an average molecular weight of about 8×10^6, yet such preparations are often somewhat degraded, and the naturally occurring molecule in the cell is probably a good deal larger. As a molecular weight of this magnitude corresponds to a helical length (on the basis of the atomic arrangement shown in Fig. 1.14) of about 50 μ this calculation implies that the DNA in a bacterial cell cannot be present as simple helix but that some super-coiling of the molecule must be present. How this occurs, and the effect of the process on the properties of DNA, are at present largely unknown.

Single-stranded DNA

Although the great majority of the DNA in a bacterial cell is double stranded and this is true of most bacteriophages (p. 223) as well, a class of phage containing single-stranded DNA does exist. These are usually very small in size and constitute some of the most simple organisms yet detected. In order to reproduce, such phages must form a typical base paired double stranded DNA from the single strand they normally carry and this process is carried out in the host cell using the host enzymes for the purpose.

Ribonucleic acids (RNA)

Superficially RNA has considerable structural similarity to DNA. The molecule has a long chain of alternating sugar and phosphate residues, although the mono-saccharide residue is *ribose* (Fig. 1.15) and not deoxyribose as in DNA. As in DNA, the sugar residues are double substituted, one bond being to the 3'-position of the molecule and the other to the 5'. Once again, the ends of an RNA strand can be distinguished, one by having a free 5' phosphate and the other by a free 3' phosphate.

In RNA the C-1 position of the sugar is substituted with purine and pyrimidine bases, with DNA, but unlike DNA the bases are adenine, guanine, cytosine, and *uracil* rather than adenine, guanine, cytosine, and *thymine.* In general the incidence of complementary base pairing is a great deal less in RNA than in DNA, and where it occurs adenine pairs with uracil rather than with thymine (see Figs. 1.18 and 1.19).

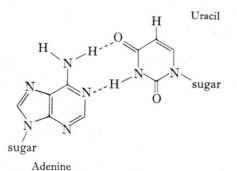

Fig. 1.19   Base pairing of adenine and uracil.

Despite a basically similar overall chemical structure, three distinct types of RNA can be distinguished on the basis of structure and function: *messenger RNA* (or m-RNA), *ribosomal RNA,* and *transfer RNA* (or t-RNA).

Messenger RNA (m-RNA)

Messenger-RNA is invariably single stranded and contains only the single purine and pyrimidine bases adenine, guanine, cytosine, and uracil: no base is further substituted (see below). A very large number of different m-RNAs exist in each cell and the variety in structure arises from the fact that the order of bases along the RNA strand is a complementary copy of a stretch of DNA (see Chap. 2). Consequently there are approximately as many different types of messenger RNA molecule as there are genes in the cell. In all cases, the molecules differ solely in their length (which is usually between about 300 and 7,000 to 10,000 residues) and in the order of the purine and pyrimidine bases in the strand.

Ribosomal RNA

About 40 per cent of the total mass of each ribosome consists of RNA. Chemical studies show that some, at least, of the material is double stranded and that some contains substituted purine and pyrimidine bases in place of the normal residues on the C-1 of the ribose residues (see discussion of the structure of t-RNA in Chap. 2). The exact structure of one component of ribosomal RNA, the so-called 5s component, has recently been determined and is shown in Fig. 1.20, but as yet no function for this molecular species of ribosomal RNA has been discovered, nor is the structure of any other part of the ribosomal RNA known in detail.

P.UGCCUGGCGGCCGUAGCGCGGUGGUCCCACCUGACCCGA...
...UGCCGAACUCAGAAGUGAAACGCCCUAGCGCCGCCGAUGG...
...UAGUGUGGGGUCUCCCCCAUGCGAGAGUAGCGAACUGCC...
...AGGCAU.OH

Fig. 1.20 The nucleotide sequence of the 5S RNA from *Escherichia coli*. (Note: in this figure and Fig. 1.21 the symbols A, G, C and U indicate the corresponding nucleotides rather than the bases as in Fig. 1.14)

Transfer RNA (t-RNA)

There are probably a hundred or more different types of transfer RNA molecule in each bacterial cell and all probably represent relatively minor but important variants of a common structure. All are very small as RNA molecules

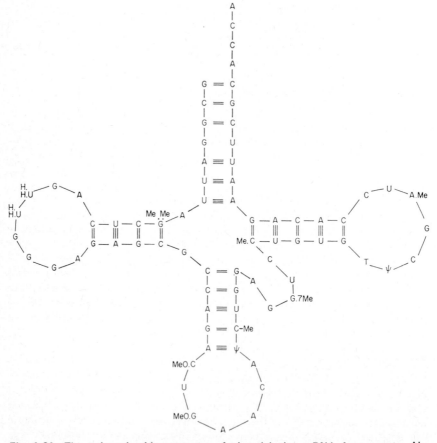

Fig. 1.21 The polynucleotide sequence of phenylalanine t-RNA from yeasts. Abbreviations as in Fig. 1.20 but in addition: 2Me.G = 2-methyl-guanine, diH.U = dihydrouracil, Me.O.G = O-methyl-guanine, Me.O.C = O-methyl-cytosine, ψ = pseudo-uridine, Me.C = methyl cytosine, G.7Me = 7-methyl-guanine, T = thymine and A.Me = methyl-adenine.

go (about 70–80 nucleotide residues seems to be an average size) and share the polyribosephosphate structure common to all RNAs. This type of RNA shows a certain limited amount of base pairing but unlike DNA (where the pairing always occurs between two distinct DNA strands, p. 26), in transfer RNA a single strand turns back on itself to allow internal base pairing to occur. However, this base pairing does not extend over the whole t-RNA molecule, nor is it uninterrupted, and the result is a key-shaped molecule of the type shown in Fig. 1.21. The detailed structure of t-RNA molecules is described in greater detail in Chap. 2 where the function of these molecules in protein synthesis is also discussed.

MUCOPEPTIDE (GLYCOPEPTIDE)

Whereas the macromolecules that have been described so far are all structures with well-defined limits, this is not so in the case of the mucopeptides that form the rigid part of the bacterial cell wall, particularly in Gram-positive bacteria. As mentioned previously these compounds consist of relatively enormous sack-shaped molecular nets with a fairly well-defined repeating composition but without any clearly defined limits.

Basically all mucopeptides consist of interlinked polysaccharide and polypeptide chains in which the polysaccharide component is composed of alternating N-acetyl-glucosamine and N-acetyl-muramic acid residues, the latter compound being a substituted sugar found only in bacterial mucopeptide and its metabolic precursors (Fig. 1.22). The linkage between the N-acetyl-glucosamine and the

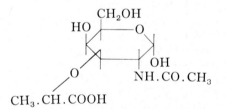

Fig. 1.22   The structure of N–acetyl–muramic acid.

N-acetyl-muramic acid is by a glycoside bond linking the C-1 of muramic acid to the C-4 of glucosamine. In addition, some muramic acid residues in some mucopeptides may also carry an O-acetyl residue on C-6, but this is not a feature common to all mucopeptides.

The polypeptide portion of bacterial mucopeptide can best be considered as two parts. One of these is a polypeptide that usually has the composition either L-alanine-D-glutamic acid-L-lysine-D-alanine. (The diaminopimelic acid, DAP; Fig. 1.23, can either be in the DD- or LL- or *meso*-stereoisomeric form depending on the bacterial species involved). In a minority of mucopeptides other amino

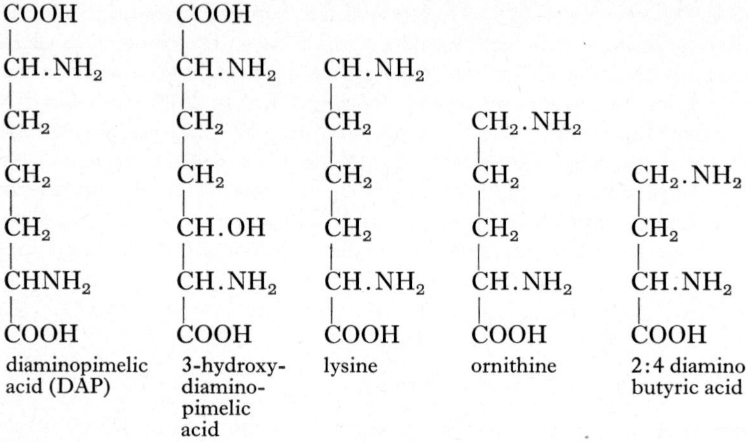

$$
\begin{array}{lllll}
\text{COOH} & \text{COOH} \\
| & | \\
\text{CH.NH}_2 & \text{CH.NH}_2 & \text{CH.NH}_2 \\
| & | & | \\
\text{CH}_2 & \text{CH}_2 & \text{CH}_2 & \text{CH}_2.\text{NH}_2 \\
| & | & | & | \\
\text{CH}_2 & \text{CH}_2 & \text{CH}_2 & \text{CH}_2 & \text{CH}_2.\text{NH}_2 \\
| & | & | & | & | \\
\text{CH}_2 & \text{CH.OH} & \text{CH}_2 & \text{CH}_2 & \text{CH}_2 \\
| & | & | & | & | \\
\text{CHNH}_2 & \text{CH.NH}_2 & \text{CH.NH}_2 & \text{CH.NH}_2 & \text{CH.NH}_2 \\
| & | & | & | & | \\
\text{COOH} & \text{COOH} & \text{COOH} & \text{COOH} & \text{COOH}
\end{array}
$$

diaminopimelic 3-hydroxy- lysine ornithine 2:4 diamino
acid (DAP) diamino-   butyric acid
    pimelic
    acid

Fig. 1.23 The structure of certain amino acid components of some bacterial muco-peptides.

acids, or substituted amino acids, are found in this part of the mucopeptide: for example some molecules contain 2,4-di-amino-butyric acid and some have ornithine or 3-hydroxy-2,6-diamino-pimelic acid (Fig. 1.23). Regardless of minor variations in composition, this part of all mucopeptides is notable for two factors: first, the polypeptide contains both D- and L-amino acids while the polypeptides found in proteins are solely made up from the L-isomers; and secondly, the D-glutamyl residues are linked in the polypeptide chain through their γ carboxyl and the α carboxyl is free.

The other part of the polypeptide of bacterial mucopeptide is one that fulfils a cross-linking role by joining the Ala-Glu-Lys-Ala side chains to one another. In *Staphylococcus aureus* the cross-bridge is Gly.Gly.Gly.Gly.Gly (i.e., penta-glycine), but in other mucopeptides the bridges can be composed of a number of other types of amino acid, e.g., serine.

Examination of the structure of all types of mucopeptide is incomplete at present, and all that is clear is that there is considerable variation in molecular arrangement from species to species. In *Staph. aureus*, for example, the inter-linking of the polysaccharide and the polypeptide components occurs as is shown in Fig. 1.24. The key bond is the peptide link between the —NH_2 group of L-alanine and the carboxyl of the $CH_3.CH(O—)COOH$ side chain of *N*-acetylmuramic acid.

At present it is unclear whether the mucopeptide in the wall is built up essentially of two-dimensional sheets of mucopeptide or whether cross-links occur linking these sheets to those above and below, as is suggested diagram-matically in Fig. 1.25. Whichever the method used, it has to allow cells to grow. Two possibilities seem feasible: either sheets of mucopeptide slide over one another or cross bridges break and reform. Little is known at present about which process operates in practice.

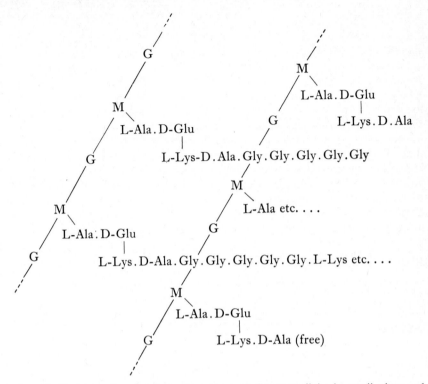

Fig. 1.24 The overall molecular arrangement of the cross links in small pieces of mucopeptide from *Staph. aureus.*

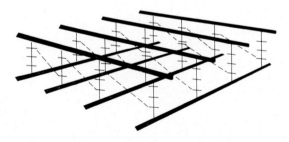

Fig. 1.25 Diagrammatic representation of the three-dimensional interchain links that may be found in bacterial mucopeptides. The solid horizontal bars represent the *N*-acetyl-muramic acid/*N*-acetyl glucosamine chains, the vertical crossed bars the polypeptide sidechains on the *N*-acetyl muramic acid residues and the dotted line the penta-glycine cross bridges. For further details see text. (After Rogers, H. J. (1965) *Symp. Soc. gen. Microbiol.*, **15**, 186.)

CONCLUSION

In summary, it should be stressed that the preceding account is a rather superficial description of some of the macromolecules that go to make up bacterial cells. But although the molecules described here, or some representatives of them, may account for up to 80 per cent of the dry weight of the cell it does not follow that relatively minor macromolecular components are unimportant to cell life and survival. Were this so, natural selection would probably have seen to it that the molecules had ceased to be produced a long time ago. Typical examples of macromolecules that have been studied a great deal, but which are not described here for lack of space, are the teichoic and teichuronic acids of bacterial cell walls, the poly-D-glutamic acid from the capsules of many *Bacillus* spp., and the complex mucopolysaccharides that are found in the capsular material of various pneumococcal strains. For details of these molecules, and many others, it will be necessary to consult a formal textbook of biochemistry or one of the monographs dealing with various classes of bacterial macromolecules.

BOOKS AND ARTICLES FOR REFERENCE AND FURTHER STUDY

An integrated list of references is given at the end of Chapter 2, p. 76.

2
Biosynthesis in Micro-organisms

INTRODUCTION

When a bacterial cell grows and divides, the final outcome of the process is that there are now two cells where there was one before. For this process to be completed, therefore, every atom and every molecule in the parent cell has to be duplicated and inserted into its correct place in the developing structure that will eventually become the mature daughter cell.

As described earlier (Chap. 1), bacterial cells are made up largely of macromolecules (proteins, nucleic acids, polysaccharides, fats, etc.) but these components are always accompanied in the cell by a proportion of low molecular weight compounds that are either destined themselves to become part of the macromolecules, or to be used catalytically in macromolecular biosynthesis, or to take a part in the energy metabolism of the cell (Chap. 3). Consequently the biochemical aspects of cell duplication can conveniently be considered under these headings:

1. Biosynthesis of the low molecular weight organic molecules.
2. Biosynthesis of the macromolecules themselves.
3. Provision of energy—in a convenient molecular form—to achieve the first two of these activities.

Certain biochemical activities that cannot easily be classified under these three headings do, of course, occur in growing cells. There are, for example, certain degradative activities that destroy unwanted molecules but, by and large, these have a relatively minor importance in the economy of the growing cell.

In this chapter therefore we will first discuss topic 1. Next the problems of macromolecular biosynthesis will be considered (topic 2) and finally how the rate of the various reactions involved in these two categories are adjusted to the requirements of the cell. The provision of energy for synthesis (topic 3), is largely dealt with in Chapter 3.

BIOSYNTHESIS OF LOW MOLECULAR WEIGHT ORGANIC MOLECULES

The chemical composition of the various types of macromolecule that are part of the bacterial cell have already been described in Chapter 1; they are built up from a relatively restricted range of simple molecules. For example, some polysaccharides consist solely of a long chain of identical monosaccharide units or else of alternating units of two different monosaccharide sugars. In others monosaccharides alternate with amino sugars or with uronic acids. Nucleic acids (both RNA and DNA) consist of a monosaccharide sugar together with four purine or pyrimidine bases and phosphate residues. Proteins, although they lie at the other end of the scale of macromolecular complexity, are still fairly simple. All are derived from the same twenty amino acids joined together by peptide bonds, the complexity being introduced by the specificity of the order of the amino acids in the polypeptide chains. Even such complex structural elements as the mucopeptide of the bacterial cell wall, that is formed from as many as 8–10 different types of component molecule, are relatively simple from the point of view of the type and range of molecular components involved.

Molecular biosynthesis vs. nutritional requirements

Although the molecular components that are needed to synthesize the various categories of macromolecule are very similar in all bacterial species, the biosynthetic load carried by the cells of different species varies widely. This wide variation reflects the enormous range of nutritional patterns found in bacteria (Chap. 5). Basically, if an organic molecule is available intact outside the cell, or the cell is equipped to take it up, then it is unnecessary to biosynthesize it. So with a complex heterotrophic organism, such as *Streptococcus faecalis*, for example, where a large range of complex molecules are taken up from the growth medium, the biosynthetic load is light; with photo- and chemo-autotrophic bacteria, on the other hand, the load may be extremely heavy; bacteria such as *Escherichia coli* that can grow well in a glucose/ammonium ion/salts medium occupy an intermediate position.

Metabolic pathways

If an organism can multiply—as many photoautotrophs can—in an illuminated medium where dissolved CO_2 (or HCO_3^- ion) is the sole carbon source, it follows that all the low molecular weight organic molecules needed for macromolecular synthesis in these cells must be derived by chemical modification of CO_2. Similarly where glucose is the sole carbon source, modification of this molecule must occur to give the required chemical end-products. However, as such chemical changes are complex, they cannot conceivably occur by a single step. All the necessary biochemical modifications therefore occur as a series of simple reactions, and these 'step-wise' modifications of a molecule are known as *pathways*. Broadly speaking such pathways are classified as *anabolic* or *catabolic* depending upon their apparent purpose. Pathways that build up the components needed for macromolecular biosynthesis are normally *anabolic* or *biosynthetic* while those that break down compounds, either to yield energy or to remove their toxic effects, are called *degradative* or *catabolic*. Sometimes it is difficult to decide whether the pathway is anabolic or catabolic, and such pathways are called *amphibolic*.

In order to consider some of the principles involved in the nature and function of pathways, it is necessary to examine a few examples in detail. This selective approach should not, however, allow one to forget that pathways exist in bacterial cells for the synthesis of all the hundreds of types of molecule necessary for cell growth and which cannot be obtained from the growth media. Figure 2.1 shows in detail the chemical steps involved in the biosynthesis of histidine from the precursors phospho-ribosyl-pyrophosphate and ATP. The pathway consists of a chain of enzymes which act together to convert the precursors to histidine in nine steps. In each case, the chemical modification introduced is small. Thus steps 1 and 2 involve removal of pyrophosphate, 3 and 6 are molecular rearrangements (isomerase action), 4 and 7 add an —NH_2 group, 8 removes phosphate, 9 removes hydrogen atoms and 5 is a ring closure reaction.

When one comes to examine the overall pattern of biosynthetic pathways in an organism, it is clear that they have evolved to give as economical an arrangement as possible. Economies fall into two classes: very often certain sections of a pathway are common to the synthesis of more than one end-product. Secondly, enzymes may occasionally act on more than one substrate, thus allowing a single enzyme to catalyse the biosynthesis of two distinct end-products.

A good example of the first of these two types of economy is provided by the biosynthetic pathway for the synthesis of aromatic amino acids (Fig. 2.2). In this pathway steps 1 to 8 are common to the synthesis of all the end-product amino acids, but after this point the pathways become specific for single products.

The best example of the second type of economy is provided by the leucine/iso-leucine/valine/pantothenate biosynthetic pathway (Fig. 2.3). Here enzymes 1–4 of the pathway are capable of acting on two distinct substrates; one enzyme is thus catalysing the formation of two end-products.

It is interesting to note that this pathway leads to the synthesis of three amino acids needed for protein synthesis (leucine, iso-leucine, and valine) and a compound (pantothenate) that ultimately becomes part of Co-enzyme A, a molecule exerting a solely co-factor activity. As a consequence of the fate of these products a great deal more leucine, iso-leucine, and valine than panthothenate must be synthesized and therefore the output of the pathways must be unequal. The way in which this is achieved is discussed later (p. 67).

Some idea of the extent to which condensation of biosynthetic pathways has

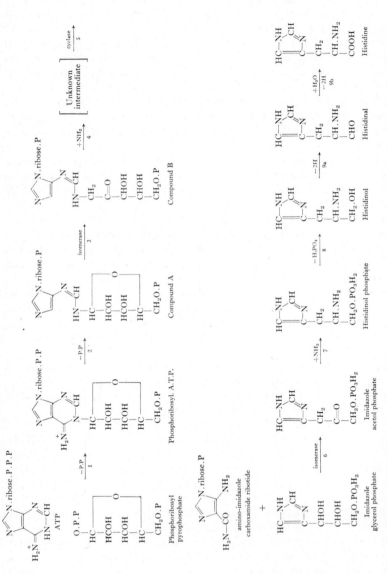

Fig. 2.1 The histidine biosynthetic pathway. P = phosphate, P.P = pyrophosphate, P.P.P = pyrophosphate. *Compound A* is phospho-*ribosyl*formimino-amino*imidazole-carboxamide ribotide. *Compound B* is phospho-*ribulosyl*-formimino-amino*imida-zole-carboxamide-ribotide.

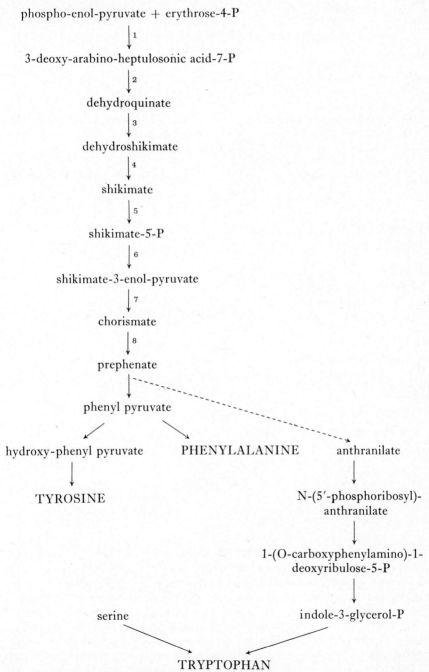

Fig. 2.2 The biosynthetic pathway leading to the aromatic amino acids—phenylalanine, tyrosine, and tryptophan.

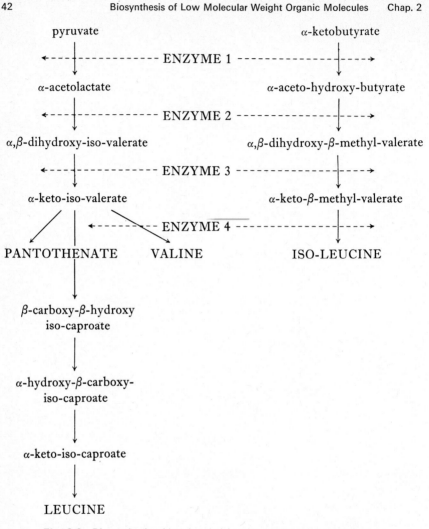

Fig. 2.3 Biosynthesis of leucine, iso-leucine, valine, and pantothenate.

occurred can be seen from the pathway shown in Fig. 2.4. In this example the biochemical interactions involved in the synthesis of a number of protein amino acids from glucose are outlined, and much of the pathway is common to all the amino acids concerned. As this pathway also has an important part to play in the energy balance of the cells it is a good example of an amphibolic one. It (Fig. 2.4) is also an example of a cyclical pathway. A number of such 'cycles' occur in bacterial metabolism and once again the situation seems to achieve economies.

Not all the small organic molecules needed by the cell become part of macro-molecules: many are needed to catalyse the transitions necessary for cell meta-

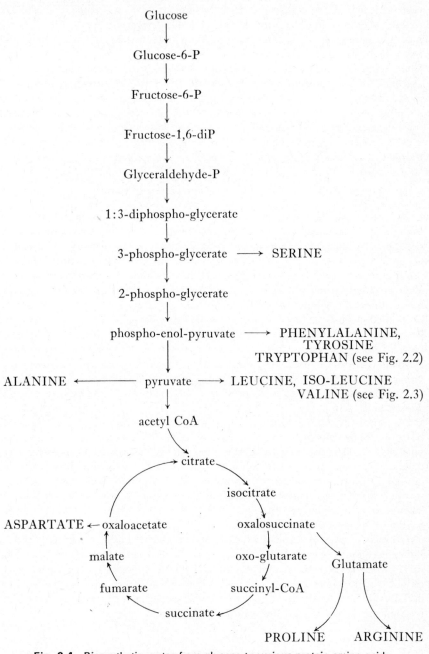

Fig. 2.4 Biosynthetic routes from glucose to various protein amino acids.

bolism in general. Some examples of such compounds are adenosine triphosphate (ATP), nicotine adenine dinucleotide (NAD), riboflavin phosphate (flavin mononucleotide, FMN), and flavin-adenine dinucleotide (FAD), which are concerned with electron transport and thus with the cell energetics, CoA (a co-factor often required for the synthesis of C—C bonds in biosynthetic pathways), and folic acid (a molecule often needed for the addition of methyl groups to molecules during biosynthesis) to mention but a few. Each of these compounds, although required in much smaller amounts than the macromolecule precursors, must nevertheless be synthesized by the cells. The biosynthesis of such molecules occurs by pathways indistinguishable in principle from those used for the synthesis of the molecular precursors of cell macromolecules.

THE PHENOTYPIC EXPRESSION OF BIOSYNTHETIC PATHWAYS

As mentioned previously, the amount of material that must be provided by each biosynthetic pathway varies greatly from cell to cell and situation to situation. For example, regardless of whether the material is synthesized endogenously via a biosynthetic pathway, or is taken up from the growth medium, a duplicating cell requires several thousand times as much of a protein amino acid, such as glutamic acid or leucine, as it does of a co-factor molecule—such as biotin, thiamine or pantothenate—where only relatively few molecules/cell will be needed. Consequently the output of the various pathways must somehow be adjusted by the cell to meet requirements. Since there is usually only one copy of the gene that specifies the formation of each protein involved in biosynthetic pathway in each cell, some method has to be found for 'setting-the-pace' of gene expression. Possible mechanisms for this process are discussed below after the mechanisms of protein and nucleic acid synthesis have been described (p. 66).

On top of the built-in and fairly inflexible 'pace-setting' of gene expression, however, many cells have sensitive and highly flexible methods of adjusting the outputs of biosynthetic pathways to changes in the amount of preformed material available exogenously. For example, even if a cell has a pathway capable of biosynthesizing histidine, the presence of this amino acid in the growth medium renders the pathway redundant, temporarily at least, and it is a great advantage to the economy of the cell to be able to shut down a pathway temporarily until exogenous material is exhausted.

In practice, the regulation of a biosynthetic pathway is achieved in two fundamentally distinct ways: either (1) the *rate of function* of the pathway is lowered, or (2) the *relative concentration* in the cell of the enzymes of the pathway is reduced by blocking further synthesis. The first of these two mechanisms is called *inhibition*—and often *feedback inhibition* because of the details of the mechanism involved—and the second is *repression*.

Feedback inhibition

In any concatenated series of biochemical reactions, such as a biosynthetic pathway, the throughput of material passing along the pathway can be influenced by blocking any one of the steps. Normally feedback inhibition of biosynthetic pathways is exercised by the *product* of the *last enzyme* of the pathway on the function of the *first* enzyme—hence the name for the phenomenon (Fig. 2.5). So to take the example of the histidine pathway (Fig. 2.1), an increase in the free histidine concentration in the cell will inhibit the conversion of phosphoribosyl-pyrophosphate + ATP to phosphoribosyl − ATP—step 1, (Fig. 2.1) in the biosynthetic pathway. Conversely, a lowering of the histidine concentration in the cell will cause a lowering of the inhibition of the first enzyme of the

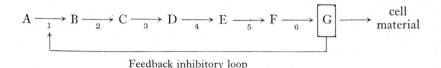

Feedback inhibitory loop

Fig. 2.5   Feedback inhibition of a hypothetical pathway.

pathway and a consequent increase in the rate of endogenous histidine biosynthesis. This method of regulation, therefore, ensures that the endogenous rate of histidine biosynthesis is controlled by the free concentration of histidine in the cell, and this in turn depends upon the balance between the rate of utilization of this material by the cell and its uptake from the growth medium. Consequently, high external histidine concentrations cause a shut-down of histidine biosynthesis without, in any way, impairing the ability of the pathway to function at a full rate if the exogenous supply of histidine subsequently fails.

Repression and Induction

Like feedback inhibition, repression is usually exerted by the product of the pathway but in this case, instead of inhibiting the *function* of a *single* enzyme, the effect is to block further *synthesis* of *all* the enzymes of the pathway. Thus whereas feedback inhibition has an immediate effect in reducing biosynthesis, repression acts more slowly (but has a more lasting effect) by reducing the total synthetic activity of the culture as the cells grow and the existing supplies of the biosynthetic enzymes are diluted out in the increasing population. A comparison of the effect of feedback inhibition and of repression on the kinetics of biosynthesis by a growing culture are shown graphically in Fig. 2.6. A fall in the endogenous concentration of the repressor molecule leads to a de-repression of synthesis of the biosynthetic enzymes—but these, of course, take a little time to reach the level in each cell that existed before repression commenced.

A rarer phenomenon that affects the expression of biosynthetic pathways is *induction*. In this case the level of expression of a pathway is increased in response to the presence of a low molecular weight molecule called an *inducer*. The inducer may be the precursor of a pathway. More often induction is a phenomenon affecting the expression of enzymes that are involved in the degradation of molecules—particularly of macro-molecules and that are not organized as part of an enzymic pathway. One of the most studied inducible enzymes in bacteria is the β-galactosidase synthesized by many strains of *Escherichia coli*. This enzyme is capable of hydrolysing many β-galactosides,

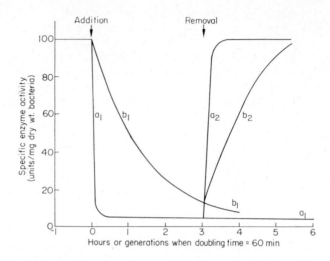

Fig. 2.6 The change of enzyme activity with time in a culture following the addition of (a_1) an inhibitor of enzyme activity, and (b_2) a repressor of enzyme formation. The second part of the graph shows the effect of removal of the compounds (a_2 and b_2).

including lactose, to yield free galactose, which is then further metabolized, in much the same manner as glucose can be by the pathways shown in Fig. 2.4, to provide energy and material for biosynthesis.

As the effect of both induction and repression is to modify the expression of genes, these types of regulation involve alterations in the rate of protein synthesis. The molecular processes underlying these types of regulation will not therefore be discussed further until the mechanism of protein synthesis has been described in detail (p. 67).

BIOSYNTHESIS OF MACROMOLECULES

General points

One feature common to the synthesis of all types of macromolecule, is that one of the reactants must provide the necessary energy for the polymerization step. Thus in the case of carbohydrate synthesis, energy must be provided for glycoside bond synthesis, with proteins the energy is needed for peptide formation, and with nucleic acids the polymerization step is the formation of a sugar-phosphate ester. In all cases the low molecular weight component of the reaction is the molecule activated to provide the energy. Thus monosaccharide nucleotides are the precursors of polysaccharides; amino acid precursors of proteins are in the form of mixed acid anhydrides with phosphoric acid; purine and pyrimidine bases for nucleic acid synthesis are in the form of nucleotide triphosphates.

The complexity of the biochemical processes underlying the synthesis of macromolecules is closely related, as may be expected, to the complexity of the macromolecules themselves. At one end of the scale, the biosynthesis of those polysaccharides that consist of a single type of monosaccharide joined together in a long chain involves, apart from the activation step, only a single enzyme acting repeatedly. Even with polysaccharides consisting of alternating monosaccharide units, biosynthesis can occur by the alternate action of two enzymes. But at the other end of the scale, the structure of proteins and nucleic acids is so complex that much more sophisticated methods of macromolecular synthesis must be used. In these cases it is not sufficient just to use the sequential action of enzymes on their own to ensure accurate synthesis of the product. Rather the enzymes have to act under the specification of a second molecule, often called a *template*, that carries information to ensure that the various enzymes operate in the correct place and in the correct order. Further details of these processes are given in the relevant sections below.

Biosynthesis of polysaccharides and related molecules

All polysaccharides are synthesized by the extension of a pre-existing polysaccharide chain by one monosaccharide unit (X) at a time, and the added monosaccharide unit enters the reaction in an activated form as a monosaccharide nucleotide, usually as the uridine-diphosphate derivative (UDP.X) but sometimes with other purine or pyrimidine nucleotides. The synthesis occurs according to the following generalized reactions:

$$...X.X.X.X.X.X.X.X.X + \text{uridine-diphosphate-}X$$
$$(n \text{ residues})$$
$$= ...X.X.X.X.X.X.X.X.X.X + \text{uridine-diphosphate}$$
$$(n+1 \text{ residues})$$

In the case of polysaccharides involving two alternating types of monosaccharide (X and Y) the generalized reaction occurs in two steps:

Step 1
$$...X.Y.X.Y.X.Y.X.Y.X.Y + \text{uridine-diphosphate-}X$$
$$= ...X.Y.X.Y.X.Y.X.Y.X.Y.X + \text{uridine-diphosphate}$$

Step 2
$$...X.Y.X.Y.X.Y.X.Y.X.Y.X + \text{uridine-diphosphate-}Y$$
$$= ...X.Y.X.Y.X.Y.X.Y.X.Y.X.Y + \text{uridine-diphosphate}$$

Not all polysaccharides consist of single unbranched chains of monosaccharide units. Some have mixtures of monosaccharides with N-acetyl-amino sugars and some of monosaccharides and uronic acids (Chap. 1). Such polymers are synthesized essentially as shown above, but the intermediates are the nucleotide derivatives of the relevant N-acetyl-amino sugars or uronic acids.

Considerable doubt exists as to the method of synthesizing branched polysaccharides. The chemical reactions involved are similar to those required for chain extension, but, as yet, there is little clear idea of how the branch points are inserted at specific points in the structure, if indeed they do occur specifically. In certain cases it has been shown that addition of a *primer molecule*, consisting of a small preformed piece of the polysaccharide to be synthesized greatly increases the rate of polysaccharide synthesis; and it may be that branch points are inserted into polysaccharides by the operation of some crude form of template, the position of the branching in the primer influencing the insertion of the branches in the new material.

Biosynthesis of lipids

Since some lipids have a more complex structure than polysaccharides, the synthetic mechanisms may be correspondingly more complex. With the synthesis of phosphatidic acids and the neutral triglycerides (Chap. 1) the situation is fairly straightforward. For these compounds the precursor of the pathway is phosphoglycerol and the synthetic steps are as follows where R_1 and R_2 are long chain fatty acids.

Step 1

$$\begin{array}{lll} CH_2OH & & CH_2O.OC.R_1 \\ \cdot & & \cdot \\ CHOH & + R_1CO.SCoA \longrightarrow & CHOH \quad + CoA.SH \\ \cdot & & \cdot \\ CH_2OPO_3H_2 & & CH_2OPO_3H_2 \end{array}$$

Step 2

$$\begin{array}{lll} CH_2O.OC.R_1 & & CH_2O.OC.R_1 \\ \cdot & & \cdot \\ CHOH & + R_2CO.SCoA \longrightarrow & CHO.OC.R_2 \ + CoA.SH \\ \cdot & & \cdot \\ CH_2OPO_3H_2 & & CH_2OPO_3H_2 \end{array}$$

Step 3

$$\begin{array}{lll} CH_2O.OC.R_1 & & CH_2O.OC.R_1 \\ \cdot & & \cdot \\ CHO.OC.R_2 & + R_1CO.SCoA \longrightarrow & CHO.OC.R_2 \ + H_3PO_4 + CoA.SH \\ \cdot & & \cdot \\ CH_2O.PO_3H_2 & & CHO.OC.R_1 \end{array}$$

In each of these reactions the energy for bond formation comes from the splitting of the —CO.SCoA bond of the substituted fatty acid.

The precursor of phosphatidyl inositol, phosphatidyl serine and phosphatidyl ethanolamine (see Chap. 1 for structures) is a phosphatidic acid of the type formed in Step 2 above. This phosphatidic acid is first converted to the cytidine-diphosphate (CDP) derivative and this is then converted to the inositol derivative by the reaction:

$$\text{CDP-diglyceride} + \text{inositol} \longrightarrow \text{phosphatidyl inositol} + \text{CMP}$$

and phosphatidyl serine is formed similarly:

$$\text{CDP-diglyceride} + \text{serine} \longrightarrow \text{phosphatidyl serine} + \text{CMP}$$

The details of the reaction leading to the formation of phosphatidyl serine are as follows:

$CH_2O.OC.R_1$

$CHO.OC.R_2$ $+ CH_2(OH).CH(NH_2).COOH \longrightarrow$

$CH_2O.P.P.$cytidine   serine
    CDP-diglyceride

$\qquad\qquad CH_2O.OC.R_1$

$\qquad\qquad CHO.OC.R_2$ $+ CMP$

$\qquad\qquad CH_2O.P.O.CH_2.CH(NH_2).COOH$
$\qquad\qquad\qquad$ phosphatidyl serine

Phosphatidyl ethanolamine is formed from phosphatidyl serine by the loss of CO_2.

Biosynthesis of nucleic acids

The biosynthesis of DNA and RNA is basically similar although some difficulties are introduced by the fact that DNA contains two complementary deoxyribonucleotide polymers while RNA is usually single stranded, at least in

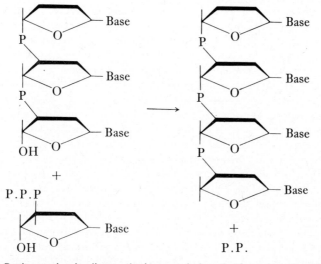

Fig. 2.7   Basic reaction leading to the increase in length of a polynucleotide chain, as found in RNA and DNA, by one unit.

the period immediately following synthesis (Chap. 1). In both molecules synthesis occurs by a step-wise extension of the molecule, the overall chemical change involved in an extension by one unit being as is shown in Fig. 2.7. In DNA the low molecular weight precursors (which also provide the energy for

the chemical bonds involved in the extension) are the *deoxyribo*nucleotide triphosphates of adenine, guanine, cytosine, and thymine, whereas *ribo*nucleotide
triphosphates of adenine, guanine, cytosine, and uracil are involved in RNA
synthesis.

Although the reactions in Fig. 2.7 show how the nucleotide backbone of RNA
and DNA can be extended, they do nothing to show the means whereby the
correct nucleotide bases are inserted in their correct positions in the growing
chain (for the sequence of purine and pyrimidine bases along the backbones of
RNA and DNA is of the greatest possible significance to the organism (see
Chap. 1 and p. 55). In practice DNA uses its double stranded nature to provide
its own information as to the order of insertion of bases, while RNA obtains this
information by using one of the two strands of DNA for the purpose. Thus
DNA provides its own template, while RNA uses a DNA template.

The insertion of the correct bases into the two growing chains of a DNA
molecule occurs as shown in Fig. 2.8. The process depends greatly on the
'base-pairing' that exists between the complementary bases in the DNA double-
strand. As the original molecule always contains an adenine residue opposite a
thymine and a cytosine opposite a guanine (Fig. 2.8a) disruption of the base
pairing and separation of the two strands allows the synthesis of two new DNA
strands using the base sequence of the old strands as a guide (Fig. 2.8b). In this
way the new strands are built up, strand *a* having complementary base pairing
with the parent strand *A*, and strand *b* with complementary base pairing to
strand *B*. When this step-wise process is complete, the end product is two
double-stranded DNA molecules each with one 'old' and one 'new' strand
(Fig. 2.8c). This type of DNA replication, the only type found *in vivo*, is called
semi-conservative replication and is carried out by an enzyme known as the
DNA-primed DNA polymerase, or often just *DNA polymerase* for short.

Synthesis of RNA bears many similarities to DNA synthesis except that the
precursor molecules are different (see above) and that the base sequence of a
DNA strand, rather than the RNA itself, provides the template for the insertion
of the correct base in the correct position in the growing RNA chain. For this
purpose, it is assumed that the base-paired double strand of DNA must open,
at least temporarily, in a manner analogous to that occurring in DNA synthesis,
and, according to this hypothesis, therefore, RNA synthesis occurs by the
mechanism shown in Fig. 2.9. Once more the ordering of the incoming bases
depends greatly upon the base-pairing properties of purine and pyrimidine
bases, although in this case the pairing is between DNA adenine and RNA
uracil, between DNA thymine and RNA adenine and between DNA or RNA
guanine and RNA or DNA cytosine (Fig. 1.19). The enzyme that carries out this
process is known as DNA-primed RNA polymerase, or just *RNA polymerase* for
short. Where double-stranded RNA molecules are required, these are synthesized as single strands in the first instance and the double-stranded structures
are formed subsequently by the formation of the appropriate base pairs. In many
cases RNA molecules are *internally* base-paired; that is the molecule bends
back on itself to form a structure in which a range of bases from one region of
the strand pair with others from a different region. As an example see the
structure of t-RNA (p. 32).

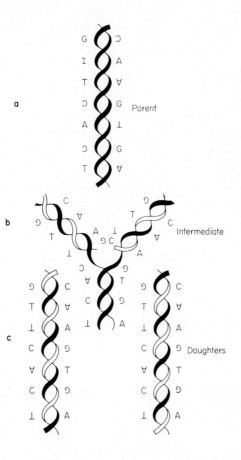

Fig. 2.8 Biosynthesis of DNA. (a) parental structure. (b) Intermediate stage with the two new strands being formed on the complementary old strands. (c) Final product consisting of two replicas of the original. Note that each replica consists of one 'new' (white) and one 'old' (black) strand.

Biosynthesis of proteins

Protein synthesis is best considered in three sections corresponding to the formation of the primary, secondary, and tertiary structure of the molecule (Chap. 1). The synthetic processes involved in these three steps are very different: the first consists of the means whereby the amino acids are activated and inserted to form a single polypeptide chain in which each component amino acid occupies its correct position in the chain. By far the majority of the experimental evidence available concerns this step. The second concerns the synthesis of the —S.S— bridges that link the different polypeptide chains together, and the third is the process whereby the newly synthesized structure is folded up to make the biologically functional molecule.

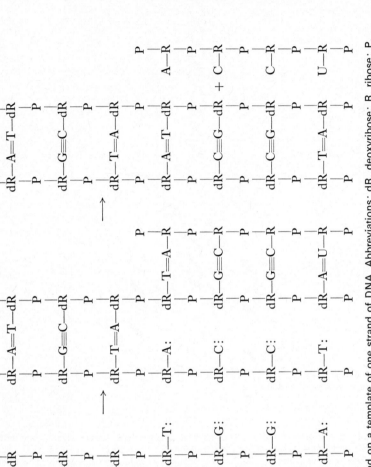

Fig. 2.9 Biosynthesis of an RNA strand on a template of one strand of DNA. Abbreviations: dR, deoxyribose; R, ribose; P, phosphate; A, adenine; G, guanine; C, cytosine; U, uracil; T, thymine. Symbols = and ≡ are hydrogen bonds. '.' and ':' are potential hydrogen bonds.

Biosynthesis of the polypeptide chains

Although up to twenty different amino acids may be found in bacterial proteins, the bond linking them all is the same—the peptide bond. Consequently, two problems exist in the formation of the primary structure of proteins:

1. How is the peptide bond synthesized?
2. How is the order of the twenty different component amino acids in the polypeptide chain specified?

As is the case with the formation of all types of bonds between two molecules in biological systems, activation of one component of the reaction mixture is necessary for bond formation to occur. In principle, *although it must be stressed that the exact details are somewhat different in practice*, the synthesis of the peptide bond takes place according to the following basic reaction:

$$...HN.CHR_1.CO.X + H_2N.CHR_2.COOH$$
$$= \cdots HN.CHR_1.CONH.CHR_2.COOH + X.OH$$

where X is a derivative that activates the carboxyl group of the amino acid, thus facilitating combination with the —NH_2 group of the second amino acid to form the peptide bond.

In practice the situation is made slightly more complex by the fact that the incoming amino acid for peptide bond synthesis is already activated for the formation of the *next* peptide bond. So the reaction sequence is as follows, where the symbol ':' indicates an activated bond.

Step 1
$$...HN.CHR_1.CO:X_1 + H_2N.CHR_2.CO:X_2$$
$$= ...HN.CHR_1.CONH.CHR_2.CO:X_2 + X_1.OH$$

Step 2
$$...HN.CHR_1.CONH.CHR_2.CO:X_2 + H_2N.CHR_3.CO:X_3$$
$$= ...HN.CHR_1.CONH.CHR_2.CONH.CHR_3.CO:X_3 + X_2.OH$$

Step 3
$$...HN.CHR_1.CONH.CHR_2.CONH.CHR_3.CO:X_3 + H_2N.CHR_4CO:X_4$$
$$= ...HN.CHR_1.CONH.CHR_2.CONH.CHR_3.CONH.CHR_4.CO:X_4$$
$$+ X_3.OH$$

etc., etc.

Thus the activation energy is derived from the activated state of the last amino acid of the growing chain and not from the activated incoming amino acid, and the chain is synthesized from the amino to the carboxyl terminus.

The step-wise reactions shown above account for the synthesis of all the peptide bonds of the polypeptide chain. There are, however, two special steps that must be considered: the first and the last. The first step—namely the formation of a dipeptide from the first amino acid—cannot occur unless the amino group of the first amino acid is substituted. In bacteria, though not in mammalian cells, the first amino acid of a growing polypeptide chain is always the substituted amino acid *N*-formyl-methionine (Fig. 2.10). In this molecule the free —NH_2 group of methionine is blocked with a —CHO residue. After the polypeptide chain has been synthesized and liberated from the synthetic site, enzymes exist that split off the whole N-formyl-methionine residue from the chain, so the ultimate N-terminal amino acid of a bacterial polypeptide chain is the second residue in the order of synthesis.

Less is known about the mechanism of chain termination. Probably all that happens is that no amino acid is presented for insertion and the chain, as synthesized up to that point, is liberated from the synthetic site but this matter will be discussed in more detail later.

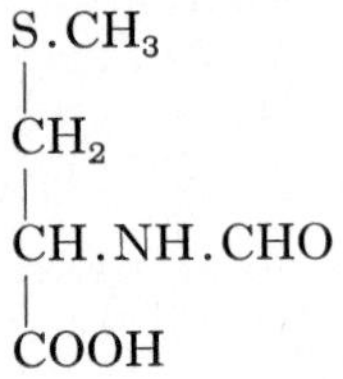

$$S.CH_3$$
$$|$$
$$CH_2$$
$$|$$
$$CH.NH.CHO$$
$$|$$
$$COOH$$

Fig. 2.10   Structure of N-formyl-methionine.

The ordering of amino acids in polypeptide synthesis

Although these reactions allow the synthesis of a polypeptide chain they do nothing to decide the order of insertion of the different amino acids in their correct position in the chain. This step is achieved by allowing polypeptide synthesis to occur on a molecular template. In the case of RNA synthesis, it may be recalled, the molecular template used was a DNA strand of complementary base sequence—but in protein synthesis the operation of the template is a great deal more complex. In this case the template molecule is a single strand of RNA, and the order of insertion of amino acids is determined by the order of the purine and pyrimidine bases along the polyribophosphate backbone of the RNA strand. Thus an essentially linear molecule, a strand of RNA, specifies the structure of a second essentially linear molecule, the polypeptide chain.

Since there are only four different types of purine and pyrimidine bases in RNA, but 20 different types of amino acid in protein, it is clear that the base sequence of individual purines and pyrimidines along the RNA strand cannot unambiguously determine the order of insertion of amino acids into the growing polypeptide chain. For specification to be unambiguous, a sequence of at least three purine or pyrimidine bases is needed for each amino acid; but on the other hand with this number there are too many possibilities and there would be more than one group of three bases available for each amino acid. In fact, if there is a 'direction of reading' of the bases (that is if the base order adenine-guanine-cytosine for example can be distinguished from the sequence cytosine-guanine-adenine) then there will be 64 distinct triplets of purines and pyrimidines available to specify the insertion of 20 amino acids.

Experimental investigation of this problem, the so-called feat of 'cracking the genetic code', shows that 61 of the 64 available triplets (or RNA *codons* as they are called) 'code' for the insertion of individual amino acids during protein synthesis, while the remaining three triplets are concerned with bringing the synthesis of a given polypeptide chain to an end in one way or another. The number of different codons that specify the insertion of each type of amino acid is different. Three of the amino acids can be specified by any six separate triplets. Similarly five of the amino acids each have four specific codons, one has three, nine have two, while two are specified by a single codon only. This variable relationship between the nature of an amino acid and the number of distinct codons that can specify its insertion into the growing polypeptide chain may account, to some extent at least, for the unequal distribution of the various amino acids in proteins. The full list of triplet attributions is given in Fig. 2.11.

2nd base	U	C	A	G	
1st base					3rd base
	Phe	Ser	Tyr	Cys	U
U	Phe	Ser	Tyr	Cys	C
	Leu	Ser	CT	CT	A
	Leu	Ser	CT	Trp	G
	Leu	Pro	His	Arg	U
C	Leu	Pro	His	Arg	C
	Leu	Pro	Gln	Arg	A
	Leu	Pro	Gln	Arg	G
	Ileu	Thr	Asn	Ser	U
A	Ileu	Thr	Asn	Ser	C
	Ileu	Thr	Lys	Arg	A
	Met	Thr	Lys	Arg	G
	Val	Ala	Asp	Gly	U
G	Val	Ala	Asp	Gly	C
	Val	Ala	Glu	Gly	A
	Val	Ala	Glu	Gly	G

Fig. 2.11 The key to the genetic code. The RNA bases adenine, guanine, cytosine and uracil are referred to as A, G, C, and U. The abbreviation CT stands for 'chain termination'—see text.

The triplet GUG is responsible for the insertion of valine unless the amino acid is the first in the chain. When this is so, the same triplet inserts N-formyl-methionine rather than valine. Similarly the codon AUG inserts methionine unless it is the first triplet in the message when N-formyl-methionine is inserted instead.

Co-linearity of DNA, RNA, and polypeptide chain

Fig. 2.12 is a diagrammatic summary of the mechanism of protein synthesis as described up to this point. The linear sequence of purine and pyrimidine bases along the RNA backbone specifies the linear order of amino acids in the polypeptide chain, three bases (known as the *RNA codon*) specifying the insertion

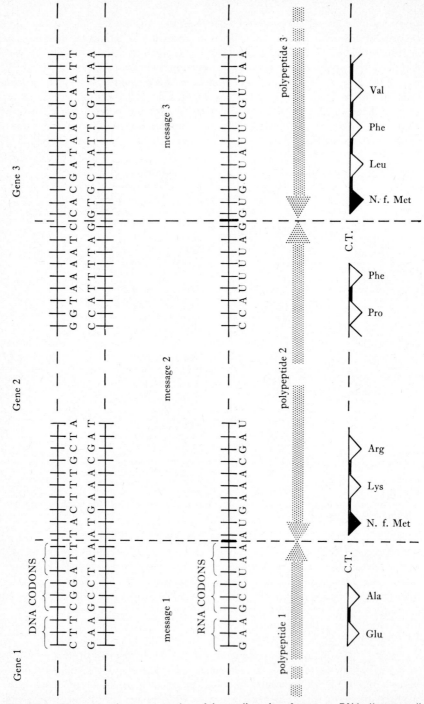

Fig. 2.12 Diagrammatic representation of the co-linearity of gene, m-RNA, ('message') and polypeptide chain. (N. f. Met = *N*-formyl-methionine; C. T. = chain termination)

of each amino acid. Fig. 2.12 also shows that a linear relationship exists between the base sequence of RNA and of one strand of DNA, since, as described earlier (p. 50), RNA is synthesized on a template of DNA. Three DNA bases (called the *DNA codon*) therefore specify the sequence of the RNA codon. However, because of the mechanisms involved, the DNA codon always has the *complementary* base sequence to that of the RNA. Thus, for example, the DNA codon AAG specifies the RNA codon UUC, which, in turn, specifies insertion of serine into the polypeptide chain.

The relationships shown in Fig. 2.12 are often summarized by saying that there is *co-linearity* between a stretch of DNA and a given polypeptide chain; or in more general terms, that there is *co-linearity between gene and enzyme*. This co-linearity of the molecules involved forms the biochemical basis of the observation that the order of mutations in a gene as determined by mapping techniques (see Chap. 4) is linearly represented by the locations of the substituted amino acids in the mutated polypeptide chain. Fig. 2.13 shows the genetic locations of certain mutations in the *trpB* gene in *Escherichia coli* and their effect on the amino acid sequence of the B-component of the tryptophan synthetase enzyme that the gene specifies.

Because of its function as a transmitter of information, the RNA molecule that acts as the intermediate between the gene and the polypeptide chain is usually known as *messenger RNA* or *m-RNA* (Fig. 2.12).

The polypeptide product of a gene is synthesized initially as a long straight-chain molecule but after synthesis this structure folds up (p. 23). Generally the polypeptide products of a single gene appear to fold up in the same way and this is a property of the amino acid sequence of the polypeptide and is not imposed by the action of another gene distinct from the one that was responsible for ordering of the amino acids in the polypeptide chain.

The role of transfer RNA

Although the steps outlined in Fig. 2.12 show how the order of amino acids in a polypeptide chain may be specified by the base sequence of the gene, they do little to explain, in molecular terms, how three purine or pyrimidine bases can specify the insertion of a single amino acid. This task is performed by a class of RNA molecules, quite distinct from messenger RNA, known as *transfer RNA* or *t-RNA*.

Structure of t-RNA's

Although a large number of types of transfer RNA molecules are found in bacterial cells (there are often more than one to every RNA codon), enough is known of their structure to be fairly certain that they are all relatively minor variants of the same basic type of structure. Fig. 2.14 shows, as an example, the structure of one of the alanine transfer RNA's from yeast, and Table 2.1 gives some details of how other t-RNA's that have been purified compare in structure. All t-RNA's have a polyribophosphate chain such as is found in other RNA's (p. 30) and in the case of the alanine t-RNA the total base sequence amounts to 77 residues. All the t-RNA's examined so far (see Table 2.1) have a

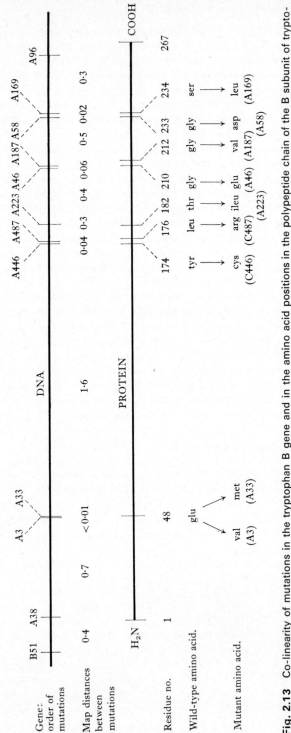

Fig. 2.13 Co-linearity of mutations in the tryptophan B gene and in the amino acid positions in the polypeptide chain of the B subunit of tryptophan synthetase. (Modified from Yanofsky. C. (1967). Gene structure and protein structure. *Scient. Am.*, **216**, No. 5, 80–94)

$$\overset{\displaystyle Me}{|}$$

P.G.G.G.C.G.U.G.U.G.G.C.G.C.G.U.A.G.U.C.G.G.U.A.G...

with H, H, H, H connections shown above: U.A.G.U / C.G.G.U

...C.G.C.G.C.U.C.C.C.U.U.I.G.C.I.Ψ.G.G.G.A.G.A.G.U...

with Me, Me above C.G.C.G and Me above U.I.G, and 36 37 38 marked below I.G.C

...C.U.C.C.G.G.T.Ψ.C.G.A.U.U.C.C.G.G.A.C.U.C.G.U.C...

...C.A.C.C.A.C.C.A.OH

A: Adenlyic acid	Di-H-U: 5,6,dihydrouridylic acid
C: Cytidylic acid	Di-Me-G: dimethyl guanylic acid
G: Guanylic acid	Me-G: 1-methyl guanosine
I: Inosinic acid (hypoxanthine ribotide)	Me-I: methyl inosinic acid
T: Thymine ribotide	Ψ: pseudo-uridylic acid
U: Uridylic acid	

Fig. 2.14   The structure of one of the alanine transfer RNAs from yeast. (See also Fig. 1.21)

Table 2.1   A comparison of the structures of t-RNAs from yeast. (A: alanine t-RNA; B: tyrosine t-RNA; C: phenylalanine t-RNA)

	A	B	C
Substrate	Alanine	Tyrosine	Phenylalanine
No. of bases	77	78	75
Type of abnormal bases			
methyl-adenine	–	1	1
thymine	1	1	1
methyl-cytosine	–	1	2
7-methyl-guanine	–	–	1
O-methyl-guanine	–	1	1
O-methyl-cytosine	–	–	1
di-methyl-guanine	1	1	1
dihydro-uracil	2	6	2
N-methyl-adenine	1	1	1
inosine	–	1	–
methyl-inosine	1	–	–
pseudo-uracil	1	–	–
Anticodon	I.G.C.	G.pseudoU.A.	O-methylG.A.A.

Note: Inosine base pairs like guanine, pseudo-uracil like uracil and O-methyl-guanine like guanine.

closely similar number of residues and, consequently t-RNA molecules are among the smaller RNA molecules encountered naturally. The terminal sequence of alanine-t-RNA is cytidyl-cytidyl-adenosine or —CpCpA.OH and this sequence is common to *all* t-RNA's yet examined.

Unlike other types of RNA, individual types of t-RNA molecule contain a variable number of 'abnormal' or 'odd' purine and pyrimidine bases such as hypoxanthine, 5,6 dihydro-uracil, dimethyl-guanine, and methylguanine, for example (see Fig. 2.15). The purpose of these substituted bases is unknown as yet. Another common constituent of t-RNA's, though not of other RNA's, is pseudouracil riboside (Fig. 2.15). This molecule seems to occur frequently at a certain point in the generalized t-RNA structure and its possible function will be discussed below.

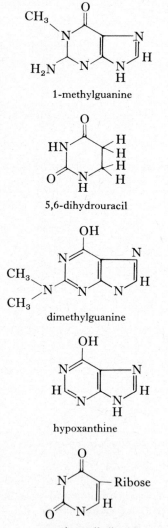

1-methylguanine

5,6-dihydrouracil

dimethylguanine

hypoxanthine

pseudouracil riboside

Fig. 2.15 The structure of a number of 'abnormal' bases found in some types of RNA.

Function of t-RNA's

As implied in the reactions shown in Fig. 2.12, the function of t-RNA molecules is to interact both with the amino acid and with the RNA codon on the messenger RNA so that the order of bases on the 'message', which is a copy of the gene, ensures that the amino acid is inserted into its correct position in the growing polypeptide chain. The two parts of the basic t-RNA structure involved in these functions can be identified with some certainty.

The part of the t-RNA molecule responsible for recognizing the RNA codon is called the *anticodon*. It consists of a sequence of three bases in the t-RNA sequence that are complementary in hydrogen bonding properties to the three bases of the RNA codon. Thus in the alanine t-RNA whose structure is shown in Fig. 2.14 the anticodon is the base sequence IGC at residues 36-38 of the structure, and these three bases interact by complementary base pairing with GCU, one of the RNA codons for alanine. The part of the t-RNA molecule that interacts with the amino acid is the final adenine residue of the —CpCpA.OH sequence at the 3′OH terminal of the molecule. The amino acid interacts with this sequence to form a mixed acid anhydride with the phosphate residue on the terminal adenine (Fig. 2.16) in all t-RNA molecules. This bond serves two

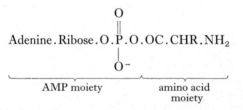

Fig. 2.16 An activated amino acid in the form of a mixed acid anhydride with AMP.

purposes: first, it attaches the amino acid to the t-RNA molecule, but secondly its nature ensures that the amino acid is held in an activated state so that a peptide bond may be formed if a suitable free —NH$_2$ group is available.

In practice the formation of the mixed acid anhydride between the amino acid and the terminal adenylic acid of the t-RNA does not occur directly but involves ATP which acts as a carrier. The reaction, therefore, proceeds in two steps, as follows, where the symbol ':' denotes an activated bond:

Step 1

$$H_2N.CHR.COOH + ATP = H_2N.CHR.CO:AMP + PP$$

Step 2

$$H_2N.CHR.CO:AMP + t\text{-RNA.cytidyl.cytidyl.adenyl.OH}$$
$$= H_2N.CHR.CO:P.adenyl.cytidine.cytidyl.t\text{-RNA} + AMP$$

The net effect of these two reactions is that the amino acid is first converted to a free adenyl derivative and then transferred from this situation to become an adenyl derivative at the 3′OH terminal of one or other of its specific t-RNA's.

Synthetase enzymes

Since all t-RNA's end in the sequence CpCpA.OH, and consequently lack specificity in this part of their structure, some means must be available to ensure that each amino acid is loaded on to its correct t-RNA. This process is carried out by a series of *synthetase enzymes*, one for each type of t-RNA, that have specificity both for the amino acid residue and for some portion of the t-RNA molecule that does not include either the anticodon, or the —CpCpA.OH terminal sequence. Investigation of the nature of this enzyme has shown that as well as ensuring that the correct amino acid is loaded on to a given t-RNA molecule, it is also the catalyst responsible for both the reactions shown in Steps 1 and 2 above. So closely integrated is the function of this enzyme in loading amino acids on to t-RNA that the intermediate amino acid adenylate ($H_2N.CHR.CO.AMP$—the product of Step 1) is never liberated from the surface of the enzyme, but is immediately used for Step 2 of the reaction.

The overall mechanism of protein synthesis

Now that the multifaceted structure of the t-RNA molecule has been described in some detail, it is possible to give a complete account of the process of poly-peptide bond formation. This is shown diagrammatically in Fig. 2.17. The process involves two essentially separate series of reactions that meet with the involvement of the t-RNA molecule. The first series achieves the activation of the amino acid in the form of a loaded t-RNA molecule, by means of the steps shown in Steps 1 and 2 on p. 53. All the 20 types of different amino acid found in protein are activated in this way and each is loaded on to one of several distinct t-RNA's specific for the amino acid in question.

The second series of reactions involved consists of the formation of messenger RNA on a template of DNA (see Figs. 2.9, 2.12 and 2.17). This process results in the mobilization of the information inherent in the base sequence of the DNA and its dispatch, in the form of a series of messenger RNA molecules, to the site of protein synthesis in the cell.

These two series of reactions impinge when the activated amino acid-t-RNA complexes are provided for the formation of the polypeptide chain, the order in which they are used being determined by the order of coding triplets in the messenger RNA (Fig. 2.17). After the loaded t-RNA has been drawn to its correct position for peptide synthesis by the interaction of codon and anticodon, and after the peptide bond has been formed by the reactions shown in Fig. 2.18 (but see also p. 53), the free t-RNA is released from the site of peptide synthesis until it has been recharged with another amino acid residue. In this way the amino acids are inserted into their correct locations in the growing polypeptide chain, starting with an *N*-formyl-methionine residue and continuing until one of the codons UAA, UAG, or UGA is encountered in the messenger RNA. At this point no amino acid is inserted and the polypeptide chain is released complete.

As far as is known, all polypeptide chains in the bacterial cell are synthesized by this means, all the information required for determining the amino acid sequence and length of the structure being stored ultimately in terms of the base sequence of the DNA.

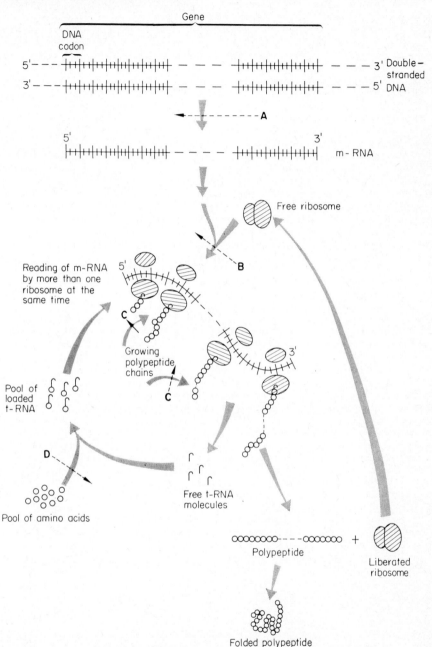

Fig. 2.17 Diagrammatic representation of the overall process of protein synthesis. The dotted arrows indicate the sites of inhibition of certain antibiotics. (**A**—actinomycin D, rifamycin; **B**—streptomycin, erythromycin, tetracycline; **C**—puromycin, chloramphenicol; **D**—some structural analogues)

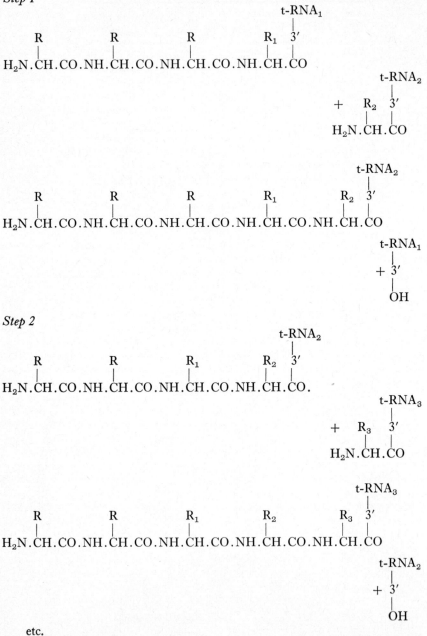

Fig. 2.18 Molecular interactions involved in the lengthening of a polypeptide chain by two units.

RIBOSOMES

As might be expected from the complexity of the reactions involved, protein synthesis does not occur in free solution in the bacterial cells but occurs on the surface of structures—they might even be thought of as organelles—known as ribosomes (see Chap. 9). Ribosomes are composed of protein and RNA (a different type from both t-RNA and m-RNA) in approximately equal proportions. Little is known, as yet, of the exact function of the many proteins and enzymes present in ribosomes, nor is the role of the RNA part of the structure clear. All that appears certain at the moment is that the RNA messenger is dispatched from its site of synthesis on the DNA template to the ribosome, and it is on the surface of these structures that the interaction with the loaded t-RNA molecules occurs. The nascent polypeptide chain remains attached to the ribosome—at least at one point on its length—until a chain-terminating triplet is encountered when the complete polypeptide is liberated free into the cytoplasm of the cell.

In practice it appears that a single messenger RNA molecule can be used a number of times before it is inactivated or destroyed. This can be shown experimentally by the kinetics of protein synthesis and also by the fact that it is possible to isolate messenger RNA molecules with more than one ribosome attached at intervals (see Fig. 2.17). Both this observation, and the size of the message in relation to the size of the ribosome, suggest that the part of the message actively involved in peptide synthesis moves in relation to the ribosome; and where there are a number of ribosomes attached to a single messenger, these structures follow one another along the RNA at regular intervals each producing a polypeptide chain. The structures in which a series of ribosomes are found attached to a thread of messenger RNA are called *polysomes*, and isolation of ribosomes from living cells often gives preparations in this form particularly if care is taken not to destroy RNA in the isolation procedure.

It is not clear, at this stage in our knowledge, what it is that determines the longevity of the messenger RNA molecule: some types of message certainly seem to survive longer than others.

FORMATION OF SECONDARY AND TERTIARY PROTEIN STRUCTURES

The reactions described in some detail in the foregoing sections are involved solely in the synthesis of the polypeptide chain using the information provided by the gene for the purpose. There still remains to be described the formation of the —S.S— bridges between cysteine residues and also the means whereby the protein molecule, with its cross-bridges formed, folds up to become the biologically active molecule.

The basic reaction involved in the cross-linking reaction is a reduction and may be written as follows:

$$R_1.SH + R_2.SH = R_1{-}S{-}S{-}R_2 + 2H$$

All such bonds are formed between pairs of cysteine residues in the polypeptide chains, although not all cysteine residues in the protein are involved in such linkages, nor are the bridges always formed between cysteine residues from different chains.

The formation of —S.S— bridges has only been studied to a limited extent in natural systems—largely because of the difficulty of getting the synthesis of whole proteins to continue for any length of time *in vitro*. Where the reaction has been investigated, an enzyme seems to be needed to form the bridge, but whether each cross bridge requires its own specific enzyme is not known as yet.

Very little is known about the folding up of proteins after their synthesis; it is usually assumed that it is spontaneous and dependent on the order and type of amino acids in the polypeptide chains. This view is supported to some extent by experiments that show that some enzymes unfold in some types of reagent (e.g. strong urea) but can refold, to recover full enzymic activity, when the reagent is removed.

Undoubtedly a further stage is important in the biosynthesis of most proteins at least when the process is examined in the framework of a growing cell. This is the means whereby the proteins are inserted into their correct positions in the molecular architecture of the cell. A simple example is the synthesis and arrangement of flagellin in bacterial flagella. Practically nothing is known about how the molecules are arranged to give the flagellum its characteristic properties and molecular pattern.

REGULATION OF THE RATE OF PROTEIN SYNTHESIS

A number of methods whereby the level of gene expression can be regulated has already been outlined on p. 44. Basically these are of two kinds: either the action is to modify the activity of existing enzymes or to influence the rate of synthesis of new ones. Only methods of affecting enzyme synthesis will be discussed here: the regulation of enzyme activity has already been dealt with in detail on p. 45.

In the preliminary discussion of enzyme regulation (see p. 44) it was pointed out that the rate of synthesis of a gene product was under two types of control—one that 'set the pace' of gene expression and which was unaffected by alterations in the composition of the growth medium and the other (induction and repression) in which alterations in the composition of the medium immediately influenced the rate of synthesis of the product of individual genes or groups of genes.

The molecular interrelationships described in Fig. 2.12 show that messenger RNA occupies a key intermediate position between the gene and its final protein product. Consequently any alteration, either in the rate of formation of this molecule or its 'translation' into protein, will, in turn, influence the rate of expression of the gene, and it is in this region of the overall protein synthetic process that the regulatory factors exert their effect rather than by limiting the availability of amino acids, or energy, or any other potentially limiting process.

'Setting the pace' of gene expression

Genetic experiments have shown that a region exists (called the *promoter* region or promoter gene) that can affect the rate of expression of a related gene or operon (for a definition of this term, see Chap. 4). The region usually seems to lie close to the structural gene (or group of genes) that it affects, but, as yet, it is unclear whether this genetic region restricts the rate of synthesis of messenger RNA or its translation into protein. One possibility is that the region determines the ease with which RNA polymerase can bind to the relevant section of DNA. In this way the region could indirectly influence the rate of synthesis of the appropriate messenger RNA. Whatever the mechanism, however, it seems that this type of region is responsible for adjusting the output of genes, in terms of their protein product, to the long-term requirements of the cell, and it is this type of mechanism that seems to be responsible for the unequal quantities of material synthesized by different genes, despite the fact that only a single copy of each gene may be present in the cell.

The molecular basis of induction and repression

The molecular processes underlying the alteration in expression of a gene when low molecular weight inducers and repressors are present in the growth medium, have been worked out in detail for a number of bacterial enzymes. In all cases the effect of the regulatory molecule is to alter the rate of synthesis of messenger RNA and hence of the synthesis of protein from a given gene or group of genes. For this process to occur, a second gene, or group of genes, known as *regulatory genes* must be present in the cell. The way in which these genes, or to be precise, their products, influence the synthesis of individual proteins was elucidated first for inducible β-galactosidase synthesis in *Escherichia coli* by the French workers Jacob and Monod and has become known as the Jacob/Monod hypothesis. Application of the tenets of this hypothesis to other bacterial inducible and repressible enzyme systems suggests that the elements of the hypothesis may apply widely (though not perhaps universally) but that the regulation of certain enzymes may be a good deal more complex than that of β-galactosidase in *Escherichia coli*.

The main element of the Jacob/Monod hypothesis, as applied to an inducible system, is that a protein product of one type of regulatory gene (called the inducibility or *i*-gene) is produced continuously and almost completely blocks the synthesis of messenger RNA by the structural gene (or genes) whose synthesis is being regulated (see Fig. 4.12). This restriction in messenger synthesis is caused by the interaction of the *i*-gene product with a second regulatory region (known as the *operator*) that lies close to the structural gene (see Fig. 4.12). Thus when the *i*-gene product is interacting with the operator, the level of expression of the structural gene is reduced to a very low level, known as the *basal level* or *basal rate* (see also p. 45). Addition of an inducer to the culture, however, leads to an immediate increase in the rate of specific protein synthesis by the structural gene. This seems to occur because the inducer combines with the *i*-gene product to prevent it complexing with the operator, thus allowing the structural gene to be expressed at its full rate (Fig. 4.12). Removal of the inducer should return the rate of enzyme synthesis to the basal rate, and this is what is observed experimentally.

In repressible systems a converse system operates. Here the regulatory gene (often called the R—or repressibility—gene) synthesizes a product that only blocks expression of the structural gene when in combination with a low molecular weight repressor molecule. In the absence of this low molecular weight repressor enzyme synthesis occurs at the full rate despite the presence of the R-gene product.

A number of relatively crucial elements in the Jacob/Monod hypothesis have only recently been verified experimentally. For example, it is now certain that the i-gene product (in the case of β-galactosidase synthesis in *Escherichia coli* at least) is a protein and that the operator region is part of the DNA of the bacterial genome that is abnormal in that it appears to make no messenger RNA or protein product. Both these conclusions were predicted by Jacob and Monod. It is still by no means clear, however, by what means the combination of the i-gene product with the operator influences the rate of messenger synthesis by the structural gene.

Although the biochemical evidence in favour of the Jacob/Monod hypothesis has appeared only recently, the original hypothesis was first put forward in 1961. In its original form, the hypothesis was based almost entirely on inferences to be drawn from purely genetical experiments. This type of evidence and its interpretation is described in detail in Chapter 4.

MUCOPEPTIDE BIOSYNTHESIS

Compared with the efforts that have been made by experimental workers in the field of protein synthesis, relatively little has been done to elucidate the mechanism of mucopeptide synthesis. Nevertheless quite a lot is known about the latter process.

As might be expected from the discussion of the mechanism of polysaccharide synthesis above, the polysaccharide chain of mucopeptide is built up by the interaction of the relevant monosaccharides after their activation to form the uridine diphosphate derivatives. Therefore this part of the synthesis is, in principle, identical with the two sets of reactions shown in Steps 1 and 2, p. 47, for the synthesis of a polysaccharide of alternating subunits. But the process of mucopeptide synthesis is unusual in that the peptide side-chain is added to the muramic acid residue before the substituted muramic acid is used for polysaccharide synthesis. And, moreover, the peptide is built up stepwise and not added to the muramic acid as a preformed peptide. As a consequence the biosynthesis of even a small part of the great sack-shaped molecule of mucopeptide (see p. 33) involves many steps, as follows, where UDP is uridine diphosphate, GlcNAc is N-acetyl-glucosamine, and MurNAc is N-acetyl-muramic acid:

Step 1   Addition of alanine.

Uridine-diphosphate-MurNAc + L-ala = Uridine-diphosphate-MurNAc

$$\text{L-ala} + H_2O$$

Step 2   Addition of glutamic acid.

$$\begin{array}{c} \text{UDP-MurNAc} \\ | \\ \text{L-ala} \end{array} + \text{D-glu} = \begin{array}{c} \text{UDP-MurNAc} \\ | \\ \text{L-ala} \\ | \\ \text{D-glu} \end{array} + H_2O$$

Step 3   Addition of lysine.

$$\begin{array}{c} \text{UDP-MurNAc} \\ | \\ \text{L-ala} \\ | \\ \text{D-glu} \end{array} + \text{L-lys} = \begin{array}{c} \text{UDP-MurNAc} \\ | \\ \text{L-ala} \\ | \\ \text{D-glu} \\ | \\ \text{L-lys} \end{array} + H_2O$$

Step 4   Addition of *two* molecules of alanine together as a dipeptide.

$$\begin{array}{c} \text{UDP-MurNAc} \\ | \\ \text{L-ala} \\ | \\ \text{D-glu} \\ | \\ \text{L-lys} \end{array} + \text{D-ala-D-ala} = \begin{array}{c} \text{UPD-MurNAc} \\ | \\ \text{L-ala} \\ | \\ \text{D-glu} \\ | \\ \text{L-lys} \\ | \\ \text{D-ala} \\ | \\ \text{D-ala} \end{array} + H_2O$$

Step 5   Formation of the C1 → C4 glycoside bond.

$$\text{...GlcNAc} + \begin{array}{c} \text{UDP-MurNAc} \\ | \\ \text{L-ala} \\ | \\ \text{D-glu} \\ | \\ \text{L-lys} \\ | \\ \text{D-ala} \\ | \\ \text{D-ala} \end{array} = \begin{array}{c} \text{...GlcNAc.MurNAc} \\ | \\ \text{L-ala} \\ | \\ \text{D-glu} \\ | \\ \text{L-lys} \\ | \\ \text{D-ala} \\ | \\ \text{D-ala} \end{array} + \text{UDP}$$

Step 6   Formation of the C4 → C1 glycoside bond.

$$\begin{array}{c} \text{...GlcNAc.MurNAc} \\ | \\ \text{L-ala} \\ | \\ \text{D-glu} \\ | \\ \text{L-lys} \\ | \\ \text{D-ala} \\ | \\ \text{D-ala} \end{array} + \text{UDP-GlcNAc} = \begin{array}{c} \text{...GlcNac.MurNAc.GlcNAc...} \\ | \\ \text{L-ala} \\ | \\ \text{D-glu} \\ | \\ \text{L-lys} \\ | \\ \text{D-ala} \\ | \\ \text{D-ala} \end{array} + \text{UDP}$$

At this stage it is worth noticing the following points:

1. As mentioned above, the UDP-MurNAc is loaded stepwise with amino acids before the glycoside bond is formed at Step 5.
2. The last two D-alanine residues are added as a dipeptide rather than singly.
3. The final polypeptide side chain on the muramic acid has the overall composition L-ala–D-glu–L-lys–D-ala–D-ala; that is with an extra D-alanine at the end of the chain when compared with the final structure in the complete molecule (see p. 35).

Step 7   Formation of the cross bridges.

At this stage two units as produced by Step 6 are joined together by a penta-glycine bridge. The bridge is formed between the ε-amino group of a lysine residue and the carboxyl group of the *penultimate* D-alanine of the muramic acid side chain. In the process the terminal D-alanine from one side chain is liberated.

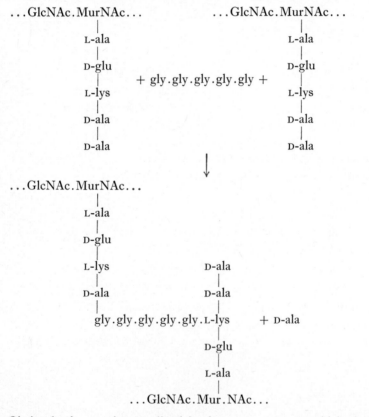

Obviously the reactions outlined in these seven steps could lead to a large molecule—particularly if the cross linking occurs between many chains to form a three-dimensional structure rather than to form a linear or sheet-like molecule. However, practically nothing is known of the detailed steps by which the prototype molecule resulting from Step 7 is interlinked to form the final macro-molecule. In particular, for example, it is not known at what stage in the overall formation of the mucopeptide that the cross-linking reaction occurs, beyond the fact that it occurs after the formation of considerable lengths of polysac-charide chain. Nor is it clear whether all the side chains to the muramic acids are cross linked by bridges to other side chains or whether some remain free. It is undoubtedly true that some of the polypeptide side chains to the muramic acid residues are free since an enzyme exists in the cell that can remove the terminal alanine residue from the chain rather than catalysing its cross linkage to another side chain. The overall reaction catalysed by this enzyme is:

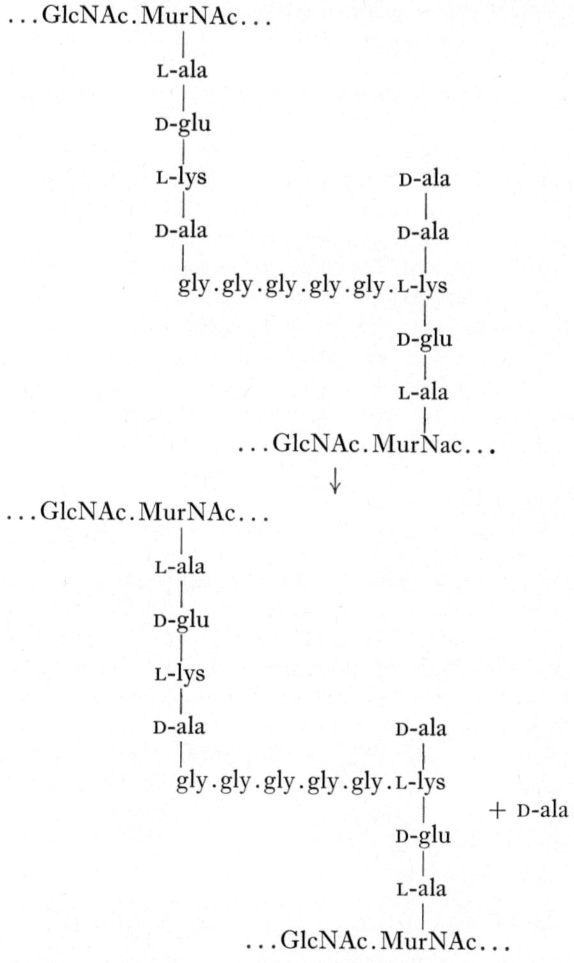

...GlcNAc.MurNAc...
     |
    L-ala
     |
    D-glu
     |
    L-lys   D-ala
     | |
    D-ala   D-ala
     | |
    gly.gly.gly.gly.gly.L-lys
                    |
                  D-glu
                   |
                  L-ala
                   |
       ...GlcNAc.MurNac...

            ↓

...GlcNAc.MurNAc...
     |
    L-ala
     |
    D-glu
     |
    L-lys
     |
    D-ala   D-ala
     | |
    gly.gly.gly.gly.gly.L-lys
                   | + D-ala
                 D-glu
                  |
                 L-ala
                  |
      ...GlcNAc.MurNAc...

Attempts have been made by some workers to gain some insight into the overall process of cell wall synthesis by using fluorescent antibodies to stain newly synthesized wall. These techniques have given conflicting evidence to date. Some species seem to form their cell walls outward as a band originating at the cell division furrow while others seem to lay down new material over the whole surface of the bacterial cell. But even if these results are taken to be a true representation of the direction of wall synthesis, the antisera are never sufficiently well fractionated to act only against a single wall component, such as mucopeptide, and consequently the fluorescent antibody approach is more likely to allow one to follow the overall synthesis of cell wall material rather than any single component within it.

MACROMOLECULAR SYNTHESIS IN THE CONTEXT OF BACTERIAL CELL DUPLICATION

In the foregoing sections we have discussed the molecular processes involved in the biosynthesis of many types of macromolecule that are important in cell duplication, but there has been little attempt to fit these reactions into their context in cell replication itself. In the following section, therefore, some attempt will be made to give an account of what is known about this aspect of macromolecular synthesis. The account will inevitably be incomplete partly because the problem has only recently become accessible to experimental attack and partly, as implied earlier, because not enough is known of the problems involved in synthesis *in vitro* to be able to study the process in a growing cell.

Of the macromolecules that have been considered so far, most is known about DNA replication in relation to cell division. When any cell divides, the problem of providing each daughter cell with one copy of the genetic complement from the parent can be considered as having two parts:

1. How are the copies produced?
2. How are the copies distributed so that one copy goes to each daughter cell?

The present hypothesis invoked to explain this process was originally proposed by Jacob, Brenner, and Cuzin. They proposed that there is a point of contact, known as the *replicator*, between the DNA and some point on the cell membrane. The replicator has the properties both of holding the DNA double thread in its correct spatial relationship with the cell as a whole but also with initiating the replication of the DNA molecule which occurs as described on p. 50. The replicator also, it is suggested, is responsible for distributing the two copies of the DNA, one copy to each daughter cell. This last step can be achieved in one of the two ways shown diagrammatically in Fig. 2.19. Either the replicator divides when the replication of the DNA double strand is complete, thus giving the pattern shown in Fig. 2.19b, or the replicator divides as DNA replication commences as shown in Fig. 2.19a. Which ever occurs, the net effect is the same: one copy of the DNA is distributed to each daughter cell.

The underlying mechanism of DNA replication and distribution as it occurs in growing cells may lead to some mutations of bizarre phenotype. One of the most dramatic is the formation of DNA-less bacteria. In this case the failure of DNA replication means that cell division procedes normally but that there is

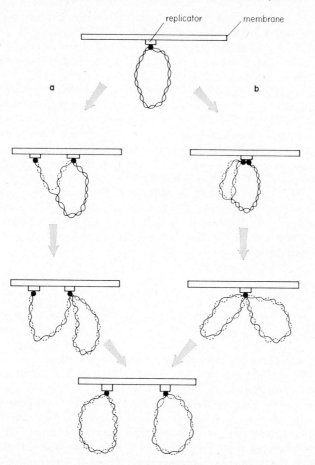

Fig. 2.19 Two methods for the replication of DNA at cell division and the distribution of the two copies to the daughter bacteria. (**a**) Replicator divides at the commencement of DNA replication. (**b**) Replicator divides at the end of DNA replication. Pre-existing DNA strands are shown as solid lines; newly synthesized DNA as dashed lines.

no duplicate copy of the DNA to pass to one of the daughter cells. Thus, in this mutant, a string of DNA-less bacteria are formed as the culture grows with a single DNA containing cell at one end of the chain. It is interesting to note in passing that the DNA-less cells are quite capable of carrying out many of the physiological functions attributed to normal cells; what they cannot do is replicate.

The other macromolecule whose synthesis has been followed to some extent in the whole cell is mucopeptide. Here two basic patterns of synthesis seem to occur. In many Gram-negative bacteria (most of the studies have been done in *Salmonella typhimurium*) the new material is inserted more or less continuously into the old wall structure so that the new material is distributed over the whole cell surface. With streptococci, however (and this may apply to all Gram-positive bacteria), the new material is laid down on each side of the division furrow, thus producing a clear line of demarcation between 'old' and 'new' mucopeptide on the cell surface.

In summary therefore it is true to say that a great deal is known of the detailed biochemical steps in the synthesis of many types of bacterial macromolecule but that our knowledge of how these synthetic processes are organized both in space and time to lead to the formation of a new bacterial cell is still rudimentary. Perhaps one of the most interesting speculations concerned with bacterial cell growth, and one that has a profound bearing on the various topics discussed here, is the possibility that the position of the various macromolecules in the cell itself provides information, in addition to that carried in the DNA, that assists in inserting newly synthesized macromolecules into their correct position in the developing cell. If this is so, then the possession of the genetic information carried in a cell is no use without a bacterial cell to interpret it. Which came first—the information or the cell?

THE ACTION OF ANTIMICROBIAL AGENTS INCLUDING ANTIBIOTICS

Having considered the biosynthesis of macromolecules, in this section the principal biosynthetic sites at which some of the clinically useful antimicrobial agents act are considered.

To be effective, antimicrobial agents have ultimately to inhibit an essential chemical reaction in the cell and the biosynthetic interconversions described in the previous section provide many excellent targets for such compounds. In practice inhibitors exist for practically every interconversion that occurs in the cell although the potency of the inhibition varies widely from compound to compound and reaction to reaction.

Broadly speaking all inhibitors can be classified under two main headings although there are a few compounds, notably those that interfere with ribosome function, that may not fall into this general scheme. The two main groups are:

1. Inhibitors that interfere with the function of enzymes by interaction at the enzyme active centre (Type A).

2. Inhibitors that interfere with nucleic acid metabolism by becoming incorporated as part of a nucleic acid molecule and thus disrupting its normal replicative activity (Type B).

In bacterial cells the majority of reactions that can be blocked effectively by antibiotics form part of either the protein/nucleic acid or cell wall biosynthetic systems. At our present stage of knowledge, the steps in the biosynthesis of lipids do not seem to be so susceptible to inhibition, but this may merely reflect the fact that we know a good deal less about the details of this process than about protein and cell wall synthesis. A variety of steps in energy metabolism of the cell are also susceptible.

The steps in protein and nucleic acid biosynthesis that may be blocked by various inhibitors are indicated in Fig. 2.17. Thus RNA polymerase is inhibited both by actinomycin D and by rifamycin, while the ability of the ribosome to catalyse the translation of m-RNA is impaired by streptomycin, tetracycline, erythromycin, and the other macrolide antibiotics. Streptomycin, tetracycline, and the macrolides all inhibit ribosome function in different ways but the detailed mechanism of their action is, as yet, unclear. Whereas tetracycline and erythromycin seem to block polypeptide synthesis completely, streptomycin (under some conditions at any rate) can allow translation of the message but many errors appear in the resulting polypeptide chain due to the insertion of incorrect amino acids as the chain grows.

Puromycin and chloramphenicol are also potent inhibitors of protein synthesis but in these it is the chemical reactions involved in polypeptide chain extension that are inhibited. Of the two antibiotics, the action of puromycin has been completely elucidated while that of chloramphenicol is not yet completely understood. Puromycin is incorporated in place of an amino acid at the growing point of the polypeptide chain and this process blocks further chain extension. It also causes the detachment of the incomplete chain from the ribosome; thus in practice the antibiotic leads to the synthesis of many incomplete, and therefore inactive, protein molecules in the inhibited cell.

Apart from the reactions shown in Fig. 2.17, a number of metabolic inter-conversions indirectly related to but essential for polypeptide synthesis can be readily inhibited. The biosynthesis of the nucleotide precursors of both RNA and DNA can be blocked very effectively by some antibiotics (e.g. azaserine and diazo-norleucine) and the loading of amino acids on to the relevant t-RNA's can sometimes be impaired severely by structural analogues of the amino acid concerned. For example, L-5-methyl tryptophan blocks the loading of L-tryptophan on to tryptophan-t-RNA in *Escherichia coli*.

Another biosynthetic process indirectly related to protein synthesis is DNA replication. This process can be effectively inhibited by a number of molecules of Type B (see above). Effective examples of such compounds are ethidium bromide, proflavine, mitomycin, and many members of the acridine series of dyes.

The other major series of biochemical reactions in bacteria inhibited by therapeutically effective antibiotics is cell wall synthesis. The overall process consists of at least seven steps (see Steps 1–7, pp. 68–71) and a number of clinically important antibiotics act by blocking one of these reactions. The penicillins and cephalosporins inhibit Step 7 (see p. 70), that is, they block the formation of the cross bridges between the two adjacent side chains on the N-acetyl-muramic acid residues. It is thought that the four membered β-lactam ring of these antibiotics plays a crucial part in their inhibitory action for the —CO—N= bond of the lactam ring opens in the course of their action to block irreversibly a key chemical group at the active centre of the enzyme catalysing Step 7.

The overall reaction is probably:

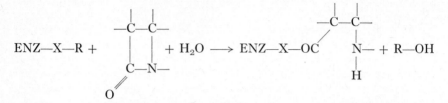

where —X—R is the key group of the enzyme blocked by the antibiotic.

In addition to penicillins and cephalosporins, the antibiotics vancomycin and bacitracin inhibit cell wall biosynthesis. In neither case has the precise mode of action of the inhibitor been worked out, but it seems likely that some reaction between steps 3 and 6 are blocked in both cases.

Just as inhibitors of the biosynthesis of the precursors of RNA and DNA may exert a lethal effect on cell growth, so inhibitors of the biosynthetic precursors for cell wall synthesis can be effective antibiotics. Step 4 in cell wall biosynthetic pathway involves the addition of D-alanyl-D-alanine to the growing side chain of the *N*-acetyl-muramic acid residues (see p. 69). The D-alanyl-D-alanine dipeptide required for this step is formed in the cell as a separate and preliminary step according to the following reaction:

$$\text{D-alanine} + \text{D-alanine} = \text{D-alanyl-D-alanine} + \text{H}_2\text{O}$$

and this reaction is sensitive to the antibiotic D-cycloserine. The antibiotic is thought to exert its action because it is a close structural analogue of D-alanine. This ensures that the inhibitor is drawn to the active centre of the enzyme that normally handles D-alanine as a substrate but, unlike D-alanine, the cyloserine interacts effectively with the enzyme and thus blocks its ability to form D-alanyl-D-alanine from its metabolic precursors.

Cell wall synthesis in many moulds may also be effectively inhibited by antibiotics. As would be expected, because of the different chemical constitution of bacterial and fungal cell wall structures, antibiotics active against bacteria do not inhibit cell wall synthesis in fungi. However, the antibiotic griseofulvin is effective. Much less is known of the biosynthesis of fungal cell walls than of their bacterial counterparts, and the precise point of griseofulvin action is not known.

The examples of antibiotic action given above are but a few of the many known; they have been selected because of their commercial production for use in chemotherapy. The therapeutic value of a bacterial inhibitor depends on many properties other than its ability to inhibit microbial growth (p. 223) and, in fact, because they lack these properties the majority of inhibitors of bacterial growth are useless for therapy.

BOOKS AND ARTICLES FOR REFERENCE AND FURTHER STUDY

Proteins

DOTY, P. (1957). Proteins. *Scient. Am.*, **197**, No. 3, 173.
PHILLIPS, D. C. (1966). The three-dimensional structure of an enzyme. *Scient. Am.*, **215**, No. 5, 32.

Nucleic acids

CRICK, F. H. C. (1954). Structure of hereditary material. *Scient. Am.*, **191**, No. 4, 54.
HOLLEY, R. W. (1966). The nucleotide sequence of a nucleic acid. *Scient. Am.*, **214**, No. 2, 30.

Ribosomes

RICH, A. (1963). Polyribosomes. *Scient. Am.*, **209**, No. 6, 44.

Genetic code

CRICK, F. H. C. (1962). The genetic code. *Scient. Am.*, **207**, No. 4, 66.
—— (1966). The genetic code. III. *Scient. Am.*, **215**, No. 4, 55.
NIRENBERG, M. W. (1963). The genetic code. II. *Scient. Am.*, **208**, No. 3, 80.

Genes and proteins

CLARK, B. F. and MARCKER, K. A. (1968). How proteins start. *Scient. Am.*, **218**, No. 1, 36.

Structure of the bacterial chromosome

CAIRNS, J. (1966). The bacterial chromosome. *Scient. Am.*, **214**, No. 1, 36.

All the above articles are incorporated in the following single volume:

HAYNES, R. H. and HANAWALT, P. C. eds. (1968). *The molecular basis of life: an introduction to molecular biology.* W. H. Freeman & Co. Ltd., Folkestone, Kent.

Mechanisms of antibiotic action

COLLINS, J. F. (1965). Antibiotics, proteins and nucleic acids. *Bri. med. Bull.*, **21**, 223.
GALE, E. F. (1967). How antibiotics work. *Sci. J.*, **3**, No. 1, 62.
NEWTON, B. A. and REYNOLDS, P. E. (1966). Biochemical studies of antimicrobial drugs. *16th Symp. Soc. for gen. Microbiol.* Cambridge University Press.

3

Bioenergetics in Micro-organisms

INTRODUCTION

The metabolic activities of a micro-organism may be classified as anabolic
(or biosynthetic) and catabolic (degradative). Anabolic metabolism (Chap. 2)
results in the synthesis of a great variety of cell constituents such as proteins,
lipids, polysaccharides and polynucleotides from fewer and much simpler
precursor molecules. These synthetic processes require energy. By contrast,
catabolic metabolism provides the energy and, in many cases, the precursors
used for biosynthetic pathways. As an alternative to catabolic metabolism,
photosynthetic mechanisms can provide the energy for biosynthetic pathways
and, by carbon dioxide fixation, the precursors for the carbon skeletons of cell
constituents.

Micro-organisms can utilize and synthesize a vast and exotic range of organic
compounds, and any attempt to describe these processes, either overall or in
detail, is beyond the scope of this chapter. However, there is a more clearly
defined central area of metabolism where both anabolic and catabolic metabolism
share a common ground where they of necessity interact. In this chapter the
coupling of catabolic and anabolic metabolism is described, with particular
reference to the underlying energy considerations.

THERMODYNAMICS OF BIOSYNTHETIC PROCESSES

Free energy changes of reactions

Systems that are not in equilibrium can be used to perform work, and in doing so, equilibrium is approached. For instance, the flow of water from one vessel to another via a connecting tube will proceed if the original levels are different. The flow from the higher to lower level could be harnessed to the performance of work by inserting a water-wheel or turbine in the connecting tube. The system is at equilibrium when the levels are equal, and work can then no longer be performed. If the system is not at equilibrium the work performed by the flow of a given mass (e.g., one mole) of water depends on the difference in water level in the two vessels, that is, the extent to which the system is away from its equilibrium position. The greater the disequilibrium, the more work can be performed. The ability of the system to perform work is known as the free energy content of the system. The free energy content cannot be measured in absolute terms since there is no convenient reference point of zero free energy. However, changes of free energy can be measured or calculated for different conditions of the system. There is a fall of free energy content when work is performed and the system moves nearer its equilibrium position. The free energy change in a work-performing process is therefore negative. Conversely, the change in free energy of a system that is being driven by the performance of work is positive.

This simple model finds its parallel in chemical reactions. Consider the reaction $A \rightleftharpoons B$. If the reaction started with A and had not reached equilibrium, then the conversion of A to B could proceed with the performance of work until equilibrium was reached. This work represents the change of free energy (abbreviated as ΔG) associated with the conversion of A into the equilibrium mixture of A and B. As with the hydrostatic model, the greater the initial disequilibrium the greater the loss of free energy in obtaining equilibrium.

Quantitative aspects are introduced by standardizing the conditions and obtaining a 'standard free energy change' (ΔG_0) for the free energy change occurring when all the reactants are maintained at unit concentration (i.e. 1 mole/litre) and 1 mole of A is converted to 1 mole of B. Let the equilibrium constant be K, which is equal to the ratio of the concentration of B and A at equilibrium:

$$K = \frac{[B]}{[A]} \qquad (1)$$

The dependence of ΔG_0 on the degree of disequilibrium of the system with its reactants at unit concentration is given by

$$\Delta G_0 = -RT \log_e K \qquad (2)$$

where R is the gas constant and T the absolute temperature. This states that ΔG_0 is proportional to the absolute temperature and also proportional to the logarithm of the equilibrium constant K. If K equals unity, then the system is already at equilibrium ($[A] = [B] = 1 = [A]/[B]$), and the value for ΔG_0 is zero. If K is large, then ΔG_0 is large and negative; work can be performed by the conversion of A to B. If K is very small (very much less than unity), then ΔG_0 is large and positive; work must be performed to convert A to B.

The value of ΔG_0 refers to the *standard* free energy change when the reactants are at unit concentration. The free energy change (ΔG) for other concentrations is given by

$$\Delta G = \Delta G_0 + RT \log_e \frac{[B]}{[A]} \tag{3}$$

or, for the general case where a number of substrates (S_1, S_2 ...) are converted to a number of products (P_1, P_2 ...)

$$\Delta G = \Delta G_0 + RT \log_e \frac{[P_1] \cdot [P_2] \cdots}{[S_1] \cdot [S_2] \cdots} \tag{4}$$

The concept of a reaction occurring at constantly maintained concentrations of reactants despite a considerable conversion of substrate to product should not be unfamiliar, since this is largely the intracellular state of affairs. Remarkably constant intracellular pools of metabolites are maintained despite the metabolic flux through the pools. An analogous case at a microbiological level is afforded by the behaviour of continuous cultures (p. 180).

When ΔG is pH dependent it is convenient to define an apparent standard free energy change, $\Delta G_0'$, that holds for a defined pH (e.g., pH 7·0) and all other concentrations at unity. The conventional units for free energy changes are calories per mole of substrate conversion.

Consecutive reactions

Such reactions are those where the product for one reaction is a substrate for the next, e.g.,

$$A \rightleftharpoons B \tag{5}$$

$$B \rightleftharpoons C \tag{6}$$

These two reactions can be added to give the overall reaction (7) from which the common reactant B apparently disappears since it occurs on both sides.

$$A + (B) \rightleftharpoons (B) + C \tag{7}$$

The equilibrium constant for the overall reaction is the product of the equilibrium constants of the part reactions. ΔG for the overall reaction is the sum of the values for the part reactions (Table 3.1).

Consecutive reactions in metabolism

Most enzyme-catalysed reactions occur in their intracellular environment as consecutive reactions of metabolic pathways. In such a series, where there is a flow of molecules from one end of the pathway to the other, ΔG for each reaction must be negative, since for any reaction where ΔG is positive the flow for that reaction would be in the opposite direction to the overall flow—a situation that is plainly impossible in a steady state. The fall in free energy occurring along

Table 3.1   Behaviour of equilibrium constants and free energy changes for consecutive reactions.

Reaction			Equilibrium constant	ΔG
Part Reaction 1	A	B	K_1	ΔG_1
Part Reaction 2	B	C	K_2	ΔG_2
Overall Reaction (1 + 2)	A	C	$K_1 \times K_2$	$\Delta G_1 + \Delta G_2$

the length of a metabolic pathway is largely associated with a minority of the individual reactions, the remaining ones exhibiting free energy changes that are not very much less than zero. This is illustrated in Fig. 3.1, where ΔG is plotted against the reaction progress of a hypothetical metabolism pathway:

$$A \longrightarrow B \longrightarrow C \longrightarrow D \longrightarrow E \longrightarrow F$$

The size of the downward step between each of the metabolites represents the free energy change associated with that reaction. There are two categories of reaction:

1. Those that have a large negative value for ΔG, and are so displaced from equilibrium that reversal of the direction of flow by altering the concentrations of reactants is impossible under biological conditions.
2. Those that have a small negative value for ΔG, and are sufficiently close to equilibrium for reversal to occur as a consequence of relatively small changes in the concentration of reactants.

The free energy changes occurring during reversal of the hypothetical pathway are also shown in Fig. 3.1. Small changes in concentrations suffice to alter the direction of flow in these reactions that are near equilibrium. The barrier formed by the effectively irreversible reactions must be scaled by the provision at that point of an alternative route which is coupled to the utilization of energy from some other source.

Energy coupling in biological reactions

Otherwise endergonic biosynthetic reactions are made thermodynamically possible by coupling with an exergonic reaction, the ΔG value for the coupled reaction being the sum of the values of the part reactions. As an example, consider the case of *Torulopsis utilis* growing aerobically with ethanol as the sole carbon source. The very process of growth provides a vivid demonstration of the *de novo* synthesis of cell constituents. Amongst these constituents will be polysaccharides containing glucose, and it is apparent that there must be a

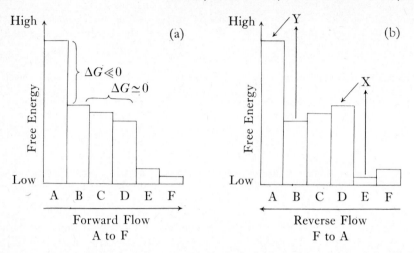

Fig. 3.1 Free energy changes in a hypothetical metabolic pathway between A and F, where B, C, D and E are the intermediates. The free energy of each compound is indicated on an arbitrary scale that runs from low to high, and changes in free energy are given by the difference between each compound. This is shown for part of the sequence in **(a)** where the pathway runs from A to F and takes the thermodynamically likely direction. In **(b)** reversal is achieved by introducing energy (e.g., from other reactions) to proceed via new intermediates X and Y, thereby overcoming the large difference in free energy between E and D, and B and A. The other reactions are near equilibrium, and their direction of flow is altered by small changes in the concentration of the reactants and products.

metabolic pathway that converts ethanol into glucose or a closely related compound. The free energy changes involved are shown below for pH 7·0.

$$\frac{1}{2} \text{ glucose}$$

$$\Delta G_0' = +26 \text{ kcal/mole ethanol}$$

$$CO_2$$

$$\text{Ethanol} \underline{\hspace{4cm}} 3O_2$$

$$\Delta G_0' = -311 \text{ kcal/mole ethanol}$$

$$2CO_2 + 3H_2O$$

The exergonic oxidation of ethanol provides energy for the endergonic synthesis of glucose. In the absence of oxygen *T. utilis* is unable to grow on ethanol, for the exergonic oxidation cannot proceed. However, growth can occur anaerobically with glucose as the carbon source, and the metabolism of glucose is directed in the energy-yielding direction of ethanol and carbon dioxide production.

Although micro-organisms utilize many diverse exergonic metabolic pathways to provide energy for driving the biosyntheses that underlie growth, the majority of mechanisms for coupling energy production to energy utilization involve ATP. At pH 7, the standard free energy of hydrolysis for the terminal phosphate of ATP is 7·4 kcal/mole. This and the $\Delta G_0'$ (pH 7·0) values for other hydrolyses of importance are shown in Table 3.2.

Table 3.2 Free energy changes for the hydrolysis of some biochemically important compounds.

Substrate	Reaction	$-\Delta G_0'$ (pH 7) kcal/mole
ATP	$ATP + H_2O \longrightarrow ADP + \text{phosphate}$	7.4
ATP	$ATP + H_2O \longrightarrow AMP + \text{pyrophosphate}$	7.6
acyl-CoA	$\text{acyl-CoA} + H_2O \longrightarrow \text{acid} + CoASH$	7.7
phosphomonoesters (e.g., glucose-6-phosphate)	$R.O\text{-phosphate} + H_2O \longrightarrow ROH + \text{phosphate}$	3.0
peptides	$R.CO.NH.R' + H_2O \longrightarrow RCOOH + R'.NH_2$	0.5
amino acid esters	$R.CH(NH_2)COOR' + H_2O \longrightarrow R.CH(NH_2)COOH + R'OH$	8.4
glycosides (e.g., starch)	$(\text{glucose})_n + H_2O \longrightarrow (\text{glucose})_{n-1} + \text{glucose}$	3.0
acyl-oxygen esters (e.g., glycerides)	$RCO.OR' + H_2O \longrightarrow RCOOH + R'.OH$	5.1
acetyl-phosphate	$CH_3.CO\text{-phosphate} + H_2O \longrightarrow CH_3.COOH + \text{phosphate}$	10.5
phosphoenolpyruvate	$CH_2{=}C(\text{O-phosphate})(COOH) + H_2O \longrightarrow CH_3{-}C({=}O)(COOH) + \text{phosphate}$	13.0

Energy production in catabolic pathways is largely concerned with the synthesis of ATP from ADP and phosphate:

$$ADP + phosphate \longrightarrow ATP + H_2O \qquad (8)$$
$$\Delta G_0' \text{ (pH 7·0)}$$
$$= +7·4 \text{ kcal/mole}$$

The manner in which this reaction (8) is driven by the exergonic reactions of catabolism is discussed later in this chapter; the way by which ATP hydrolysis can drive biosynthetic reactions is considered here. Two specific examples can be used to indicate the more general features of energy coupling in biosynthetic pathways.

Chain extension in fatty acid synthesis

Escherichia coli may be grown with acetate as the sole carbon source, and amongst the cell components that must be synthesized for growth are various long-chain fatty acids. The carbon skeletons of these fatty acids are derived from an end-to-end condensation of a number of acetate units, e.g.,

$$CH_3.CO.SCoA + CH_3.CO.SCoA \longrightarrow$$
acetyl-CoA   acetyl-CoA
$$CH_3.CO.CH_2.CO.SCoA + CoASH \quad (9)$$
acetoacetyl-CoA

$\Delta G_0'$ (pH 7·0) for reaction (9) is $+6·6$ kcal/mole acetoacetyl-CoA, but by coupling to the hydrolysis of ATP the $\Delta G_0'$ value can be dropped to less than unity. This is demonstrated by uniting the two reactions (8 and 9), and then adding them to obtain the overall reaction:

		$\Delta G_0'$ (pH 7) kcal/mole	
2 acetyl-CoA	$\longrightarrow$ acetoacetyl-CoA + CoASH	$+6·6$	(9)
ATP + H$_2$O	$\longrightarrow$ ADP + phosphate	$-7·4$	(8)
Sum: 2 acetyl-CoA $+$ ATP $+$ H$_2$O	$\longrightarrow$ acetoacetyl-CoA + ADP $+$ phosphate + CoASH	$-1·2$	(10)

The thermodynamic sums themselves do not imply that the reaction *will* happen, but only that it *can*. A suitable reaction pathway is needed to couple the two reactions via a common intermediate, and an appropriate enzyme is almost invariably required to catalyse the reaction in the direction of its thermodynamic fate. Reactions (11) and (12) demonstrate the mechanism whereby ATP hydrolysis is coupled to the condensation of carbon skeletons; the trick is to proceed via an intermediate common to both reactions, and this is achieved by means of the energy-dependent carboxylation of acetyl-CoA to malonyl-CoA, the common intermediate. The sum of reactions (11) (carboxylation) and (12) (condensation) is the desired reaction, namely, (10). The common intermediate appears on both sides of the summed reaction and therefore vanishes from the sum.

$$
\begin{aligned}
\text{ATP} + \text{CO}_2 & \longrightarrow \text{ADP} + \text{phosphate} \\
+ \text{ acetyl-CoA} + \text{H}_2\text{O} & \qquad + \text{ malonyl-CoA} \qquad\qquad (11)
\end{aligned}
$$

$$
\begin{aligned}
\text{Malonyl-CoA} & \longrightarrow \text{acetoacetyl-CoA} \\
+ \text{ acetyl-CoA} & \qquad + \text{CO}_2 + \text{CoASH} \qquad\qquad (12)
\end{aligned}
$$

$$
\begin{aligned}
\text{Sum: } 2\text{ acetyl-CoA} + \text{ATP} & \longrightarrow \text{acetoacetyl-CoA} + \text{ADP} \\
+ \text{H}_2\text{O} & \qquad + \text{ phosphate} + \text{CoASH} \qquad (10)
\end{aligned}
$$

A convenient means of displaying this type of coupling is drawn below, where the contact between the arrows of the two reactions represents by analogy the coupling mechanism as if in a friction drive between wheels:

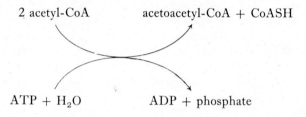

$$
\begin{array}{ccc}
2\text{ acetyl-CoA} & & \text{acetoacetyl-CoA} + \text{CoASH} \\
\\
\\
\text{ATP} + \text{H}_2\text{O} & & \text{ADP} + \text{phosphate}
\end{array}
$$

The synthesis of acyl-thioesters

It follows from the above section that an essential metabolic step for *E. coli* growing on acetate is the conversion of acetate into acetyl-CoA, a reaction that is endergonic (reaction *13*). This apparent impasse is overcome by coupling acetyl-CoA synthesis to ATP hydrolysis. In this particular instance it is the other pyrophosphate linkage of ATP that is hydrolysed, yielding AMP and pyrophosphate (reaction *14*).

$$
\begin{array}{lll}
& & \Delta G_0' \text{ (pH 7)} \\
& & \text{kcal/mole} \\
\text{acetate}^- + \text{H}^+ \longrightarrow \text{acetyl-CoA} + \text{H}_2\text{O} & +7\cdot7 & (13) \\
+ \text{ CoASH} & & \\
\\
\text{ATP} + \text{H}_2\text{O} \longrightarrow \text{AMP} + \text{pyrophosphate} & -7\cdot6 & (14) \\
\hline
\text{Sum: acetate}^- + \text{CoASH} \longrightarrow \text{acetyl-CoA} + \text{AMP} & & \\
+ \text{ ATP} + \text{H}^+ \qquad\qquad + \text{ pyrophosphate} & +0\cdot1 & (15)
\end{array}
$$

The common intermediate involved in the coupling mechanism is acetyl-adenylate which remains enzyme bound (reactions *16* and *17*).

$$
\text{H}^+ + \text{acetate}^- + \text{ATP} \longrightarrow \text{acetyl-adenylate} + \text{pyrophosphate} \quad (16)
$$

$$
\text{acetyl-adenylate} + \text{CoASH} \longrightarrow \text{acetyl-CoA} + \text{AMP} + \text{H}^+ \quad (17)
$$

The sum is the same as that of reaction (*15*).

This section can be concluded by summarizing the three major features of energy coupling in biosynthetic processes:

(a) Biosynthetic pathways contain one or more reactions that are thermo-dynamically unfavourable unless energy is supplied.

(b) The most common source of energy for this purpose is that obtained from the hydrolysis of ATP.

(c) Coupling of ATP hydrolysis to the energy-dependent biosynthetic reaction is achieved by the introduction of a new reaction that contains an intermediate common to both the biosynthetic and ATP-hydrolytic reactions. The free energy changes of the two reactions, biosynthetic and ATP-hydrolytic, are added to give the energy change for the overall reaction.

Oxidation–reduction potentials in biological systems

Autoxidizable compounds are those which rapidly and reversibly react with each other in oxidoreduction reactions. Hydrogen is not autoxidizable, but in the presence of platinum catalyst it reacts with the oxidized forms of dyes (e.g., methylene blue), with quinones, and with ferric ions. The oxidized and reduced forms of each reactant (e.g., H^+ and H_2; Fe^{3+} and Fe^{2+}, quinone and hydroquinone) form a redox pair or couple, and the equilibrium and free energy changes involved in reactions between different couples are treated in a similar manner to that for other reactions (p. 80). However, an additional concept to be introduced is that of the electron-donating activity of a redox couple. Although this property, like free energy, can be measured only in a relative rather than absolute sense, it is convenient to choose an arbitrary zero point, and for this purpose the electron-donating activity of the standard hydrogen couple is defined as zero volts. This is the oxidoreduction potential of a hydrogen/proton couple at 1 atmosphere pressure of hydrogen and pH 0. Standard or mid-potentials of other couples refer to their difference from the standard hydrogen couple at the 'mid-point' when the ratio of oxidized to reduced form is unity. The abbreviation for mid-potential is E_s and, at defined pH values other than zero, E_s'. The units are volts and the value may be positive or negative according to whether the couple concerned is reducible or oxidizable by the standard hydrogen couple.

The electrical analogy is straightforward. The work obtainable from an electrical charge is proportional to the voltage across which the charge is allowed to flow. The work is also proportional to the size of the charge. Likewise, for an oxidoreduction involving the transfer of n electrons between couples whose mid-potentials differ by ΔE_s volt

$$\Delta G_0 \propto n \Delta E_s \qquad (18)$$

The proportionality constant is the Faraday, F, the caloric yield of 1 electron equivalent traversing a rise in potential of 1 volt (23,000 kcal.), and equation (18) becomes

$$\Delta G_0 = n F \Delta E_s \qquad (19)$$

or

$$\Delta G_0 = 23,000 n E_s \, \text{kcal.} \qquad (20)$$

E_s' values referring to pH 7·0 are more convenient for biological work than E_s values at pH zero. Table 3.3 lists the E_s' (pH 7·0) values for a number of important couples. E_s and E_s' values refer to the standard or mid-potentials when the couple is half reduced, half oxidized. When the ratio of oxidized to reduced form is other than unity, the redox potential E differs from E_s by an amount that is proportional to the logarithm of that ratio:

$$E = E_s' + \frac{0.06}{n} \log_{10} \frac{\text{oxidized}}{\text{reduced}} \qquad (21)$$

It follows from equation (21) that for a couple where $n = 1$ a ten-fold change in the ratio oxidized/reduced changes the redox potential by 0·06 V (Table 3.4).

Table 3.3 Oxidoreduction potentials at pH 7·0 for some biologically important couples.

Couple (ox/red)	E_s' (pH 7·0) volt
Acetate + CO_2/pyruvate	−0·70
acetate/acetaldehyde	−0·60
H^+/H_2	−0·42
Ferredoxin	−0·41
3-phosphoglycerate/3-phosphoglyceraldehyde	−0·37
CO_2 + pyruvate/malate	−0·33
CO_2 + acetyl-CoA/pyruvate	−0·32
NAD^+/NADH	−0·32
$NADP^+$/NADPH	−0·32
CO_2 + 2-oxoglutarate/isocitrate	−0·28
NH_4^+ + 2-oxoglutarate/glutamate	−0·14
NH_4^+ + pyruvate/alanine	−0·13
fumarate/succinate	+0·03
ubiquinone/reduced ubiquinone	+0·10
cytochrome b_2 (ferric/ferrous)	+0·12
2,3-enoyl-CoA/acyl-CoA	+0·19
cytochrome c (ferric/ferrous)	+0·22
cytochrome aa_3 (ferric/ferrous)	+0·29
oxygen/water	+0·81

Relationships between equilibrium constants, ΔG, and redox potentials

Two autoxidizable redox couples will react with each other until equilibrium is reached, at which point their potentials are equal and ΔG is zero. During the reaction electron flow is from negative to positive; therefore the couple with the more negative potential will reduce (i.e., donate electrons to) the couple with the more positive potential. The same is true for non-autoxidizable redox couples, provided that suitable catalysts (e.g., enzymes) are also present. Thus the difference in redox potential of two couples is a measure of the extent of disequilibrium between them, and also of the work that can be performed by oxidizing one couple at the expense of the other. The equations, symbols and units and inter-relationships for equilibrium constants, free energy changes and redox potential are summarized in Table 3.4.

All three methods of expressing the disequilibrium of a reaction are essentially equivalent, since they all describe the same underlying phenomenon. The use of different methods of saying the same thing is a convenience for use in appropriate contexts, and should not obscure their interconvertibility. To illustrate this, one may consider the transfer of energy from a catabolic pathway

Table 3.4 Inter-relationships between equilibrium constant, free energy change and oxidoreduction potential.

Conditions	Equilibrium constant	Free Energy Change (kcal/mole)	Oxidoreduction Potential (volts)
At equilibrium	$K = \dfrac{[P_1][P_2]\cdots}{[S_1][S_2]\cdots}$	ΔG = zero	Couples at the same potential
All reactants at unit activity i.e., standard conditions	Disequilibrium unless K equals unity	$\Delta G_0 = -RT \log_e K$ which, at 25°C, becomes $\Delta G_0 = -1.363 \log_{10} K$	Couples at their respective standard potentials, ΔE_0, which differ by ΔE_0 $$\Delta E_0 = \frac{-\Delta G_0}{nF}$$
Reactants at other concentrations	Disequilibrium	$\Delta G = \Delta G_0 + RT \log_e \dfrac{[P_1][P_2]\cdots}{[S_1][S_2]\cdots}$ which, at 25°C, becomes $\Delta G = \Delta G_0 + 1.363 \log_{10} \dfrac{[P_1][P_2]\cdots}{[S_1][S_2]\cdots}$	For each couple $$E = E_0 - 2.3 \frac{RT}{nF} \log_{10} \frac{[red]}{[ox]}$$ At 25°C, the value of $2.3\, RT/F$ is 0.06 volt

to a biosynthesis. Initially, the transfer of electrons from oxidoreduction carriers of low potential to higher potential (volts) is utilized for the synthesis of ATP, and the conserved energy is expressed as a free energy change (calories). The subsequent involvement of ATP in a metabolic pathway can then profoundly alter the equilibrium position of an otherwise unfavourable reaction.

Oxidoreduction potentials and biosynthetic processes

The synthesis of more complex cell components from simpler precursors may require not only the condensation of small carbon skeletons into longer ones, but also reduction reactions to yield products that are of a lower oxido-reduction potential than the precursors. Examples of this are seen in the conversion of acetate to long-chain fatty acids, carbohydrates and the bases of nucleotides. It follows that reductive biosyntheses require an electron or hydrogen donor that possesses a redox potential sufficiently low for the reaction under consideration to proceed in the desired direction. Inspection of Table 3.3 will demonstrate that certain couples (e.g., cytochromes) have redox potentials that are too high for driving reductive biosyntheses such as glutamate synthesis, whereas others (e.g., NAD) have suitably low potentials for this particular purpose.

Reducing equivalents arise primarily either from photosynthesis or a catabolic pathway, and mechanisms are required to couple the generation of reductants to their re-oxidation in biosyntheses. From the thermodynamic viewpoint it is possible to classify three general ways for transferring reducing equivalents from the reaction where they become available to the reaction where they are utilized.

1. Direct involvement of a common oxidoreduction carrier

The reductive biosynthesis of oxoglutarate and ammonium to glutamate is important in nitrogen fixation, and the mid-potential of this system (-0.14 V at pH 7) is such that the reduction is readily driven by the NADH/NAD$^+$ couple (-0.32 V at pH 7) which can in its turn be reduced by the oxidation of pyruvate, isocitrate and 2-oxoglutarate, e.g.,

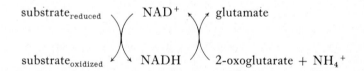

2. Prior alteration of the acceptor potential

A yeast such as *Torulopsis utilis* growing on lactate as the sole carbon source must synthesize dicarboxylic acids such as malate in order that the many functions of the Krebs cycle can be fulfilled. The immediate metabolic product of lactate is pyruvate, and reductive carboxylation to malate could occur.

$$\text{pyruvate} + CO_2 + NADPH + H^+ \longrightarrow \text{malate} + NADP^+ \qquad (22)$$

At pH 7·0, E_s' for the pyruvate/malate couple is only 0·01 V more negative than that for NADP$^+$/NADPH. However, the intracellular concentrations of reactants do not favour the reductive carboxylation, and an alternative route is utilized that results in the synthesis of an acceptor with a higher redox potential:

$$\text{pyruvate} + CO_2 + \text{ATP} \longrightarrow \text{oxaloacetate} + \text{ADP} + Pi \qquad (23)$$

$$\text{oxaloacetate} + NADH_2 \longrightarrow \text{malate} + NAD \qquad (24)$$

Sum: pyruvate + CO_2 + $NADH_2$
      + ATP $\longrightarrow$ malate + NAD + ADP + Pi
where Pi = phosphate.

E_s' (pH 7) for the oxaloacetate/malate is -0.19 V, and the couple is readily reduced by $NADH_2$. Thus the energy available from the hydrolysis of ATP has been utilized to drive an otherwise unlikely biosynthetic reduction by increasing the oxidoreduction potential of the acceptor couple.

3. Prior alteration of the donor potential

As an alternative to increasing the oxidoreduction potential of the acceptor couple, it would seem reasonable that the potential of the donor could be lowered to overcome an otherwise unfavourable equilibrium. For instance, cytochrome b has a mid-potential of about 0.1 V at pH 7.0 and is unable to act as a reductant in glutamate biosynthesis. This difficulty may be overcome by effecting an energy-dependent reduction of NAD according to reaction (25).

$$2 \text{ cyt.} b^{++} + NAD^+ + H^+ + ATP \longrightarrow$$
$$2 \text{ cyt.} b^{+++} + NADH + ADP + Pi \quad (25)$$

This reaction demonstrates the interconvertibility of redox potential and free energy.

THERMODYNAMICS OF CATABOLIC PROCESSES

The necessary energy for biosynthetic processes is derived either from photosynthetic mechanisms or by coupling to catabolic reactions that have large and negative free energy changes. In the case where the coupling involves solely oxidoreduction reactions, the mechanism is fairly straightforward; the oxidation of a catabolite effects the reduction of a biosynthetic intermediate and the transfer of reducing equivalents is made by an intermediate oxidoreduction carrier such as NAD or NADP, e.g.,

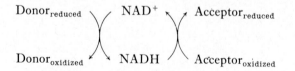

The other major case is where energy transfer between catabolic pathways and biosynthetic pathways is via the synthesis and hydrolysis of ATP, e.g.,

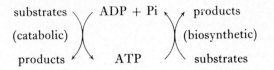

substrates ⟍ ⟋ ADP + Pi ⟍ ⟋ products
(catabolic)   ⟩ ⟨   ⟩ ⟨ (biosynthetic)
products ⟋ ⟍   ATP   ⟋ ⟍ substrates

The utilization of ATP in biosynthetic routes has been discussed above, and the remainder of this section deals with the production of ATP by catabolic processes.

Substrate level phosphorylation

A reaction with a large and negative free energy change can be coupled to ATP synthesis through a common intermediate that is involved in both the reactions. Such a mechanism is exactly the reverse of that used in coupling ATP hydrolysis to a biosynthetic reaction through a common intermediate (see p. 85). In at least one case, that of glucose metabolism, a reaction that *utilizes* ATP in the production of glucose from simple precursors (e.g., lactate) is identical with one that *produces* ATP during glycolysis, i.e., the interconversion of glyceraldehyde-3-phosphate and 3-phosphoglycerate:

$$\text{glyceraldehyde-3-phosphate} + ADP + Pi + NAD \rightleftharpoons$$
$$\text{3-phosphoglycerate} + NADH_2 + ATP \qquad (26)$$

where Pi = phosphate.

The direction of flow in this reaction in the intracellular environment is set by the relative concentration of reactants and products.

It is reasonable to expect that some biological advantage can be attached to the efficient use of ATP in biosynthetic reactions, where 'efficient' refers to the relative proportion of the free energy of hydrolysis of ATP that is conserved during a biosynthetic reaction rather than lost as heat. In short, the biologically 'ideal' free energy change for an energy-coupling reaction would be approximately zero. Given that this can be achieved, then it is a simple matter to use the same reaction under different conditions for ATP production rather than biosynthesis.

Oxidative phosphorylation

The transfer of electrons from a couple of relatively lower oxidoreduction potential to one of higher potential is associated with a fall in free energy which may appear as heat. In many biological systems there are mechanisms which conserve this energy and couple it to the synthesis of ATP, e.g.,

$$ADP + Pi \longrightarrow ATP \qquad (8)$$
$$\Delta G_0{}' \text{ (pH 7)} = 7 \cdot 4 \text{ kcal/mole}$$

For an oxidoreduction to drive ATP synthesis it is necessary that the free energy change for the oxidoreduction, given by $\Delta G = nF\Delta E$, is also about 7·4 kcal/mole but negative. At pH 7·0 and standard (i.e., 1M) concentration of ADP, ATP and phosphate, an oxidoreduction involving the transfer of 2 electrons must cover a span of +152 mV from reductant to oxidant to obtain a free energy charge of

-7.4 kcal/mole. At equal concentration of ATP and ADP, and 10^{-2}M phosphate, which are likely intracellular concentrations, the necessary ΔE value rises to $+210$ mV.

The oxidoreductions most commonly used for ATP synthesis in non-photosynthetic organisms are those occurring in the respiratory chain of carriers which catalyses the final stages of the oxidation of catabolic intermediates for oxygen. The general features of this process can be illustrated by referring to the respiratory chain of *Torulopsis utilis*. In this diagram (Fig. 3.2), the various substrates and respiratory carriers are plotted against an ordinate which indicates their $E_s{}'$ (pH 7) values, while displacement along the abscissa represents progress from the input of reduced substrate towards oxygen, the final acceptor.

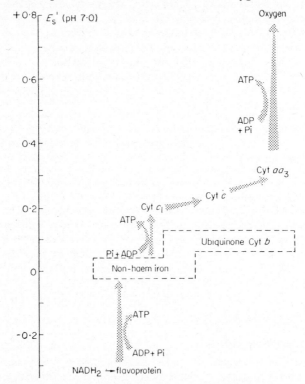

Fig. 3.2 Mid-potentials of the respiratory chain carriers. The vertical scale indicates the mid-potential of the carriers of mammalian mitochondria, and these are probably similar to those for *Torulopsis utilis*. The flow arrows indicate the direction of electron flow from $NADH_2$ towards oxygen. The placing of a group of carriers (ubiquinone, non-haem iron and cytochrome *b*) within a box implies uncertainty about their actual position with respect to each other in terms of electron flow.

It is apparent from this representation that there are three spans each of sufficient magnitude for the synthesis of ATP coupled to the transfer of two electrons; from NADH and its dehydrogenase to the group that includes ubiquinone and cytochrome *b*, from this group to cytochrome c_1 and *c*; and from cytochrome *a* and a_3 to oxygen. The first two spans are such that it could be anticipated that the electron flow would readily be reversed (i.e., positive to negative) if driven by the hydrolysis of ATP. The role of such a reversal in biosynthetic reductions has already been mentioned.

The exact mechanism of oxidative phosphorylation is unknown. By analogy with substrate level phosphorylations, it might be expected that there are common intermediates which couple electron flow to ATP synthesis. However, no such intermediates have been detected and more recent attempts to understand the coupling mechanism involve a greater appreciation of the importance of the membrane that is invariably the site of the respiratory chain and its associated energy conservation mechanism.

The individual features of the respiratory chain vary considerably amongst micro-organisms, and the sequence of carriers given for *T. utilis* is not widely applicable. Nevertheless, the thermodynamic rules are the same for all species, and it is likely that common molecular mechanisms underlie the same biological phenomena.

BACTERIAL PHOTOSYNTHESIS

Photosynthetic bacteria use light energy for the formation of ATP and the reduction of pyridine nucleotides, trapping light with pigments that are similar to those found in higher plants. The ATP and reducing power derived from the photochemical reaction may then be used to reduce CO_2 in a series of 'dark' reactions, probably following the action of ribulose diphosphate carboxylase, as in the green plant system described by Calvin and Bassham.

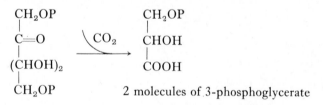

Ribulose 1,5-diphosphate

The complete photosynthetic carbon reduction cycle, as worked out for green plants, is shown in Fig. 3.3. Many of the key enzymes have been found not only in photosynthetic bacteria but also in other autotrophic bacteria and the mechanism shown in Fig. 3.3 may be of very general use. In chemoautotrophic bacteria, the energy required to produce reduced pyridine nucleotides and ATP needed in the Calvin–Bassham cycle is obtained from the metabolism of inorganic electron donors (e.g., the oxidation of Fe^{2+} by iron bacteria or of NO_2^- by *Nitrobacter*). The ATP produced by the mechanism of electron transfer phosphorylation already described (p. 91) can be used to form reduced pyridine nucleotides by a reversal of electron flow (see p. 90) where the substrates are not sufficiently reducing to reduce NAD directly. The photosynthetic bacteria have elaborated specialized membranes containing chlorophylls and carotenoids that utilize light energy to initiate electron transfer reactions resulting in the

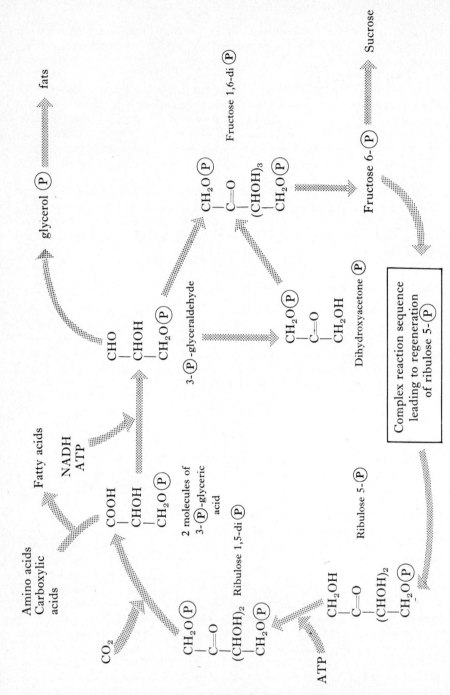

Fig. 3.3 A simplified representation of the photosynthetic carbon reduction cycle.

formation of ATP; in some cases this photosynthetically formed ATP is used in the reduction of pyridine nucleotides by reversed electron flow.

Classification of photosynthetic bacteria

There are three groups of photosynthetic bacteria that differ in the nature of their major chlorophyll and the substrate that is utilized. The general reaction of photosynthesis may be represented thus:

$$CO_2 + H_2A \xrightarrow{\text{light}} CH_2O + H_2O + 2A$$

The nature of the electron donor, H_2A, varies. In green plants H_2A is H_2O and oxygen is evolved; the photosynthetic bacteria are *anaerobic* and do not evolve oxygen. Sulphur bacteria use H_2S as electron donor whereas the non-sulphur bacteria (Athiorhodaceae) normally use organic materials such as succinate or malate as donors. The three groups of photosynthetic bacteria, shown in Table 3.5, have chlorophylls similar to those of green plants, differing

Table 3.5   Classification of photosynthetic bacteria.

Organism or group	*Main Chlorophyll pigment*	*Substrate Utilized* (H_2A)
1. Purple sulphur bacteria (Thiorhodaceae) e.g., *Chromatium*	Bacteriochlorophyll *a*	H_2S
2. Purple nonsulphur bacteria Athiorhodaceae e.g., *Rhodospirillum*, *Rhodopseudomonas*	Bacteriochlorophyll *a* or Bacteriochlorophyll *b*	Organic e.g., succinate
3. Green sulphur bacteria Chlorobacteriaceae e.g., *Chlorobium*, *Chloropseudomonas*	Chlorobium chlorophyll	H_2S

in the substituents on the tetrapyrrole rings, and in the case of bacteriochlorophylls, having absorption bands much further to the red region, so that whereas in green plants the photoreactive centre absorbs light at 700 mμ in *Rhodospirillum rubrum* the reaction centre absorbs at 890 mμ.

Electron flow in bacterial photosynthesis

Light energy absorbed by chlorophyll or carotenoids is transferred rapidly to a special form of chlorophyll at the reaction centre (called P890 in *R. rubrum*). This energy raises the chlorophyll to an excited state and an electron is donated to an acceptor molecule and is replaced by an electron from a donor (see Fig. 3.4). The donor is almost certainly a cytochrome of the *c* type and the ultimate acceptor is probably a quinone, but an intermediate carrier, Z, of low potential may be involved. Certainly the reduction of a quinone is one of the very early changes that follow illumination of particles from the photosynthetic bacteria, and in some bacteria (*Chromatium*) the oxidation of a cytochrome *c* takes place on illumination, even at 77°K, showing that the cytochrome is intimately associated with the reaction centre. The reduced quinone is reoxidized and the cytochrome *c* reduced through a chain of carriers that differs in different bacteria. During this electron flow from low potential carrier to higher potential cytochrome *c* phosphorylation occurs, a process known as cyclic photophosphorylation, and in *R. rubrum* the ATP or its precursor $(X \sim I)$ may be used to form NADH by reversed electron flow. The importance of this mechanism can be shown in whole cells where NADH formation in light is abolished by un-coupling agents. Electrons that are removed from the cyclic system during the synthesis of reduced pyridine nucleotides are replaced by electrons from the reduced substrate H_2A (e.g., H_2S for sulphur bacteria, succinate for non-sulphur bacteria). It is possible that in some bacteria (e.g., *Chlorobium*) NAD is reduced directly by reduced ferredoxin, as shown in Fig. 3.4 by a dotted pathway. Ferredoxin, a low potential electron carrier, is a non-haem iron protein that is not present in all photosynthetic bacteria, but is found in *Chlorobium.*

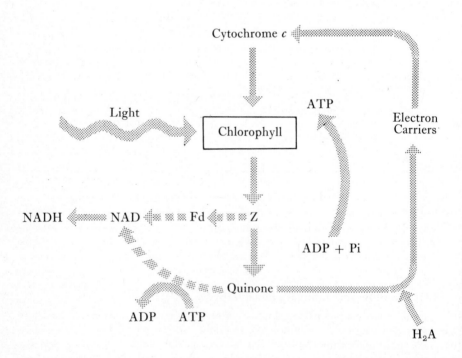

Fig. 3.4 Diagrammatic representation of electron flow in bacterial photosynthesis. The 'electron carriers' differ in different organisms but are usually cytochromes of either the *b* or *c* types. Fd is ferredoxin, Z is a hypothetical intermediate.

BOOKS AND ARTICLES FOR REFERENCE AND FURTHER STUDY

The abbreviations used throughout this chapter are in common biochemical use and both these and the relevant metabolic background can be obtained from either of the following texts.

MAHLER, H. R. and CORDES, E. H. eds. (1967). *Biological Chemistry*, Harper and Row, and John Weatherill, Inc., 872 pp.

—— and —— eds. (1968). *Basic Biological Chemistry*, Harper and Row, and John Weatherill, Inc., 527 pp.

4

Genetics and Gene Action

INTRODUCTION

Genetics concerns the study of variation and inheritance. It has been clear
ever since Koch's introduction of pure culture study on solid media in 1881
(p. 6) that bacteria divide to produce more bacteria of similar kind, and therefore that they must have some hereditary apparatus; likewise, they have long
been known to give rise to genetically stable variants ('sports'). Yet only in
recent decades has it been found that bacteria have mating systems which allow
us to construct linkage maps (p. 110) analogous to those of fungi. Thus the twin
areas of genetical research—mutation and recombination—are both open to the
bacterial geneticist.

Bacteria have many advantages for genetical study. Consider, for example, *Escherichia coli*, perhaps the most thoroughly investigated of all bacteria. It exists in the form of single cells, with all the genes in a single linkage group (the genetical equivalent of a *chromosome*), although usually more than one copy of this linkage group is present in each cell; sometimes, as an exception to this, a few genes are in one or more distinct linkage groups called *episomes* or *plasmids* (p. 119). It is easily cultured under standard conditions, and it has an extremely short generation time (*c.* 25 min.; *cf.* 25 years in man). Most important of all, huge numbers of cells can be studied, and thus mutants can be selected with ease. These properties have enabled bacteria, and the viruses that parasitize them, to be the subject of more fundamental research than is possible with any other group of living things. The exciting discoveries that have emerged from this work, and from some parallel studies on the human haemoglobins, have drawn together *genetics*, which deals with the characters of organisms (*phenotype*) and their relation to the heritable factors determining those characters (*genotype*), and *biochemistry*, which deals with nucleic acids and enzymes. The result has been a virtual collapse of the barriers between the two sciences, and a wholly new science, *molecular biology*, has appeared.

The purpose of this chapter is to survey the classical genetics of eukaryotes (which will be restricted to fungi) and to see how these genetical principles have worked out in bacteria and viruses. We have not hesitated to introduce specialized words since these are often useful in mastering complex ideas. The reading list contains a number of accessible books and articles, at various levels, in which these ideas are more fully expounded, and in which more adequate explanations of topics which we have been obliged to treat here only briefly will be found.

NATURE AND REPRODUCTION OF GENETIC MATERIAL

The genome: nucleic acid or protein?

Although it has long been evident that cellular metabolism is controlled by the nucleus, the physico-chemical nature of the genetic material (*genome*) was not defined until the last decade. The only parts of the eukaryote nucleus that persist through cell division are the chromosomes, which consist of deoxyribonucleic acid (DNA) and a complex basic protein (histone). The interphase nucleus also contains varying amounts of ribonucleic acid (RNA), but this cannot be a candidate for the role of genetic material because it is largely confined to the nucleolus, which disappears during cell division.

Because DNA and histone replicate and divide together it is technically difficult to determine which is the genetic material. The first clue was indirect: it was found that a simple protein—protamine—replaces histone in sperm, thus pointing to DNA as the only substance with genetic continuity. This was confirmed when it was discovered (p. 118) that pure DNA could transfer information from one bacterial strain to another; since then refined experiments, using isotopically labelled DNA and protein precursors, have shown beyond doubt that only the DNA is transferred intact from one generation to the next in organisms as far removed as bacteriophage and man.

Division of the genome

Viruses

The amount of genetic information carried by viruses is insufficient for them to initiate and complete their own replication. Because of this, they take control of the infected cell's metabolic system to energize their replication. For example, when a bacteriophage infects a bacterium, synthesis of bacterial DNA, RNA, and protein stops within five minutes; then the bacterial system begins to reproduce the phage DNA and later to form phage coat protein. Similarly, when a tobacco cell is infected with tobacco mosaic virus (TMV), it first synthesizes virus RNA and then the virus coat protein. In other words, viruses do not divide—they force other systems to copy them.

Bacteria

Electron microscopy has shown that the DNA of *E. coli* exists as a closed ring; DNA replicates by sequential division of the two single strands in this ring (p. 139). Circular DNA has not been observed in *Bacillus subtilis*, although it is possible that the DNA of this organism is in circular form during replication. DNA replication in bacteria does not occur at the same time as cell division, so that there may be more than one copy of the chromosome in a cell; the mean number of chromosomes per bacterium in a culture of *E. coli* varies between one (in resting populations) and four (during exponential growth).

Mitosis

We have seen that genome replication in prokaryotes is a continuous process which may be out of phase with cell division. In the more complex cells of higher organisms, where the DNA is organized into more than one chromosome, there is usually a regular sequence of replication—the *mitotic cycle*. Immediately after the mitotic division there is usually a period (G1) when RNA and protein, etc. are synthesized. Then in the synthetic (S) phase the DNA and histone of the chromosomes doubles. Following S there is another growth phase, G2, and only then is the cell capable of a further mitosis.

In mitotic *prophase* the chromosomes become visible as double strands (*chromatids*) which are attached to each other only at a point of constriction, the *centromere*. At the end of prophase the nuclear membrane breaks down and a spindle is formed. The centromeres become attached to the spindle fibres and come to lie on the equator of the cell at *metaphase*. *Anaphase* is marked by the sudden division of the centromeres, whose daughters, with their attached chromatids, travel or are pulled towards opposite poles of the spindle. At *telophase* the daughter chromatids coalesce and nuclear membranes are formed around the two new nuclei. In this way identical genomes are passed to the two daughter cells.

Meiosis

Underlying the many different life cycles of eukaryotes there is a basic alternation of *haploid* cells containing one set of genetic information and *diploid* cells containing two sets. *Meiosis* is the process by which one diploid nucleus gives rise to four haploid nuclei (fusion of two haploid nuclei results in formation of a *zygote* and restores the diploid condition).

The meiotic process differs from mitosis in two ways—there are two successive nuclear divisions without intervening synthesis of DNA, and the nuclei produced are not identical to each other or to the parent nucleus. Because of the complexity of meiotic prophase it is convenient to divide it into a number of stages. When

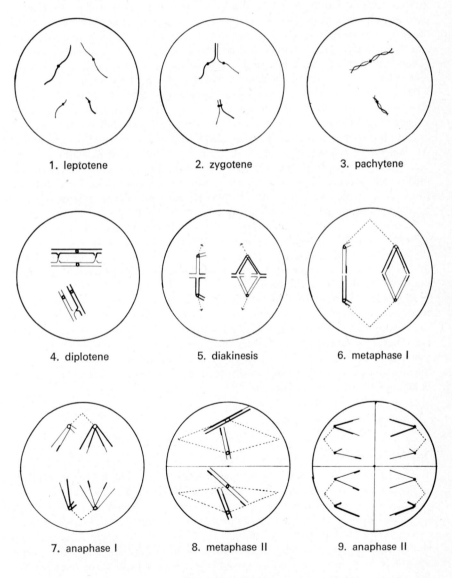

1. leptotene   2. zygotene   3. pachytene

4. diplotene   5. diakinesis   6. metaphase I

7. anaphase I   8. metaphase II   9. anaphase II

Fig. 4.1 Meiosis.

the chromosomes first become visible in *leptotene*, they appear as single strands (*cf.* mitosis) although DNA synthesis is complete. During *zygotene*, homologous chromosomes from the two parental strains (shown as heavy and light lines in Fig. 4.1) come to lie together, twining around each other so that in *pachytene*, when pairing is complete, it is impossible to distinguish them. These pairs of homologous chromosomes are now referred to as *bivalents*. In *diplotene* the bivalents open out and each chromosome is seen to consist of two chromatids; at various points along the bivalent pairs of chromatids are seen to have exchanged partners. These exchanges or *chiasmata* (sing. *chiasma*) are the products of breakage and reunion of chromatids during pachytene, at which time there is a small amount of DNA synthesis, and they are the cytological expression of genetic *crossing-over* (p. 110). The homologous undivided centromeres repel each other and but for the chiasmata would pull the bivalents apart.

In *diakinesis* the nuclear membrane breaks down and a spindle is organized. The centromeres become attached to spindle fibres and jockey for position until at *metaphase I* the individual bivalents are positioned with their chiasmata more or less on the equator and their centromeres randomly orientated on either side. At *anaphase I* the attraction between paired chromatids lapses and the still undivided centromeres pass to opposite poles of the spindle to give two *telophase I* nuclei each with the haploid number of undivided chromosomes. Note that as a result of crossing-over the sister chromatids of these chromosomes (or half bivalents) are not genetically identical.

At this stage there may be a short *interphase*, but more usually the chromosomes pass directly to *prophase II* and two new spindles are organized. This time the centromeres come to lie on the equator of the spindle by *metaphase II* and divide normally at *anaphase II*. The daughter chromatids pass to the four poles to form four haploid nuclei.

Although superficially akin to mitosis, this second nuclear division differs from it fundamentally because of the random orientation of centromeres and the crossing-over which took place during prophase I. These two factors ensure that all four products of meiosis are different. In the Euascomycetes (p. 397), which have been important in many genetic investigations, the *tetrad* is arranged linearly and each member undergoes a further genuine mitosis to form an ascus containing a row of eight ascospores.

MUTATION

Nuclear and extra-nuclear mutation

The phenomenon of *mutation* is fundamental to genetics; it is a prerequisite of the evolution and, thence, the recognition of new genes. A mutation can be defined as any permanent change in the quality or quantity of an organism's heritable information. Not all this information is confined to the chromosomes; for example, both plastids and mitochondria are known to have genetic continuity and contain their own DNA. The existence of nuclear genes controlling plastid development, however, indicates that this specific DNA is not sufficient for plastid autonomy.

Since the genome is replicated and divided exactly between daughter cells, chromosome mutation can be detected almost immediately in haploid organisms and will appear in specific proportions in diploids following meiosis and fertilization (p. 100). On the other hand, if one plastid mutates in a cell containing twenty, a number of cell generations must elapse before a cell is formed which by chance contains none but mutant plastids. There is no reason to suspect that the mutational events in extra-nuclear particles are in any way different to those within the chromosomes, but the technical difficulties of their isolation and characterization have so far led to concentration of research on nuclear genes.

Most spontaneous mutations occur at a frequency of the order of 10^{-7}–10^{-9} per cell generation, and their behaviour can usually be interpreted as a change in the nucleotide sequence specifying the phenotype studied. Some genes, however, mutate at much higher frequencies, and it is sometimes necessary to invoke ingenious explanations for these *unstable genes*. The following account is restricted to the better-understood mechanisms of nucleic acid mutation and of structural changes in chromosomes.

Nucleic acid mutation

Substitution

Substitution mutations are of two types: *transition* mutations involve the replacement of a purine by a purine or of a pyrimidine by a pyrimidine, e.g., when A is replaced by G in DNA; *transversion* mutations are those in which a purine is replaced by a pyrimidine or vice-versa (e.g., A by T). Both types usually lead to a simple change of one amino acid in the gene product (*mis-sense* mutations, Fig. 4.2); if, however, a single base change generates a nonsense triplet (e.g., CAG to UAG; see p. 55), translation of m-RNA is terminated at the site of the mutation and the resultant polypeptide is shorter than its wild-type analogue (*nonsense* mutations).

DNA (sense strand)	C A T C G T T T A	C A T C A*T T T A
RNA (messenger)	G U A G C A A A U	G U A G U*A A A U
Polypeptide	– Val – Ala – Asn –	– Val – Val* – Asn –

Fig. 4.2   Replacement of alanine by valine following a G to A transition(*) : mis-sense mutation.

Addition or deletion

Unlike the single amino acid changes induced by substitution, deletion or addition of a single nucleotide leads to a one-base shift in the frame of reading of the code. Such a *frameshift* mutation (Fig. 4.3) produces a completely new polypeptide and is therefore likely to alter the phenotype (p. 99) much more profoundly than a simple mis-sense mutation. Sometimes a more drastic type of mutation occurs in which more than one nucleotide is deleted in a single mutational event; these *multisite* or *deletion* mutants, in which the base

C deleted

```
DNA   A G C  G C G  A T G  A C G  A        A G C  G G A  T G A  C G A
RNA   U C G  C G C  U A C  U G C  U        U C G  C C U  A C U  G C U
Poly-  –  Ser  –  Arg  –  Tyr  –  Cys  –    –  Ser  –  Pro*  –  Thr*  –  Ala*  –
peptide
```

Fig. 4.3 New polypeptide chain(*) produced by deleting a single base from a triplet coding for arginine: frameshift mutation. It is assumed in this figure and in Fig. 4.2 that the m-RNA is read from 'left' to 'right'.

sequence corresponding to an entire polypeptide is occasionally lost, are extremely useful in genetic mapping (p. 121).

Chromosome mutation

It is not desirable in this brief account to describe all the types of chromosome aberration that can occur; inversions, duplications, deletions, and translocations of chromosome segments have been found. They can all be placed in one of two categories: those which affect the viability of the cell's immediate progeny, and those with a delayed effect on the organism's fertility. The molecular mechanisms of mutation are almost certainly similar in the chromosomes of bacteria and those of higher organisms, but the complex structure of the latter makes it difficult to recognize any alteration apart from gross structural changes.

Viability

A break in a chromosome has no effect on the cell containing it so long as no division takes place. At division, however, those chromosome fragments which lack a centromere usually fail to reach the telophase nuclei, which therefore lose some genetic information. In a diploid nucleus, the other homologue may be able to supply the lost information and there may be no immediate effect, but in haploids loss of information can lead to early death.

Fertility

Exchange of chromosome parts within or between chromosomes usually has no immediate physiological effect on the cell and its mitotic progeny since there is no change in the total information content. At meiosis, however, crossing-over between paired aberrant chromosomes frequently leads to the formation of unbalanced gametes with consequent reduction of fertility.

Chemical mutagens

Chemical mutagens are perhaps of greater intrinsic interest than physical mutagens because it is sometimes possible to predict their effects from a knowledge of their properties. They fall into three groups, not always mutually exclusive: nucleotide analogues; products which attack nucleic acid or chromosomes directly; and those products which, through their interference with various metabolic processes, indirectly affect the stability, synthesis, or division of DNA or chromosomes.

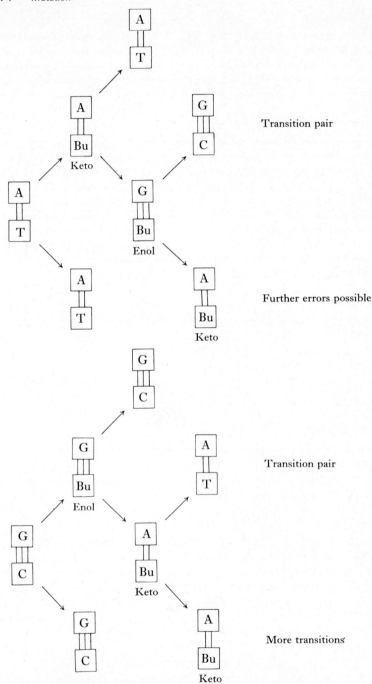

Fig. 4.4 Transitions induced by 5-bromo-uracil.

Nucleotide analogues

Fig. 4.4 shows how incorporation of a typical analogue, 5-bromo-uracil (BU), which in its normal keto form can pair with adenine and in its rare enol form with guanine, can lead to AT → GC transitions. Similarly 2-amino-purine (AP), which normally pairs with thymine, can occasionally pair with cytosine without changing its tautomeric state, again leading to AT → GC transitions.

Direct action

Nitrous acid also leads to transition mutations, by deaminating A, C, and G to give respectively hypoxanthine (pairs with C instead of T), uracil (pairs with A instead of G) and xanthine (pairs with C, like G; hence no effect). *Alkylating agents* induce transitions by upsetting base-pairing relationships (e.g., methylated guanine can pair with T instead of C), but they can also cause loss of bases from the sugar phosphate backbone. Base deletion may be followed by replacement by either a purine or a pyrimidine (i.e., both transitions and transversions are induced), or it may be followed by rupture of the phosphate backbone leading to chromosome breakage. Another important mutagen is *proflavin*, which is thought to cause both deletions of and short additions to DNA.

Indirect action

As well as including the indirect effects of mutagens which fall into the first two groups, this group includes drugs, hormones, respiratory inhibitors, excess O_2, pH changes, and all the other factors which may upset the cell or its metabolism in any way.

Physical mutagens

Ionizing radiations

Ionizing rays and particles can produce all the changes outlined above and are therefore much less specific than chemicals. The variations in the effects of different types of radiation depend very largely on their penetration and on the density of the induced track of excited radicals.

Other radiation

Ultra-violet light of wavelength 250–270 mμ is absorbed by DNA and is frequently used to induce mutations in unicellular organisms, bacterial and fungal spores, and pollen grains, etc. UV has little effect on multicellular organisms, however, because of its lack of penetration. Infra-red and visible light may also be mutagenic under rather special conditions; the latter can induce chromosome aberrations in the presence of oxygen if the chromosomes have been pre-treated with acridine orange in the dark.

Miscellaneous

Ultrasonic treatment and centrifugation have also been used to upset cell division and produce chromosome aberrations.

Spontaneous mutation

If large numbers of progeny from a single source are examined it is always possible to find spontaneous mutants, but one cannot be certain that these have not resulted from mutagenic effects in the background. Since it is possible to select strains with high mutation rates, and since some genes are more likely to mutate than others, this is clearly not a simple situation. Apart from the mutagenic background, there are cellular control mechanisms, such as DNA repair enzymes, which could affect mutability in different strains.

BIOCHEMICAL MUTANTS

Prototrophs and auxotrophs

Prototrophic (effectively = wild-type, symbol +) micro-organisms can often be grown on simple media. *Chlamydomonas* requires only inorganic salts to grow normally in liquid culture or on agar plates in the presence of light. *E. coli* will grow on a similar medium supplemented with an energy source such as glucose, while *Neurospora crassa* requires the vitamin biotin in addition.

Auxotrophic mutants are unable to grow on the minimal media which will support prototrophs because they cannot synthesize particular compounds. For example, *Neurospora* mutants which cannot synthesize arginine will only grow on arginine-supplemented medium. Such mutants are given the symbol *arg*; there are similar abbreviations to designate alleles conferring requirements for other amino acids and vitamins (p. 109).

Auxotroph selection

It is simple to pick out a prototrophic mutant in an auxotrophic population—only the mutant will be able to produce a colony on a suitable selective medium. It is less easy, however, to select auxotrophic mutants from prototrophs because the prototrophs will grow just as well as the mutants, if not better, on supplemented medium.

One way of solving this problem depends on the fact that penicillin kills only growing bacteria. If a suspension of *E. coli* prototrophs is treated with a mutagen and then grown in minimal medium containing a lethal concentration of penicillin, all the cells which are capable of growth will be killed. After a suitable delay the remaining bacteria, including any auxotrophs present which cannot grow because of lack of various supplements, are centrifuged down, washed, resuspended in penicillin-free medium, and poured on to supplemented agar plates. The resultant colonies are tested and classified as requiring various amino acids or vitamins for normal growth.

Similarly, mutants of mycelial organisms such as *Neurospora* can be isolated by the *filtration technique*. Treated spores are suspended in liquid minimal medium and allowed to grow for a short time. The suspension is then filtered; the ungerminated auxotrophs pass through the filter, while the germinated prototrophs are retained and discarded.

Auxotroph classification

The original investigations of biochemical mutants, which led to the formulation of Beadle and Tatum's 'one gene—one enzyme' hypothesis (p. 128), were carried out by laboriously examining the growth requirements of thousands of unselected *N. crassa* spores. Lederberg, however, developed a *replica plating* technique; this was originally used to demonstrate the spontaneous nature of mutation of *E. coli* to phage resistance in the *absence* of phage, but it also provides an easy way of classifying mutants of organisms which grow as discrete colonies and it has been used with modification for *Neurospora* too.

For replica plating (Fig. 4.5), a suspension of treated bacterial or fungal cells is poured on to a master plate containing medium supplemented with those

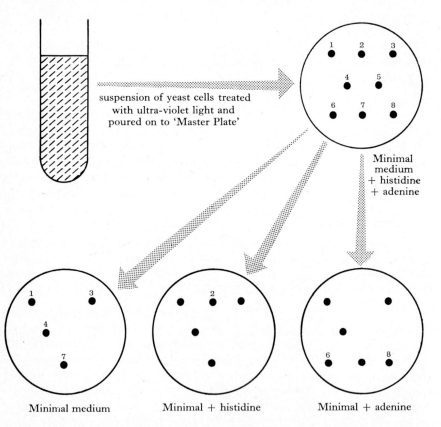

Fig. 4.5 Selection of adenine- and histidine-requiring mutants by replica-plating.

growth factors which will allow the desired mutants to grow. After incubation, a piece of sterile velvet mounted on the end of a cylindrical block is pressed gently on to the plate; this 'replicator' picks up a few cells from each colony on the master plate, and it is pressed successively on to a series of plates containing minimal medium alone and with various supplements. After further incubation, colonies appear on these replica plates in the same pattern as on the master plate; absence of a particular colony from a plate is presumptive evidence that it requires the supplement missing from that plate.

Biochemical pathways

One of the first biosynthetic pathways to be worked out was that of arginine. Arginine-requiring mutants of *N. crassa* were found to fall into a number of groups including:

1. Those which grew on ornithine, citrulline, or arginine.
2. Those which grew on citrulline or arginine, but not on ornithine.
3. Those which grew only on arginine.

This suggested that arginine is the product of the sequence:

$$\text{Precursor} \xrightarrow{\ 1\ } \text{Ornithine} \xrightarrow{\ 2\ } \text{Citrulline} \xrightarrow{\ 3\ } \text{Arginine,}$$

and that the three groups of mutants are deficient in the enzymes mediating reactions 1, 2, and 3 respectively. The enzymes were subsequently isolated and shown to catalyse these reactions. Some mutants can be shown to accumulate excess precursors before the blocked reaction—for example, adenine-requiring mutants of yeast produce a red pigment which distinguishes them from the colourless wild-type colonies. Numerous biochemical pathways have since been elucidated in this way.

GENETIC RECOMBINATION AND MAPPING IN EUKARYOTES

Meiosis and Mendel

The principle of segregation

When two haploid yeast cells fuse, one containing a chromosome bearing the gene *A* mediating, say, the citrulline → arginine reaction and the other with its homologous chromosome bearing the mutant *allele a*, the diploid (*Aa*) will be able to grow on minimal medium unsupplemented with arginine. Thus in the presence of the *dominant* gene *A* we can no longer detect the presence of the *recessive* allele *a*. After meiosis, however, two of the haploid cells produced will contain *A* and the other two *a*, so that in the haploid generation only half the cells will be able to grow on minimal medium. In other words *A* and *a*, although associated together in the diploid, segregate independently and unchanged into the haploid cells produced by meiosis.

Mendel originally deduced this principle of segregation, in 1865, by observing the products of fertilization in pea plants, which are diploid; we can observe it directly in yeast, however, which has a haploid generation. Following random fusion of haploid yeast cells carrying the two genes, yielding diploids, we can distinguish between the *genotypic* ratio of 1 *AA* : 2 *Aa* : 1 *aa* and the *phenotypic* ratio (p. 99) of three prototrophs to one auxotroph.

The principle of independent assortment

If we now consider a second gene pair B and b, mediating, say, biotin production versus requirement, carried on a *second* pair of homologous chromosomes, we again get the genotypic ratio of 1 BB : 2 Bb : 1 bb and the phenotypic ratio of three biotin producers to one biotin requirer, following random fusion of cells carrying either B or b. When we observe the inheritance of *both* sets of genes, however, we find that the haploid progeny of the $AaBb$ diploids includes four types of cell in approximately equal numbers:

1. Grow on minimal medium; hence genotype AB.
2. Require biotin; hence genotype Ab.
3. Require arginine; hence genotype aB.
4. Require both arginine and biotin; hence genotype ab.

This indicates that the two factors A and B segregate at random independently of each other.

Chromosome mapping

The exception: linkage

We have considered the inheritance of two pairs of genes carried on different pairs of chromosomes. If we look at the haploid progeny of an $AaBb$ diploid in which both pairs of genes are carried on the *same* pair of chromosomes, however, then we find that the principle of independent assortment no longer holds good. A and B still segregate from a and b, but no longer at random.

If A and B were originally on one of the parental chromosomes and a and b on the other, then Ab and aB, the *recombinants*, will only be formed if crossing-over takes place between A and B. In Fig. 4.6, where A and B have been placed at equal distances along the chromosomes, this means that if there is only a single chiasma per bivalent (p. 102) then only $\frac{1}{3}$ of the tetrads will include the recombinant haploids Ab and aB. There will therefore be an excess of the parental types (AB and ab derived from the original haploids) and the two genes are said to be *linked*. Further, if a number of genes are found to be linked in this way, they are said to constitute a *linkage group*, implying that they all belong to the same chromosome. Note that the two recombinant classes, Ab and aB, will be present in approximately equal numbers, since both parents contribute similar haploid chromosome complements to the cross.

Random spore analysis

The information gained from the study of linked genes can be used to determine the order and degree of separation of the genes along the chromosome. Thus in Fig. 4.6 one-third of the tetrads are 1 AB : 1 Ab : 1 aB : 1 ab while the other two-thirds total 4 AB : 4 ab; that is, *one-sixth* of the haploids are recombinant. This $\frac{1}{6}$ expressed as a percentage gives a map distance of 16·67 units (known as *centimorgans*) between A and B. If we then find a third gene C to be 12 units from B and 29 units from A (i.e., 58 per cent of tetrads include the results of chiasmata between A and C), we can deduce that the linear order of the genes is:

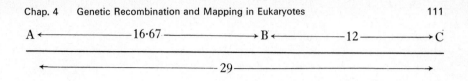

Gene order

Although this method of genetic mapping is slightly more complex than we have made it appear, it is possible to build up a linkage map of all the genes in a chromosome in this way, using the so-called *three-point test cross*, on the principle that the probability of recombination between any two genes is proportional to the distance between them.

Interference

The three-point test cross reveals occasional double cross-overs, which are the products of independent chiasmata occurring in the two regions *AB* and *BC*. Now, the chance of the two occurring together in the same bivalent is the product of the chances of either event occurring singly—16·67 per cent × 12 per cent, i.e., 2 per cent, in this example. But the actual number of double cross-overs observed is often significantly lower than the expected number as

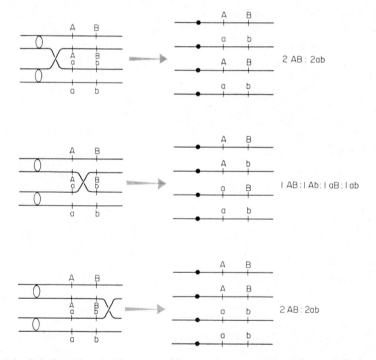

Fig. 4.6 Relationship of chiasma position to the segregation of two linked genes, *A* and *B*.

calculated in this way, since a chiasma in one region is found to reduce the chance of another occurring in an adjacent region. This phenomenon, known as *interference*, is explained by the mechanical restraints imposed on crossing-over by the rigidity of the chromatids.

Interference is usually expressed as the *coincidence*, i.e., the ratio of expected to observed double cross-overs. As we should expect, the degree of interference usually increases as the distance between cross-overs decreases. For a discussion of the *negative* interference observed in mapping very closely linked genes the student should consult an advanced textbook; see also p. 124.

Tetrad analysis

So far, the mapping procedures described have relied on analysis of the geno-types of random spores. In ascomycetes such as *Neurospora* and *Sordaria* where *linear* tetrads are formed the products of individual meioses can be isolated mechanically and analysed. When spore colour mutants are studied this analysis can be carried out by direct microscopic observation of the tetrads and we can use these observations to determine the map distance of the gene from the centro-mere. For example, the *asco* mutant of *Neurospora* has colourless instead of black

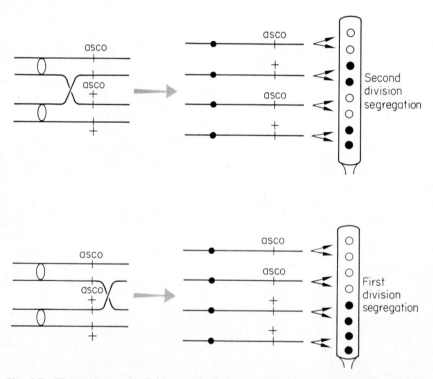

Fig. 4.7 First and second division asci of the *asco*/+ heterozygote of *Neurospora crassa*.

spores, and the distribution of black and colourless spores in the asci of a heterozygote depends on whether or not a chiasma has occurred between the gene and the centromere. Thus if we find that 20 second division asci occur for every 80 first division asci, we know that a chiasma occurs between the centromere and the gene in 20 per cent of the asci (Fig. 4.7).

Note that in tetrad analysis we are counting crossover asci, *not* crossover spores containing crossover chromatids. This is important to remember because in a second division ascus only *half* the spores contain chromosomes derived from the crossover chromatids. Because of this, if we wish to equate gene-centromere distances obtained in this way with the map distances obtained by random spore analysis, we must divide the gene-centromere distance by two. In this example the map distance between the gene and the centromere is:

$$\frac{\text{Second division asci}}{\text{Total asci}} \times \frac{100}{2} = \frac{20}{100} \times \frac{100}{2} = 10 \text{ units.}$$

Similarly, if we obtain linkage data from tetrad analysis, the scores must always be divided by two to equate them with those obtained from random spore analyses.

Mitotic crossing-over

Although meiosis has not been observed in the imperfect fungus *Aspergillus niger*, Pontecorvo has discovered a parasexual system in this species which allows gene recombination. If strains of the same mating type are allowed to grow together, they form hybrid mycelia (*heterokaryons*) in which the nuclei of both strains co-exist in the same cytoplasm; the haploid nuclei, however, occasionally fuse to form diploids which continue to divide with the haploids in the same mycelium. By chance sectors arise containing only the heterozygous diploid nuclei, and sometimes non-meiotic haploidization occurs ($c.$ 10^{-3} per mitosis). Some of the haploid conidia derived from these sectors turn out on investigation to contain both marker genes, thus proving that mitotic crossing-over must have taken place in the heterozygous diploid nuclei at some stage. Since this demonstration, mitotic crossing-over has turned out to be an important tool in genetic studies in both ascomycetes and basidiomycetes, and it has even been suggested that it may be of use in human tissue culture investigations.

MATING SYSTEMS IN BACTERIA

Partial genetic transfer

We have seen that in eukaryotes both parents in a cross contribute similar chromosome complements to the progeny. The various mechanisms of mating in bacteria all differ from those in higher organisms in that one parent (the 'female' or recipient) contributes all its genetic material, while the contribution of the other (the 'male' or donor) is usually much smaller; three systems are known:

1. Conjugation: a tube is formed between donor and recipient cells through which DNA is transferred.
2. Transduction: a small piece of DNA is carried from donor to recipient by a bacteriophage (p. 263) which acts as vector.
3. Transformation: a small piece of donor DNA is adsorbed to the recipient cell (without the mediation of a phage vector).

Conjugation

The F factor

Conjugation in *E. coli* was first studied by Lederberg and Tatum in 1946. As an illustration, consider a strain requiring the three amino acids A, B, and C (genotype $a^- b^- c^- d^+ e^+ f^+$); this is mixed with a second strain requiring amino acids D, E, and F ($a^+ b^+ c^+ d^- e^- f^-$) and after an interval the mixture is plated on minimal medium, which selects prototrophs ($a^+ b^+ c^+ d^+ e^+ f^+$). Under these conditions Lederberg and Tatum recovered about one prototroph from every 10^7 cells plated. This could not be accounted for by simultaneous back-mutation of all three mutant genes in either of the parents (e.g., a^- to a^+), and yet the prototrophs could be propagated indefinitely on minimal medium. Cell-to-cell contact between the two strains was necessary, and also, if both were distinguished by character differences which were not involved in selecting the prototroph (e.g., phage-resistance or sensitivity), these *unselected* markers behaved as if they were linked to the selected markers. In this way the various genetic markers were accommodated, at first, in a single linear linkage group.

Reciprocal recombinant classes were not found in these early experiments, as would be expected if the two parents had formed gametes by a meiotic process (p. 100). A finding of great importance was that genetic transfer was a one-way process; one of the parents donated genetic material, but the viability of the *other* parent alone was essential for recombinant formation. Finally, the order in which the markers were linked depended on which particular strains were crossed.

The experiments which resolved this paradoxical situation are described by Hayes (1968), who proposed in 1953 that transfer of genes was mediated by an agent present in donor cells called the F (fertility) or *sex* factor. He showed that the donor (F$^+$) character could be transmitted to up to 95 per cent of recipient (F$^-$) cells when fresh cultures were mixed, although prototrophs were formed only at low frequency. The F factor is therefore an infectious particle, quite independently of its ability to transfer bacterial genes; it is now known that cells carrying F synthesize a specific organ on the cell surface, the *sex pilus* or *fimbria* (p. 305, Fig. 9.15), which can interact with recipient (F$^-$) cells and mediate the passage of genetic material (probably the donor genes pass *through* the sex pilus).

How is possession of this F factor related to the ability to transfer genes? The rare recombinants inherit few of their genes from the donor (F$^+$) strain; they more closely resemble the recipient (F$^-$). Thus the donor transfers only part of its genome to the recipient. This was elucidated when it was found that occasional donors, when mixed with recipients, could yield recombinants at a frequency about 1,000 times higher than the F$^+$ strain from which they originated; these high frequency (Hfr) donors differ from F^+ donors in some other respects, and a comparison of the two types follows:

1. Conjugal pairs are formed between donor and recipient cells at high frequency in both types of cross ($F^+ \times F^-$; Hfr $\times$ F^-); these can be seen by electron microscopy.
2. $F^+ \times F^-$ yields few recombinants, whereas Hfr $\times$ F^- yields recombinants at high frequency.
3. When F^+ is mixed with F^-, the F factor itself is usually transferred to F^-; in contrast, an Hfr donor very rarely transfers its F factor.
4. Genetic analysis of $F^+ \times F^-$ recombinants shows that F is not linked to any particular gene. All the cells of a particular Hfr strain, however, exhibit *oriented transfer* of donor genetic material; the donor genes incorporated by the recipient depend on the amount of donor chromosome transferred. The F factor is linked to different donor genes in different Hfr strains.
5. Hfr and F^+ donors are interconvertible, and both can revert to the F^- state. F^- strains, however, never revert spontaneously to F^+ or Hfr; they can only acquire the F factor by infection from another F^+ culture.

We can now propose an explanation for all these findings. Since different Hfr strains can be isolated which exhibit different orders and orientations of transfer of genes, there must be a single *circular* linkage group in *E. coli*. To form an Hfr donor, the F factor enters this circular structure and breaks it at the point of entry; it then behaves as if attached to one end of the (now linear) linkage group and can mobilize the transfer of the *other* end through the conjugation tube formed by the agency of the sex pilus. Each Hfr strain consists of a population of cells each of which can transfer its chromosome in the same manner, such that one end always enters the recipient cell first. The recipient cell becomes diploid (p. 100) for the region transferred, and after recombination occurs it segregates normal haploid daughter cells. In a F^+ donor, the F factor is not attached to the chromosome, but it can cause the formation of a conjugation tube and move through it.

There is good evidence for this interpretation; we describe here only the *interrupted mating experiment* which is widely used in genetical analysis. Consider an Hfr donor which transfers a group of linked genes (shown here as a^+ b^+ c^+ d^+ e^+) to recipients (a^- b^- c^- d^- e^-) at high frequency, and let us suppose that when a^+ is selected the other four genes appear among the recombinants at these frequencies: 90 per cent b^+, 75 per cent c^+, 40 per cent d^+, and 25 per cent e^+. Thus this Hfr donor transfers its genes in the order a^+ b^+ c^+ d^+ e^+. Fresh cultures of Hfr and F^- are mixed, and samples are removed at intervals and agitated at high speed, a process calculated to break any conjugation tubes but not to affect cell viability. It is found that samples agitated at intervals up to, say, 8 min. after mixing yield no recombinants; at 8 min., a^+ recombinants appear; at 9 min., a^+ b^+; at 10 min., a^+ b^+ c^+; at 17 min., a^+ b^+ c^+ d^+; and at 25 min., a^+ b^+ c^+ d^+ e^+. Therefore the agitation procedure must break the donor DNA during its transfer to recipient cells; the time between mixing and observation of a donor gene in recombinants is the time taken for the donor to transfer that gene, which is a measure of the distance of the gene from the end of the chromosome penetrating the recipient.

Conjugation in genetic analysis

Many Hfr strains have been isolated, all with different points of entry of the F factor into the chromosome and thus different origins of transfer, but the 'time interval' between any two genes is always the same. It is possible to draw a complete map of the *E. coli* linkage group 'calibrated' in terms of times of transfer. The whole chromosome takes about 89 min. to be transferred, the last marker to enter the recipient being F itself; the chance of the conjugation tube breaking during this time is, however, quite high, and hence the rarity of transfer of F in Hfr $\times$ F$^-$ crosses.

Substituted F factors

We must mention one other type of donor. The F factor, which is made of DNA, occasionally recombines with the chromosome of the cell which carries it; when this happens, it picks up a small region of the chromosome, into which may be inserted a corresponding piece of DNA from F. The importance of donors carrying these *substituted* F factors (F', F prime) is that they transfer only a small piece of the donor chromosome (*sexduction*); the recipient cell therefore becomes diploid for this region and complementation tests (p. 130) can be carried out.

Conjugation in other bacteria

The F factor can be transferred from *E. coli* to *Shigella* and *Salmonella*, and conjugation systems have been described in other bacteria.

Transduction

Unrestricted transduction

The first observation of transduction—the transfer of genetic material from donor to recipient by a phage vector—was made by Zinder and Lederberg in *Salmonella typhimurium* in 1952. They prepared a U-tube at the bottom of which was a filter impermeable to bacteria; on one side of the filter was placed a histidine-requiring culture, and on the other a tryptophan-requiring culture that was lysogenized (p. 263) with phage P22. The tube was incubated, gentle suction being applied to each side alternately to allow mixing of the culture supernatants through the filter. Prototrophs (*his$^+$ trp$^+$*) appeared at a frequency of about 10^{-7} on the side initially containing the tryptophan-requiring lysogen, showing that phage P22 particles had migrated to the side containing the histidine-requiring strain, picked up *trp$^+$* markers from it, migrated back, and converted the *trp$^-$* strain to *trp$^+$*.

It has since been shown that phage P22 lysates can transduce almost any marker from a *S. typhimurium* donor on mixing with a suitable recipient (*unrestricted* or *generalized transduction*). It is unusual, however, for a recombinant to receive more than a very few adjacent donor genes; only if two markers are closely linked can they be transduced by a single phage particle (*joint trans-*

duction). Thus two differences between conjugation and transduction are evident:

1. Conjugation requires intimate contact between donor and recipient cells; for transduction, it is sufficient to treat the recipient with a suspension of transducing phage grown on the donor culture.
2. An Hfr donor can transfer a variable and sometimes large piece of its chromosome to a recipient during conjugation, while the phage vector in transduction carries only a small piece of donor DNA. This is understandable when it is remembered that the amount of DNA in a phage particle is only about 1 per cent of that in the bacterial chromosome.

The mechanism of transduction is probably as follows. A phage particle infects a cell of the donor culture; the chromosome of this cell breaks up, and part of it (the *transduction fragment*) is incorporated into a progeny phage particle. This particle is released in the cell lysate, and it infects a cell of the recipient culture; the transduction fragment is then incorporated by recombination into the recipient cell's chromosome. It is important to note that as a result of acquiring some bacterial DNA the phage vector loses its own viral genetic material; because of this it cannot replicate in and lyse the recipient cell. Transducing phage particles are therefore *defective* in the functions of replication.

Restricted transduction

We can refer to P22 as an unrestricted transducing phage since it can transduce almost any donor gene in *S. typhimurium*; phage P1 performs a similar role in *E. coli*. A different type of transduction in *E. coli* is mediated by phage λ (restricted transduction). If a *gal*$^+$ (galactose-fermenting) strain is infected with λ, lysogens are formed which we may designate (λ) *gal*$^+$; these lysogens can be *induced* with UV light (p. 124), whereupon λ is enabled to replicate and lyse its host cells. When this lysate is mixed with a *gal*$^-$ strain, a few *gal*$^+$ transductants appear (*c.* 10^{-6}); when, however, one of these *gal*$^+$ transductants is induced and the lysate is used to infect a fresh *gal*$^-$ culture, *gal*$^+$ transductants now appear at a frequency as high as 10^{-3}—about half the λ particles released can transduce *gal*$^+$.

The transduction frequency here is about 1,000 times greater than in the *S. typhimurium*—phage P22 system, and only the galactose and other closely linked genes can be transduced. The explanation is that λ associates with the chromosome just beside the *gal* region (p. 125); in a (λ) *gal*$^+$ lysogen, a λ particle occasionally picks up a *gal*$^+$ fragment and transduces it to a *gal*$^-$ recipient. This recipient is converted to *gal*$^+$, but instead of the transduced *gal*$^+$ fragment recombining with the *gal*$^-$ chromosome, which is what would happen in P22-mediated (generalized) transduction, it is *reversibly* inserted (p. 126) adjacent to the galactose region. If this transductant is also infected with a *non*-transducing (lysogenizing or *helper*) λ particle, which is almost certain to occur if there is a high proportion of phage particles to recipient bacteria, and if the transductant is induced, the λ *gal*$^+$ particle is enabled to replicate because the functions in which it is deficient are supplied by the non-transducing particle. So the lysate contains about half normal λ and half transducing (λ *gal* $^+$) particles—and hence the large number of transductants when a further *gal* $^-$ stock is infected.

Transduction in genetic analysis

Transduction has been reported in many bacterial species; its value is that it allows fine structure analysis to be carried out within a small region of the genes, corresponding to a single transduction fragment, where the time intervals in conjugational transfer would be too short to be accurately measurable by interrupted mating experiments; it is probable that recombination between two adjacent nucleotides can be detected. The principle is that if two strains carry mutations at different sites within a small region, and a transduction is carried out between them, the number of recombinants should be proportional to the distance between the mutational sites (two-factor crosses). This method has various sources of error which limit its application, but some of these can be eliminated if the sites being studied happen to be in a gene which can be jointly transduced with a second gene; the number of recombinants arising in different crosses can be standardized by examining how many incorporate a marker from this linked gene.

Transformation

Transforming DNA

The discovery of transformation precedes that of the other two systems by several years. Griffith, in 1928, injected mice with live non-capsulated and heat-killed capsulated pneumococci (p. 301); neither of these proved fatal when injected separately, but when both were injected together, some of the mice died and living capsulated bacteria were recovered from their blood. This same conversion of non-capsulated to capsulated pneumococci was later demonstrated *in vitro*, and it was found that a cell-free extract of the capsulated strain could carry out the transformation. Griffith at first supposed that the agent (*transforming principle*) that transformed the non-capsulated strain was the capsular polysaccharide of the killed culture, but later work has identified it unequivocally with DNA. Transformation has since been demonstrated in many bacteria, although work has been concentrated on the pneumococcus, *Haemophilus influenzae*, and *Bacillus subtilis*; it is also found that the transforming principle can carry any donor gene.

In contrast to transduction, the size of the DNA fragment transferred depends on how much the donor chromosome is broken down during extraction; in general it is probably about 1/200 the size of the chromosome. Joint transformation of two donor markers (*cf.* joint transduction) occurs only when they are very close together. Labelling experiments show that DNA penetrates the recipient cells, but the mechanism is not entirely clear; it is probable that only one strand of the incoming DNA recombines with that of the recipient. Not all the cells in a population are physiologically capable of taking up DNA (*competent*); if transformation is carried out in synchronously dividing cells, it is found that cells are competent during only a small period of the growth cycle.

Once the transforming DNA is inside the cell, it may not be incorporated into the recipient's DNA for 2 or 3 generations, during which time it does not divide with the recipient chromosome. Incorporation is probably by breakage and reunion of the two pieces of DNA; there is then a lag of a few generations before expression of the incorporated material can be detected.

Transformation in genetic analysis

A DNA fragment of similar size is transferred in transduction and in transformation, and both processes can therefore be employed in fine structure analysis; transformation has the added advantage that it can be used to study bacteria in which no transducing phage is available.

Extra-chromosomal elements in bacteria

Episomes and plasmids

The F factor can exist in two states in the cell: either *autonomous* (independent of the chromosome, as in F$^+$) or *integrated* (as part of the chromosome—Hfr). This property is also shown by temperate phages (p. 263)—e.g., λ can be autonomous (vegetative phage) or integrated (prophage). Although the F factor and temperate phages have such different physiological functions, both have about 1 per cent as much DNA as the bacterial chromosome, both can be lost and re-acquired by infection, and both can pick up pieces of chromosomal material and carry these to other cells (F$'$ factors and λ *gal*$^+$ particles; see above).

Jacob and Wollman proposed in 1958 that genetical elements able to exist in these alternative states should be called *episomes*; an episome can exist either in an autonomous state, in which it replicates independently of (often faster than) the chromosome, or in an integrated state in which it replicates synchronously with the chromosome and behaves essentially as a small region of it. An episome in the autonomous state is rather like a small chromosome. We can distinguish episomes from *plasmids*: the latter term is used for genetically similar bodies such as plastids and mitochondria (p. 365) which have *not* been observed to associate with the chromosome/s, although recent work has shown that this distinction may have little foundation when applied to bacterial elements.

Many temperate phages are known besides λ; we mention now two groups of F-like factors and also the staphylococcal penicillinase plasmid.

Colicinogenic factors

Some *E. coli* strains possess extrachromosomal particles which enable them to produce proteins called *colicins* which are lethal to other strains of the same species. These *colicinogenic* (Col) factors, like F, can be transferred to non-colicinogenic strains by cell contact, but they have not been shown unequivocally to be linked to any chromosomal gene. Many different Col factors can be distinguished, according to the particular colicin produced, but there are probably two classes as judged by the sex pilus which mediates their transfer: some are transferred by means of a pilus resembling that synthesized by F, and others by means of a distinct (I) pilus. Col factors can promote the transfer of chromosomal genes, and they confer immunity to the action of the colicin they produce and related colicins.

R *factors*

The property of *transferable drug resistance* (p. 582) in the *Enterobacteriaceae* is determined by a unit which can be transferred from cell to cell independently of chromosomal genes. This unit is the R factor; it consists of a *resistance transfer factor* (RTF) to which a variable number of genes conferring drug resistance are attached. R factors, like F and Col, can promote chromosomal transfer, and they confer immunity to superinfection (p. 126) like temperate phage. Two types are distinguished according to whether or not they interfere with an F factor present in the same cell: fi^+ R factors repress F functions (*fertility inhibition*) and fi^- do not; again, some use an F-like and some an I-like pilus.

Penicillinase plasmids

The agent determining inducible penicillinase synthesis in the staphylococcus is a plasmid. It differs from the F, Col, and R factors in that it is transferred not through a conjugation tube but by transduction (a matter which perhaps concerns the genus of the host rather than the factor itself); it can replicate autonomously. The plasmid carries both regulator and structural genes (p. 134) for penicillinase synthesis, as well sometimes as genes conferring resistance to erythromycin and to various inorganic ions.

VIRAL GENETICS

Genetic analysis

The bacteriophage as an ideal virus

The study of a virus is inseparable from that of its host cell. It is therefore not surprising, when the simple genetic apparatus of bacteria is compared to that of eukaryotes, to find that far more is known of the genetics of phage than of other viruses. Also, despite the superficially complex structure which enables phages to penetrate the bacterial cell wall, the phage life cycle is much better understood than that of any other group of viruses (p. 259). This section is thus mainly concerned with phages, among which most is known of the double-stranded DNA coliphages T2, T4, and λ.

Mutants

A phage is detected by its ability to form a plaque on a sensitive host strain layered on solid medium; thus the easiest type of mutant to observe is one with altered plaque morphology. Phage T4 forms small and fairly clear plaques on *E. coli* B; mutant strains may produce large clear plaques (*r*, rapid lysis), very small plaques (*m*, minute) and plaques with turbid rings (*tu*). Another type of mutation alters the range of strains on which the phage can form plaques (*h*, host-range).

Other mutants are detected by examination of artificial lysates under the electron microscope; some mutations prevent production of phage heads and infective DNA, and others affect components of phage tails or block lysozyme formation and thus prevent release of mature particles. These *defective* mutants are either completely unable to mature, or they are *conditional* defectives, i.e., they mature only in suitable conditions. Two types of conditional defective mutant are known:

1. Temperature-sensitive, *ts*. Wild-type phage T4 forms similar numbers of plaques at 25° or 42°; *ts* mutants form plaques at 25° only.
2. 'Amber', *am*, These cannot form plaques on *E. coli* B, but they grow on a strain (CR 63) which has a *suppressor gene* (p. 132) for the amber mutation; amber is one of the three nonsense triplets (UAG; p. 55).

Many other mutants are known, some of which will be mentioned later. The value of conditional defective mutations is that they can be sited in *any* phage gene, unlike, e.g., plaque-type mutations. Let us examine how they are plotted.

Phage crosses

The principle of phage crosses is that two genetically marked stocks are allowed to infect one host strain at multiplicities such that most cells receive at least one particle from each stock. It sometimes happens that when a cell receives one particle, a second *superinfecting* particle is broken down and thus unable to participate in recombination; before phage is added, therefore, the host cells are treated with KCN, which allows phage adsorption but prevents host metabolism. After allowing a few minutes for adsorption, anti-phage serum is added to remove unadsorbed phage; the infected cells are plated out at suitable dilutions with a sensitive host, which effectively removes the KCN and allows phage development to proceed synchronously in all the cells, and parental and recombinant classes can be scored.

A genetic map can be built up by carrying out two-factor crosses (p. 118) in this way. Benzer studied rapid lysis (*r*) mutants of phage T4; these fall into three groups, *r*I, *r*II, and *r*III, according to the hosts on which they form large plaques, and all three are located in discrete regions of the map. Some 2,400 *r*II mutants were collected by Benzer; since it would require no less than $2,400^2$ two-factor crosses to plot all these, he devised an elegant short-cut known as *deletion mapping*. The principle is that if a known region of a mutant phage is deleted (p. 103), a second mutant can give wild-type recombinants when crossed with this deletion mutant only if its mutational defect does not lie in the region corresponding to the deletion in the latter. A large number of deletion mutants have been accurately delineated, which saves many two-factor crosses when plotting new mutations.

Organization of genetic material

Linkage group versus chromosome in phage T4

When phage T4 mutants are mapped as described above, the resulting linkage map shows two features of interest:

1. All the markers can be accommodated in a single circular linkage group.
2. The genes are grouped in clusters related to their functions in phage development. Beginning at an arbitrary point in the circular map, we find, sequentially, genes involved in DNA synthesis, tail fibre production, head protein, tail components, and finally lysozyme (required for liberation of mature phage).

It is necessary to distinguish clearly the terms *linkage group* (or *map*) and *chromosome*. We have seen from conjugation studies that *E. coli* has a single circular linkage group (p. 115) and it will also be recalled that microscopy reveals a single circular chromosome (p. 100). There is little doubt that the genetic map corresponds to the cytological chromosome. But these two approaches reveal quite different entities in phage T4, where the circular linkage map cannot be correlated with a circular chromosome. Streisinger has found that the chromosome populations of phage T4 (and also T2) lysates show two curiosities:

1. The chromosomes behave as if they were linear structures resulting from breakage of the circular linkage group at random points, i.e. the genes are *circularly permuted.*
2. The sequence of about 1,000 base pairs at one end of each chromosome (double helix) is repeated at the other end (*terminal redundancy*).

These apparent complexities point to quite a simple explanation for the maturation of T4 DNA. Let us assume that the genome of T4 is divided into 10 units (designated a–j) and that the region of terminal redundancy in each chromosome extends for two of these units. Because of circular permutation of the genes, a chromosome population might consist of a b c d e f g h i j *a b*, e f g h i j a b c d *e f*, h i j a b c d e f g *h i*, etc. (note: each chromosome here has ten units plus two 'redundant' units—in italics—at the end). This suggests that each 12-unit chromosome is cut from a long DNA polymer consisting of repeated phage genomes (a b c d e f g h i j a b c d e f g h i j a b c d . . .), and that the length of DNA in each chromosome is determined by the amount that can be accommodated in a phage head. In support of this, Frankel has observed a long-chain intermediate (*concatenate*) in T4 replication. Also, it is known that if part of the T4 genome is deleted (p. 103), the length of the region of terminal redundancy is increased accordingly; if, in our representation above, units c and d were deleted, but the phage head still accommodated 12 units, we should find chromosomes of type a b e f g h i j *a b e f*, g h i j a b e f *g h i j*, etc.—i.e. there is now a terminal redundancy (again in italics) of *four* units.

Phage lambda

The genes of the temperate phage λ resemble those of T4 in being found in clusters associated with different functions in development. The chromosome, however, has no terminal redundancy, nor are the genes circularly permuted; although the chromosome is found in ring form during replication, it is linear when in the phage head. λ has one peculiarity not found in T4: about 16–18 nucleotides protrude from the 5′-end (p. 25) of each strand of the double helix, and these single-stranded ends are found to be complementary to one another in base composition. On annealing λ DNA *in vitro*, the ends bond together and the molecule is converted into a ring.

Viral nucleic acids

The features of viral DNA revealed by T4 and λ are by no means restricted to these alone. In particular, many viruses (polyoma, SV40, papilloma, and the single-stranded DNA phage ϕX174) and other virus-like bodies (extra-chromo-somal elements, p. 119) contain DNA in ring form. The DNA of the T-even phages (T2, T4, T6) is unusual in containing 5-hydroxymethylcytosine instead of cytosine, and other chemical differences are known. Many viruses differ from other life forms in that their genetic information is encoded in RNA instead of DNA; it has been shown beyond doubt that these RNA viruses specify the products necessary for their own synthesis, and that RNA replication does not require the formation of a DNA intermediate. Some (RNA bacteriophages, poliovirus, TMV) contain their RNA as a single polynucleotide strand, and others (reovirus, wound tumour virus, rice dwarf virus) as a double strand. Intermediates have been detected in the replication of these viruses, but the details are in doubt.

Gene expression during phage development

We have emphasized that the genes of T4 and λ are grouped in clusters related to their functions in development. These clusters in T4 are linked in the order in which their functions are expressed; those involved in DNA synthesis (early functions) lie in a different part of the map to those specifying structural features and lysozyme formation (late functions). It has long been known that mutants which cannot make DNA are also unable to form any structural components or lysozyme, although all the genes except the mutated one are intact. Conversely, in the normal development of wild-type phage, the early genes are no longer expressed once the late functions begin to appear. This implies the presence of a *control* mechanism (p. 133); some step associated with DNA synthesis is required to 'switch on' the late genes, and a late function in turn must 'switch off' the early genes. There is some evidence that the λ chromosome is divided into two parts which are *transcribed* (p. 137) as separate units for this purpose.

Recombination in bacteriophage

Participants in phage crosses

The technique for crossing phage strains was described on p. 121. If, for example, we cross a rapid lysis host-range mutant ($h^-\ r^-$) with the wild-type strain ($h^+\ r^+$), it should be possible to recover parental-type ($h^-\ r^-$, $h^+\ r^+$) and recombinant ($h^-\ r^+$, $h^+\ r^-$) progeny. In contrast to the three bacterial mating systems, both parent phages contribute their entire chromosomes to the cross.

Since in mating experiments each host cell is infected with about ten particles and ultimately releases a burst of some two hundred, clearly we cannot study the outcome of *single* mating events. During phage maturation (p. 262), a DNA pool is formed from which chromosomes are incorporated into mature particles, and recombination occurs among the genomes in the pool. Studies on premature and delayed lysis of infected cells support the view that mating occurs at random among the replicating genomes, and that recombinants can themselves take part in further recombination events. Visconti and Delbrück showed that the results could be explained if each chromosome underwent an average of five rounds of mating; their approach was based on assumptions which are not all valid, but probably it is essentially correct. Let us note some curiosities of phage crosses.

Evidence from genetical experiments

1. Triparental Matings. If bacteria are infected with *three* genetically marked particles, recombinants incorporating all three markers can be recovered.
2. Non-reciprocity. So far we have envisaged recombination as a reciprocal process, e.g., a cross between h^+r^+ and h^-r^- should yield h^+r^- and h^-r^+ recombinants in equal numbers. But single-burst phage experiments do *not* always give this result; there is evidence that *single* recombination events do not yield both recombinant types, but the significance of this is not known.
3. Negative Interference. This is the tendency of a recombination event to *increase* the probability of a second recombination occurring nearby (*cf.* p. 111).
4. Phage Heterozygotes. If *E. coli* is infected with r^+ and r^- strains of T4, about 2 per cent of the liberated progeny particles produce mottled plaques. These particles possess both the r^+ and r^- alleles and are therefore *heterozygous*; on further propagation, heterozygotes continue to appear at a frequency of 2 per cent, the remaining 98 per cent being segregants containing r^+ or r^- only. The heterozygous region may extend over a few loci in any part of the T4 genome, and there is some evidence that one source of heterozygosity lies in the region of terminal redundancy (p. 122) described above.

Conclusions: the evidence from labelling experiments

The fate of isotopically labelled components during recombination has been studied in phage λ, which unlike the T-even phages undergoes an average of only 0·5–1 instead of 5 rounds of mating during the growth cycle; recombinants incorporate label from both parents, so that the process must involve *breakage and reunion* of parental molecules. This mechanism is also supported by work on the T-even phages. More data must be obtained, however, before all the genetic findings can also be explained; perhaps the study of strains deficient in recombination capacity (*rec* mutants) will shed some light on the problem.

Lysogeny

Prophage and the lysogenic cell

The phenomenon of lysogeny (p. 263) has been studied most thoroughly in λ, the prophage of which occupies a specific region in the bacterial genome (see below); while integrated at this position, the prophage replicates synchronously with the host chromosome. In general, phage arising from a lysogenic cell is genetically indistinguishable from the phage used to establish lysogeny, and prophage can mutate and recombine with other prophages; thus we can regard prophage as being similar to any other genetic material. Its presence confers certain properties on the lysogenic cell:

1. The cell can release a burst of free phage if the prophage enters the vegetative state. This occurs at a low frequency which can sometimes be increased, especially with λ, by *induction* with agents such as ultra-violet light which break down the mechanism restraining prophage from lytic development.
2. The cell is *immune to superinfection* by the same or closely related phages (p. 126).

3. The cell may acquire new antigens or other factors which result from expression of prophage genes (*lysogenic* or *phage conversion*). In a sense, property 2 is an example of lysogenic conversion, but the term is usually reserved for physiological changes less obviously associated with the presence of prophage; two well-known examples are:
   a. *Corynebacterium diphtheriae* only produces toxin if it is lysogenized with phage β; non-lysogenic strains are invariably atoxigenic.
   b. Various antigenic patterns among group E salmonellae are associated with the presence of ϵ prophages.

It is quite possible that lysogenic conversion is very similar to transduction, and it is hard to lay down a rigid distinction between them. In general, however, a phenotypic change (p. 99) is recognized as resulting from lysogenic conversion only if it is referable to an active prophage, in contrast to the defective fragment usually carried in a transducing particle (p. 117).

The relation between the prophage and the bacterial chromosome

It has been shown that the prophage of phage λ is associated with the *E. coli* chromosome near the galactose region. If a lysogenic strain is crossed with a non-lysogenic one, the lysogenic character segregates among recombinants and is correlated with inheritance of galactose genes; further, different λ mutants are closely linked to their parental *gal* alleles. Jacob and Wollman have also identified specific chromosomal locations for many other prophages.

Whatever the mode of association with the bacterial chromosome, it is reversible: a lysogenic cell occasionally loses its prophage (without vegetative development of the latter), and the resulting *cured* cell then becomes sensitive to infection by further phage and can again be lysogenized by it. We can therefore propose two possible interpretations of the mode of association (Fig. 4.8):

1. The prophage is *synapsed* to the chromosome and lies alongside it.
2. The prophage is *inserted* into the chromosome and becomes continuous with it.

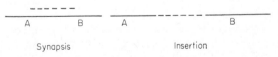

Fig. 4.8 Two possible modes of association of a temperate phage chromosome (------) with a region of the bacterial chromosome (———). A and B are bacterial genetic markers. Redrawn by kind permission of Blackwell Scientific Publications Ltd.

None of the early experiments designed to distinguish between these alternatives decisively ruled out either model; recently, however, evidence has accumulated in support of the insertion hypothesis, at any rate in λ and related phages, and in 1962 Campbell proposed a mechanism for insertion and excision of prophage with which this evidence is in good accord. It will be recalled that although the λ chromosome is linear when in the mature particle, it is in ring form when intracellular (p. 122). Campbell envisages that this circular phage chromosome has a 'recognition region' corresponding to its region of insertion into the bacterial chromosome; recombination occurs between these regions, as a result of which the phage chromosome unloops and becomes continuous with the chromosome of its host. If this process is reversed, the prophage is excised.

The best evidence for Campbell's idea of 'unlooping' is that the order of the genes in vegetative phage is found to be a permutation of that in prophage (shown in Fig. 4.9). The model has the further merit that it suggests a simple mode of formation of restricted transducing particles (p. 117) by occasional errors in prophage excision.

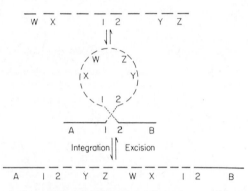

Fig. 4.9 The Campbell model for insertion of prophage (–––) into the bacterial chromosome (———). A and B are bacterial markers; W, X, Y, and Z are phage markers, and 1 and 2 represent the 'recognition regions' (see text) where recombination (✕) occurs. Redrawn by kind permission of Professor A. Campbell and Acad. Press Ltd.

Superinfection immunity

If a cell carries a λ prophage and is then superinfected with a second λ particle, the latter injects its DNA, but usually this neither replicates nor is incorporated as further prophage; this *superinfection immunity* also extends to closely related (homo-immune) phages. There are two possible explanations: the prophage initially present either blocks a region of the host chromosome that is required for phage replication or prophage insertion, or it produces a cytoplasmic *immunity substance* which prevents further development of the superinfecting particle. The second of these can be shown to be correct by carrying out conjugational crosses using Hfr donors (p. 115). If the recipient is non-lysogenic and the donor lysogenic, it is found that as soon as the λ prophage enters the cell it immediately leaves the donor chromosome, replicates, and lyses the recipient, producing a burst of mature phage. This so-called *zygotic induction* does not occur if the cross is carried out the other way round (i.e., if the Hfr donor is non-lysogenic and the recipient lysogenic), nor does it take place if both donor and recipient harbour λ prophages carrying different genetic markers.

Since zygotic induction only occurs when a prophage enters the cytoplasm of a non-immune cell, it follows that phage replication requires the absence of an immunity substance (*repressor*) in the cytoplasm.

The attachment specificity of a prophage—i.e. its specificity for the region of the host chromosome where it is inserted—must be distinguished from its immunity specificity. We can make hybrids between λ and related phages, e.g., $\phi434$, which derive their immunity specificity from one parent and their attachment specificity from the other.

The immunity substance not only inhibits superinfecting phage; it is also the agent which prevents prophage from expressing most of its genes and replicating as vegetative phage. The mechanism of immunity has been elucidated by the isolation and mapping of phage mutants:

1. Clear Plaque (*c*) Mutants. These mutants cannot form prophage; they can only develop vegetatively, and hence form clear plaques (whereas temperate phages normally form plaques with turbid centres; see Chap. 8). Complementation tests (p. 130) show that they plot in three genes, c_1, c_2, and c_3, the first of which is now known to be the gene which produces the immunity substance. In mixed infections with c_1 and wild-type ($c_1{}^+$) particles, lysogenic clones which carry both prophages can be recovered; these clones can segregate $c_1{}^+$ lysogens, but never c_1, consistent with the idea that $c_1{}^+$ is dominant to c_1 and produces a repressor of phage development which is absent in c_1. Similar mutants have been isolated from the temperate phage P22.
2. Non-inducible (*ind*) Mutants. These are insensitive to induction (p. 124); they plot within the c_1 gene, and if double lysogens are made (as above) it turns out that ind^- (non-inducible, mutant) is dominant to ind^+ (inducible, wild-type). ind^- mutants therefore behave as if the cytoplasmic immunity substance were unusual in not being antagonized by ultraviolet light.

This interpretation of immunity has been confirmed with the isolation of the immunity substance itself, which has been shown to be a protein and to bind to λ DNA but not to the DNA of a λ mutant of different immunity specificity.

Conclusions

Little is known of the sequence of events which determines, when a cell is infected, whether the genome of a temperate phage will be incorporated as prophage or develop vegetatively. This section gives a simplified account of our knowledge of lysogeny; we shall see later (p. 141) that there are striking parallels between lysogeny and the mechanism of regulation of enzyme production.

Animal and plant viruses

Most of what is known of the genetics of non-bacterial viruses derives from cytological studies. Thus, it is known that some plant virus infections (p. 244) result in the production of more than one nucleoprotein component of which only one is fully infectious (squash mosaic virus, tobacco necrosis virus); tobacco rattle virus also produces two components, one of which seems to carry the genetic information for replication and the other for coat protein. Among animal viruses (p. 239), patterns of dependence and interference are found, especially among viruses of the leucosis complex, which are reminiscent of phage behaviour. The genetical analysis of animal viruses is difficult mainly for technical reasons: few virus particles in a suspension may succeed in establishing infection, and differences in plaque character are much less reliable than in phage. Some virus genomes may be as small as 1 per cent of that of phage T4, lowering the chance of recombinants being detected accordingly. Recombination, however, has been observed in various animal viruses, among which influenza virus has an unexpectedly high rate. The recent isolation of temperature-sensitive mutants among these viruses offers the most promising hope of constructing genetic maps.

An agent has been described, causative of Scrapie (p. 244) in sheep, which is extremely stable to heat and is inactivated neither by nucleases nor by ultra-violet light at the wavelength specifically absorbed by DNA. It is too early to conclude that the genetic material of the Scrapie agent is something other than a nucleic acid, but if this turns out to be so, it will be the first example of an entirely new basis of heredity.

GENE EXPRESSION

The gene and its product

One gene—one enzyme

The concept of 'one gene—one enzyme' has been considered in Chapter 2 (p. 57), and reference to that chapter indicates that the concept sometimes has to be modified to take account of some types of regulatory gene that appear to have no protein product. It is more accurate (see below) to consider the gene as specifying a *polypeptide*. The folding of the polypeptide product of a gene is determined by the sequence of amino acids in the polypeptide chain, and is hence dictated by the structural gene (p. 134) and not by any additional gene.

The colinearity of gene and polypeptide

The idea that the gene is translated into a sequence of amino acids is strongly supported by the finding that the two are *colinear*: the site of a gene mutation corresponds to the position of the altered amino acid in the chain. This was originally shown in haemoglobin, but it is also demonstrated by Yanofsky's studies on *E. coli* tryptophan synthetase. The reaction mediated by fully-aggregated tryptophan synthetase, AB, is conversion of indoleglycerol phosphate to tryptophan, but A and B can catalyse the reaction independently via indole, although at a much lower rate than the conversion mediated by AB. The A and B proteins can be extracted from mutants defective in one or other subunit. Yanofsky has shown that mutations affecting A protein activity usually act by changing one amino acid (mis-sense mutations, p. 108). He was able to locate the position of the altered amino acid in each mutant by proteolytic degradation, and this always correlated with the site of the mutation in the genetic map.

Similar studies have been carried out with amber (nonsense, p. 103) mutants of phage T4 head protein. The site of each amber mutation in the genetic map can be correlated with the position at which polypeptide chain synthesis is interrupted.

The cistron: unit of function

The cis-trans test in phage T4

Genes were once thought of as indivisible units of the chromosome ('beads on a string'); this became untenable when it was shown that crossing-over can occur within a gene, and we now know that the minimal units of recombination and mutation are far smaller than the gene itself—both probably correspond to a single base pair. For the unit of function, however, we must look again at Benzer's rapid lysis (*r*) mutants of phage T4 (p. 121). Of the three groups of *r* mutants, *r*II is characterized by inability to grow on a particular strain of *E. coli*, K12 (λ); Benzer found that the *r*II mutants could be subdivided into two further groups, *r*IIA and *r*IIB. The distinction was based on whether or not plaques were formed when K12 (λ) was infected by *two r*II mutants at the same time; if the two mutants were both *r*IIA, or both *r*IIB, no plaques appeared, but if one was *r*IIA and the other *r*IIB then plaques were formed. Further, *r*IIA and *r*IIB mutations plot in different parts of the *r*II region of the genetic map.

The explanation (Fig. 4.10) is that *r*IIA and *r*IIB are two different *units of*

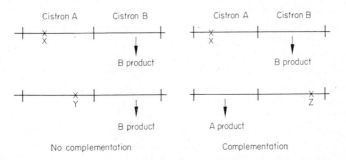

Fig. 4.10   The cis-trans test in the *r*II region of phage T4 (only the trans arrangement is shown). X and Y are *r*IIA mutants; they can not complement one another since the cell they infect has no cistron A product. Z is an *r*IIB mutant; it can complement X as the infected cell has the products of an intact A and an intact B cistron. Redrawn by kind permission of Blackwell Scientific Publications Ltd.

function, and so long as a K12 (λ) cell is infected by two phages mutant in different units of function it contains all the genetic information for phage replication—i.e., a plaque of wild-type phenotype can be formed in which each *individual* particle is still, nevertheless, an *r*II mutant. When the two mutations are carried by different phage particles, they are said to be in the *trans* position; if a phage is constructed which carries both mutations in the same chromosome they are in the *cis* position. Hence the name of *cis-trans test*: if two mutations in the *trans* position allow formation of a phenotypically wild-type plaque they are in different units of function, and if not, they must be in the same unit of function. The *cis* part of the test is a control to confirm that both mutations are recessive; clearly if a cell is co-infected with a double mutant and a non-mutant phage, the latter should form a wild-type plaque anyway. In practice, it is usually adequate to do the *trans* test only, and this is called the *complementation test.*

The cis-trans test, therefore, defines the unit of function (the *cistron*); the *r*IIA and *r*IIB cistrons are said to *complement* each other since they supply different phage functions. The cistron is the unit which, when transcribed into messenger RNA, specifies the production of a single polypeptide chain.

Complementation tests in bacteria and fungi

The value of the complementation test is that it does not merely define two particular cistrons in phage T4; it can be extended to other functions in *any* organism as long as two provisions are made:

1. Two genomes, each carrying a single mutation, must be introduced into the same cell; we have seen that this is achieved in phage by co-infection of one host cell by two mutant particles.
2. Recombination must *not* be allowed to occur between the two genomes. When it is recalled that recombination can probably occur between any two different mutant sites, it is obvious that a non-mutant genome could be formed and the *trans* arrangement of the two mutations would become *cis*— which would invalidate the test. This can be checked in the *r*II situation described above by confirming that the plaques formed on K12 (λ) contain *r*II, and not wild-type (r^+), particles.

Two methods satisfy these provisions in bacteria. First, we can set up partial diploids using F′ factors (sexduction, p. 116), which incorporate a small piece of the bacterial chromosome into their structure. If, say, we isolate an F′ carrying part of the lactose region (F-*lac*) from a strain carrying the mutation *lac-1*, and infect a second strain carrying *lac-2*, the recipient becomes diploid for the lactose genes; thus if *lac-1* and *lac-2* are in different cistrons, the cell will ferment lactose. Essentially the same approach is to use defective λ transducing particles (λ *gal*; see also p. 117), although these are restricted to the galactose genes. The second method relies on the fact that when a transduction is carried out between two mutant stocks, the transduction fragment is not always incorporated into the recipient's chromosome. When this happens, the fragment does not divide with the chromosome, but remains in the cytoplasm (*abortive transduction*); such a cell is again a partial diploid, and the complementation test is carried out as follows. Transducing phage is grown on, say, a mutant *leu-1* and then used to infect a second mutant *leu-2*; the mixture is plated on medium selective for leucine-independence, and a few large colonies (the normal *complete* trans-ductants) appear as a result of successful recombination between the transduction fragment and the chromosome. If *leu-1* and *leu-2* are in different cistrons, any abortive transductants that are formed will contain all the non-mutant cistrons required for leucine synthesis—*but*, because the abortive transduction fragment does not divide with the bacterial chromosome, leucine synthesis is restricted to one cell alone. Thus the colony resulting from an abortive transduction is very small since it has only a limited degree of independence of leucine. The finding of such very small colonies, which may in practice be far more numerous than the complete transductants, indicates that *leu-1* and *leu-2* are in different cistrons. Abortive transductants are distinguished from, e.g., slow-growing mutants by streaking them gently on the surface of the medium; because of their genetic constitution, they give rise to only a single progeny colony each.

The principle of complementation tests in *Neurospora* and *Aspergillus* is to try to form a *heterokaryon* (p. 113), in which nuclei from different mutant strains share the same cytoplasm. If a mixed inoculum of two compatible strains is placed on a suitable selective medium, a heterokaryon should form if the strains are of the same mating type, in order to prevent ascus formation, and if they complement each other. Good heterokaryon formation under these conditions shows that the mutants are functionally distinct, but absence of growth does not necessarily mean absence of complementation; it is possible to resort to the *forced heterokaryon technique* (see Fincham and Day, 1963) in this event.

Intracistronic complementation

So far we have used complementation tests to define different cistrons, but a complication sometimes arises which at first seems to invalidate the test. Some pairs of mutants are found to map in the same cistron, and yet they exhibit partially wild-type phenotype when tested for complementation. This is known variously as *intracistronic, intragenic, interallelic,* and *partial* complementation, and it is distinct from the normal *intercistronic* (or *intergenic*) complementation

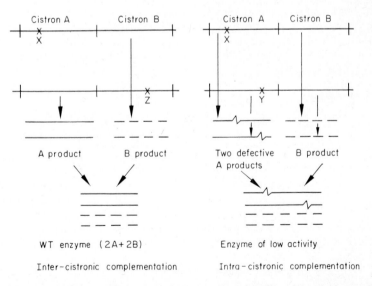

Fig. 4.11   Inter-cistronic and intra-cistronic complementation; X and Y are cistron A mutants, and Z is a cistron B mutant. (See text for explanation)

described above. The reason is that each cistron produces one polypeptide, but the enzyme on which phenotype is judged may be a polymer in which any one polypeptide is represented more than once. Fig. 4.11 shows a comparison between inter- and intra-cistronic complementation involving such an enzyme (e.g., tryptophan synthetase, p. 128). In intercistronic complementation, some wild-type enzyme is formed as the cell still has the products of both wild-type cistrons; in intracistronic complementation, wild-type enzyme can never be formed, but the defects in the two polypeptides *partially* compensate for each other and an enzyme of low activity is formed. It must be emphasized, however, that it is unusual for two mutants to specify defective polypeptides that happen to be mutually corrective, and the enzyme so formed has very low activity—sometimes barely detectable.

Conclusions

We have seen that the complementation test defines the unit of genetic function (or translation), the cistron; the potential complexity introduced by intracistronic complementation does not invalidate the catchphrase 'one gene—one enzyme' if this is re-phrased 'one cistron—one polypeptide'—the cistron is a convenient definition of the gene.

Suppressors

Isolation and characterization

When an auxotrophic culture is plated on medium selective for prototrophs, a few *revertant* colonies often grow, some at wild-type rate (probably back-mutants to wild-type) and some more slowly. Any slow-growing colony probably results from a *suppressor* mutation: a mutation at a site distinct from that of the primary mutation (and often in a different gene) which restores, at least partially, wild-type phenotype. Suppressor mutations are distinguished according to the level at which they suppress the phenotype conferred by the gene carrying the primary mutation.

'Indirect' suppressors

These act by altering the physiological conditions in the cell; three types are known:

1. The primary mutation may cause production of an enzyme which is inactive because it is sensitive to some ion in the cell, e.g., Zn^{2+}. If a second mutation reduced the zinc concentration, the enzyme would become active and the effect of the primary mutation would therefore be suppressed; this has been found in *N. crassa*.
2. The suppressor mutation may occur within a gene whose altered product can substitute for the defective enzyme specified by the gene carrying the primary mutation.
3. It is possible in some biochemical pathways, e.g., proline, for a suppressor mutation to open an alternative pathway so that the cell can by-pass the original metabolic block.

Intragenic suppressors

These act by altering the actual mode of expression of the gene carrying the primary mutation; they are of particular interest because they give information on the transcription and translation processes. One type is known to affect the translation of m-RNA by causing a change in the amino acid or codon specificity (p. 61) of t-RNA. Yanofsky studied a mutant in which a glycine residue was replaced by arginine at one position in the *E. coli* tryptophan synthetase A protein; a suppressor mutation which partly restored wild-type phenotype was then isolated. He found that the gene for the A protein (the *A* gene) in this suppressed strain produced two polypeptides—one (the majority product) had the arginine residue characteristic of the primary mutant, and a small proportion had a glycine residue (wild-type polypeptide). This can be understood when it is remembered that each amino acid has a specific t-RNA which directs its incorporation into growing polypeptide chains; the suppressor mutation probably acts by altering glycine t-RNA to a form which sometimes recognizes the codon for arginine, so that glycine is sometimes built into polypeptide in place of arginine. Since the original mutation causes the *A* gene to insert arginine in place of glycine, 'two wrongs make a right' and the suppressor mutation allows synthesis of a small amount of wild-type A protein. The same result could be obtained if other components of the translation apparatus were altered. Altered t-RNA has also been found to be responsible for suppression of nonsense mutants (p. 103).

One type of suppression results from interference with protein synthesis at the level of the ribosomes. Gorini has described *E. coli* mutants that normally require a growth factor but can grow without this factor if streptomycin is supplied (*conditional streptomycin-dependent* mutants); streptomycin is known to affect ribosomes (p. 75), and it causes misreading of the genetic code in these mutants so that, once again, two wrongs make a right and a 'correct' amino acid (i.e. characteristic of wild-type polypeptide) is sometimes inserted in place of the 'incorrect' one specified by the mutant gene.

Suppression and gene expression

It is possible to think of other ways in which a suppressor might intervene in the sequence of events between the gene and its product. We have tried to show, however, that if suppressor genes sometimes contravene the principle of 'one cistron—one polypeptide', they supply a number of insights into the genetic and biochemistry of the microbial cell.

THE REGULATION OF GENE EXPRESSION AND DNA REPLICATION

The biochemical reactions involved in gene expression have been described in Chapter 2 (p. 66). As pointed out there, the experimental evidence for the molecular processes described is largely genetic, and it is only recently that biochemical support for the original interpretations has been forthcoming. The genetic evidence is presented in the following section, and the extent to which it supports the biochemical models outlined in Chapter 2 will be apparent.

Historically, much of the elucidation of enzyme regulation came from a study of the lactose enzymes, and this system will be discussed in some detail.

Enzyme induction and repression: the Jacob–Monod model

Kinetics of induction of the lactose enzymes

When *E. coli* is grown in a medium containing glucose as carbon and energy source, it produces virtually no enzymes for breaking down the disaccharide lactose. If the cells are then washed and resuspended in lactose medium, three enzymes which degrade lactose are formed after a few minutes; these are:

1. β-galactosidase; hydrolyses β-galactosides (product of z cistron).
2. β-galactoside permease; concentrates β-galactosides in cell (product of y cistron).
3. β-galactoside transacetylase; acetylates β-galactosides (metabolic role not clear; product of a cistron).

If the cells are again washed and transferred back to *glucose* medium, synthesis of these enzymes ceases and they are diluted out by cell division. Lactose is therefore an *inducer* of the enzymes required for its own degradation; when it is added, the enzyme levels rise at a rate proportional to the increase of total protein in the culture.

Inducers

The lactose enzymes are not induced by lactose alone, but also by many molecules structurally similar to lactose. These other inducers are not necessarily degraded or even bound by the lactose enzymes, and some of them can induce higher enzyme levels than does lactose itself. Whatever the inducer, however, the enzymes induced are always the same, they are always induced synchronously, and the amount of any one enzyme induced is always the same relative to the other two. Thus the property of inducibility is independent of the enzymes induced.

The Jacob–Monod model

The property of inducibility can be examined genetically:

1. Mutants are found which form no lactose enzymes on addition of inducer, although when crossed they can be shown to possess intact z, y, and a cistrons; other mutants (*constitutive*) produce all three enzymes whether or not inducer is present.
2. The dominance relationships of these mutants, and the ability of the regions involved to complement one another, can be studied by making partial diploids using F′ factors carrying fragments of the lactose region (F-*lac*; p. 116).

This work is described in the classic paper of Jacob and Monod (1961), whose model for induction of the lactose enzymes is widely accepted. It is convenient to outline the current form of this model before discussing the evidence in its favour; it comprises a *regulator gene* (symbol i), three *structural genes* (z, y, a;

see above), and an *operator* (*o*) and *promoter* (*p*) closely linked to the structural genes (Fig. 4.12). The regulator gene produces a substance known as *apo-repressor* which can diffuse through the cytoplasm; in the absence of lactose, the

a. Molecular interaction in the uninduced state.

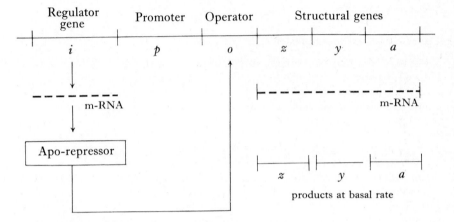

b. Interaction in the induced state.

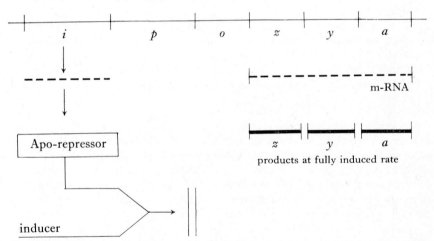

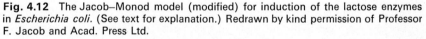

Fig. 4.12 The Jacob–Monod model (modified) for induction of the lactose enzymes in *Escherichia coli*. (See text for explanation.) Redrawn by kind permission of Professor F. Jacob and Acad. Press Ltd.

apo-repressor is able to interact with the operator, which in turn prevents the structural genes from being read into m-RNA (and thus into protein) so long as it is complexed with apo-repressor. When lactose is present, however, the apo-repressor forms a complex with the *inducer* (lactose), and this apo-repressor plus inducer complex is *not* able to interact with the operator; thus the structural genes can then be expressed.

It must be emphasized that the model, which at present we have deliberately left rather vague, is intended to account solely for the genetical findings. These inevitably generate some physiological predictions (for instance, we should expect to be able to isolate apo-repressor molecules), but several physiological alternatives remain tenable (see p. 138).

The regulator gene and regulator mutations

We have seen that the property of inducibility is independent of the enzymes induced; the genetic evidence for a regulator gene rests on the finding of mutations which affect inducibility and yet are not located in the structural genes whose expression is affected. The wild-type regulator allele is written i^+ (i.e., inducible), and among the many i mutants known the following are important:

1. i^-. The phenotype of i^- mutants is constitutive, i.e., enzymes are produced even in absence of lactose. If a partial diploid of the type i^+z^-/Fi^-z^+ is set up (i.e., i^+z^- in the chromosome and i^-z^+ in the F′ factor), it turns out that β-galactosidase is still inducible; thus i^- is recessive to i^+. This is consistent with the idea that the apo-repressor in i^- mutants is altered in such a way that it can no longer interact with the operator; as the partial diploid also contains i^+ gene product, the operators of both the chromosomal and episomal lactose regions can be prevented from 'switching on' the lactose genes adjacent to them.

2. i^s. i^s mutants are permanently non-inducible—no lactose enzymes are produced whether or not lactose is present, and i^s is dominant to both i^+ and i^-. The apo-repressor in such mutants can no longer be antagonized by lactose (*super*-repressor); it can interact with the operator and prevent it from 'switching on' the lactose genes even if the partial diploid also contains i^+ or i^- gene products.

What kind of molecule is the apo-repressor? The finding of a third type of mutant, i^t, which is inducible at low but constitutive at high temperatures, suggests that the i gene product can undergo the kind of temperature-sensitive modification characteristic of protein, and indeed the highly specific interaction between a small molecule such as lactose and the apo-repressor is hard to explain otherwise. In spite of this and other circumstantial evidence, it is only recently that the i product has been isolated—and it has the physical properties of protein.

The operator and operator mutations

The apo-repressor molecule must be able to interact specifically not only with the inducer, but also with some element which can 'switch' the z, y, and a cistrons on and off co-ordinately. This element is the *operator*, and the operator plus the three cistrons it controls together form the *operon*. Operator mutants can be isolated in which the operator region can no longer interact with apo-repressor molecules, so that the structural genes are expressed whether or not lactose is present; the phenotype of these operator-constitutive (o^c) mutants therefore resembles that of i^- mutants (see above).

How are the operator mutants distinguished from i mutants? Firstly, they plot in a different part of the map (see Fig. 4.12). A general feature of inducible (and repressible) systems is that the operator is always adjacent to the genes it controls; the regulator is not necessarily linked, although it happens to be so in the lactose system. Secondly, when the dominance relationships of o mutants are examined in partial diploids, it is found that in contrast to i alleles they act in the *cis* position only. To illustrate this, a diploid of genotype $o^c z^+ y^-/Fo^+ z^- y^+$ forms z gene product (β-galactosidase) constitutively and y gene product (β-galactoside permease) inducibly—in other words, an o allele affects only those structural genes that are on the same fragment (chromosomal or episomal) as itself. This is an important piece of evidence for the proposal that whereas i forms a cytoplasmic product (the apo-repressor), o does *not* form a product but is an integral part of the genetic region $o\ z\ y\ a$.

The operon: a unit of transcription

To understand how the apo-repressor prevents expression of the structural genes, recall that it has two properties—it recognizes a region (the operator) adjacent to zya, and it lowers the level of expression of all three structural genes co-ordinately. It is now thought that the zya region is transcribed sequentially, beginning at the operator end, into a single m-RNA molecule (*multicistronic messenger RNA*), and that this m-RNA contains nonsense codons (p. 103) which enable it to be translated into the necessary three polypeptides (we assume here that all three lactose enzymes consist of only one species of polypeptide each, although there is some evidence that this may not be so). Thus the genetic segment adjacent to zya probably consists of two regions: a recognition region for the apo-repressor (*operator*), and a region at which transcription is initiated, the *promoter*, probably by attachment of an RNA polymerase molecule. Recent work has shown that the promoter maps between i and the operator; perhaps the presence of apo-repressor blocks the progress of RNA polymerase from the promoter to the structural genes. Mutations which confer non-inducible phenotype, some of which were formerly thought to lie in the operator, are now believed to be of two types: promoter mutations (preventing transcription), and mutations at the operator-proximal end of the z gene (preventing translation of the multicistronic m-RNA).

We have seen that the cistron is the unit of translation of m-RNA; we can now add that the operon is the unit of transcription of DNA.

Repression of the histidine biosynthetic enzymes

An example of a repressible system is furnished by the histidine pathway. In absence of histidine, *S. typhimurium* produces some ten enzymes which catalyse steps in the pathway; when histidine is added, however, the synthesis of all these enzymes stops abruptly. In the words of Jacob and Monod, enzyme repression is 'very closely analogous, albeit symmetrically opposed, to the induction effect'. The histidine genes are all clustered together in an operon, but the difference to the lactose system is this: the histidine regulator gene (R) produces an apo-repressor which in absence of histidine does *not* interact with the histidine operator, while when histidine is present it forms a histidine–apo-repressor complex which interacts with the histidine operator and prevents expression of the genes in the operon. Histidine acts here as a *co-repressor* (*cf. lactose* as an *inducer*); the prefix *co-* distinguishes the exogenous repressor substance from the product of the regulator gene, the *apo*-repressor.

How widely does the Jacob–Monod model apply?

Both inducible and repressible enzyme systems are widespread among micro-organisms; in general, enzyme pathways for the degradation of exogenous substances are inducible by the molecules they degrade (β-galactosidase, etc., in *E. coli* by lactose, penicillinase in *B. licheniformis* by penicillin, histidase in *S. typhimurium* by histidine), whereas biosynthetic pathways are typically repressed by their end products (histidine enzymes in *S. typhimurium* by histidine, tryptophan enzymes in *E. coli* by tryptophan). There is little doubt that the Jacob–Monod model applies, at least in outline, to all these systems, but it is uncertain, for instance, whether apo-repressor molecules always act by preventing transcription; perhaps they block translation of m-RNA in some systems. There may be genetic differences, too. The dominance relationships of regulator gene mutants in the lactose system led to the idea of an apo-repressor, i.e., the regulator exerts *negative* control over the lactose operon. This negative control is found in inducible and repressible pathways generally, but the regulator gene exerts *positive* control in at least one inducible system—the arabinose pathway in *E. coli*, where the regulator gene product activates the arabinose operon rather than repressing it. We shall need to invoke this concept again for the regulation of DNA synthesis (see opposite).

Enzyme regulation is also known among fungi, but the histidine genes in *Neurospora* and yeast, instead of being clustered in one operon as in *Salmonella*, are scattered around the genome, although the pathways are identical. Only three of the ten genes for galactoside utilization in yeast are tightly linked, while the expression of all ten is co-ordinately controlled. The concept of the operon may be characteristic of bacteria rather than fungi, but even in bacteria a few exceptions are known: the arginine genes in *E. coli* are co-ordinately repressed by arginine, yet only four of them are clustered, while four more (plus the regulator gene) are unlinked.

Finally, although the rate of synthesis of many degrading enzymes, such as β-galactosidase, is controlled by specific induction mechanisms, there is an additional control known as *metabolite repression* or the *glucose effect*: in presence of glucose (or indeed of many carbon sources when in excess) the synthesis of β-galactosidase and many other enzymes is repressed *even in high concentrations of inducer*. There is evidence that at least one component of this effect is alteration of the *i* gene product, and it is known that the metabolite repressor is not glucose itself but a related compound. Inducible enzyme pathways in bacteria are therefore regulated not only by their specific inducers but also by a non-specific repression system; the latter probably ensures that the cell uses glucose before degrading metabolites which may be less readily available.

The regulation of DNA replication

The states of DNA in the bacterial cell

It is evident that just as there are control mechanisms in protein synthesis, so there must be controls in the replication of the genetic material; DNA synthesis occurs throughout the growth cycle of the bacterial cell, and yet division of the chromosome and of any extra-chromosomal elements present is usually co-ordinated with cell division. The enzyme requirements for DNA synthesis are understood only in outline, but it is known that although protein synthesis is needed for *initiating* replication, it is not required to maintain the process once started.

The genetic material of *E. coli* is a ring of DNA (p. 100); replication begins at some point in this ring, and proceeds in one direction around the chromosome until two rings are made, each consisting of one old and one new polynucleotide strand and attached at the starting point. Then the two rings detach, and each can begin a new replication cycle. To understand how chromosome replication is co-ordinated with that of episomes, which are probably themselves small DNA rings, consider the three states in which the genetic material of one episome, temperate bacteriophage, can occur in the cell (p. 119):

1. Autonomous. During lytic infection of the host cell, phage replication is rapid and is controlled by the phage genome, not by that of the cell.
2. Integrated. When the phage chromosome is reduced to prophage it becomes essentially a part of the host chromosome (p. 125); thereafter its rate of replication is synchronized with that of the chromosome.
3. Superinfecting. A lysogenic cell is immune to homologous superinfecting phage (p. 126); the latter cannot replicate, either on its own or as part of the chromosome, and it is therefore diluted out (and perhaps broken down) during cell division.

We can conclude that a bacterial chromosome or autonomous episome carries genes which cause it to behave as a unit of replication. An integrated episome (e.g., prophage, state 2 above) behaves as part of the host unit of replication.

The replicon hypothesis of Jacob and Brenner

Jacob and Brenner sought to explain these findings in 1963 by proposing a model similar to the Jacob-Monod hypothesis for regulation of protein synthesis. They postulate two elements, analogous to the regulator and operator respectively:

1. A regulator of replication (the *initiator*) which determines the synthesis of a cytoplasmic initiator substance.
2. An operator of replication (the *replicator*) which is recognized by its specific initiator substance and permits replication of the DNA ring (unit of replication) of which it is a part.

To complete the analogy, the replicator plus the unit of replication to which it belongs is termed a *replicon* (*cf.* operon, p. 136).

Let us interpret the three states of temperate phage DNA on this basis. When a phage replicon enters the lytic cycle, it determines the synthesis of its own initiator which acts on the phage replicator; thus the phage DNA divides (state 1). When it lysogenizes, it becomes a part of the host replicon, and only replicates when the host initiator allows replication of the latter (state 2). Finally, if it superinfects a lysogenic cell, it is repressed by the immunity system (p. 126) and so cannot produce its own initiator, but because it does not become part of the host replicon it cannot respond to host initiator—so it cannot replicate at all (state 3).

The evidence for the Jacob–Brenner model is admittedly circumstantial. It is possible to isolate phage mutants which can replicate as prophage but not autonomously, as if the phage initiator gene were defective. Temperature-sensitive F mutants which replicate autonomously at 30° but not at 42° are known, although when integrated (as in Hfr, p. 115) they replicate with the host chromosome at either temperature; again, some chromosomal mutants cannot initiate DNA replication at high temperature, although they can complete a cycle at high temperature once it has begun at a lower temperature. All these mutants are suggestive of defects in initiator genes, and at least two proteins which have a role in regulating DNA replication have been detected.

The mesosome

Finally, we have to account for the finding that all the replicons in a cell—chromosome and autonomous episomes—are usually passed to daughter cells. Jacob and his colleagues propose that the replicons are all attached to a single membranous structure, the *mesosome* (p. 286), and that initiator is produced on a signal from this which causes each replicon to divide (each DNA ring perhaps rotates *through* its point of attachment to the mesosome as it replicates); the mesosome itself then divides, drawing one copy of each replicon to the daughter cells. The electron microscopic evidence in *B. subtilis* is consistent with this, and if chromosomal and episomal DNA are suitably labelled, they can be shown to remain together for several cell divisions. Although one should be careful in extrapolating from bacteria to higher organisms, the idea of all replicons being attached to a cellular 'backbone' which draws them apart after replication is obviously reminiscent of the arrangement of chromosomes on a mitotic spindle (p. 100).

The operator-regulator relationship

The Jacob–Monod model for regulation of protein synthesis and the Jacob–Brenner model for regulation of DNA replication both comprise a regulator element and an operator element. Very similar models have been invoked to explain patterns of gene instability in maize and also in bacteria (p. 103), although the operator and regulator elements in these instability systems are probably episome-like units (p. 119) which associate reversibly with regions of the genome, rather than being parts of the maize or bacterial chromosomes themselves. It is tempting, although irrelevant to the subject of this book, to speculate on the role of these systems in the complex regulatory relationships involved in cell differentiation in higher organisms.

We can now interpret lysogeny (p. 124) a little more fully. When a phage genome enters the prophage state, it confers immunity to homologous super-infecting phage on its host cell; the immunity substance (p. 127) is analogous to Jacob and Monod's apo-repressor, and c and ind λ mutants can be regarded as mutants of the regulator gene of immunity (*cf. i^-* and i^s mutants respectively; single-step 'operator-constitutive' mutants have not so far been isolated, how-ever). We can distinguish a further regulatory system concerned with integration and excision of prophage: there is an operator (*att*) and a regulator (*int*) of integration. Finally, the immunity system of the prophage requires it to remain part of the host replicon (p. 139) instead of synthesizing its own initiator and replicating autonomously. Even in this simplified picture of lysogeny it is necessary to envisage no less than three operator-regulator systems—for immunity, prophage integration, and replication.

L'ENVOYE

We have tried, in this brief account of microbial genetics, to show how the ideas of the classical geneticist have been harnessed to the resources of the bio-chemist to elucidate heredity at the molecular level. This approach has had some tremendous successes, chiefly in the 'cracking' of the genetic code and in under-standing the control of gene expression; the solution of other problems, such as the mechanism of recombination, remains less complete at present. These advances have been achieved at a price: we know a great deal about the genetic behaviour of *E. coli* and a few other bacteria, but there is little idea of how widely this behaviour applies; it may be that there are other organisms, in which mating systems have not so far been detected, which hold the clue to some of the curious instability systems (p. 103) which occasionally come to light in the *Entero-bacteriaceae*.

Two points remain. Bacteria were once treated as 'bags of enzymes'; when bacterial genetics became a respectable field of study, they were regarded as bags of genes, and they now tend to be looked on as bags of operons (p. 136). It may be suspected that many ideas about control systems, at present fashionable, will turn out to be over-simplifications when examined in the context of different levels of organization in the cell. Secondly, we have already indicated that some tenets of microbial genetics are proving useful in understanding the processes involved in cell differentiation, which is one of the last unconquered areas in biology. Perhaps it is in this fascinating study that microbial genetics will find its apotheosis.

BOOKS AND ARTICLES FOR REFERENCE AND FURTHER STUDY

Important original papers

BEADLE, G. W. and TATUM, E. L. (1941). Genetic control of biochemical reactions in *Neurospora*. *Proc. nat. Acad. Sci., Wash.*, **27**, 499–506.
JACOB, F. and MONOD, J. (1961). Genetic regulatory mechanisms in the synthesis of proteins. *J. mol. Biol.*, **3**, 318–356.

Reference books

BRAUN, W. (1965). *Bacterial Genetics*. 2nd Edn. W. B. Saunders Co., Philadelphia, 380 pp.

ESSER, K. and KUENEN, R. (1967). *Genetics of Fungi*. Springer-Verlag, New York, 500 pp. Up-to-date and comprehensive.

FINCHAM, J. R. S. (1965). *Microbial and Molecular Genetics*. English Universities Press, London, 149 pp. Simply written introductory text.

―――― and DAY, P. R. (1963). *Fungal Genetics*. Blackwell Scientific Publications, Oxford, 300 pp. Standard text at intermediate level.

HAYES, W. (1968). *The Genetics of Bacteria and their Viruses*. 2nd Edn. Blackwell Scientific Publications, Oxford, 925 pp. First-class text at introductory and more advanced levels, containing (despite its title) much background material on fungi.

HARTMAN, P. E. and SUSKIND, S. R. (1969). *Gene Action*. 2nd Edn. Prentice-Hall, Englewood Cliffs, New Jersey, 260 pp.

HERSKOWITZ, I. H. (1968). *Basic Principles of Molecular Genetics*. Nelson, London, 302 pp.

SRB, A. M., OWEN, R. D. and EDGAR, R. S. (1965). *General Genetics*. W. H. Freeman, London, 557 pp. 2nd Edn. An introduction to the basic principles of genetics.

TAYLOR, J. H. ed. (1965). *Selected Papers on Molecular Genetics*. Academic Press, New York, 649 pp. A reprinting of many important original papers, including the two quoted above, with introductory articles.

WOESE, C. R. (1967). *The Genetic Code*. Harper and Row, New York, 200 pp.

Articles from 'Scientific American'. All short and easy to read; among many that have appeared in the last few years the best are perhaps:

JACOB, F. and WOLLMAN, E. L. (1961). Viruses and genes. **204**, No. 6, 92–107.

BENZER, S. (1962). The fine structure of the gene. **206**, No. 1, 70–84.

CRICK, F. H. C. (1962). The genetic code. **207**, No. 4, 66–74.

NIRENBERG, M. W. (1963). The genetic code: II. **208**, No. 3, 80–94.

EDGAR, R. S. and EPSTEIN, R. H. (1965). The genetics of a bacterial virus. **212**, No. 2, 70–78.

CAIRNS, J. (1966). The bacterial chromosome. **214**, No. 1, 36–44.

CRICK, F. H. C. (1966). The genetic code: III. **215**, No. 4, 55–62.

YANOFSKY, C. (1967). Gene structure and protein structure. **216**, No. 5, 80–94.

WOOD, W. B. and EDGAR, R. S. (1967). Building a bacterial virus. **217**, No. 1, 60–74.

WATANABE, T. (1967). Infectious drug resistance. **217**, No. 6, 19–27.

5

Microbial Nutrition and the Influence of Environmental Factors on Microbial Growth and other Activities

INTRODUCTION

In this chapter the main types of nutrition found amongst micro-organisms are outlined and the effects of environmental factors on microbial growth and reproduction are discussed in general terms. Exceptions to the generalizations will inevitably occur and reference should be made to Chapters 9 and 10 for more precise information on particular groups of organisms.

Though the various environmental factors are discussed individually it is important to realize that they are frequently interrelated. Thus the nutrients required may depend amongst other factors on the pH and temperature of the environment, and for photosynthetic organisms on whether light is available. Most of the investigations on the effects of environmental factors relate to one or only a few factors, and their interrelationship is often not sufficiently taken into account. Similarly there are interactions between nutritional factors, the requirement for a particular substance depending on other constituents of the medium (see, for example, under vitamins, p. 151).

Organisms grown axenically (i.e., cultures of a single species of micro-organisms) on a simple sterilized medium are unlikely to behave in the same way as they would in a complex environment such as the soil, where other organisms are present and may further complicate the situation by competing for available nutrients and by producing various toxins and antibiotics. Sterilization of artificial media may further complicate interpretation by chemically altering the medium.

In addition to these problems of assessing the effect of environment and nutrition on micro-organisms there is a basic difficulty of measurement of growth and/or reproduction. The absolute measure of growth is dry weight increase, but it may be difficult to separate organisms from the medium in which they are growing. Dry weight measurement is also destructive, though replicate cultures may be used for further observations. Efforts to avoid these problems by measuring some other parameter, e.g., volume growth, linear growth, or increase in optical density of suspensions have the disadvantage that the value measured may be water uptake, vacuolation, or pigment production rather than true growth (Chap. 7). All these interactions, difficulties, and short-comings have to be taken into account when the results of laboratory experiments are evaluated or applied to the natural environment.

For the isolation of micro-organisms and the study of their gross physiology, relatively crude culture media are often used. These media are based on natural products and represent attempts to reproduce the natural environment of the organism to be studied. Autotrophs, which do not require organic nutrients, can best be grown on simple solutions of inorganic salts supplemented with soil extracts or similar materials when vitamins and minor elements are required. Among media widely used for the culture of heterotrophs (see below) are those containing meat extract, blood, milk, peptones, hydrolysed casein, yeast extract, malt extract, or potato extract; all are, to varying degrees, crude sources of vitamins, amino acids, and bases and are sometimes supplemented with sugars and mineral salts. These may be used as fluid culture media or as 'solid' media when gelled with gelatin or, more usually, agar. Solid media are widely used for the isolation and maintenance of cultures but they contribute little to our knowledge of the precise nutritional requirements of organisms.

For the precise investigation of nutritional requirements completely synthetic culture media must be used. The composition of a minimal culture medium may be determined in one of two ways. Attempts may be made to grow the organism in various mixtures of pure chemicals, usually sugars, amino acids, bases, vitamins, minerals, etc. Alternatively, an attempt may be made progressively to replace components of a crude culture medium with pure chemicals; if necessary, the natural products are analysed chemically and their constituents tested by addition to the culture medium singly and in combination. With more fastidious organisms these procedures may be extremely laborious, especially as other factors, such as pH, gaseous atmosphere, etc., may need to be precisely controlled to permit growth. Despite these difficulties, the investigation of nutrition has led to many fundamental discoveries in metabolism and enzymology.

NUTRITIONAL TYPES

All organisms require for growth the elements necessary for synthesis of their cellular constituents, and an energy-generating system which consists essentially of an electron donor and an electron acceptor. If an organism is growing under aerobic conditions oxygen will be the terminal electron acceptor, but with anaerobic organisms a variety of compounds are capable of replacing oxygen. Differences in the source of the energy, that is the electron source, are used to distinguish between various groups of micro-organisms. Autotrophs are able to grow using either the energy liberated by the oxidation of certain inorganic compounds or the energy from light (see Chap. 3). The former are chemosynthetic autotrophs and the latter photosynthetic autotrophs; both can utilize inorganic sources of carbon and nitrogen. Heterotrophs use organic compounds as energy and carbon sources; in many instances the same compound serves both purposes.

Fungi, most protozoa, and many bacteria are heterotrophs. The complexity of their nutritional requirements varies enormously according to the synthetic ability of the species or even the particular strain. Many fungi can grow and sporulate satisfactorily with a simple sugar, an inorganic or simple organic nitrogen source, and mineral elements; a small number of vitamins are frequently required. Other fungi have been cultured only on chemically undefined plant or animal extracts which contain a wide range of organic compounds. The obligate parasites of plants, such as most of the rusts and smuts (see Chap. 19), have unknown, presumably very complex, requirements and have not been cultured axenically.

Protozoa generally have complex nutritional requirements and many have not been grown in axenic culture or in chemically defined media. Those that have been investigated commonly require a range of amino acids and vitamins in addition to carbohydrates and in this they resemble the metazoa. A complication arises in the close relationship between protozoa and some of the green algal flagellates, the phytoflagellates. In dark conditions and/or in the presence of certain chemicals in a suitable nutrient medium some of these algae may lose their chlorophyll and change from photosynthetic autotrophs to heterotrophs. This change may be the result of a mutation or, alternatively, a phenotypic response. Other algae are also capable of heterotrophic growth in dark conditions but the range of substrates used is more restricted than that metabolized by true heterotrophs. The majority of algae are photosynthetic autotrophs.

Bacteria are nutritionally the most versatile micro-organisms, ranging from photosynthetic and chemosynthetic autotrophs to heterotrophs with requirements of varying complexity. Some of the heterotrophs can grow on a single simple sugar and an ammonium salt; others, such as *Lactobacillus* spp., require a wide range of amino acids and vitamins. Many parasitic bacteria have only been grown in complex undefined media whilst *Mycobacterium leprae* has never been grown in artificial culture. Considerable use is made of the nutritional requirements of bacteria for taxonomic purposes (see Chap. 11). However, less use is made of this approach in dealing with fungi, algae, and protozoa because of the availability of morphological characters for classification.

NUTRITIONAL REQUIREMENTS

Quantitatively the most important elements required by living cells are carbon, hydrogen, oxygen, nitrogen, sulphur, and phosphorus. A wide range of other elements and compounds is also required in much smaller amounts.

The ability of an organism to utilize any particular compound depends on many factors. Enzyme systems capable of metabolizing the compound must be present and there must be a means of getting the substance into the cell. Many large polymers, such as some oligosaccharides and some proteins, are not utilized by some micro-organisms because they are not taken up, though within the cells enzymes are present which can degrade the polymers. However, many fungi and bacteria release extra-cellular enzymes which break down the polymers into transportable units. These extra-cellular enzymes and even whole series of enzymes involved in particular metabolic pathways in the organism may be inducible, that is, they are produced only in the presence of the substrate or closely related compounds (Chaps. 2 and 4). The ability to induce enzymes is a measure of an organism's latent adaptability to a changing environment.

Even if enzyme systems for a particular compound are present it may still not be utilized because of the deficiency in membrane transport systems. Uptake of substances (p. 165) frequently involves carriage across cell membranes by transport proteins. These 'permeases', like enzymes, are inactivated by, for example, extreme pH values and this may prevent utilization of various nutrients. Furthermore the permease systems are specific for particular compounds or group of compounds and may be differentially affected by such factors as pH. If a variety of compounds is available only one may be taken up though all are utilizable; an order of preference for various nutrients is thus established.

These problems of assaying the use of nutrients may be increased by changes which the organism makes to its own environment. Toxic substances resulting from metabolism may accumulate; for example, ammonia may be produced, particularly when amino acids are being used as a carbon source, and under conditions of poor aeration this may be toxic. Various organic acids may drastically alter the pH of the medium. Such effects may prevent further growth, or even cause death of the organisms.

The response of fungi to variations in the environment frequently depends on the particular stage in the life cycle which is studied. The requirements for reproduction, particularly sexual reproduction, are generally more exacting than for growth; thus the range of compounds used and the range of concentrations which are suitable may be less, and extra vitamins, minerals and trace elements may be required.

When considering the nutritional requirements of a micro-organism or the variety of compounds it uses it must be remembered that the observed effect on growth or reproduction is a reflection of a large number of environmental, intracellular, and genetic factors, any of which may be limiting at a particular time.

Carbon sources

Carbon is the basic structural unit of all organic compounds and is therefore needed in comparatively large quantities. Autotrophs can synthesize organic compounds from carbon dioxide, or the bicarbonate ion, and no other source of carbon is required. Heterotrophs need an organic carbon compound as an energy source which also serves to provide most of the carbon required by the cell. Studies with radioactive carbon have shown that many heterotrophs require in addition small quantities of carbon dioxide, although they cannot use it as their sole source of carbon since it cannot be further oxidized to produce energy.

The range of carbon compounds used is enormous, varying from gaseous carbon dioxide through simple sugars and amino acids to lipids, polysaccharides, and proteins.

The heterotrophs—fungi, most protozoa, and many bacteria—commonly, and often preferentially, use the simple carbohydrates, particularly D-glucose. Disaccharides, monocarboxylic acids, amino acids, lipids, alcohols, and polymers such as starch and cellulose are less readily utilized approximately in the

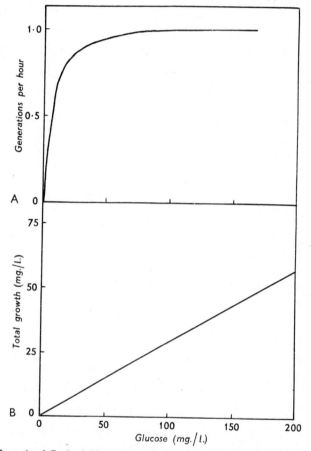

Fig. 5.1   Growth of *Escherichia coli* in a defined medium with glucose as the carbon source. (**A**) The effect of glucose concentration on growth rate. (**B**) The effect of glucose concentration on total growth. (Monod, J. (1942). *Recherches sur la croissance des cultures bactériennes*, Hermann et Cie, Paris)

order given. There is a great variation between organisms in the range of compounds used. Many fungi and bacteria have a large range of catabolic and anabolic enzymes and can synthesize all the substance they require from a single carbon source or, conversely, can use a large selection of substances; mixtures may be stimulatory but are not essential. Other fungi and bacteria, and protozoa generally, are more exacting; they may require a specific, and often complex, selection of carbon sources because, being metabolically deficient, they lack the enzymes to convert a single source to all the compounds that they require. These complex requirements may include several carbohydrates, a selection of organic acids and, with the protozoa, various steroids, phospholipids, purines, and pyrimidines. The response of micro-organisms to different concentrations of carbon depends largely on the type of compound. Firstly, if the substance is readily used then the growth rate increases with increasing concentration up to an optimum value (Fig. 5.1); then excessive osmotic pressure or limited availability of some other nutrient prevents further increase or reduces the rate. When the carbon compound is less readily used the pattern of response is similar, except that both the maximum growth rate and the rate at a given concentration are less. If a compound is not used to any appreciable extent then very little growth occurs at any concentration.

The effect of carbon source and concentration on the sexual sporulation of fungi is different from their effects on the growth response. Provided that other environmental and nutritional factors are favourable, sporulation is often induced by a reduction in carbon concentration or by the provision of a less readily utilizable source, i.e., by conditions less than optimal for growth.

Nitrogen sources

The main role of nitrogen in cells is as a constituent of proteins, which occur as enzymes as well as structural polymers. It is also a constituent of the purines and pyrimidines which, in turn, are constituents of nucleic acids and some growth factors. Within the cell the basic units of nitrogen metabolism are amino acids, which are obtained from the environment or formed by combination of ammonium ions and some organic acids. The utilization of inorganic nitrogen has a great effect on the pH of the medium (p. 161). Uptake of ammonium ions involves the release of hydrogen ions and the medium therefore becomes acidic, whereas the use of nitrate tends to cause alkalinity. These changes, by affecting enzyme systems and the solubility of various ions, may affect the ability to use many different nutrients.

Some bacteria can 'fix' atmospheric nitrogen; that is, they can convert nitrogen gas into organic nitrogen. *Azotobacter* spp. and *Clostridium* spp., together with the symbiotic organism *Rhizobium* spp. (p. 319), are perhaps the best known of these. However, recent work using isotopically labelled nitrogen has shown that the ability to fix this gas is much more widespread than was once thought. Many blue-green algae and some species of bacteria, such as *Achromobacter*, *Nocardia*, *Pseudomonas* and *Aerobacter* are able to do this.

Inorganic nitrogen is used by many fungi, algae and bacteria and a few protozoa. Ammonium salts are usually most favourable, with nitrates less readily used and nitrites used by only a few organisms since they are frequently toxic. There is a requirement by protozoa and some bacteria for more complex nitrogen sources, particularly amino acids. Gram-positive bacteria are more exacting in this respect that gram-negative ones; many protozoa and some bacteria require a selection of 10 to 15 different amino acids. Apart from this absolute requirement, many micro-organisms may grow better when supplied with a mixture of amino acids, proteins or protein hydrolysate, than with a single amino acid or an inorganic source. The mixture of amino acids may supply the organism with one which it can synthesize only slowly and which, in the absence of an exogenous supply, limits growth. In addition to amino acids it is possible that some protozoa require peptones or natural proteins, although some of the evidence for this may be otherwise interpreted; for example, traces of growth substances (p. 151) may be contained in poorly purified proteins; proteins may also chemically combine with toxins or stimulate pinocytosis (p. 166), all of which give an apparent response to added protein.

The effect of nitrogen concentration on growth has not been extensively investigated, but it appears that the pattern of response is similar to that for carbon concentration provided that toxic levels are not reached and the pH is controlled. The importance of the carbon/nitrogen ratio seems to be variable. It is claimed that a high ratio is best for the growth of some fungi and a lower one for sporulation; other micro-organisms may show no response to variation in the ratio. The interpretation of experimental results may be complicated because the provision of organic nitrogen source may involve the simultaneous addition of available carbon. However, it is probable that the particular carbon and nitrogen sources and their actual concentration are more important than the ratio between the two elements.

Other elements

Elements other than carbon and nitrogen required by micro-organisms are supplied usually as inorganic compounds, but some organisms require certain sulphur- and phosphorus-containing organic compounds. The elements needed can be divided into two classes according to the concentration at which they are required. The macronutrients such as phosphorus, potassium, sulphur, magnesium, calcium, iron, and sodium are usually required at relatively high levels (10^{-3} to 10^{-4}M) and hence it is easy to demonstrate that growth does not occur in their absence. Micronutrients or trace elements such as manganese, copper, cobalt, zinc, and molybdenum are required only at very low concentrations (10^{-6} to 10^{-8}M) and it is an exacting task to demonstrate a requirement for them. The usual methods for removing macronutrients from a medium leave a sufficient level of trace elements as contamination to satisfy the need for them and in many instances a requirement for one micronutrient can be satisfied by another one. The distinction between macro- and micronutrients is not clear cut; certain elements such as iron lie between the two classes and may be considered to belong to either. The position of sodium is also anomalous (p. 151). Generally speaking, it is true to say that the macronutrient elements are incorporated into compounds of structural importance or have a physiological role such as osmoregulation and are thus required in large quantities and that the micronutrients are co-factors for enzymes.

The availability of radicals in the medium is greatly influenced by the pH and the presence of other chemicals. Care must therefore be taken when formulating a medium to ensure that each element is present in the correct amount and that it is available in a form which can be taken up by the organism. Other constituents of the medium may react with trace elements to give insoluble salts. Iron, for example, may be removed in this way and some organic compounds can chelate, that is form weak chemical bonds with, metal ions such as copper and zinc.

Since all these elements are intimately involved in basic metabolism their requirement is fairly general amongst all groups of micro-organisms, though the amounts required vary. There are special cases where inorganic ions play a special part in the metabolism of the cell; for example, the iron and sulphur bacteria metabolize large quantities of some compounds containing these elements.

Deficiencies in the supply of minerals and trace elements may cause both qualitative and quantitative effects. Lack of macronutrient elements reduces growth, sometimes severely, and may prevent sporulation in fungi. A deficiency in trace elements may also reduce growth and/or sporulation but can produce very specific effects depending on the particular enzyme system affected by the loss of co-factors; for example, chlorosis in algae, lack of pigmentation and unusual growth forms may result. Organisms with trace element deficiencies may also accumulate metabolic products owing to the inactivation of particular enzymes in a reaction series. This fact can be put to good use in certain industrial processes (Chap. 22).

The form in which macronutrients are supplied in culture and some of their uses within the cell are summarized below.

Phosphorus is usually supplied in artificial media as an inorganic phosphate, though phosphorus-containing organic compounds which may satisfy this requirement under natural conditions may also be required *in vitro*; phosphate salts may serve the additional function of buffering the medium. Phosphorus is of major importance in metabolism and is required for the production of nucleic acids and certain lipids. Chemical bonds between phosphate groups and other compounds, so-called 'high energy bonds', are important in energy storage and transfer; they exist, for example, in nucleoside phosphate sugars, ADP and ATP (p. 83). Phosphate is also an integral part of some co-enzymes such as NAD^+ and $NADP^+$.

Potassium, often supplied as one of the phosphate salts in artificial media, is important in enzyme activation and in maintaining osmotic and electrical potentials within the cell, being accumulated by active transport (p. 161) across the cell membrane. In the control of electrical potential it is closely linked with sodium, but the latter seems not to be an essential nutrient for all organisms unless potassium is in limited supply when sodium may take its place.

Sulphur is usually supplied in artificial media as sulphate ions, but some organisms have lost the ability to reduce the sulphate ion and require a supply in the form of hydrogen sulphide or the amino acids cystine or methionine which are the chief products of sulphate reduction in non-exacting organisms. Cystine is an important structural constituent of proteins and it also serves as a precursor for the synthesis of several other important sulphur-containing molecules, such as biotin and thiamine.

Magnesium is assimilated from solutions of its salts. It is an essential co-factor for several enzymes, is an integral part of the chlorophyll molecule and is involved in bacterial cell wall metabolism.

Calcium is deposited as calcium carbonate or oxalate by some algae and fungi (e.g., *Mucor* spp. and *Myxomycetes*) and certain of the 'higher' bacteria; it is essential for the stability of some extracellular enzymes and for sporulation in the Bacillaceae and in certain fungi (e.g., *Chaetomium*). It is taken up as the free ion.

Iron is required for the synthesis of cytochromes, the chlorophyll of purple bacteria and prodigiosin, produced by *Serratia marcescens*, and affects the synthesis of some of the bacterial toxins (e.g., diphtheria toxin). A few organisms (e.g., *Haemophilus* spp.) are unable to synthesize the iron-containing haemins and are nutritionally exacting with respect to these compounds.

Sodium is not generally essential for growth although it is taken up by most organisms. However, a specific requirement for it exists in some marine bacteria and blue-green algae.

The uses of micronutrients are less well known than those of the elements dealt with so far. The function of some is unknown, even though a definite requirement for them has been demonstrated in at least some micro-organisms. Thus boron is needed by algae, phytoflagellates, and some bacteria, and vanadium, tungsten, chromium, and gallium are sometimes stimulatory to fungi. Copper is essential for melanin synthesis and for nitrate-reducing enzymes. The latter also require manganese and molybdenum. Zinc plays a part in alcohol dehydrogenase activity and in the synthesis of cytochrome c, and cobalt is a constituent of vitamin B_{12} complex.

Vitamin and growth factor requirements

There is often confusion between the terms 'vitamin' and 'growth factor'. In this book 'vitamin' is restricted to mean a complex organic compound required in small quantities by living organisms for the normal functioning of their physiological processes; vitamins are constituents or precursors of co-enzymes. The very low levels of vitamins required, usually only a few micro-grams per litre of culture medium, is explained by their function. Organic compounds required for growth and reproduction but not occurring in co-enzymes are regarded as growth factors. Some are required in relatively high concentrations for structural purposes and are not distinct from other nutrients (e.g., amino acids) which are specifically required by an organism unable to synthesize them.

The basic metabolic pathways and therefore the co-enzyme requirements of most organisms are very similar. There is, however, a very wide variation in the ability of micro-organisms to synthesize vitamins and consequently a variation in those that have to be supplied. Thus there are some fungi, bacteria, algae, and protozoa that are able to synthesize all the compounds of this type that they require, while others do not have the synthetic machinery for one or several vitamins and these have to be supplied. Usually protozoa have a limited synthetic power and often require an external supply of a wide range of vitamins. In bacteria and fungi the reverse is true; most species requiring an exogenous source need only one or two vitamins. The inability to synthesize vitamins may not reflect a complete absence of the appropriate pathway and in some cases suitable precursors satisfy the requirement as readily as the vitamin itself.

The situation is complicated by two other factors. Firstly, there may be a partial requirement for a vitamin where the organism can synthesize sufficient for survival but an exogenous supply is necessary for vigorous growth. Thus a fungus may grow on a vitamin-free medium, but if the vitamin is supplied growth is better and sporulation may occur. Secondly, the demand for vitamins may vary with the environmental factors and other constituents of the medium, which determine the metabolic pathways capable of operating. Thus the vitamin requirement may vary with the carbon source available and, conversely, the ability to use the carbon source may depend on the presence of a particular vitamin. The presence or the concentration of a vitamin may therefore affect morphogenesis determining, for example, whether reproduction is possible.

Vitamins

The commonest vitamins required by micro-organisms are as follows:

Thiamin (vitamin B_1: Fig. 5.2A). This is a co-enzyme, metabolically active as its pyrophosphate derivative and used for reactions involving breaking carbon-carbon linkages, e.g., some decarboxylases, transaldolases, and transketolases. All micro-organisms examined can phosphorylate thiamin (usually supplied in culture as thiamin HCl) and convert it to the active diphospho-form but may be unable to synthesize it except from the two moieties pyrimidine and thiazol.

Biotin (Fig. 5.2B). Biotin is a co-enzyme in a wide variety of reactions including carboxylations, fatty acid metabolism, deamination of some amino acids and in the urea cycle. This large number of activities may explain why a partial requirement (p. 151) is frequently reported for this vitamin; when in short supply the activity of some pathways may be reduced but others still function.

Riboflavin (vitamin B_2: Fig. 5.2C). This is one of the precursors of a large group of compounds, the flavoproteins, which are very important in metabolism. The common members of this group are flavin mononucleotide (FMN) and flavin adenine dinucleotide (FAD) to which riboflavin is easily converted. They are concerned with many oxidation/reduction processes and the cytochrome electron transport system in aerobic organisms. Riboflavin may also function as a photoreceptor in growth and phototropism of some fungi.

Pyridoxin (vitamin B_6: Fig. 5.2H). The vitamin B_6 complex contains several other related compounds such as pyridoxal and pyridoxamine. Particular variants may be more active for certain species and sometimes the metabolically active pyridoxal phosphate is required. The latter is a co-enzyme for transaminases, amino acid decarboxylases and some amino acid racemases.

Pantothenic acid (Fig. 5.2G). Pantothenate is part of co-enzyme A which is concerned with carbohydrate, lipid, and amino acid metabolism. Some micro-organisms require pantothenate itself; others need only one or both of the moieties, pantoic acid and β-alanine.

Nicotinamide (Fig. 5.2D). This is part of the pyridine nucleotides (NAD$^+$, NADP$^+$) which are concerned with almost all major oxidation and reduction pathways in metabolism. Some *Haemophilus* strains require the nucleotides themselves.

Vitamin B_{12} Group (cyanocobalamine). This is a group of closely related and very complex compounds. They are concerned with thymidine synthesis, transmethylations, and isomerizations of some organic acids.

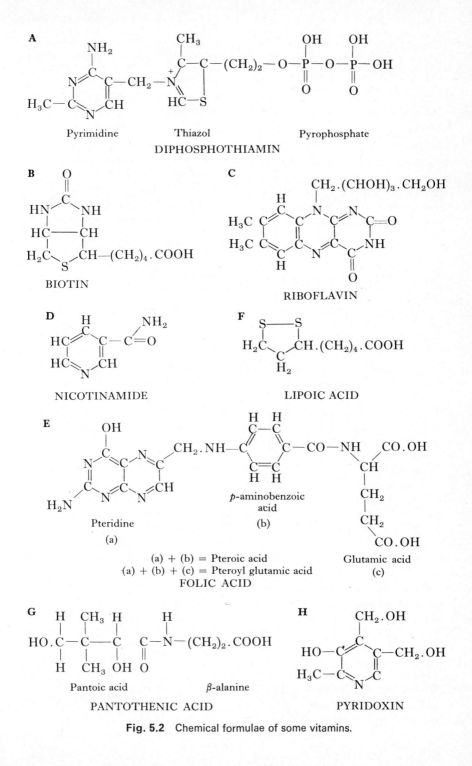

Fig. 5.2 Chemical formulae of some vitamins.

Para-aminobenzoic acid group. This is part of folic acid (Fig. 5.2E) which is a precursor of tetrahydrofolic acid, a co-enzyme involved in the transfer of residues containing single carbon atoms.

Lipoic acid (dithiooctanic acid: Fig. 5.2F), takes part in the carriage of hydrogen and acyl groups.

Growth factors

Micro-organisms, particularly protozoa, are known to require several compounds which are involved in membrane formation. Inositol, for example, is required by many fungi, particularly yeasts, but *Actinomyces israelii* is apparently the only bacterium known to need this compound. Absence of inositol from culture media of exacting fungi often causes the formation of morphologically peculiar cells. The high level at which inositol is required, as compared with most other growth factors, is also in keeping with a structural role. Choline, too, is required at levels similar to inositol by some organisms (e.g., pneumococci) and is also probably involved in membrane formation. Inositol and choline are both required by many *Mycoplasma* species which, unlike bacteria but like some protozoa, also require sterols, which are incorporated in the cell membrane.

There are many instances among bacteria, mycoplasmas and protozoa, and a few among fungi, where a requirement for long-chain fatty acids has been demonstrated. For example, oleic acid is essential to several *C. diphtheriae* strains, though the requirement is not often specific and can be satisfied by other unsaturated fatty acids. These factors are required in concentrations greater than that of vitamins but levels in excess of that required are frequently found to be inhibitory unless a detoxicant such as serum albumin or Tween is present, and in some instances the stimulating effect of fatty acids is seen only in the presence of such an agent.

Several amines are known to be necessary growth factors for some organisms although their role is as yet uncertain. Putrescine, spermidine, and permine are needed by *Haemophilus parainfluenzae* and a mutant strain of *Aspergillus nidulans*, for example, and are required at levels of 0·2–10 µg/ml. They are thought to be involved in membrane stability but polyamines such as spermine and spermidine occur also in association with ribosomes. Glutamine and asparagine are respectively required by some *Neisseria gonorrhoeae* and *Pediococcus* strains at levels which suggest that they do not have a structural role.

Purines and pyrimidines are found to be only stimulatory to the growth of many bacteria but are essential for others. The stimulatory effect is usually not highly specific since compounds causing the stimulation can frequently be replaced by another substance of the same type. *Lactobacillus arabinosus* is stimulated by the purine guanine but the other three purine bases, adenine, xanthine, and hypoxanthine, may be effectively substituted for it. However, *Shigella boydii* has a specific requirement for adenine, as has *L. bifidus* for adenine, guanine, xanthine and the pyrimidine uracil. The latter compound also stimulates the growth of some *Lactobacillus* species and can be replaced, in most instances, by the related compounds uridine, uridylic acid, cytidine and cytidylic acid and sometimes by cytosine. But again, instances are known of organisms having specific requirements such as *Staphylococcus aureus*, when growing anaerobically, for uracil and *L. bulgaricus* for orotic acid. Purines and pyrimidines are used to synthesize nucleic acids (p. 28) and it is therefore not surprising to find that some organisms such as *Lactobacillus gayonii* are stimulated not by purines, nor purine nucleosides, but by purine nucleotides.

There are several other compounds known to be required as growth factors including amino acids, peptides and proteins and such compounds as hematin and biopterin but, in addition, there may be further unknown compounds which are required by organisms such as those bacteria which can be cultivated only on complex undefined media although it is possible that there is another explanation for their reluctance to grow on relatively simple media.

Microbiological assay of nutrients

The fact that certain nutrients are necessary for the growth of some micro-organisms and determine the amount of growth that can take place, makes it possible to estimate the amounts of these nutrients in materials containing them. In general, the microbiological assay of a particular nutrient is based on a comparison of growth of a micro-organism in media containing various known amounts of the nutrient with that in media containing various amounts of the test material; it is assumed that equal growth will result when equal amounts of the nutrient are present. The culture medium used in this procedure must be carefully selected to contain all the nutrients necessary for vigorous growth with the exception of the one to be assayed; substances known to be only slightly stimulatory to growth must be included as otherwise such substances present in the material might lead to spurious results.

Among the substances which have been estimated by this method are amino acids, vitamins, and minor elements. Some vitamins (e.g., vitamins A, B_1, and C) and some amino acids (e.g., tryptophan and hydroxy-proline) are more conveniently, and usually, estimated chemically. Others, particularly the B group vitamins, when present in small quantities, are assayed microbiologically using lactic acid bacteria mainly, but the yeasts *Schizosaccharomyces pombe* and *Kloeckera brevis* are used for inositol and the vitamin B_6 complex respectively. Certain strains of *Saccharomyces carlsbergensis* have been used for the assay of inositol, pantothenic acid, and pyridoxin.

Water

Most micro-organisms contain about 90 per cent water by weight. Spores and other structures of this type contain considerably less than this amount. Water is the solvent in which all metabolic reactions occur and through which, in some instances, all exchange of substances with the environment takes place.

The amount of water available to micro-organisms cannot be measured simply as the total amount in the system since some will be associated with solutes and other molecules in such a way as to be unavailable. The amount of water bound in this way affects the vapour pressure (V.P.) of a solution and the ratio of this to the V.P. of pure water gives a measure of the amount of water in the solution available to organisms:

$$a_w = \frac{P}{P_0} = \frac{\text{Relative humidity}}{100}$$

where a_w = available water

        P = vapour pressure of the solution

       $P_0 =$,,   of pure water

Micro-organisms grow in the range $a_w = 0 \cdot 63 – 0 \cdot 99$. Bacteria, algae, and protozoa are active only at the top end of this range, and only a few xerophilic
fungi (i.e., those able to grow at very low available water levels) are able to grow
at a_w values of down to $0 \cdot 63$. Although no micro-organisms can grow without
relatively high levels of available water many have stages in their life cycle,
when spores or cysts are formed, which enable them to survive long periods of
desiccation.

Fungal mycelium and bacterial cells growing on solid surfaces need a fairly
high humidity to prevent desiccation and *in vitro* this need is met by the high
water content of solid nutritional media. The spore-bearing structures of fungi
are frequently raised above the substrate and hence are exposed to a greater
risk of desiccation; asexual structures such as conidiophores and sporangiophores require 95–100 per cent relative humidity, though sometimes a saturated
atmosphere is necessary for the completion of spore formation. When more
than one type of spore is produced the relative abundance of the two types may
be affected by the humidity as may the production and morphology of the more
complex sexual fruit bodies of the higher fungi.

Osmotic pressure

Micro-organisms have a higher internal concentration of solutes than the
surrounding aqueous medium and, therefore, have a higher osmotic pressure;
for example, Gram-positive bacteria have an internal osmotic pressure of about
20 atmospheres and Gram-negative bacteria of 5–10 atmospheres. The internal
osmotic pressure may be reduced to a minimum by converting all low molecular
weight substances not immediately involved in metabolism into insoluble storage
products such as lipids and polymeric carbohydrates. However, the excess
osmotic pressure over that of the environment causes water to enter the cell
which would swell and burst if no mechanism of osmo-regulation existed. In
algae, fungi and most bacteria a more or less rigid cell wall exists surrounding
the protoplast which prevents its bursting. In filamentous fungi however the
hyphal tip has only a very thin wall where it is being laid down and a decrease
in the osmotic pressure of the environment can cause bursting of this tip. There
is evidence that in fungi and bacteria the osmotic pressure of the cytoplasm
is adjusted according to the environmental osmotic pressure in such a manner
that a very large pressure difference between the two is prevented. In protozoa,
because there is no rigid cell wall to contain the protoplast, water is continuously
withdrawn from the cytoplasm and is periodically expelled into the environment by the contractile vacuole or similar mechanism. This removal of water
against the osmotic gradient requires metabolic energy. The solute uptake by
some protozoa may be limited so that excessive differences in osmotic pressure

between the cell and the environment do not occur, and the solute concentration is often less in protozoa than in bacteria and fungi.

The effects of changes in the external osmotic pressure of micro-organisms have not been extensively investigated but a distinction must be made between immediate effects and the long term effects on growth at a high osmotic pressure. Protozoa normally live in an environment with a very low osmotic pressure, an increase of which causes plasmolysis. Bacteria (Fig. 9.2) and fungi are similarly plasmolyzed if transferred to a medium of high concentration but the protoplast does not always completely separate from the cell wall. In all these instances, unless the change is drastic, normal growth follows a return to solutions of normal osmotic pressure.

When a fungus is placed in a solution of a higher osmotic pressure than that in which it was previously growing but which does not cause bursting of the hyphal tip, apical growth is stopped either temporarily or permanently and lateral branches are produced below the apex. Similarly, growth of bacteria in high osmotic pressure environments may lead to morphological variation; changes have been observed in both directions between bacillary and coccal forms when grown under conditions of high osmotic pressure.

Hydrostatic pressure

Some protozoa, fungi, and bacteria can live under conditions of considerable hydrostatic pressure, for example, in the depths of oceans, either in suspension or in the sediments, and in oil wells. Viable bacteria have been recovered from environments where the depth of water exceeds 10,000 metres and the pressure is up to 1,140 atmospheres. It is not certain whether these organisms are active at these depths but it is known that not all of these isolates are barophilic (i.e., not adversely affected by high pressures). It is suggested that some isolates grow at 1 atmosphere but not at the high pressures of the environment from which they were isolated. Various bacteria, yeasts, and viruses can withstand pressures up to 12,000 atmospheres for at least a few minutes but not even deep-sea bacteria can reproduce at 1,500 atmospheres and prolonged exposure even to pressure of 100–600 atmospheres inhibits the normal growth of most bacteria. The effect of growth at just tolerable pressures is generally to increase cell size and especially to cause the formation of long, bizarre filamentous cells. The influence of pressure has been only slightly studied at a metabolic level but it is known to affect bacterial luminescence, several enzyme reactions, sulphate reduction, and pseudopodium formation in amoebae. The reasons for the effects of pressure are far from clear and no coherent pattern of response is apparent. In many instances insufficient consideration has been given to related effects, particularly of temperature and other environmental and nutritional factors.

The response of organisms to water and aqueous solutions is of practical importance. Desiccation provides an efficient means of preventing biodeterioration and of keeping organisms in culture collections (p. 194). High sugar or salt concentrations are used in food preservation, although some organisms such as the osmophilic fungi, particularly some yeasts, and halophilic bacteria and fungi which grow in solutions with high solute concentration may still cause spoilage (Chap. 20).

Oxygen and carbon dioxide

Obligate aerobic micro-organisms use molecular oxygen as their terminal electron acceptor. Obligate anaerobes, which fail to grow but may survive in the presence of oxygen, use some alternative substance. The reason why obligate anaerobes cannot grow in the presence of oxygen is obscure. The evidence put forward to support the two main theories, that they require a low redox potential for growth and that inhibitory substances are formed in the presence of oxygen, is often contradictory and it may be that different obligate anaerobes are prevented from growing by different mechanisms. Facultative anaerobic organisms are capable of growing in the presence or absence of air.

The microaerophiles require low concentrations of oxygen but some organisms often included in this group are really misnamed and, in fact, need a high carbon dioxide concentration rather than a low oxygen tension, e.g., *Brucella abortus*. Carbon dioxide requirement is probably a universal feature of micro-organisms and is not confined to autotrophs and a few others as was once thought. This requirement is often difficult to demonstrate as the quantity needed may be so low that it can be supplied by metabolically derived carbon dioxide, but it has been shown that the removal of this gas from bacterial cultures inhibits growth and that at least some bacteria and protozoa die when deprived of it for too long. In contrast, too high a concentration of carbon dioxide adversely affects growth. In the fungi, for example, some normally filamentous forms become yeast-like (an aspect of dimorphism) and some aquatic Phycomycetes form resistant sporangia.

Algae are normally aerobic organisms but a few species, such as *Scenedesmus*, are known to be capable of anaerobic autotrophic, but apparently not heterotrophic, growth. Thus under both aerobic and anaerobic conditions, these algae require carbon dioxide. Fungi too are mainly aerobes though some, such as the yeasts, are facultative anaerobes. Vegetative growth and asexual reproduction can take place under anaerobic conditions but sexual reproduction almost invariably requires at least some oxygen. Bacteria and protozoa show much the same range of diversity with respect to gas requirement. They range from obligate aerobes such as *Pseudomonas* and the protozoa which inhabit the better aerated bodies of water through the facultative protozoa and bacteria such as *Escherichia coli*, the microaerophilic bacteria such as *Lactobacillus casei*, the heterotrophic bacteria with high carbon dioxide requirements (capneic bacteria), such as *Brucella abortus*, to the obligate anaerobes such as *Clostridium* and the protozoa which occur in the gut of metazoa. In an environment unfavourable with respect to gases, growth of these organisms slows and ceases and motile bacteria become non-motile, a point of practical importance when examining for motility. Aerobic spore-forming bacteria generally initiate spores only in the presence of relatively high oxygen concentrations but once sporogenesis has begun it continues under lower oxygen tensions. Conversely clostridia sporulate better in the complete absence of oxygen.

In liquid cultures of aerobic organisms oxygen frequently becomes the growth limiting factor. This is mainly due to its relatively low solubility in water. To aerate a culture fully, vigorous mechanical agitation or rapid bubbling of sterile air or oxygen through the culture is required. If oxygen is used care must be taken to ensure that carbon dioxide deficiency does not result.

Movements in response to chemical substances

Two different kinds of response are possible; if the whole organism moves then *chemotaxis* is exhibited. A change in direction of growth may occur in filamentous fungi and algae which are not motile; such a response is called *chemotropism*. Motile micro-organisms may move towards or away from chemicals, positive- or topo-chemotaxis and negative- or phobo-chemotaxis respectively. Generally compounds with inhibitory or deleterious properties cause a phobo-tactic response, but if a substance inhibits motility it may cause an accumulation of cells within its effective area without actually attracting the organisms. Compounds which cause a topo-chemotactic response are sometimes nutrients but chemicals not required for nutrition are involved in the initiation of aggregation of myxobacteria (p. 339) and Acrasiales (p. 426) and in the attraction of the motile gametes of some micro-organisms. Similar chemotropic responses may occur along chemical gradients in hyphal anastomoses or prior to the fusion of progametangia in Phycomycetes.

The chemotactic response of motile bacteria to the concentration of oxygen in the environment, by arranging themselves at a point on a concentration gradient appropriate to their requirement for this gas, is a special instance of this type of tactic movement and is called *aerotaxis*. The arrangement of cells in this way is a result of a response to insufficient or excess oxygen rather than a positive response to the optimal gas concentration and is of the phobotactic type which is often exhibited by bacteria.

PHYSICAL FACTORS

Temperature

Temperature affects the rate of all processes occurring in micro-organisms and it may determine the type of reproduction, the morphology of the organism and the nutrients required. For any particular organism, four important temperatures may be defined. The thermal death point (p. 199) and the minimum, optimum, and maximum temperatures. The minimum and maximum are the lowest and highest temperatures respectively at which growth occurs, and the optimum temperature is that at which the growth rate (or any other process) is the greatest. The minimum, optimum, and maximum temperatures are known as the cardinal temperatures or points of the particular organism (Table 5.1). They are not invariable since their values vary between different strains of a single species and with the stage in the life cycle, the age of cultures and the nutritional status of the medium. Nevertheless, these points are general guides to the response of organisms to temperature which are useful in many practical considerations.

In general, the minimum temperature for growth of micro-organisms is 10–15°C, the optimum 24–40°C and the maximum 35–45°C. The saprophytic bacteria, fungi, algae, and free-living protozoa tend to occupy the lower part

Table 5.1 Cardinal temperatures (°C) for vegetative growth of representative organisms.

	Minimum	Optimum	Maximum
Thermophils			
Bacillus stearothermophilus	30	55	75
Clostridium thermocellum	50	60	68
Mesophils			
Escherichia coli	10	37	45
Streptococcus faecalis	0	37	44
Neisseria gonorrhoeae	30	36	38
Bacillus cereus	10	30	46
Mycobacterium tuberculosis			
var. human	30	37	40
var. avian	30	42	50
Saccharomyces cerevisiae	1	29	40
Chilomonas paramecium	10	28	35
Psychrophils			
Pseudomonas fluorescens	−8	20	37
Candida scottii	0	4–15	15
Cladosporium herbarum	−6	?	20

of these temperature ranges and the bacterial and protozoal parasites the upper
parts. For any particular group the range of temperature between the maximum
and minimum is generally less for reproduction than for growth. The optimum
temperature does not lie midway between the minimum and maximum temperatures but is nearer to the upper limit of the range. This is due to the effect
of temperature on living cells and on their metabolic processes. The latter obey
normal chemical laws, the rate of reaction increasing with increasing temperatures, until a point where denaturation of the thermo-labile enzymes involved
in catalysis of the reactions begins to have a significant effect on the rate. The
optimum temperature is the result of a balance between the effect of high temperatures on reaction rates and the denaturation of the labile proteins. The
determination of optimum temperatures at elevated temperatures may be complicated by a time factor; initially the growth rate may be rapid but subsequently
it may decrease as proteins are slowly denatured. Thus, the optimum temperature
should be considered as the temperature which allows the best sustained growth.
Enzymes vary in the temperature at which they are denatured. Thus, as the
environmental temperature is raised, enzyme systems may be inactivated one
by one. Those systems necessary for reproduction may be the most sensitive
and therefore inhibited before those required for growth, thus explaining the
wider temperature range for vegetative growth than for sporulation.

It is common practice to assign organisms to one of three categories according to their temperature relationships. There is considerable overlap between
these classes and differences of opinion as to the characteristic temperatures.
The majority of micro-organisms are *mesophils* and grow at moderate temperatures as discussed above. Among all the main groups of micro-organisms there
are, however, those that are able to grow at, or actually have an obligatory

requirement for, elevated temperatures. These organisms are called *thermo-phils*. The range of temperature over which they grow is not wider than that characteristic of mesophils but the three cardinal points are all higher, for example, the optimum may be 55–60°C and the minimum to maximum range 40–80°C. Conversely, there are examples of organisms with optima of approximately 15°C and a range from $-5°C$ to 30°C. Such low-temperature organisms are called *psychrophils*. How thermophils deal with the problem of denaturation of enzymes and how psycrophils maintain a sufficiently rapid metabolic rate to grow is uncertain. It has been suggested that not only do thermophils produce enzymes more rapidly than mesophils, thus countering thermal denaturation, but their enzyme proteins are particularly thermostable. Isolated enzymes from psychrophils do not show differences from comparable enzymes from meso-phils and it has been suggested that the ability of psychrophils to transport solutes across the cell membrane efficiently at low temperatures accounts for their ability to grow under these conditions.

Hydrogen ion concentration

There are many ways in which the hydrogen ion concentration or, as it is usually expressed, pH $(= -\log_{10} [H^+])$ of the environment may affect micro-organisms and some of these have been mentioned earlier in the chapter. The net effect of pH acting on these various factors is expressed by the resulting growth and reproduction of the micro-organisms.

Micro-organisms grow over a pH range usually with a fairly well defined optimum which, in contrast to the temperature optimum, lies approximately at the middle of the range permitting growth. The minimum pH for growth is generally about pH 2·5 and the maximum about pH 8–9; the optimum varies widely between and within the various groups of micro-organisms but is frequently between pH 5 and pH 7·5. Bacteria such as *Thiobacillus thiooxidans* and *Acetobacter* spp. and some imperfect fungi such as species of *Penicillium* and *Aspergillus* are capable of growth at the very low pH values of between pH 0 and pH 2 and some *Bacillus* spp. can grow at pH 11. The pH range for reproduction of fungi is often less than that for growth.

Living cells are very well buffered internally against pH changes and environmental values have to be extreme before the intracellular pH is much affected. If in extremely acidic or alkaline conditions the intracellular pH is much changed intracellular enzymes may cease to operate. Extracellular enzymes are of course directly influenced by the environmental pH.

It is the nutritional status of the environment which is most markedly affected by pH both by altering the solubility of ions (p. 148) and the dissociation of molecules thus determining their availability to the micro-organisms and suitability for transport across the cell membrane (p. 166). In artificial culture growth limiting pH conditions frequently arise as a result of the metabolic activities of the micro-organisms (p. 148).

Light

Light of particular wavelengths from the range between the near ultra-violet (250 mμ) to the near infra-red (1100 mμ) affects micro-organisms in four main ways. It provides the energy for photosynthesis, it causes oriented response (tropic and tactic movements), it may be required for or may stimulate sporulation in fungi, and it can have deleterious or lethal effects. Light responses may be mediated by pigments which are transformed by absorption of light energy; this effect is translated into the observed response by light-independent reactions ('dark' reactions). Of the four responses mentioned, photosynthesis, the tropic and tactic movements, and fungal sporulation are dealt with here. The deleterious or lethal effects are discussed in Chapter 7.

The pigment primarily involved in photosynthesis is chlorophyll but carotenoids and phycobilins, accessory pigments, can also absorb light and transfer this energy to chlorophyll. The structures of the pigments are not the same in

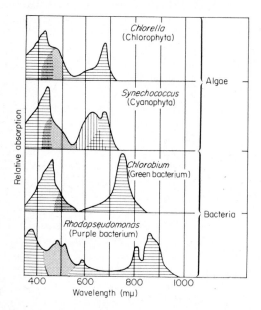

Fig. 5.3 The absorption spectra of some representative phototrophic micro-organisms, measured on whole cells with the aid of opal glass, except that of *Chlorobium*. Spectra of *Chlorella*, and *Synechococcus* taken from Shibata *et al.* (1954); spectrum of *Chlorobium* taken from Larsen (1953), and corrected for light scattering. The contributions by the various classes of photosynthetic pigments are approximately indicated as follows: chlorophylls, horizontal hatching; carotenoids, stippling; phycobilins, vertical hatching. (After Stanier, R. Y. and Cohen-bazire, G. (1957). The Role of Light in the Microbial World. In *Microbial Ecology*, 7th Symp. Soc. Gen. Microbiol. Camb. Univ. Press. 66 pp.)

all micro-organisms and the different types have different absorption spectra. Thus the wavelength of light supplied will determine which of a mixture of organisms can photosynthesize. In Fig. 5.3 it can be seen that between them phototrophic micro-organisms absorb light between 350 mμ and 950 mμ. The differences in the absorption maxima correlate with the ecological distribution of the various organisms (Chap. 13). In some photosynthetic organisms, light may induce effects not directly attributable to photosynthesis, for example sterol content of *Scenedesmus* is raised in high compared with low light conditions.

For most non-photosynthetic organisms light is unnecessary or deleterious but in some fungi certain metabolic processes, including sporulation, are dependent on it. Sporulation of fungi, both sexual and asexual, is often initiated or increased by radiation in the ultra-violet or blue spectral regions. Thus zonation of a colony may occur when it is grown in alternating periods of light and dark; bands of dense sporulation develop as a result of the periods of illumination. Light may also influence pigment production itself. Some normally pigmented fungi and mycobacteria fail to produce any pigment or produce reduced amounts if grown in the dark.

The most thoroughly studied effects of light are those concerned with movement of whole organisms or their parts. The phototactic behaviour of the purple bacterium *Rhodospirillum rubrum* in fact results from a response to lack of light. If light is projected on to part of a slide on which there is a suspension of these organisms they accumulate in the light zone because organisms which have crossed from the light to the dark zone stop and reverse their direction, whereas those crossing from a dark to a light zone carry on. The reason for the reversal of direction of movement is not certain but it is known that the action spectra for phototaxis and photosynthesis are the same, indicating that the photosynthetic pigments are involved. The first evidence of the identity of action spectra for these two processes was obtained by projecting a spectrum on to a slide on which there was a suspension of bacteria. The organisms accumulated in the infra-red and blue-green regions where bacteriochlorophyll and the carotenoids respectively absorbed. Later experiments with narrow wavebands of light have confirmed that the peaks of the action spectra for the two processes are coincident.

Phytoflagellates also respond tactically to light, accumulating in regions of light provided the intensity is not too great. In *Euglena* the photoreceptor is not the photosynthetic pigment since the action spectra for photosynthesis and phototaxis do not correspond; only the light from the blue region of the spectrum is effective in inducing movement. The photoreceptor for phototaxis is in the front end of the organisms but may not be the carotenoid pigment of the 'eye-spot' as was once thought, since other flagellates which are photoresponsive do not possess this structure. The photoreceptor in *Euglena* has been suggested to be a thickening near the base of the flagellum. *Euglena* responds to light of lower intensity than do phototactic bacteria and responds both positively and negatively to light. Both these types of response occur also in the blue-green algae and the diatoms depending on the wavelength and intensity of the light.

Photosynthetic rate, and hence growth rate in autotrophs, is roughly proportional to light intensity, until saturation point is reached, after which further increase in light intensity does not affect the rate. The light intensity which is saturating depends on other environmental factors such as temperature and mineral nutrition (Fig. 5.4). The growth rates of *Chlorella pyrenoidosa* and *Scenedesmus costulatus* are much less when they are growing heterotrophically in darkness than in light. One strain of *Euglena gracilis* behaves similarly; another grows equally well in light or in a nutrient solution in the dark.

Spore-bearing structures in fungi often respond to unilateral illumination by directional growth or increase in growth rate, that is to say they are phototropic. They usually grow towards light, positively phototropic, which in the natural environment will increase the likelihood of spores being released away from the substrate. Thus stipes of Agaricaceae, asci of some Ascomycetes, and the necks

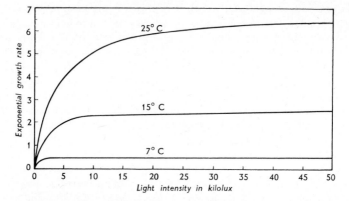

Fig. 5.4 Growth of the alga *Chlorella ellipsoides* in liquid culture, illustrating the inter-action of light and temperature in determining exponential growth rate (figures arbi-trary). The saturation point, that is the light intensity beyond which further increase does not result in an increased growth rate, is about 3 kilolux at 7°C, 10 kilolux at 15°C, and 50 kilolux at 25°C. (After Burlew, J. S. (1953). *Alga Culture*, Fig. 7, Chap. 16, Carnegie Institution, Washington, D.C.)

of the perithecia of some others, sporangia of many Phycomycetes and the conidiophores of a few Deuteromycetes are all positively phototropic.

Gravity

The response to gravity generally results in the organism being in the favour-able position for growth, reproduction or dispersal.

In fungi geotropism is demonstrated mainly by spore-bearing structures, though it may be modified by phototropic effects (see above), which are usually dominant. Thus sporangia of some Zygomycetes are negatively geotropic (i.e., grow away from the earth). The higher Basidiomycetes orientate in response to gravity very precisely so that the gills or pores (p. 399) are vertical, allowing the spores borne on them to fall freely and be efficiently dispersed.

The reactions of protozoa to gravity have not been fully investigated. It has been suggested that some free-living amoeboid forms may respond positively, maintaining their position in the bottom sediments of natural water, whilst free-swimming forms may react negatively remaining in the surface layers of water. The position of protozoa within the environment is, however, controlled by many other factors such as light and chemical gradients (see above) and many show no reaction to gravity.

Both positive and negative geotaxis has been observed in algae, too, but this problem has been little studied in these organisms.

Mechanical stimuli

Mechanical stimuli may include anything from contact on a microscopic level between organisms and something in their environment to the considerable damage caused by such processes as scraping or cutting a fungus colony. Bacteria are too small to be affected by gross processes; grinding with fine abrasives, ultrasonic treatment, or extrusion of a frozen cell mass through a small aperture have to be used to break the cells (p. 206).

Motile bacteria with flagella arranged at one end of the cell, ciliate protozoa and phytoflagellates exhibit a complex avoidance behaviour on contacting an object in their path; a series of movements involving backing off, turning at an angle and proceeding again is repeated until the organism has moved around the obstruction. In contrast bacteria which have peritrichous or amphitrichous flagella (p. 303) merely reverse their direction of motion on coming into contact with an obstruction. Algae, too, usually have negative thigmotactic responses.

In fungi few tactile responses are fully documented owing to the difficulty in demonstrating them conclusively. However, sporulation may sometimes be induced by the fungus growing up against some barrier although in this case changes in the nutrient distribution at the obstruction may complicate the interpretation. In natural conditions one example of the tactile response is fruit body formation induced in some hypogeous (underground) fungi by contact with plant roots. The formation of appressoria in some plant pathogens (e.g., *Botrytis*, p. 631) may be stimulated by contact with the host plant surface. Tactile responses are also thought to play a part in the penetration of some antheridia into the oogonium during fertilization (e.g., *Achyla* sp., p. 392). Some Deuteromycetes may be stimulated to sporulate by very drastic tactile stimuli such as cutting or scraping the mycelium, though whether this is a response to touch or to metabolites released into the medium from killed cells is in doubt.

Some bacteria align themselves according to the orientation of the molecules in the substrate. Probably the best known examples of this phenomenon, elasticotaxis, is the alignment of *Cytophaga* cells with the cellulose micelles in cellulose fibres. Other myxobacteria are known to arrange themselves along lines of substrate orientation induced by stress in agar.

UPTAKE AND TRANSLOCATION OF NUTRIENTS

The rate and method of nutrient uptake depends on the nature of the nutrient, the structure and properties of the cell envelope, and the environment of the cell. The cytoplasmic membrane is essentially composed of lipoprotein and its structure is discussed in Chapters 9 and 10. The cell wall of actively growing organisms is generally regarded as being freely permeable to small molecules but the walls of spores and cysts are almost impermeable. Movement of molecules across membranes may occur as a result of diffusion gradients, electrical potential gradients, or by 'active transport' mechanisms which move substances against such gradients and bring about their accumulation in the cytoplasm.

The diffusion of substances from a high to a low concentration takes place in both directions across the membrane. It is most common in fat-soluble substances, this property being necessary for easy unassisted passage across the membrane. However, water also crosses the membrane very freely. The most common substances moving on concentration gradients are gases such as carbon dioxide and oxygen. The movement of gases is not generally controlled by the cell, except inasmuch as the metabolism is responsible for establishing the gradients. The movement of water is generally considered in terms of osmosis, that is, water moves from dilute solutions to more concentrated ones; it is under cellular control by adjustment of the amount of solute rather than any positive control of water movement itself (pp. 156–7).

The movement of ions, most minerals, and trace elements in and out of cells is complex and not fully understood. Firstly, ions will follow concentration gradients, diffusing in the normal way, and secondly they will follow gradients in electrical potential, moving towards solutions of opposite net charges to themselves. This type of movement is usually not just a matter of mineral ion and trace element distribution, for the situation is complicated by various charged groups on proteins, amino- and organic-acids, and bases within the cell. Superimposed on these gradients are the 'active transport' systems which move ions against concentration and electrical potential gradients. Such systems require metabolic energy to be expended, often exhibit requirements for trace elements, and have kinetics which indicate that the substance being transported forms a complex at some stage with another substance from which it is subsequently released on the other side of the membrane. The complex formation and disruption is probably enzyme mediated and various specific 'permeases' have been proposed. This type of transport system for ions operates in both directions across the membrane though the affinity for different ions may be different in the two directions, e.g., potassium ions are transported into cells in preference to sodium or hydrogen ions, but the reverse preference is shown in their outward movement. This system constitutes the so-called potassium–sodium pump and results in an intracellular accumulation of potassium ions. These slowly leak out of the cell down the concentration and, possibly, potential gradients. Phosphate, or rather $H_2PO_4^-$, is actively transported, provided the cell has adequate potassium-ion concentration, though the transport systems are independent. Uptake of heavy metal ions, particularly manganese and magnesium, requires phosphate for their transport. These examples illustrate the interdependence of nutritional factors. Furthermore, all these transport systems for ions, since they depend on the correct dissociation of the ion, and in some cases on the movement of hydrogen and hydroxyl ions, are very sensitive to pH changes both in the environment and within the cells.

Active transport, or at least systems dependent on chemical carriers to cross membranes, are also involved in the uptake of sugars and amino acids. Disaccharides may cross the membrane as such or may be hydrolysed first, as are polysaccharides, the constituent sugars then being taken up by their own transport systems. Many model transport systems have been suggested to account for the observed kinetics of uptake. Some of these are quite complex involving a series of carriers and their associated enzymes. As mentioned earlier, the preferential use of various nutrients in a complex medium may reflect the affinity of the active transport system for the various compounds, or the control of the synthesis of carrier systems.

Further methods by which nutrient substances cross the cell membrane are pino- and phago-cytosis. These are particularly important in some heterotrophic protozoa, myxomycetes and Acrasiales (p. 426). When a food particle comes into contact with the cell membrane the latter invaginates, surrounding the particle; this process is called phagocytosis. Eventually the invagination is occluded and the food vacuole so formed moves into the cytoplasm. Pinocytosis is a similar phenomena involving the enclosing of liquids rather than particles. Enzymes are secreted into the food vacuole and the products of digestion presumably cross the vacuole membrane by one of the systems indicated above. Some protozoa show high specificity, phagocytosis occurring only with certain types of food particle, while others will ingest almost anything, even latex particles, provided they are of a suitable size. The mechanism of the specificity and the stimulus required for this process is not precisely known, though various inducers have been suggested.

The control mechanisms for transport are not well understood. Availability and diffusion gradients account for the movement of some substances outside direct cellular control and many of the active transport systems are induced by the presence of the substrate to be transported. Transport deficient (i.e., cryptic) mutants are known where some of the permease or carrier systems for a substrate are lacking, though the enzymes for its metabolism are present.

Fungi may translocate nutrients and elaborated food material through the mycelium. Functionally, the hyphae consist of open tubes. If septa are present they normally have at least one pore in them and thus the cross walls are not thought to limit translocation under normal conditions. Movement occurs towards growing and sporulating regions from the older parts of the colony. The rate of movement is too great for diffusion to be solely responsible and the main translocation method is protoplasmic streaming, though the mechanism of this is obscure.

BOOKS AND ARTICLES FOR REFERENCE AND FURTHER STUDY

AINSWORTH, G. C. and SUSSMAN, A. S. eds. (1965). *The Fungi*, an advanced treatise, Vol. 1. The Fungal Cell. Academic Press, N.Y., 748 pp.
——— ——— eds. (1966). *The Fungi*, an advanced treatise, Vol. II. The Fungal Organism, Academic Press, N.Y., 805 pp.
COCHRANE, V. W. (1958). *Physiology of Fungi*, John Wiley & Sons Inc., N.Y., pp. 524.
FARRELL, J. and ROSE, A. (1967). Temperature Effects on Micro-organisms. *A. Rev. Microbiol.*, **21**, 101.
FOGG, G. E. (1953). *The Metabolism of Algae*, Methuen & Son Ltd., London. John Wiley & Son Inc., N.Y., 149 pp.
FRUTON, J. S. and SIMMONDS, S. (1953). *General Biochemistry*, 2nd edn. John Wiley & Sons Inc., N.Y., 1077 pp.
GUIRAD, B. M. and SNELL, E. E. (1962). *Nutritional Requirements of Microorganisms* in *The Bacteria* IV. ed. Gunsalus, I. C. and Stanier, R. Y., Academic Press, New York and London, pp. 33–93.
HALL, R. P. (1967). Nutrition and Growth of Protozoa. In *Research in Protozoology*, Vol. I. ed. Chen, T-T, Pergammon Press, Oxford, pp. 388–404.
HAWKER, L. E. (1957). *The Physiology of Reproduction in Fungi*, Cambridge Univ. Press, 128 pp.

HOLZ, G. G. (1964). Nutrition and Metabolism of Ciliates. In *Biochemistry and Physiology of Protozoa*, Vol. III, ed. Hutner, S. H., Academic Press, N.Y., pp. 199–242.

KITCHING, J. A. (1957). Some factors in the life of free living Protozoa. In *Microbial Ecology*; 7th Symp. Soc. Gen. Microbiol., Cambridge Univ. Press, pp. 259–286.

—————— (1967). Contractile vacuoles, ionic regulation and excretion. In *Research in Protozoology*, Vol. I. ed. Chen, T-T. Pergamon Press, Oxford, pp. 307–336.

KOSER, S. A. (1968). *Vitamin Requirements of Bacteria and Yeast*, Charles C. Thomas, Springfield, Illinois, U.S.A., 663 pp.

LEWIN, R. A. ed. (1962). *Physiology and Biochemistry of Algae*, Academic Press, N.Y. and London, 929 pp.

LILLY, V. G. and BARNETT, H. L. (1951). *Physiology of the Fungi*, McGraw-Hill Book Co. Inc., N.Y., 464 pp.

WERKMAN, C. H. and WILSON, P. W. eds. (1951). *Bacteriol Physiology*, Academic Press, N.Y., 707 pp.

WILLIAMS, R. E. O. and SPICER, C. C. eds. (1957). *Microbial Ecology*; 7th Symp. Soc. gen. Microbiol., Cambridge Univ. Press, 388 pp.

6

Growth of Micro-organisms in Artificial Culture

INTRODUCTION

The word 'growth' in common usage is not a precise term (see Oxford Dictionary for various meanings). An object or an organism or a population is said to 'grow' when it increases in size. Such a popular use of the word includes increase in size of an amorphous mass, such as a cloud; the more orderly increase in size of a crystal, which 'grows' by the addition to its lattice structure of solute molecules from the surrounding solution; and the highly complex 'growth' and differentiation of a living organism.

Growth of a biological system or of a living organism or part of one may, however, be defined as *an increase in mass or size (in any direction) accompanied by the synthesis of macromolecules, leading to the production of new organized structure.* Increase in size of a macromolecule in a living cell by the addition of molecular groups from the neighbouring pool of intermediate substances as a result of the activity of cell enzymes is growth. When such macromolecules are integrated into the cell organelles, thus producing new organized structure, the organelles grow in size. Organelles which are foreign to the normal cell, may be produced when extraneous DNA is introduced into the cell, as in infection by a virus. Virus organelles may then grow at the expense of host cell material and result in either symbiotic growth with the host cell or in its destruction.

Growth of a whole living cell, tissue, or organism is a more complex process. The osmotic intake of water by a cell when it is transferred from a concentrated to a dilute solution may cause it to swell (i.e., to increase in size), but this is *not* growth since no synthesis of material is involved. Swelling of a cell may, however, be accompanied by such synthesis and this *is* growth even if this synthesis is at the expense of the original cell reserves and results in a net loss of weight.

The germination of a fungus spore (Figs. 10.13 and 14) may be taken as an example. The endogenous metabolism of a mature spore is at a low level (p. 381). If a spore potentially capable of germination, is placed in a suitable nutrient solution and other conditions are favourable, it will sooner or later begin to germinate. The first visible sign of germination is usually swelling of the spore. Electron microscopy has revealed that this is accompanied by internal structural changes, such as an increase in number of mitochondria, nuclear division and, in some species, the formation of a new cell wall layer. Here swelling reflects true growth. The spores of some species of fungi will reach this stage if placed in distilled water, i.e., in the absence of an external supply of food. Here the synthesis of macromolecules and structural changes take place entirely at the expense of the original cell material and may then result in a loss of weight. Nevertheless this is still growth. The next stage in germination is the emergence of one or more germ-tubes, which are capable of growing indefinitely to form a new mycelium as long as suitable nutrients are available and if no other factor, either external or internal, becomes limiting. Here both size and mass increase and synthesis of macromolecules results in the production of new organized structure, often highly differentiated. If, however, external nutrients are initially absent or become exhausted growth ceases when the reserve material in the organism has been exhausted. Weight is reduced before growth actually ceases.

A living tissue grows by increase in size of its component cells, which may be accompanied by an increase in their number by cell division, and/or, in some animal tissues, by the inclusion of organized interstitial cellular products. As with a tissue, a colony of unicellular organisms on an agar plate grows by increase in size of the individual cells, by their replication (which may not involve increase in size) and/or by the accumulation of complex extracellular products, such as capsular material. All these processes involve the synthesis of macromolecules and the production of new organized structure. Populations of organisms, whether or not they form colonies, may also be said to grow because increase in number of organisms by replication is accompanied by synthesis and new organized structure is produced. As with the growth of a fungus germ tube when external nutrients are used up, localized growth and replication within a colony or population may continue for a short period but with a net loss of mass.

In multicellular organisms, such as the higher fungi, some algae and higher plants and animals, growth may include differentiation to produce cells or organs of particular form and performing particular functions.

Thus growth, whether of a single macromolecule in a living cell or of a complex plant or animal, produces order out of chaos, materials being incorporated in orderly sequence under the control of the genetic material of the particular organism.

Because the word 'growth' is applied to a number of analagous biological phenomena at different levels of organization, it is often advisable to use it preceded by a qualifying descriptive word, such as population, mass, volume, linear, localized, etc., to avoid ambiguity.

GROWTH OF INDIVIDUAL ORGANISMS

Growth of unicellular organisms

Many unicellular organisms, including the true bacteria, the fission yeasts (*Schizosaccharomyces* spp.) and many algae multiply vegetatively by binary fission (p. 373). Since, under standard conditions, the size of mature cells is remarkably uniform for a particular species, it is clear that each daughter cell must approximately double its size before it itself divides.

The quantitive aspect of the growth of single bacterial cells has to some extent been studied though the small size of bacteria presents a serious obstacle to such investigations. Growth of a bacterial cell is normally followed by the division of the nuclear body, the formation of a transverse septum and the separation of the resulting daughter cells (Fig. 9.16) but imperfect synchronization of these processes sometimes occurs and gives multinucleate cells or multicellular rods and cocci (p. 289). In extreme instances, growth may occur in the absence of division, with the result that long forms are produced (p. 305).

The scatter of interdivision times of bacteria has been considered statistically. Observations on cells dividing on permeable membranes placed over nutrient medium have shown that the frequency of interdivision times is scattered in an asymmetrical manner, some cells having prolonged division times compared to the doubling time of the population. Growth inhibitory substances at concentrations that have little effect on the most rapidly dividing organisms often increase this scatter of interdivision time.

More exact data are available for the growth of yeast cells. Recent advances in the use of the interference microscope have made it possible to measure the dry weight of single yeast cells. The most notable and unexpected finding, obtained with both the fission yeast *Schizosaccharomyces pombe* and the budding yeast *Saccharomyces cerevisiae*, is that increase in the dry weight of a single cell with time is linear, and not exponential, as it would be were the rate of increase proportional to the size already attained. Hence it is suggested that synthesis is controlled by cytoplasmic particles which remain constant in number and activity throughout the life of the cell and which double at cell division, being shared equally between the daughter cells. The other notable finding is that increase in cell volume does not closely parallel the increase in dry weight and that in the two species studied the pattern of increase differs considerably. Since the pattern of growth and division is rather different in fission and budding yeasts, it appears that the linear increase in weight may be a widespread feature of living cells and that the volume increase is a more variable character.

Little data are available concerning the growth of other unicellular fungi, such as the chytrids, or of unicellular algae. The diatoms are a special case since, as already described (p. 451) the box-like nature of the rigid cell wall ensures that at each division one daughter cell is slightly smaller than the parent. A return to the original size is brought about periodically by the formation of auxospores and the development of new cells from them (p. 451).

In a study of the increase in weight, volume, and protein content of the protozoan *Amoeba proteus*, increase in weight was at first linear with time but then fell off gradually and ceased four hours before cell division occurred. Change in protein content in sychronously dividing cultures of a few dozen individuals, which was measured by a micro-chemical method, closely paralleled increase in weight. After the cell had ceased to increase in size the volume of the nucleus increased rapidly prior to nuclear division. Similar results have been obtained with the ciliate *Tetrahymena*.

Growth of filamentous organisms

Some filamentous organisms, such as the filamentous or thread-forming iron and sulphur bacteria, some green algae (e.g., *Spirogyra* and other members of the Zygnemaceae) and some blue-green algae (e.g., *Nostoc* and *Anabaena*) are little more than chains of individual organisms. Any cell of such a filament is potentially capable of binary fission and hence the growth in length of the chain is merely the sum of the growth of the individual component cells.

Fungal hyphae, the filaments of *Streptomyces* and related genera, and those of many algae, increase in length only by the extension of a zone just behind the tip of the hypha. This has been most studied in the fungi. In such a growth pattern under constant conditions new primary wall material is constantly being produced at the tip of the hypha and soon after formation this primary wall loses its extensibility. Although growth in *length* of the hypha takes place only over a narrow subapical zone it is clear that the cytoplasm in a much longer part of the hypha must increase (grow) in volume in order to supply the advancing tip. For *Neurospora* (p. 398) it has been calculated that at least a 12 mm portion of the hypha must be involved in such duplication of the cytoplasm. Translocation of nutrients and organelles is thus essential to apical growth of a hypha.

It has been generally accepted that under constant conditions, established fungal hyphae grow at a uniform rate, although a few apparent exceptions have been claimed. Branches occur by extension of areas of wall which have either retained or regained their plasticity. The pattern and frequency of branches is controlled by the genetic constitution of the species, by environmental factors, and by the phenomenon of apical dominance. The mechanism of the latter is not fully understood, but the result is to favour growth of the parent hypha and to limit that of the branches.

Secondary growth of the individual cells of more complex organisms

Young cells are usually hyaline, thin-walled, and non-vacuolate. Vacuoles develop and granules of reserve foods and drops of oil commonly become more conspicuous as the cell ages. The structure of the cell wall is also subject to change. The wall of a fungal hypha is thin and elastic at the growing tip but becomes thickened and inextensible behind the zone of elongation, through the deposition of secondary wall material. In spores and in the individual cells of the complex fruit-bodies of the higher fungi, secondary thickening of the wall may not take place uniformly all over the cell; this may result in extension of the cell in some directions and not in others, often giving complex shapes. A study of the stages in development of the basidia and basidiospores of certain Basidiomycetes showed that the final shapes could be explained as the result of the combined effects of the pressure of the surrounding cells during development, and of the rate and site of hardening of the cell wall. Cells, such as the asci of most hymenial Ascomycetes, which develop in a closely packed single layer or hymenium and are thus subjected to equal lateral pressure, tend to be cylindrical or clavate; those developing under conditions of equal pressure from all directions, such as the irregularly distributed asci of the Plectascales, are usually globose.

Secondary thickening of fungal cell walls is most striking in many spores, and in the peripheral cells of many sclerotia, rhizomorphs, and fruit-bodies. This is often associated with deposition of a black pigment, usually assumed to be melanin. Some spores remain thin-walled and are usually viable for only a limited time. Others develop thick walls and may become dormant. An entirely new layer or layers may be laid down inside the original spore wall from which it may differ in structure and presumably also in chemical composition. In some spores, e.g., the zygospores of *Rhizopus* and allied genera, the original wall may be sloughed off and replaced by several distinct layers differing from it and from each other. The cell walls (or particular wall layers) of fungus spores are usually inextensible, largely impermeable to water and dissolved salts and often differ markedly in fine structure from the walls of vegetative hyphae. On germination these inextensible walls break to allow emergence of the germ tube, the wall of which is continuous either with the innermost layer of the spore wall or with a new inner layer formed just prior to germ-tube emergence. The ornamentation which is such a conspicuous feature of the spores of many fungi may arise in a number of ways according to the species, but always involves deposition of secondary material.

Pigments develop in the cytoplasm or wall of many micro-organisms, notably in spores and reproductive structures. All fungal spores are colourless when first formed but many become pigmented later. The basidiospores of some species of agarics do not become fully pigmented until some hours after attaining full size and are not shed until pigmentation is completed. Maturation thus continues after full size is attained.

GROWTH OF POPULATIONS OF UNICELLULAR ORGANISMS

The following account, although based largely on studies with bacteria, is in general applicable to other groups of unicellular micro-organisms.

Measurement of populations

Populations of unicellular organisms may be measured in terms either of the number of individual cells or of their mass. The former is sometimes referred to as the *cell concentration* and is defined as the number of individual cells per unit volume, whilst cell mass or *cell density* is defined as the weight of cells per unit volume. The methods used may be calibrated against each other so that comparative values can then be converted into absolute measurements.

Measurement of cell concentration may include either both living and dead cells, or only living cells and therefore may be expressed as the *total count* and *viable count* respectively.

Total counts

Total counts may be obtained by a number of techniques. These include (1) direct counting under the microscope in counting chambers with known area rulings and a known distance between coverslip and rulings; (2) direct counting by a device that counts the cells in a measured volume of suspension as they pass one by one through a minute aperture whose electrical conductivity they disturb, as in the instrument called the Coulter counter; (3) direct counts on dried stained smears of known volumes of a suspension of organisms spread over a known area or relative counts of bacterial cells mixed with a known number of other cells; (4) opacity of suspensions measured by visual comparison with known standards or electrophotometric measurement of turbidity either with an absorptiometer (which measures the residual amount of light after passing a beam of known intensity through a standard volume of the cell suspension) or with a nephelometer (which measures the amount of light from a standard source which is scattered by the cell suspension); (5) separating cells from a suspension by filtration or centrifugation and weighing either wet or dry after standard treatment; and (6) chemical assay of elements (e.g., nitrogen, which is proportional to the amount of protein present in the cells) or other compounds.

Viable counts

Viable counts are methods of determining the number of organisms capable of multiplying to form colonies on solid media or cultures in particular media under specified conditions (e.g., p. 584). With pure cultures conditions are usually chosen to give optimal growth, thus allowing almost all living cells to be counted; with mixed cultures a single set of culture conditions will not suit all organisms. With different media, temperatures, pH's, oxygen and carbon dioxide supplies, etc., different counts will be obtained according to the ability of the organisms to grow under the particular conditions chosen.

Methods used for determining viable counts include (1) inoculating appropriate dilutions of the cell suspension on to the surface of plates of media or incorporating the suspensions into the medium when this is poured, and incubating to produce visible colonies for counting; (2) diluting the suspension so that 50 per cent of appropriate sample volumes may be expected to contain no viable organisms: when such samples are inoculated into a large number of tubes of suitable medium and incubated, the observed proportion of tubes remaining sterile allows statistical calculation of the average number of bacteria per sample volume; (3) special methods are used to determine viable counts in air (p. 530).

Difficulties arise with all these methods. Cells may clump together, grow in chains or fail to separate on replication. Clumps of organisms can be broken up by agitation in a shaking machine and the counts are thereby increased. The number of cells in a chain, or multinucleate cells which have failed to divide can be determined by visual observation on appropriately stained smears. Opacity measurements can yield erroneous results if foreign particles are present in the suspensions and the use of control blanks and comparison between different dilutions are necessary to interpret results.

Phases of growth

In batch culture, in which the food supply is limited, growth proceeds through a series of phases. The shape of the growth curve which includes these phases is given in Fig. 6.1. The different parts of the growth curve may, however, be altered in length or slope according to the conditions prevailing in the culture. Temperature has a marked effect on the rates of chemical reactions involved in metabolism and hence on growth rate. Increase in temperature results in increase in rates of growth up to an optimum (p. 159, Fig. 6.2A and B), above which a relatively small further increase in temperature leads to cessation of growth. Many practical features in the control of bacterial populations are elucidated if the factors affecting each part of this growth curve are fully appreciated.

The exponential or log phase of growth

If bacteria are present in a medium which is fully satisfactory for their growth and if they are in the suitable state the population doubles in a definite and constant length of time. This is called the *population doubling time* to avoid the ambiguity inherent in the old term 'generation time'. Within a population the time taken for each cell to divide and form two daughter cells, i.e., the *interdivision time** or *cell doubling time*, varies considerably. The average of the individual cells' doubling times may be greater than the doubling time of the populations in which a few cells may divide very slowly in contrast to the rest of the population. However, the population doubling time during the exponential phase is constant and the size of the population increases in geometrical progression with time. If the logarithm of the number of organisms is plotted against time a straight line is therefore obtained (Fig. 6.1B).

Lag phase

When a medium is inoculated with organisms the conditions at first are usually different from those of the previous environment from which they were

---

* To avoid confusion the term 'interdivision time' has been used for individual cell replication time. The mean interdivision time is given by

$$\bar{\tau} = (\log_e 2)/k + (k\sigma^2)/2$$

where $(\log_e 2)/k$ is the doubling time (T_d) and σ is the standard deviation of the interdivision times (τ) (see Painter and Marr, 1967). It is obvious from the equation that τ is equal to T_d only if σ^2 is zero, i.e., if all cells have the same interdivision time. The interdivision times of related cells are not subject only to chance variations but have positive correlations between sister cells and negative correlations between mother and daughter cells.

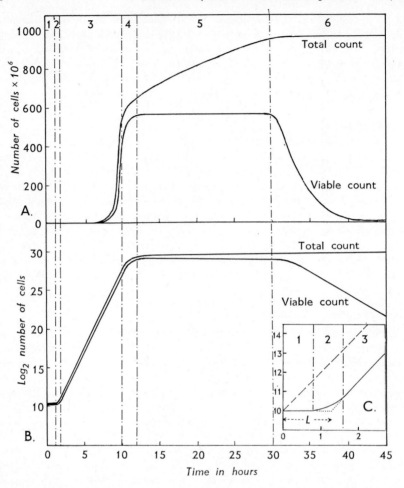

Fig. 6.1 Phases of growth. (**A**) A typical growth curve of *Escherichia coli* in a static 5 ml. volume of nutrient broth at 30°C. Both total and viable counts are plotted on an arithmetic scale against time and the phases of growth are indicated by numbers, viz.: 1. Lag phase; 2. Acceleration phase; 3. Exponential phase; 4. Retardation phase; 5. Stationary phase; 6. Phase of decline. (**B**) The same data as in A are redrawn by plotting numbers of cells as logarithms to the base 2, against time. This graph clearly demonstrates the exponential nature of phase 3 of the growth curve. A unit increase on the logarithm to the base 2 axis indicates a doubling of the population. The doubling time is the time required for this unit increase. (**C**) Graphical derivation of lag time (data as in B). The solid line represents the observed viable growth curve; the broken line the 'ideal' curve if no lag in growth occurred. Lag time, *L*, can be obtained by extrapolation of the exponential portion of the growth curve. It may be seen that the lag time, derived in this way, is of longer duration than the lag phase (Phase 1).

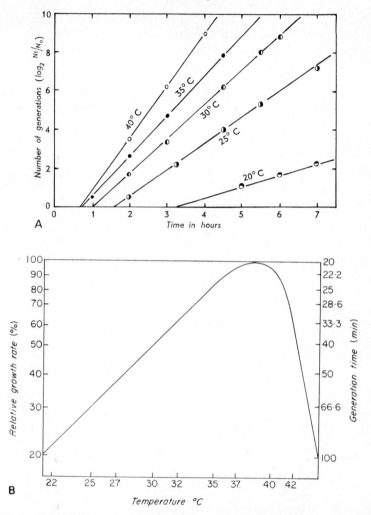

Fig. 6.2 Temperature and growth. (**A**) The influence of temperature on the Lag Time and Growth Rate of *Klebsiella pneumoniae* growing in a static culture of 5 ml. of nutrient broth. The number of generations are plotted against time and the slopes of the graphs indicate the growth rate at each temperature. The lag time is shown by the point of intersection of each curve with the time axis (A. H. Linton). (**B**) The relative growth rate (percentage of the maximum) plotted against temperature (on a scale of the reciprocal of the absolute temperature). (Data from Cooper, K. E., Gillespie, W. A., Linton, A. H., and Sehgal, S. N. in Kavangh, F. (1963) *Analytical Microbiology*, p. 43)

taken. The organisms may not be adapted to their new environment, and may be even in an unhealthy condition. Immediately on inoculation they may die or fail to commence to metabolize until either toxic substances have diffused away or essential foodstuffs have accumulated in the bacterial cell to an adequate concentration. Thus before the logarithmic phase of growth begins there will be a lag period in which the number of organisms produced by replication may be only a little greater or may be less than the number destroyed.

The length of the lag period is defined in different ways for different purposes. The lag in viable count for instance, is greater than the lag in metabolic rate of some of the chemical reactions involved in growth. The length of the lag period varies with changes in the environment, for instance, a rise in temperature within the range permitting growth leads to a decrease in lag time (Fig. 6.2).

Stationary phase and stage of decline

As the number of organisms increases, the concentrations of foodstuffs decrease and the concentrations of waste products of bacterial metabolism increase. This results in either (a) starvation, or (b) poisoning of organisms. The result is first inhibition of growth and finally death of organisms. When the number of organisms destroyed is equal to the number produced the stationary phase is reached. The number of viable cells present in the stationary phase is remarkably constant for a specific organism under similar conditions. This is known as the maximum population.

With further cultivation the viable count decreases, and the growth curve demonstrates the phase of decline. The total count ceases to increase when multiplication ends; it can only decrease if lysis of cells occurs.

Phases of acceleration and retardation

The beginning and end of the logarithmic phase have transition stages linking it with the earlier lag phase and the later stationary phase. These are spoken of as phases of acceleration and retardation. For mathematical purposes it is often convenient to regard these as complexes of an extrapolated log phase to the size of the initial inoculum on the one hand, or the count of the stationary phase on the other. The point at which the extrapolated log phase equals the size of the original inoculum can be referred to as *lag time* to distinguish it from true lag (see inset C, Fig. 6.1).

Calculation of the growth rate

Batch culture

During the exponential phase in which no death occurs the population doubles at regular intervals of time, i.e., the population doubling time (T_d). In an ideal system the course of events may be presented as follows:

Number of Doubling Times

$$0 \qquad 1 \qquad 2 \qquad 3 \qquad 4 \qquad \ldots \; n$$

Where an inoculum of one cell is used the number of cells (N) in each population for corresponding times is

$$N = 1 \qquad 2 \qquad 4 \qquad 8 \qquad 16 \qquad \ldots \; 2^n$$

Where an inoculum of N_0 cells is used the populations for each doubling time may be expressed in the following way:

$$N = N_0 \qquad 2N_0 \qquad 4N_0 \qquad 8N_0 \qquad 16N_0 \qquad \ldots \qquad N_0 2^n$$

The total growth time (T), measured from the time of inoculation, is the sum of the duration of the Lag Phase (L) and the number of increments of the doubling time (T_d) which have occurred. Thus the corresponding times for each stage of the population growth will be

$$T = L \qquad L + T_d \quad L + 2T_d \quad L + 3T_d \quad L + 4T_d \quad \ldots \quad L + nT_d$$

If the inoculum consisted initially of N_0 viable cells and n is the number of doubling times after the end of the Lag Phase, or from the time of inoculation if there is no lag, then the number of viable cells (N) at time T may be expressed as follows:

$$N = N_0 2^n \tag{1}$$

but the time required for n doubling events to have occurred has been shown above to be

$$T = L + nT_d \tag{2}$$

This equation may be re-arranged:

$$n = (T - L)/T_d$$

Therefore, from equations (1) and (2)

$$N = N_0 2^{(T - L)/T_d} \tag{3}$$

This expression can be simplified by taking logarithms to the base 2 of each side of the equation, when

$$\log_2 N = (T - L)/T_d + \log_2 N_0 \tag{3a}$$

Thus a plot of $\log_2 N$ against time gives a straight line from the point where the population is equal to the inoculum size (i.e., $N = N_0$) at the end of the Lag Phase (i.e., $T = L$). The slope of this line is called the exponential growth rate by many biologists and equals $1/T_d$. It is approximately equal to the number of generations per unit time, though more accurately it is the number of population doubling times (see footnote, p. 175).

Plotting in units of $\log_2$ is particularly valuable since an increase of one division on the ordinate scale of the graph represents doubling of the population. Tables of $\log_2$ are published but the values may be obtained by multiplying common logarithms (i.e., $\log_{10}$) by 3·322.

Exponential growth when considered in terms of the synthesis of molecules instead of cell divisions is best described in terms of natural logarithms (i.e., logarithms to the base e) when the rate of the controlling chemical reaction is given by the specific rate constant k:

$$m = m_0 e^{kT} \tag{4}$$

and

$$\log_e m = \log_e m_0 + kT \tag{4a}$$

The value of k is given by the slope of the graph ($\log_e m$ against time) where m and m_0 represent the number of molecules at time T and time 0 respectively if there is no metabolic lag. Where a metabolic lag (L'') occurs then T must be replaced by $T - L''$. A lag in viable count may occur when a metabolic lag does not.

This kind of consideration is relevant when growth is measured in terms of dry weight (W). The corresponding formulae would then be

$$W = W_0 e^{k(T - L'')} \tag{5}$$

and

$$\log_e (W/W_0) = k(T - L'') \tag{5a}$$

The slope of the line obtained by plotting $\log_e W$ against T is spoken of as the *instantaneous specific growth rate* since

$$\frac{d(\log_e W)}{dt} = \frac{1}{W}\frac{dW}{dt} = k \tag{6}$$

If we consider the time taken to double the dry weight of organisms, then

$$W = 2W_0$$

and

$$\log_e (W/W_0) = \log_e 2 = k(T - L'') \tag{7}$$

and either the metabolic lag (L'') may equal 0 or else time is measured from the end of L''. From expression (7) the doubling time (T_d) is therefore given by

$$T_d = (\log_e 2)/k \tag{8}$$

If the dry weight of cells (W_1 and W_2) is determined for two times (T_1 and T_2), both on the exponential curve, no lag period is involved and this term is eliminated from the equation, viz:

$$\log_e (W_2/W_1) = k(T_2 - T_1) \tag{9}$$

Continuous culture

Continuous culture is a system of cell growth in which nutrients are being continuously added and spent culture media is being continuously removed from the culture vessel. In the steady state which can be thus achieved the viable population of cells can be kept constant, the removal of dead and viable cells being balanced by new cell growth. Open systems of continuous culture are found in many natural circumstances, such as in lakes as part of river courses, in the alimentary tract of man and animals, and in infected body cavities during the course of disease which are supplied with nutrients from the body tissues and may be drained naturally or artificially.

Industrial and laboratory processes have been developed for the continuous production of micro-organisms or their products such as, for example, the production of vinegar from alcohol. The theory of continuous culture has recently been developed and specialized apparatus designed with automatic control for laboratory research. Continuous culture under steady-state conditions is now possible for long periods of time, over a wide range of controlled conditions.

The growth rate, the density of the microbial population, and the concentration of products may be kept constant by maintaining a constant flow of nutrient medium, which limits the multiplication rate by the concentration of one of its constituents. The flow may be kept constant by pump or gravity under a constant head, and by flow valves affecting the inflow of the medium to a vessel which is made to contain a constant volume of culture by overflow at an exit. The ratio of the flow rate, f, to the culture volume, v, is called the dilution rate, D, and this gives the number of complete volume changes in the culture vessel per hour (hr^{-1}). Thus

$$D = f/v \qquad\qquad (10)$$

The reciprocal of D is a measure of the average time a particle remains in the vessel, provided that a constant, immediate and vigorous agitation is maintained. A diagram of a typical apparatus, modified from Herbert (1958), is shown in Fig. 6.3.

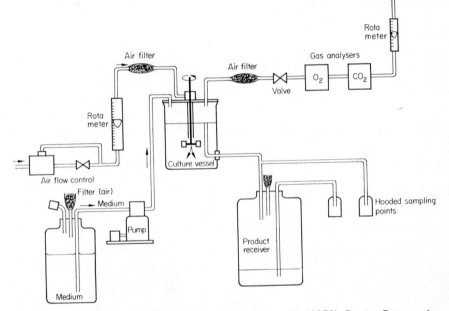

Fig. 6.3 Continuous culture apparatus. (After Herbert, D. (1958) *Recent Progress in Microbiology*, 382 pp. VIIth International Congress for Microbiology. Almqvist and Wiksell, Stockholm)

Flow rate may be adjusted in response to substrate concentration in various forms of apparatus called *chemostats*, or to the opacity of the organisms in the culture vessel as in *turbidostats*, or to the outflow of some particular product. Growth may take place under aerobic conditions, when adequate oxygen supply and agitation has to be maintained, or under anaerobic conditions in controlled gas flow. In all cases very rapid mixing by mechanical stirring and agitation is necessary if theoretically predicted results are to be obtained.

We have seen that with batch cultures during the logarithmic phase of growth the doubling time $T_d = \log_e 2/k$ (equation 8). The rate of change of W (dry weight) is $dW/dt = kW$, from which it follows that $k = 1/W \cdot dW/dt$ (equation 6) and k has been called the *specific growth rate* (also known as the exponential or instantaneous growth rate).

With continuous culture, during the steady-state conditions of constant limiting nutrient supply, a constant growth rate is produced by maintaining a constant dilution rate (within certain maximum and minimum limits). That growth rate depends on the concentration of the limiting substrate (S) in the culture was shown by Monod (1942) who demonstrated the relation

$$k = k_m\left(\frac{S}{S_k + S}\right) \qquad (11)$$

where S_k is a saturation constant equal to the substrate concentration which produces half the maximum possible growth rate k_m (Fig. 6.4).

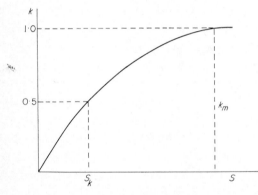

Fig. 6.4 Diagram showing the dependence of growth rate on the concentration of the limiting substrate in continuous culture. S_k is the concentration of the substrate which produces half the maximum possible growth rate k_m.

When the limiting nutrient is carbohydrate or amino acid S_k has low values and most of the foodstuff is utilized. The efficiency with which the growing organisms use a substrate is expressed by the term *yield constant, Y*.

$$Y = \frac{\text{weight of organism formed}}{\text{weight of substrate used}} \text{ (for a finite growth period)}$$

$$= -\frac{dW}{dS} = \frac{dW/dt}{-dS/dt} \qquad (12)$$

In *batch cultures* the growth rate $dW/dt = kW$ (equation 6); in *continuous culture* the rate at which cells are washed out is equal to the cell density multiplied by the dilution rate (equation 10). Thus

$$-dW/dt = DW \qquad (13)$$

The rate of increase of cells in the continuous culture vessel is therefore, from equations (6) and (13),

$$\text{increase} = \text{growth} - \text{output}$$

$$\frac{dW}{dt} = kW \quad - DW$$

$$= W\left[k_m \frac{S}{S_k + S} - D\right] \tag{14}$$

The rate of substrate utilization, from equations (12) and (14) is

$$\frac{dS}{dt} = \frac{dW}{dt} \cdot \frac{dS}{dW} = [(k - D)W]\left[-\frac{1}{Y}\right]$$

$$= \frac{W}{Y}[D - k]$$

and, where S is the level of substrate *in the culture,*

$$\frac{dS}{dt} = \frac{W}{Y}\left[D - k_m \frac{S}{S_k + S}\right] \tag{15}$$

It can be seen from equation (12) that, if during a finite growth period the weight of organisms formed is W and the weight of substrate utilized is $S_R - S$ (i.e., the difference between the amount in the entering medium S_R, and that remaining in the culture vessel S), then

$$Y = W/(S_R - S)$$

or

$$S_R - S = W/Y \tag{16}$$

Equation (15) may therefore be derived in another way. In the culture vessel

$$\text{rate of increase in substrate} = \text{input} - \text{output} - \text{consumption}$$

$$\frac{dS}{dt} = DS_R \quad - DS \quad - k(W/Y)$$

$$= D(S_R - S) - \frac{W}{Y}\left[k_m \frac{S}{S_k + S}\right]$$

and, from equation (16) this becomes

$$\frac{dS}{dt} = \frac{W}{Y}\left[D - k_m \frac{S}{S_k + S}\right] \tag{15}$$

Under steady state conditions in the culture vessel $dS/dt = 0$ and $dW/dt = 0$, since this is what is meant by 'steady state' conditions. If $\tilde{W}$ is the cell concentration at the steady state when $S = \tilde{S}$, the limiting concentration, from equation (16) is

$$\tilde{W} = Y(S_R - \tilde{S}) \tag{17}$$

From equation (15)

$$D = k_m \frac{\tilde{S}}{S_k + \tilde{S}}$$

or, by rearrangement

$$\tilde{S} = S_k \left(\frac{D}{k_m - D} \right) \tag{18}$$

so that the dry weight of cells maintained in the steady state is

$$\tilde{W} = Y \left[S_R - S_k \frac{D}{k_m - D} \right] \tag{19}$$

and W is therefore determined by the absorption constants of the organism for the limiting foodstuff, and its yield efficiency, by the concentration supplied to the culture vessel and by the dilution rate.

With a particular organism and growth limitation by a particular nutrient, the whole process will achieve a steady state controlled by (a) the concentration of the foodstuff and (b) the dilution rate. If, however, the dilution rate is too great for the *maximum* possible growth rate, the organisms will be completely washed out of the apparatus. If, on the other hand, the flow rate is too small to supply sufficient nutrients to support the *minimum* possible growth rate, the organisms will fail to multiply and will then be washed out. Within these limits a steady cell concentration will be maintained in the culture vessel such

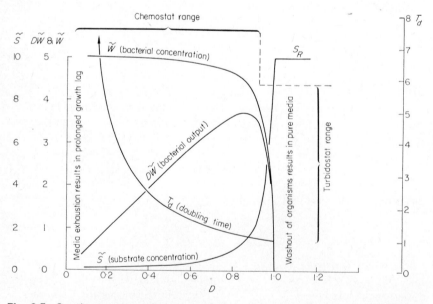

Fig. 6.5 Steady-state theoretical curves. The constants used are: Y (yield constant) = 0·5; S_k (substrate concentration) = 0·2; K_m (maximum growth rate) = 1·0; S_R (substrate entering the culture medium) = 10·0; all units are g/l/hr^{-1}.

that (a) *the specific growth rate* (k) is equal to the *dilution rate* (D) (see equations *8*, *14*, and *18*), i.e.,

$$k = \frac{\log_e 2}{T_d} = k_m \left[\frac{\tilde{S}}{S_k + \tilde{S}} \right] = D \qquad (20)$$

and (b) the *cell densities*, expressed as dry weight of cells $(\tilde{W})$, is given by equation *(19)*. The theoretical steady state conditions for given constants are shown in Fig. 6.5.

COLONIAL GROWTH

Growth of bacteria on solid media

The growth of bacterial colonies on solid media is often characteristic of the species or even the strain of organism (Fig. 9.18). The conditions affecting the shape of the colony and its rate of growth are obviously complex. For instance, the limitation of available nutrients diffusing from the medium to and into the colony are complicated by conditions such as the solid media aqueous surface film and atmosphere, as well as the thickness and diameter of the colony. With prolonged growth the proximity of neighbouring colonies, the thickness of the medium and the availability of oxygen and carbon dioxide are important factors. Death and autolysis of cells in the centre of an older colony or dispersed throughout it, temperature, pH, and osmotic conditions will all influence colonial growth. Mathematical expressions for viable counts of colonies, or their size (diameter or thickness) are of limited validity.

Multiplication of bacteria on surfaces like talc, powdered glass, chalk or precipitates can take place when they are suspended in water containing very low concentrations of nutrients, (e.g., dissolved ammonia and carbon dioxide), owing to the absorption of solutes on to these surfaces producing an increased local concentration.

Growth of colonies of filamentous organisms

Growth of fungal colonies and solid media

If a plate of a suitable agar medium is inoculated at the centre with fungal spores or mycelium, a colony of circular outline is produced within a few days and will continue to grow at a rate dependent on the isolate used, the nature of the medium, and environmental factors. Three phases in the growth of such a colony may be distinguished, a lag phase, a phase of linear growth, and a phase of staling. The final phase may not occur in some circumstances (Fig. 6.6).

Owing to the complexity of growth of fungal mycelium (p. 374) no satisfactory general mathematical treatment of growth on a solid medium has been achieved, although such treatment has been applied to particular species (Mandels, 1956).

LAG PHASE. When spores are inoculated on to a solid medium some hours elapse before germination is completed. If a mycelial inoculum is used, the regeneration of broken and damaged hyphae will take a comparable period. There will

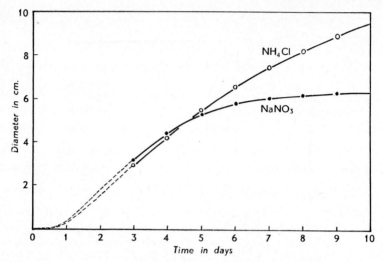

Fig. 6.6 Growth of the fungus *Sclerotinia fructigena* on agar media in petri dish cultures. There is a lag of about one day, after which approximately linear growth occurs. When the nitrogen source is ammonium chloride, no marked staling occurs and the petri dish margin is reached in about 10 days; with sodium nitrate as the nitrogen source, however, there is a strong staling, beginning at about 5 days. The broken lines indicate growth during the first three days, which in this particular experiment was estimated but not measured. (M. J. Carlile)

then be a further interval before a maximum rate of metabolism and hence of growth is attained, this latter interval being analogous to the combined lag and acceleration phases of growth or populations of unicellular organisms. Lag in fungus cultures has been studied less than for bacterial growth; but it is clear that, as with bacteria, the duration is influenced by the size, age, and nature of the inoculum as well as by the composition of the medium employed.

PHASE OF LINEAR GROWTH. The lag phase is followed by one in which the diameter of the colony increases linearly with time, a constant rate of growth being maintained at the margin of the colony. This increase in diameter of a colony is readily measured in petri dish cultures and hence is much used in studies of fungus growth. Such measurements relate to the growth rate at the margin of the colony, and may bear little relation to the increase in weight. Thus a colony on agar without added nutrients commonly increases in diameter much more rapidly than one on a nutrient medium. Such a colony will, however, consist of a thin network of hyphae in contrast to the thicker mat of mycelium produced by one on a rich medium, and has clearly a far smaller dry weight than that of the latter. Provided that these limitations are appreciated, the method is a valuable one, as it permits repeated observations on the same culture. It is particularly applicable to the measurement of the effect of temperature or other physical factors on media of similar composition.

The rate of linear growth is influenced by the composition and concentration of the medium, its pH and osmotic pressure, and the temperature of incubation. The oxygen and carbon dioxide content and relative humidity of the atmosphere within the petri dish may also be of importance. Linear growth may sometimes continue until the margin of the petri dish is reached but the growth rate may decline before this. The growth rate of fast-growing fungi can conveniently be measured in long horizontal tubes half filled with medium and inoculated at one end. The linear phase of growth can be maintained for several days in these growth tubes.

STALING PHASE. The growth rate of a fungus colony usually declines as the margin of the petri dish is approached but it may do so some distance from such a mechanical barrier. The decline in growth rate is then usually due to the harmful effects of the colony's own metabolic products and is hence referred to as *staling*. It is far commoner on rich than on poor media and is frequently associated with an unfavourable change in the pH of the medium. In addition to a reduction of growth rate, staling generally causes an uneven growth leading to the production of an irregular or even strongly lobed colony outline instead of the smooth one characteristic of the linear phase of growth. In some species staling is accompanied by the autolysis of the mycelium at the centre of the colony. Staling is more usual at supra-optimal than sub-optimal temperatures, no doubt owing to the increased rate of metabolic processes at high temperatures.

Growth of fungal colonies in liquid media

For many purposes, including the study of nutritional factors, measurements of colony diameter on a solid medium are unsuitable, as already pointed out. Dry weight of mycelium gives a more informative result but can seldom be estimated satisfactorily with solid media. Moreover, the solidifying agents are difficult to purify. For critical metabolic studies, the use of a liquid medium of known composition is essential.

Unfortunately most fungi, other than aquatic species, do not grow normally when submerged in liquid. Attempts to support the mycelium on glass beads or other inert substances have been unsatisfactory owing to the difficulty of separating mycelium and substrate. Shake cultures or those aerated by other means are an improvement on static ones but present many problems (see Campbell, in Ainsworth & Sussman, 1965). Aquatic fungi have been much used for metabolic studies, but are not necessarily representative of the fungi as a whole.

Many fungi produce a thick mat of hyphae at the surface of a static liquid medium. In such circumstances conditions will not be uniform throughout the mycelium and in particular there will be a gradient from an anaerobic or partially anaerobic to an aerobic state. Shaken or aerated cultures achieve homogeneity at any one time. Those continuously irrigated with fresh medium maintain chemically stable conditions over the duration of the experiment. Under such conditions, some fungi grow as an amorphous filamentous colony; others, including the species of *Penicillium* employed in the production of penicillin, produce compact spherical bodies or pellets.

The factors determining pellet formation are not fully understood. The pellets consist of an inner mass of non-growing and presumably inactive hyphae and an outer zone of actively growing ones. The thickness of this outer layer is limited by the rate of the diffusion of oxygen and nutrients. The mathematical analysis of pellet growth has been discussed by Pirt (1966).

FACTORS INFLUENCING GROWTH

Growth both of individuals or of populations is controlled by the genetical constitution of the species or strain and by a number of internal and external factors. Under natural conditions these factors interact to influence growth and development but in the laboratory the effects of altering one external factor can be studied while others are kept constant. In the study of each factor it is important to distinguish between its effects on the various phases of growth of an organism. The influence on growth of a number of external factors has been considered in Chapter 5; these include nutrition, temperature, pH, light, and gravity. The following paragraphs discuss several other important factors.

Genetic constitution

Species differ greatly in the rate of growth and cell division attained under optimal conditions (Table 6.1), and within a species there may be marked differences between strains. Lengths of lag period and total growth are also influenced by genetic constitution.

Table 6.1 Doubling times and growth rates of representative organisms under optimum conditions.

	Doubling time
Escherichia coli	20 minutes
Mycobacterium tuberculosis	18 hours
Chilomonas paramecium	$6\frac{1}{4}$,,
Amoeba proteus	24 ,,
Chlorella pyrenoidosa	$8\frac{1}{4}$,,
Anagaena cylindrica	24 ,,
Saccharomyces cerevisiae	2 ,,
Schizosaccharomyces pombe	4 ,,
	Radial growth rate of colony
Neurospora crassa	12 cm./day
Sclerotinia fructigena	0·7 cm./day

Age of cells

Cells taken from old cultures of those unicellular micro-organisms, which divide by binary fission, are often erroneously referred to as old cells. The age of a single cell cannot be greater than the time between two divisions. A cell, when first formed by cell division, although often called a young cell, contains materials both genetic and protoplasmic which pre-existed in the parent cell. The effect of 'age' is, however, apparent in batch cultures which have been growing for a long time. Cells from old batch cultures differ in many ways from those of young cultures because they have been modified by a changing environment. Sub-culture from old batch cultures to fresh medium results therefore in a long lag period during which adaptation to the new conditions takes place. In contrast, the lag phase can be virtually abolished by subculturing cells from an actively growing culture in the early logarithmic phase to a medium of identical composition and under the same physical conditions if mechanical damage is avoided by the act of inoculation.

Changes in the growth medium

Transferring an inoculum of micro-organisms to a different medium usually leads to the establishment of a new growth rate. Other enzymes and chemical reactions have to be utilized and the control mechanism regulating the synthesis

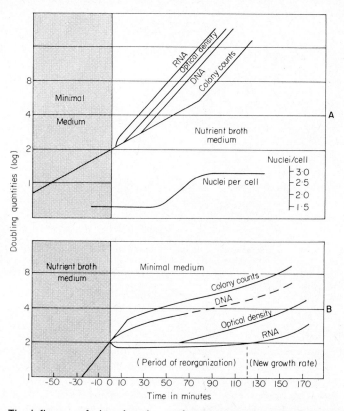

Fig. 6.7 The influence of changing the nutrient content of medium on growth rate and biosynthesis. (**A**) Change to greater growth rate. *Salmonella typhimurium*, cultured at 37°C transferred from a minimal medium to a nutrient broth. The curves are fitted to coincide. The inset graph shows numbers of nuclei per cell determined with stained preparations. Transfer to a medium of higher growth rate shows differing lag periods in synthesis and cell division. (**B**) As in **A** above but in which the culture was transferred from nutrient broth to a minimal medium. (After Kjeldgaard, N. O., Maaloe, O. and Schaechter, M. (1958) *J. gen. Microbiol.* **19**, 609)

of various cell components brought into step with the new growth rate. If the medium is richer and the temperature the same, the rate of synthesis of RNA is first affected; this is followed by an increase in protein and DNA synthesis and finally cell division occurs. The number of organelles, such as nuclei and ribosomes per cell, is also changed (Fig. 6.7A). Transfer from a nutrient broth to a minimal synthetic medium, producing a reduced growth rate, affects these syntheses in the reverse order though there is a considerable lag period for re-organization and turnover of cell constituents (Fig. 6.7B).

In particular situations when a medium contains two alternative foodstuffs, one may be used preferentially and subsequently the initial rate of growth slows as this foodstuff is exhausted. After a period of adaptation, during which the cell synthesizes new enzymes required to utilize the alternative foodstuff, the lag is followed by a new growth rate. This phenomenon was called 'diauxie' by Monod (Fig. 6.8).

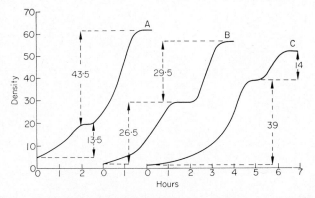

Fig. 6.8 Growth of *Escherichia coli* in a mixture of glucose and sorbite, ratios A, 1/3; B, 2/2; C, 3/1. The figures show that the total growth corresponds to each 'cycle of growth'. (After Monod, J. (1942) *Recherches sur la Croissance des Cultures Bactér-iennes*, Hermann et Cie, Paris)

Competition

In the natural habitat the growth of a population of micro-organisms is influenced by the activities of other species. A study of population growth as influenced by competition for the same food source was made with the ciliates *Tetrahymena pyriformis* and *Chilomonas paramecium* (Fig. 6.9). *Chilomonas* reached a population of about 80,000 in a week, and apart from minor fluctuations remained approximately at that value until the experiment was terminated after 5 weeks. Under similar conditions *Tetrahymena* reached about 45,000 in ten days and also remained roughly constant until the end of the experiment. Results of simultaneous inoculation were unexpected. Both species at first grew more rapidly than when alone, *Chilomonas* reaching over 130,000 in a week. Then, however, the *Chilomonas* died out completely while *Tetrahymena* in-

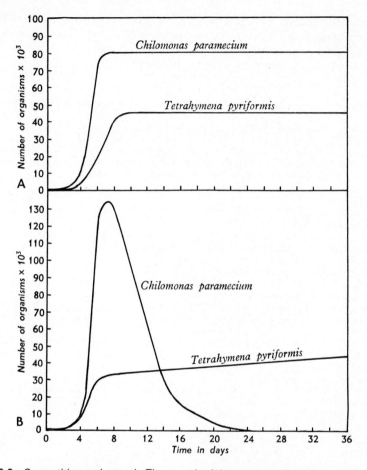

Fig. 6.9 Competition and growth. The growth of the protozoa, *Tetrahymena pyriformis* and *Chilomonas paramecium*, cultured separately (**A**) and together (**B**). (Data from Mučibabić, S. (1957) *J. gen. Microbiol.* **16**, 561)

creased slowly to reach its 'single-culture' value. Not only did growth in numbers differ in pure culture and in mixed culture, but the size and form of the organisms were affected. The prey-predator relationships, such as one protozoan species feeding on another or on bacteria or yeasts have also been studied. Such experiments may end with the extinction of the prey followed by that of the predator. Alternatively, the prey may be reduced to low numbers; the predator then dies out owing to insufficient food and the prey re-establishes itself. In a study of competition between *Paramecium aurelia* and *P. caudatum*, both preying on *Pseudomonas aeruginosa*, *P. aurelia* was normally the more successful, but when yeast was used for food *P. caudatum* prevailed.

Mixed populations of bacteria, such as occur in sewage or unhygienically produced raw milk (p. 654), produce a succession of high populations of different bacteria types. As each type exhausts the foodstuffs which it is most capable of utilizing the population of this type falls and the succeeding type rises by utilizing different foodstuffs or even the products of an earlier population.

BOOKS AND ARTICLES FOR REFERENCE AND FURTHER STUDY

AINSWORTH, G. C. and SUSSMAN, A. S. eds. (1965). *The Fungi*, Vol. 1, Academic Press, London and New York, in particular chapters 25–28, i.e., Mandels, G. R., Kinetics of Fungal Growth; Robertson, N. F., The Mechanism of Cellular Extension and Branching; Jerebzoff, S., Growth of Rhythm; Campbell, A., Special Growth Techniques (Synchrony Chemostasis).

BRODY, S. (1945). *Bioenergetics and Growth*, Reinhold Publishing Co., New York, 1023 pp.

BURNETT, J. H. (1968). *Fundamentals of Mycology*, Edward Arnold, London, Chapters 3 and 4.

DEAN, A. R. C. and HINSHELWOOD, C. (1966). *Growth, Function and Regulation in Bacterial Cells*, Clarendon Press, Oxford, 439 pp.

ERRINGTON, F. P., POWELL, E. O., and THOMPSON, N. (1965). Growth Characteristics of Some Gram-negative Bacteria. *J. gen. Microbiol.*, **39**, 109–123.

GUNSALUS, I. C. and STANIER, R. Y. *The Bacteria*, Vol. IV. *The Physiology of Growth*, Academic Press, New York, 459 pp.

HERBERT, D. (1958). Some Principles of Continuous Culture, in *Recent Progress in Microbiology*, VIIth International Congress for Microbiology, pp. 389–396. Almqvist & Wiksell, Stockholm.

MANDELSTAM, J. and MCQUILLEN, K. (1968). *Biochemistry of Bacterial Growth*, Blackwell Sci. Pub., Oxford and Edinburgh, 540 pp.

MEADOW, P. and PIRT, S. J. eds. (1969). *Microbial Growth*. 19th Symp. Soc. gen. Microbiol., Cambridge University Press, 450 pp.

MONOD, J. (1942). *Recherches sur la Croissance des Cultures Bactériennes*, Hermann et Cie, Paris, 210 pp.

—— (1949). The Growth of Bacterial Cultures. *Ann. Rev. Microbiol.*, **3**, 371.

NOVICK, A. (1955). Growth of Bacteria. *Ann. Rev. Microbiol.*, **9**, 97–110.

PAINTER, P. R. and MARR, A. G. (1967). Inequality of Mean Interdivision Time and Doubling Time. *J. gen. Microbiol.*, **48**, 155–159.

PIRT, S. J. (1966). A Theory of Growth of Fungi in the Form of Pellets in Submerged Culture. *Proc. Roy. Soc.* (London) **B.166**, 369–373.

POWELL, E. O. (1956). Growth Rate and Generation Time of Bacteria with Special Reference to Continuous Culture. *J. gen. Microbiol.*, **15**, 492–511.

TEMPEST, D. W. and HERBERT, D. (1965). Effect of Dilution Rate and Growth-Uniting Substrate on the Metabolic Activity of *Torula utilis* Cultures. *J. gen. Microbiol.*, **41**, 143–150.

7

Influence of External Factors on Viability of Micro-organisms

SURVIVAL IN NATURE

In nature a wide variety of agents may bring about the death of micro-organisms, and many of these are employed by man for the destruction of harmful species.

Organisms are able to grow only within certain limits of pH, temperature, osmotic pressure, and other environmental conditions (p. 155). Outside these limits they may survive but be unable to grow and reproduce; further departure from optimal conditions may lead to death. Death may result also from desiccation or starvation, or may be caused by predators, parasites or by competitors which produce antibiotics or other toxic products. An organism may cause its own destruction by producing toxic metabolites, a process known in mycology as 'staling', or by inducing other unfavourable changes in the environment (p. 188).

Micro-organisms differ widely in their ability to survive unfavourable conditions, and different phases of the same species vary greatly in their sensitivity. A state of reduced metabolic activity (hypobiosis), whether brought about artificially or occurring naturally, greatly increases an organism's ability to survive unfavourable conditions.

Hypobiosis

Hypobiosis can be brought about artificially by low temperatures, lack of oxygen, loss of water, and exposure to high salt concentrations, or by a combination of these. With the exception of low temperature, these factors are probably also involved in the natural hypobiosis occurring in such resting bodies as the endospores of bacteria, the sclerotia and chlamydospores of fungi, and the cysts of protozoa. In most instances thick cell walls help to maintain appropriate concentrations of oxygen and water.

In a state of hypobiosis, organisms are able to survive for long periods in the absence of nutrients, and have, moreover, a greatly increased resistance to lethal agents. For example, the amoeboid phase of *Naegleria gruberi* is destroyed by drying for 15 seconds, freezing to $-30°C$ for 45 minutes or heating to 50°C for 2 minutes. Cysts of this protozoan, however, will survive drying for 23 months, freezing for $3\frac{1}{2}$ months, or heating to 50°C for 30 minutes. Spores of *Bacillus anthracis* have been shown to be viable after 31 years in the dry state and those of *Clostridium sporogenes* after 46 years in alcohol. Dried fruit-bodies of the fungus *Schizophyllum commune* have survived 34 years in a vacuum and retained the ability to resume spore production when moistened.

The circumstances that determine the production of resting bodies are not fully understood. Starvation is a frequent cause, as with many protozoa which produce cysts when the supply of bacteria on which they feed is exhausted. In contrast, bacteria may produce endospores on rich media under certain circumstances (p. 293). The germination of protozoan cysts has been found to be induced by a wide variety of agents. Germination of cysts of *Schizopyrenus russelli*, a soil amoeba, is induced by the presence of various amino acids; that of *Colpoda duodenaria*, a ciliate, is brought about by a variety of factors which are thought to stimulate activity of the tricarboxylic acid cycle.

Hypobiosis has been subdivided into *dormancy*, in which metabolic activity although greatly reduced is detectable, and *cryptobiosis* (sometimes *anabiosis* or *latent life*) in which metabolic activity is absent or undetectable. There is no doubt that a wide variety of organisms can survive a complete cessation of metabolism. For example, many algae, spores of bacteria and of some fungi, and fragments of lichens have survived exposure for 2 hours to a temperature a fraction of a degree above absolute zero. It has been calculated that at $-200°C$ simple chemical reactions would be eight million times slower than at 20°C so at even lower temperatures the complex reactions involved in metabolism cease completely. The absence of chemical activity at these temperatures, however, ensures the maintenance of structural integrity and hence the resumption of metabolism on return to normal conditions. One method of inducing hypobiosis, namely freeze-drying, is now widely employed for preserving cultures of bacteria for long periods. Cell suspensions are rapidly frozen, by means of 'dry ice' (solid carbon dioxide) in alcohol and are then dried under high vacuum. It is usually necessary to include in the suspension protective agents such as glucose, ascorbic acid, gelatin, serum, or glutamate. The way in which these substances act is still not clear. The dried suspensions are maintained under vacuum or in an inert gas such as nitrogen to avoid oxidation to which they are very susceptible in the dried state.

Senescence and natural death

The life of most higher organisms, if other mishaps are avoided, is terminated by a gradual decline in metabolic activity (*senescence*) ending in death. This, however, is unusual in micro-organisms, in which the individual commonly divides into two daughter cells. In this, micro-organisms resemble the unspecialized cells of higher organisms under favourable conditions (such as in tissue culture or in an actively growing tissue) rather than individual multicellular organisms.

In fungi a common example of natural death is provided by Basidiomycete fruit-bodies which die on completion of spore discharge, i.e., when their function is fulfilled. In *Neurospora crassa* a 'natural death' mutant is known in which the vegetative growth of a colony ultimately ceases; rejuvenation can be obtained by crossing or by heterocaryosis with a normal strain. In the suctorian protozoan *Tokophyra infusionum* which reproduces by budding, a definite limited life-span occurs, the length of which may be modified by nutritional factors, excess food in general shortening life.

LETHAL AGENTS

A wide variety of physical and chemical agents bring about the death of micro-organisms. Differences in efficiency are found between the various agents and each is influenced by (1) many environmental factors and (2) the nature of the micro-organisms exposed to them.

Factors influencing the activity of lethal agents

Certain environmental factors influence the activity of both physical and chemical agents. Since the rate of killing is approximately logarithmic (p. 197) the time taken to kill all organisms is dependent upon the initial load of organisms present. The pH of the environment also influences the efficiency of many lethal agents. Heat, for instance, is more effective at acid pH, a property which makes it possible to sterilize acid fruit juices at lower temperatures than neutral solutions (p. 648). The activity of many acid and basic agents is strongly dependent upon pH since, in general, unionized molecules enter cells more readily than do ions; consequently the concentration attained within the cell is greatest when the pH outside the cell is one at which the agent is in an undissociated state. Some chemical agents, such as mercuric chloride, whose action depends on the ionized mercuric ion, are more effective in aqueous than in alcoholic solution. The presence of other salts may similarly affect their activity by suppressing ionization of the molecule.

The rate of killing of chemical agents is dependent on the concentration of the agent and the temperature, the activity usually being greater as the temperature is raised (Fig. 7.1B). A population of *Salmonella paratyphi*, for instance, is killed by exposure to 0·01 per cent mercuric chloride for $2\frac{1}{2}$ minutes at 31°C, 36 minutes at 14°C, or 101 minutes at 0°C.

Many chemicals are most effective in a pure or inorganic aqueous environment. Chlorine, for instance, is effective as a disinfectant for the purification of drinking water when present in a few parts per million (p. 584). The activity of such chemicals, however, may be greatly reduced in the presence of organic matter such as serum, pus, milk, faeces, etc. Usually the concentration of the disinfectant is lowered by chemical interaction with organic matter; sometimes the latter protects micro-organisms against the lethal agent. Lethal chemicals which are not readily inactivated by organic matter and have the additional property of low toxicity for mammalian tissue, are used in the treatment of infectious diseases (p. 211).

Under normal conditions, most non-sporing micro-organisms are susceptible to physical and chemical agents. A few notable exceptions occur; for instance, whilst mycobacteria are as sensitive to heat as other non-sporing micro-organisms, they are much more resistant to chemical agents by virtue of their protective waxy envelope. This selective resistance to chemicals facilitates their isolation from mixed floras in pathological materials. Fungal spores are usually more resistant than non-sporing micro-organisms but bacterial endospores are the most resistant forms of life under natural conditions. High temperatures are required to kill endospores and these structures, together with mycobacteria, are susceptible to only a few chemical agents, including formaldehyde, glutaraldehyde and halogens, and then only under certain conditions. Few disinfectants are active against viruses; halogens and phenol are generally active and some viruses are inactivated by quaternary ammonium salts and xylenols (p. 209). A number of compounds able to inactivate viruses *in vitro* have little action on them in tissue cells because of the close association of viruses with host cells. In outbreaks of foot and mouth disease more reliance is placed on cleansing with sodium carbonate solutions than on viricidal action. Fungi are not necessarily killed by substances lethal to bacteria but many chemicals are fungicidal and some may be specifically active against particular groups of fungi (p. 211). Griseofulvin, for example, is selectively active against dermatophyte infections of animals (p. 76) and is used in treatment of these.

The choice of a lethal agent for use in a particular situation therefore depends on the materials to be treated, the environmental factors and whether or not sterility is required. No single lethal agent is suitable for every purpose. The use of heat sterilization, whilst being the most readily controlled, is limited to heat stable materials; chemicals, on the other hand, cannot be used for the sterilization of food or culture media because of the residual chemical's adverse effects.

Definition of terms

The antimicrobial agents discussed in this chapter are used in a number of ways and the terminology to describe these processes requires careful definition. *Sterilization* is an absolute term indicating the complete destruction or removal of all micro-organisms, including the most resistant bacterial spores. *Disinfection*, usually used of chemical agents, is the killing of a population of harmful micro-organism without necessarily achieving sterility. In practice, disinfectants by killing important pathogenic organisms often render certain objects safe for use, e.g., clinical thermometers. Chemicals which are less toxic for mammalian tissue but kill micro-organisms are termed *antiseptics*. They act usually more slowly than disinfectants. These are applied to the skin and mucous membranes without producing toxic tissue effects. Antimicrobial chemicals which possess sufficiently low toxicity are used as *chemotherapeutic* agents in the treatment of infectious disease in animals and plants; these include the important group of naturally produced antibiotics.

Some antimicrobial agents are mainly employed as killing agents whereas others inhibit growth and multiplication. These antibacterial agents are described as *bactericidal* (*germicidal*) or *bacteriostatic* according to whether they kill bacteria or merely stop their multiplication. Killing agents are essential for sterilization, but many of the useful chemotherapeutic agents are bacteriostatic. These inhibit multiplication of the infective agent and thereby enable the body's defences to destroy the micro-organisms.

Dynamics of disinfection

Exposure of a population of micro-organisms to a lethal agent does not normally result in instantaneous death but an exponential decrease in the number of viable cells in the population, that is, in each successive unit of time a constant fraction of the micro-organisms survives. Plotting the logarithm of the number of surviving organisms against time often results in a straight line graph (Fig. 7.1A) the curve being similar in form to that expressing the course of a unimolecular reaction. This, however, is true only under certain limited conditions such as when a disinfectant is present in excess and the rate of disinfection is sufficiently rapid. Departure from the exponential relationship can be caused by a variety of factors; for instance, the death-rate may be initially slow but soon accelerates, reaching a constant value which is maintained until most organisms are killed with the exception of a few organisms which may persist and die much more slowly than the majority. Plotting values under these conditions of survival against time result in sigmoid curves (Fig. 7.1B) rather than straight lines.

Two measures of the effectiveness of a lethal agent are commonly used. These are the LD_{50} (i.e., the dose needed to kill 50 per cent of the individuals) or, alternatively, the dose which kills the entire population. The LD_{50} is particularly useful for comparing the potency of different lethal agents since it can be determined with ease and accuracy and allows for differences in suceptibility of individual cells. However, in practice it is often necessary to ascertain the dosage needed to kill all the organisms, a measurement that is of less scientific value, since it is dependent on the incidence of abnormally resistant individuals, but of great medical importance.

PHYSICAL AGENTS

Many physical agents have an adverse effect upon the viability of micro-organisms. Some lethal agents, e.g., heat and certain radiations, are widely used in sterilization procedures. Others, such as cold, desiccation and those producing cell

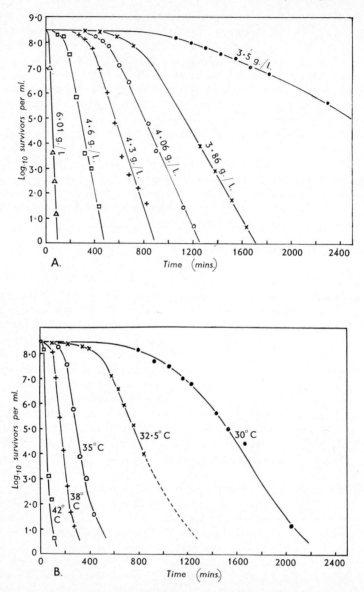

Fig. 7.1 The effect of phenol on *Escherichia coli*. (**A**) The relationship between the logarithms of survivors and time, for *E. coli* when exposed to various concentrations of phenol at 35°C. (After Jordon, R. C. and Jacobs, S. E. (1944) *J. Hyg., Camb.*, **43**, 275) (**B**) The relationship between the logarithms of survivors and time, for *E. coli* when exposed to 4·6 g phenol per litre at various temperatures. (After Jordon, R. C. and Jacobs, S. E. (1945) *J. Hyg., Camb.*, **44**, 210)

damage, kill a proportion of a population of cells but cannot be relied upon to produce sterility. The mechanical process of filtration is also used in sterilizing fluids.

Heat

High temperature is one of the most reliable and widely used techniques available for killing micro-organisms. Unlike disinfection by chemicals no toxic residue remains when sterilization is completed and its use is only limited where materials to be sterilized may be damaged by the temperatures used or adversely affected by moisture.

The temperature that is lethal for a particular micro-organism depends on the time of exposure (Fig. 7.2). Estimates of the 'thermal death point', generally defined as the lowest temperature at which a suspension of bacteria or other organisms in an aqueous medium is killed in 10 minutes, are of little

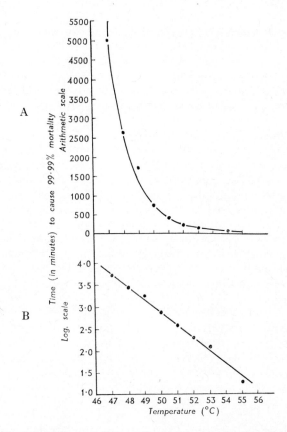

Fig. 7.2 The lethal effect of different temperatures on *Escherichia coli* at pH 7. In **A**, the exposure time needed to cause death of 99·9 per cent of a population is indicated; in **B**, log₁₀ of this is plotted. (After Jordon, R. C., Jacobs, S. E. and Davies, H. E. F. (1947) *J. Hyg., Camb.*, **45**, 333)

value. A more practical term is the 'thermal death time' which is the time required to kill a culture at a specified temperature. In a moist environment the thermal death time for non-sporing bacteria ranges from several hours at 47°, one hour at 60°C, to 5 minutes at 70°C, none being able to survive more than a few minutes at 80°C. In practice heating a culture at 80°C for 10 minutes in the presence of water, kills all non-sporing bacteria and the majority of fungal spores. More resistant spores, such as the ascospores, found in certain strains of *Byssochlamys fulva*, resist 86–88°C for 30 minutes. Some bacterial endospores are killed by this temperature but the majority are killed in the same time only by temperatures above 100°C. The resistance of bacterial spores constitutes the major problem in heat sterilization.

The majority of viruses are inactivated by exposure to temperatures between 50 and 60°C for 20 minutes but a few, e.g., poliovirus and the infective hepatitis virus, are inactivated only by higher temperatures such as 75°C for 30 minutes.

Moist heat

It is important to distinguish between moist and dry heat processes of sterilization. Moist heat is the more efficient lethal agent, sterilizing at a lower temperature and in a shorter time. Death of micro-organisms is the outcome of coagulation and denaturation of the structural proteins and cellular enzymes and this occurs more readily in the hydrated than in the dry state. Irrespective of the material being processed moist heat sterilization takes place in an aqueous environment.

Exposure to temperatures of the order of 63°C for 30 mins. or 72°C for 20 secs. are used in processes of pasteurization, as for milk (p. 590) to kill non-sporing pathogens and beverages (p. 647) to terminate fermentations. Raising the temperature to boiling point also kills many bacterial spores but several hours exposure are required to kill the more heat resistant spores and sterility cannot be ensured. However, boiling is widely used to render instruments safe for use in some medical and dental work but since sterility is not guaranteed, this treatment is inadequate for surgical instruments. The addition of 2 per cent sodium carbonate to the boiling water gives as effective killing in 10 minutes as boiling in pure water for several hours. The corrosive action of the alkali on plated metal (e.g., surgical instruments), however, limits the use of this technique.

Steam is a particularly effective lethal agent since it penetrates into all exposed surfaces of the treated material as it contracts upon condensing on the colder object and in the process transfers heat by giving up its latent heat of vaporization. This highly efficient process of heat transfer continues until temperature equilibrium is reached. It is essential that pure moist steam is used, i.e., all air must be discharged from the equipment (Fig. 7.3) and the steam must not be superheated else the process of heating becomes virtually a less efficient dry heat process.

Sterilization by steam is usually carried out in an autoclave. Using pure steam the autoclave pressures correspond to a range of temperatures and the time of exposure required to achieve sterilization depends on the working temperature used (Table 7.1). The sterilizing times in this table are recommended for high

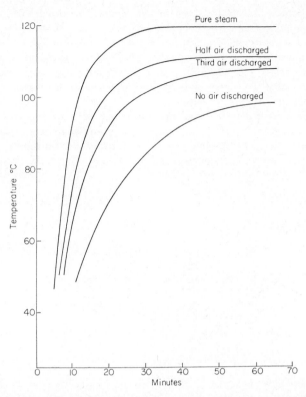

Fig. 7.3 Temperatures of steam–air mixtures in an autoclave. Steam at 15 lb. per square inch was introduced into a chamber completely or partially evacuated and the temperatures within the chamber were determined at intervals. (Data, from Underwood, W. B. (1941) *A textbook of sterilization*, American Sterilizer Co, Erie Pa.)

Table 7.1   Equivalent minimum sterilizing time/temperatures with moist and dry heat.

	Moist heat			Dry heat	
lb./sq. in. (*pure steam*)	Temperature °C	Sterilizing time		Temperature °C	Sterilizing time
0	100	20·0 hours			
5	109	2·5 ,,			
10	115	50·0 minutes			
15	121	15·0 ,,		120	8·0 hours
20	126	10·0 ,,			
30	134	3·0 ,,		140	2·5 ,,
				160	1·0 ,,
				170	40·0 minutes
				180	20·0 ,,

vacuum autoclaves and include the minimum time required for sterilization plus a safety margin (see M.R.C. Report, 1959). In autoclaves where air is removed by downward displacement, longer times are required especially when permeable materials are being sterilized. Under these conditions exposure for 25 minutes at 121°C (i.e., at 15 lb. per sq. in.) is adequate; with bacteriologically clean and heat-labile materials, shorter times (e.g., 20 minutes) or, lower temperatures (e.g., 109°C) are used.

The sterilization of contaminated heat labile materials, such as woollen blankets and medical equipment containing plastic components, etc., has proved a problem. Recently the combined use of steam at sub-atmospheric pressures (Fig. 7.4) and a bactericide, such as formalin, has proved effective. At sub-

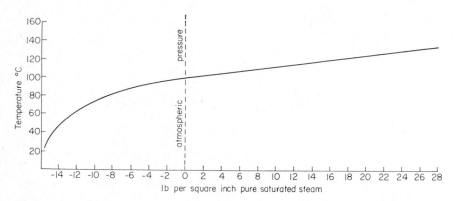

Fig. 7.4   Temperatures of saturated steam at sub-atmospheric and positive pressures.

atmospheric pressures steam is produced at temperatures well below 100°C (e.g., 80°C at 17 mm. Hg). The penetration and efficiency of heat transfer by steam makes the technique superior to others at the same temperature. Formalin vapour is injected into the apparatus during the heating process and this is carried by the steam into all exposed parts of the equipment and all spores are destroyed after 2 hours. The formalin is finally removed under vacuum and residual traces neutralized with ammonia.

Dry heat

Sterilization by dry heat has limited application. The higher temperatures and longer exposure times (Table 7.1) are often damaging to many materials. It is used mainly for the sterilization of materials which must be kept dry, e.g., anhydrous fats, oils, certain surgical instruments, powders, certain glassware, etc. Usually a hot air oven is used, the heating of which takes place by heating elements in the base or sides, the air being circulated by a fan. The process of heat transfer to the objects is by convection, radiation and conduction. Time must be allowed (1) for heat penetration to bring all objects to the required temperature, (2) for the object to be held at a standard temperature to kill all spores, and (3) for the oven to cool before opening. The entire process takes several hours and is therefore considerably longer than the moist heat processes.

Non-sporing organisms are killed by exposure to 100° dry heat for 1 hour, fungal spores at 115°C for 1 hour and the majority of bacterial spores (excluding a few highly resistant spores) at 160°C for one hour.

Infra-red radiation

Infra-red radiation is used for the sterilization of particular materials such as all-glass syringes. This is a form of dry-heat sterilization. The materials to be sterilized are exposed to infra-red radiation as they pass on a belt through the equipment. Temperatures of 180°C are rapidly attained with greater certainty than in a hot air oven and exposure for not less than 10 minutes is adequate to sterilize. Under vacuum, infra-red radiation can attain temperatures of 280°C and subsequent introduction of nitrogen avoids oxidation of the materials on cooling.

Cold

Chilling and freezing bacterial cultures brings about the death of a proportion of the cells, this proportion varying with the cooling procedures used. Rapid freezing to low temperatures, however, often results in a state of hypobiosis (p. 194) leading to prolonged survival. The death rate during *cooling* or *chilling* depends on several factors including the rate of cooling, the nature of the suspending medium and the growth phase of the organism. For instance, cold shock caused by the rapid cooling of young cultures of *Escherichia coli* from 37° to 4° produces a 95 per cent reduction in viable cells whereas similar cultures slowly cooled to 4° over a period of 30 minutes show no loss in viability. The greatest loss in viability occurs with actively metabolizing cells.

Freezing causes further physical and metabolic damage to the cells. The death rate is found to be greatest at temperatures just below freezing point compared with still lower temperatures. In addition to mechanical damage from ice crystals, forming in the cell, the major effect is the precipitation of coagulable cellular proteins which is much greater at −2° than at −20° or lower. More organisms survive sub-freezing temperatures if they are protected by organic substances such as peptone, glucose, sucrose, glycerol, sodium glutamate, etc. Dependent upon the presence or absence of these substances the number of survivors under frozen conditions varies with the time of storage.

Some moulds and yeasts are more resistant to freezing than are most bacteria but bacterial spores are virtually unaffected. Some viruses are inactivated at low temperature; but others including bacteriophage, are preserved by holding at −60°. A few protozoa can be preserved by freezing.

Desiccation

When dried some loss in viability occurs with all micro-organisms. Bacterial spores are least affected but vegetative cells demonstrate rapid and considerable loss in viability during the drying process followed by reduced loss under storage. Drying is less destructive under vacuum than in air, as is also storage in an inert atmosphere which guards against the lethal effects of oxidation. A residual water content of 30–40 per cent is considered to be the most harmful.

The period of survival after natural drying varies between organisms. Some delicate pathogens survive for a few hours only; others remain viable for weeks or months especially if they are protected by body fluids, such as serum, and dried on fabrics, as blankets and clothing.

Spray-drying used for the preparation of dried milk and egg powders is not a means of sterilization and both pathogenic and non-pathogenic organisms survive. Care must be taken therefore to safeguard against bacterial multiplication of these products after reconstitution prior to cooking.

Visible and ultra-violet radiation

Light can bring about chemical change, and hence biological damage, only if it is absorbed; light that passes through the cell without being absorbed has no effect. Visible light, that is, electromagnetic radiation of wavelength 400–750 mμ, is absorbed by relatively few of the compounds present in non-photosynthetic organisms, and therefore has little harmful effect, and the same is true of ultra-violet radiation of wavelength (300–400 mμ). Ultra-violet radiation of wavelength less than 300 mμ, on the other hand, is strongly absorbed by proteins and nucleic acids, in which it brings about chemical change, with the result that relatively small doses of such radiation will cause chromosome damage,* genetic mutation or death. Higher doses are required to cause inactivation of enzymes.

If micro-organisms are treated with various dyes (e.g., erythrosin) they become sensitive to damage by visible light. Such dyes are said to possess photodynamic action. The germicidal effect of sunlight is due largely to the ultra-violet component, which is only a few per cent of the total. The lower limit of wavelength of the ultraviolet received at the earth's surface is about 290–300 mμ, but is dependent on the clarity of the atmosphere and on the altitude and latitude. It will be appreciated, therefore, that the germicidal power of sunlight varies greatly.

In some micro-organisms the harmful consequences of exposure to short wavelength ultra-violet radiation can be partly averted by prompt exposure of treated organisms to visible light, an effect known as photoreactivation. Such micro-organisms, e.g., *E. coli* and baker's yeast, possess photoreactivating enzyme. In its simplest terms this repair of U.V. damage can be pictured in terms of the Michaelis–Menten reaction scheme;

$$\begin{array}{ccc} \text{PR Enzyme} & \text{Enzyme} & \text{PR Enzyme} \\ + & \rightleftharpoons \text{Substrate} \longrightarrow & + \\ \text{Substrate} & \text{Complex} & \text{Product} \end{array}$$

* Quite high doses are required to cause *breaks* in DNA. The dose required to reduce the molecular weight of *H. influenzae* DNA by 50 per cent, i.e., 1 double-strand break, is $2\cdot5 \times 10^5$ ergs/mm^2 which is well beyond the biological dose range.

The photoreactivating (PR) enzyme combines mechanically with the irradiated DNA (but not with unirradiated DNA), the U.V. lesion acting as the substrate. The complex absorbs light, producing repaired DNA with subsequent liberation of the PR enzyme. The enzyme–substrate complex does not dissociate in the dark.

The main U.V. induced lethal photoproduct formed in DNA is the pyrimidine dimer in which adjacent pyrimidine bases (thymine or cytosine) in the same DNA strand dimerize. The hydrogen bonds to the purine bases on the opposite strand are thus broken and the DNA becomes distorted in that region. As a result DNA replication past the dimer is slowed down to the point where the presence of a few dimers is lethal to the cell. Such dimers are the only known substrate for PR enzyme, although for binding of the enzyme and the dimer, the latter has to be associated with a piece of double stranded DNA at least 9 bases long. Following absorption of light the dimer is monomerized *in situ*. It has recently been calculated that *E. coli* possesses about 25 PR enzyme molecules per cell.

In the laboratory, cabinets to be used for carrying out inoculations can be conveniently sterilized with ultra-violet radiation from a quartz mercury vapour lamp. Similar sources are used for the sterilization of indoor atmospheres (p. 594).

Atomic radiation

There are two distinct types of atomic radiation; electromagnetic waves of the same class as light waves, but having much shorter wavelengths and far higher energies, and subatomic particles travelling at very high velocities. Electromagnetic waves produced by high voltage apparatus are referred to as X-rays; if produced by the breakdown of radioactive elements such as radium, they are known as γ-rays. The high speed subatomic particles, released by radioactive decay or accelerated by means of such instruments as the cyclotron, include charged particles such as β-rays (high speed electrons), protons, α-particles (*helium nuclei*) and uncharged particles such as neutrons. The effect of all these atomic radiations on matter, living or otherwise, is to cause ionization. An electron is knocked out of one molecule and is gained by another. As a result, both molecules become ions of high chemical reactivity capable of causing biological damage. The different types of atomic radiation differ in the number and distribution of the ionization they cause, and in their penetrating power. A major cause of radiation damage in living tissue is probably the production of free hydroxyl and hydrogen radicles, an indirect result of the ionization of water.

There are vast differences in sensitivity of different organisms to atomic radiation, micro-organisms being much more resistant than most higher organisms. Thus the LD_{50} for most mammals, including man, is less than 1,000 roentgens (r),* whereas for *Escherichia coli* it is 10,000 r, yeast 30,000 r, *Amoeba* 100,000 r, *Paramecium* 300,000 r, and for *Micrococcus radiodurans*, the most

* The roentgen is a unit of dosage (intensity of radiation × time of exposure) employed in radiobiological work.

resistant vegetative bacterium known, 750,000 r. The severity of radiation damage depends to a considerable extent on conditions at the time of irradiation. Thus the sensitivity of cells is considerably reduced by anaerobic conditions at the time of irradiation. Recently it has been found that pre-treatment with certain chemicals such as β-mercapto-ethanolamine, gives some protection against subsequent irradiation. The effects of irradiation damage may be modified by post-irradiation treatment. For example, if irradiated *E. coli* cultures are subsequently kept at 18°C far more bacteria survive than at higher or lower temperatures. It is thought that at this temperature the processes which repair radiation damage are more effective than those extending the damage.

Since vegetative bacteria present a more or less identical target to radiation, the amount of radiation damage caused does not vary much from species to species. Thus differences in radiation sensitivity, which vary enormously from species to species, must be due to differences in the ability of each bacterial species to modify such damage, and in fact, can be explained largely by the absence or presence of a DNA dark-repair mechanism, and if present, of its efficiency. The dark-repair mechanism operates by excising damaged single-stranded regions of DNA and repolymerizing the missing bases. Unlike photo-reactivation which operates only for U.V. damage, dark-repair mechanism will repair DNA damaged by U.V. and ionizing radiation, nitrogen mustards, and alkylating agents, etc.

In dealing with the effects of both short-wavelength ultra-violet and atomic radiation it is important to distinguish between genetic and physiological effects. If severe, both may cause death, but whereas a threshold value exists for physiological damage, there is apparently no minimum dose required to cause mutation.

Radiation, as a means of sterilization, is finding an increasing number of applications. Package irradiation of syringes, catheters, petri dishes, surgical gloves, and sutures is widely used. To avoid risk of subsequent contamination through fine cracks in the package materials meticulous care in handling is essential. The use of radiation in the sterilization of foods is limited by their effect on its appearance and flavour.

Cellular disintegration

Micro-organisms can be broken up by ultrasonic vibrations or mechanical agitation. Ultrasound, which includes inaudible sound waves with vibrations of the order of 700,000 cycles per second, is only effective in liquid suspension and, apart from heating effects, the lethal action is attributed almost entirely to the physical destruction of cells. Efficiency is related more to the intensity of the wave emission than to its frequency. Damage to cells is attributed mainly to 'gaseous cavitation'; it is presumed that the cell walls are weakened by the creation of minute cavities resulting from the formation of small gas bubbles in the fluid under the alternating pressures produced by the sound field. As with other lethal agents the order of destruction in general follows an exponential rate (Fig. 7.5).

All actively living cells including micro-organisms are susceptible to ultrasound. Gram-negative bacteria are generally more susceptible than Gram-posi-

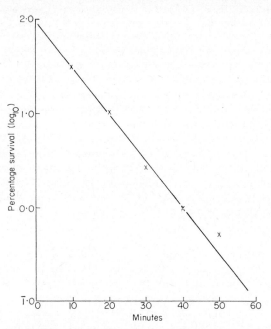

Fig. 7.5 The rate of killing of *Klebsiella pneumoniae* by ultrasonic waves, 700,000 cycles per second. (Data from Hamre, D. (1949) *J. Bact.* **57**, 279)

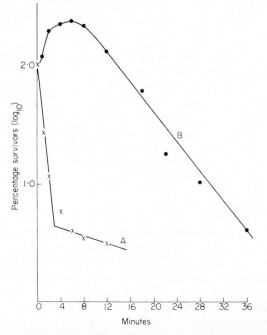

Fig. 7.6 The rates of inactivation of vaccinia virus by ultrasonic waves, 20,000 cycles per second. (**A**) the virus is suspended in buffered saline; (**B**) the virus is suspended in 20 per cent inactivated rabbit serum. (T. J. Hill)

tive bacteria and rod-shaped organisms more than cocci. Bacterial spores are highly resistant. Some workers report viruses to be resistant to ultrasound whilst others consider them to be susceptible (Fig. 7.6). The different findings may in part be related to the size and shape of the virus particles. Ultrasound is used in virology to disrupt tissue cells to release intracellular viruses and to break up aggregates of virus; this may explain the initial increase in pox forming units in the serum suspension in Fig. 7.6.

Cellular disintegration can also be brought about by mechanical agitation in the presence of abrasives (e.g., carborundum), sand or glass beads. Alternatively shearing forces developed by forcing a fluid or frozen suspension through a narrow opening bring about cell disintegration. These techniques are often used to produce cell-free enzymes for biochemical studies since the preparations are not subject to heating as with ultrasonics.

Filtration

Filtration, not itself a lethal agent, is used widely in industry and laboratory for the sterilization of heat-labile fluids. By this process organisms are removed from fluids and may be separated from their soluble products of metabolism, e.g., exotoxins. Filters are made from many materials, including unglazed porcelain, diatomaceous earth (kieselguhr), asbestos fibres, sintered glass, and synthetic membranes of cellulose nitrate (collodion) and cellulose acetate. All types are available in a range of grades of different pore size and are selected according to the nature of the fluid to be filtered and the size of particle which is to be removed. In addition to the mechanical action of filters other factors must be considered such as the effect of pressure exerted to force the fluid through and the electric charges on the filter and on the suspended particles which may result in their electrical adsorption to the filter.

In the past, filters were used to remove micro-organisms of bacterial size or larger, and 'sterilization by filtration' was regarded as rendering a fluid bacteria-free and not virus-free. Membrane filters are now made in a wide range of pore size including many much smaller than the smallest bacterium. The finer ones are used for ultra-filtration and measure from 5 mμ in size. These can be used for filtering fluids free of viruses and by a choice of different membranes of known pore size it is possible to estimate with reasonable accuracy the virus particle size according to which membrane just fails to retain the virus.

CHEMICAL AGENTS

Disinfection by chemicals is a much more complex subject than sterilization by physical agents. Many chemicals can cause the death of micro-organisms but their modes and rates of action are extremely diverse. Some are general protoplasmic poisons damaging all living cells; others, in particular the antibiotics (p. 213), are highly specific, being relatively non-toxic to mammalian tissues and some micro-organisms but active in high dilution against specific ones, a feature termed selective toxicity. Chemical disinfectants may be lethal to micro-organisms at certain concentrations (i.e., bactericidal) but inhibit multiplication without lethal effects (i.e., bacteriostatic) at high dilution. These differing activities are considered to be the result of quite separate and unrelated modes of action and not simply of dilution. Thus mercuric ions in low concentrations interfere with enzyme action by combining with —SH groups, a reaction which can be reversed or prevented by adding other —SH containing compounds such as glutathione; in high concentration mercuric ions irreversibly denature cell proteins.

The majority of disinfectants are general poisons, their lethal action being the result of their capacity to coagulate, precipitate or otherwise denature both structural and essential enzymes of tissue and micro-organismal cells. Groups of chemicals which do this are considered below. Because of their general toxicity some are useful only as general disinfectants; others, with milder toxicity, can be applied as antiseptics to living tissue.

Halogens

Halogens or compounds which release halogens, e.g., hypochlorites, chloramines, hypobromites, are strongly bactericidal by the oxidation of proteins and similar substances. In consequence of their vigorous chemical activity, solutions are readily inactivated by combination with organic matter, and this limits their usefulness (p. 196). An alcoholic solution of iodine (1 per cent) is a useful skin disinfectant but may cause reactions in sensitized individuals.

Alkylating reagents

Heterocyclic compounds, e.g., ethylene oxide, ethylene imide and ethylene sulphide, as well as methyl bromide and lactones, e.g., β-propiolactone, are effective bactericides through their ability to alkylate bacterial structures. The gas ethylene oxide effectively kills bacterial spores and can be used for the sterilization of heat-labile materials and equipment in hospital practice. To avoid explosion, the gas is mixed with carbon dioxide under pressure at a standard humidity. β-propiolactone also rapidly kills spores; it is unstable in aqueous solution being readily hydrolysed to non-toxic residues, a property which makes it a useful disinfectant to be added to serum (to inactivate the virus of infective hepatitis) and to sterilize tissue grafts.

Phenolic compounds

This group of disinfectants which includes phenol, lysol (a solution of cresols in soap), and xylenols are intensely toxic by virtue of protein denaturation. Proteins are precipitated by 1 to 2 per cent phenol. Their activity is not greatly reduced by organic matter but none are capable of killing bacterial spores. Thus most vegetative cells of bacteria are killed by 1 per cent phenol in 5–10 minutes at 20°C, but anthrax spores survive 24 hours in 5 per cent phenol. Newer phenolic preparations such as 'clear soluble phenolics' are as efficient as lysol but much safer.

A group of halogenated phenols, which include 'Dettol', a chloroxylenol, are much less toxic than the phenols but their activity is severely depressed by organic matter since the halogen contributes to the activity. Hexachlorophene is widely used against staphyloccoal infections of the skin. It has little action against Gram-negative organisms.

Chlorhexidine ('Hibitane') is a fairly powerful disinfectant with a wide spectrum of activity and very low tissue toxicity. It is used as an alcoholic solution for skin disinfection and, in aqueous solution, as a general antiseptic, either alone or mixed with cetrimide (a quaternary ammonium compound).

Aldehydes

Formaldehyde, which is usually marketed as 40 per cent aqueous solution (formalin), even when diluted to 1 per cent, is able to kill bacterial spores and tubercle bacilli. In contrast to most disinfectants it can therefore produce complete sterility but is highly toxic. As a gas or solution its penetration into organic matter and crevices is slow but it is highly efficient for sterilizing smooth surfaces. Suitable dilutions have been used to treat footwear of persons suffering with fungal infections, such as athlete's foot, and to treat wool prior to sorting to kill any spores of anthrax bacilli which may be present.

Other useful aldehydes include a 2 per cent aqueous solution of glutaraldehyde; it is less irritant than formalin but more rapidly bactericidal.

Alcohols

Alcohols kill vegetative bacteria rapidly but have no action on spores. Their activity requires the presence of water and 70 per cent gives optimal activity, absolute alcohol being relatively inactive. Isopropyl alcohol is slightly more effective than ethyl alcohol and is preferred as a skin disinfectant.

Acids and alkalis

Mineral acids and alkalis produce their main lethal effects through their hydrogen and hydroxyl ions respectively although their characteristic ions may also contribute. The more highly dissociated molecules therefore produce the greatest effect. Hydrogen ions are more effective than hydroxyl ions. Organic acids often owe their activity to the undissociated molecules.

Heavy metals and their salts

All the heavy metals are antibacterial and antifungal to some degree. Silver and copper exhibit these properties at minute concentrations, a property referred to as oligodynamic activity. The salts and organic complexes of mercury, tin, silver and, to a lesser degree, copper are all actively lethal agents. When ionized in aqueous solution the metal ions combine with and precipitate cell proteins.

Surface active compounds

These include cationic, non-ionic, and anionic compounds of which the former include the quaternary ammonium compounds, the largest and most important antibacterials of these compounds. Their activity has been variously attributed to denaturation of cell protein, inactivation of enzymes and disruption of the cell membrane as a result of their surface-active properties. It is likely that inactivation of enzymes concerned with energy-producing metabolism is the most important mechanism of inhibition. They are not highly bactericidal but by virtue of their 'surface wetting' properties they are sometimes used in combination with antiseptics, e.g., chlorhexidine ('Hibitane'), to make the latter more effective. Gram-positive organisms are generally more sensitive than Gram-negative ones and this may be due to the differences in type and content of phospholipids in the two groups of bacteria. Quaternary ammonium salts are not active against *Mycobacterium tuberculosis*, bacterial spores, viruses or fungi.

Dyes

Prior to the introduction of sulphonamides and antibiotics, the triphenyl-methane dyes (e.g., methyl violet, brilliant green, etc.) and the acridines (e.g., acriflavine and proflavine), because of their high antiseptic action and low toxicity, were widely used in the treatment of wounds and burns. An outstanding feature of the acridines is their sustained activity in the presence of serum, a property which, until recently, was unique among antiseptics. Dyes are more active against the Gram-positive than Gram-negative bacteria and crystal violet, also fungistatic, was used against mycotic skin infections. The discoloration of the skin by these compounds has resulted in their being superseded by more acceptable antiseptics.

SELECTIVE TOXICITY

Although many antiseptics kill vegetative bacteria some have a relatively restricted spectrum of activity. Certain antiseptics (e.g., triphenylmethane dyes, hexachlorophene, and acriflavines) are selectively active against Gram-positive organisms. Those which also have an action against Gram-negative bacteria inhibit these organisms only at relatively high concentrations. Some Gram-negative bacteria, e.g., *Pseudomonas aeruginosa*, are susceptible to very few of the commonly used antiseptics. The wider the spectrum of activity the greater the usefulness of an antiseptic.

The property of selective toxicity is most marked among the antibiotics; each has its own spectrum of activity (Table 16.4). Consequently the range of microbial diseases which can be treated by each is essentially restricted.

In addition to the importance of selective toxicity between micro-organisms the relative toxicity of an antibacterial agent for the host compared with its toxicity for the micro-organism is of great practical importance. This is considered below.

Chemotherapy

This term was coined by Paul Ehrlich who defined it as 'the use of drugs to injure invading organisms without injury to the host'. Toxicity to the host is usually the limiting factor which determines whether or not a chemical can be used as a therapeutic agent. Frequently a substance which is harmful to micro-organisms is harmful to the cells of the host and therefore the choice of a chemotherapeutic agent depends on the difference in degree between these two actions. The ratio

$$\frac{\text{minimum curative dose}}{\text{maximum tolerated dose}}$$

devised by Ehrlich to evaluate the usefulness of a therapeutic agent, was called the chemotherapeutic index; the higher the ratio, the more likely is the drug to be useful, all other factors being equal.

The anti-malarial drug quinine and a few other effective chemotherapeutic agents were known for centuries, but a rational approach to the discovery of new agents was not possible until it was appreciated that infectious disease was caused by micro-organisms. In the late nineteenth century the differential action of some dyes in staining bacteria without staining the surrounding tissues suggested to Ehrlich the value of searching for toxic compounds having a greater affinity for micro-organisms than for mammalian tissues. Ehrlich undertook a systematic study of organic arsenical compounds in the hope of finding one which retained the effectiveness of inorganic arsenic in killing the causal organism of syphilis, *Treponema pallidum*, while being less toxic to man. He finally succeeded, discovering a compound, arsphenamine, named Salvarsan or 606 (since it was the 606th compound he tested) which was used for many years in the treatment of syphilis.

The two principal approaches now employed in the search for new chemotherapeutic agents are the synthesis of chemical analogues of known metabolites, and the screening of the many naturally produced antibiotics of soil organisms for their suitability as chemotherapeutic agents.

Metabolic analogues

Essential metabolites, whether substrates or co-enzymes, if present in small amounts in a cell or tissue, can be antagonized by substances known as metabolic analogues. Each analogue exerts its antagonism by occupying and blocking the enzyme sites used by the metabolite. The similarity must be in the chemical configuration, dimensions and electron distribution since the active sites on the enzymes are known to be highly polarized. Many examples of metabolic analogues are known but one example only will be considered here.

THE SULPHONAMIDES AND FOLID ACID ANTAGONISTS. The discovery by Domagk (1935) that certain azo dyes containing an aromatic sulphonamide group had a better chemotherapeutic index against streptococcal septicaemia in mice than any previously known, was destined to alter the whole outlook in chemotherapy. Tréouel, Nitti, and Bovet (1935) showed that the reduction product of the dye, *p*-aminobenzene sulphonamide, was active both *in vitro* and *in vivo*. Since then a large number of sulphonamide compounds have proved to be effective against virulent streptococci, pneumococci, neisseria and, to a lesser extent, salmonellae and shigellae.

Woods (1940) observed the specific antagonism of *p*-aminobenzoic acid (PABA) to sulphonamide action and Fildes (1940) proposed the theory that sulphonamides inhibit bacterial growth by blocking an enzyme system associated with the use of PABA as an essential metabolite. Many other antagonists to sulphonamides exist and not all can be explained by this theory. Nevertheless, the primary and most important effect on most susceptible micro-organisms appears to be that of interference with the incorporation of PABA into folic acid during its enzymic synthesis. The structural similarity and absolute size of the sulphonamide molecule to that of PABA is one of the best examples of a structural analogue interfering with an enzyme controlled synthesis (Fig. 7.7).

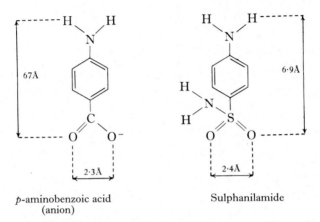

p-aminobenzoic acid
(anion)

Sulphanilamide

Fig. 7.7 The structural similarity between p-aminobenzene sulphonamide and p-aminobenzoic acid.

Some microbial parasites cannot use external sources of folic or folinic acid since these compounds cannot penetrate the cell wall; they must be therefore synthesized within the cell from PABA, pteridine, and glutamic acid (Fig. 5.2). Sulphonamides, by virtue of their structural similarity to PABA act as competitive inhibitors of this synthesis and therefore have powerful bacteriostatic effects on organisms dependent on PABA. These include many pathogenic bacteria, toxoplasma, and plasmodia. In the cell folic acid is converted to folinic acid by the bacterial enzyme dihydrofolate reductase. This conversion can be blocked by a number of diaminopyrimidines (e.g., trimethoprim and pyrimethamine) which have an affinity for this enzyme. Consequently these compounds are effective chemotherapeutic agents against certain bacteria and protozoa. The function of folinic acid is to act as a co-factor for enzymes concerned in the synthesis of purines and pyrimidines used in the nucleic acid synthesis but none of these later stages are sensitive to sulphonamides (Fig. 7.8). Organisms which can absorb folic acid from an external source are insensitive to sulphonamide.

Most vertebrates can satisfy their requirements for folic acid either from their diet or from organisms in their intestinal tract which are able to synthesize it; alternatively they may be supplied with folinic acid in their diet. Any slight effect of sulphonamides on mammalian enzymes is counteracted by excess folic and folinic acids supplied to the cell.

Antibiotics

Antibiotics are chemical agents produced by many micro-organisms which are harmful to other microbial species. The clinically useful antibiotics are produced by certain actinomycetes and by some species of *Penicillium* and *Bacillus*. Many

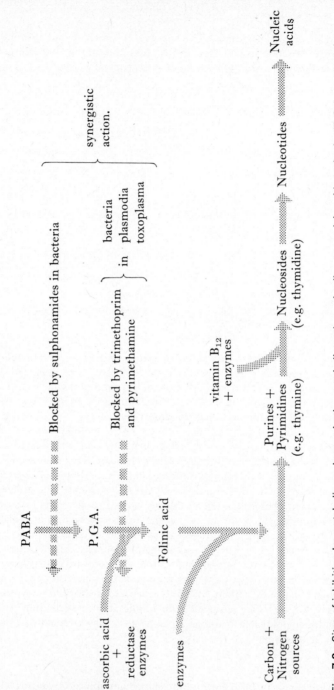

Fig. 7.8 Sites of inhibition by metabolic analogues in the metabolic pathway leading to nucleic acid synthesis. (P.G.A. = pteroyl glutamic acid)

hundreds of antibiotics have been isolated but relatively few are clinically useful because of an unsatisfactory antimicrobial spectrum of activity, or excessive toxicity *in vivo*. The mechanisms of some antibiotics are considered in Chapter 2 (p. 74), the spectra of activity of clinically useful ones in Chapter 16 (p. 581) and their commercial production in Chapter 22 (p. 681).

METHODS OF TESTING DISINFECTANTS AND ANTIBIOTICS

The testing of disinfectants

The function of a disinfectant is to kill micro-organisms but, as already noted, many environmental factors influence the activity of a disinfectant. Chemical assay thus gives incomplete information and microbiological assessment is essential. Many techniques have been developed but no completely satisfactory method of standardization of disinfectants has been devised to take into account the effect of organic matter, time/temperature relationships, toxicity for the host tissues, etc.

Frequently the activity of a disinfectant against a pure culture of a test organism is compared with that of phenol over a limited time range (e.g., 5 to $7\frac{1}{2}$ minutes). Modifications include the addition of organic matter (e.g., dried yeasts) and extension of the test period (e.g., 30 minutes). These tests are only valid with phenolic compounds which have a similar mechanism of action. In hospitals, disinfectants are often assessed in the situation for which they are required, by 'in use' tests; samples of the disinfectant are withdrawn periodically to determine if the disinfectant is continuing to be active against the bacterial flora in any particular situation.

The assay of antibiotics

The potency of samples of antibiotics is assessed by bio-assay. As with the bio-assay of nutrients (p. 155), the activity of the sample of unknown potency is compared with that of a standard preparation against a test organism known to be sensitive to the antibiotic under test. In contrast to bio-assays of nutrients which measure the amount of growth, those of antibiotics measure inhibition of growth. The bio-assay of antibiotics is usually carried out by serial dilution or agar diffusion techniques.

Serial dilution methods

A series of dilutions of both the standard antibiotic preparation and the test sample are prepared in a fluid nutrient medium. The tubes are inoculated with a standard suspension of the test organism and incubated. The greatest dilutions of the standard and of the sample which completely inhibit growth is noted and from these the potency of the test sample is estimated. This measures the bacteriostatic concentration of the drug; the bactericidal level may be determined by sub-culturing from each tube and determining which concentration of the drug kills the test organism. Defects of the test are that it is laborious, requires strictly aseptic conditions (a chance contaminant may completely invalidate results) and

is liable to be influenced by the growth of resistant mutants which have a selective advantage in the presence of the antibiotic. Serial dilutions are now mainly
used for the assay of antibiotics which diffuse too slowly for agar diffusion
methods to be applicable.

Agar diffusion methods

In this bio-assay the antibiotic diffuses from a confined source through
nutrient agar seeded with the test organism whilst the plate is incubated. A

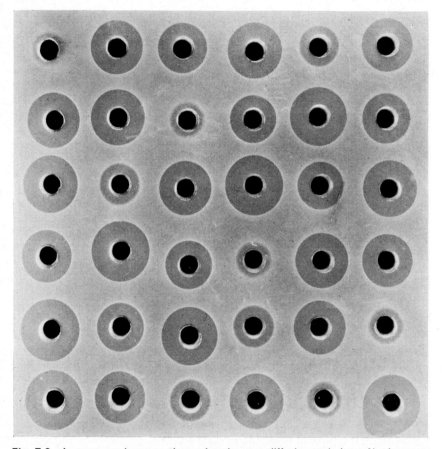

Fig. 7.9 A streptomycin assay plate using the agar diffusion technique. Nutrient agar
is seeded with the test organism (*Klebsiella pneumoniae*) and the antibiotic is placed
in the cup, cut in the agar with a sterile cork borer, from which it diffuses. The size of
each zone of inhibition is proportional to the concentration of the antibiotic in the
cup. In the assay plate demonstrated, six different dilutions are used, three of which are
standard solutions of known concentration, the other three are dilutions of the unknown
solution being assayed. These are randomly distributed throughout the plate, each row
and each column including all six dilutions. The average of six zone measurements for
each dilution is determined and this cancels variations which may occur in different
parts of the plate. (A. H. Linton)

concentration gradient of the antibiotic is thus set up and this results in a zone of inhibition being formed around the antibiotic reservoir. Under constant conditions of temperature, medium, and inoculum size, the size of the zone of inhibition is proportional to the concentration of the antibiotic at the source (Fig. 7.9). In this bio-assay the size of zones produced by a range of dilutions of a standard preparation are plotted and the concentrations of antibiotic in various dilutions of the sample which give zones of comparable size are read from the graph. With antibiotics of small molecular size the relationship between the square of the zone diameter and the logarithm of the antibiotic concentration is strictly linear (Fig. 7.10).

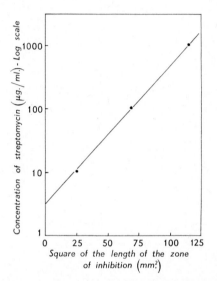

Fig. 7.10 A standard assay curve obtained with *Klebsiella pneumoniae* and strepto-mycin by the agar diffusion technique. A straight line relationship results by plotting logarithms of the antibiotic concentrations against the corresponding values of the square of the inhibition-zone diameters. The assay of antibiotic solutions of unknown potency can be determined by reading from the standard curve the concentrations correspond-ing to the square of the inhibition-zone diameters produced by these solutions. (After Linton, A. H. (1958) *J. Bact.* **76**, 94)

BOOKS AND ARTICLES FOR REFERENCE AND FURTHER STUDY

ALBERT, A. (1968). *Selective toxicity*, Methuen and Co. Ltd., London, 531 pp.
COWAN, S. T. and ROWATT, E. eds. (1965). The Strategy of Chemotherapy. *8th Symp. Soc. gen. Microbiol.* Cambridge Univ. Press, 360 pp.
CRUICKSHANK, R. ed. (1965). *Medical Microbiology—a guide to Laboratory Diagnosis and Control of Infection*, E. & S. Livingstone Ltd., Edinburgh, 1067 pp.
KAVANAGH, F. (1963). *Analytical Microbiology*, Academic Press, New York, 707 pp.
KELSEY, J. C. and MAURER, I. M. (1966). An in-use test for hospital disinfectants. *Mon. Bull. Min. Hlth.*, **25**, 180–184.
NEWTON, B. A. and REYNOLDS, P. E. eds. (1966). Biochemical Studies of Anti-microbial Drugs. *16th Symp. Soc. gen. Microbiol.* Cambridge Univ. Press, 349 pp.

Report, M.R.C. (1959). *Sterilization by steam under increased pressure*, Lancet *i*, 425.

SYKES, G. (1964). *Disinfection and Sterilization*, E. and F. N. Spon, London, 486 pp.

WILLIAMS, R. E. O., BLOWER, R., GARROD, L. P., and SHOOTER, R. A. (1966). *Hospital infection—causes and prevention*, Lloyd-Luke Ltd., London, 386 pp.

8

Structure, Biology, and Classification of Viruses

GENERAL INTRODUCTION

The discovery of viruses

By the beginning of the twentieth century much was known about the causes of the infectious diseases of animals and plants. Many parasitic bacteria and fungi had been isolated and their life cycles and relations with their hosts studied. It was felt to be only a matter of time before these types of micro-organisms would be shown to be the causal agents of all infectious diseases. This state of optimism was soon shattered and the causes of many important diseases, e.g., smallpox, foot and mouth, influenza, tobacco mosaic, remained unknown. Iwanowski in 1892 showed that the symptoms of tobacco mosaic (p. 244) could be reproduced in healthy plants by rubbing their leaves with juice from infected ones, even after the juice had been passed through a filter capable of holding back particles of bacterial size. Unwittingly, Iwanowski had demonstrated the existence of an infectious agent entirely different from any other known at that time. However, he was sceptical of his results and clung to the idea of a bacterial aetiology of tobacco mosaic. Six years later Beijerinck confirmed these filtration experiments and realized that an infectious agent different from bacteria was responsible. However, his idea of a '*contagium vivum fluidum*' implied the idea of a non-particulate agent and he used the term 'virus' derived from the Latin meaning of a 'poison'. In the same year as Beijerinck's work on tobacco mosaic, Loeffler and Frosch showed that foot and mouth disease of cattle could be induced by bacteria-free filtrates of the fluid from vesicles characteristic of the disease, and Sanarelli obtained comparable results with myxomatosis of rabbits. Other plant and animal diseases were soon added to the list of those caused by agents able to pass through bacterial filters.

These agents were termed 'filterable' viruses but this description is no longer used since it is now possible to make filters fine enough to retain the smallest virus particles.

It also became clear that viruses propagate only in living cells, i.e., they are obligate intracellular parasites. This explained the repeated failure of the early microbiologists to grow viruses in normal bacteriological media.

Twort (1915) and D'Herelle (1917) respectively described cultures of staphylococci and dysentery bacilli showing clear areas due to lysis of bacterial cells. From such cultures bacteria-free filtrates were obtained which were able to produce the effects in new cultures. It was concluded that this lysis was due to a virus, now termed bacteriophage (often abbreviated to phage or ϕ). The actinomycetes are also attacked by viruses (actinophages) and viruses have recently been discovered in blue-green algae (p. 268) and fungi (p. 268). Indeed it seems that most, if not all, types of organisms are attacked by viruses.

The nature of viruses and their place as micro-organisms

Following the early, rather nebulous ideas on the nature of viruses, the development in the 1930's of several physical techniques (filtration with filters of graded pore size, ultracentrifugation and electron microscopy) made it clear that viruses were particulate, virus particles being termed virions.

In the same period, Stanley (1935) was able to purify and crystallize tobacco mosaic virus (TMV). In the following year Bawden and Pirie showed that this virus is a nucleo-protein. Many viruses have since been purified by the use of similar techniques. The ability of some viruses to crystallize emphasizes their simplicity and regularity of structure.

As knowledge about viruses increased, it became clear that they differed radically in many respects from other infectious agents. Indeed, the ability of some viruses to crystallize, called into question the concept of viruses as living micro-organisms. Subsequently the uniqueness of viruses was further emphasized when, by using gentle extraction methods, it was found possible to separate the protein and nucleic acid components of some simple plant and animal viruses (e.g., TMV and poliovirus) and to show that the nucleic acid alone was able to initiate infection.

The debate on the nature of viruses as living organisms continues although most of the disputes are purely semantic and revolve around usage of the words 'living' and 'organism'. It is not proposed here to outline the pros and cons of considering viruses as living organisms although this may be a useful exercise for the student of microbiology to undertake.

Clearly the protagonists of the 'infectious chemical' attitude to viruses may question their inclusion within a text book of micro-*organisms*. However, it is indisputable that virology grew out of bacteriology and that owing to their ultra-parasitic nature, viruses are intimately connected with a wide range of living organisms. Moreover, the study of bacteriophage has done much to further our knowledge of genetics both of the phages themselves and of their host bacteria (p. 120). It is clear too that viruses unlike, say, the transforming DNA of some bacteria (p. 118), have developed distinct means of transferring their genome from one host cell to another.

Virus-like micro-organisms

Considering the great diversity of micro-organisms and the varying degrees of parasitism that have developed during the course of evolution, it seems likely that there may exist organisms which resemble viruses in some ways and bacteria in other. Such organisms are now recognised as belonging to two groups, namely the Rickettsiae (p. 345) and the *Bedsoniae*, also known as the Chlamydozoaceae, Miyagawanella or psittacosis-lymphogranuloma-venereum agents. Organisms belonging to both these groups are small (approximately 300 mμ in diameter—a size only a little larger than the largest of the viruses) and, like the viruses, they are obligate intracellular parasites. For these reasons, they were classified for many years as viruses and many virological techniques, e.g., methods of culture, have proved useful in their study. However, the Rickettsiae and *Bedsoniae* agents are now known to possess properties more closely related to the bacteria. The bacterium-like properties of the Rickettsiae and *Bedsoniae* are as follows: (1) particles contain both RNA and DNA, whereas viruses contain only one of these, (2) multiplication occurs by binary fission or budding, (3) particles possess bacterial cell wall constituents and some of their associated metabolic pathways (consequent to this is the susceptibility of these agents to antibiotics like penicillin). A summary of the main differences between viruses and some other micro-organisms is given in Table 8.1.

For the above reasons the Rickettsiae and *Bedsoniae* are no longer considered as viruses but as very small bacteria which have developed an ultraparasitic mode of life.

Structure and chemical composition of viruses

Clearly, in determining both the chemical and physical characteristics of viruses, efficient methods of preparing virus particles free of host cell constituents are necessary. The development of such purification methods (e.g.,

Table 8.1 A comparison of the properties of viruses with some other micro-organisms.

	Bacteria		Mycoplasmas	Rickettsiae	Bedsoniae	Viruses
	Gram-positive	Gram-negative				
Size	c. 1 μ diam.	c. 1 μ diam.	Some have reproductive units < 150 mμ	c. 300 mμ	c. 300 mμ	Largest are 300 mμ Majority < 150 mμ
Nucleic Acids	DNA and RNA	DNA and RNA	DNA and RNA	DNA and RNA	DNA and RNA	Either DNA or RNA
Growth on artificial media	+	+	+	-	-	-
Intracellular replication	-[1]	-[2]	+	+	+	+
Mode of replication	Binary fission	Binary fission	Binary fission and budding	Binary fission	Binary fission and ? budding	Intracellular assembly of constituent parts (nucleic acid and protein subunits, etc.)
Muramic acid	+	+++	-	+	+	-
Sensitivity to antibiotics used in clinical treatment[3]	++	++	++	++	++	-

[1] Some will replicate intracellularly, e.g., Mycobacterium tuberculosis.
[2] Some will replicate intracellularly, e.g., Brucella abortus.
[3] See also Table 16.4.

chromatography, electrophoresis, density gradient centrifugation), has greatly facilitated the study of virus structure and composition. To these techniques must be added electron microscopy and X-ray diffraction which have added considerably to our knowledge of virus structure.

The simplest of the viruses (e.g., poliovirus, tobacco mosaic virus), possess two main chemical constituents, protein and nucleic acid. Viruses, unlike other micro-organisms, are characterized by the presence of only one type of nucleic acid in the infectious particle. Thus viruses may be of two types, containing either RNA or DNA. Until recently all higher plant viruses examined were RNA-containing but a DNA-containing virus infecting cauliflowers has now been described. The viruses of other organisms may be of either type.

The extremely small size of viruses means that the particles are capable of containing only minute amounts of nucleic acid (Table 8.2). The extreme smallness of viral genomes has considerable influence upon viral fine structure.

Table 8.2   The approximate composition of some viruses.

Virus	% RNA	% DNA	% Protein	Molecular weight of Nucleic acid	Molecular weight of Virion
Poliovirus	25–30	0	70–75	$2 \cdot 2 \times 10^6$	$3 \cdot 6 \times 10^6$
Influenza[1]	0·8–1	0	60–75	$2 \cdot 0 \times 10^6$	280×10^6
TMV	6	0	94	$2 \cdot 1 \times 10^6$	40×10^6
Turnip yellow mosaic	40	0	60	$1 \cdot 7 \times 10^6$	5×10^6
Tomato bushy stunt	17	0	83	$1 \cdot 7 \times 10^6$	11×10^6
Vaccinia[2]	0	5	89	150×10^6	2000×10^6
Adenovirus	0	13	87	23×10^6	177×10^6
Coliphage T2	0	50	50	120×10^6	300×10^6
Coliphage ϕX174	0	26	74	$1 \cdot 7 \times 10^6$	$6 \cdot 2 \times 10^6$

[1] Also contains approximately 30% lipid and 6% non-nucleic acid carbohydrate.
[2] Also contains approximately 6% lipid.

It is now well established that DNA and RNA generally occur as double- and single-stranded molecules respectively. The nucleic acids of some viruses, however, are atypical. For example, although most DNA viruses contain the double-stranded molecule, some phages (e.g., ϕX174, S13), have single-stranded DNA. In RNA viruses the nucleic acid is usually single stranded but double-stranded RNA has been demonstrated in the reoviruses (a group of animal viruses), the fungal viruses and the wound tumour virus of plants.

Most of the proteins of the virion occur as a coat, termed a capsid, surrounding the nucleic acid. Such proteins have two main functions. They protect the viral genome from adverse environmental conditions and also aid in the penetration of the virus into the host cell. The latter function may be merely the attachment of the capsid on to the host cell membrane (as with many of the animal viruses) or it may be concerned with active injection of the viral nucleic acid through the host cell wall (as with some bacteriophages). Beside structural proteins, essential for the morphological integrity of the particle, some viruses also contain proteins with enzyme functions, e.g., the phages contain the enzyme lysozyme which facilitates entry of the phage nucleic acid through the host cell wall; the myxoviruses (a group of animal viruses including influenza virus) contain the mucolytic enzyme neuraminidase which may be involved in penetration of the virus into the host cell or release of progeny virus after infection. The total amount of protein present in viruses differs from virus to virus (Table 8.2).

Besides the two basic constituents, protein and nucleic acid, some viruses also contain lipids, e.g., the myxoviruses, herpes viruses (including herpes simplex virus of man), the arboviruses (including the virus of yellow fever), the poxviruses (including the virus of smallpox) and the phage PM2 which infects a species of *Pseudomonas*. The amount of lipid present varies from approximately 30 per cent in the case of the myxoviruses to 6 per cent in the case of the poxviruses. The origin and function of this lipid component is discussed later.

Some viruses also contain a small percentage of non-nucleic acid carbohydrate e.g., influenza virus contains approximately 6 per cent. The function of this component is unknown.

Finally, some viruses, viz., phages and plant viruses, are known to contain polyamines. The basic properties of these compounds serve to neutralize the acidic phosphate groups of the nucleic acids. It is thought that this facilitates

Table 8.3 The main features of the size and structure of some animal, plant, and bacterial viruses.

Virus	Nucleic acid	Symmetry of capsid	Presence of envelope	Size of virion
Vaccinia	DNA	?	+	230 × 300 mμ
Herpes simplex	DNA	icosahedral	+	non-enveloped 110 mμ enveloped 180–250 mμ
Tipula iridescent virus (infects larvae of the crane fly), *Tipula paludosa*	DNA	icosahedral	–	130 mμ
Adeno	DNA	icosahedral	–	75 mμ
T4 phage	DNA	icosahedral & helical	–	head: 95 × 65 mμ tail: 110 × 25 mμ
φX174	DNA	icosahedral	–	30 mμ
Tobacco mosaic	RNA	helical	–	rigid rod 300 mμ long
Potato X	RNA	helical	–	flexible rod 500 mμ long
Influenza	RNA	helical	+	80–200 mμ
Polio	RNA	icosahedral	–	28 mμ
Turnip yellow mosaic	RNA	icosahedral	–	28 mμ
MS2 (phage)	RNA	icosahedral	–	24 mμ

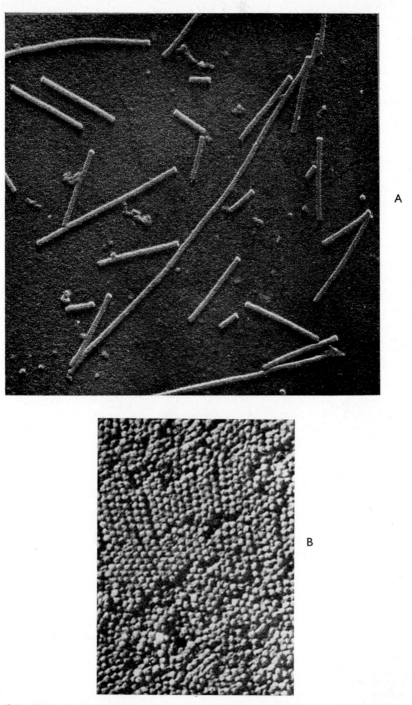

Fig. 8.1 Electron micrographs of two plant viruses illustrating the metal shadowing technique. (**A**) the rod shaped virus TMV (×33,000). (**B**) the isometric virus of tomato bushy stunt. (×72,000. H. L. Nixon)

the folding up of the nucleic acid molecule within the small volume of the virion. Similar substances may occur in animal viruses.

For many years after the discovery of viruses, little was known about their fine structure. Techniques such as filtration and ultracentrifugation gave estimates of the sizes of virions (Table 8.3) but viral fine structure only began to be elucidated following the development of electron microscopy and X-ray diffraction.

Electron microscopy, with the aid of metal shadowing (p. 9) demonstrated clearly that some viruses like TMV are rod shaped (Fig. 8.1) while others like poliovirus and tomato bushy stunt virus (Fig. 8.1) are nearly spherical (isometric). However, it was X-ray diffraction which first demonstrated the structure of viruses at the molecular level. This technique can only be applied fruitfully to those viruses which, when purified, form crystals or paracrystals

Fig. 8.2 Diagram showing approximately 1/20th the length of the rod-shaped TMV virion. Note the helical arrangement of both the protein structure units (large shoe-shaped structures) and the inner RNA (beaded structure). The last turns of the protein helix have been omitted to show the position of the RNA more clearly. (After Klug and Caspar *et al.* (1962) *Cold Spring Harbor Symp. quan. Biol.* **27**, 49)

e.g., poliovirus and TMV respectively. The virus, first and most extensively investigated by X-ray diffraction, was TMV. From such studies it is now clear that the capsid of TMV is a hollow rod consisting of identical protein molecules (structure units) arranged in the form of a helix around the long axis of the virion (Fig. 8.2). Furthermore, the RNA of the virion also exists as a helix, within and closely associated with, the helix of protein subunits. The virions

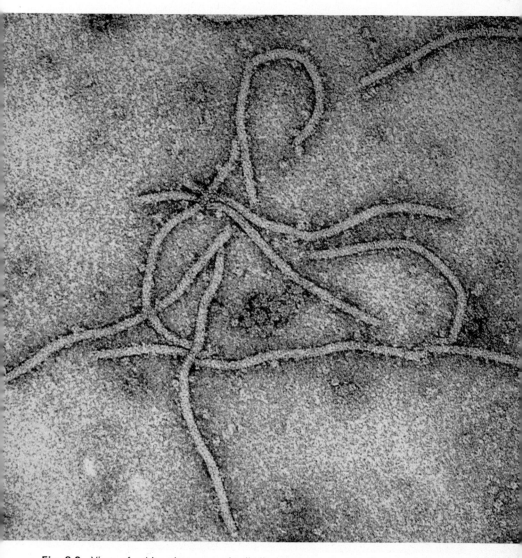

Fig. 8.3 Virus of white clover mosaic (helical symmetry), negatively stained with uranyl acetate. Note the flexuous nature of the particles. ($\times$174,000. R. G. Milne, Rothamsted Experimental Station)

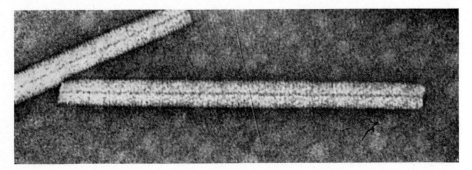

Fig. 8.4 Virions of TMV (helical symmetry) negatively stained with phosphotung-state. Note: **(a)** the stain has filled the central hole and **(b)** the groove between the helix of structure units is visible as cross striations.

of other rod-shaped plant viruses (e.g., white clover mosaic virus, Fig. 8.3), although less rigid than those of TMV, have a similar helical organization.

Viruses with isometric (nearly spherical) capsids were found to have a somewhat different structure. The protein subunits forming the capsids of these viruses are arranged as if located on the surface of a special class of deltahedron viz., an icosadeltahedron. This is a polyhedron whose faces are all equilateral triangles and with the symmetry of an icosahedron (a twenty sided solid). Such solids are said to have cubic or icosahedral symmetry.

The introduction of negative staining to electron microscopy (p. 9) gave confirmation of the results obtained by X-ray diffraction. Electron micrographs of viruses with helical and icosahedral symmetry are shown in Figs. 8.3, 8.4, 8.5 and 8.6 respectively. The morphological subunits revealed on the surface of virions by negative staining are termed capsomeres. The number of capsomeres present varies greatly between different viruses, e.g., virions of the phage ϕX 174 have 12 capsomeres whereas the virion of herpes simplex (a human virus) has 162. It seems probable that in most cases a single capsomere consists

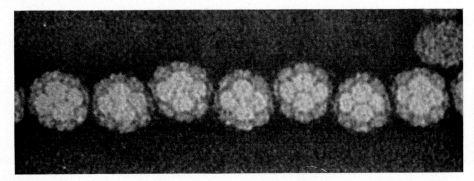

Fig. 8.5 A row of turnip yellow mosaic virions (icosahedral), negatively stained with uranyl acetate ($\times$ 500,000). Note the clearly visible capsomeres on the surface of the virions. (J. T. Finch)

Fig. 8.6 (A) A virion of adenovirus negatively stained with phosphotungstate. (Valentine, R. C. and Pereira, H. G. (1965) *J. Mol. Biol.* **13**, 13–20) **(B)** A model of the capsid of adenovirus. Note the distinctive structure of the capsomeres (coloured black) at the vertices of the icosahedral structure. Such capsomeres have five neighbours and are called 'pentons'; the remaining capsomeres with six neighbours are called 'hexons'. (Valentine and Pereira)

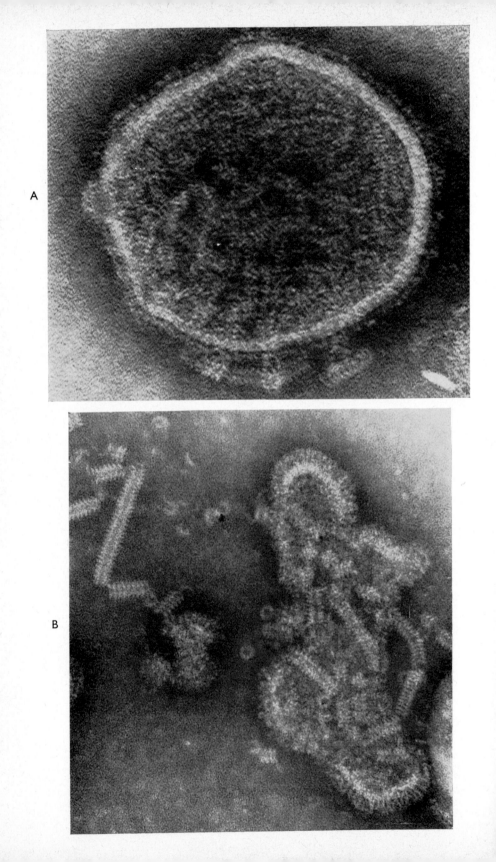

A

B

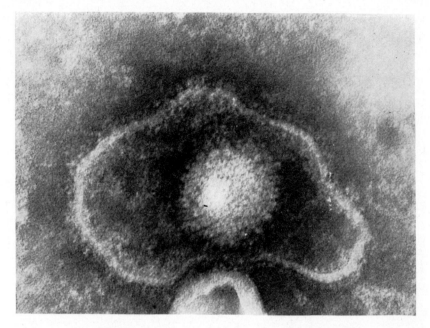

Fig. 8.8 Herpes simplex virion negatively stained with phosphotungstate (×300,000) Note the loose outer envelope and the inner capsid with the regularly arranged capsomeres. (Mrs. J. D. Almeida)

of more than one structure unit (protein molecule) the latter being the subunits demonstrated by X-ray diffraction. The arrangement of the nucleic acid molecule within the capsid of icosahedral virions is not yet clear.

Negative staining has also shown that some animal viruses, although more or less spherical, nevertheless have capsids with helical symmetry. In such viruses (e.g., influenza virus, Newcastle disease virus), the helical capsid is wound up like a ball of wool inside an envelope of lipoprotein. The capsid can be released from its lipoprotein envelope by treatment with lipid solvent (Figs. 8.7A and B), hence revealing the helical symmetry (analogous to the tubular TMV virion).

As mentioned previously, several groups of viruses contain lipids. Such viruses may have particles of helical or icosahedral symmetry, e.g., the myxoviruses and herpesviruses respectively. In all cases the lipid is found as a lipoprotein envelope surrounding the nucleocapsid (nucleic acid core plus protein capsid, Fig. 8.8). The viral lipoprotein component has an interesting origin. There is strong evidence to suggest that the lipoprotein is derived directly from the host cell membranes. For example, in the case of herpes simplex virus the virion

Fig. 8.7 **(A)** Virion of Newcastle disease virus (NDV) negatively stained with phosphotungstate (×300,000). Note the distinct fringed outer envelope and the less distinct inner nucleoprotein capsid. **(B)** Disrupted NDV virion negatively stained with phosphotungstate (×300,000). The structure of the helical capsid is clearly visible. Note the similarity with the structure of TMV (Fig. 8.4). (Mrs. J. D. Almeida)

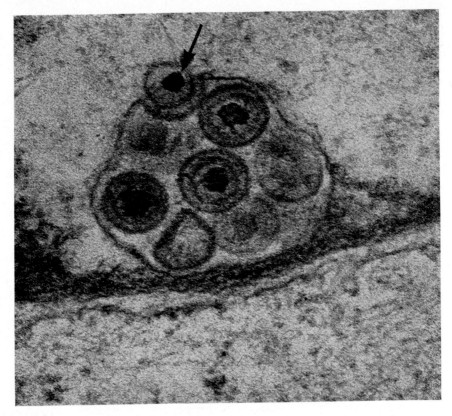

Fig. 8.9 A portion of a nucleus (above) with marginated chromatin and pale nucleo-plasm, from a cell infected with a herpes virus. The nuclear envelope, with its double membranes bounding the perinuclear space, crosses the field and extends as a mem-brane-bounded invagination into the nucleoplasm. The invagination contains several mature particles each enveloped by its characteristic outer membrane and with a dia-meter of about 130 mμ. An immature particle is maturing by budding into the invagina-tion (arrow) and acquiring the outer membrane of maturity as it passes through. ($\times$ 136,000. From Epstein, Achong, Churchill, and Biggs (1968) *J. Nat. Cancer Inst.* **41**, 805)

acquires an envelope of host nuclear membrane as it passes from the nucleus into the cytoplasm (Fig. 8.9). In the myxoviruses, e.g., influenza, the envelope is acquired from the cytoplasmic membrane as the virions 'bud off' from the cell surface. Hence viral envelopes are of host cell origin although they may be modi-fied by the inclusion of viral antigens. In the animal viruses, the envelope is probably of importance in facilitating entry of the virion into the host cell; the virus envelope fusing with the cell membrane, thereby allowing the capsid direct access to the cell interior.

In summary, most viruses can be grouped according to their structure, hav-ing either icosahedral or helical symmetry; the virions in some groups also

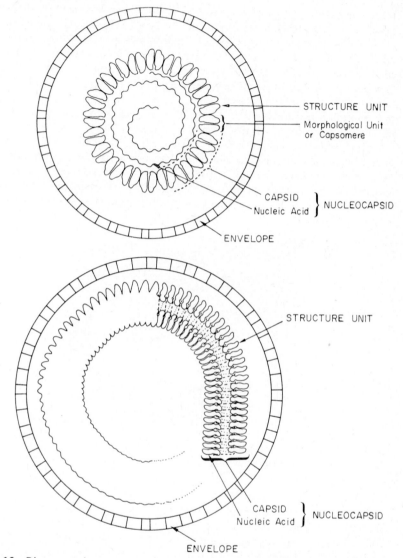

STRUCTURE UNIT

Morphological Unit
or Capsomere

CAPSID } NUCLEOCAPSID
Nucleic Acid

ENVELOPE

STRUCTURE UNIT

CAPSID } NUCLEOCAPSID
Nucleic Acid

ENVELOPE

Fig. 8.10 Diagrammatic representation of sections through (**A**) an enveloped virus with icosahedral symmetry, (**B**) an enveloped virus with helical symmetry. (After Caspar, 1962)

possess a lipoprotein envelope. The two types of virion architecture are illustrated diagrammatically in Fig. 8.10A and B.

In 1956–7, Crick and Watson proposed the theoretical basis for virus structure. It was suggested that the extreme smallness of viral genomes was the main factor influencing their structure. For example, a virion of poliovirus (RNA-containing) has approximately 6×10^3 nucleotides which, assuming a code of three nucleotides per amino acid, may code for approximately 2000 amino acids. This is equivalent to approximately 10–15 proteins and these must include not only structural proteins but also enzymes involved in the intracellular replication of the virus. The most efficient method of constructing a virion with such limited information and materials is to construct a capsid of identical 'building blocks' (protein molecules); a fact which is amply confirmed by X-ray diffraction and electron microscopy. Moreover, in the construction of such a protein shell it is desirable that the structure should be in the most stable state, i.e., in a state of minimum free energy. It can be shown that such a state exists when the constituent structure units make identical bonds with their neighbours. Such a situation can be achieved most closely by arranging the structure units in either a helix or a sphere (or a close approximation to a sphere, i.e., an icosadeltahedron). Again these theoretical predictions fit in with the observations made on virus fine structure by X-ray diffraction and electron microscopy.

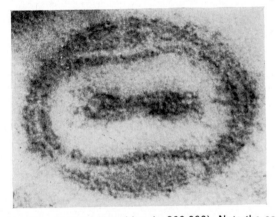

Fig. 8.11   Section of a vaccinia virus virion (× 200,000). Note the complex structure of membranes surrounding the inner DNA-containing nucleoid. (M. A. Epstein)

Although most viruses appear to fit neatly into the helical or icosahedral categories, there are at least two virus groups which have a more complex structure. The poxviruses (a group including smallpox virus) are large brick shaped or ovoid viruses, the virion being composed of a central DNA containing body (the nucleoid) surrounded by a series of protein and lipid membranes (Fig. 8.11). Many of the larger phages have a complex structure which is clearly correlated with the mechanism by which the viral nucleic acid is injected into the host cell. These phages possess a head and tail (Fig. 8.12A), the head contains a tightly packed core of nucleic acid (usually DNA) within a thin protein coat, the latter structure being similar to the capsid of the simpler isometric viruses. The head is spherical, ellipsoidal or hexagonal in shape and sizes range from 25 to 100 mμ in diameter. Structural distinction between the nucleic acid and protein components is well illustrated when phage heads are emptied of

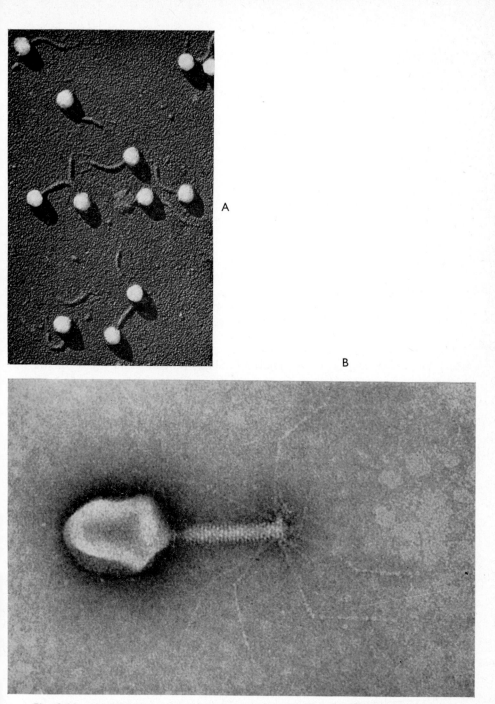

Fig. 8.12 (A) T.5 phage metal shadowed. The head and tail of the phage virion are clearly visible (×60,000). (B) T.4 phage negatively stained with phosphotungstate (×280,000). Note the tail fibres radiating from the tail base plate, the striations corresponding to the contractile outer sheath of the tail and the elongated head. (E. Kellenberger)

their DNA by osmotic shock, leaving hollow-headed protein ghosts. The phage tail is narrower than the head and may be up to 100 mμ in length. It consists of several distinct proteins but its structure varies in complexity with different phages. In the T-even phages (T₂, T₄, T₆, a group of phages attacking *Escherichia coli*), the tail consists of an inner tube (possibly having helical symmetry) and an outer contractile sheath. At the head end the sheath is attached to a collar and at the opposite end, to a hexagonal plate (Fig. 8.12B). Fibres radiating from this terminal plate serve to attach the phase to the host bacterium prior to infection (p. 260). It is thought that these phages represent a combination of icosahedral and helical symmetry in the structure of the head and tail respectively. Of all viruses, it is perhaps in the structure of phages, the T-even in particular, that structure is seen to be correlated with function most elegantly.

Virus replication

As pointed out earlier, viruses replicate only in living cells. The use of the term 'replicate' infers that the process of virus multiplication is different from that of micro-organisms and tissue cells which divide by binary fission with or without mitotic division of their genetic components. Whilst the mode of entering the host cell varies from virus to virus the mode of replication is considered to be similar for all and has been most completely worked out for bacteriophage (for a detailed description see p. 259). The viral nucleic acid upon entering the cell takes over control of the cellular metabolic processes and codes for the separate synthesis of viral nucleic acid and protein which later combine to form the mature virus particle. The virus yield from a cell infected with a single virus particle varies widely but often ranges from 10 to 100 particles.

The host cell must be capable of supporting this sequence of steps in viral replication. Many viruses have a single or limited host cell requirement; others may replicate in a range of different host cells but the quantity of virus produced in each cell type may differ widely.

Viruses may be propagated in susceptible animals, plants or micro-organisms, or in tissue cultures made from animal or plant tissues. When using animals it is necessary to consider: (1) their natural susceptibility to infection (Table 8.4) or immune status to the virus; (2) the possibility of latent infection with the same or other virus (often the challenge of another virus stimulates a latent virus to become active, as occurs with herpes simplex in man, the cause of the common cold sore on the lips which often erupts when the patient is challenged by a common cold virus); (3) the most suitable route of inoculation which is usually related to the affinity of the virus for particular tissues (Table 8.4). Infection is recognized by characteristic signs and symptoms of disease.

A widely used host for animal virus propagation is the developing chick embryo. The tissues and associated membranes support the replication not only of fowl viruses but also of mammalian viruses to which the hatched chick and adult bird are not susceptible. Whilst maternal antibodies against fowl viruses may be present in small amounts in the yolk of the fertile egg the chick embryo does not possess an antibody-forming mechanism. Evidence of virus replication may be indicated by pock formation in the chorioallantoic membrane (Fig. 8.13), by lesions and inclusion bodies (p. 240) in particular tissues, death of the embryo,

Table 8.4 The natural and experimental hosts of some common animal viruses.

Virus	Natural host	Experimental host	
		Animal	Route of inoculation
Arthropod-borne Encephalitides (Arboviruses)	Man and other mammals	Mouse	Cerebral
Canine Distemper	Dog	Ferret	Nasal
Foot-and-Mouth	Cloven-hoofed animals	Guinea pigs Suckling Mice	Hind footpads Cerebral
Influenza	Man	Chimpanzee, Ferret	Nasal
Mumps virus	Man	Monkey	Parotid gland
Poliovirus	Man	Monkey	Cerebral, occasionally peritoneal
Rabies	Dog, Man, etc.	Mouse	Cerebral

or the demonstration of the presence of viral antigen by specific serological tests (p. 242). Since one pock on the chorioallantoic membrane can arise from infection with one viral particle it is possible to assay for the number of pock-forming units in a virus suspension by this techique.

Laboratory cultures of tissue cells provide the most useful means of propagating viruses both for production of virus antigens and for diagnosis of virus diseases. A variety of tissue cultures are available but those including actively multiplying tissue cells are the most useful for the propagation of viruses. Monolayer cultures are widely used for the study of particular virus-cell interactions

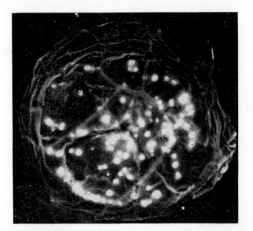

Fig. 8.13 The chorioallantoic membrane of a developing chick embryo infected with vaccinia (nat. size). After incubation for two days the membrane was removed to formal saline and examined for lesions. Plaques of necrosis resulting from virus activity are clearly seen. (D. B. Peacock)

A

B

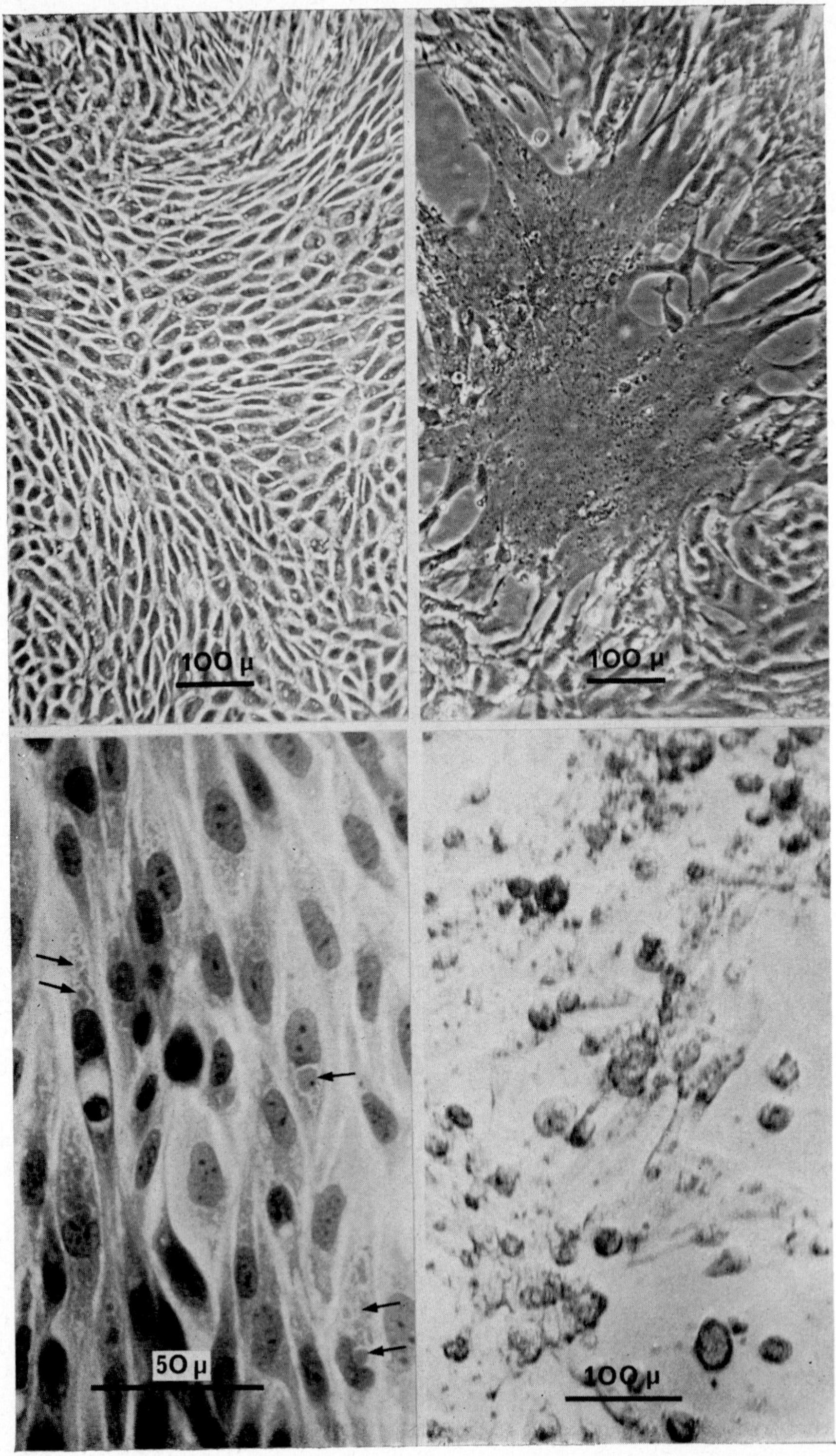

C

D

100 μ

100 μ

50 μ

100 μ

and cytopathic effects (p. 238). Tissues freshly taken from a living or recently dead animal are treated with enzymes (e.g., trypsin), a process which, whilst maintaining cell viability, causes disaggregation of the cells. These cells are seeded into culture tubes or bottles containing physiologically and nutritionally balanced fluids plus appropriate antibiotics to inhibit bacterial and fungal contaminants. The cells adhere to the lower surface of the vessel when incubated at 37°C, multiply, and form a confluent tightly-packed sheet of cells within 2–4 days; the term monolayer is used to describe the cell sheet which is rarely more than one cell thick (Fig. 8.14). Latent viruses, when present in the animal tissue from which the cultures were prepared, may be stimulated into active replication by either the enzymic disaggregation process or the subsequent infection of the tissue culture with another virus. Despite this possibility, tissue culture constitutes a highly efficient *in vitro* system for the propagation of viruses, with the outstanding advantage that cultures can be prepared from most cells or tissues known to be susceptible to particular viruses.

VIRUS DISEASES OF VERTEBRATE ANIMALS

General considerations of virus infection

Unlike the majority of plant virus diseases (p. 244) most animal virus diseases cannot be diagnosed solely on their signs and symptoms. Viruses do not produce exotoxins and the diseases they cause are the direct result of their primary and secondary replication cycles within the various tissue cells of the animal body. An understanding of how this replication occurs in tissue cells provides an understanding of the disease processes taking place in the animal body as a whole. Replication is studied in *in vitro* systems of animal tissue cultures.

The replication of viruses in tissue cells leads to the biological malfunctioning of those cells and if large numbers of cells are involved malfunctioning of the organ generally follows. This may result in death of the animal.

Some animal viruses replicate in a limited range of tissue cells, e.g., influenza virus replicates only in cells of the respiratory tract, while others replicate in a wide variety of tissue cells, e.g., smallpox virus in cells of the skin, lungs, and other internal tissues. The latter category of viruses can therefore spread to other susceptible tissues by blood-borne dissemination from a primary site of infection. The presence of virus in the blood stream is not necessarily an indication of wide tissue susceptibility since viruses with a limited tissue range may 'spill-over' into the blood stream following replication in the susceptible tissues.

Fig. 8.14 Virus cytopathic effect in tissue culture. (**A**) Normal monolayer of dog kidney cells. (**B**) Large, central giant cell (polykaryocyte) formed by fusion of many dog kidney cells when infected with canine distemper virus. (**C**) Intra-cytoplasmic inclusions (some arrowed) in dog kidney cells infected with canine distemper virus. (**D**) Extensive rounding-off and degeneration of monkey kidney cells infected with poliovirus. (**A**, **B**, and **C**: G. H. Poste; **D**: L. W. Greenham)

As with other microbial diseases (p. 557) the severity of a virus disease depends upon the 'size' of the infecting dose, the state of health of the animal, its age, sex, and degree of immunity (Chap. 16). The various aspects of the epidemiology of animal virus diseases are considered elsewhere (p. 592).

In vitro infections

Cell susceptibility

There is for each virus a range of susceptible (sensitive) tissue cells in which it will readily replicate. Other types of cells, in which replication is difficult or impossible, constitute the resistant (insensitive) cells. Cell susceptibility is related to such factors as the ability of the virus to adsorb to the cell plasma membrane, which, in turn, depends on the presence of virus 'receptors' on the membrane, as well as the provision by the cell of the appropriate biosynthetic processes for the support of replication of the invading virus.

Replication sequence

Tissue culture studies have revealed the nature of the replication sequence which occurs in the infected host cell. This sequence includes: (1) *adsorption* of the virion to the cell plasma membrane; (2) *penetration* of the virion into the cell through active pinocytosis (i.e., engulfment in a manner similar to phagocytosis but on a smaller scale) by the host cell itself or by fusion of the virus envelope with the cell membrane; (3) *eclipse* of the virion, due to enzymically controlled decapsidation processes which release the nucleic acid 'core', intact virus being no longer found in the cell; (4) *the latent period*, when the biosynthesis of new virus components is undertaken; (5) *the assembly period*, when early, immature virus forms are first appearing in the cell; (6) *maturation* of progeny virions from the early immature forms; and (7) *release* of the progeny virions from the infected cells by either budding-off processes (e.g., influenza virus) or cell lysis (e.g., poliovirus). With some animal viruses the complete sequence from (1) to (7) may take as little as 7–8 hours.

Cytopathic effects

Following virus replication within the host cell observable cytopathic effects (CPE) may result. These include: (1) *hyperplasia*, or cellular proliferation (e.g., herpes simplex virus); (2) *polykaryocytosis*, or giant cell formation (e.g., measles virus and ectromelia virus); (3) *inclusion body formation*, when inclusions are formed either intracytoplasmically (e.g., poxviruses) or intranuclearly (e.g., herpesviruses) but only occasionally at both sites at the same time (e.g., measles, distemper and rinderpest viruses); (4) *malignant transformation*, when the host cells acquire tumour cell-like characters and may give rise to tumours *in vivo* (e.g., papova viruses); (5) *vacuolation*, of either the cytoplasm (e.g., SV (simian virus) and VA (vacuolating agent) groups of viruses) or the nucleus (e.g., pig pox virus); (6) *necrosis*, which has as its initial signs a 'rounding-off' (e.g., polioviruses) or a 'tailing off' (e.g., adenoviruses) of the host cells and as its terminal signs a shrinking and a contraction in cell volume with heavy nuclear staining (pyknosis), observable in stained preparations (e.g., polioviruses). With some viruses a single CPE is observed in the host cell; with others more than one are found. When necrosis is one of these it is always the terminal cytopathic effect.

In vivo infections

With a few exceptions the signs and symptoms of virus diseases in animals are insufficient to be diagnostic. However, they may be sufficient to permit a presumptive diagnosis to be made, while awaiting the laboratory identification of the causal virus.

Signs and symptoms

The signs and symptoms presented by a particular virus disease reflect closely the primary and secondary replication sites in the body and the ensuing CPEs. Two examples illustrate this. *Poliovirus* infects man principally via the oral route and is insensitive to the low pH of the stomach. The primary replication sites are the tonsils and the various lymph nodes of the digestive tract. During this stage, virus is found in abundance in the faeces and diarrhoea may be present. Both virulent and occasionally avirulent strains pass via the intestinal wall into the blood stream (viraemia) by which the virus is disseminated throughout the tissues. The viraemia results in an elevated temperature and may induce an invasion of the central nervous system (CNS). Secondary replication sites next become infected including the CNS, lymphatic structures and brown fat and, at this stage of the disease, overt paralysis is observed. Since the virus induces necrosis in its host cells it follows that widespread necrosis of neurons in the CNS, a tissue system with a poor regenerative capacity, results in a permanent paralysis which varies in degree with the extent of the necrosis.

Smallpox virus in man usually infects via the respiratory route, although infection via the dermal tissues may sometimes occur. The enteric route is much less likely because of the susceptibility of the virus to a low pH, such as would be encountered in the stomach. Replication of the virus may produce transient signs of respiratory disease. From this primary site the virus passes to the regional lymph nodes and thence, via the blood stream, to the liver, spleen, and other tissues where the secondary replication takes place. From these tissues the virus proceeds via the blood stream to the skin and mucous membranes where the typical focal rash is formed. Both viraemias will produce temperature elevations, the second being most severe (*ca.* 104–105°F). During the second viraemia the patient suffers a toxaemic illness of about 4 days' duration with fever, systemic symptoms and prostration, often with accompanying headache, backache, pains in the limbs, and vomiting. The typical focal rash is seen on the third or fourth day, involving the buccal and pharyngeal mucosa and the skin, principally of the face, forearms, and lower legs, although the trunk is involved. The rash, at first macular, soon becomes papular, vesicular, and finally pustular (8–9 days) when crusts commence to form.

From these two examples it is evident that viruses are individualistic in respect of their replication sequence and CPE in the animal body. Similar signs, symptoms, and lesions are, however, not necessarily evidence for common patterns of replication. For example, both smallpox and papilloma (wart) viruses give lesions in the skin, but whereas smallpox virus replicates in internal tissues the papilloma viruses are confined entirely to the skin tissues. Also lesions in the CNS are encountered in poliomyelitis, rabies, and the various encephalomyelitides, although each virus has a different pattern of replication in the body.

Laboratory diagnosis

Laboratory investigations are usually necessary for the diagnosis of virus diseases. These depend upon sampling the right clinical specimens from the patient at the optimum stage of the disease, and upon their transport to the laboratory under conditions which will ensure maximum survival of the virus. Clinical specimens may be used for any one of three purposes: (1) for virus isolation: (2) for serological examination; (3) for direct examination.

VIRUS ISOLATION: the specimens, or suitable extracts of them, are inoculated into appropriate tissue cultures and/or susceptible animals. With fast multiplying and adaptable viruses (e.g., poliovirus and some myxoviruses) only 4–10 days may be needed for their isolation and identification; with slow multiplying and less adaptable viruses (e.g., adenoviruses, measles virus) up to 1–2 months may be required. Sometimes one or more 'blind' subcultivations will be required in the tissue cultures before a CPE is obtained. Positive isolation enables some form of definitive diagnosis to be made; negative isolation, however, does not necessarily indicate the absence of a virus. Upon isolation it is usually possible to make a presumptive identification based on the observed CPE of the tissue culture cells. Confirmation of this identification rests on specific serological tests, e.g., complement fixation, neutralization, haemagglutination-inhibition, haemadsorption-inhibition, interference-inhibition, and plaque-reduction tests (p. 569).

SEROLOGICAL EXAMINATIONS require two samples of serum, an acute serum at the time of onset of the disease and a convalescent serum at least 14 days later. These are treated by suitable serological tests (*vide supra*), using appropriate standard virus preparations as antigens, and the level of circulating antibodies in each serum is determined. A fourfold rise in antibody titre in the convalescent compared with the acute serum is usually accepted as significant and diagnostic. A single serum sample is worthless for diagnosis by serological means, although it may be of use in epidemiological surveys which set out to determine mean antibody levels to a particular virus.

DIRECT EXAMINATION is made of specimens obtained from localized lesions. For instance, the scab overlying an old cutaneous lesion or preferably, because of the difficulty in softening a dried scab, scrapings from beneath it, are usually a rich source of virus material. These specimens are smeared on microscopic slides, air dried and methyl alcohol fixed. The smear is then ready for staining and microscopical examination. Conjunctival scrapings from herpetic and inclusion conjunctivitis can be similarly tested. Emulsified scab material is often screened under the electron microscope for viruses of the poxvirus or herpesvirus groups. By these examinations attempts are made to identify either virus particles or virus inclusions.

At postmortem, various tissue specimens may be collected for direct examination for evidence of a virus CPE. Unlike the lesions produced by bacteria, little or no pus is formed in virus lesions. The tissue reaction is therefore never as vigorous, and may be limited to infiltration of the affected site by mononuclear cells or other leucocytes of the blood. Tissue specimens are examined by conventional histopathological means; the CPEs most usually observed are inclusion body formation, hyperplasia, and cellular necrosis.

Inclusions, when found, are so characteristic as to be diagnostic. Their importance in diagnosis was so widely recognized that it became the custom at one time to name them after their discoverer (e.g., Negri bodies in the rabid brain). They are acidophilic in staining reaction and most are homogeneous in composition. Inclusions most frequently indicate the original 'factory sites' of virus replication within the cell and are not usually composed of aggregates of virions. The characters of virus inclusions are generally similar but specifically different; details can be found in standard virological textbooks.

Hyperplasia is the characteristic CPE of oncogenic (i.e., tumour-forming) viruses. Cellular proliferation may be so rapid as to give tumour-like masses in the body, e.g., the Shope papilloma and the Shope fibroma of rabbits, and the Rous sarcoma of chickens. Less localized proliferation may give rise to the diffuse swellings seen in rabbit myxomatosis. More recently, a herpesvirus has been observed in association with the proliferating lymphoid cells of Burkitt's lymphoma.

Cellular necrosis, when observed, is usually focal in nature. Pus is never observed at these sites except when there is secondary bacterial infection, e.g., smallpox pocks on the skin. Necrosis in a vital organ or system, e.g., liver, kidney, or CNS, may lead to malfunction. For example, the characteristic jaundice of yellow fever is the result of malfunctioning of the liver.

The role of interferons in cytoimmunity

In contrast to other microbial infections, antibody immunity (p. 562) is not now considered to be a major factor in recovery from virus infections; whilst formed against virus antigens, antibodies are ineffective intracellularly and do not, therefore, inhibit virus multiplication. There is increasing evidence that interferon is one of the major factors in this recovery. Interferons are soluble proteins, with a molecular weight of about 30,000 and their production in tissue cells is induced by many viruses and some non-viral agents. Inactivated or avirulent viruses are usually more active inducers than infective, virulent viruses since the latter rapidly inhibit cellular protein and RNA synthesis, both of which are required for interferon production. Interferons exhibit species specificity and inhibit the multiplication of a variety of viruses only in the same cells or cells from the same animal species.

There is little uptake of interferon by treated cells and it has no effect on the virus extracellularly. It does not interfere with virus adsorption, penetration or assembly of progeny virus particles. Inside the cell interferon induces the formation of an inhibitor protein via DNA-dependent RNA synthesis and the inhibitor protein interferes with transcription of the information on the virus genome by preventing polysome formation. The infected cell still dies but no progeny virus is synthesized or released.

Attempts to utilize interferons for the prevention and treatment of virus infections have not proved fruitful but studies on the mode of action of interferons have yielded considerable information on virus-cell interactions.

'Slow viruses'

Of recent years certain agents, called 'slow viruses', have been described which produce chronic infections. These include CHronic Infectious Neuropathic Agents (termed CHINA in the abbreviated form) of which scrapie in sheep is one example. Another slow virus is the agent of Aleutian disease of mink. They share in common a long initial period of incubation, often months or years, and a protracted course once clinical symptoms arise. Many of the 'slow viruses' have very unusual properties, e.g., the scrapie agent is less than 25 mμ in size, is extremely resistant to heat and chemicals (e.g., formalin) and contains little or no nucleic acid. Such agents clearly pose problems when compared with more conventional forms of viruses. Their mode of replication may involve some as yet undiscovered mechanisms of molecular biology.

VIRUS DISEASES OF PLANTS

Despite recent advances, we are not yet in a position to identify all viruses by their structure. They are usually distinguished either by the visible symptoms they produce in the host or by serological methods (p. 569). Viruses causing disease in animals induce the formation of antibodies in the host just as bacterial pathogens do. In plants such a response does not occur, but specific antibodies are produced when plant viruses are injected into animals. The reaction between antisera produced in this way and sap from plants infected with the same virus is useful in providing a rapid means of detecting and identifying some plant viruses. The symptoms of virus diseases are, however, of primary importance in the identification of viruses as well as being a measure of the harm sustained by the host.

Symptoms of virus diseases of plants

External symptoms

A virus causing a plant disease may be systemic, that is distributed throughout the plant, or localized, producing only local lesions (necrotic spots) on leaves or other parts of the host. The same virus may produce a systemic infection on

Fig. 8.15 Virus diseases of plants. (**A**) Tobacco plant (*Nicotiana tabacum* var. White Burley) systemically infected with tobacco mosaic virus. (**B**) Leaf of *N. glutinosa* showing necrotic local lesions caused by tobacco mosaic virus. Left half of leaf inoculated with 10^{-3} mg/litre suspension of tobacco mosaic virus; right with 10^{-2} mg/litre suspension. (**C**) Leaf of lettuce systemically infected with lettuce mosaic virus. (**D and E**) Enations (outgrowths) on undersides of cucumber leaves caused by tobacco ringspot virus. (**A–E** Rothamsted Experimental Station, Copyright)

one host and only local lesions on another; strains of a virus may give markedly different symptoms on a particular host. Thus the strains of tobacco mosaic virus (TMV) from tobacco (*Nicotiana tabacum*) are systemic in this host but produce only local lesions on leaves of *Nicotiana glutinosa* (Figs. 8.15 A and B). Strains from tomato, however, produce only local effects on the tobacco plant. When infection is systemic the whole plant may show the characteristic symptoms of the disease or these may be restricted to certain parts, as with tobacco mosaic which infects tobacco systemically but produces the typical mosaic symptoms only on young shoots developing after primary infection has taken place.

The most common symptom of plant virus disease is a reduction in size and cropping power of the plant or part of it, as in 'reversion' of blackcurrant where the berries are abnormally small, lettuce mosaic where the diseased plants are smaller than healthy ones, or potato virus diseases where the tubers are few and small (Fig. 8.16A). Such reduction in size is usually accompanied or preceded by other symptoms, notably mosaic, mottling, striping or spotting of the leaves (Fig. 8.15A and C), wrinkling or curling of leaves (Fig. 8.15A) or a general chlorosis of the green parts of the plant. Wilting and premature leaf fall may occur, as with potato Y virus (Fig. 8.16B). The shape of leaves may be altered, as with reversion of blackcurrant and the 'fern' leaf effect produced by cucumber mosaic virus in tomato leaves. Outgrowths (enations) from the leaf surface also occur (Fig. 8.15D and E). Striping or mottling of petals is frequent, as with the well-known 'broken' or Rembrandt tulips. The fruit may be distorted or altered, as in swollen shoot of cacao where the pods of diseased trees are abnormally short, in cucumber mosaic where warty outgrowths occur on the fruit, or certain virus diseases of tomato which cause spotting or cankering of the fruit. Shoot elongation may be inhibited by the presence of a virus as with nettlehead of the hop, proliferation of lateral buds may cause a 'witches' broom' effect, or local excess formation of xylem may give rise to canker as in swollen shoot of cacao. Roots may also be affected, often as a result of damage to the phloem as in elm phloem necrosis, or they may die from starvation resulting from reduction in leaf area or chlorophyll content of the shoot.

The recognition and identification of a virus disease is complicated by the production of different symptoms on the same host by a particular virus under different conditions, by the existence of strains of a particular virus often differing in virulence and consequently in effects, and by the fact that the same virus may attack a range of often unrelated hosts producing strikingly different symptoms.

Some diseases once thought to be due to viruses are now considered to be caused by mycoplasmas (p. 344).

Fig. 8.16 Virus diseases of plants. (A) Leaf roll of potato. Left, healthy plant and tubers produced by it. Right, diseased plant showing stunting, curled leaves and reduced crop. (W. C. Moore) (B) Leaf drop streak of potato caused by potato virus Y; showing spotting of younger leaves, wilting and leaf drop of older ones. (W. C. Moore) (C) Leaf of potato (var. Majestic) infected with potato virus Y showing necrotic local lesions characteristic of the leaf drop streak (B). (Rothamsted Experimental Station, Copyright) (D) Leaf of White Burley Tobacco systemically infected with tomato spotted wilt virus. (Rothamsted Experimental Station, Copyright)

Effect of environment on symptoms

Factors such as temperature, light and the water supply available to the host, profoundly influence the symptoms of virus diseases. It is well known that, in glasshouse experiments, the severity of a virus disease may alter according to the season, some diseases being more severe in summer and others in winter. The former are mainly diseases causing the accumulation of the products of photosynthesis in the leaves, as with leaf roll of potato, and hence the effect is increased with greater light intensity.

High temperature reduces the severity of the symptoms of a number of virus diseases but the limiting temperature varies for different viruses. Thus, while a short period at 20°C suppresses symptoms of crinkle and mosaic of potato, 35°C is necessary to suppress symptoms of tobacco mosaic. Low temperature also may mask the presence of a virus, as with tobacco mosaic which produces typical symptoms only over a range of 10–30°C, and strawberry yellow edge which does not show at temperatures below 16°C although susceptible host plants may become infected at temperatures too low for the development of visible symptoms. Over the range permitting the development of typical symptoms an increase in temperature or in light intensity usually reduces the incubation period, i.e., time from inoculation to the appearance of visible symptoms. The intensity of the response may also increase.

The effect of plant nutrients and water supply on the symptoms of virus diseases are usually correlated with their effects on rate of growth of the host, symptoms being generally more severe when growth is rapid.

Internal symptoms

Virus diseases often cause the destruction or modification of tissues or cell contents. Phloem necrosis is common in virus-infected plants and leads to abnormal accumulation of carbohydrates in the leaves. In mosaic diseases chloroplasts are altered or destroyed. Some viruses induce abnormalities in meiosis leading to failure to produce viable seeds.

Mature particles of certain viruses may be demonstrated in the host tissues with the aid of the electron microscope (Fig. 8.17).

Intracellular inclusions have long been noted in the cytoplasm of cells infected with certain plant viruses These may be amorphous (X-bodies) or may consist of crystals of characteristic shape. Electron micrographs suggest that both types may be aggregates of the virus. One virus, tobacco etch, causes the formation of intranuclear crystalline inclusions in tobacco.

Transmission of plant viruses

Natural transmissions

A large proportion of plant viruses are carried from a diseased plant to a healthy one by vectors (most commonly insects) (p. 609). Blackcurrant reversion is carried by a mite (itself the cause of 'big bud' of blackcurrant); nematodes have also been shown to act as vectors of some viruses; the chytrid (p. 391) *Olpidium brassicae* is able to transmit tobacco necrosis virus and big vein virus

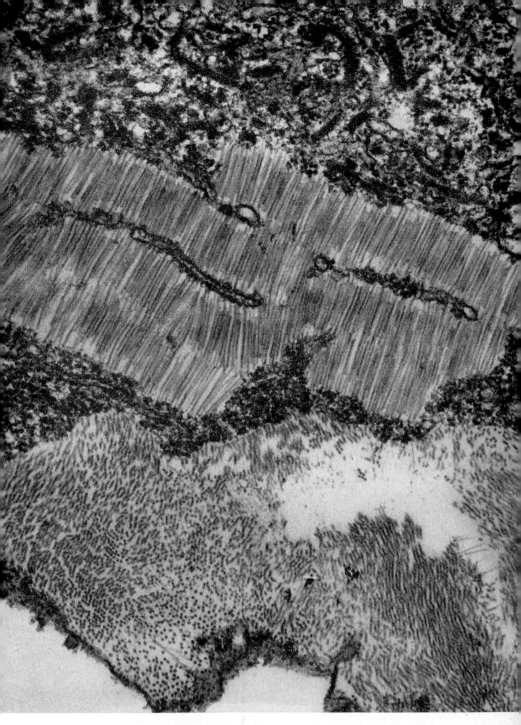

Fig. 8.17 Electron micrograph of TMV particles in a cell of *Chenopodium amaranticolor*. The virus causes local lesions on leaves of this plant. The rod-shaped particles of TMV are seen cut transversely or obliquely at bottom of plate and as whole particles in a band across centre of plate. Section stained by uranyl acetate and lead citrate. (Electron micrograph by R. G. Milne, Rothamsted Experimental Station)

to lettuce plants; mop top virus of potato is transmitted by *Spongospora sub-
terranea* (a member of the Plasmodiophorales, p. 392). Some viruses can be
transmitted only by a specific vector, some may be transmitted by a vector and
also by other means, some cannot be insect transmitted. Many viruses are transmitted mechanically by diseased plants rubbing against healthy ones in windy
weather (e.g., potato X virus) and others by root contact or through the soil.
The more resistant viruses (e.g., TMV) may survive in the dead remains of diseased plants in the soil, others (e.g., *Arabis* mosaic) are known to survive in
weed seeds. Crop plants may become infected from these sources. Virus diseases of crop plants may be spread by common horticultural practices such as
grafting, pruning, disbudding, or picking flowers. Viruses are not generally
carried by true seeds and it is possible that infected seeds are commonly rendered
non-viable and thus are incapable of spreading the disease. However, a few
virus diseases, such as lettuce mosaic and *Arabis* mosaic are seed-borne. Vegetative propagating units, such as bulbs, corms, rhizomes or tubers frequently
carry viruses. A notable example is the spread of potato viruses by the use of
infected 'seed' tubers. Since the tubers of virus-infected plants are smaller than
those of healthy ones they tend to be selected for planting. Pollen from diseased
barley plants carries barley false stripe virus to healthy ones.

Experimental transmission

Inoculation of healthy plants with juice from one suspected of containing a
virus is one method of identifying viruses. Inoculations are made by grafting,
by mechanical means or by the use of suitable vectors (p. 609).

Where it is possible to graft a shoot of a diseased plant on to a healthy one, or
vice versa, transmission usually follows. Only fairly closely related species, however, can be grafted to one another. A parasitic plant, the dodder, has been used
to transmit viruses from one host plant to another but again this method is
limited in application.

The simplest method of transmission is by rubbing infected sap on to a healthy
plant but many viruses cannot be transmitted in this way. Even with those which
can be so transmitted care must be taken that the amount of damage to the plant
tissues is at a minimum or the virus will not become established. Carborundum
powder or other abrasives may be added to the virus extract to produce uniform
slight abrasions of the leaf when rubbed; buffers should also be added, to control the pH of the plant sap.

Insects are bred experimentally and used for transmission of those viruses
which cannot be successfully established by other means. This involves a
special technique of handling insects and host plants and the use of cages to
confine the insects on the experimental plants.

Control of plant virus diseases

A plant systemically infected with virus can seldom be cured. This can sometimes be done by heat treatment, however, where there is a sufficient favourable
margin between the temperature at which the virus is destroyed and that which
damages the plant. The most successful form of heat treatment is that used for

the control of the flower-distorting virus (Aspermy virus) in chrysanthemums. This is carried out commercially by a number of cutting producers and has resulted in a great improvement in chrysanthemum flower quality. Other examples are 'phony' disease of peach which is eradicated if infected trees are exposed to a temperature of 35°C for a short period (obviously practicable only with trees growing under glass) and some sugar cane viruses which can be eliminated from cuttings by treatment in a hot water bath at 50°C.

Control of the vectors of virus diseases is difficult but effective, and modification of agricultural techniques sometimes prevents spread of disease.

Recently virus-free clones of certain crop plants, e.g., potato, have been raised from tissue cultures of the meristems. Growing points of shoots are usually free of virus and cultures from these thus give disease-free plants.

Control of most plant virus diseases must, however, be sought through the breeding of resistant varieties, general plant hygiene, removal of source of infection, and the prevention of spread of a virus within a crop.

INSECT VIRUSES

Introduction

Two major types of virus disease of insects are the inclusion body diseases and non-inclusion diseases. In inclusion body diseases the virus particles are occluded in polyhedral protein crystals (polyhedra), in granules (capsules), or in spindle-shaped bodies. These constitute the inclusion bodies that are to be seen under the light microscope in the tissues of the diseased insects. Polyhedral inclusion bodies may lie either in the nuclei or in the cytoplasm of the host's cells. In relation to this the corresponding viruses are divided into those that cause nuclear and cytoplasmic polyhedroses respectively. These two sorts of virus differ not only in their sites of multiplication but also in their morphology and in the symptoms they produce. The non-inclusion viruses differ strikingly from the foregoing types in that they lie quite free in the tissues of the hosts.

Viruses of invertebrates have for long been regarded as a separate and rather special group of viruses. This is not only because their hosts are distinctive but also because viruses that have polyhedral and granular inclusion bodies are not known in vertebrates or plants. The gap between insect virology and general virology has, however, narrowed greatly with the discovery in recent years of an increasing number of non-inclusion viruses in insects and other invertebrates, and with the recognition of two other distinct and important categories of insect viruses. These are the arboviruses (arthropod-borne viruses) that infect both vertebrates and blood sucking arthropods, and an analogous group of viruses that multiply in both plants and the insects that serve as vectors.

Survey of different groups of insect viruses

Viruses that cause nuclear polyhedroses

Nuclear polyhedroses occur most commonly in larvae of Lepidoptera (the moths and butterflies) but are known also in Hymenoptera and Diptera. The viruses multiply within nuclei in chiefly the skin, blood, fat-body, and tracheae.

The virus particles contain DNA and are rod-shaped (*ca.* 20–50 mμ diameter, 200–400 mμ long), and become enclosed either singly or in bundles within paracrystalline protein which forms microscopically visible polyhedra. These enlarge and rupture the nuclei in which they form, and ultimately the cell bursts. As the disease progresses the tissues of the host disintegrate and liquefy and the skin, which by now has become very fragile, ruptures. The liberated fluid contains myriads of virus-containing polyhedra. These are insoluble in water and resistant to ordinary bacterial decay. If diseased insects are left to decay in water, the polyhedra eventually form a layer at the bottom of the vessel. In nature, the polyhedra become ingested with food plants and dissolve in the alkaline juices of the gut of susceptible insect species. Though they are insoluble in water, polyhedra may dissolve if the pH falls below 5 or rises above 8·5. It appears that their dissolution in the gut and the consequent liberation of occluded virus particles is not enzymic. Analyses of polyhedra reveal traces of silicon (e.g., 0·1–0·3 per cent). Solution of silicates incorporated in the structure of the polyhedra by alkaline agents in the gut may account for dissolution of the polyhedra. From the gut the viruses enter susceptible tissues and disease develops usually 10–12 days after ingestion of the virus. In addition to infection via the mouth there is much evidence to suggest that nuclear polyhedrosis viruses may be transmitted through the eggs. This may be possible because infection does not always lead to acute disease and viable eggs may be laid by diseased females.

Viruses that cause cytoplasmic polyhedroses

Many cytoplasmic polyhedroses are known. Most affect only Lepidoptera. As in nuclear polyhedroses the virus particles are occluded in polyhedral protein crystals, but these do not dissolve completely in weak alkali. A rather sponge-like residue remains which is pitted with the sockets in which the nearly spherical RNA-containing virus particles were formerly located. The individual virus particles are regular twenty-sided (icosahedral) bodies 60–65 mμ diameter.

The development of cytoplasmic polyhedroses is different from that of nuclear polyhedroses. The viruses multiply only in cells of the gut, usually the epithelial cells of the mid-gut, from which they spread to fore and hind guts. The skin is not attacked, and as a result the body does not become fragile and burst. The polyhedra appear in the cytoplasm of infected cells. They vary much in size, even within the one insect, but are clearly visible with the light microscope. Though the polyhedra themselves are located in the cytoplasm, recent autoradiographic work has indicated that the viral RNA may in fact be synthesized within the nucleoli of infected cells. From thence it presumably passes into the cytoplasm where the protein coat of the virus particle is formed. The polyhedra in the gut wall frequently show through the dorsal integument of the host as pale yellow or whitish areas. Infected larvae develop more slowly than do healthy ones, and thus are smaller; they have reduced appetites, and sometimes disproportionately large heads or long bristles. When the disease is far advanced large numbers of polyhedra are liberated in the lumen of the gut from which they may be regurgitated or voided in faeces. Besides spread in this way there is almost certainly hereditary transmission of virus via the egg.

Viruses that cause granuloses

Granuloses have so far been encountered only in larvae, and very occasionally pupae, of Lepidoptera. The diseases are characterized by the appearance in the tissues of millions of granules, so small (*ca.* 300–500 mμ diameter) as to be barely visible beneath the light microscope. Each granule is a capsule that invests one or occasionally two rod-shaped virus particles which contain DNA. The granules occur in both nuclei and cytoplasm of the host's cells of which those of the fat body seem to be the main site of virus development. As the disease progresses other tissues are affected, and normally these include the skin. In this event the marked liquefaction of the internal tissues after death, usually 4–20 or so days after infection, is followed by rupture of the fragile body wall, and the symptoms of the granulosis then closely resemble those of nuclear polyhedroses.

Like polyhedra the granules consist of protein arranged in a paracrystalline lattice, and usually within each capsule a solitary virus particle lies inside two membranes. The capsule, which is resistant to ordinary putrefactive processes, is soluble in various acids and alkalis. The function of the capsule appears to be the same as that of polyhedra, namely the protection of the otherwise labile virus particle(s) after release from the host body.

Granulosis viruses in general appear to remain infective for up to two years and even may survive in the soil for this time with little deterioration. Percolating water does not readily remove them from the soil. A granulosis virus has been found to remain infective on leaves for up to four months. Spread of granuloses in nature among individuals of the same generation is mainly by ingestion of contaminated food material, but transmission from one generation to the next can be achieved by way of the eggs.

Spindle viruses

Viruses of a new type—the spindle viruses—have recently been discovered in larvae of the cockchafer *Melolontha melolontha*. They are characterized by the formation in the cytoplasm in susceptible host cells of inclusion bodies of which some are spindle-shaped and 0·5–12 μ long and others ovoid and up to 20 μ long.

Viruses that cause non-inclusion diseases

The number of non-inclusion or free viruses known in insects and other invertebrates has increased rapidly in recent years, and is now rather more than a dozen. The majority of these are located in the cytoplasm of host cells. Only two—densonucleosis of *Galleria mellonella* and the flaccidity virus of *Antheraea eucalypti*—are known to have an affinity with the nucleus. Other free viruses include several that infect bees, for example acute (ABPV) and chronic (CBPV) paralysis viruses, sacbrood virus (SBV), and one concerned in the causation of European foulbrood; transparency or 'wassersucht' virus of beetles; paralysis virus of crickets; the virus causing 'clear heads' disease of silkworms; lethargy virus in cockchafer; Malaya disease of Indian rhinoceros beetles; a virus disease of *Cirphis unipuncta*; 'iridescent viruses' in *Tipula* (TIV), *Chilo* (CIV), *Sericesthis* (SIV), and mosquitoes (MIV); and hereditary sigma virus of *Drosophila*. Paralysis diseases of crabs and mites are also caused by free viruses.

Non-inclusion viruses are a rather heterogeneous group and vary among other respects in the shape of the virus particles and the sort of nucleic acid contained. The majority are more or less spherical, but CBPV particles are slightly elongated and variable in size, and the Malaya disease virus is rod-shaped in its mature form but spherical at an earlier stage of organization of the particles. Densonucleosis virus, TIV, SIV, and CIV contain DNA, but transparency, *Antheraea* flaccidity, CBPV, ABPV, sacbrood, and European red mite viruses contain RNA. The sort of nucleic acid has not been determined for all.

TIV has received much study. It causes a disease of larvae of the crane fly (*Tipula paludosa*) in which the body fluid becomes brilliantly iridescent owing to the presence of microcrystals that are built up of crystalline arrays of rather large (130 mμ diameter) virus particles. The latter in addition to DNA contain lipid and chiefly protein. Though DNA-containing, they lie in the cytoplasm of the host's cells. At a late stage of the disease a quarter of the dry weight of the insect may be virus. This is the highest proportion known for an animal disease. Cross inoculation tests with this distinctive virus have shown it will infect other Diptera as well as Lepidoptera and Coleoptera.

Drosophila flies infected with Sigma virus become permanently paralysed if exposed to carbon dioxide in doses that merely anaesthetize uninfected flies. This virus is passed on from generation to generation.

Arboviruses

Arboviruses are acquired by blood-sucking arthropods (mosquitoes, sand-flies, certain gnats and ticks) when they feed on vertebrates in whose blood there is a sufficiently high level of virus. The viruses multiply, become established in the salivary gland, and subsequently are transmitted back to susceptible vertebrates, mainly wild animals including rodents, birds and reptiles. 204 arboviruses had been catalogued up to 1967, and of these about 75 cause disease in man. The nucleic acid in all arboviruses for which information is available is RNA, and the shape of the majority of arboviruses is spherical. Their antigenic relationships allow about three-quarters of known arboviruses to be distributed among a number of antigenic groups, more than half of which include human pathogens. The human pathogens are found mainly in groups A and B.

Typical arboviruses are generally believed to produce no cellular damage in the arthropod host. In susceptible vertebrates the symptoms of infection range from influenza-like ailments to severe disorders of the central nervous system and from mild to severe haemorrhagic disease. Arbovirus diseases in man include St. Louis encephalitis, Western encephalitis, Venezuelan equine encephalitis (that also affects horses), dengue, and yellow fever. Epidemics of arbovirus diseases sometimes cause considerable public concern. Though vaccination has proved effective in controlling yellow fever, it is possible that for arbovirus diseases generally, the best means of control is reduction of populations of the appropriate arthropods. The effective reservoirs of arboviruses are not generally known, but might include ticks, in which life-long infections may occur, and in temperate regions certain vertebrates such as hibernating rodents.

Plant pathogenic viruses in insects

Insects serve as vectors for many viruses that cause disease in higher plants. Some insects act in this way merely by becoming contaminated about the mouth parts whilst biting the plant. Others act by introducing virus-containing fluid into the plant. Some of these insects merely transmit the virus particles that they have acquired from infected plants on which they have fed, but some leaf hoppers and aphids actually become infected by the viruses that cause disease in plants and within their bodies the viruses multiply. The viruses of wound tumour, maize mosaic, maize rough dwarf, and pea enation mosaic are examples. In all these, virus particles have actually been seen in cells of the insect with the aid of the electron microscope. In a number of instances the infections have been shown to cause cytopathic changes, or to reduce the fecundity or longevity of the insect.

Some general aspects of the biology of viruses that infect insects

Two different viruses may concurrently infect the one insect. It has been found in the laboratory that administration of even heat-inactivated granulosis viruses may enhance the virulence of nuclear polyhedrosis viruses, though no marked effect has been found with the converse arrangement. The feeding to an insect of a virus not known to occur in it can stimulate a latent virus to activity. This can be a complicating factor in experiments on cross infection. Latent viruses are extremely frequent in insects, and a variety of factors ('stressors') which can induce a state of stress in the insect are effective in activating them. Such stressors include unfavourable temperatures or humidities, overcrowding, unsuitable food, feeding with certain chemicals, and inoculation with a foreign virus. Stressors probably act indirectly through the metabolism of the insect rather than directly upon the latent virus. The physical state of the latent virus—for example whether it is in a provirus or fully formed condition—is unknown. Viruses which normally form inclusion bodies probably do not do so in their latent state.

Natural and applied control of insect pests by virus diseases

Spontaneous virus diseases are commonly responsible for severe outbreaks of disease in insect populations in the field and their recurrence each season may contribute to pest control. Though outbreaks may result from acquisition of new infections it is possible that latent infections that become activated by stress are of more importance.

There are several examples of the successful use by man of viruses to control insect pests, chiefly those which feed openly on the host plant. A number of circumstances serve to heighten the chance of success in any given instance. The virus used should be highly virulent and should kill the insect quickly enough to prevent it doing much damage before it dies. Because nuclear polyhedroses and most granuloses result in liquefaction of the body contents and disintegration of the skin, they offer the best chance of further liberation and spread of the virus. It is advantageous if the insect to be controlled is one which is gregarious and feeds openly on the foliage. As in all infectious diseases, a high host population density favours spread of the disease. Inclusion body viruses offer the advantage that the inclusion body protects the virus and so allows its longer storage and extends its retention of infectivity once distributed in the environment. Virus preparations have been applied as dusts or sprays. Administration of a number of insect viruses to man and other mammals have revealed no harmful effects, but in view of the host specificity known among viruses that infect mammals it would be unwise to generalize from the limited data available. The very existence of arboviruses argues for caution. Although some insect viruses have been successfully produced in tissue culture, current technology virtually requires that viruses for control purposes must be produced by rearing and infecting host populations in the laboratory.

BACTERIOPHAGE

Introduction

Enormous numbers of different bacteriophages (phages) are known since almost all species of bacteria, including actinomycetes, are susceptible to infection by specific ones. Bacteriophages exhibit high specificity of action, each being able to infect only one group of closely related organisms, usually a single species (although the T phages of *Escherichia coli* will also attack *Shigella sonnei*), or occasionally only a few strains of a single species. This highly selective activity provides a very useful method for typing bacterial isolates, and phage-typing methods are widely used in modern epidemiological studies.

Phages together with their host bacteria provide a model which is convenient for the intensive study of host–parasite relationships and virus multiplication. In particular extensive studies of the T-even coli phages (T2, T4 and T6) have contributed greatly to present-day knowledge of the genetics of micro-organisms (p. 120).

Demonstration of bacteriophages

Bacteriophages can be isolated from many natural sources such as faeces, sewage, and polluted water. Like all viruses, phages multiply only within living cells and since they can be seen directly only by electron microscopy, phage activity must usually be demonstrated by indirect means. When a drop of a bacteria-free filtrate of a fluid containing phage particles is added to a young broth culture of susceptible bacteria, the culture will clear visibly within a few hours due to lysis of the bacterial cells. The bacteria-free filtrate can be used to infect a second broth culture and this process may be continued indefinitely.

The lytic activity of phage was first demonstrated on solid media where, on a culture plate seeded or carpeted with susceptible bacteria, phage growth and activity is indicated by cleared areas known as plaques (Fig. 8.18). These plaques behave like colonies; they may show differences in size and general morphology and can be 'picked off' by touching with an inoculating needle and transferred

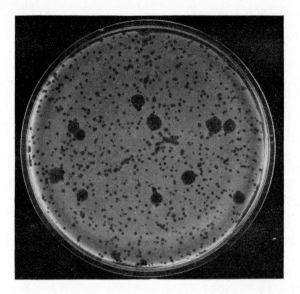

Fig. 8.18 Plaques of different sizes caused by the growth of at least two bacterio-phages on a carpet of actively multiplying bacteria ($\times \frac{1}{2}$). Each plaque arises from the growth of a single phage particle initially inoculated on to the culture. (A. H. Linton)

to a fresh culture of susceptible bacteria. Moreover, a single phage particle produces one plaque by multiplication and liberation of many phages which attack and lyse neighbouring bacteria. Hence, fluids containing phage particles can be titrated by plating measured volumes of suitable dilutions of the fluid with sensitive bacteria. The plaque count is analogous to a colony count in bacterial counting methods.

To study the physical and chemical properties of a phage, the preparation needs to be as free as possible from host cell material. Phages are usually concentrated and purified by growth in liquid cultures of host bacteria; the virus particles are then separated from bacterial debris by differential centrifugation and repeated washing.

However, the presence of phage is not always evident since latent infection may occur without bacteriolysis. Accordingly, phages are subdivided into two main groups by the relationship they establish with their bacterial hosts. *Virulent* or *lytic* phages always produce lysis of the host cells. *Temperate* or *symbiotic* phages may lyse the host, but sometimes their genetic material becomes associated with that of the host cell as *prophage* and is transferred to daughter cells on division. Thereafter a small proportion (10^{-2}–10^{-7} per cell generation) of the progeny cells lyse to release free phage. Hence bacterial cultures which carry temperate phages are termed *lysogenic* and the process by which the phage enters the prophage state is known as *lysogenization*.

Characteristics of bacteriophage

Although spherical and filamentous phages are known, most larger phage particles are tadpole shaped, consisting of a head and a tail (p. 235).

The proteins which make up the head membrane, tail core, sheath, and fibrils are all distinct from each other and are antigenic (p. 562). Phage preparations exhibit high serological specificity without cross-reactions with host antigens. Complement-fixing antigens (p. 572) have been recognized on the head protein but antibody to this protein does not neutralize the infectivity of phage. Some antigens have been demonstrated in the tail, which produce neutralizing antibodies that are thought to inactivate the phage irreversibly by blocking a specific site on the phage tail, but they do not destroy the phage.

Phage particles are organized structures, exhibiting general stability of type, but since their properties are controlled by genes, they are liable to change through mutation. Phage mutation may yield mutants which differ from the parent in host range, plaque type, and many other properties.

When a bacterium is infected by two closely related phages, *genetic recombination* (p. 123) can occur in which some of the new phage particles liberated at the end of the reproductive cycle may possess the combined properties of the two original phages. These properties will then subsequently be passed on to new phages during further reproductive cycles. A detailed consideration of phage genetics is given elsewhere (p. 120).

Virulent bacteriophage

If a very large number of phage particles attack a susceptible bacterial culture, there may be so many points of attack on each cell that the cells may disrupt, the contents bursting out through the numerous holes in the cell wall, made by the phages. This is described as 'lysis from without'. When the number of phages is few, they are able to infect the bacterial cells without causing immediate lysis. Then, instead of the bacterial cell synthesizing its own DNA, its metabolic machinery produces components for phage DNA and protein structures. After a short interval of 15–30 minutes, many new phage particles are assembled inside the cell and are released by the cell bursting; this is 'lysis from within'. The stages in this infective process are similar to the multiplication cycles of other viruses, and may be considered as follows: (1) adsorption of free phage; (2) penetration of phage nucleic acid; (3) intracellular development; (4) maturation and release of new phage (Fig. 8.19).

Study of the characteristic sequence of events during a viral multiplication cycle entails making various determinations at precise times after initial infection. To do this, the infection must be *synchronized*; otherwise infection of different cells is spread over a long period of time during which new virus particles, released when some infected cells lyse, will infect other cells. Synchronization can be achieved by allowing adsorption of virus to bacteria for only a few minutes and then removing the unadsorbed virus. When it is desired to study the multiplication cycle in cells infected by single virus particles, 'one-step conditions' are employed; once adsorption is complete, the virus cell mixture is diluted into a large volume of nutrient broth so that progeny viruses released

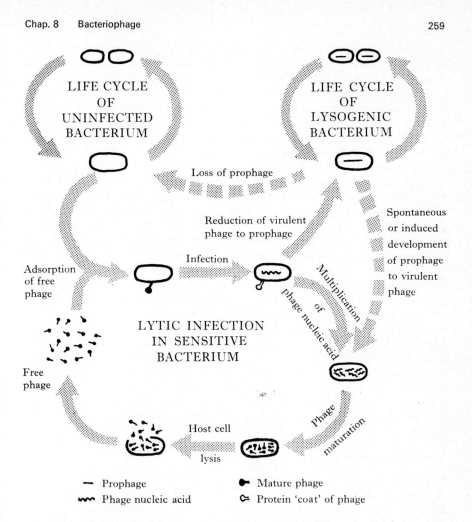

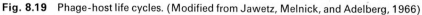

Fig. 8.19   Phage-host life cycles. (Modified from Jawetz, Melnick, and Adelberg, 1966)

later are very unlikely to become adsorbed to the uninfected cells because they are so sparse.

Adsorption of free phage

For adsorption to take place, the phage particles must first collide with the cell, then bonds must be established between the two surfaces. Under optimum conditions, this process is very rapid, almost all phages being adsorbed within a few minutes. Adsorption is usually irreversible but can sometimes be reversed by varying conditions such as pH and temperature. Frequency of collision between phages and bacteria depends mainly on the concentrations of virus particles and cells. When there is an excess of phage particles, as many as 300 may adsorb to one bacterial cell. Frequency of collision is also affected by the size and mobility of the virus and the mutual attraction or repulsion brought about by electrostatic forces. The formation of bonds after collision is dependent on the chemical structure of the two surfaces. As with the union between an antigen and antibody, the presence of complementary chemical groups on the two surfaces is essential, e.g., with coli T1 amino groups on the viral surface bond with receptors containing acidic groups on the cell surface. The receptor sites of the bacterial cell render it highly specific for a small range of phages and cannot be found in the walls of bacteria resistant to those phages. The receptors may be situated in different layers of the wall, e.g., receptors for phages T2 and T6 exist in the lipoprotein layer of the wall whereas phages T3 and T4 adsorb to receptors in the lipopolysaccharide layer. Receptor substances have been separated from cell wall preparations by gentle chemical extraction and have been shown to combine specifically with phage active against the whole cell. The phage surface must also be modified before adsorption can occur; this may be by attachment of positively charged cations (e.g., Mg^{++} or in some cases, L-tryptophane. The extent of modification varies from one phage to another but each phage is specific with regard to the co-factor required for adsorption.

Some phages (T-even phages) are equipped with fibrils which extend from the base plate at the tip of the tail. These phages adsorb to receptors on the bacterial cell by attachment of their fibrils to the cell surface, leaving the virus particle in a characteristic position at right angles to the cell wall and with the head sticking out (Fig. 8.20A). The receptors for the very small phages f1 and f2 are only found on the surface of F^+ cells; these phages attach to the specific pili which confer the male character on the F^+ cells.

Bacterial mutation can change the nature of the chemical groups on the receptors and by so doing prevent phage adsorption, rendering the cells phage-resistant.

Penetration of phage nucleic acid

Phage nucleic acid is injected into the bacterial cell usually by insertion of the tail core in the cell wall and contraction of the sheath, leaving the empty protein coat outside. The T-even phages are known to hydrolyse the mucopeptide of the bacterial cell wall by means of a phage lysozyme which is attached to the tip of their tails. The dissolved hole may help in the insertion of the tail core in readiness for injection of nucleic acid, but such a hole is apparently not essential since some other phages which possess no lysozyme are still able to insert their tail core. Holes produced by phage lysozyme may cause a leakage of cell contents; normally the holes are repaired fairly quickly but if a very large number of phages produce holes in the same cell within a short time, there may be so much leakage that the cell lyses without virus multiplication.

Fig. 8.20 (A) Electron micrograph of T2 bacteriophage adsorbed on to empty cell walls of *Escherichia coli* (× 40,000). Each phage particle is adsorbed by its tail, the end of which has been introduced through the cell wall (E. Kellenberger). (B) Electron micrograph of a thin section of *Escherichia coli* infected with T2 bacteriophage thirty minutes before fixing (× 50,000). Mature phage particles are visible, their polyhedral heads being well preserved. The bacterial cell wall, detached from the cytoplasm, is clearly visible. (Kellenberger, E. *et al.* (1958) *J. Biophys. Biochem. Cyt.* **4**, 671)

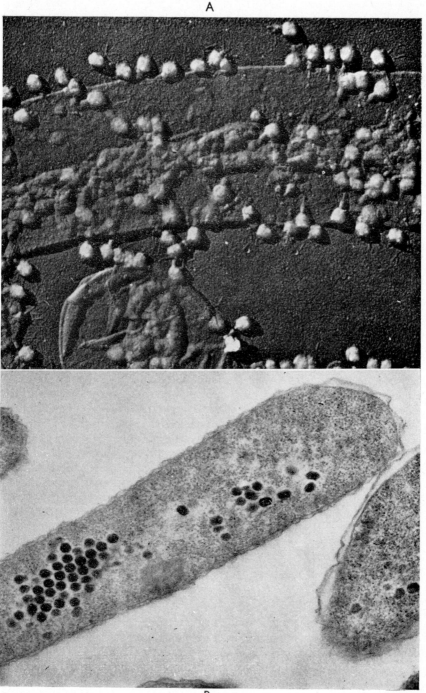

A

B

Electron microscopy of T-even phages has shown that irreversible adsorption of a phage always results in contraction of the tail sheath. The base plate and fibrils are held firmly against the cell surface and contraction of the sheath pulls the head towards the cell, causing the hollow tube or core to penetrate the cell wall reaching through to the cell membrane but probably not piercing this. Rather like primitive muscle, the sheath contains the source of energy for inserting the tube through the wall and does not depend on an energy supply from the bacterium. Other phages which are not equipped with contractile sheaths are still able to effect penetration with their tube but at a much slower rate than the T-even phages and are dependent on the host cell for the energy required to achieve entry of their tube or, in the case of the tiny spherical phages, of their capsid.

Once the cell membrane is reached, the release of phage DNA is triggered off and the contents of the head are expressed on the membrane through which the nucleic acid molecules are able to pass into the cell. The empty protein head and tail structures remain attached to the bacterial cell wall but they take no further part in the infective process.

Intracellular development

If the bacterial cells are disrupted soon after the injection of phage nucleic acid, the presence of phage cannot be demonstrated by plaque formation on a susceptible plate culture. This short period before phage can once more be demonstrated is called the *eclipse phase.*

Immediately after injection of phage DNA into the host cell, synthesis of a new protein required by the phage for multiplication commences. The first products, often referred to as 'early proteins', include certain enzymes which are required for later construction of the complicated molecules peculiar to the phage, e.g., kinases for the formation of nucleotide triphosphates, and a phage DNA polymerase.

Subsequently during the eclipse phase, 'late proteins' appear which include the protein sub-units for the phage head and tail structures. Construction of new phage DNA is accomplished with some components from the medium and some from degraded host DNA. Genetic recombination, in which phage particles undergo random exchange of genetic material can occur at this stage (p. 116).

Maturation and release of new phages

The assembly of the newly-synthesized components to form a complete or mature, infectious virus particle is known as maturation. Maturation of phages involves assembly of many components and therefore occurs in several steps. Viral DNA molecules are first condensed into large particles surrounded by a membrane and resembling phage heads. These are recognizable before maturation is complete in thin sections of infected bacteria (Fig. 8.20B). Phage heads are completed by further condensation of capsid monomers around the DNA condensate, but the heads are not yet stable since premature lysis of the cells at this stage by artificial means results in the DNA escaping from the protein coat of the new head. The hollow tube and base plate appear to be assembled next, followed by the contractile sheath and finally the fibrils are added. Each step of the phage assembly is controlled by genes in the phage nucleic acid, maturation being a sequential process in which each step recognizes or requires the previous step. The assembly can be achieved *in vitro* by adding the components in the correct combination.

For almost all phages, synthesis of phage components and assembly of mature phages continues until the bacterial cell bursts. The cell wall is weakened by the action of phage lysozyme and the cell finally bursts releasing the newly formed infective phages into the surrounding medium. One exception to this method of release is the tiny phage M13 which appears to leak out of the cells by an unknown mechanism, without killing the host cells.

Temperate bacteriophages

Unlike virulent phages, temperate phages enter into a symbiotic relationship with their hosts and do not usually produce lysis (Fig. 8.19). They gain entry to bacterial cells in the same way as virulent phages but thereafter do not interfere with the synthesis of bacterial components nor do they usually produce any visible changes in the culture.

Although temperate phages do not lyse their hosts, their presence can often be demonstrated because the occasional phage particle takes on a virulent role and as a result a very small number of cells in a culture may lyse. By plating an excess of sensitive indicator strain for the phage with a small inoculum of the lysogenic culture, any release of phage in virulent form will be clearly shown by plaques in the carpet of sensitive cells around the affected colonies of the lysogenic culture.

The prophage (p. 257) exists as one or a few particles per cell, it behaves like a host gene and is reproduced synchronously with the host nucleus, to which it is bound. The mechanism of attachment has been studied in the case of phage λ where it has been found that the small circular phage chromosome is integrated into the much larger circular chromosome of the bacterium at a point where a short sequence of homologous nucleotide pairs occurs in the two chromosomes. One prophage may be substituted for another in a lysogenic culture following infection with a second temperate phage.

Infection by temperate phages confers resistance on the host bacterium to superinfection with the same or related phages. This immunity is clearly different from the resistance of certain bacteria to virulent phages through failure to adsorb these. The occasional spontaneous change in prophage which allows the production of virulent phage particles can be greatly accelerated by the use of various agents, e.g. ultra-violet light.

The phenomenon of *transduction* depends on the activities of temperate phages. When a temperate phage from one lysogenic culture is allowed to infect a second bacterial culture, it is possible for a small number of closely linked genes to be transferred from one bacterium to another. The transferred material remains as a stable feature in the recipient bacterium; properties such as resistance to antibiotics can be transferred in this way.

Some temperate phages can confer new properties on the host cell and this is known as *phage conversion*. The prophage behaves as a host gene, each cell receiving the prophage acquiring the new property. For example, virulent toxin-producing strains of *Corynebacterium diphtheriae* have been isolated from cultures of avirulent non toxin-producing strains as a result of infection with specific phages.

Medical applications

Phage-typing

The success of phage-typing depends on the very high specificity of phages for their host bacteria. Phage-typing methods were first introduced in 1925 to distinguish strains of salmonellae. As a result of subsequent development of the methods, the type distribution of *Salmonella typhi* is now known for most of the civilized world and routine typing contributes greatly to our knowledge of the epidemiology of typhoid fever. The phages employed for *S. typhi* are active only against freshly-isolated strains possessing the Vi-antigens and the method is termed 'Vi phage typing'.

Phage-typing is widely used for distinguishing strains of *Staph. aureus*. This enables the sources of a particular phage type of staphylococcus responsible for an outbreak of sepsis, to be traced and eliminated. Staphylococcal phage-typing presents quite different problems from the typing of salmonellae. Whereas there are only a few hundred typhoid carriers in England, there are many millions of staphylococcal carriers and thus a much greater variety of types are found among staphylococci than typhoid bacilli. Despite the fact that staphylococcal phages are not as stable as those of *S. typhi*, the method does permit a fairly reliable identification of individual strains which is very valuable in short-term investigations.

Phage-typing of *Staph. aureus* is technically more difficult than the typing of salmonellae; the staphylococcal phages are derived from lysogenic strains of *Staph. aureus* and selected on the basis of their range of host specificities. Twenty-two phages are included in the basic set of typing phages; the phages are numbered by international agreement and are divisible into four groups of antigenically related phages. Almost all strains derived from human sources belong to groups I, II, or III. Any one strain of *Staph. aureus* is likely to be susceptible to several phages, usually of the same antigenic group. Some strains of *Staph. aureus* are resistant to all the phages used and cannot be typed but new phages are added to the basic set from year to year, thus increasing the range and value of the typing method.

The phages are prepared for use from filtrates of infected susceptible host cultures known as *propagating strains*. The filtrates are purified and then titrated to determine the dilution required to just produce confluent lysis when spotted on to a culture plate seeded with the host bacterium; this dilution is referred to as *Routine Test Dilution* (or R.T.D.) and is the dilution normally employed in the typing method.

A culture plate is seeded with the staphylococcus to be typed, then each of the phages of the standard set is dropped on to the agar surface in a standard pattern, often within a grid marked on the culture plate (Fig. 8.21). After drying, the plates are incubated, usually at 30°C. The phage type of the test strain is then designated by the numbers of the phages which have produced lysis, under the standardized conditions of the test, e.g., phage type 84/85 is a strain of *Staph. aureus* which has recently caused several serious outbreaks of hospital infection, and which is lysed only by phages 84 and 85. Some strains which are not lysed by phages at their routine test dilution, may be typable when more concentrated phage suspensions are employed (usually 1000 × R.T.D.).

Phage-typing is a very valuable method but since it is an empirical procedure, it is open to many sources of error. For example, if phage lysates are too concentrated, non-specific lysis may occur. Moreover, since most staphylococcal propagating strains are lysogenic, some harbouring as many as five different temperate phages at a time, concentrated lysates are likely to contain both the

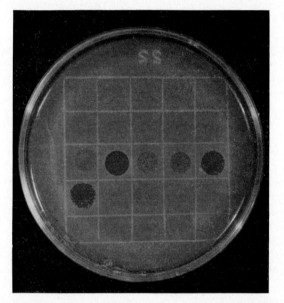

Fig. 8.21 Phage-typing of a strain of Staphylococcus (× ½). The strain to be typed is carpeted on to the surface of a nutrient agar medium and specific phages are spotted on to the centre of the appropriate squares. The phage-type is determined by the pattern of sensitivity to the various phages. (K. B. Linton)

propagated typing phage and some of the temperate phages. These temperate phages may themselves become lytic or they may alter the susceptibility of the test organism to the typing phage.

Phage therapy

Because phages kill bacteria but are harmless to man, d'Herelle regarded the presence of phages as an important factor in recovery from infectious disease and urged its therapeutic administration in a variety of infectious processes. Phage therapy has been attempted on many occasions, e.g., oral administration in enteric infections, instillation into the urinary bladder in cases of cystitis, local application on boils, etc., but the results have been disappointing. The great variety of phage-specific strains of bacteria within a species and the rapidity with which phage-resistant forms can develop are probably important limiting factors. Hence, with the possible exception of cholera, it appears that phage therapy is of little value.

BACTERIOCINS

The bacteriocins are a class of bacterial products which exercise a very specific bactericidal activity; they have some relationship with bacteriophages and possibly also with sex factors (p. 114). They were discovered and first studied in the intestinal Gram-negative organisms but many other bacteria, both Gram-negative and Gram-positive, produce them. They take their specific names from the organisms producing them, e.g., colicins from *Escherichia coli* (Fig. 8.22), pyocins from *Pseudomonas pyocyanea* (*aeruginosa*), megacins from *Bacillus megaterium*.

All true bacteriocins are proteins but some may become associated with the lipopolysaccharide of the cell wall of the producer and form a complex which can be split only by drastic methods. Some bacteriocins, particularly pyocins, in electron micrographs resemble bacteriophage tail structures. They may be considered as products of defective phage genomes, able to code for only part of the phage structure. The initiation of bacteriocin synthesis kills the producer cell.

Like bacteriophages they are adsorbed to specific receptors on, or possibly under, the bacterial surface from which they exercise a lethal action by a mechanism which has not, so far, been elucidated. The target of the lethal effect varies between bacteriocins. They may degrade the cell's DNA, or inhibit DNA and RNA synthesis, or interfere with protein synthesis. Unlike bacteriophages they are not regenerated in the cells to which they have been adsorbed and, as a rule, they do not lyse them in the process of killing.

While adsorption of the bacteriocin 'particles' takes place very rapidly, there is an interval between adsorption and the irreversible bactericidal action during which the cell may be rescued by inactivation of the bacteriocin already adsorbed to the cell with trypsin.

Most, though not all, bacteriocins are good antigens and evoke in rabbits the formation of neutralizing antibodies (p. 574). Precipitating antibodies are produced also, but it is not entirely clear whether they react with the bacteriocin itself or with an associated component.

The potential or actual synthesis of bacteriocin is due to genetic determinants, the bacteriocinogenic factors, which resemble prophages but are probably not integrated into the bacterial chromosome. Cells carrying such factors are immune to the effect of the corresponding bacteriocin.

As with a prophage, the synthesis of the lethal product specified by a bacteriocinogenic factor is repressed in most cells harbouring the factor and is active in only a few of them; but the synthesis of bacteriocin can be induced, like the synthesis of phage components, by U.V. irradiation, mitomycin C and other agents. The massive amount of bacteriocin produced after induction is due

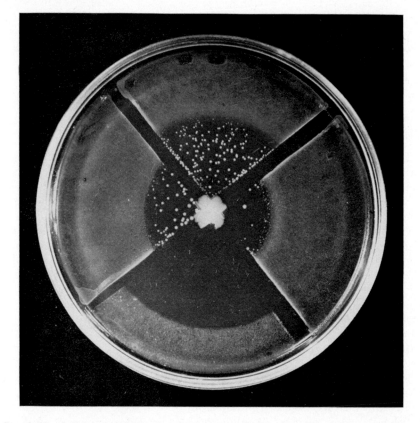

Fig. 8.22 Colicin-sensitivity of strains of *Escherichia coli*. The macro-colony of *E. coli* in the centre of the plate is an active colicin-producer, the colicin diffusing out from the colony into the medium as it is produced. This colony was grown for twenty-four hours after which each of the four quarters of the plate was seeded with strains of *E. coli* of varied degrees of sensitivity to the colicin. After incubation, zones of inhibition proportional to the sensitivity of each strain developed. Within the zones, growth of colicin-resistant colonies can be seen. (K. B. Linton)

mainly to a great increase in the proportion of organisms producing it rather than to an increase in the amount produced by individual cells.

Some colicinogenic factors (genetic determinants for colicins) can be transferred from producer to non-producer strains. The factor for colicin Ib (col I b), acts as a sex factor directing the formation of pili (p. 114) which allow cells to conjugate and thus to pass on the col factor. Other col factors are not able to cause formation of their own pili, but rely for transmission on col I b, the F factor, or on transduction by bacteriophages. Whether there is any mechanism for the transfer of bacteriocinogenic factors other than col factors is not known.

The specificity of action of bacteriocins has led to their use for bacterial typing. Whereas most phage typing schemes are based on the sensitivity of the test strain to phage, most bacteriocin typing schemes are based on the activity spectra of bacteriocins produced by the test strain. Typing of *E. coli*, *Sh. sonnei*, *Proteus* spp., and *Pseudomonas aeruginosa* has been done with encouraging results, but only occasionally have quantitative controls, as for instance, those analogous to the determination of the critical test dilution in bacteriophage typing, been rigorously applied.

VIRUS DISEASES OF FUNGI

Until the early 1960s no disease of a fungus had been proved to be due to a virus, although viruses had been suggested as the cause of certain abnormalities in fungus fruit-bodies and mycelium. Indeed several authors had cited the supposed lack of virus diseases as a character distinguishing fungi from actinomycetes.

Mushroom growers, however, had long been concerned about poor cropping and the production of abnormal fruit-bodies under conditions of intensive cultivation. It has now been shown conclusively that a number of well-known disorders of cultivated mushrooms, such as 'watery stipe' and 'die-back', are due to a virus or a complex of viruses. Electron microscopy shows that virus particles of at least two and probably three distinct types (viz., small and larger spherical particles and elongated ones) may be present in watery stipe fruit-bodies. These particles are morphologically comparable with those of certain plant viruses and are distinct from those of bacteriophages. Serological tests and density-gradient centrifugation gave similar results for all the types of virus particle observed, but improved techniques of virus isolation are desirable. Spread of the virus along infected hyphae and through anastomoses between diseased and healthy ones has been demonstrated.

Polyhedral virus particles, 25–30 mμ in diameter and containing RNA, have been isolated from a slow-growing strain of *Penicillium stoloniferum*. The viral RNA from this fungus stimulates the production of interferon (p. 243) in animals. Cultures of *P. stoloniferum* free of virus were obtained from heat-treated conidia of the infected strain. There is strong evidence that some other examples of slow growth and abnormal form of moulds is due to virus infection.

ALGAL VIRUSES

Virus infections of algal cells have only recently been recognized and are best known in the Cyanophyta (p. 346).

The blue-green algal virus (BGAV) was first isolated in 1963 from infected cells of *Lyngbya* cultured from a waste stabilization pond. It also infects species of *Plectonema* and *Phormidium* and hence has been coded as LPP-1. This virus is stable over a pH range from 5 to 11 which is similar to that within which the host *Plectonema boryanum* grows. In culture, plaques vary in size between 0·1 to 8 mm in diameter and this is considered to be due to strain variation. The virus contains double-stranded DNA with a molecular base composition of 55 per cent guanine plus cytosine.

The BGAV virus causes a lateral displacement of the photosynthetic lamellae and the virus particles remain attached to the lamellae after the whole cell has lysed. The technique of growing the virus within algae growing on agar plates revealed that the virus particles had long 'tails'. It is suggested that the virus may develop in 'virogenic stroma' in the nucleoplasm or associated with 'polyhedral bodies'. 'Ghost' virus particles can be seen attached to the cell wall after infection.

Lysis of cells occurs randomly along the filaments of *Plectonema boryanum* and eventually the filaments fragment into separate cells. The LPP-1 virus exhibits a one-step growth curve typical of bacterial viruses. The growth cycle is as follows:—a virus particle attaches to the outer wall of the host cell by its tail; remains on the outside but injects its DNA into the cell as in some phages; then multiplies in the nucleoplasm where fine threads are observable in infected cells but not in uninfected ones, and is then thought to migrate to the photosynthetic lamellae. As virus formation continues, empty spaces appear and from these the helices move into the virogenic stroma formed where the photosynthetic lamellae are displaced. Here the helices become enclosed in a protein coat and then compressed into the mature particles.

It is possible that algal blooms (p. 572) could be controlled by virus pesticides. The naturally occurring virus is widespread, having been found in 11 out of 12 waste stabilization ponds investigated in the U.S.A.; it has also been recorded in freshwater ponds in Israel.

Lysis of *Chlorella* strains has been recorded and virus particles reported and designated as 'chlorellophages' by Russian workers. The active lysis was only possible in the presence of bacteria although these had no harmful effect on the growth of the *Chlorella*.

THE CLASSIFICATION OF VIRUSES

Introduction

Early attempts at viral taxonomy were thwarted by the paucity of fundamental information about viruses. Such systems of virus classification were based largely on the more easily observed features of virus diseases, e.g., tissue tropisms, host range, symptomatology, and pathology, rather than characteristics of the virus particles, such as chemical composition and structure, which are more difficult to determine.

In more recent years the ever-increasing tempo of virological research has produced a concomitant expansion in fundamental knowledge about viruses. Hence, the more recently proposed systems of viral classification have been centred on the more basic and immutable characters of the virion, e.g., type of nucleic acid and fine structure.

An internationally agreed system of virus classification and nomenclature has yet to be formulated. Proposals presently under consideration include a system which embraces all types of viruses (plant, bacterial, vertebrate, and invertebrate) and in which viruses are given latinized binomial names; the genera, etc., being defined by a non-phylogenetic hierarchy of characters (type of nucleic acid, virion fine structure, etc.). In opposition to these proposals, others have suggested a binomial nomenclature based on the present vernacular names plus an informative cryptogram which for any particular virus could be amended as more facts are obtained. This 'vernacular name and cryptogram' nomenclature was designed to be non-prejudicial with respect to virus classification but its proponents favour the eventual adoption of the Adansonian type of classification (p. 464).

Which, if either, of these systems of classification will be chosen is a matter for speculation. Meanwhile, as the debate proceeds, virologists must continue to use the existing nomenclature (largely based on the vernacular names of virus diseases), together with the systems of virus grouping which have evolved out of necessity over the past few years.

Viruses can be conveniently divided into two main sections, the DNA and RNA viruses respectively. The former includes most bacteriophages and the larger animal viruses; the latter most plant viruses so far investigated (p. 223), together with the smaller animal ones and some phages. Other characters currently used in working classifications are capsid symmetry, presence of envelope, site of replication in the cell, reaction to ether treatment, number of capsomeres (for viruses with icosahedral symmetry), diameter of helix (for viruses with helical symmetry), particle size. Individual viruses are more often identified by serological means (p. 569).

The following groups of animal viruses have been defined using the above criteria.

DNA-containing animal viruses

Parvovirus group (also known as picodnaviruses). A recently recognized group of small (Latin, *parvus* = small), non-enveloped icosahedral viruses with particles approximately 22 mμ in diameter. Members of the group include several viruses isolated from rodents, e.g., Kilham's rat virus, and the virus of feline panleucopenia.

Papovavirus group. Named after three of the group members: the *pa*pilloma viruses (causing warts), *po*lyoma virus (causing tumours in mice) and *va*cuolating agent (found in 'normal' monkey kidneys). The virions are non-enveloped, icosahedral and 40–55 mμ in diameter. Nearly all members of the group are oncogenic (tumour producing).

Adenovirus group. First isolated from the adenoidal glands of man. Members of the group are non-enveloped, icosahedral viruses, 70–80 mμ in diameter. Most contain a common group antigen as well as a type specific antigen. Adenoviruses are generally associated with respiratory diseases and often produce latent infections. Man, ox, dog, mouse, and chicken are among the species from which adenoviruses have been isolated. Some of these isolates produce tumours when inoculated into newborn hamsters.

Herpesvirus group. The icosahedral virions of this group are approximately 110 mμ in diameter and are readily recognized by their surrounding lipoprotein envelope (Fig. 8.8). The integrity of this envelope appears to be essential for infectivity. The latter is therefore destroyed by ether treatment. The group takes its name from the virus causing herpes simplex (cold sores) in man. Other herpesviruses include those of varicella-zoster (chicken pox-shingles), pseudorabies, bovine rhinotracheitis, and infectious laryngotracheitis (fowl). Many herpes infections show latency.

Poxvirus group. These are the largest (230 × 300 mμ.) and most complex (Fig. 8.11) of the vertebrate viruses. Replication occurs in the cytoplasm of the host cell; this feature is unique since all other DNA viruses replicate in the nucleus. The various poxvirus subgroups depend largely on serological relationships e.g., viruses of the vaccinia subgroup (vaccinia, cowpox, mousepox, smallpox, etc.), and are sufficiently closely related to give common neutralizing antibodies. Some poxviruses produce tumours, e.g., rabbit fibroma and myxoma, others either produce local skin lesions (pocks) or more generalized infections with a rash, e.g., smallpox.

RNA-containing animal viruses

Picornavirus group. These small (pico = small), non-enveloped, icosahedral viruses may be subdivided according to their acid sensitivity. Acid stable picornaviruses include poliovirus, the coxsackieviruses and the echoviruses (*E*nteric *Cy*topathic *H*uman *O*rphan viruses). Acid labile members include the rhinoviruses (common cold viruses) and the virus of foot and mouth disease. Infections with picornaviruses are often inapparent (hence the term 'orphan' viruses—viruses without a disease). Diseases produced by different members of the group include infections of the gut (sometimes spreading to the central nervous system as in poliomyelitis), upper respiratory tract, meninges, skin, and heart.

Reovirus group. Non-enveloped, icosahedral viruses, 70–75 mμ. in diameter. Reoviruses have been isolated from the respiratory and enteric tract of man and other animals (*R*espiratory *E*nteric *O*rphan viruses). As their name suggests they are not definitely associated with any known diseases.

Arbovirus group. As presently defined this group is somewhat unsatisfactory since it contains a large number of viruses, many of which have different physicochemical properties. However, some members are known to have icosahedral symmetry and possess some form of envelope. The group derives its name from the fact that members are *A*rthropod-*bo*rne, i.e., arboviruses can multiply in arthropods and then be transmitted by bite to vertebrates (including man, domestic animals, birds, bats, and reptiles. The group contains more than 150 viruses including the equine encephalitis viruses and the virus of yellow fever.

Myxovirus group. The group name is derived from the fact that members have receptors for certain mucins. Hence, myxoviruses are able to agglutinate red blood cells (haemagglutination). The particles are roughly spherical, 80–200 mμ. in diameter and have a coiled-up tubular nucleocapsid (80–90 Å in diameter) with helical symmetry. Surrounding the latter is an envelope containing haemagglutinin. The particles also contain the enzyme neuraminidase which causes elution of virus from red blood cells.

There are three main serological subgroups; type A viruses cause disease in man, birds, horses, and swine, while type B and C have been recovered only from man.

Paramyxovirus group. These viruses are similar in some respects to the myxo-viruses, e.g., some produce haemagglutination and contain neuraminidase. However, the particles and helical nucleocapsids are somewhat larger, 100–300 mμ and 18 mμ in diameter respectively (Fig. 8.7). This group includes the parainfluenza viruses, mumps virus, and Newcastle disease virus. The sero-logically related viruses of measles, distemper, and rinderpest have been in-cluded largely on morphological grounds. Some members of the group produce mild infections of the upper respiratory tract while others produce more severe generalized diseases with rashes.

Rhabdovirus group. A recently recognized group of bullet-shaped viruses (Greek, rhabdos = a rod) 60 × 225 mμ in size. The virions are enveloped and have helical symmetry. Members of the group include the viruses of rabies and vesicular stomatitis (a bovine disease).

Other viruses

Other viruses (plant, bacterial, and insect), have been grouped according to principles similar to those used above for the vertebrate viruses (see Lwoff, A. *et al.* (1962); Smith, K. (1965)).

While no generally accepted group names have been established, some of the plant viruses can be grouped according to clearly defined structural and bio-logical characters, e.g., the tobacco mosaic (TMV) group distinguished by the rod-shaped virion, its relative stability outside the host plant, serological charac-teristics, and absence of transmission by vectors; the turnip yellow mosaic (TYMV) group, differing from TMV in shape of particle, transmission, and serology. Much more work is required, however, before all plant viruses can be satisfactorily grouped in this manner.

BOOKS AND ARTICLES FOR REFERENCE AND FURTHER STUDY

General Virology

ANDREWS, C. H. (1967). *The Natural History of Viruses.* The World Naturalist Series, Weidenfeld & Nicolson, London, 237 pp.
CASPAR, D. L. D. (1965). Design principles in virus particle construction, in *Viral and Rickettsial Infections of Man*, edited by Horsefall, F. L. and Tamm, I. 4th edn. Pitman Medical Publishing Co., Ltd., London and J. B. Lippincott Co., Philadelphia, pp. 51–93.
CRAWFORD, L. V. and STOKER, M. G. P. eds. (1968). The Molecular Biology of Viruses, *18th Symp. Soc. gen. Microbiol.* Cambridge Univ. Press, 372 pp.
FRAENKEL CONRAT, H. ed. (1968). *The Molecular Basis of Virology*, Amer. Chem. Soc. Monograph, No. 164, Reinhold, New York, 642 pp.
HAHON, N. ed. (1964). *Selected Papers on Virology*, Prentice Hall, Inc., Englewood Cliffs, N.J. (a collection of papers from the 18th Century Work of Jenner onwards, illustrating the development of virology), 313 pp.
HORNE, R. W. (1963). The structure of viruses, *Scient. Am.*, **209**, No. 1, 48.
KELLENBERGER, E. (1966). The genetic control of the shape of a virus. *Scient. Am.*, **215**, No. 6, 32.
NEWTON, A. A. (1970). The requirements of a virus, in *Organization and Control in Prokaryotic and Eukaryotic Cells*, Charles, H. P. and Knight, B. C. J. G., eds. 20th Symp. Soc. gen. Microbiol., Camb. Univ. Press, pp. 323–358.

WOOD, W. B. and EDGAR, R. S. (1967). Building a bacterial virus. *Scient. Am.*, **217**, No. 1, 60.

Animal Viruses

BELL, T. M. (1965). *An Introduction to General Virology*. William Heinemann Medical Books Ltd., London, 284 pp.
BROWN, F. (1968). The foot and mouth virus, *Sci. J.*, **4**, No. 2, 32.
BRUNER, D. W. and GILLESPIE, J. H. (1966). *Hagan's Infectious Diseases of Domestic Animals*, 5th ed. Baillière, Tindall & Cassell, London, 1105 pp.
BURNET, F. M. (1960). *Principles of Animal Virology*, 2nd edn. Academic Press, New York and London, 490 pp.
COHEN, D. (1966). Epidemiology of virus diseases. Chapter 8, pp. 185–206, in *Basic Medical Virology*, Prier, J. E. ed., The Williams and Wilkins Co. Baltimore.
FINTER, N. B. (1966). *Interferons*. Research Monographs, Frontiers of Biology, **2**. North Holland Publishing Co., Amsterdam, 340 pp.
PATTISON, J. H. (1967). Scrapie, *Sci. J.*, **3**, No. 3, 75.
WILSON, G. S. and MILES, A. A. (1964). *Topley and Wilson's Principles of Bacteriology and Immunity*, 5th. edn., pp. 1476 *et seq.* Edward Arnold Ltd., London.

Plant Viruses

BAWDEN, F. C. (1964). *Plant Viruses and Virus Diseases*. Cambridge Univ. Press, 4th. edn., 361 pp.
SMITH, K. M. (1966). *The Biology of Viruses*, Home Univ. Lib., London, 162 pp.
——— (1968). *Plant Viruses*. 4th edn., Methuen, London, 166 pp.

Insect Viruses

MARAMOROSCH, K. ed. (1968). Insect viruses. In *Current Topics in Microbiology and Immunology*, Vol. **42**. Springer Verlag, Berlin, Heidelberg, New York.
SMITH, K. M. (1967). *Insect Virology*, Academic Press, New York and London, 256 pp.
STEINHAUS, E. A. ed. (1963). *Insect Pathology*, Vol. 1, Chapters 12–16. Academic Press, New York, pp. 389–575.
VAGO, C. and BERGOIN, M. (1968). Viruses of invertebrates. *Adv. Virus Res.*, **13**, 247.

Bacteriophage and Bacteriocins

BRADLEY, D. E. (1967). Ultrastructure of bacteriophages and bacteriocins. *Bact. Rev.*, **31**, 230–314.
DAVIS, B. D., DULBECCO, R., EISEN, H. N., GINSBERG, H. S., and WOOD, W. B. (1968). *Microbiology*, Chapter 42, pp. 1055–1098. Harper & Row, New York and London.
EDGAR, R. S. and EPSTEIN, R. H. (1965). The Genetics of a Bacterial Virus. *Scient. Am.*, **212**, 70.
OZEKI, H. (1968). Methods for the study of colicine and colicinogeny, in *Methods in Virology*, edited by Maramorosch, K. and Koprowski, H. Vol. IV. Academic Press, New York and London, pp. 565–591.
REEVES, P. (1965). The bacteriocins, *Bact. Rev.* **29**, 24–45.
STENT, G. S. (1963). *Molecular Biology of Bacterial Viruses*, W. H. Freeman & Co., San Francisco and London, 414 pp.

Fungal Viruses

BANKS, G. T. *et al.* (1968). Viruses in fungi and interferon stimulation. *Nature,*
   Lond., **218**, 542–545.
HOLLINGS, M., GANDY, D. G. and LAST, F. T. (1963). A virus disease of a fungus;
   Die-back of cultivated mushroom. *Endeavour*, **22**, 112.
LAST, F. T. (1966). Physiology of parasitism I. *J. Indian bot. Soc.*, **45**, 185–198.

Algal Viruses

SMITH, K. M., BROWN, R. M., WALNE, P. L. and GOLDSTEIN, D. A. (1966). Electron
   microscopy of the infection process of the blue-green alga virus. *Virology*, **30**,
   182–192.

Classification

ANDREWS, C. H. and PEREIRA, H. G. (1967). *Viruses of Vertebrates*, 2nd edn. Baillière,
   Tindall and Cassell, London, pp. 432.
GIBBS, A. (1969). Plant virus classification. *Adv. Virus Res.*, **14**, 263–327.
LWOFF, A., HORNE, R. and TOURNIER, P. (1962). A system of viruses. *Cold Spring
   Harbor Symp. Quant. Biol.*, **27**, 51–55.

9

Structure, Biology, and Classification of Prokaryotic Micro-organisms

THE FINE STRUCTURE OF THE PROKARYOTIC CELL

The prokaryotic cell is a more complex structure than viruses but is generally less complex than the eukaryotic cell. It differs from the latter not only by lacking a perforated nuclear membrane surrounding the nucleoplasm, as already pointed out (p. 3), but also in the lack of complex organelles, such as mitochondria (p. 365), the absence of cytoplasmic streaming and in the structure and composition of cell walls, flagella, and chromatophores where present. None of these characters offers such a satisfactory and precise means of differentiating between the two groups as does nuclear structure. Nevertheless, considered together and in conjunction with the latter they afford impressive support for this concept of two fundamentally different groups of organisms. The existence of a definite reproductive cycle in most, but not all, eukaryotes and the general absence of any such precise cycle among prokaryotes is additional evidence for the more primitive nature of the latter groups.

Distinctive aspects of the fine structure of various groups and sub-groups of prokaryotes will be considered with their other characters in the systematic account which follows. It will be seen that differences in fine structures frequently confirm systems of classification based on other characters but occasionally lead to an entirely new treatment of a particular group. The prokaryotes include the large and varied group of the bacteria together with the blue-green algae.

BACTERIA

INTRODUCTION

Bacteria include a great variety of micro-organisms of different shape and size. A few are as large as 5 μ in diameter but the majority fall into the range of 0·15 to 1·5 μ in diameter; most measure only a few microns in length but a few are known measuring several hundred microns. These dimensions lie between those of the fungi and the viruses.

Bacteria were first described and drawn as early as 1682 by the Dutch microscopist, van Leeuwenhoek, using simple lenses of his own grinding, but a systematic study of these small organisms did not begin until early in the nineteenth century when attempts were made to classify those now recognized as bacteria. Ehrenberg (1838) subdivided them into a number of genera, including the genus *Bacterium*, in which were placed rod-like organisms dividing transversely. None of these genera can be recognized with certainty today. In 1857 Nägeli, although uncertain whether bacteria were plant or animal in nature, placed them in a new group, the Schizomycetes (fission fungi). Despite the facts that the name implies relationship with the fungi (which is now known to be remote) and that all members divide only by fission (whereas a few reproduce by budding), this term is still retained for the class which includes all orders of bacteria.

The first critical classification of bacteria was by the botanist Cohn (1872). His division of the group into six genera based on accurate morphological observation formed the starting point which established bacteriology as a science. He concluded that bacteria form a fairly closely-linked group of organisms not related to the fungi but with close affinities to certain blue-green algae. The study of fine structure has justified this view (p. 346).

A distinction was gradually established between a homogeneous group of simple forms often termed the 'true bacteria' (Pseudomonadales and Eubacteriales) and a number of widely differing groups frequently referred to as the 'higher bacteria'. Recent work has shown that this distinction is untenable, since so-called higher bacteria are not necessarily more complex than the true bacteria and many of them are more closely related to the latter than each other.

As in other branches of biology, no one system of classification is universally accepted and many attempts have been made to arrange bacteria in natural groups. A comprehensive classification of the bacteria is given by *Bergey's Manual of Determinative Bacteriology*, 7th Edn. (1957) and, with important exceptions such as the exclusion of viruses and other modifications, his general groupings will be followed here. Eleven orders are recognized.

PSEUDOMONADALES AND EUBACTERIALES

These two orders will be considered first since they have been most thoroughly investigated. Some structural features which also apply to other orders of the Schizomycetes will be indicated by cross reference.

Habitat

Bacteria have exploited every environment where an energy-supplying substrate is available. They occur under both aerobic and anaerobic conditions, are abundant in the soil and are found in both fresh and salt water. Some species are essential for the maintenance of life through their ability to break down plant and animal remains, others parasitize living plants and animals, some causing serious disease by interference with normal form and function.

Nutrition

The range of habitats occupied by bacteria is paralleled by the modes of nutrition found in each group. A few species are autotrophs, utilizing carbon dioxide as their carbon source and deriving energy either from light radiation (photosynthetic autotrophs) or from the oxidation of inorganic compounds (chemosynthetic autotrophs, p. 145). The majority are heterotrophs and include relatively non-exacting species (mostly saprophytes) and exacting forms, which cannot yet be cultured on a chemically defined medium. A few species, such as *Mycobacterium leprae*, have not yet been grown in artificial culture. Growth of bacteria in pure culture, that is in the absence of other organisms, on solid or in liquid media, is extensively employed in their study. The subject of nutrition is dealt with at greater length elsewhere (Chap. 5).

Bacterial anatomy

Methods of examination

Bacteria, owing to their minute size, can be examined under the light microscope only with objectives of high resolving power. With visible light, objects can be magnified up to 1500 diameters, but no further detail can be seen above this magnification because particles less than half the wavelength of the light used cannot be resolved. In practice only objects of diameter 300 mμ or larger can be distinguished as separate particles.

Most bacteria are transparent and have a refractive index similar to that of the aqueous fluids in which they are suspended. When they are examined in transmitted light, with the ordinary optical microscope, little detail can be seen. Staining with aniline dyes, after killing and fixing, was introduced during the later half of the nineteenth century and yielded some information on internal structure. This is still a useful method of examining bacteria. More recently observation of living cells by phase contrast microscopy (p. 7) has been used making it possible to study structures previously seen only in fixed and stained preparations. Further magnification was achieved by using ultra-violet light as illuminant, but far more information on cell structure has been obtained by electron microscopy. Our present knowledge of bacterial anatomy is derived from a combination of all these methods of investigation and Figs. 9.1A and B diagrammatically present the main anatomical features which are recognized.

EXAMINATION OF STAINED BACTERIA. Owing to the small size of bacteria little structural detail can be seen with the ordinary light microscope unless the organisms are stained. The majority stain readily with aniline dyes. Some staining techniques, such as the Gram and Ziehl Neelsen (p. 327) stains, although of great diagnostic value because of their differential staining properties for specific bacteria, reveal little internal structure. Others, such as the Feulgen stain for nuclear bodies, demonstrate specific structures. Because of its importance, the Gram stain is considered in some detail; other important stains are described in appropriate sections of the text.

GRAM'S STAIN. This stain was developed empirically by Christian Gram (1884) to differentiate bacteria in tissue sections. Bacteria are subdivided by their reaction to this stain into those which retain it, termed Gram-positive, and those which are decolourized, termed Gram-negative.

The organisms are stained with a basic dye of the tri-phenyl-methane group (e.g., crystal violet) at slightly alkaline pH, followed by mordanting, usually with iodine in a solution of potassium iodide, but sometimes with other agents such as picric acid. The crystal violet combines with the iodine to form a complex. At this stage of the process there is no difference in appearance between Gram-positive and Gram-negative organisms, the amounts of crystal violet and iodine taken up being similar. Subsequent washing of the stained preparation with a neutral (organic) solvent (usually ethanol or acetone) causes the crystal violet–iodine complex to be eluted from the Gram-negative species whilst the Gram-positive organisms remain fully stained. The difference is rendered more striking by counter-staining with another dye of contrasting colour (e.g., dilute fuchsin).

The main difference in behaviour between Gram-positive and Gram-negative organisms is thus seen to reside in the ease with which the Gram complex can be eluted. This may be due either to its being bound to specific organismal components or to differences in permeability of the cell wall of Gram-positive and Gram-negative organisms.

Many workers have looked for specific cell wall constituents which could account for the binding of the Gram complex, but none of the major constituents of the wall of bacteria is exclusive to Gram-positive organisms and it is thought that this factor alone cannot fully account for the Gram reaction. Quantitative differences in Gram staining may depend, however, on the presence of varying amounts of mucopeptide predominant in the wall of Gram-positive organisms or of lipids predominant in the wall of Gram-negative organisms.

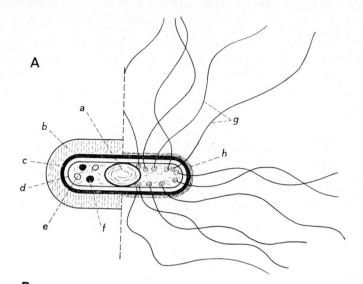

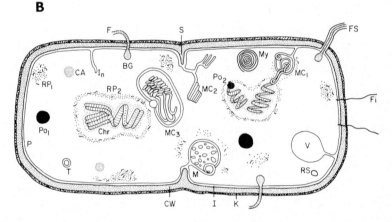

Fig. 9.1 Diagrams of typical anatomical structures common to many bacteria.

(**A**) Structures visible under the light microscope: a capsulated, non-flagellated bacillus is illustrated to the left of the dotted line and a flagellate but non-capsulate bacillus to the right. (**a**) endospore (oval and central); (**b**) capsule; (**c**) cell wall; (**d**) cytoplasmic membrane; (**e**) oil droplet or sulphur granule; (**f**) volutin granule; (**g**) flagella, each arising from within the cytoplasm, and (**h**) slime layer. (A. H. Linton)

(**B**) Structures revealed by the electron microscope in thin sections of a Gram-positive bacterium: BG—basal granule; CA—cytoplasmic aggregate; CW—cell wall; Chr—chromosome; F—flagellum; Fi—fimbria; FS—flagellar sheath; K—capsule; I—intermediate layer between cell wall and plasma membrane; In—invagination of plasma membrane; M—mesosome; MC_1—membrane complex attached to chromosome; MC_2—membrane complex in form of stacked tubules; MC_3—membrane complex containing tubules; My—myelin body; P—plasma membrane; PO_1—polyphosphate body; PO_2—polyphosphate body at end of chromosome; RP_1—ribosomes and polysomes in cytoplasm; RP_2—ribosomes and polysomes surrounding the nuclear area; RS—reserve substance; S—developing septum; T—tubule free in cytoplasm; V—vacuole. (Modified from *Bildatlas path. Mikroorganismen*, Bd. III, G. Fischer, Stuttgart)

A more likely explanation is that Gram-positivity may be the result of dehydration of the cell wall by alcohol, with consequent decrease in pore size thus
impeding the passage of the crystal violet–iodine complex and thereby making
its extraction more difficult. In Gram-positive organisms the mucopeptides
in the cell wall occur as thick continuous sheets, but in Gram-negative
organisms they are present in thin layers. This difference may be an important
factor in the Gram reaction. The integrity of the cell wall is essential for Gram-
positivity since the reaction is lost by cells damaged mechanically or treated
with lysozyme.

The ability of cells to retain the Gram stain is not a property of all living
matter, but is confined almost entirely to yeasts and some bacteria. Moulds stain
somewhat irregularly and only certain granules present in the hyphae tend to
retain the stain. Old cultures of Gram-positive bacteria often give a feeble or
negative reaction; hence Gram staining is of diagnostic value only with young
cultures.

The cell wall

The cell envelope usually consists of two components, an outer cell wall and
beneath it a cytoplasmic or plasma membrane which is actually the outermost
layer of the protoplast (p. 285). A few bacteria, including the L-forms (p. 310)
and the Mycoplasma (p. 344), have no cell wall.

The cell wall is a rigid structure which gives shape to the bacterium. Its
presence can be demonstrated by plasmolysis (Fig. 9.2B), by electron microscopy of cell wall residues, which retain their shape when freed of their contents,
and by the fact that the cellular contents assume a spherical shape when the
cell wall is enzymically dissolved (p. 284). The wall is chemically fairly inert
and does not stain by normal techniques but special cell wall stains are available (Fig. 9.2A and C). The cell walls of a representative bacterium, *Staphylococcus aureus*, account for 20 per cent of the dry weight of the cell.

Pure preparations of bacterial cell walls can be obtained by mechanical disintegration of the organism, followed by differential centrifugation. Autolysis,
osmotic lysis, and heat shock have also been used for some cell wall preparations. Chemical and enzymic methods are not used because these alter the

---

Fig. 9.2   Bacterial cell walls and associated structures.

   (**A**) *Bacillus megaterium* (× 2,700) stained with Victoria blue to reveal the cell walls
and cross-wall septa. The cytoplasm is everywhere in close contact with the cell wall
and the latter cannot be distinguished from the cytoplasmic membrane. (C. F. Robinow)

   (**B**) *Bacillus megaterium* (× 2,700) plasmolyzed in 5 per cent KNO₃ and stained as in
(**A**). The cytoplasm has separated from the cell walls thus revealing their rigid nature.
(Robinow, C. F. (1953) *Exp. Cell Res.* **4**, 392)

   (**C**) *Tetracoccus canadensis* (× 2,700) stained by Hale's boundary stain to reveal
the cell walls and cross-wall septa. The cocci show cross-walls growing in from the
periphery in two planes. (C. F. Robinow)

   (**D**) *Spirillum* sp. (× 70,000). Electron micrograph of the wall of a crushed cell. The
cell wall is shown curling over at the bottom of the photograph thus revealing the outer
face of the wall. The top of the photograph shows the inner face of the cell wall from
which part of the inner membrane has come off exposing the inner face of the outer
membrane. This consists of spherical particles arranged hexagonally in a single layer.
(Houwink, A. L. (1953) *Biochim. biophys. Acta*, **10**, 360)

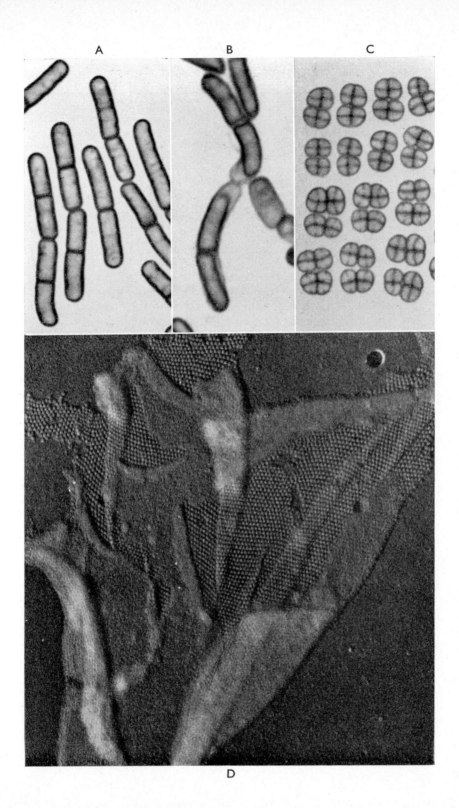

A B C

D

Table 9.1 Chemical constitution of cell walls of Gram-positive and Gram-negative bacteria.

	Gram-positive	Gram-negative
Amino acids	3 or 4 including alanine, glutamic acid, and lysine or diaminopimelic acid. (no aromatic or sulphur-containing amino acids)	Most of the amino acids commonly found in proteins; with diaminopimelic acid in addition.
Muramic acid	Present	Present
Lipids	0–2%	10–20%
Polysaccharides	35–60%	15–20%
Teichoic acids	up to 50%	none

chemical composition. Traces of contaminating protein and nucleic acids are removed by washing or by treatment with appropriate enzymes such as trypsin or ribonuclease. The preparations may be examined with the electron microscope or subjected to chemical analysis. A typical analysis is shown in Table 9.1. As already stated (p. 278), the cell walls of Gram-positive and Gram-negative bacteria differ mainly in their amino acid and lipid content.

The rigidity of the cell wall of Gram-positive bacteria is attributed to a mucocomplex, a polymer consisting of repeating units of the amino sugars N-acetyl-glucosamine and N-acetyl-muramic acid (Latin: murus, a wall) linked together by 1–4 β linkages (Figs. 1.22–24; p. 33). To each carboxyl group on the muramic acid residues is linked a peptide chain of alanine, glutamic acid, and either diaminopimelic acid (absent from the walls of Gram-positive cocci and related organisms) or lysine. The polysaccharide chains are linked together by peptide branches and these are themselves further inter-linked by peptide bridges, the latter being more extensive in the walls of Gram-positive then in those of Gram-negative bacteria. This results in a three-dimensional, rigid, multi-layered network around the organism, for which several possible macromolecular structures have been proposed (Fig. 1.25).

Within this open meshwork are found a number of matrix materials, chiefly polysaccharides. They include the immunologically active substances which determine the antigenic specificity of certain organisms, e.g., the C polysaccharide and M protein of species of *Streptococcus* (p. 322). One group of these matrix materials is the teichoic (Greek: teichos, wall) and teichuronic acids. These are highly acidic substances present only in the cell walls of Gram-positive bacteria in which they may account for up to 50 per cent of the dry weight of the cell wall. They consist of either ribitol or glycerol units joined by phosphodiester linkages. Contrary to earlier views these substances are now thought to be specifically fixed to the structural components of the cell wall by covalent bonds.

Many monosaccharides have been identified in cell wall preparations following acid hydrolysis. The distribution of these sugars varies from group to group but is constant within a particular group (Table 9.2). Rhamnose, for instance, is almost unique to species of *Streptococcus* (Lancefield groups A to G).

Table 9.2   Cell wall components as an aid to bacterial taxonomy.

	Arabinose	Rhamnose	Galactose	Glucose	Mannose	Glucosamine	Galactosamine	Muramic acid	Alanine	Glutamic acid	Lysine	Diaminopimelic acid
Actinomyces bovis	−	−	+	−	−	+	−	+	+	+	+	−
Nocardia asteroides	+	−	+	+	−	+	−	+	+	+	−	+
Mycobacterium tuberculosis	+	−	+	−	−	+	−	+	+	+	−	+
Corynebacterium diphtheriae	+	−	+	−	+	+	−	+	+	+	−	+
C. pyogenes	−	+	−	+	±	+	+	+	+	+	+	−
Streptococcus pyogenes	−	+	−	−	−	+	±	+	+	+	+	−
Lactobacillus acidophilus	−	−	−	±	−	+	−	+	+	+	+	−
Staphylococcus aureus	−	−	−	−	−	+	−	+	+	+	+	−

(+ present; ± present in some species; − absent)

The similarity between cell wall components of the genus *Nocardia, Mycobacterium,* and *Corynebacterium* suggests that they are closely related. *C.pyogenes,* although classified with the *Corynebacterium* differs from this genus both in cell wall composition and many other biological features, but closely resembles the *Streptococcus. Lactobacillus* and *Streptococcus,* genera of the family Lactobacillaceae, contain lysine but not diaminopimelic acid.

The chemical constitution of cell walls is therefore a useful tool in bacterial taxonomy and its study has revealed relationships not previously recognized as, for instance, the close relationship between *Corynebacterium, Mycobacterium,* and *Nocardia,* the first of which is usually placed in the Eubacteriales and the other two in the Actinomycetales (p. 325).

Gram-negative bacterial cell walls are chemically more complex than those of Gram-positive bacteria. They contain mucopeptides, lipids, and a full range of amino acids indicating the presence of protein, but less amino sugars than do the cell walls of Gram-positive bacteria. Certain polysaccharides of the cell walls of Gram-negative bacteria have been correlated with antigenic specificities (p. 576).

The distinction between Gram-positive and Gram-negative species is based, not only on differences in chemical composition but also on the fine structure of their walls.

Walls of Gram-negative bacteria are relatively thin, those of Gram-positive bacteria are thicker and usually amorphous. Cells walls of *Escherichia coli* are about 8 mμ thick; those of Gram-positive bacteria vary in thickness, e.g., 15 mμ in *Staphylococcus aureus,* 22 mμ in *Bacillus megaterium* and 80 mμ in *Lactobacillus acidophilus.* In shadowed preparations of *Spirillum* spp. the inner surface consists of spherical macromolecules ranging from 10 to 14 mμ in diameter, arranged in a hexagonal pattern (Fig. 9.2D). Such a regular macromolecular pattern is absent in the walls of all but a few of the Gram-negative bacteria examined.

Electron microscopy of thin sections of cell walls reveals various layers. Three layers have been demonstrated in *E. coli*, *Spirillum serpens*, and *Bacteroides* sp., in which an electron-transparent layer is enclosed on either side by an electron-dense layer. Three layers have also been demonstrated in the Gram-positive organism *Bacillus megaterium* but only in wall sections fixed with osmium or permanganate and not stained with uranyl acetate. In *Thiobacillus thiooxidans* five layers of alternating levels of density are found in a wall of 20 mμ overall thickness.

The halophilic bacteria possess a non-rigid structure external to the plasma membrane but this lacks both muramic acid and diaminopimelic acid and is therefore not a typical bacteria cell wall. The cytoplasmic membrane of these organisms is probably not subjected to differences in osmotic pressure in their natural environment as are non-halophilic bacteria and a rigid cell wall may not therefore be essential to their stability. The walls of these organisms disintegrate when suspended in pure water, suggesting that the structure is bonded by hydrogen- rather than covalent-bonds.

The protoplast

The chemical and morphological individuality of the cell wall makes it possible to remove it without damaging the underlying structures. Viable protoplasts may be obtained by completely dissolving the cell wall with the enzyme lysozyme in the presence of a stabilizing solution such as 0·2 M sucrose which protects the naked protoplast from osmotic shock. It is also possible to prevent wall formation altogether by inclusion of penicillin in the medium or, with auxotrophic strains, by depriving the organisms of the essential diaminopimelic acid. When part of the cell wall remains attached to the protoplast the whole structure is referred to as a spheroplast. Naked protoplasts always assume a spherical form even when derived from rod-shaped organisms (Fig. 9.3), indicating that the cell wall is responsible for the maintenance of the characteristic shape of the organism. The wall also prevents osmotic rupture of the protoplast in the dilute solutions in which most bacteria usually live and may protect them to some extent against harmful external agents. Isolated protoplasts are not susceptible to bacteriophage infection (p. 259) the cell wall being essential for this. However, protoplasts prepared from cells already infected with phage are able to support phage replication. Similarly protoplasts prepared from cells about to produce endospores (p. 291) proceed to sporulation.

The protoplast accounts for between 60–80 per cent of the dry weight of the whole cell; it does not contain diaminopimelic acid, a substance which is found in many cell walls. Any flagella (Fig. 9.13) which were present on the original cell are retained by the separated protoplast although the latter itself is non-motile. Isolated protoplasts are capable of metabolizing many oxidizable substrates and increase in size with time although at a slower rate than do whole cells. Their ability to reproduce or regenerate cell walls is still in doubt. Sphero-

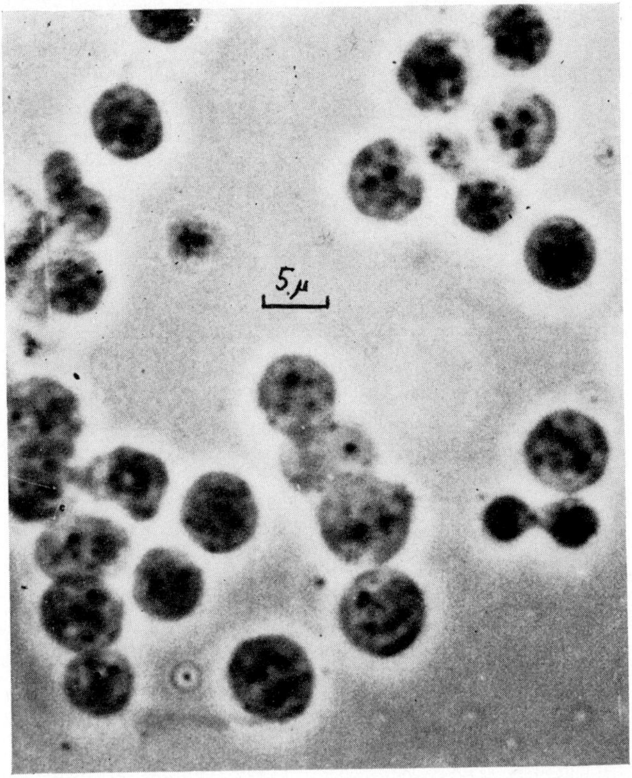

Fig. 9.3 Bacterial Protoplasts. Phase contrast photomicrographs of protoplasts of *Bacillus megaterium*. The cell walls have been dissolved away by lysozyme in the presence of a stabilizing agent (sucrose) leaving the living protoplasts which assume a spherical shape. (K. McQuillen)

plasts on the other hand are able to synthesize cell walls, suggesting that the residual cell wall material either has its own synthetic mechanisms or may contain a primer essential for cell wall synthesis.

THE PLASMA MEMBRANE. The protoplast is bounded by a plasma membrane which is distinct from the enveloping cell wall, as shown by its shrinkage from the latter under high osmotic pressure (Fig. 9.2B). It is largely lipoprotein in nature and has marked affinity for dyes.

In electron micrographs, the plasma membrane in most bacteria appears as two dense lines between the cytoplasm and the cell wall. This suggests that the plasma membrane consists of three layers, outer and inner electron-dense layers each about 2 to 4 mμ thick and an inner electron-transparent space about 3 to 5 mμ thick. This composite structure is called a 'unit membrane' and resembles similar membranes of eukaryotes, but it may not be the only type of membrane structure to be found in bacteria cells.

The plasma membrane is the effective permeability barrier of the cell regulating the inflow and outflow of metabolites to and from the protoplast. Soluble substances must first pass through the pores of the cell wall. This is not solely a function of their molecular size since large molecules may pass and smaller ones be retained. Certain components of the cell wall, such as the teichoic acids, may act as an ion exchange resin regulating the passage of positively charged ions. The passage of soluble substances through the plasma membrane is more complex. A few substances of low molecular weight can cross the plasma membrane by diffusion but the majority depend on specific proteins (permeases) associated with the plasma membrane. The solute is picked up by the carrier molecule, transported across the membrane and released into the cytoplasm. Some of these carrier mechanisms require metabolic energy for transport.

Relatively pure membrane fractions can be obtained by the lysis of protoplasts. Their chemical composition is similar to that of mammalian cell membranes and includes 40 per cent lipid, 60 per cent protein, and small amounts of carbohydrate. Sterols, however, are absent.

CYTOPLASM. The cytoplasm stains homogeneously with basic stains in young cultures but become increasingly granular with age. Unlike that of many eukaryotes, bacterial cytoplasm is relatively immobile, i.e., it does not show streaming.

The fine structure of the cytoplasm of prokaryotes reveals important differences from that of eukaryotes. The cytoplasm consists of a complicated three-dimensional network of fine fibrils stretching from the plasma membrane to the nuclear mass, but lacks the elaborate endoplasmic reticulum characteristic of the eukaryotic cell (p. 363).

In *Bacillus* species, typical of Gram-positive bacteria, structures continuous with the plasma membrane penetrate into the cytoplasm as tubular or vesicular invaginations (Fig. 9.4). These are called mesosomes or chondrioids. Evidence at present available ascribes to them a number of possible functions. They have been shown to be the principal sites of the cellular respiratory enzymes and probably play a similar role in bacteria to that of the mitochondria of eukaryotes (p. 365). In intact bacteria they are often situated at the site of the synthesis of cross-wall septa and are also attached to the chromatin material (Figs. 9.4 and 9.18). They are therefore suitably located to perform a co-ordinating role for nuclear division and cell growth under genetic control. Mesosomes have not been seen in those Gram-negative bacteria which have been examined under the electron microscope but rosette-like structures, including rods and lamellar structures without a limiting membrane, have been located near the plasma membrane. In the absence of typical mesosomes these rosettes are considered to be the sites of respiratory enzymes, but they may also be concerned with

Fig. 9.4 Electron micrograph of a thin section of *Bacillus subtilis* ($\times$170,000). The triple-layered plasma membrane is seen beneath the thicker densely stained cell wall; within the cytoplasm the pale areas are the nuclear bodies, one of which is associated with the globular mesosome which is continuous with the plasma membrane at the point of septum growth, an early stage in cell division. (W. van Iterson (1965))

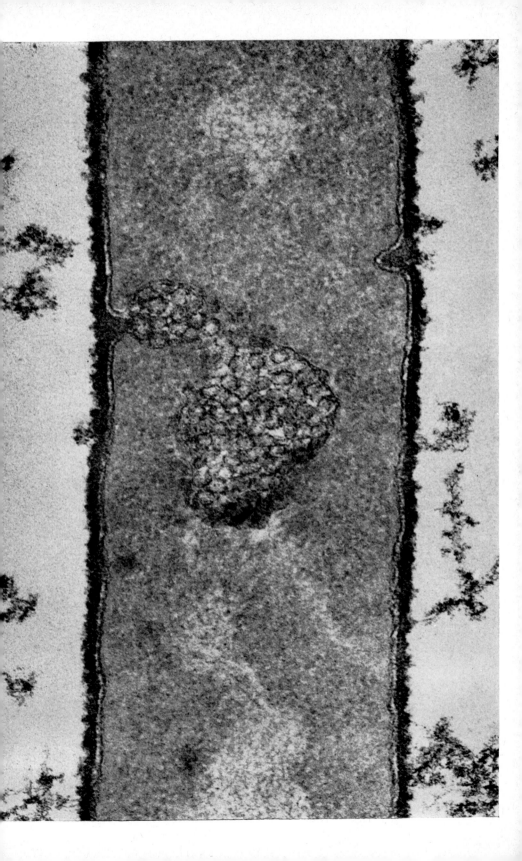

flagella synthesis. Isolated protoplasts and L-forms lack mesosomes. If bacteria are treated with lysozyme the mesosomes originally present are soon extruded into the space between the cell wall and the plasma membrane and are later released along with, but separate from, the bulk of the protoplast and the cell wall fragments. In such an isolated protoplast the synthetic capacity continues, but the cell has lost its ability to synthesize new cell wall material and to divide, properties which presumably depend on the presence of mesosomes.

In photosynthetic bacteria, the pigment-containing structures, known as chromatophores, are associated with membrane structures similar to mesosomes. Under the light microscope the pigments appear to be uniformly distributed through the cell but the electron microscope shows them to be present in distinct chromatophores, which are, however, less complex than the plastids of eukaryotes. In some species the chromatophores are homogeneous particles (e.g., *Rhodospirillum rubrum*, where they measure 60 mμ in diameter), in a few (e.g., *Rhodospirillum vannielli*) lamellar structures are present and resemble those of the blue-green algae (p. 346). Some of these lamellar structures are arranged in pairs at the periphery of the cell; others are in clusters. The pigments in photosynthetic bacteria differ from those of the blue-green algae and from the chlorophyll of plants, but are able to catalyse photophosphorylation (p. 96).

The chief centres of protein synthesis in the cell are the ribosomes (p. 65). By ultracentrifugation, particles with a sedimentation rate constant of 70 have been separated from disrupted cells of *E. coli*. These have been calculated to measure 14–17 mμ in width by 19 mμ in length, and since they account for 30 per cent of the cell's dry weight they would be expected to form a considerable part of the visible fine structure of the cytoplasm. Attempts have been made to correlate these physical properties with structures detectable in electron micrographs. Usually individual particles are not distinguishable; instead the ribosomes are seen to be integrated in long linear complexes or polysomes, closely associated with the complicated fibrillar structures of the cytoplasm. These complexes are often joined together and are possibly continuous with the plasma membrane and with mesosomes if present.

Nuclear apparatus

As already pointed out bacteria lack an organized nucleus of the type seen in cells of higher organisms but they possess DNA: a body of electron density less than that of the surrounding cytoplasm which corresponds to a nucleus; this is commonly referred to as a nuclear apparatus. Chromosomes of the type seen in cells of higher organisms are not found in bacteria but a single large molecules of DNA constitutes a single bacterial chromosome which contains a complete genome (p. 99). This is generally considered to be a continuous molecule in the form of a circle (p. 139) which may open under certain circumstances to become a simple double-ended molecule. In addition some bacterial cells possess a short length of extrachromosomal DNA in the form of a plasmid or episome. This is readily transmissible to recipient cells by conjugation or transduction and codes for characters such as drug resistance, colicin production and toxin production (p. 119).

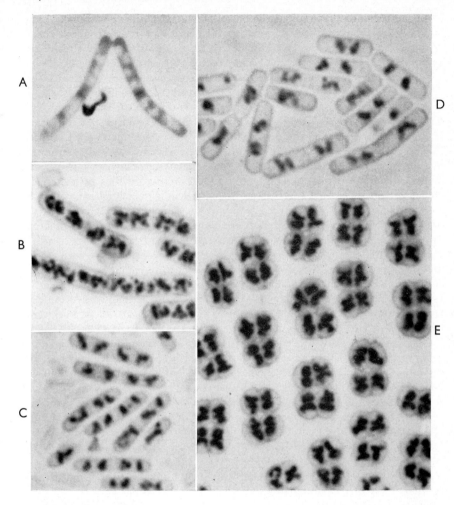

Fig. 9.5 Nuclear bodies.

(A) *Bacillus subtilis* (×3,900). In this phase-contrast micrograph, the light patches within the bacilli correspond to chromatin bodies. The dark object lying outside one of the cells is a discarded spore coat

(B) *Bacillus megaterium* (×3,600). The osmium-fixed acid-hydrolysed preparation is stained by a modified Feulgen stain in which the fuchsin of the Schiff reagent is replaced by azure A. This is a selective nuclear stain and shows the chromatin bodies within the bacilli. The spore coats are still attached to several of the bacilli. (Robinow, C. F. (1956). Bacterial Anatomy, *6th Symp. Soc. gen. Microbiol.*, p. 181)

(C) *Bacillus subtilis* (×3,600). Stained as in (B) to show chromatin bodies. The shadowy shapes in the background are the discarded spore coats

(D) *Bacillus cereus* (×3,600). A feulgen-stained preparation from a culture grown for ten hours on 3 per cent salt medium

(E) *Tetracoccus canadensis* (×3,600). Feulgen-stained preparation to show chromatin bodies. (C. F. Robinow)

In most stained preparations the bacterial cytoplasm, being extremely baso-philic, masks the presence of chromatin which also stains readily by basic stains. Chromatin, however, can be differentially stained by the Feulgen technique (Fig. 9.5). This stain is a specific colour test for DNA in higher cells of animals and plants and for bacterial chromatin. The dye, basic fuchsin, is first decolor-ized, by reduction to the leuco-form, in the presence of sulphurous acid, to give Schiff's reagent, which is used to detect the presence of aldehyde groups. Hydrolysis of smears of bacteria with hydrochloric acid removes the cytoplasmic RNA and liberates aldehyde groups on the DNA which reacts with reduced Schiff's reagent to give a purplish colour to the chromatin material. Chromatin bodies may also be demonstrated by phase contrast microscopy (Fig. 9.5), and by electron microscopy (Figs. 9.4 and 19) of thin sections of bacteria in which the nuclear region is less electron dense than is the cytoplasm.

The fine structure of the nuclear body in thin sections is characterized by the absence of cell components such as ribosomes (p. 65). The appearance of DNA in the nuclear body depends on the techniques of fixing and staining used; in some preparations numerous small aggregates of DNA fibres are seen whilst, in others, the nuclear body appears almost homogeneous; here the DNA fibres are too fine to be resolved (diameter of DNA helix is 20 Å). The fibres of DNA have been shown to be orientated in the direction of the long axis of the cell; cross-sections of cells reveal that aggregates of these fibres are folded backwards and forwards into bundles which may be twisted around each other.

The shape and size of the nuclear body varies according to the physiological state of the cell. In resting cells it is spherical, oval or dumbbell-shaped and usually only one body is present in a single cell. In the non-resting cell it is almost permanently in a state of division, changing its shape within minutes from a compact to a more tape-like appearance prior to division. Division is not by mitosis with spindle formation as in eukaryotes but by controlled replication of the DNA (p. 73). The migration and anchoring of the divided DNA as two separate masses at either end of the dividing cell is considered to be one of the functions of mesosomes (Fig. 9.4). Division of the nuclear body occurs more or less synchronously with cell division but not infrequently more rapid division results in two or four chromatin bodies being present in one cell.

Inclusion bodies

These are non-living bodies deposited in the cytoplasm of bacteria. They are found in specific organisms and their presence usually depends on the growth conditions of the culture. Volutin, i.e., metachromatic granules, are strongly basophilic, measure up to 0·5 μ in diameter and are widely distributed among bacteria and yeasts. Their presence is used as a basis for the rapid laboratory diagnosis of *Corynebacterium diphtheriae*, the causal pathogen of diphtheria (Fig. 9.6). Metaphosphate is usually but not exclusively present in the granules; the volutin of *Aerobacter aerogenes*, for instance, occupies 20 per cent of the total cell volume of which only 1 per cent is metaphosphate.

In many species polysaccharide (including glycogen and bacterial starch) may be detected by staining with iodine. Glycogen stains red-brown and starch,

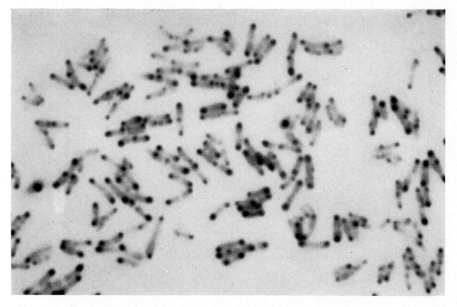

Fig. 9.6 Neisser's stained preparation of *Corynebacterium diphtheriae* showing volutin granules. (× 3,600. C. F. Robinow)

blue. Poly-β-hydroxybutyric acid granules are found in a large number of different types of bacteria (Fig. 9.7A). This substance can be detected with Sudan Black and acts as a storage material, being used by the cell for endogenous metabolism under appropriate conditions. Protein crystals are sometimes found, a notable example being in sporing cultures of *Bacillus thuringiensis* where a single large crystal is present in each cell. Vacuoles containing gas occur in halophilic species, such as *Halobacterium halobium* and a few other genera (Fig. 9.7B). Intracellular sulphur is frequently deposited by photosynthetic and chemosynthetic bacteria able to utilize hydrogen sulphide.

Endospores and cysts

Endospores are produced by members of the family Bacillaceae and some species of *Vibrio*, *Spirillum*, *Metabacterium*, and *Sporosarcina*. With the exception of *Metabacterium polyspora* and certain gut bacilli, one endospore is formed in each bacterial cell (Fig. 9.8), which ultimately germinates to give rise to a single vegetative cell; the process is not therefore one of multiplication. Endospores may be examined as highly refractile bodies in unstained preparations (Fig. 9.8.3), by special staining techniques, in which hot concentrated stains, such as malachite green, are used to penetrate the highly resistant spore envelope, or by electron microscopy.

The spore position in the vegetative cell, its shape and size which may or may not distend the sporangium, are characteristic of particular species of the aerobic genus *Bacillus* and the anaerobic *Clostridium*. The spores of *Cl. tetani*, for

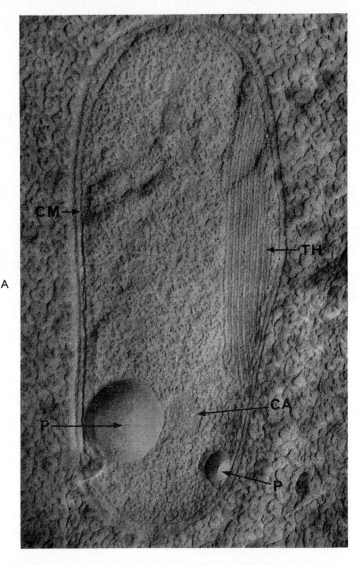

Fig. 9.7 Cellular inclusions in bacteria. (**A**) Freeze-etching preparation of cells of *Rhodopseudomonas viridis* showing variously-sized bodies of the reserve substance poly-β-hydroxybutyric acid. A thylakoid stack of lamella structures which carry the photosynthetic pigments is clearly visible. CM—cytoplasmic membrane; CA—cytoplasmic aggregate; TH—thylakoid stack; W—cell wall; P—poly-β-hydroxybutyric acid ($\times$95,000. *Bildatlas path. Mikroorganismen*, Bd. III, G. Fischer, Stuttgart). (**B**) Gas vacuoles in a thin section of *Rhodopseudomonas spheroides*. ($\times$110,000. *Bildatlas path. Mikroorganismen*, Bd. III, G. Fischer, Stuttgart)

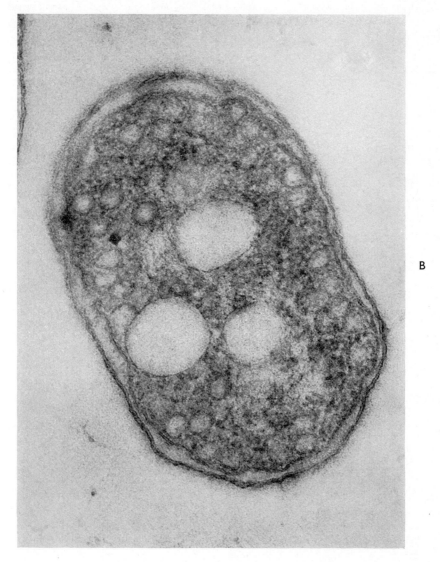

B

instance, are spherical and terminal (Fig. 15.4E); those of *Cl. septicum*, oval and sub-terminal. In both species the spores distend the vegetative cells.

ENDOSPORE FORMATION. Spore formation is initiated and completed in culture within a relatively short time from inoculation. Whilst the exact conditions inducing spore formation are not known it generally occurs in physiologically vigorous organisms containing a full complement of enzymes after a period of rapid vegetative growth and consequent depletion of exogenous nutrients. This depletion may produce changes from exogenous to endogenous metabolism

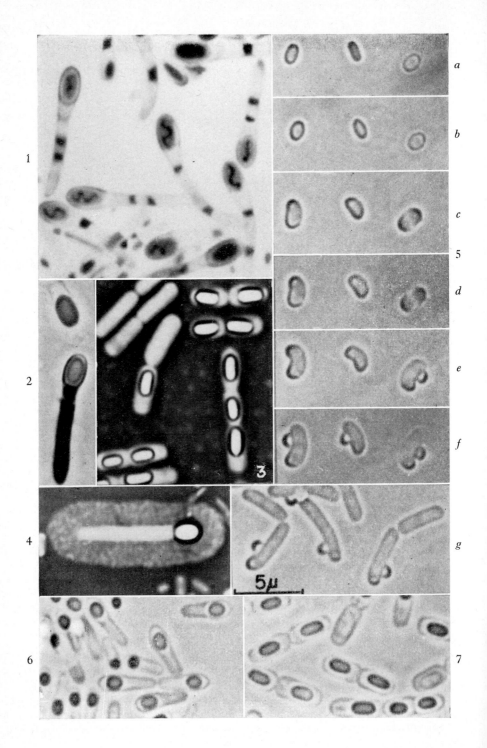

which may be the stimulus for spore formation. Sporulation involves the synthesis of new materials from low molecular weight compounds, which must be available to the cell, as well as the condensation of existing cell materials. For instance, dipicolinic acid (2,6-pyridinedicarboxylic acid) is unique to spores. Other substances, antigenically distinct from the vegetative cell, also may be found in spores. The materials necessary for spore synthesis may be provided, at least in part, by the autolytic breakdown of vegetative cell components which accompanies sporulation and which may occur before visible degeneration of the vegetative cell. Other events leading to sporulation, such as an increased requirement for cations, also take place before visible changes can be observed. Once the cell is committed to sporulation it cannot revert to vegetative proliferation.

The first visible evidence of spore formation is the appearance of a faint, clear area, later to become the spore. This gradually becomes more opaque to form the forespore, and at this stage, stains intensely. Finally the spore envelope is laid down after which the spore no longer stains readily, owing to the impermeable nature of the wall, and becomes highly refractile. The sporangium continues to be metabolically active for a time but eventually loses its Gram-positivity, autolyses and liberates the spore.

Electron microscopy reveals the stages of the formation of the spore envelope which differs in detail between species. In *B. megaterium* the electron dense forespore is separated from the cytoplasm by a less electron dense ring. Later distinct layers are formed. The central core which contains the nuclear body is surrounded by the spore wall, a delicate membrane from which the cell wall

---

Fig. 9.8 Bacterial endospores.

(**1**) *Clostridium pectinovorum* ($\times$3,600). Giemsa stained after acid hydrolysis, to reveal chromatin bodies which, in the developing spores of this organism, are in the shape of a helix. Cells with uniformly stained spores are earlier stages. The single bacillus lying vertically on the left of the photograph has a delicate transverse line or 'collar' across the shaft just below the spore ; this is a reinforcement of the cell wall and comes away with the spore

(**2**) *Clostridium pectinovorum* ($\times$3,600). Stained by mounting in Lugol's iodine to demonstrate starch-like materials in the cells. At the top is a mature spore with 'collar' attached

(**3**) *Bacillus cereus* ($\times$3,600). The preparation has been made in nigrosin, dried and mounted in air and photographed from the reverse side—a technique which emphasizes the high refractivity of the spores.

(**4**) *Clostridium pectinovorum* ($\times$3,700). Prepared and mounted as in (**3**). The polysaccharide capsule is revealed around the spore-bearing bacillus. The small rods are contaminants

(**5**) *Bacillus subtilis* ($\times$3,600). Spore germination serially photographed at the following times after inoculation on to nutrient medium: (**a**) 7, (**b**) 22, (**c**) 60, (**d**) 105, (**e**) 165, (**f**) 176, and (**g**) 200 minutes

(**6**) *Bacillus sphaericus* ($\times$3,600). Fixed with osmium tetroxide vapour and photographed wet between coverslip and agar surface. This aerobic spore bearer shows bulging of the bacillus around the spore, a feature which is unusual among aerobes and more common among anaerobes

(**7**) *Bacillus cereus* ($\times$3,600). Prepared and examined as in (**6**). This organism shows the more characteristic appearance of aerobic spore-bearing organisms. (Figs. 1–7 by C. F. Robinow)

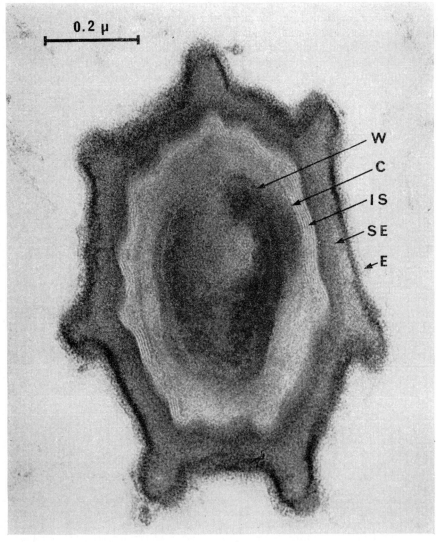

A

Fig. 9.9 Electron micrographs of thin sections of bacterial endospores. **(A)** Resting spore of *Bacillus circulans* showing sculptured ridges in the outer spore envelope (SE). IS—layered inner spore envelope; C—cortex; W—cell wall of spore; E—exosporium (×130,000). **(B)** Swelling of spore of *B. megaterium* has caused the spore envelope (SE) to burst open followed by growth of the young cell. M—membrane body; Chr—chromosome. (×145,000. *Bildatlas path. Mikroorganismen*, Bd. III, G. Fischer, Stuttgart)

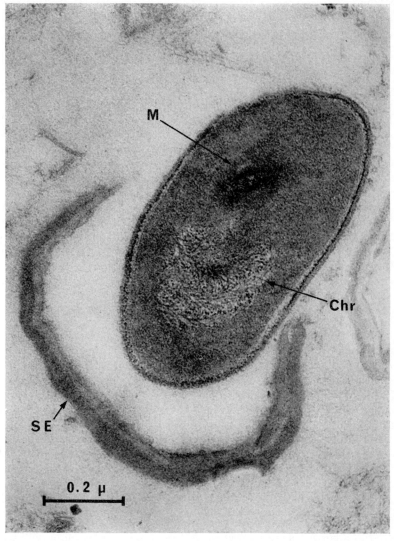

B

of the future vegetative organism will develop. Outside this is the cortex, a thicker but less electron-dense layer which, in turn, is enclosed by several layers forming the spore coat, the structure which gives the spore its high refractivity. The spore coat may be smooth, grooved or raised in geometrically-patterned ridges (Fig. 9.9). In some species an additional outer structure called the exosporium surrounds the spore.

Nuclear changes accompany spore formation. In the earlier stages a part of the nuclear material, in the form of an axial filament, becomes attached by a mesosome to the adjacent plasma membrane at the site of the spore formation.

Subsequently the mesosome guides or accompanies this chromatin material into the future spore. The outer part of the nuclear material is anchored at the opposite end of the cell and ultimately decays with the parent cell or sporangium.

ENDOSPORE GERMINATION. Usually endospores do not germinate in the mother culture and it has been suggested that exogenous energy-supplying nutrients are essential. Germination occurs a few hours after endospores are transferred to media conducive to vegetative growth. In some species the spore first loses its optical density, refractivity, and heat resistance whilst demonstrating an increase in stainability and respiration. It next swells to about twice its volume, the rate of respiration increases, and the spore wall gradually disappears leaving only a thin outer spore coat. Eventually the germ cell breaks through this spore coat, usually at one pole (Fig. 9.8.5 and 9.8.2). In other species little swelling and loss of refractivity occurs, the germ cell rupturing the spore coat either at one pole or along the equatorial plane leaving the parts of the coat attached to the germ cell. Finally respiration increases to the rate characteristic of the vegetative state and the germ cell elongates to the full size of the vegetative cell prior to fission.

Upon germination the nuclear material, which is not easy to demonstrate in the endospore, once more assumes a normal structure. Early in germination mesosomes develop in the germ cell, DNA replication commences and later the nuclear body is seen to be attached to the plasma membrane at the site of the first transverse septum.

ENDOSPORE RESISTANCE. Bacterial endospores are among the most resistant forms of life known but within the same species considerable differences to lethal agents, such as high temperature, are found (p. 200). They are extremely resistant to desiccation and thus are able to survive for long periods and to germinate when conditions are once more favourable for vegetative growth. Their resistance to high temperature and most disinfectants enables them to survive heat sterilization and disinfection processes respectively which are adequate to kill vegetative cells (p. 200). This resistance may be attributed to a lower water content, a higher calcium content, and an altered protein composition compared with the parent cell. The amino acid, dipicolinic acid, is unique to bacterial spores and is thought to play a role in enzyme inactivation. Some enzymes are active in the spore but most are dormant.

CYSTS. Some species of *Azotobacter* are able to form cysts. An entire cell rounds off and develops a thick-walled, highly refractile cell (Fig. 9.10). They have a low endogenous metabolism and have a resistance to environmental influences greater than that of vegetative cells but not as great as that of bacterial spores. Like the latter, a cyst can remain dormant for a long period and subsequently germinate to give rise to a single vegetative cell.

Slime layers and capsules

Most bacteria produce a non-living secretion of viscid material around the external surface of their cell wall. It may be removed without disturbing the viability or metabolism of the cell but its presence protects the organism against

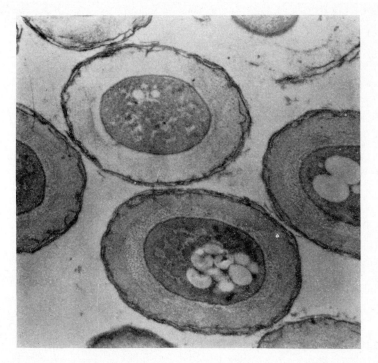

Fig. 9.10 Cysts of *Azotobacter*. Electron micrographs of a thin section. The hollow areas are thought to be lipid globules embedded in a chromatin body which occupies all the central area of the cyst. (× 23,000. Orville Wyss)

certain external factors. When this layer is sharply defined it is often referred to as a capsule but no limiting membrane can be demonstrated. Frequently a chain of organisms are seen embedded in a continuous mass of capsular substance. Capsules may be demonstrated in wet preparations of the organisms mounted in indian ink (Fig. 9.11A), or in stained preparations. With the exception of Alcian blue, capsular stains usually require copper salts as mordants. Capsules may also be rendered visible by treatment with specific anticapsular antiserum which increases their refractivity (e.g., the 'Quellung reaction', Fig. 9.11B). No structural detail is revealed by electron micrographs.

Slime layers usually, but not always, consist of a single chemical component, the nature of which varies with the species. For instance, the slime layers in *Diplococcus pneumoniae* and species of *Streptococcus* consist of polysaccharide containing glucosamine and hyaluronic acid respectively, in *Leuconostoc mesenteroides* of dextran (i.e., α 1–6 linked polymer of glucose), and in *Acetobacter xylinum* of cellulose (i.e., β 1–4 linked polymer of glucose). Some bacteria produce capsules of polypeptides (e.g., *Bacillus anthracis*, composed of 1-D-glutamic acid), of protein (e.g., *Pasteurella pestis*), or DNA (in certain halophilic species grown in sub-optimal salt concentrations). Occasionally more than one component may be produced by the same species under different growth conditions.

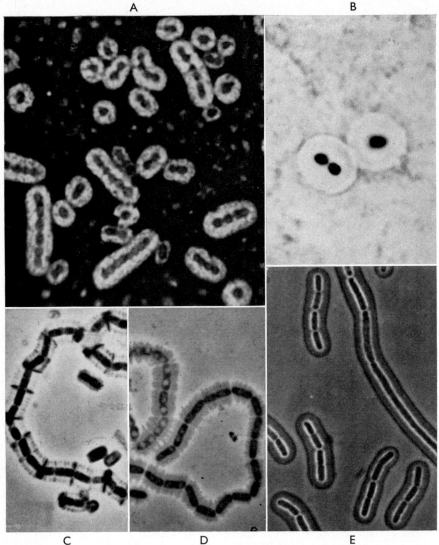

A

B

C **D** **E**

Fig. 9.11 Bacterial capsules. (**A**) Smear of capsulated pneumococci of an uncentri-fuged spinal fluid from a fatal case of meningitis. The capsules are demonstrated by negative staining using indian ink. ($\times 3{,}600$. C. F. Robinow.) (**B**) Capsule swelling test performed on sputum from a pneumonic patient—see text. ($\times 3{,}600$. C. F. Robinow.) (**C–E**) Phase-contrast photographs of *Bacillus megaterium* ($\times 2{,}500$) showing the influence of nutrition on the nature of the capsule. (**C**) Culture grown in carbohydrate-rich broth and treated with polysaccharide antibody. The polysaccharide of the capsule with which the antibody reacts, is shown to be concentrated at the ends of the organ-isms and at transverse septa across the capsular region corresponding to the points of cell division. (**D**) Culture grown in carbohydrate-rich broth and treated with polypep-tide antibody. This reacts with polypeptide capsular material and shows that this is dis-tributed uniformly throughout the capsule except for the limited spaces occupied by polysaccharide. (**E**) Culture grown in a carbohydrate-free medium and treated with polypeptide antibody. The uniform distribution of polypeptide materials throughout the capsule indicates the absence of polysaccharide under these conditions. (Tomcsik, J. (1956) *Ergebn. med. Grundlag. forsch.*, **1**)

Thus in *Streptococcus salivarius* the slime layer consists of a levan and an anti-genically distinct polysaccharide; in *Bacillus megaterium*, polysaccharide is laid down in disks surrounding the sites of cell division and polypeptide material between these disks (Fig. 9.11C, D and E). Differences in the chemical nature of capsular polysaccharides in strains of *Diplococcus pneumoniae* are respon-sible for serological differences of which 70 antigenically different types are recognized.

Capsule production, although primarily under genetic control, is commonly also determined by the constitution of the growth medium. For instance, when grown in the presence of sucrose, *Streptococcus salivarius* produces a dextran of high molecular weight, the quantity of the polymer being proportional, within certain limits, to the concentration of sucrose present.

Capsule formation may be responsible for considerable economic loss is dairy and other food industries (p. 654). Carbohydrate-containing materials become 'ropy' when encapsulated organisms grow in them. However, some organisms, e.g., *Leuconostoc* species, are employed commercially for the produc-tion of dextran as a plasma 'extender' used in the treatment of shock resulting from blood loss.

The presence of a capsule is often associated with the virulence of pathogenic organisms. When these organisms enter the animal body they are able to resist phagocytosis by the white cells of the blood, i.e., they are not readily engulfed by these scavenger cells or, if they do become engulfed, they resist digestion by the intracellular enzymes and continue to multiply. Loss of capsule, which sometimes occurs following a mutation, not only leads to loss of virulence but is often accompanied by other changes, such as alteration of the colony structure on the surface of solid media. In some organisms, e.g., *B. anthracis*, capsula-tion is only one of many factors involved in the pathogenicity of the organism (p. 552).

Flagella

Not all true bacteria are motile but motility, when it does occur, is effected by means of flagella (singular flagellum; from Latin meaning 'whip'). Flagella are unbranched, sinuous filaments of uniform thicknsss throughout their length which may be many times greater than that of the bacterium (Figs. 9.12 and 13). Their small size precludes their being seen in their natural state under the light microscope but considerable thickening by means of metal deposition or special staining techniques, such as fluorescent dyes, renders them visible. Their motion can be observed by dark field microscopy or by the agitation of indian ink par-ticles in the vicinity of flagellate bacteria mounted in soft agar to slow down their rate of movement. Dark field examinations can be greatly improved by suspend-ing organisms in 0·5 per cent methyl cellulose which not only reduces the rate of movement but also induces the aggregation of the flagella in bundles. In stained preparations the wavelength and amplitude of each wave of an individual flagellum can be determined (Fig. 9.13). Flagella of different wavelength may be present on the same organism.

The presence or absence of flagella and their arrangement and number is a species characteristic. A single bacterium may possess from one to a hundred

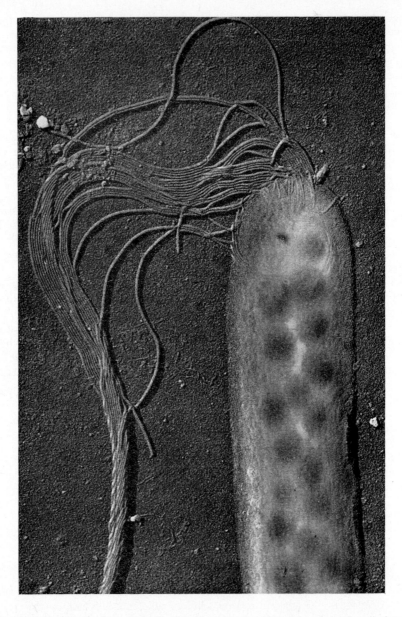

Fig. 9.12 *Spirillum* sp. (×40,000). Electron micrograph showing polar flagella inter-twined to form a thick bundle. The dark areas may be vacuoles. The cell wall structure is just visible. (Houwink, A. L. (1953) *Biochim. biophys. Acta*, **10**, 360)

flagella. Vibrios (p. 319) have a single flagellum (monotrichate—from Greek trichos meaning 'hair') (Fig. 9.17, xiii), *Spirillum* (p. 319) a tuft of flagella at one or both ends (lophotrichate) (Fig. 9.13.3, 9.17. xiv). In rod-shaped organisms the flagella may be single or numerous (Fig. 9.17, xii) distributed around the whole organism (peritrichate) or near one or both poles. Flagella can aggregate into a bundle and function as a single propulsive unit (Figs. 9.12 and 9.14). Only a few spherical bacteria are flagellate (Fig. 9.13.9).

Electron micrographs of specimens shadowed by heavy metals or negatively stained with phosphotungstate demonstrate that a single flagellum consists of several fibrils, usually three, arranged in a helix (Fig. 9.16). Flagella have an average diameter of 12 mμ. They are thus of the same order of magnitude as a single protein molecule. They are thought to arise from a basal cytoplasmic granule, and to pass through the cell membrane and the cell wall. The size of the granule depends on the type of bacterium; that of *Vibrio comma* is between 150 and 200 mμ. Some workers have expressed doubt about the size of these granules in various organisms and possibly even their existence, suggesting that they may be artifacts. In protoplasts of *Bacillus stearothermophilus* the flagella are attached by means of hooks to infoldings of the cytoplasmic membrane. As already stated (p. 284), flagella are retained by protoplasts prepared from motile organisms (Fig. 9.13) but under these conditions lose their power to produce motility. Flagella grow very rapidly, at approximately the rate of 0·5 μ per minute, attaining full length in 10 to 20 minutes. They are thought to propel the organism, not by a lashing action, but by waves of contraction passing from base to tip, imparting a spiral motion to the flagellum and an opposite rotation to the whole cell. The speed of cell movement is very high but varies with the species; speeds of 25 μ per second for peritrichous flagellate cells and 200 μ per second for the monoflagellate *Vibrio cholerae* have been recorded.

Pure preparations of flagella can be obtained by prolonged shaking of bacterial suspensions followed by differential centrifugation. Analysis shows that at least 98 per cent of the total weight is a protein called flagellin and X-ray diffraction patterns indicate a marked resemblance of this to the contractile protein of animal muscle fibres. Amino acid analysis of a number of flagellins reveals that cysteine is absent, the content of aromatic amino acids is low and that of glutamic acid and aspartic acid high. Flagella proteins constitute the H-antigens of motile bacteria (p. 576) the specificity of which may well be determined by differences in the arrangement of amino acids in the flagellar proteins.

Pili (from Latin for 'hair') or Fimbriae (from Latin for 'fringe')

These are very fine, hair-like, surface filaments much smaller than flagella (0·5 μ in length and less than 10 mμ in diameter), found only in certain Gram-negative rods. Pili are best demonstrated in metal-shadowed specimens by electron microscopy (Fig. 9.15). They are frequently found in freshly isolated strains; not all cells in the same culture are uniformly piliate; the number per cell varying from 1 to 400, and a single cell may have one or more morphologically different pili in addition to flagella and capsule. At least eight different types of pili are recognized; all are able to grow out through the cell wall without lysing the cell.

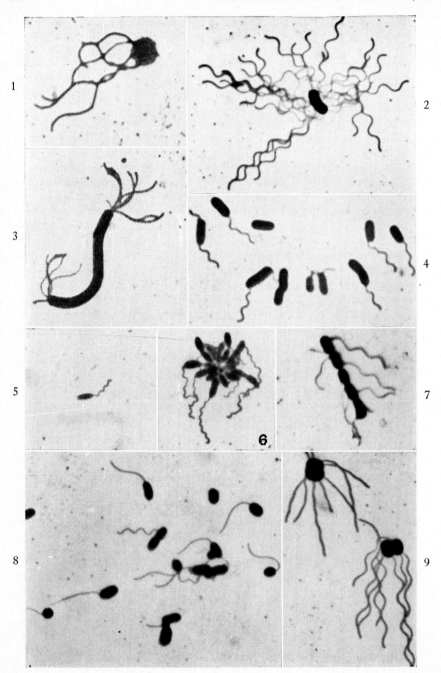

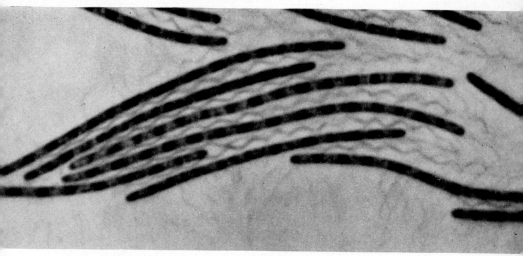

Fig. 9.14 *Proteus vulgaris;* a swarming culture fixed through the agar to the slide with Bouin's fluid. The flagella between the filamentous organisms have intertwined to form thick bundles. The white patches are chromatin bodies. (× 3,600. C. F. Robinow)

Pili cause bacteria to adhere both to one another and to foreign cells, such as red blood cells. Piliate colonies may be detected by their more compact colony form, but more readily by their ability to haemagglutinate red blood cells when suspensions of these are poured over a plate culture. Antigenic differences occur; for instance, all strains of piliate *Shigella flexneri* are found to have pili of the same antigenic composition; in contrast, at least five antigenically distinct sero-types of piliate *Salmonella* spp. are known.

Apart from their adhesive properties the function of the majority of pili is not known. Most work has been done on sex pili which are found only on strains of certain Gram-negative bacteria, and are capable of transferring self-replicating molecules of DNA (F, colicinogenic, and R factors) from donor to recipient bacteria, p. 114). Only 1 to 4 of these pili occur on each cell corresponding to the number of factors present in the cell. The donor cell synthesizes the units

----

Fig. 9.13 The shape and arrangement of bacterial flagella in stained preparations. **(1)** Protoplast of *Bacillus megaterium* (× 3,000) showing attached flagella after the cell wall has been dissolved away. (Weibull, C. (1953) *J. Bact.*, **66**, 688.) **(2)** *Clostridium parabotulinum* (× 2,000) with peritrichous flagella. **(3)** *Spirillum* sp. (× 2,000) with lophotrichous flagella. **(4)** *Pseudomonas aeruginosa* (× 2,000) with single polar flagellum. (Leifson, E. (1951) *J. Bact.*, **62**, 377.) **(5)** *Cellvibrio* sp. (× 2,000) with single polar flagellum. **(6)** *Caulobacter vibrioides* (× 2,000) showing a typical rosette of organisms each with a single flagellum. **(7)** *Streptococcus* sp. (× 2,000) showing a flagellated chain of cocci; flagella are rare in this genus. **(8)** Marine organism of un-known identity (× 2,000) with single polar flagellum; on some bacilli the flagella are straight whilst on others they have a distinct wave. **(9)** *Sarcina* sp. (× 2,000). One cell, shows waves of large amplitude, the other waves of smaller amplitude. **(2–9**, E. Leifson)

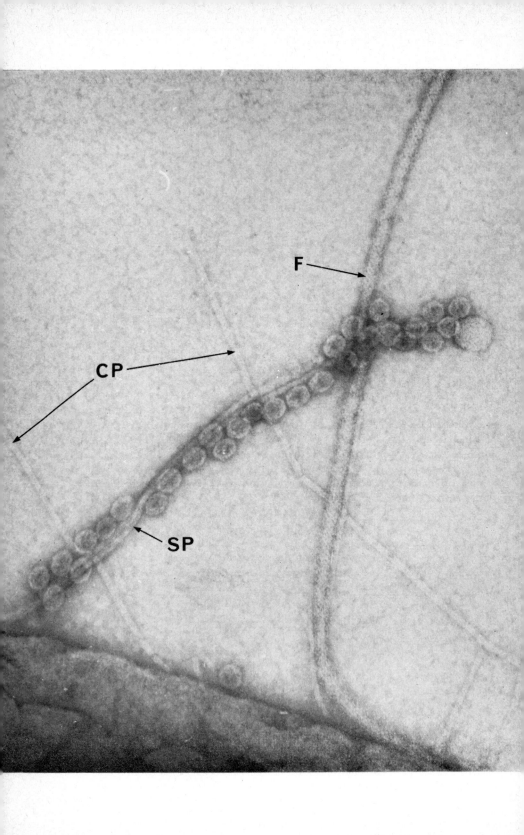

of pilus protein, called pilin; these are assembled into hollow, hair-like structures with an axial hole of 2·5 mμ diameter, and this structure is thought to be laid down around the DNA molecule being transferred. This complex of pilus and DNA passes through the cell wall. The distal end makes contact with, and attached itself to, a recipient cell which receives the DNA through the hollow pilus.

Sex pili also act as sites for male-specific phages (Fig. 9.15; p. 224). The male specific RNA phages attach to the outside of the projecting pilus, enter through the pilus wall and pass along the hollow canal. The male-specific, rod-shaped DNA phages attach to the pilus end to end, the pilus again acting as a canal along which the nucleic acid enters the cell.

Stalked bacteria

Several genera of bacteria are stalked but the stalk varies considerably both in its size and nature. *Caulobacter* has a slender, relatively short protoplasmic stalk arising from the pole of the rod shaped cell. The distal end of the stalk has an adhesive holdfast by which the organism attaches itself to the substratum, either other micro-organisms or inanimate substrates, or to the holdfasts of other *Caulobacter* cells, thereby forming rosettes of cells (Fig. 9.13.6). The stalk of *Gallionella* is usually extremely long, often spirally twisted, and has been variously considered to consist entirely of ferric hydroxide, or of organic matter impregnated with ferric hydroxide or even to be the basic structure of the cell on which inorganic material is deposited. The first of these alternatives is most unlikely and there is dispute as to which of the other suggestions is correct.

External morphology

Three morphological types are recognized among true bacteria—the coccal, bacillary, and spirillar (Fig. 9.17). Within a population arising from a single cell, variations in form and size of individual cells may be found. These variations are particularly characteristic of certain genera, and were termed pleomorphism by early bacteriologists. In addition under certain circumstances many true bacteria may change their form to produce involution forms and L-forms. These are considered later.

The coccal form

Most cocci (Greek and Latin for 'berry') approximate to true spheres but variations on this basic structure occur (Fig. 9.17, i–vi). *Staphylococcus aureus* and *Streptococcus pyogenes* are spherical, *Str. faecalis* ellipsoidal and *Neisseria gonorrhoeae* kidney shaped. Cocci may divide irregularly in one, two, or three

---

Fig. 9.15 Electron micrograph (approximately × 200,000) of appendages of *Escherichia coli*. F = flagellum (p. 301); CP = common pili (fimbriae; p. 303); SP = sex pilus (p. 305), about 2 μ long, to which are attached particles of the RNA bacteriophage MS2. Note the knob at the end of the sex pilus, a feature frequently seen on sex pili but never on common pili. (Kindly provided by Dr. A. M. Lawn, Lister Institute of Preventive Medicine)

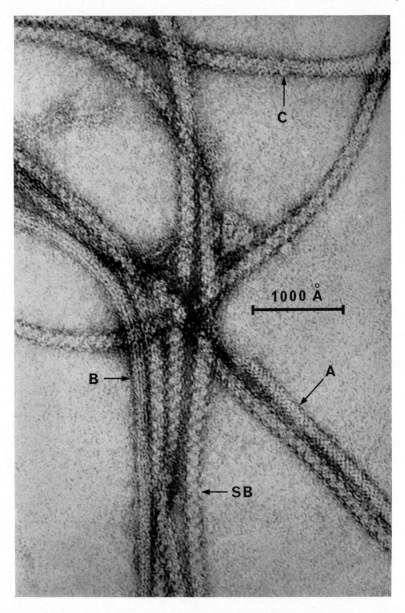

Fig. 9.16 Molecular architecture of flagella. Bacterial flagella are built of globular flagellin molecules. These are strung together like beads and may be arranged in two different functional forms; one shaped like a helix (**A**), the other with a predominantly longitudinal arrangement (**B**). Also intermediate stages may occur (**C**). SB is a flagellum probably surrounded by a sheath. (*Pseudomonas rhodos* ×215,000. Lowy, J. and Hanson, J. (1965) *J. Mol. Biol.*, **11**, 293)

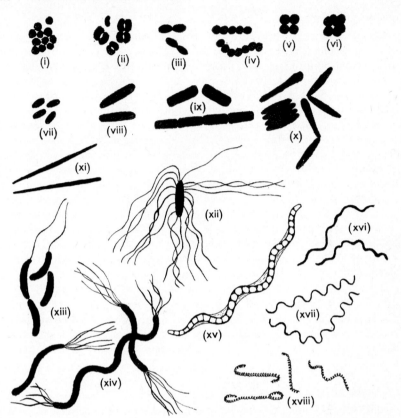

Fig. 9.17 Various morphological forms of typical bacteria drawn to approximately the same size. **(i)–(vi)** various forms of spherical bacteria. **(i)** *Staphylococcus*; **(ii)** diplococci (e.g., *Neisseria*); **(iii)** diplococci (e.g., *Diplococcus pneumoniae*); **(iv)** *Streptococcus*; **(v)** micrococci (e.g., *Micrococcus tetragenus*); **(vi)** *Sarcina*

(vii)–(xii) various forms of rod-shaped organisms. **(vii)** small rods (e.g., *Serratia marcescens*); **(viii)** larger, round-ended rods (e.g., *Escherichia coli*); **(ix)** flat-ended rods both single and in chains (e.g., *Bacillus anthracis*); **(x)** clubbed bacilli arranged singly, in palisades and at various angles to one another (e.g., *Corynebacterium diphtheriae*); **(xi)** fusiform bacilli (e.g., *Fusobacterium fusiforme*; **(xii)** flagellate rod (e.g., *Salmonella typhi*)

(xiii)–(xviii) various forms of spirillar organisms. **(xiii)** vibrios with single polar flagellum (e.g., *Vibrio cholerae*); **(xiv)** spirilla with tufts of polar flagella (e.g., *Spirillum*); **(xv)** *Cristispira*; **(xvi)** *Borrelia*; **(xvii)** *Treponema*; **(xviii)** *Leptospira*

planes. The daughter cells separate or remain attached, the resulting cell-arrangements being taxonomically useful. Division in one plane gives rise to pairs of organisms (diplococci), e.g., *Diplococcus pneumoniae*, or to chains, e.g., *Str. pyogenes*. Regular division in two planes at right angles to each other result in tetrads (e.g., *Micrococcus tetragenus*) and regular division in three planes gives rise to cubic packets of eight cocci as in the genus *Sarcina*. Irregular division in two or three planes gives the grape-like clusters characteristic of *Staph. aureus*. Diameters of cocci range from 0·5 to 1·25 μ.

The bacillary form

Bacilli (Latin for 'stick') are straight or slightly curved cylinders. The ratio of length to diameter varies greatly and in the cocco-bacillary forms is not much greater than unity (Fig. 9.17, vii–xii). For example, *Brucella melitensis* is 0·5–0·7 μ in diameter and 0·6–1·2 μ long, and *Haemophilus influenzae* 0·3–0·4 × 1–1·5 μ. In contrast, *Bacillus anthracis* is 1–1·2 μ wide × 3–8 μ long. The end walls may be flat (truncate) or slightly concave (e.g., *B. anthracis*, Fig. 15.4F), tapering (e.g., *Fusobacterium fusiforme*) or convex (e.g., *Corynebacterium diphtheriae*) (Fig. 9.6). Some species of *Corynebacterium* are clubbed or swollen at one end, hence the generic name, meaning clubbed bacillus. Certain species of bacilli invariably occur as single cells, others, especially species which produce rough colonies (p. 313), remain attached in chains. In the animal body *B. anthracis* occurs as single cells or short chains of two or three bacilli; in culture it invariably forms long filamentous chains. Post-fission movements may give rise to characteristic arrangements as, for example, in *Corynebacterium* in which transverse fission is followed by a snapping movement, the cells remaining attached at one point, the daughter cells consequently producing angular or palisade arrangements (Fig. 9.17, x).

The spirillar form

These are curved, twisted rods (Fig. 9.17, xiii–xviii). *Spirillum* spp. are long, and cork-screw-shaped; *Vibrio* spp. are short and appear comma-shaped. *Vibrio cholerae* is 0·2–0·4 μ × 1·5–4 μ long; *Spirillum* spp. may be much longer, up to 50 μ.

Involution forms

The action of autolytic enzymes, normally restricted in actively metabolizing cells, may be responsible for the frequent occurrence in old cultures of organisms of abnormal morphology often characteristic of the bacterial species which produce them, e.g., *Pasteurella pestis* and *Neisseria* species. These are known as involution forms. Such cells are mostly dead, but those still viable, give rise to normal cells again when introduced into favourable medium.

Large bodies and L-forms

Many bacteria, under certain conditions such as cold shock, the presence of specific antiserum, bacteriophage attack or the presence of antibiotics, such as the penicillins which interfere with cell wall synthesis, tend to develop an unusual form of pleomorphism involving the formation of large spherical or distorted cells, the so-called 'large bodies'. The cells revert to the normal bacillary form if the inducing stimulus is removed after only a brief exposure. If the stimulus persists for a long time, the cells either perish or may become stabilized in the so-called L-form. L-forms consist of minute filterable bodies and non-rigid, fragile, large globules of cytoplasm and resemble the pleuropneumonia and the pleuropneumonia-like organisms (p. 344). They break up readily when touched but may be examined *in situ*, by dark field microscopy from fluid cultures or in impression smears stained by Giemsa. The smears are prepared by allowing a fixative, such as Bouin's fluid, to diffuse through the agar medium, thereby fixing the colonies to coverslips placed over them. Usually serum or blood is required in the medium for their growth. They may then breed true for many generations. L-forms are probably spheroplasts (p. 284) of bacteria which have lost their ability to produce typical cell walls or in which cell wall production is inhibited. They were originally discovered in *Streptobacillus moniliformis*, and have been subsequently reported in many other organisms including *Proteus vulgaris*, *Escherichia coli*, and various species of *Haemophilus*, *Salmonella*, and *Streptococcus*.

Cell division

The normal mode of cell division in bacteria is by binary fission into two similar daughter cells. This is intimately associated with synthesis of new cell wall material, cell growth, and DNA replication. The cell first elongates beyond its normal length prior to division. During this elongation, new cell wall material is laid down in an equatorial zone at the sites of future septum formation in Gram positive bacteria; in Gram-negative species the addition of new cell wall material is not limited to one site but is built into any part of the original wall. Division of the cell is initiated by an ingrowth of the cytoplasmic membrane. This is followed by ingrowth of the cell wall like a slowly closing diaphragm, which gradually closes the aperture to form a complete transverse septum which is thicker than the normal cell wall. Both processes seem to be under the direct influence of associated mesosomes (Figs. 9.18 and 19). Finally separation is achieved by a median splitting of the cross wall usually from the periphery inwards. Nothing is known about the enzymic regulation of this cleavage.

Before cross-wall septa are produced, the nuclear body divides. When both these processes are synchronized, the resulting daughter cells each contain a single nuclear body. Nuclear division often occurs faster than cell division and multinucleate cells are then produced; when ingrowth of cross-wall septa takes place faster than separation of daughter cells, multicellular rods and cocci result (Fig. 9.18).

Differences in detail in the process of separation of daughter cells is responsible for the characteristic post-fission shape and arrangement of cells in individual genera and species. When separation is incomplete, chains of cocci or filaments of bacilli result. The production of cross wall septa at different planes in regular sequence accounts for groups of four or eight cocci, as in *Micrococcus tetragenus* and *Sarcina* species respectively; irregular division results in clusters as in *Staphylococcus* species.

Complex life-cycles have been described for some bacteria. While it is undoubtedly true that in some species remarkable pleomorphism occurs, evidence for a true life cycle is incomplete. The accepted usage of the term life cycle implies the occurrence of a cyclic series of phases in an obligatory sequence and that the completion of this sequence is essential for the fulfilment of some important activity, such as reproduction, genetic recombination or dispersal. Such cycles are common in higher organisms but it is most unlikely that they occur in bacteria.

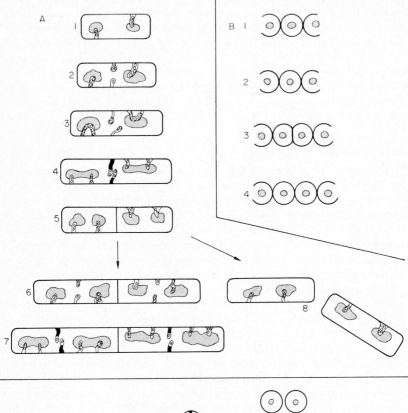

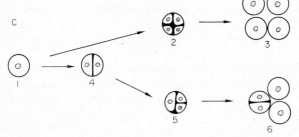

Fig. 9.18 Cell division in bacteria. **(A–C)** Diagrammatic representation of binary fission in different morphological forms of bacteria. The cell wall and cytoplasmic membrane are represented by the thick outer line, the chromatin bodies by the shaded mass in the centre of the organism and, in **A** the mesosomes by the vesicular structures. It seems that in most of the bacilli, the nucleus is in a constant contiguity with a mesosome. This contact corresponds to a real attachment between nucleus and mesosome which persists through the whole division cycle

(A) Cell division in bacilli. 1. Mature cell ready to divide, each nucleus is connected with a mesosome. 2. Early stage of division: new mesosomes appear in the region where the septum of division will form. 3. While the septum begins to form, the nuclear volume increases and each nucleus is connected with two mesosomes. 4. The two nuclear meso-

Bacterial colonies

When bacteria are inoculated on to gelled nutrient media, single organisms, by repeated division, grow into a population of progeny producing colonies visible to the naked eye on the surface of the medium. The colonial form is usually fairly constant for a given species and is therefore of taxonomic value (Fig. 9.21). Colony diameter is however readily affected by crowding, the size of the colonies then being limited by the metabolic activities of others. More constant characteristics are form (e.g., whether raised, flat, dome shaped, or swarming, as a continuous film over the surface of the medium), colour, surface appearance (e.g., smooth or rough, matt or glossy) and the nature of the colony margin (e.g., entire, lobed or crenated).

Colony form may alter with the nature of the nutrients available or as a result of a genetic change (p. 553). A frequent type of mutation is from a smooth colony, with an almost circular outline and a glistening appearance, to a rough form, with an irregular or toothed edge and a dull wrinkled surface (Fig. 9.21C). Such a change is known as a Smooth to Rough or S → R variation. It is often accompanied by other changes such as loss of the cell surface antigens characteristic of the smooth form, partial or complete loss of virulence, altered sensitivity to bacteriophages (p. 260), or agglutination of suspended cells in the presence of electrolytes and certain dyes, such as acriflavine. Thus the visible change from smooth to rough colony form is a valuable indicator of other fundamental changes. The opposite change from rough to smooth is less common. Loss of capsules often results in loss of the smooth character. For instance, capsulated (virulent) strains of pneumococci produce smooth colonies, and non-capsulated (avirulent) strains rough ones.

The precise mode of cell division is of great importance in determining the form of many colonies; the slightly asymmetric division of *Bacillus cereus* var. *mycoides* results in long twisted chains of cells turning to left or right, giving the

---

somes move farther and farther apart, while the nuclei begin to divide and the septum grows. 5. The nuclei have divided into two daughter nuclei, each of which is connected to one of the two mesosomes. The septum has completely grown across the bacillus and the two daughter cells are formed, although still remaining attached. 6. Daughter cells have grown to full size but remain attached to form a chain. 7. Mature cells begin to divide (as stage 1–5) but remain attached to form a chain. Such cells, when stained by nuclear stains, reveal a multinucleate appearance. 8. Many bacilli do not remain in chains but separate after division (stage 5), grow to full size and then divide. (After A. Ryter)

(**B**) Cell division in streptococci. 1. Mature cells about to divide. 2. The chromatin body has divided. The cross wall is beginning to grow inwards at right angles to the plane of the streptococcal chain. 3. Cross wall complete. 4. Daughter cells round off to form two spherical cocci remaining attached by their cell walls, to form a chain of cocci.

(**C**) Cell division in *Micrococci* (1–3) and *Staphylococci* (1 and 4–5). 1. Mature coccus. 2. Simultaneous ingrowth of cross walls in two planes at right angles to each other resulting in division of the coccus into four equal parts. 3. Four cells of equal size and maturity result to form a tetrad. 4. In staphylococci, a cross wall occurs in a single plane at first. 5. A further cross wall, at right angles to the first, divides one of the daughter cells only. 6. Unequal stages of maturity and irregular arrangements result, owing to the non-synchronized division in two planes.

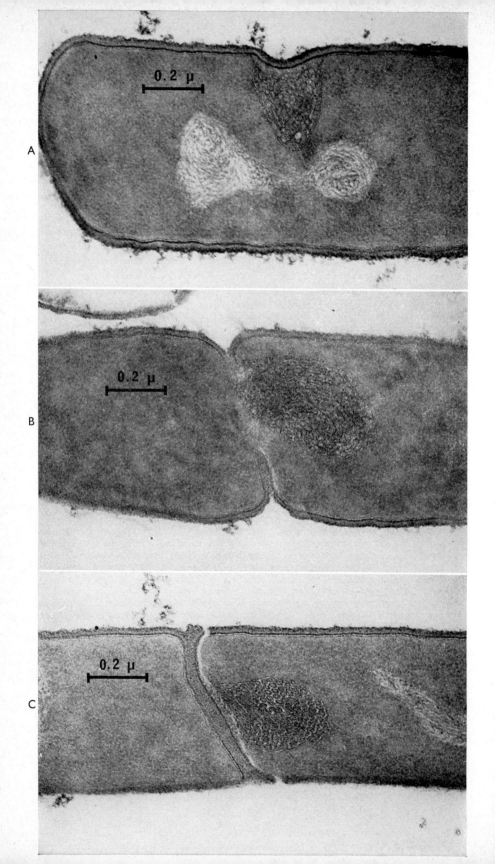

colony a rhizoidal structure (Fig. 9.20). A few motile bacteria of the genus *Bacillus* give rise to colonies that migrate across the surface of solid media leaving a well-defined track of growth.

Fig. 9.20 *Bacillus cereus* var. *mycoides* (× 2). Colony grown for forty-eight hours at 37°C. The organism spreads over the surface of the medium in long twisted chains of cells turning to the right—the dextral form. (A. H. Linton)

Bacterial variation

Extensive variation within bacterial species occurs. Most of these have a genetic basis and are considered in Chapter 4.

Classification

The following is an outline of the two orders Pseudomonadales and Eubacteriales, modified from Bergey.

A. *Pseudomonadales*

Straight, curved or spiral, rigid, rod-shaped bacteria; usually polar-flagellate at one or both ends of the cell; all Gram-negative.

Fig. 9.19 Formation of the transverse wall by membrane bodies in *Bacillus megaterium*. The formation of the transverse wall takes place with the participation of membrane bodies (mesosomes). After the cytoplasmic membrane has undergone folding, vesicular and tubular structures are formed. (**A**) At first the mesosome lies near to the wall in which a 'dent' has commenced to form. (**B**) Local synthesis of cell wall material causes a transverse septum to grow inwards like a closing iris diaphragm. (**C**) The mesosome lies along side the synthesized septum after this is complete. (× 85,000. *Bildatlas path. Mikroorganismen*, Bd. III, G. Fischer, Stuttgart)

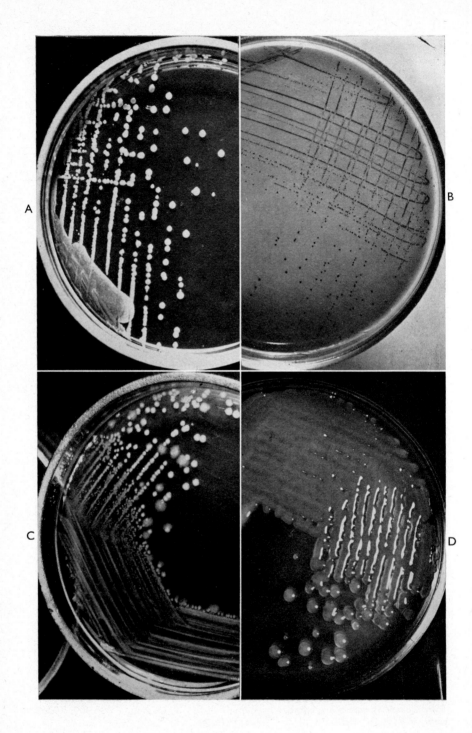

1. Rhodobacteriineae

Photosynthetic bacteria; able to grow strictly anaerobically in the presence of light; include coccal, bacillary, and spirillar forms. Three families recognized.

(A) THIORHODACEAE. Purple and red sulphur bacteria containing bacteriochlorophyll and carotenoids; preferentially use hydrogen sulphide as hydrogen donor in photosynthesis, sulphur being deposited intra-cellularly as globules; probably not a taxonomically homogeneous group.

Thiospirillum. Brownish to purplish red spiral cells, motile by means of polar flagella; in mud or stagnant water.

Chromatium. Similar to *Thiospirillum*, but short rods or vibrios; marine or in stagnant ponds.

Thiocystis. Spherical or ovoid, purplish red cells, forming colonies embedded in a gelatinous matrix; in mud or stagnant water.

(B) ATHIORHODACEAE. Non-sulphur, purple and brown bacteria; contain photosynthetic pigments as above but use organic hydrogen donors and hence do not deposit sulphur; cells motile by means of polar flagella.

Rhodopseudomonas (cells spherical or rod-shaped) and *Rhodospirillum* (cells spiral); in mud and stagnant water.

(C) CHLOROBACTERIACEAE. Green sulphur bacteria; contain 'chlorobium chlorophyll', which has different absorption characteristics to higher plant chlorophyll and bacteriochlorophyll; hydrogen sulphide (or some other inorganic sulphur compound) is the hydrogen donor in photosynthesis; do not usually deposit sulphur intracellularly, but frequently extracellularly; in sulphur springs, and stagnant water.

Chlorobium. Spherical or rod shaped cells, singly or in chains, non-motile; in marine or fresh water mud.

2. Pseudomonadineae

Non-photosynthetic organisms which may produce pigments some of which are water-soluble; most species occur in salt or fresh water or soil and a few are pathogenic to vertebrates. Seven families recognized.

(A) NITROBACTERACEAE. Autotrophs; mainly rod-shaped; different species derive energy by oxidation of ammonia to nitrite, or nitrite to nitrate; common in soil or fresh water.

The only genera which can be regarded as conclusively established are *Nitrosomonas* (rods with single polar flagellum) and *Nitrobacter* (motile or non-motile rods). The former oxidizes ammonia to nitrite, the latter nitrite to nitrate. Both are strict autotrophs and do not grow on organic media.

Fig. 9.21 Colonies of different bacteria grown for twenty-four hours at 37°C (normal size). (**A**) *Staphylococcus albus* on nutrient agar. (**B**) *Streptococcus faecalis* on bile lactose agar with neutral red added as indicator. The acid produced by fermentation of the lactose results in the indicator changing to a strong pink colour which is taken up by the growing colonies. Most streptococci produce small colonies irrespective of the media on which they are grown. (**C**) *Shigella sonnei* on nutrient agar showing smooth (S) and rough (R) colonies. A rapid rate of mutation from S → R is typical of this organism. (**D**) *Klebsiella pneumoniae* on lactose nutrient agar. Vigorous production of capsular polysaccharides results in mucoid colonies. (A. H. Linton)

(B) METHANOMONADACEAE. Rod-shaped organisms deriving energy by oxidation of simple compounds of hydrogen and carbon (e.g., H_2, CO, CH_4); occur in soil or water.

Hydrogenomonas oxidizes hydrogen, *Methanomonas* oxidizes methane and *Carboxydomonas* oxidizes carbon monoxide as can also some *Hydrogenomonas*, which raises doubt about the value of the species *Carboxydomonas*.

(C) THIOBACTERIACEAE. Straight, rod-shaped bacteria which oxidize sulphur compounds and may deposit granules of sulphur outside the cell; described often as colourless sulphur bacteria; never filamentous; in soil and water; species adequately studied are all obligate autotrophs.

Thiobacillus is the only well-established genus but a number of inadequately studied genera (including cocci and curved rods) are included by some authorities.

(D) PSEUDOMONADACEAE. Straight, rod-shaped, Gram-negative aerobic; usually motile by means of polar flagella, single or in tufts; may produce pigments either water-soluble diffusing into the surrounding medium or non-water soluble; most saprophytic in soil or water (including sea water and brines), some plant and a few animal pathogens.

Pseudomonas. Usually motile by monotrichous or lophotrichous flagella frequently produce fluorescent, diffusible pigments (green, blue, violet, or yellow); many species in soil and water (fresh, salt, and brines); numerous plant pathogens. *Ps. aeruginosa* (*pyocyanea*), an 'opportunist' organism producing 'blue pus' in many sites in a wide range of animal hosts.

Xanthomonas. Cells usually monotrichous; produce non-water-soluble pigments on agar; proteins are usually digested; pectin liquefied by some species; many are pathogenic to plants causing necrosis, e.g., *X. hyacinthi* causes yellow rot of hyacinth bulbs.

Acetobacter. Ellipsoidal or rod-shaped, occurring singly or in short or long chains; usually motile by polar flagella; oxidize various organic compounds to organic acids (e.g., acetic acid) or further; widely distributed in nature, abundant in fermenting plant materials and important in vinegar making (p. 690). *A. xylinum* synthesizes cellulose.

Aeromonas. Short, rod-shaped cells, usually motile by monotrichous flagella; heterotrophic; majority aquatic, many pathogenic to reptiles and fish. *A. hydrophila* produces a lethal septicaemia in frogs; *A. salmonicida* causes furunculosis, i.e., abscess-like lesions, of fresh water fish.

Photobacterium. Motile (monotrichate) or non-motile, short rods; luminescent, especially when grown on salt water agar, or in sea water, on dead fish or other salt water animals.

Zymomonas. Ellipsoidal to rod-shaped; motile by lophotrichous flagella; under anaerobic conditions, ferments glucose producing CO_2, ethyl alcohol, and lactic acid; found in fermenting beverages.

Halobacterium. Rod-shaped, but highly pleomorphic, obligately halophilic requiring at least 12 per cent salt for growth; frequently chromogenic, producing non-water-soluble carotenoid pigments (orange to red); occurring in tidal pools, salt lakes, salted fish and meat.

(E) CAULOBACTERIACEAE. Usually aquatic; non-filamentous, rod shaped bacteria usually attached to a substratum by a stalk, or, in the absence of a stalk, motile or free-floating. This family is heterogenous in that the two best-known genera *Caulobacter* and *Gallionella* differ substantially in the nature of the stalk (p. 307).

Caulobacter. Protoplasmic stalk arises from the pole of a rod-shaped or vibroid cell and adhesive material is secreted by the distal end; daughter cells arising by division of the stalked cell are non-stalked and motile with a single polar flagellum; may produce yellow or orange pigments; known to occur in fresh water, sea water, soil, and the intestine of millipedes.

Gallionella. Stalk, often spirally twisted on which feric hydroxide is deposited; associated with the stalk are kidney shaped cells which may be motile; occur in iron-bearing fresh water and sea water.

(F) SIDEROCAPSACEAE. Spherical, ellipsoidal or rod-shaped bacteria; commonly embedded in mucilaginous capsules in which iron or manganese compounds may be deposited; occur in water, on surface films and on submerged objects; this family contains several genera which have been inadequately studied.

(G) SPIRILLACEAE. Rigid, curved or spirally twisted cells, frequently attached after division thus forming chains, motile by tuft of polar flagella in *Spirillum* and single flagellum in other genera; usually aerobic but a few anaerobic strains known (e.g., *Desulphovibrio*); mainly water organisms but some parasites of higher animals.

Vibrio. Cells single, comma-shaped or remaining attached to form spirals; usually grow rapidly on laboratory media; widely distributed as saprophytes in salt and fresh water and soil, some parasitic in animals, including a few important pathogens, e.g., *V. cholerae*, cause of asiatic cholera in man; *V. fetus*, one cause of infertility and abortion in cattle and sheep; *V. coli*, cause of swine dysentery; and *V. jejuni*, cause of dysentery in cattle.

Desulphovibrio. Cells, single, curved or joined in short chains; strictly anaerobic; reduce sulphates to hydrogen sulphide; occurring in sea water, marine mud, fresh water, and soil.

Spirillum. Long screw-shaped cells, motile by tufts of flagella at one or both ends; volutin granules usually present; aerobes; in fresh and salt water; one species, *Sp. minus*, causes rat-bite fever in man.

B. Eubacteriales

Simple undifferentiated, rigid cells; spherical or straight rods; if motile by peritrichous flagella; no photosynthetic pigments; stain easily by aniline dyes; not acid-fast. Thirteen families recognized.

(A) AZOTOBACTERACEAE. Large, oval pleomorphic Gram-negative cells, sometimes yeast-like; obligate aerobes, fix free nitrogen in the presence of carbohydrates or other energy source; free living in soil and water. *Azotobacter* (p. 501).

(B) RHIZOBIACEAE. Cells rod-shaped, usually Gram-negative; aerobic; saprophytes, symbionts, and plant pathogens forming abnormal growth on roots and stems.

Rhizobium fixes nitrogen when grown symbiotically in root nodules of leguminous plants (p. 501).

Chromobacter produce a violet pigment which is soluble in alcohol but not in water or chloroform; saprophytic soil and water organisms.

(C) ACHROMOBACTERIACEAE. Gram-negative, small to medium size rods; motile by means of peritrichous flagella or non-motile; some produce water-insoluble yellow, orange, brown, or red pigments; mostly non-pathogenic; usually found in water or soil.

Alcaligenes. Not usually pigmented on ordinary media; when chromogenic are yellow or yellowish; do not attack polysaccharides or produce acid from carbohydrates; occur in the intestinal tracts of vertebrates or in dairy products.

Achromobacter. Similar to *Alcaligenes* but small amounts of acid produced from hexoses; occur in salt and fresh water and in soil where they are one of the predominant genera.

Flavobacterium. Yellow, orange, red, or brown pigmented on ordinary media; do not attack polysaccharides; occur in water and in soil where they are one of the predominant genera.

Agarbacterium. Attack polysaccharides, characteristically found on decomposing seaweed and in sea water, but also in fresh water and soil.

Beneckea. Differentiated from the previous genus mainly by the ability of these organisms to decompose chitin; occur in salt and fresh water and in soil.

(D) ENTEROBACTERIACEAE. Gram-negative, aerobic rods; active fermenters of carbohydrates, often producing CO_2 and H_2; many genera serologically related; many species are saprophytes; others inhabit the gut of animals, frequently causing intestinal disturbance; some are parasitic on plants, causing blights and soft rots.

Escherichia. Ferment many sugars, including lactose, to produce acids and gases (H_2 and CO_2, ratio 1:1); acetyl-methyl-carbinol not produced, methyl red test positive. *E. coli* is a parasite in the gut of many animals where it is generally non-pathogenic, consequently its presence in water supplies is used as an indicator of faecal pollution (p. 585). It occasionally causes cystitis and peritonitis when gaining access to the bladder and peritoneum respectively; specific serotypes cause enteritis and septicaemia in newborn calves, bowel oedema in piglets and enteritis in babies. Other species, e.g., *E. freundii*, may also be found in the gut of animals, but also have a wider distribution in water and soil.

Aerobacter. Ferments many sugars, including lactose, to produce acids and gases (H_2 and CO_2, ratio 1:2); produces acetyl-methyl-carbinol but methyl red test is negative; can utilize citrate as sole carbon source; widely distributed in nature, including the gut of man and other warm-blooded animals.

Klebsiella. Sugar fermentation variable, but usually a number of sugars, including lactose, are fermented; non-motile; acetyl-methyl-carbinol may or may not be produced; grows on certain media as mucoid colonies owing to excessive polysaccharide-capsule production; one cause of lobar pneumonia and infection of the urinogenital tract in man; one cause of mastitis in cattle and metritis in mares.

Erwinia. Motile rods which normally do not require organic nitrogen compounds for growth; plant pathogens causing dry necroses, galls or wilts, e.g., *E. amylovora*, the cause of fireblight or blossom-blight of pear and twig-blight of apple; unable to liquefy pectate gels or produce gas from sugars.

Pectobacterium. Motile rods, normally not requiring organic nitrogen compounds for growth; plant pathogens causing soft rots, e.g., *P. carotivorum* causing soft rot of the fleshy (parenchyma) including tissues of carrots, turnips, potatoes, bulbs, and mushrooms; able to liquefy pectate gels and many can produce gas from sugars.

Serratia. Small rods; produce various amounts of red pigment (prodigiosin), often depending upon the fermentable carbohydrate present. *S. marcescens* (syn. *Chromobacterium prodigiosum*) readily produces white variants; occurs in water, soil and milk; in the latter produces blood-like discoloration.

Proteus. Motile, straight rods with many flagella; some species swarm over the surface of solid media devoid of bile salts; pleomorphism is marked; glucose invariably fermented and usually other sugars but not lactose; urea usually decomposed; phenylalanine oxidatively deaminated to phenyl pyruvic acid; occurs in faeces or putrefying materials; often present in wounds.

Salmonella. Motile rods (except *S. gallinarum* and *S. pullorum*); many sugars fermented, but not lactose; gas is usually formed except by *S. typhi* and *S. gallinarum*); indole not produced, urea not split, acetyl-methyl-carbinol not produced and methyl red test positive; pathogenic to man and animals. *S. typhi* and *S. paratyphi* cause enteric fevers in man; other species cause infective food poisoning in man, and paratyphoid and chronic infections in animals and birds, animals may be symptomless carriers. Genus subdivided serologically, somatic antigens being characteristic of sub-groups and certain flagella antigens being species specific.

Shigella. Non-motile rods; sugars fermented without gas formation, lactose may be fermented late; possess distinctive antigenic structures; some cause bacillary dysentery in man and occasionally in primates; a few species are non-pathogenic inhabitants of the gut of warm-blooded animals.

(E) BRUCELLACEAE. Small, coccoid to rod-shaped Gram-negative cells; filamentous and pleomorphic forms occasionally found; aerobic or facultatively anaerobic; fastidious in their growth requirements and hence obligate parasites, often pathogens of warm-blooded animals.

Pasteurella. Small ellipsoidal rods showing bipolar staining by special techniques; *Past. pseudotuberculosis* the only motile species, the flagella growing only at temperatures 18–26°C; most ferment various sugars (rarely lactose) to produce acids but no gas; pathogens include *Past. pestis*, causes bubonic and pneumonic plague in man and rodents, *Past. multocida* (*septica*), causes fowl cholera, swine plague, and haemorrhagic septicaemia in many warm-blooded animals; *Past. pseudotuberculosis*, produces tuberculous-like lesions in rodents and birds, may cause abortion in sheep; *Past. tularensis* (sometimes included in *Brucella*) causes tularaemia in man, is vector-transmitted from rodents.

Bordetella. Minute coccobacilli which do not ferment sugars; usually require complex blood media for growth; *Bord. pertussis*, causes whooping cough in humans, other related species cause similar human respiratory infections; *Bord. bronchisepticus* a motile, less fastidious organism, often causes bronchopneumonia in rodents and dogs.

Brucella. Small, non-motile coccobacillary rods, no sugars fermented; urea decomposed; *Br. abortus* requires additional CO_2 for growth; the various species cause undulant fever in man, usually contracted by the drinking of raw milk from infected animals, and contagious abortion or infertility in cattle, goats and pigs; chronic infections sometimes occur in other animals, e.g., the horse.

Haemophilus. Minute rod-shaped, non-motile organisms, each species differing in its dependence upon certain growth factors in blood (phosphopyridine nucleotide or V-factor and haemin or X-factor); some such as *H. influenzae* affect the respiratory tract, conjunctiva, and meninges of man and some other warm-blooded animals and may be associated with the influenza virus; others, such as *H. ducreyi* affect the genital region producing soft chancre in man.

Actinobacillus. Non-motile, small to medium-sized rods; some sugars fermented to produce acids only; *A. lignieresi*, causes actinobacillosis in cattle, commonly in the form of granulomatous lesions in the soft tissues, usually of the head, as in 'wooden tongue', small granules present in pus from these lesions, contain colonies of the growing organism.

(F) BACTERIODACEAE. Pleomorphic, usually Gram-negative rods, with rounded or pointed ends, varying in length to long filamentous forms; normally strict anaerobes; found primarily in the intestinal tract and in the oral cavity of warm-blooded animals; sometimes pathogenic, often associated with gangrenous conditions in man and other animals.

Bacteroides fragilis, has been isolated from cases of acute appendicitis, pulmonary gangrene and abscesses of the urinary tract in man; *Sphaerophorus necrophorous* is the cause of calf diphtheria; related organisms cause liver abscesses in cattle, labial necrosis in rabbits and kangaroos and foot rot in sheep.

(G) MICROCOCCACEAE. Gram-positive, spherical cocci arranged in irregular clusters, tetrads or cubical packets; catalase-positive; free living, saprophytic, parasitic or pathogenic.

Micrococcus. Cocci in irregular clusters or tetrads; some produce yellow, orange, or red pigments; saprophytes in milk and dairy products, soil, salt and fresh water or parasitic but never pathogenic.

Staphylococcus. Single cocci or in irregular clusters; the pathogenic strain, *Staph. aureus*, is usually pigmented (golden yellow), haemolytic (to red blood blood cells of ox, sheep, and rabbit, but not horse) and invariably produces the enzyme coagulase (mainly responsible for the fibrin-containing wall of the characteristic abscesses produced); a few strains produce an enterotoxin in food which results in toxic food-poisoning when ingested by man (p. 587); mastitis in cattle and pigs is frequently caused by staphylococci; *Staph. aureus* and the non-pathogenic *Staph. albus* are frequently found on the skin, in skin glands and on the nasal mucous membranes of warm-blooded animals.

Sarcina. Division in three planes results in the formation of cubical packets of cells; saprophytes.

(H) NEISSERIACEAE. Gram-negative, kidney-shaped cocci occurring characteristically in pairs; some non-pathogens will grow on nutrient agar but pathogenic species require media enriched with blood or serum; some species commensal in the nasopharynx of warm-blooded animals, others pathogenic for man.

Neisseria gonorrhoeae is the causal organism of gonorrhoea; *N. meningitidis* is one of the causal organisms of meningitis.

(I) BREVIBACTERIACEAE. Gram-positive, non-sporing rods, some almost coccal, others filamentous; many are pigmented; carbohydrates may or may not be fermented; aerobic and facultatively anaerobic species occur; in dairy products, soil, decomposing materials, salt and fresh water.

(J) LACTOBACILLACEAE. Gram-positive rods or cocci, which divide in one plane to produce chains; carbohydrates, essential for growth, are fermented to lactic acid and other products; microaerophilic (i.e., optimum growth at reduced oxygen tension) to anaerobic; catalase negative.

Streptococcus. Cells spherical or ovoid, in pairs or chains of various lengths; some possessing a group-specific capsular polysaccharide by which they are serologically subdivided into Lancefield Groups (A to S); each group produces its own range of pathogenicity in specific animal host; Group A are haemolytic streptococci (*Str. pyogenes*) causing scarlet fever, tonsillitis, puerperal fever, and erysipelas in man; streptococci causing mastitis in cattle include *Str. agalactiae* (Group B) and *Str. dysgalactiae* (Group C); a third mastitis streptococcus, *Str. uberis*, does not fall into a Lancefield group; other species in Group C principally cause pyogenic infections in warm-blooded animals (e.g., *Str. equi*, causes strangles in the horse); Group D includes the enterococci, e.g., *Str. faecalis*, normal inhabitant of the intestinal tract but able to cause cystitis and sub-acute bacterial endocarditis; Group G includes species mainly pathogenic to dogs, e.g., *Str. canis*; Group N includes *Str. lactis*, found in milk, an important starter in cheese manufacture. Other streptococci, which do not possess a specific polysaccharide and therefore cannot be classified into Lancefield groups are commonly commensal in the mouth (e.g., *Str. salivarius*) but when introduced into the blood during dental procedures are able to settle on heart valves previously damaged by rheumatic fever, to cause sub-acute bacterial endocarditis. These are often termed viridans streptococci by virtue of their ability to discolour blood pigments.

Diplococcus pneumoniae (pneumococci). Lancet-shaped cells, usually in pairs; surrounded by a distinct polysaccharide capsule; share with viridans streptococci the property of partial haemolysis, but are bile soluble; occur in the respiratory tract of man and animals causing lobar pneumonia, meningitis, and other infections.

Peptostreptococcus. Strictly anaerobic, non-haemolytic streptococci; found in septic and gangrenous lesions of man and animals.

Leuconostoc. Spherical cocci, which grow in solutions of carbohydrates (especially sucrose) or media containing carbohydrates (e.g., milk or plant juices) producing characteristic polysaccharide slime: *Leuc. mesenteroides* found in fermenting plant and vegetable materials, such as crude sugar in refineries or silage.

Lactobacillus. Slender, non-motile rods; many workers recognize three main subdivisions of the genus according to their growth temperatures and end-products of carbohydrate fermentation; *Thermobacterium* and *Streptobacterium* are homo-fermentative (i.e., carbohydrates are fermented to lactic acid only); species of the former grow at 45°C but not at 15°C, species of the latter grow at 15°C but rarely at 45°C; *Betabacterium* are heterofermentative (i.e., carbohydrates are fermented to lactic acid and volatile acids and CO_2); commonly present in the intestine of warm-blooded animals (particularly in breast-fed infants), fermented milk and milk products and some plant products. In consequence of their requirements, of specific vitamins for growth, certain species have been used for the biological assay of vitamins.

(K) PROPIONIBACTERIACEAE. Gram-positive irregular rods, ferment carbohydrates, and often organic acids (e.g., lactic acid) and alcohols with the production of aliphatic acids (e.g., propionic, butyric, acetic acids and ethanol); grow best anaerobically; occur in the intestinal tract of animals and in other suitable substrates, (e.g., dairy products); important in the breakdown of the products of acid-producing flora in cheeses, producing propionic and acetic acids together with CO_2, which give aroma and texture to many Swiss cheeses.

(L) CORYNEBACTERIACEAE. Gram-positive, usually non-motile rods, often barred, some containing metachromatic granules; some species pleomorphic; mainly aerobic; occurring in dairy products, soil and as parasites and pathogens of animals and plants.

Corynebacterium. Stain irregularly with methylene blue to produce characteristic barring; metachromatic granules present in many species; often club-shaped; following transverse division the daughter cells remain attached but orientate themselves into angular or palisade arrangements; non-motile except for a few plant pathogens; *C. diphtheriae* produces a powerful exotoxin causing diphtheria in man; *C. renale, C. equi, C. pyogenes*, and *C. ovis* are animal pathogens causing pyelonephritis in cows, pneumonia in foals, general pyogenic infections in many animals and chronic lesions resembling tuberculosis in sheep respectively.

Listeria monocytogenes (the only species), small rods producing characteristic tumbling motility; catalase positive; produces cerebral lesions in sheep and cattle resulting in inco-ordinated movement referred to as 'circling disease'; produces monocytosis (i.e., an increase in blood monocytes) in rabbits.

Erysipelothrix. Non-motile rods with a tendency to mass together giving the appearance of a network of cells; catalase negative; *Ery. insidiosa (syn: rhusiopathiae)* causes swine erysipelas but also infects birds and man.

Microbacterium. Unbranched, non-motile, rod-shaped organisms; thermoduric saprophytes occurring in dairy products.

Cellulomonas. Small pleomorphic rods or cocci, usually motile; commonly attacking cellulose; of soil or plant origin.

Arthrobacter. Usually rods in young culture becoming coccoid later; usually non-motile; occur in soil, where they form an important part of the autochthonous flora, on plants, in cheese, and in activated sludge.

(M) BACILLACEAE. Gram-positive, large rods, sometimes in chains; the chief endospore-producing family; frequently motile, by peritrichous flagella; either aerobic or anaerobic, some species being facultatively anaerobic or aerotolerant; active in decomposition of animal tissues and residues; mostly saprophytes commonly found in soil, some are pathogens of animals including man.

Bacillus. Aerobic spore-bearing bacilli, spores are usually no wider than the vegetative cells; frequently produce rough colonies sometimes with a hair-like periphery resulting from long chain formation; active producers of enzymes such as penicillinase, collagenase, chitinase; widespread; many thermophils, e.g., *B. stearothermophilus*, cause spoilage of canned foods; others are pathogenic to insects, e.g., *B. larvae*, the cause of American foul brood of bees and one species, *B. anthracis*, is the cause of anthrax in many warm-blooded animals.

Clostridium. Anaerobic or aerotolerant, spore-bearing rods; cells usually swollen by the spore which may be either spherical or oval, terminal, sub-terminal or equatorial; commonly found in soil and in the intestinal tract; powerful exotoxins are produced by many pathogenic species, e.g., *Cl. tetani* and *Cl. botulinum*, the causal organisms of tetanus and botulism respectively; in addition to producing exotoxins other pathogens are saccharolytic and proteolytic, e.g., *Cl. perfringens (welchii)*, the causal organism of gas gangrene in man, lamb dysentery and enterotoxaemia in sheep, *Cl. chauvoei* the causal organism of blackleg in cattle, *Cl. septicum* the causal organism of gas gangrene in man, malignant oedema in cattle and braxy in sheep; saprophytic species may be thermophilic and cause spoilage of canned foods, e.g., *Cl. thermosacharolyticum*, others are of particular importance in industrial processes, e.g., *Cl. acetobutylicum* used in the manufacture of butyl alcohol from carbohydrates. Several species are active in the fixation of nitrogen in soil, e.g., *Cl. pasteurianum*.

ACTINOMYCETALES

Introduction

The actinomycetes are a group of Gram-positive bacteria with a capacity for mycelial growth. For long it was uncertain whether their phylogenetic affinities lay with the bacteria or the true fungi. However, their prokaryotic cellular organization, the chemistry of their cell walls, their nitrogen metabolism and their sensitivity to antibiotics and phages, more specifically termed actinophages, leave little doubt that they are indeed bacteria.

The production of a very fine mycelium is the primary characteristic of actinomycetes. The individual hyphae are usually not more than *ca.* 1 μ in diameter. The hyphae branch and possibly in all species become septate by the inward growth of rings of wall material. Within each compartment occur one or more chromatin bodies, often of irregular form. The mycelium of some species lies wholly within or in contact with the substrate, but in others there are aerial parts too.

Many actinomycetes produce spores which differ from the endospores of bacilli not only in method of formation (p. 291) but in being only mildly resistant to heat. When they germinate they form one or more germ-tubes whose walls are continuous with the spore's own walls. The combination of aerial growth and sporulation usually confers a cottony or powdery texture on the surface of the colony, while colonies which lack aerial mycelium are either glossy or matt.

Spores may be formed on special hyphae (sporophores) either singly or in groups of small numbers (e.g., 2 or 4) or in chains in indefinite numbers. Conidial chains of indefinite length are characteristic of the genus *Streptomyces* (Fig. 9.22, F–H). The sporophore wall first becomes two-layered, and then extensions of the inner layer grow inwards to become the septa which delimit the spores. The outer layer forms a common sheath until it ruptures and frees the spores. The aerial hyphae of 'streptomycetes' have been reported to contain a variety of configurations of the chromatin material. During sporulation, elongated chromatin masses in the young aerial hyphae appear to subdivide in a series of stages to yield a number of round chromatin bodies of which one is included in each spore. Several experimental approaches have indicated that each spore contains only a single haploid genome. Reports that the sporophores of *Streptomyces* originate by the germination of special initial cells found in tangled nests of hyphae in the substrate mycelium lack substantiation as also do reports that dumbbell-shaped configurations of chromatin material seen in immature spores represent stages in nuclear fusion.

The sporophores in streptomycetes may sometimes be grouped to form structures resembling the coremia (p. 401) or pycnidia (p. 402) of true fungi. One family of actinomycetes (the Actinoplanaceae) is characterized by the formation of 'sporangia' (Fig. 9. 22, I, J) which, however, are quite unlike those of the fungi. A sporangium is formed as a terminal vesicle on a hyphae, and into it grows one or more filaments which then segment into chains of spores which in some species become flagellate and motile. In the genus *Dermatophilus* the mycelium breaks up into motile cocci (Fig. 9.22D), and motility has also been

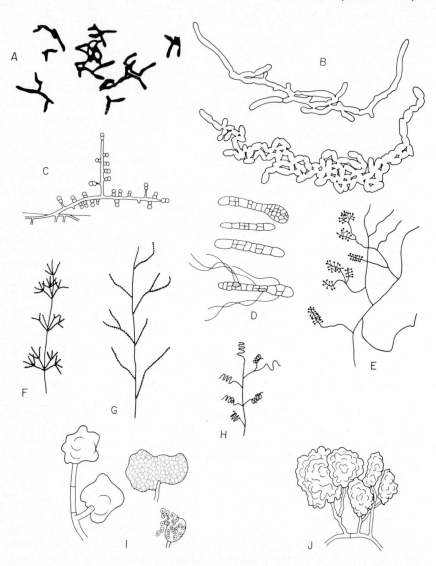

Fig. 9.22 Actinomycetes. (**A**) *Actinomyces israeli* (×1400). Branching elements from culture. (Drawn from photograph by Pine, L., Howell, A. and Watson, S. J. (1960) *J. gen. Microbiol.*, **23**, 403.) (**B**) *Nocardia corallina* (×1,900). A young colony already beginning to fragment; and below, the same colony nearly nineteen hours later and showing extensive fragmentation. (Drawn from photograph by Brown, O. and Clark, J. B. (1966) *J. gen. Microbiol.*, **45**, 525.) (**C**) *Microbispora rosea*. Schematic diagram of typical aerial hypha. (After Lechevalier, M. P. and Lechevalier, H. (1957) *J. gen. Microbiol.*, **17**, 104.) (**D**) *Dermatophilus dermatonomus*. Filaments with transverse and longitudinal septa, and one producing motile cocci. (After Thompson, R. E. M. and Bisset, K. A. (1957) *Nature, Lond.*, **179**, 590.) (**E**) *Micromonospora* sp. Fine mycelium bearing single spores. (After Skinner, C. E., Emmons, C. W. and Tsuchiya, H. M. (1947)

recorded for a number of species of *Nocardia*, but is lacking in most actinomycetes.

Most actinomycetes stain relatively easily with basic dyes and are Gram-positive. A few, principally the pathogenic mycobacteria, stain only with difficulty but once stained, resist decolorization by mineral acid. This property of acid-fastness varies between species and depends to some extent upon the conditions of growth. It is attributed to a cellular lipid component (mycolic acid) in the morphologically intact cell, the degree of acid-fastness being proportional to the amount of mycolic acid present. The Ziehl Neelsen technique is used for staining these organisms. The preparation is stained for 10 minutes with a hot, concentrated solution of fuchsin in 5 per cent phenol (carbol-fuchsin). Prolonged decolorization with 20 per cent mineral acid (or a mixture of acid and alcohol) removes the dye from all but acid-fast micro-organisms, non-acid-fast ones being subsequently stained by a dye of contrasting colour.

In *Streptomyces* the nuclear bodies stain with Feulgen's stain (p. 290) throughout all stages in their division, and in no actinomycetes has a spindle apparatus been recognized. Electron microscope studies reveal that the cytoplasm contains a particularly extensive membrane system and many ribosomes. Most of the membranes are orientated randomly, but adjacent to septa they are ordered into discrete bodies which connect with the plasma membrane adjoining the developing septa. They are thought to be concerned with septum formation as in the true bacteria (p. 286).

The cell wall composition of actinomycetes differs from that of fungi and resembles that of Gram-positive bacteria. The walls are thin (*ca.* 10–20 mμ) and contain neither chitin nor cellulose, but appear to be complexes of sugars, amino-sugars, and a few amino acids. Two features particularly characteristic of actinomycetes are the presence of muramic acid (p. 282) and the narrow range of amino acids present. The walls contain major amounts of *either* 2,6-diaminopimelic acid (DAP) *or* lysine. The walls of actinomycetes, as with Gram-positive bacteria in general (p. 282), are conspicuously lacking in aromatic and sulphur-containing amino acids, in proline, in histidine, and in arginine.

Actinomycetes have a considerable economic importance. In nature they contribute to the mineralization of organic residues. Only a few are plant pathogens but at least one of these causes a plant disease of great importance, i.e., common scab of potato. A number cause debilitating or even lethal infections of animals and man. On the credit side, they produce the majority of antibiotics (p. 682) of medical importance, including substances active against bacteria, fungi, protozoa, rickettsiae, larger viruses, and neoplasms.

Henrici's Molds, Yeasts and Actinomycetes, 2nd. edn. John Wiley & Sons, Inc., N.Y.) (**F**) *Streptomyces phaeochromogenes.* Straight or slightly wavy sporophores arising on side branches of aerial hypha. (After Ettlinger, L., Corbaz, R. and Hütter, R. (1958) *Arch. Mikrobiol.,* **31**, 336.) (**G**) *Streptomyces reticuli.* Straight, branching sporophores borne in verticils. (Source as F.) (**H**) *Streptomyces violaceoniger.* Spiral spore chains. (Source as F.) (**I**) *Actinoplanes utahensis* (× 1,000). Sporangia, one showing spores within, one dehiscing. (After Couch, J. N. (1963) *J. Elisha Mitchell Sci. Soc., 79,* 53.) (**J**) *Amorphosporangium auranticolor* (× 1,000). Group of sporangia. (Source as I)

Natural occurrence

Most actinomycetes are soil inhabitants. Indeed the characteristic odour of soil after it is ploughed or wetted by rain is largely due to these organisms, many of which produce strong earthy smells in pure culture. The numbers of actinomycetes in soils as revealed by the dilution plate method are second only to the numbers of bacteria, but there is in fact no very accurate way of estimating their abundance in the soil. The large count on dilution plates suggests that actinomycetes are present there very largely as spores. Actinomycetes are worldwide in their distribution. Their numbers generally increase with the warmth of the climate and decrease with depth in the soil. The difficulty with which streptomycete spores are wetted may enable them to be dispersed in the air-water interfaces in the upper layers of the soil rather than to sink to lower regions. Like most other bacteria their growth is favoured by alkaline conditions. Cultivation practices which improve soil aeration often lead to increased numbers of streptomycetes which are the commonest of the soil-inhabiting actinomycetes. In acid and waterlogged soils they are relatively scarce.

The important and unique role of actinomycetes in the soil is attributable to their ability to flourish after faster growing micro-organisms have transformed organic residues into a dark homogeneous mass, rich in lignin, certain hemicelluloses, and proteins, which is no longer favourable for the continued growth of these other organisms. Actinomycetes compete effectively with the latter only through their ability to break down residual nutrients remaining in the soil.

In view of the large numbers of soil actinomycetes that produce antibiotics under laboratory conditions one might expect these substances to be of importance in nature. However, the formation of antibiotics by actinomycetes actually living in the soil is hard to demonstrate, and occurs, if at all, to only a slight extent owing to the unsuitable nutritional conditions. Further, any antibiotic effect would be limited to the vicinity of the colony responsible because antibiotics are inactivated by adsorption on to such surfaces as clay minerals and are destroyed by non-biological and biological reactions. However, antibiotic-producing streptomycetes may contribute to the fungistatic properties of soil (p. 511).

If compost is allowed to become heated by microbial action, the surviving microflora which continues the decomposition contains many thermophilic actinomycetes. Several species of *Streptomyces* and *Micromonospora* are thermophiles, as are all species of *Thermoactinomyces* and *Thermomonospora*. Species of *Micromonospora* sometimes form a lime-like coating on hot decomposing masses of plant material. The inhalation of dust from mouldy hay which has spontaneously heated during its maturation may produce an allergic response known as 'Farmer's Lung' in persons who have become hypersensitive to antigens in the dust. Spores of thermophilic actinomycetes, notably *Thermopolyspora polyspora* and *Micromonospora vulgaris*, are the principal sources of this antigen. Because these spores are small they can penetrate deeply into the respiratory tract, in the peripheral parts of which the allergic symptoms develop.

A number of actinomycetes are responsible for infectious diseases of man and animals (p. 550). Leprosy and tuberculosis are caused by species of *Myco-*

bacterium (p. 549); certain species of *Actinomyces, Nocardia, Dermatophilus,* and *Streptomyces* cause superficial or even deep seated lesions (p. 550).

Although actinomycetes are so common in the soil, few cause disease in plants. *Streptomyces scabies* causes the extremely important scab disease of potato, and other streptomycetes cause scab of mangel and sugar beet, and pox or soft rot of sweet potato. Nevertheless other relationships with higher organisms may exist. Actinomycetes are probably the endosymbionts within the 'nodule clusters' (rhizothamnia) that are found on the roots of a variety of dicotyledons (e.g., members of the Casuarinaceae, Myricaceae, Betulaceae, Elaeagnaceae, Coriariaceae, and Rhamnaceae) and of which a number have been shown to fix atmospheric nitrogen. Although actinomycetes have been isolated from such nodules, and electron microscope studies reveal a morphological resemblance of the endosymbionts to actinomycetes, there are no conclusive reports of successful reinfection of symbiont-free plants with the isolated organism. Undoubted actinomycetes have been found living symbiotically with insects. Usually arthropods contain symbiotic micro-organisms only if they feed on nutritionally incomplete diets during their *entire* life cycle. Certain bloodsucking hemipterans, for example, contain *Nocardia* in crypts produced by folding of the midgut epithelium. It is suspected that such symbionts furnish their insect partners with vitamins. Organisms interpreted as actinomycetes are present even inside cells of certain insects.

Spores of streptomycetes are common in the air (p. 537), and strains of *Streptomyces, Micromonospora,* and *Actinoplanes* have been demonstrated in fresh and salt water. The activity of actinomycetes in reservoirs may impart unpleasant odours to water supplies. The structure of those members of the family Actinoplanaceae that produce motile sporangiospores is clearly suited to an aquatic existence, and soil water is a rich source of these organisms.

Physiology

The germinating spore that initiates the new actinomycete colony normally swells and then puts out one or two germ-tubes. Some spores fail to germinate if they are washed unless provided with dextrose. The germ-tube typically develops into a system of filaments which grow apically and branch monopodially. In *Actinomyces* and *Nocardia* the mycelium soon fragments into pieces which grow at both ends. Such fragmentation in *Nocardia corallina* (Fig. 9.22B) appears to be induced by agents which accumulate in the culture medium. In *Dermatophilus* the filaments become septate both transversely and longitudinally and separate into motile coccoid spores. In other genera the substrate mycelium typically remains intact, though in the Actinoplanaceae it is prone to fragment when grown on certain media which include potato dextrose agar. The mycelium of colonies which do not fragment grows almost entirely in the apical regions, the older parts either remaining quiescent or dying. The mycelium at first contains an optically homogeneous cytoplasm but this becomes vacuolated as it matures and fat and volutin granules may be found in the vacuoles. Anastomosis of filaments has been claimed but the evidence for this is equivocal. In *Streptomyces* and some other genera the substrate mycelium

soon gives rise to aerial mycelium on which the spores are formed. The onset
of sporulation in streptomycetes is related to a depletion of nutrients in the
medium or to other unfavourable circumstances. The ability to form aerial
mycelium implies a capacity for growth at a distance from the nutritive substrate and hence for an internal translocation of nutrients. This is a fundamentally different biological organization from that encountered in other
bacteria. Growth in well-aerated submerged culture is rapid and is followed by
lysis of the mycelium and even of the spores. In many streptomycetes the lytic
phase is accompanied by increased liberation of antibiotics. Although some
species are thermophilic and grow best at 50–56°C the majority of actinomycetes
are mesophilic. Isolates from soil and water grow best at about 25–30°C;
those from warm-blooded host animals grow best at about 37°C.

The basic nutrition of the majority of saprophytic actinomycetes is simple
and resembles that of a large number of other bacteria and a majority of fungi.
There is a requirement for an organic source of carbon and energy, but beyond
this all other nutritional needs can generally be met by inorganic salts, such as
nitrate or ammonium salts for nitrogen, and salts of the necessary mineral
nutrients among which phosphorus, potassium, and sulphur are needed in
greatest amounts. Some actinomycetes have additional requirements for particular vitamins, or for particular amino acids. Special nutrients may also be
necessary to sustain particular metabolic activities especially in industrial applications of actinomycetes. For example, cobalt needs to be present for synthesis
of vitamin B_{12} (cobalamin) and chlorine for the synthesis of chloramphenicol.
Although nitrogen fixation by free-living actinomycetes has sometimes been
claimed there is still much uncertainty as to whether this really does occur. The
general consensus of opinion is that it does not. Although the minimal nutrition
of actinomycetes is generally simple, the total pattern of nutrition is complicated firstly by the diversity of complex macromolecular nutrients which can
be digested by extracellular enzyme action; secondly by the breadth of the range
of organic nutrients that can enter the metabolic pathways of the cell; and thirdly
by the development in some species of considerable degrees of specificity for
particular organic nutrients. The pathogenic mycobacteria require very complex media for artificial culture. *Mycobacterium tuberculosis* (p. 549), grows only
in the presence of serum or coagulated egg to which amino acids, particularly
asparagine, are added. At blood temperature, cultures take up to 4 weeks to
develop under aerobic conditions. *Mycobacterium johnei* (p. 549) requires, in
addition, extracts of mycobacteria, which contain substances similar to vitamin
K and even on such complex media grows more slowly than do tubercle bacilli.

The majority of actinomycetes inhabit the soil and these as a group are
characterized by an extensive ability to digest complex materials and to use
unusual molecules. This is the principal reason for their success as soil inhabitants. Enzymes able to degrade dextrins, starch, cellulose, hemicelluloses, such
as mannans and xylans, pectin, chitin, fats, peptones, proteins, including kera-
tin, humic acids, and possibly even lignin are commonly produced. Laminarin,
alginates, and even agar may be digested by some species. Substances which
lyse fungal hyphae and bacteria are also formed. Many soil actinomycetes, and
Nocardia species in particular, utilize unusual molecules as sources of energy

and carbon. These include long-chain fatty acids, hydrocarbons, benzene, various aromatic compounds, steroids, and heterocyclic nitrogen compounds. Species of *Streptomyces, Nocardia,* and *Mycobacterium* able to oxidize ethane have been isolated from various soils. Most have no perceptible specificity for hydrocarbons for they grow well on a variety of other media. *Mycobacterium paraffinicum,* however, grows poorly on the usual protein and carbohydrate media and grows well only with gaseous hydrocarbons such as ethane, propane, or butane. Some strains will utilize only ethane. For this reason, *M. paraffinicum* is considered to have potential value for petroleum prospecting (p. 675). Among the simpler nutrients utilized as sources of carbon and energy are glucose, maltose, lactose, cellobiose, glycerol, alcohols, and various organic acids including fatty and amino acids. The latter are also particularly suitable nitrogen sources. A minority of actinomycetes utilize sucrose. The use of oxalic acid by some species of *Streptomyces* and *Nocardia* suggests they may have an important natural role in the detoxification of calcium oxalate in plant residues.

Most actinomycetes are aerobic organisms but obligate anaerobes occur in the genera *Actinomyces* and *Micromonospora.* Anaerobic actinomycetes possess a fermentative metabolism which yields acid but not gas from a variety of carbohydrates. Even aerobic forms may produce considerable quantities of acid—largely lactic—but the acid is later consumed. Indeed, in well-aerated cultures as much as 25–30 per cent of the substrate carbon may be converted to cellular material. The intermediary metabolism of actinomycetes is still not well known, but in several streptomycetes enzymes characteristic of the tricarboxylic acid cycle have been demonstrated. A feature of the metabolism of nitrogen in those actinomycetes in which it has been studied is that lysine is synthesized through the α,ϵ-diaminopimelic acid pathway as it is in other bacteria, and not through the alternative α-aminoadipic acid pathway. Actinomycetes, like other bacteria, do not contain sterols.

The smells that are characteristic of many actinomycetes are attributable to a variety of metabolites which includes acetic acid, acetaldehyde, ethanol, isobutanol, and isobutyl acetate. An earthy smelling neutral hydrocarbon oil, named geosmin, has been isolated from several species including *Streptomyces griseus.* Actinomycetes generate a great variety of substances in their environment, and these include vitamins, pigments, and antibiotics. The biosynthesis of some of these has been developed on an industrial scale. For example vitamin B_{12} is produced by *Streptomyces aureofaciens* and other species as a by-product of the aureomycin fermentation. Aureomycin, streptomycin, and neomycin (produced by *S. aureofaciens, S. griseus,* and *S. fradiae* respectively) are just three examples of the many medically important antibiotics pr oduced Up to 1962, approaching five hundred different antibiotics that are formed by actinomycetes had been described; only a few are of clinical value. The vast majority were from species of *Streptomyces,* while a few were derived from species of *Nocardia, Micromonospora, Thermoactinomyces,* and certain of the Actinoplanaceae. Antibiotic production appears to be a manifestation of the secondary metabolism of the organism. Given the appropriate precursors and enzymes systems an actinomycete begins to synthesize secondary metabolites in response to a slowing down of primary metabolic activity at the end of its growth phase or in response to derangement of normal growth activity. Antibiotics are thus produced only when the organism is grown under certain specific physical and chemical conditions. Antibiotic production is maximal near the stage of maximum growth of the actinomycete, but good growth can proceed without antibiotic production because antibiotics are not inevitable by-products of growth. The ability to produce particular antibiotics is not correlated with taxonomy, for the same substance may be produced by members of different species or different genera, while sometimes different strains of the one species or even the one organism produce several different antibiotics. Further, a strain may lose or gain the ability to produce an antibiotic as a result of mutation.

Genetical change in actinomycetes may result in the development of sectors in colonies, in changes in nutrition such as the acquisition of special organic requirements, or in alterations in pigment and antibiotic production, or pathogenicity. The ultimate source is mutation. Mutagenic agents have sometimes been used to obtain strains of altered antibiotic activity. Genetical variation by mutation is augmented by the occurrence of a form of heterokaryosis (p. 113) and genetical recombination. Heterokaryosis implies the transfer of a complete 'nucleus' from one cell to another in the manner familiar in fungi, though of course actinomycetes are not eukaryotic organisms, and the active migration of nuclei commonly occurring in fungi would not be expected to occur. Such a modified heterokaryosis has been reported in a number of streptomycetes. The simplest mechanism that may be invoked to account for the establishment of heterokaryons is that nuclear bodies are transferred by way of occasional hyphal fusions in the substrate, but the existence of such anastomoses has not been proven. Genetical recombination occurs in many streptomycetes apparently without the participation of any specialized sexual structures. It is not known whether this is a regular part of the life cycle. In *S. coelicolor* the possible sequence of events commences with the transfer of probably incomplete nuclear bodies by a process of conjugation between substrate hyphae. This yields heterozygous 'nuclei' which multiply and undergo some chromosomal loss and terminal deletion, and possibly mitotic crossing over. Eventually by crossing over and reduction, haploid recombinants emerge. In *S. coelicolor* the genes are located on two linkage groups, of which the simplest physical counterpart would be a pair of chromosomes. This, if true, would contrast with the single chromosome that exists in *Escherichia coli*. Recombination has even been reported as occurring between different species.

Claims that genetical change has been effected by transduction and transformation require substantiation. Cytoplasmic inheritance has been reported in respect of the ability of *Streptomyces scabies* to produce tyrosine.

Particular viruses (p. 256) attack actinomycetes, and so far have been found in several streptomycetes, *Actinomyces bovis*, and *Nocardia farcinica*. Such viruses—'actinophages'—are abundant in the soil. It has been possible experimentally to separate the life-cycle of actinophage into phases of attachment, insertion, growth, maturation, and lysis. Phages attack mycelium only at certain stages of its development. Aerial mycelium and spores are not lysed. The harbouring of phage without lysis occurring ('lysogeny') has also been demonstrated. Some actinophages possess a tail, which in several at least is unlike that of the T-even coli phages in that it lacks a visible contractile sheath, tail fibres, and a base plate. A terminal knob or group of prongs possibly assumes the function of the base plate, though the mechanism of attachment is obscure. In other actinophages there is no tail at all.

Classification

Early attempts to classify Actinomycetes on morphological characters resulted in confusion owing to the tendency for these organisms to alter during periods of artificial culture. The recent trend towards the increased use of the chemistry of the organism as a source of taxonomic criteria promises to give a more satisfactory system, but at present reliance must still be placed on morphological characteristics.

Table 9.3 contains a key to the major taxa and notes on these are added below. Extensive keys for the identification of actinomycetes have been published by Waksman (1961).

Mycobacteriaceae contains two genera, *Mycobacterium* and *Mycococcus*, the members of which range from saprophytes to obligate parasites. They are characterized by the capacity to form a rudimentary mycelium under at least some conditions of cultivation. This may consist of branched vegetative cells or may approach the level of development seen in *Nocardia* (in Nocardiaceae p. 335). Morphology, the presence in the cell walls of large amounts of amorphous lipid (possibly lipoprotein) together with arabinose, galactose, and *meso*-DAP, and serological resemblances suggest a relationship not only with *Nocardia* but with *Corynebacterium* (p. 324) normally classified in the Eubacteria but sometimes included here. *Mycobacterium* has been distinguished on the property of acid-fastness but this is not a sound distinction since many species of *Corynebacterium* and *Nocardia* are acid-fast under particular nutritional conditions.

Mycobacterium: acid-fast, cells usually rod-shaped but tending to branch. *M. tuberculosis* (Fig. 15.5B) exists in a number of strains causing tuberculosis in man, cattle, pigs, birds, or fish; *M. leprae* and *M. lepraemurium* cause leprosy in man and rats respectively; *M. johnei* causes chronic enteritis in cattle and sheep; *M. rubiacearum* is the N-fixing bacterium causing leaf-nodules in some shrubby members of the Rubiaceae (p. 620). *Mycococcus*: non-acid-fast, cells usually spherical, but branched and resting cells may develop; widely distributed soil saprophytes.

Actinomycetaceae, now usually restricted to the single genus *Actinomyces*. *Actinomyces*: true mycelium which readily fragments; microaerophilic or anaerobic; cell wall contains lysine in place of the DAP characteristic of all other actinomycetes and lacks arabinose; best-known species are parasitic but serologically related ones have been isolated from soil water. Actinomycosis, a chronic disease of animals, including man, is caused by *A. israeli* and *A. bovis*. *A. israeli* is frequently isolated from the crypts of the normal human tonsil, where it can survive in the microaerophilic conditions prevailing, and from this site is thought to invade mouth wounds, e.g., those caused by tooth extraction. The lungs may also be infected. *A. bovis* causes 'lumpy jaw' of cattle, in which the jawbone is invaded and enlarged by the formation of new bone.

Table 9.3   A key to the major taxa of Actinomycetes.

Mycelium rudimentary or absent, no spores . .   MYCOBACTERIACEAE
                                                (*Mycobacterium, Mycococcus*)

True mycelium, fragmenting into bacillary
  elements; anaerobic or microaerophilic   . .   ACTINOMYCETACEAE
                                                (*Actinomyces*)

True mycelium, dividing in all planes to
  produce motile cocci   . . . .   DERMATOPHILACEAE
                                                (*Dermatophilus*)

True mycelium, dividing only transversely:
  Sporangia formed, sometimes conidia also   .   ACTINOPLANACEAE
                                                (*Microellobosporia, Strepto-*
                                                sporangium, Spirillospora,
                                                Actinoplanes, Ampullariella,
                                                Amorphosporangium)

  No sporangia:
    Aerial mycelium absent at ordinary tempera-
      tures; conidia borne singly on lateral
      branches of fine substrate mycelium;
      glycine and *meso*-DAP are major con-
      stituents of walls . . . .   MICROMONOSPORACEAE
                                                (*Micromonospora*)

    Aerial mycelium bearing chains of conidia;
      glycine and L-DAP are major consti-
      tuents of walls . . . .   STREPTOMYCETACEAE
                                                (*Streptomyces*)

    With or without aerial mycelium;
      mycelium sometimes fragmenting;
      conidia absent, or single, or paired,
      or in chains; contain *meso*-DAP, but lack
      major amounts of glycine in walls . . .   NOCARDIACEAE

        Mycelium fragmenting into bacillary
          elements, aerobic · . . . .   *Nocardia, Micropolyspora*

        Mycelium not fragmenting; aerial
          mycelium present; conidia borne—
          Singly . . . .   *Thermomonospora, Thermoactino-*
                                                myces

          or in pairs   . . . .   *Microbispora*

          or in chains . . . .   *Pseudonocardia*

Other tissues, including the udder, may be invaded and a similar infection sometimes occurs in sows. *A. baudetii* is present in certain lesions in cats and dogs.

  In tissues affected by *Actinomyces* and in the resulting pus, small yellow granules (the so-called 'sulphur granules') occur. These are compact calcified colonies of the organism, consisting of a central Gram-positive mass of mycelium
with rods and cocci derived from fragmentation and a peripheral zone of radiating Gram-negative club-shaped structures (Fig. 15.5C and D). These latter
which give the name 'ray fungus' to these organisms, consist mainly of lipid
substances and are thought to be a host reaction to invasion. Actinomycoses
are susceptible to penicillin treatment.

Nocardiaceae contains several genera, some of which were formerly included in Actinomycetaceae but differing from *Actinomyces* in cell wall composition, plus others formerly placed in Micromonosporaceae (see below). *Nocardia*, mycelium at first non-septate, later becoming septate and fragmenting into rod-shaped cells (occasionally motile by one to four polar or lateral flagella); aerobic; potentially acid-fast; cell walls containing lipids (as in *Mycobacterium* but differing from those of that genus); *meso*-DAP (in contrast to the L-DAP of the streptomycetes) and, in most species, arabinose and galactose. Most species are soil saprophytes but a number cause diseases (nocardioses) of animals including man (Fig. 15.5A). *N. asteroides* causes mammary and cutaneous lesions in cattle; dogs and man may also be affected. *N. farcinica* attacks lungs and lymph nodes in cattle, causing a chronic disease known as bovine farcy, inducing the development of nodules in usually one limb and resulting in death; *N. madurae* is the principal cause of madura foot, a tropical or subtropical disease of man. The organism gains entry through abrasions, caused by the habit of walking barefoot on soil, and induces swelling and malformation of the foot. Small granules of varied colour develop in the pus.

Other genera differ from *Nocardia* in details of cell wall composition and in the degree of fragmentation of the hyphae. Further subdivisions of the family may follow further investigations.

Dermatophilaceae, single genus *Dermatophilus* (syn. *Polysepta*); aerobic parasites; slender hyphae (1–2 μ diam.) giving rise to wider ones (3–5 μ) which divide transversely into narrow segments and then longitudinally two or three times (Fig. 9.22D), the individual cells then separating as motile cocci each with a tuft of long flagella; cell walls contain major amounts of *meso*-DAP (cf. *Nocardia*). *D. congolensis* and *D. dermatonomus* cause acute local infections of the skin of cattle, sheep and other animals, following damage from other causes, and sometimes prove fatal. The damage caused to hides or wool is of economic importance. *D. pedis* is thought to cause strawberry foot rot, an extensive inflammation of the feet of sheep. A recently described genus *Geodermatophilus* has been assigned to this family to include organisms resembling *Dermatophilus* in their complex life cycles and wall composition, but differing in having only rudimentary mycelium and in being soil inhabitants.

Micromonosporaceae contains single genus, *Micromonospora*, slow growing; mycelium extremely fine (0·3–0·6 μ diam.) profusely branched, aerial mycelium lacking at ordinary temperatures; aerial mycelium is, however, abundant on hot decomposing masses of plant material and is normally produced by the thermophilic *M. vulgaris* at 60°C; bearing spores singly at apices of short lateral branches (Fig. 9.22E); cell walls contain major amounts of glycine and *meso*-DAP. L-DAP has also been reported.

Streptomycetaceae contains the well-known genus *Streptomyces* and a few other less known ones (often of relatively complex morphology, e.g., forming sclerotia or producing spores on verticillate branches or in 'pycnidia'). *Streptomyces*, aerobic; abundant and widespread in soils; aerial mycelium giving rise to chains of spores; in *S. griseus* and allies chains arise on substrate mycelium also); cell walls containing major amounts of glycine and L-DAP (distinguished from those of *Nocardia* by absence of arabinose and *meso*-DAP). Sporulating streptomycetes can be identified on such characteristics as random or whorled arrangement of sporophores and whether spore chains are straight or coiled (Fig. 9.22, F, G, H).

Electron microscope studies reveal that the spores are smooth, warty, spiny, or hairy, and that the ornamentation of a given species is more or less independent of conditions of cultivation. The smooth form is commonest. Species with straight or flexuous spore chains are smooth spored. Species with ornamented spores always produce spiral spore chains, but the converse is not always true. Strains with blue to blue-green spore masses seem always to have spiny spores. The ornamentation promises to be a useful diagnostic feature in certain sections of the genus. Production of hydrogen sulphide in culture is another useful diagnostic feature. A codified system of taxonomic criteria has been proposed to simplify the identification of streptomycetes by providing a working classification which characterizes isolates which may not have been fully named. It is based on capacity to pigment the medium darkly, spore ornamentation, form and colour of aerial mycelium, and utilization of selected carbon sources. *S. scabies* causes common scab of potato (p. 625). Many saprophytic species are active producers of antibiotics (including several clinically useful ones, p. 682).

Actinoplanaceae contains a few genera all of which produce spores in sporangia. These organisms are common and widespread in soil from which they are isolated by baiting with pollen of *Liquidambar* and similar materials. The spores of *Actinoplanes*, *Ampullariella*, and *Spirillospora* are motile, those of other genera are not. Genera are distinguished by form of sporangia and spores (Fig. 9.22, I); some produce chains of conidia in addition to sporangia. In culture most species produce brilliant pigments, usually orange-coloured, and usually lack aerial mycelium.

Observed differences in wall composition suggest that this family may not be a natural group and may well be split up as a result of further study.

CHLAMYDOBACTERIALES

The order Chlamydobacteriales includes several organisms which have not been isolated and studied thoroughly and whose role in nature is in many instances far from clear. Failure to take sufficient account of the fact that the appearance of the Chlamydobacteriales varies to a large extent with age and the conditions in which they are grown has led to considerable confusion. Original descriptions made from natural materials have in some instances been rejected but sometimes have been subsequently resurrected after more extensive morphological and physiological studies. The group has the common characteristics of being aquatic and the mature structure consisting of many cells. Whether these organisms are multicellular or whether they are separate but contiguous single cells, is in doubt in some instances. It is certain, however, that any appearance of branching is false and is due to adherence of liberated single cells to an existing trichome (uniseriate arrangement of cells which may or may not have a sheath) and formation of a new chain thus giving the appearance of a branch.

The best-known genera of this group are *Sphaerotilus* and *Leptothrix*. The former are motile rod-shaped cells 2 μ by 3–6 μ which commonly contain poly-β-hydroxybutyrate granules (Fig. 9.23) but which also rapidly deposit sulphur internally when exposed to hydrogen sulphide. Chains are formed when motile cells fix themselves to a surface with secreted gum, usually with their long axes in a vertical position. Transverse division then occurs and further gum is secreted. This gum is carbohydrate and distinct from the protein-polysaccharide-lipid complex which forms the substance of the sheath laid down subsequently. As the filaments age, the cells in the basal region disappear leaving a rigid but brittle empty sheath (Fig. 9.23).

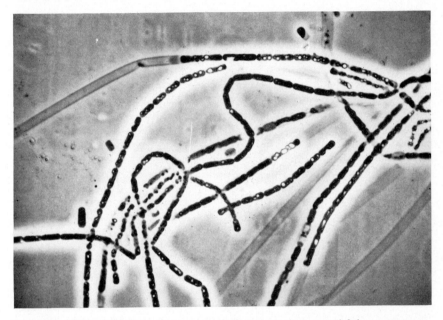

Fig. 9.23 *Sphaerotilus natans*, grown on 0·1 per cent glucose and 0·1 per cent peptone. Empty sheaths and inclusion globules of poly-β-hydroxybutyric acid can be seen. (× 1,620. Mulder, E. G. (1964) *J. appl. Bact.*, **27**, 151)

Leptothrix is morphologically very similar and often indistinguishable from *Sphaerotilus* and is considered by some authors to be the same organism. However, others believe they are not identical since *Leptothrix* oxidizes manganous salts and deposits manganese in its sheath, whereas *Sphaerotilus* has not been shown to carry out this oxidation.

Toxothrix is basically very similar to *Sphaerotilus* and *Leptothrix* but characteristically the trichomes lie loosely, spirally twisted together and an arched, fan-shaped arrangement of the sheathed chains of cells sometimes develops. Iron is deposited in the sheath.

The individual cells of *Peloploca* and *Pelonema* are 2 μ or less in width and contain a gas vacuole. The trichomes are often one to several hundred microns long and occur in bundles in *Peloploca* and singly in *Pelonema*. The organisms occur in fresh-water habitats where algae are decomposing and the dissolved oxygen content is low but they do not oxidize sulphur or deposit iron oxides.

Crenothrix, Phragmidiothrix, and *Clonothrix*, in contrast to other organisms in this order, are all wider at the base of the trichome than at the tip. Unfortunately none of these three has been isolated and thoroughly studied. The cells of *Crenothrix* and *Phragmidiothrix* divide longitudinally and transversely near the tip and form a mass of non-motile coccoid cells whereas the cells near the tip of the trichome of *Clonothrix* divide only transversely producing a single row of coccoid elements. In these organisms false branching is frequently seen. The sheath of a *Crenothrix* and *Clonothrix* may be impregnated with manganese or iron when the organism is growing in waters containing salts of these elements. *Phragmidiothrix* does not, so far as is known, deposit these elements.

The bacteria of this group are of considerable nuisance in certain situations. In sewage works using the activated sludge process (p. 692) masses of *Sphaerotilus natans* may cause 'bulking' (prevention of settling of the sludge). *Crenothrix polyspora* sometimes develops in reservoirs to such an extent that it makes the water unpalatable and it may also cause blockage of water pipes. In contrast some of these organisms may increase the efficiency of sand filters at water works and by removing iron and manganese by oxidation and precipitation improve the quality of the water produced. *Sphaerotilus* is thought to play an important part in the degradation of organic material in natural waters contaminated with sewage and in the effluent from paper mills and dairies.

BEGGIATOALES

In most of the genera included in this order the cells are arranged in trichomes but in a few species they occur singly. The organisms are non-flagellate and when motile move by a gliding, jerking or rolling motion. Those occurring as trichomes are very similar to blue-green algae (p. 346) the major difference being that they lack the pigments occurring in the latter. The morphological similarity is so close as to suggest that some of the Beggiatoales are more closely related to blue-green algae than to bacteria. They are found in aquatic and terrestrial environments usually in association with decaying organic material, often of algal origin, where hydrogen sulphide is present.

In the past, *Beggiatoa* has been regarded as an autotroph deriving its energy from oxidation of sulphides to sulphur which is deposited as granules within the cell. However, this genus has now been shown to be heterotrophic and does not have an obligatory requirement for H_2S although it is stimulated by this substance. The possession of intracellular sulphur granules is an important difference between *Beggiatoa* and *Vitreoscilla* but confusion has been caused by failure of some authors to mention the presence or absence of these granules. In addition to sulphur granules poly-β-hydroxybutyrate granules have been identified in *Beggiatoa*. The trichomes of *Beggiatoa* are fragile structures, but no suggestion has been made that individual cells from different trichomes can exchange with one another as has been claimed for *Vitreoscilla* when the paths of two gliding trichomes cross.

Leucothrix in contrast is a very robust organism, usually with a basal hold-fast. It liberates gliding gonidia from the trichome trip and when cultured in a rich medium the trichome tends to form knots of varying complexity. Another difference between *Leucothrix* and other members of the Beggiatoales is that the former deposits sulphur externally when sulphides are oxidized.

Achromatium is a single celled organism larger than 2 μ wide and some strains described, but not isolated in pure culture, are said to be 35–100 μ in diameter. The cells divide by constriction and under appropriate conditions contain substantial intracellular calcium carbonate sphaerules and sulphur granules.

All these organisms occur in terrestrial and aquatic environments where there is organic material and/or hydrogen sulphide. Their presence tends to prevent the hydrogen sulphide concentration building up.

HYPHOMICROBIALES

The bacteria included in this order are peculiar in that the products of the division are two obviously unequal cells which arise as a result of budding. It should be noted here than the division of *Caulobacter* cells (p. 319) also produces unequivalent cells, but as a result of fission and not of budding. The order is subdivided on whether the buds are borne on the tips of slender hyphae or whether they are sessile. In *Hyphomicrobium* a chemoheterotrophic organism and in the anaerobic photoheterotrophic *Rhodmicrobium* sp. the buds are borne on filaments whereas in *Pasteuria* and *Blastocaulis* they are sessile. Several new species have been recently described including one, *Pedomicrobium*, which deposits iron and manganese salts on the cell surface but the relationships of this and other organisms in the order are far from clear.

Hyphomicrobium occurs in soil and fresh, salt and brackish water, but in normal culture media is quickly overgrown by other species unless high concentrations of chemicals unfavourable to most bacteria but to which hyphomicrobia are remarkably indifferent, are present. It is also able to grow oligocarbophilically, i.e., in the absence of an added carbon source. Many other budding bacteria can grow in a nutritionally poor environment and this may be associated with their preference for growing on surfaces on which nutrients are known to be concentrated by adsorption.

MYXOBACTERIALES

In 1892 Thaxter described several species of bacteria which exist alternately either as individual unicellular vegetative cells or aggregated to form differentiated multicellular fruiting bodies visible to the naked eye (Fig. 9.25D). He placed these organisms in a new order, the Myxobacteriales. Since then other organisms have been added which resemble those described by Thaxter in their gliding mode of locomotion, their lack of rigid cell walls, their cell division by constriction, the manner of their colony growth and in their formation of microcysts by shortening and thickening of the wall of the vegetative rod. However, the added species, *Cytophaga* and *Sporocytophaga*, do not even form fruiting bodies. Thus like so many groups of the 'higher bacteria' the order Myxobacteriales is probably not sufficiently homogeneous to be satisfactory. The guanine and cytosine contents of the DNA of the fruiting myxobacteria have been shown to be 67–71 per cent and that of the non-fruiting types 33–42 per cent. This adds weight to the suggestion that *Cytophaga* and *Sporocytophaga*, together with certain of the Beggiatoales, nutritionally unlike the majority of organisms in that order, and with the gliding, flexible, nutritionally unspecialized *Flexibacter*, should be placed in a new order Flexibacteriales.

Both myxobacterial types have been isolated from water and are widespread and abundant in soil particularly in association with decaying plant residues, including seaweed, and dung. They are very efficient in breaking down high molecular weight compounds. The fruiting types produce antibiotic compounds active against Gram-positive and possibly also Gram-negative bacteria. They liberate extracellular enzyme mixtures which are poorly characterized but which have been shown to contain proteinases, polysaccharidases, lipases, and amidases attacking the cell wall mucopeptide and which cause lysis of bacterial cells. The products of digestion by these enzymes are used as nutrients and the association of myxobacteria with dung is probably because of the large number of eubacteria contained in this material. The cytophagas, in contrast, are thought to require contact between their surface and the substrates they attack because, among other things, they do not cause zones of clearing in cellulose-agar. They are nevertheless very efficient in hydrolysing cellulose. One *Cytophaga* species has been shown to oxidize cellulose only slightly more slowly than glucose and at about the same rate as cellobiose.

The vegetative cells of myxobacteria are either slender with round or blunt ends or fusiform. They are of low refractility and their walls are not demonstrable by staining. They may be motile when in contact with a solid surface or a surface film of a liquid: linear locomotion, swinging, and flexing movements have been observed but flagella are absent and in suspension the cells are non-motile.

The fruiting body is initiated by the formation of a heap of motile vegetative cells. A distinct fruiting body is then formed and this may be sessile as in *Myxococcus* (Fig. 9.24) or borne on a stalk as in *Synangium*. Within these fruiting bodies shortening of the vegetative rod leads to the formation of rod-shaped, spherical or ellipsoid microcysts. The microcysts may be contained in a cyst or be embedded in slime but those of *Sporocytophaga* which does not form a fruiting body are, of course, entirely free. *Cytophaga* does not form microcysts although in old cultures coccoid forms occur. The microcysts may be dispersed by wind and rain and are more resistant to heat and desiccation than vegetative cells but are not as resistant as eubacterial endospores. The first signs of germination are swelling of the microcyst and loss of refractility; the vegetative rods then emerge leaving an empty sheath behind.

The control of the developmental cycle is not understood although certain nutritional conditions affecting it have been established. Phenylalanine and tryptophan inhibit microcyst formation above certain concentrations although they are necessary for growth of the vegetative cells. The fruiting myxobacteria, like the slime moulds which have a superficially similar life cycle, provide fascinating material for the study of morphogenesis in a relatively simple system.

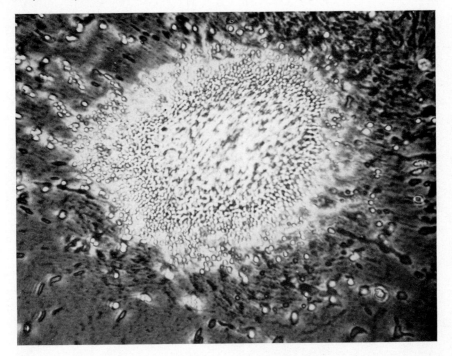

Fig. 9.24 *Myxococcus xanthus,* grown on a slide of *Escherichia coli* agar at 30°C for seventy-two hours. The mature fruiting body, with microcysts embedded in the wall, can be seen. The preparation is viewed from below and focused on the basal region of the fruiting body. (× 720. A. Pinches)

SPIROCHAETALES—THE SPIROCHAETES

The spirochaetes (Fig. 9.17, xv–xviii) are long, slender, motile organisms, coiled about their long axis in the form of a helix. They differ from spiral Eubacteria, such as *Spirillum,* in being flexible, in lacking true flagella and, with the exception of *Borrelia,* in being difficult to stain with aniline dyes. They are best examined in the living state by dark-field illumination, by negative staining with nigrosin after fixing or by electron microscopy (Fig. 9.25, A). Pathogens may be demonstrated in tissue sections after staining by Levaditi's technique which involves the deposition of silver on the organism's surface. They are divided into two families; the Spirochaetaceae which are relatively large, varying in length from 30 to 500 μ, often showing cross striations and various inclusion bodies, and the smaller Treponemataceae which rarely exceeds 16 μ in length and in which the protoplasm shows no obvious structural features.

Habitat

The Spirochaetaceae are mainly free-living saprophytes, and the parasitic members cause little damage to their hosts. Free-living strains are found in

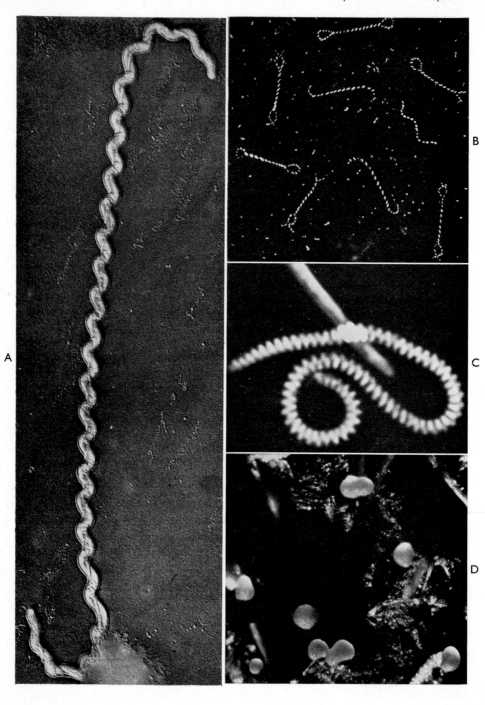

fresh water, marine slime, sewage, and stagnant waters. Others are parasitic on fresh water and marine shellfish including oysters and other molluscs.

The Treponemataceae are mainly parasites of birds and mammals, including man, and cause a number of serious diseases. *Treponema pallidum*, the cause of syphilis, can survive for only a short time outside the host and hence non-venereal transmission is a very rare occurrence. The primary lesion is usually confined to the genitalia; secondary lesions are found at other sites including the mouth. In the tertiary phase vital organs, particularly of the circulatory and central nervous system, are attacked. Yaws, an ulcerative disease of man in tropical countries, is caused by the serologically related species, *Tr. pertenue*; *Tr. cuniculi* is transmitted venereally in rabbits; other species of *Treponema*, apparently harmless, have been found in the human mouth and in the gut of termites, cockroaches, and toads. Species of *Borrelia* are also pathogenic for a number of hosts. *Bor. anserina* is responsible for a septicaemic disease of fowls, commonly known as fowl spirochaetosis. It is mechanically transmitted by the chicken mite *Dermanyssus gallinae* and possibly also by ingestion of faecal material from infected ticks. *Bor. recurrentis*, the causal organism of relapsing fever in man, is also a blood-borne infection transmitted from person to person by the body louse, *Pediculus humanus* var. *corporis*. Species of *Borrelia* are frequently found in association with fusiform bacilli in infections of the gums of the human mouth.

The genus *Leptospira* includes many species pathogenic to various hosts and other harmless species, e.g., *L. biflexa*, living in fresh water, which are often morphologically indistinguishable from the pathogens. Pathogenic species of *Leptospira* (Fig. 9.25) differ from those of *Treponema* and *Borrelia* in being able to survive much longer outside their hosts and in having a much wider host range. These organisms are blood-borne but concentrate in the kidneys and are excreted in the urine. Three serotypes are known to be of particular significance to human and animal health: *L. icterohaemorrhagiae*, a natural infection of rats causes leptospiral jaundice or Weil's disease in man; *L. canicola*, primarily causes leptospirosis in dogs but is transmissible to man; *L. pomona* causes leptospirosis in pigs from which man can contract swine-herd's disease. Many other serotypes have been isolated from both wild and domestic animals.

Nutrition

A few saprophytic and parasitic species have been grown in pure culture on complex media. Leptospirae are grown relatively easily in fluid media containing salts and fresh rabbit serum. Anaerobic conditions are needed for *Treponema*

Fig. 9.25 The morphology of various spirochaetes. (**A**) *Leptospira canicola* (× 21,875). Electron micrograph of a freshly fixed preparation, shadowed with gold-palladium alloy. (Swain, R. H. A. (1957) *J. Bact. Path.*, **73**, 155.) (**B**) Diagram of *Leptospira* sp. as seen under dark ground illumination (× 2,000 approx.). The fine primary coils show as bright dots throughout the length of the organisms. (A. H. Linton.) (**C**) An unknown spirochaete with finely turned coils (× 3,600). Nigrosin preparation. (C. F. Robinow (1958). In *General Microbiology*, Stanier *et al.*, Macmillan, London.) (**D**) *Myxobacteria*. Fruiting bodies of *Myxococcus virescens* on sterilized dung. (× 44. B. N. Singh (1947) *J. gen. Microbiol.*, **1**, 1)

and *Borrelia*; the other genus is aerobic. It is generally agreed that the pathogen *Tr. pallidum* has not been grown in artificial media but it can be maintained in the testes of male rabbits. Many species of Treponemataceae, owing to their slender form, will pass through bacterial filters. This property is utilized in obtaining cultures free of other bacteria.

Structure of spirochaetes

Electron micrographs show *Tr. pallidum* to be covered by an outer periplast. If this is removed by digestion with pepsin or trypsin three fine filaments, about 10 mμ in diameter, are revealed, twisted around the organism following the contours of the coils. Damage to the filaments results in loss of coil. It is thought that they are contractile in nature and concerned in the maintenance of the shape of the organism. Loose filaments were at one time mistaken for flagella. Spirochaetes are flexible and also able to rotate and move in a corkscrew like manner.

Leptospirae are slender cells, being only 0·1 μ in diameter and coiled throughout their length and are best seen under dark-ground illumination. In addition, 'secondary' coils may be seen at either or both ends giving the appearance of hooks when the cells are rotating rapidly. The cytoplasm is wound about a single straight axial filament, the whole being contained within the cell wall. *Borrelia* are thicker than *Treponema* and *Leptospira* and consist of loose, open coils.

Genera of the family Spirochaetaceae have characteristic structures. In *Spirochaeta* and *Saprospira* the centre of the helix is also occupied by an axial filament and in *Cristispira* a membrane, the crista, spirals along the organism.

CARYOPHANALES

This small order is characterized by the possession of nuclear bodies clearly visible in living cells, and by the organization of the cells into unsheathed filaments. Numbers occur in water containing decomposing organic matter, and in the gut of animals.

MYCOPLASMATALES

These organisms are similar to the L-forms of true bacteria (p. 310) and it is thought, but without conclusive evidence, that they are derived from unknown bacterial parents, having lost the ability to revert. In the absence of knowledge of their precise bacterial origin they have been placed in a separate order Mycoplasmatales which contains the single genus *Mycoplasma*. The first species to be investigated was that causing bovine pleuropneumonia; hence the causal organism is often referred to as the pleuropneumonia organism (PPO) or the group to which it belongs the pleuropneumonia-like organisms (PPLO). Both parasitic and saprophytic PPLO's are now known. Pathogenic ones include *Mycoplasma mycoides* the cause of pleuropneumonia in cattle, *M. agalactiae*, the cause of contagious agalactia in sheep and goats, *M. gallisepticum*, the cause of chronic respiratory disease in fowls, and several species responsible for arthritis in different hosts. Parasitic strains of doubtful pathogenicity have been isolated from the normal respiratory and genital tracts of cattle and man. Saprophytic strains have been isolated from sewage, soil, compost, and manure. Identification is based on biological characters, serological specificity and, with parasites, the host from which they are isolated. A rigid cell wall is lacking, the cell envelope resembling the cytoplasmic membrane of mammalian cells.

The form of the organisms is variable, including rings, globules, filaments, and small elementary bodies. Structures as small as 125 mμ may be separated by differential filtration and are capable of growth in artificial media. Parasitic species require a medium rich in proteins, usually supplied as serum protein, unsaturated fatty acids, and cholesterol or a related sterol. On solid media they grow as delicate colonies in which the central part grows down into the medium making them difficult to remove. They easily break up when touched but can be fixed to coverslips and stained by Giemsa for microscopical examination. In fluid media, growth results in faint turbidity, which is best examined by dark ground illumination. Most grow aerobically, but occasionally a particular strain grows better under increased CO_2 tension and one obligate anaerobe has been described.

Most mycoplasma are resistant to many antibacterial agents including penicillin, thallium acetate, and sulphathiazole but are moderately susceptible to the tetracyclines.

A number of plant diseases formerly thought to be caused by viruses are now known to be due to mycoplasma.

RICKETTSIALES

The rickettsias are a group of small bacteria-like prokaryotic organisms that occur naturally in the tissues of arthropods and can be transmitted to mammals in which they cause disease. They have properties which are intermediate between those of bacteria and viruses. Although generally smaller than bacteria they can be resolved under the light microscope. They do not stain readily by ordinary technique but Giemsa stain is satisfactory. Rickettisas may be spherical or rod-shaped and range from 0·2 μ to 0·6 μ in diameter. Their rigid cell envelope is chemically and structurally similar to the walls of Gram-negative bacteria and they contain muramic acid which strongly supports a phylogenic relationship with bacteria. Reproduction is usually by binary fission.

They resemble the viruses in their dependence upon host cells for growth except for one species, *Rickettsia quintana*, which can be cultivated on non-living media. Unlike viruses, however, they exhibit some degree of metabolic activity independent of that of the parasitized host cell from which they derive essential nutrients. Unlike viruses rickettsias contain both DNA and RNA (Table 8.1) and most are susceptible to broad-spectrum antibiotics, such as tetracyclines, a property which they share with the psittacosis lymphogranuloma group of viruses.

Rickettsias have been successfully grown in tissue culture and in the developing chick embryo, the best site for propagation being the yolk sac where large quantities of *R. prowazekii* are grown in the preparation of vaccines against typhus fever. Most form a reservoir of infection in wild animals, especially rodents, and are transmitted to man and other animals by lice, ticks, mites or fleas. Some workers classify them according to the diseases they produce in man and the vector responsible for transmission but this is an artificial scheme since only a small number cause human disease, many more being symbionts in a variety of arthropods.

Bergey recognizes four families, based on morphology, serology, their mode of transmission between hosts and whether they occur in the cytoplasm or nucleus of the cells they parasitize. Organisms of the family Rickettsiaceae usually parasitize the gut cells of arthropods which transmit them to other hosts. Those that affect mammals attack the capillary endothelium producing red skin rashes. Members of the family Chlamydiaceae are usually transmitted by non-vector means such as on dried infectious excreta and are inhaled on dust by susceptible hosts. Members of Bartonellaceae and Anaplasmataceae primarily parasitize mammalian erythrocytes. Transmission between mammalian hosts is dependent on intermediate arthropod hosts.

BLUE-GREEN ALGAE

CYANOPHYTA

The cells of the Cyanophyta (Myxophyceae) are considerably larger than those of other prokaryotic groups and contain chlorophyll α, β-carotene, and the phycobilin pigments phycocyanin and phycoerythrin. Thus they rank alongside the algae in size and the ability to photosynthesize via chlorophyll α with the liberation of oxygen, but possess the bacterial characteristics of a prokaryotic nucleus and absence of mitochondria. The method of cell division, the copious production of mucilage and the absence of any sexual stages in the life history are all features drawing the group closer to bacteria than to algae. There has been considerable discussion regarding their position in the plant kingdom but irrespective of this they appear to be a very ancient group playing an important role in the microbiological economy.

The morphology of Cyanophytes is basically that of unicells or aggregations of these to form colonies and filaments (Fig. 9.26). Under the microscope they appear a bluish-green colour but distinct chromatophores are absent, the pigments appearing to diffuse throughout the cells. Small granules of a polysaccharide can sometimes be seen and in some species larger blackish vacuoles occur; these are the gas vacuoles containing mainly nitrogen and present in only a small number of species. The cell walls are indistinct but often there is a prominent sheath of mucilage around the cell or filament and this is frequently stained brown by iron or manganese deposits and occasionally covered by crystals of carbonates. Species of Cyanophyta are very important constituents of the thermophilic flora and are also found on and in the surface of rocks, tufa, etc., where they build up laminated deposits. In the past the protoplast has been thought to consist of a central region, the nucleoplasm—containing the nuclear material without any bounding membrane and an outer chromatoplasm in which the pigments are dispersed; it is not at all easy to define these areas with the

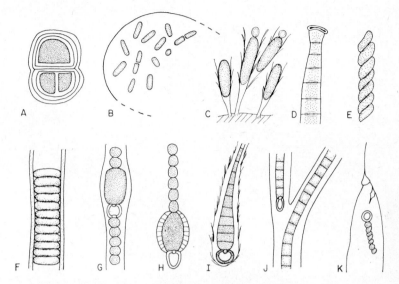

Fig. 9.26 Cyanophyta. (**A**) *Chroococcus*, (**B**) *Aphanothece*, (**C**) *Chamaesiphon*, (**D**) *Oscillatoria*, (**E**) *Spirulina*, (**F**) *Lyngbya*, (**G**) *Anabaena* with heterocyst and akinete. (**H**) *Cylindrospermum* with basal heterocyst and akinete above. (**I**) *Calothrix*, (**J**) *Tolypothrix* with heterocyst above point of branch, (**K**) *Richelia* in the diatom *Rhizo-solenia*.

light microscope. Under the electron microscope sections of Cyanophycean cells reveal a series of lamellae (thylakoids, Fig. 9.27) mainly in the peripheral regions of the cell but also penetrating into the centre. Although these are not bounded by any enclosing membrane they are almost certainly the sites where photosynthesis is proceeding. The nuclear material appears as structureless masses and within the cell are bodies of various kinds whose structure and function are only suspected. Even under the light microscope the cell wall had been defined as at least a bipartite structure (inner and outer investment) but clearly from Fig. 9.27 there are at least three regions to the inner investment and an outer somewhat fibrous sheath. These are not unit membranes as described by Robertson (1959). The ingrowth of iris-diaphragm-like cross walls is character-istic of many filamentous Cyanophyta and can be seen in the light microscope. Several cross walls can be initiated within one cell and the precision with which these are formed midway between older cross walls is striking.

Whilst some filamentous Cyanophytes are relatively immobile within their mucilaginous sheaths others, e.g., *Oscillatoria*, *Spirulina*, and *Lyngbya*, glide rapidly and at the same time rotate about the longitudinal axis. Although it is not yet known precisely how this movement is motivated, electron microscope observations have revealed rows of pores on either side of the cross walls at least in *Symploca*. The sheath material is often left behind as the filament moves and the pores in the inner investment may be involved in this gliding through the sheath. Fine pores or plasmodesmata also penetrate the cross walls whilst

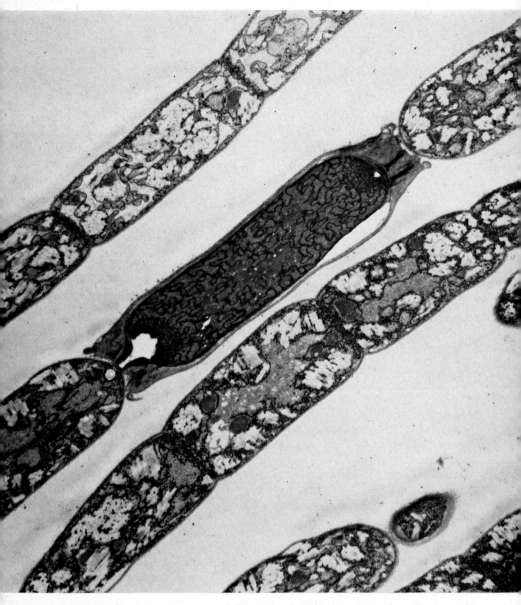

Fig. 9.27 Electron micrograph of filaments of *Aphanizomenon*. One filament with elongate heterocyst in which the thylakoids are prominent and aggregating towards the poles. (×12,500 approx. Photograph kindly supplied by Dr. R. V. Smith)

in the most advanced group, the Stigonematinales, there are more pronounced central pit-connections between the cells.

Some, but not all, planktonic Cyanophyta are buoyant, and under calm conditions float to the surface of lakes and ponds forming the so-called water blooms. This buoyancy is associated with the presence in the cells of pseudo-vacuoles or gas vacuoles. Ultra-structural studies have revealed a series of tubes (or gas cylinders) within each gas vacuole (Fig. 9.28).

Vegetative growth is the only known method of increase in cell numbers of many Cyanophytes especially of the coccoid series (Chroococcales). A small number of genera form exospores (e.g., *Chamaesiphon*, Fig. 9.26, C), some filamentous genera differentiate small segments of the filaments (hormogonia) which are more actively motile and form a vegetative means of propagation. These small segments are sometimes surrounded by a thicker wall and are then incapable of movement—the so-called hormocysts. Both types eventually germinate by the simple outgrowth of the filament. Enlarged cells (with dense contents), sometimes with inflated walls, occur singly or in rows in *Anabaena*, *Cylindrospermum*, etc., and act as akinetes (Fig. 9.26, G). These are almost certainly resting spores which germinate to form a new filament.

An apparently empty slightly enlarged cell is often associated with the akinete but may also occur elsewhere in the filament—this is the enigmatic heterocyst (Fig. 9.27). Recent electron microscope studies (Lang, 1968) have shown that the thick wall of the heterocyst is laid down between the three-layered inner investment and the outer sheath. As the wall thickens the photosynthetic lamellae become contorted and pack towards the poles of the heterocyst and the polar thickenings develop *inside* the plasma membrane and are not therefore part of the thickening of the wall as light-microscope observations often suggest. Thus the apparently empty heterocyst as seen in the light microscope still contains cytoplasmic constituents, a feature observed by some early light microscopists, and in line with the rare observations that under certain conditions heterocysts do germinate to form filaments. In some genera (e.g., *Tolypothrix*) branching occurs beneath the heterocyst (Fig. 9.26, J).

The Cyanophyta are very widely distributed organisms—they occur in hot sulphur springs at temperatures between 40° and 70°C and in all soils and most aquatic environments. They are particularly abundant in the tropics. They are not, however, very abundant in oceanic plankton (represented by *Trichodesmium* only) though the interesting symbiont *Richelia* occurs in the cells of the diatom *Rhizosolenia* (Fig. 9.26, K). Symbiotic relationships are quite common amongst Cyanophyta, e.g., *Anabaena* species in the fern *Azolla*, liverwort *Anthoceros*, gymnosperm *Cycas*, and angiosperm *Gunnera*. Aberrant Cyanophyta also occur in the algae *Glaucocystis* and *Cyanophora* whilst they are common components of lichens (p. 617). The aggregation of cells and filaments in mucilaginous masses often yields macroscopic colonies which lie on lake sediments, (e.g., *Aphanothece*), float freely in the water (e.g., *Microcystis*), rest on soil (e.g., *Nostoc*) or form blackish clusters on rock faces (e.g., *Stigonema*). Colonies of *Calothrix* often form a conspicuous zone on rocks in the upper inter-tidal zone, in some areas associated with other blue-green algae which actively bore into the rock; Cyanophyta also grow on and in mollusc shells. Other species

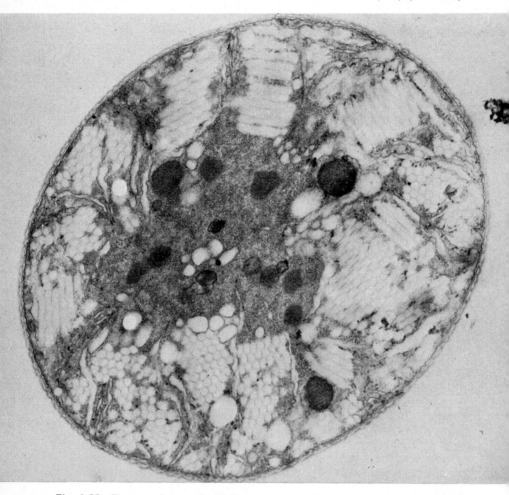

Fig. 9.28 Electron micrograph of *Microcystis* cell cut in transverse section showing the ill-defined central nuclear material and numerous gas vacuoles, each consisting of a bundle of gas cylinders. Some gas vacuoles are cut transversely, others obliquely or longitudinally. Note the very sparse thylakoids. (×30,000 approx. Photograph kindly supplied by Dr. R. V. Smith)

build up deposits by precipitation of lime within the mucilage sheaths and many of the species living around thermal springs are involved in the precipitation of lime to form travertine. The black streaks down the sides of concrete buildings in wet tropical and even temperate regions are formed by a community of Cyanophyta. Blackish patches on soils, lawns, etc., are produced by the aggregation of filamentous species, e.g., of *Anabaena*, *Phormidium*, etc. Filamentous species of *Anabaena*, *Oscillatoria*, and *Spirulina* are frequently found both floating in water (planktonic) or moving freely over the surface of under-water sediments (epipelic) and great concentrations of filaments appear bluish-green or even black.

Two notable features of the metabolism of the Cyanophyta are the ability of some species of the Nostocales to fix atmospheric nitrogen (a property which they share with certain bacteria (p. 319) and the formation by some species of very potent toxins (e.g., by *Microcystis, Anabaena,* and *Aphanizomenon*) which have been responsible for death of cattle from drinking water containing these genera.

BOOKS AND ARTICLES FOR REFERENCE AND FURTHER STUDY

General References including Pseudomonadales and Eubacteriales

BREED, R. S., MURRAY, E. G. D., and SMITH, N. R., eds. (1957). *Bergey's Manual of Determinative Bacteriology*. 7th Edn. The Williams & Wilkins Co., Baltimore, 1094 pp.

BRIEGER, F. M. (1963). *Structure and Ultrastructure of Micro-organisms*. Academic Press, New York and London, 327 pp.

DAVIS, B. D., DULBECCO, R., EISEN, H. N., GINSBERG, H. S., and WOOD, W. B. (1968). *Microbiology*. Harper & Row, New York and London, 1464 pp.

FUHS, G. W., VAN ITERSON, W., COLE, R. M., and SHOCKMAN, G. D. (1965). Symposium on the fine structure and replication of bacteria and their parts. *Bact. Rev.,* **29**, 277–345.

GUNSALUS, I. C. and STANIER, R. Y. (1960). *The Bacteria, Vol. 1. Structure*, Academic Press, New York and London, 573 pp.

HAYFLICK, L. and CHANNOCK, R. M. (1965). Mycoplasma species of man. *Bact. Rev.,* **29**, 185–221.

KLIENEBERGER-NOBEL, E. (1962). *Pleuropneumonia-like Organisms (PPLO), Mycoplasmataceae*, Academic Press, New York and London, 157 pp.

POINDEXTER, J. S. (1964). Biological properties and classification of the Caulobacter group. *Bact. Rev.,* **28**, 231–296.

POLLOCK, M. R. and RICHMOND, M. H. (1965). Function and structure in microorganisms. *15th Symp. Soc. gen. Microbiol.,* Cambridge Univ. Press, 405 pp.

ROBINOW, C. F. (1956). The chromatin bodies of bacteria. *Bact. Rev.,* **20**, 207–242.

SALTON, M. R. J. (1963). *The Bacterial Cell Wall*, Elsevier Pub. Co., Amsterdam and London, 293 pp.

SKERMAN, V. B. D. (1967). *A Guide to the Identification of the Genera of Bacteria*, 2nd Edn., Williams & Wilkins Co., Baltimore, 303 pp.

SPOONER, E. T. C. and STOCKER, B. A. D., eds. (1956). Bacterial anatomy, *6th Symp. Soc. gen. Microbiol.,* Cambridge Univ. Press, 360 pp.

VAN ITERSON, W. (1958). *Gallionella Ferrugina Ehrenberg in a Different Light*, N.V. Noord-Hollandsche Uitgevers Maatschappij, Amsterdam.

VAN NIEL, C. B. (1955). Classification and taxonomy of the bacteria and blue-green algae. In *A Century of Progess in the Natural Sciences*, California Acad. of Sci., San Francisco.

Actinomycetales

CROSS, T. and MCIVER, A. M. (1966). An alternative approach to the identification of *Streptomyces* species, a working system. In *Identification Methods for Microbiologists*, edited by B. M. Gibbs and F. A. Skinner, Part A, pp. 103–110, Academic Press, New York and London.

KUSTER, E. (1967). The actinomycetes. In *Soil Biology*, edited by A. Burges and F. Raw, Chapter 4, pp. 111–127, Academic Press, New York and London.

LECHEVALIER, H. A. and LECHEVALIER, M. P. (1965). Classification des actinomycetes aérobies basée sur leur morphologie et leur composition chimique. *Annls Inst. Pasteur*, Paris, **108**, 662–673.

——— and ——— (1967). Biology of actinomycetes. *A. Rev. Microbiol.*, **21**, 71.

SERMONTI, G. and HOPWOOD, D. A. (1964). Genetic recombination in *Streptomyces*. In *The Bacteria*, Vol. 5, *Heredity*, edited by I. C. Gunsalus and R. Y. Stainier, Chapter 5, pp. 223–251, Academic Press, New York and London.

WAKSMAN, S. A. (1961). *The Actinomycetes*, Vol. 2, *Classification, Identification and Descriptions of Genera and Species*, Baillière, Tindall and Cox, Ltd., London.

——— (1967). *The Actinomycetes, A Summary of Current Knowledge*, Ronald Press, New York.

Other Orders

DWORKIN, M. (1966). Biology of myxobacteria. *A. Rev. Microbiol.*, **20**, 75–106.

ECHLIN, P. (1966). The blue-green algae. *Sci. Amer.*, **214**, 74–83.

HOLM-HANSEN, O. (1968). Ecology, physiology and biochemistry of the blue-green algae. *A. Rev. Microbiol.*, **22**, 47–70.

LANG, N. J. (1968). The fine structure of blue-green algae. *A. Rev. Microbiol.*, **20**, 75–106.

MULDER, E. G. (1964). Iron bacteria, particularly those of the *Sphaerotilus-Leptothrix* group, and industrial problems. *J. appl. Bact.*, **27**, 151–173.

STANIER, R. Y. (1942). The cytophaga group: a contribution to the biology of myxobacteria. *Bact. Rev.*, **6**, 143–196.

STARR, M. P. and SKERMAN, V. B. D. (1965). Bacterial diversity: the natural history of selected morphologically unusual bacteria. *A. Rev. Microbiol.*, **19**, 407–454.

10
Eukaryotes

FINE STRUCTURE OF THE EUKARYOTIC CELL

THE FUNGI

PROTOZOA

THE SLIME MOULDS (MYXOMYCOTA)

THE ALGAE

FINE STRUCTURE OF THE EUKARYOTIC CELL

As already pointed out (p. 3) eukaryotic cells are characterized by the possession of one or more discrete nuclei each of which is enclosed by a perforated nuclear membrane. In this they differ from prokaryotic cells, the nucleoplasm of which has no enclosing membrane (p. 275). In addition to this important difference eukaryotic cells are more complex than prokaryotic ones in a number of other ways, notably in the possession of mitochondria and vacuoles. Fungi, algae (with the exception of the blue-green algae, p. 346), protozoa and slime moulds are all eukaryotic and although the cells of some of them may be less complex than those of higher plants and animals, considerable differentiation may occur in others. Special ultrastructural features of the cells of the various groups of eukaryotic micro-organisms will be considered in the sections dealing with those groups, but first a general survey of the structures involved is necessary.

CELL ULTRASTRUCTURE

For the purpose of this account of the ultrastructure of eukaryotic protists in general, it is convenient to consider the different groups of organisms simply in terms of cells. Thus protozoa, slime mould swarm cells, fungal and algal zoo-spores may be defined as unicells while the vegetative phases of fungi and many algae are multicellular. Although a large number of organelles are common to all cells, some are found in only a few. It is, therefore, also convenient to illustrate the range of differentiation by means of a hypothetical eukaryotic cell. It must be emphasized that such a diagram (Fig. 10.1) cannot represent what is essentially a dynamic structure, or even the fixed image of an existing organism. However, with this in mind it is hoped that it will serve to summarize the morphology and spatial arrangement of some of the structures to be described in this section.

Cell walls and the plasmalemma

Cells of the thalli of algae and fungi possess a true cell wall (Figs. 10.1, 10.2A), which is more or less rigid like that of bacteria (p. 280) but is of different chemical composition and architectural structure. Generally the wall consists of a reticulate microfibrillar skeleton or framework embedded in an amorphous matrix (Fig. 10.2B). Some of the most extensive studies have been made on certain genera

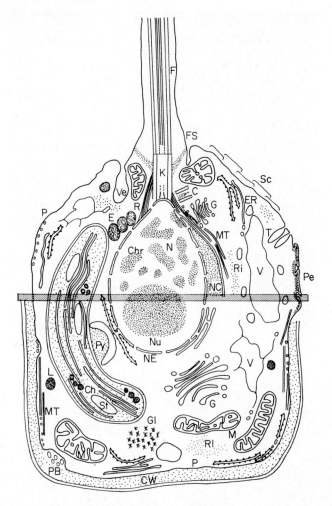

Fig. 10.1 Diagrammatic representation of ultrastructure of eukaryotic protist cells. Upper portion includes features found in flagellated cells (flagellate protozoa, zoospores of algae and fungi, myxomycete swarm cells). Lower portion shows structures found in vegetative cells of algae and/or fungi. The organelles shown do not occur in all cells, but are included to illustrate their spatial arrangement (e.g., chloroplasts occur in algae and green flagellates but are absent from all fungi, myxomycetes, and most protozoa). Key to lettering: C = Centriole, E = Eyespot, F = Flagellum, G = Golgi Apparatus, K = Kinetosome, L = Lipid drop, M = Mitochondria, N = Nucleus, P = Plasmalemma, R = Rhizoplast, T = Trichocyst, V = Vacuole, Ch = Chloroplast, Chr = Chromosome, ER = Endoplasmic Reticulum, FS = Flagellar Sheath, Gl = Glycogen, MT = Microtubules, NE = Nuclear Envelope, NC = Nuclear Cap, PB = Paramural Body, Py = Pyrenoid, Pe = Pellicle, Ri = Ribosomes, Sc = Scales, Ve = Vesicle.

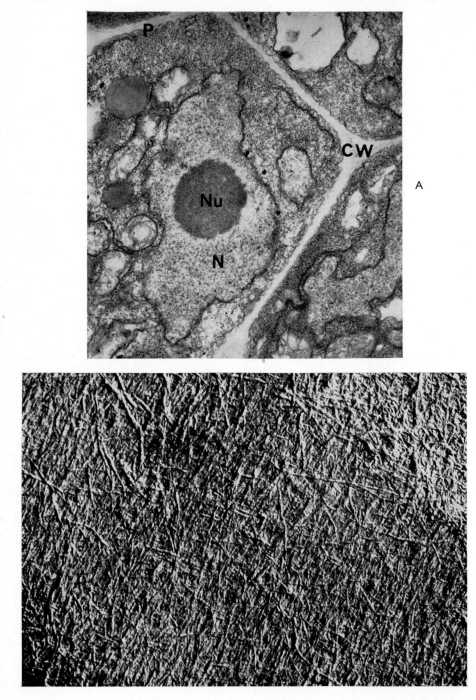

A

B

of green algae where cellulose has been found to constitute the microfibril layer (p. 437).

In most fungi the fibrillar layer is composed of chitin, though cellulose does occur in some groups (p. 369). Considerable variation is found in the wall structure of cells from different stages in the life cycle of an organism. For example some fungal spore walls possess complex ornamentations of several layers, and are often pigmented. Similarly some algal groups have what might be termed a 'wall layer' composed of scales (p. 437). Scales and other inorganic deposits are common in many protozoa and more typical cell walls are found enclosing the oocyst stages of some of them. Slime layers and filamentous layers are formed at the cell surface in some protozoa but such layers are far removed from the rigid structures of fungi and algae described above.

Within the cell wall the living protoplast is bound by a membrane, the plasmamembrane or plasmalemma (Figs. 10.1, 10.2A). Many protist cells do not possess a true cell wall and are merely enclosed by the plasmalemma. Such cells are often referred to as naked cells.

The electron microscope has shown the chemically fixed plasmalemma to be a tripartite structure composed of two electron dense layers separated by an electron transparent layer. Each layer measures approximately 25–30 Å in width and the total width of the plasmalemma is 75–100 Å. This tripartite structure which has been found to occur in many biological membranes is termed a unit membrane. Numerous pieces of indirect evidence led to the concept of the plasmalemma as a bimolecular leaflet of lipid on to the outer side of which is adsorbed an extended layer of protein.

The bimolecular lipid phase is considered by many workers to consist of molecules with polar and non-polar ends, the former facing outwards and therefore associated with the protein phase.

Recent detailed studies employing electron microscopy of both chemically fixed and freeze-etched material, together with a consideration of the physical and chemical properties of biological membranes have resulted in a more dynamic view of their structure, as being a micellar subunit complex. It now seems likely that not only are various membranes constructed differently but that any one membrane may undergo structural changes with time and according to function. A rigid presentation of the ultrastructure of the plasmalemma or any other cell membrane may be meaningless in the light of our present knowledge.

The plasmalemma is usually addressed to the cell wall, when present, but may at some stages of development of certain organisms or under certain conditions become undulating or invaginated.

Fig. 10.2   (A) A section through parts of the closely packed ascogenous hyphal cells of the fungus *Podospora curvicolla*, showing the true cell wall (CW), plasmalemma (P), nucleus (N), and large dense nucleolus (Nu). (× 18,000. A. Beckett, University of Bristol.) (B) Piece of hyphal wall of *Allomyces macrogynous* showing outer layer (top right) of randomly arranged microfibrils and an inner layer (centre and bottom) of orientated microfibrils (approximately parallel to longitudinal axis of hypha) shadowed with Pd–Au. (× 13,800. From J. M. Aronson and R. D. Preston (1960) *Proc. R. Soc., B.* **152**, 346)

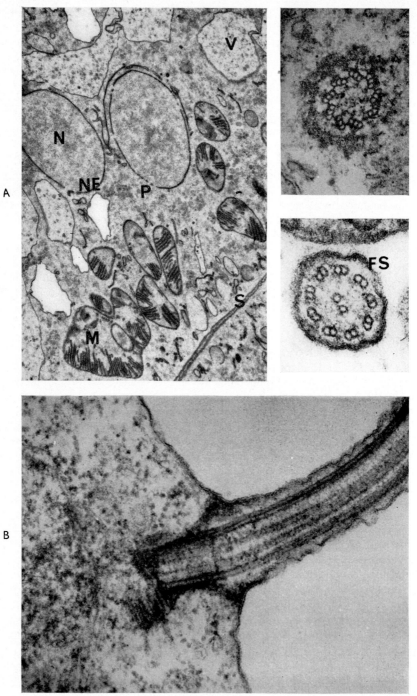

Membranes or vesicular structures associated with the plasmalemma have recently been termed paramural bodies. These in turn may be subdivided according to their origin into lomasomes or plasmalemmasomes, the former originating from cytoplasmic membranes while the latter are derived entirely from the plasmalemma. It has been suggested that these paramural bodies may be associated with cell wall synthesis either as a means of incorporating wall material or as sites of enzyme incorporation for extracellular synthesis. Although these structures have been found in a number of fungi, a few algae, and some higher plants, no conclusive evidence regarding their function has been obtained. It is interesting to note that in some cases the incorporation of cytoplasmic membrane bound vesicles into the plasmalemma involves the fusion of the two membrane systems and suggests an interrelationship between them. Such associations occur between membranes of the Golgi system and the plasmalemma.

The reverse process whereby substances enter the cell from the exterior and pass into the cytoplasm within membrane bound vesicles is termed pinocytosis and is particularly common in amoebae and myxamoebae.

The plasmalemma in euglenoids is regularly folded around the underlying pellicle while in many fungal zoospores it forms a more or less smooth boundary to the cell.

Attempts have been made to demonstrate a continuity between the plasmalemma and various internal cytomembrane systems but results have been inconclusive. Microdensitometry of the plasmalemma in chemically fixed cells has shown it to possess a polarity, such that one of the electron-dense layers is thicker than the other. However, there is little apparent regularity of this feature since the plasmalemma of some fungi has an outer layer less dense than the inner one, while in higher plants, for example, the reverse is true.

Nuclei

The nuclear envelope

The nuclear envelope of most eukaryotic protist cells consists of two membranes. These are separated by a space termed the perinuclear space. Each membrane measures 70–90 Å thick and is perforated by a number of pores

---

Fig. 10.3 (A) A longitudinal section through part of the zygospore of the fungus *Rhizopus sexualis* showing nuclei (N), vacuoles (V), mitochondria (M), and part of a developing septum (S). The nuclear envelope (NE), is perforated by several pores (P). (× 15,000. M. A. Gooday, Bristol University.) (B) A longitudinal section through part of a zoospore of the chytrid fungus *Phlyctochytrium* showing the flagellum enclosed within the sheath which is continuous with the plasmalemma. The kinetosome, and vestigial kinetosome lie at the base of the flagellum. (× 80,000. L. W. Olson and M. S. Fuller, University of California, Berkeley.) (C) A transverse section through a centriole of the aquatic fungus *Catenaria anguillulae*. Each of the fibres is composed of a triplet of sub-fibres and each triplet is radially tilted anticlockwise. This pattern is also seen in parts of the kinetosome. (× 120,000. M. S. Fuller, University of Georgia.) (D) A transverse section through the flagellum of *Coelomomyces indicus*, a fungus parasite of mosquitoes. The flagellar sheath (FS) encloses the peripheral ring of nine doublet sub-fibres which together with the central pair constitute the axoneme. (× 92,000. A. Beckett, Bristol University)

at the edges of which the two membranes are continuous (Figs. 10.1, 10.3A, 10.4C).

Additional membranes are found inside the nuclear envelope in some amoebae.

Freeze etched preparations of yeast nuclei show both open and sealed pores in the envelope and it has been suggested that a regulatory mechanism may exist to control the flow of specific nuclear substances from the nucleoplasm to the cytoplasm.

In some fungal zoospores a double membrane continuous with the outer membrane of the nuclear envelope gives rise to an enclosed structure surrounding part of the nucleus. This structure is the nuclear cap (Fig. 10.1), within which densely packed ribosomes are found. Pores are absent from the nuclear cap membrane.

Chromosomes

The electron microscope has in general been unsuccessful in contributing to our knowledge of chromosome fine structure. Following glutaraldehyde or osmium tetroxide fixation chromosomes are usually seen as rather amorphous electron-dense structures. In some of the better preparations obtained so far, fine fibrillar components have been observed. In *Amoeba proteus* and *Pelomyxa* the nucleus contains fine threads in the form of regular helices. However, the dimensions of these structures are considerably greater than the double helix of deoxyribose nucleic acid (DNA) (p. 26).

Recent studies on the fine structure of chromosomes in the dinoflagellate *Woloszynskia micra* have shown them to be electron-dense banded structures. This banding is apparently due to the arrangement of the DNA fibrils. An interesting feature of these chromosomes is that division into chromatids begins at one end and proceeds along the length of the chromosome with the inter- mediate formation of V or Y-shaped configurations. In the fungi so far in- vestigated chromosomes appear as irregular dense masses.

Nucleoli

Nucleoli are seen in suitably prepared cells usually as compact, granular bodies consisting of electron dense ribosomes (p. 363, Fig. 10.2A).

Spindles and centrioles

Nuclear division in eukaryotic protist cells usually involves a spindle apparatus composed of tubular structures measuring 200–240 Å in diameter. These microtubules are often enclosed within a persistent nuclear envelope which does not break down until the telophase separation of daughter nuclei. This is the case in many fungi (p. 370), slime moulds, and certain flagellate and ciliate protozoa.

Little or no information is so far available as to what degree or how these microtubules are involved with the separation of chromosomes. In many instances where they occur, no connections with the chromosomes can be detected, while in others, some microtubules appear to terminate at or within the chromosome. Usually continuous microtubules are found passing apparently between the chromosomes and linking the polar ends of the nucleus. The mitotic nucleus of many eukaryotic cells is associated at its polar regions with an astral ray-centriole complex. Astral rays are microtubular elements often seen radiating from the cylindrical centriole. All centrioles have a system of fibres running along their length, which, when viewed in transverse section, are seen to be arranged in a peripheral ring of nine, each composed of a triplet of subfibres. Each triplet is radially tilted in a clockwise direction when viewed from the proximal end (Fig. 10.3C). The cylinder of triplets is a structure characteristic of at least part of the kinetosome at the base of flagella in motile cells (p. 358).

Although as mentioned above many eukaryotic cells possess centrioles or their structural equivalents, in some, these organelles are not apparently functionally involved in the polarization of nuclei during mitosis. In addition, other organisms have what is termed anastral and acentric mitosis where neither astral fibres nor centrioles are present at all. However, when centrioles do occur and are actively involved in mitosis, they undergo division at some stage in the mitotic cycle.

In the aquatic fungus *Catenaria anguillulae* a single centriole lies at either end of the prophase nucleus. These then divide some time before metaphase so that a pair of centrioles can be seen at the poles. One of each pair lies at right angles to its partner. Although little is known of the mechanism involved in replication, DNA has been reported to occur in centrioles. It has been suggested that the centriole in certain protozoa, which is associated with complex fibrous structures, may be the organizing centre for protein molecules that are synthesized in the cytoplasm.

Flagella and associated organelles

The electron microscope has shown that in all eukaryotic protists possessing motile cell stages, flagella are composed of an axial core, enclosed within a membranous sheath continuous with the plasmalemma (Figs. 10.1, 10.3B). With a few possible exceptions the axial core, or axoneme, consists of two sets of fibres, a peripheral ring of nine and a central pair. Transverse sections show the peripheral fibres to consist of a doublet of sub-fibres, and both these and the central fibres have a dense outer layer enclosing a less dense core (Fig. 10.3D).

Cilia of protozoa have an axial component identical to the one described above for flagella.

In some protists, for example dinoflagellates and euglenoids, varying degrees of additional differentiation within the sheath have been found. Dinoflagellates possess two flagella, one posterior, the other transverse and often situated within a groove running round the cell. Electron microscopy has shown the transverse flagellum to be composed of a fibrous strand of periodic structure in addition to the usual 9 + 2 axoneme and linked with it by dense packing material.

Euglenoid flagella in transverse section have their diameters considerably increased by the inclusion of an amorphous or crystalline paraflagellar material within the sheath. An electron dense body at the base of one of the flagella and within the sheath membrane is also found in euglenoid flagellates. This is the flagellar swelling, considered by many to be associated in some way with photoreception.

Electron microscope studies on whole cells shadowed with heavy metals have shown that a variety of ornamentations may occur along the length of some flagella. Thus it has been possible to distinguish between the so-called whiplash flagellum (Peitschengeissel), which is devoid of appendage, sand the hairy flagellum (Flimmergeissel) possessing lateral hairs. These lateral hairs have been found to be bipartite in that the distal end is very much more slender than the proximal region.

Some flagellates possess both types of flagella on the one cell. In such cases the posterior flagellum is of the whiplash type, while the hairy flagellum is situated anteriorly. This arrangement is termed heterokont flagellation.

Scales are found covering the flagella of some algal cells and flagellar spines occur on the motile cells of some brown algae.

A further variation to the structures described above is found in the algal flagellates belonging to the family Haptophyceae (see Christensen, 1962). Here a short filamentous projection arises from a position within the cell close to the base of the true flagella and associated with it by striated fibrous connections. This structure, the haptonema, is a tube of three concentric membranes surrounding an inner core of seven fibres. Bridges occur locally between the central core and the haptonema wall. Transverse sections of the basal region of the haptonema show an increase in the number of central core fibres to nine, though in view of the fact that the whole structure grows out from the base, then developmentally there are nine original fibres which decrease with growth until only seven remain.

Associated with flagella and cilia is a number of organelles which lie inside the cell proper, usually at or near the base of these motile appendages. The basal body, or kinetosome (Fig. 10.1, 10.3B), is a cylinder composed of a ring of nine fibres, each fibre being made up of three sub-fibres. As mentioned previously this structure is consistent with that of the centriole (p. 361). However, the kinetosome does not retain this regular structure throughout its length, since a central 'cartwheel' is present usually towards the lower end of the cylinder, while in the kinetosome-flagellum transition region a complex stellate fibre pattern is often seen. In some fungal zoospores this region is marked by the gradual transition from the peripheral ring of nine doublet fibres (as in the axoneme) to nine triplet fibres. The third sub-fibre arising as an outgrowth of one of the doublet sub-fibres. Distally the kinetosome is delimited from the flagellum or cilium by a 'crosswall' of dense material, the terminal plate (Fig. 10.1).

Bands of fibrous material are closely associated with the kinetosome in many motile cells. A network of fibres linking the basal bodies is found in ciliates. This network, the kinetodesma, has been extensively studied in *Paramecium*. In many flagellates, fibres, termed rhizoplasts, link the kinetosome with the nucleus; others, known as parabasal fibres, link the kinetosome with the Golgi apparatus.

Cytoplasmic membranes

Rough and smooth membranes

A variety of membranes is found in most eukaryotic cells. Some of these membranes are studded with dense granules (ribosomes) similar to those found in the nucleolus, while others are smooth. These are commonly referred to as rough and smooth endoplasmic reticulum respectively. The concept of the endoplasmic reticulum (ER) is held by many cell biologists to infer a continuity and interrelationship between all the internal cell membranes. Connections found between the outer membrane of the nuclear envelope and both types of ER suggest that these membranes are part of one interrelated system. This concept has recently been questioned with the suggestion that in fact different types of membranes are distinct entities.

Vacuoles

Eukaryotic cells in contrast to those of prokaryotes possess a number of vacuoles. These are usually bound by a single unit membrane termed the tonoplast. Vacuoles of varying shapes and sizes perform a number of functions according to cell type and state. In some fungal and algal cells they serve to localize storage products; in others they contain water. Many protozoa possess a more or less complex vacuole system, the contractile vacuole, which is sometimes connected to a system of small vesicles by tubular ducts.

Food vacuoles are formed by invagination of the plasmalemma; the process of pinocytosis (p. 403) also results in the production of vesicles within the cytoplasm. Such processes are often associated with considerable membrane synthesis.

The Golgi apparatus

The Golgi apparatus is the term given to the complex system of smooth membranes present in a wide range of eukaryotes and seen with the electron microscope as stacks of elongated vesicles from the ends of which smaller spherical vesicles may be budded (Figs. 10.1, 10.4A). Electron microscope studies on a number of organisms possessing a Golgi system have shown that the predominant function is probably secretory. In certain algal cells vesicles arising from the Golgi system have been found to contain scales which are subsequently deposited around the outside of the cell (p. 436). In many eukaryotes the Golgi body exhibits a polarity and is associated with the nuclear envelope, ER, or with fibres of the parabasal filament.

Since a large number of organisms apparently lack Golgi systems (most higher fungi and ciliates) they cannot be essential cell components. It is conceivable that a simpler system of isolating secretory products within membrane packets may operate in these groups, and the so-called multivesicular bodies found in many eukaryotes may be part of such a system.

Ribosomes

Ribosomes are small electron-dense granules preserved in cells fixed with glutaraldehyde or osmium tetroxide. They measure approximately 170–230 Å in diameter and consist of two sub-units which in the functional ribosome must be joined together. By means of fractional centrifugation, ribosomes can be isolated from the cell and subsequent chemical analysis has shown them to contain about 40 per cent protein and 60 per cent ribonucleic acid (RNA).

Ribosomes of both prokaryotes and eukaryotes are the site of protein synthesis (p. 65), and their occurrence and number may be related to the rate of synthesis within the cell.

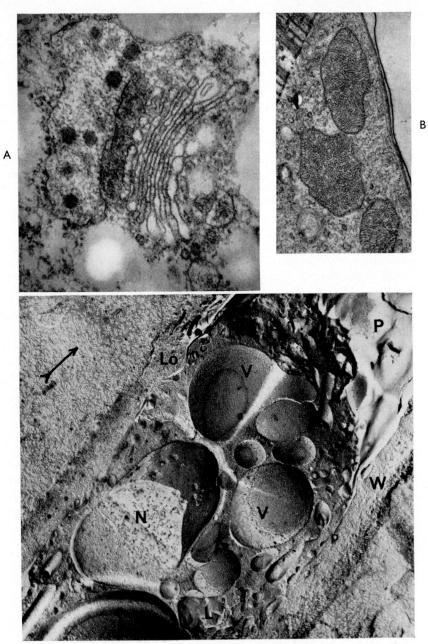

Eukaryotic ribosomes are larger than those of prokaryotes and are situated along the surface of rough ER, the nuclear membrane, inside the nucleus and free in the cytoplasm. Ribosomes are also found in chloroplasts, where they appear to be of the smaller 'prokaryotic type'.

Mitochondria

Typical mitochondria are characterized by a double delimiting membrane and an internal differentiation in the form of flattened membranous sacs (cristae) or tubules (tubuli) (Figs. 10.3A, 10.4B).

It is within these organelles that the enzymes of the respiratory system occur. Morphologically mitochondria undergo considerable changes reflecting not only their functional state but also the metabolic state of the cell as a whole.

In many motile cells mitochondria are often intimately associated with the flagellar apparatus and in the alga *Prymnesium parvum* not only is there a relationship between the haptonema base and a mitochondrion, but also the cristae within this mitochondrion exhibit a marked polarity.

Mitochondria are apparently absent from some eukaryotes living in anaerobic environments.

Many ideas have been suggested regarding the ontogeny of mitochondria, perhaps the most plausible being that they arise from membrane bound initials derived from the nuclear membrane and containing the necessary genetic information for their subsequent morphogenesis.

Plastids, pyrenoids, and eyespots

Plastids are found in algal thallus cells and their motile stages (phytoflagellates). These plastids may be differentiated into chloroplasts, chromoplasts, or amyloplasts containing chlorophyll, carotenoids, and starch respectively. Details of their structure and distribution are given on p. 432.

Chloroplasts in phytoflagellates appear to replicate by division of pre-existing ones rather than to develop from undifferentiated proplastids as in higher plant cells. The pattern of replication is probably controlled to some extent by existing chloroplasts. However, there is some evidence to suggest that the initial formation is from the nuclear membrane, during which process the genetic information necessary for the synthesis of RNA and DNA (known to occur within chloroplasts) is included.

Fig. 10.4   (A) A section through part of the Golgi apparatus of *Selenidium terebellae*, an archigregarine gut parasite of marine worms. Some of the elongated vesicles are inflated at their ends forming the peripheral Golgi vesicles. (× 60,000. Mrs. G. Dorey, Bristol University.) **(B)** A section through part of the ciliate *Euplotes eurystomus* showing mitochondria with typical tubuli. (× 17,500. R. Gliddon, Bristol University.) **(C)** A section from a freeze-etched hyphal cell of the fungus *Alternaria brassicicola*. The following features are shown: The hyphal wall (W) in section; the plasmalemma in surface view (P), and in section (p), with lomasomes (Lo) adjacent to the cell wall; both inner and outer surfaces of vacuoles (V) and a lipid droplet (L). The nucleus has been fractured through part of the nucleoplasm (N) and the inner surface of part of the nuclear envelope, with pores, is also shown. The direction of shadowing is indicated by (→). (× 20,000. R. Campbell, Bristol University)

Pyrenoids (Fig. 10.1) are regions of granular, dense material embedded within the chloroplast or arising as an outgrowth of it, and usually associated with the deposition of reserve materials derived from photosynthates (p. 433). In many chloroplasts, discrete membrane-bound starch granules are found between the lamellae. However, in euglenoids, the reserve material paramylon is formed outside the chloroplasts.

Following aldehyde or osmium tetroxide fixation, eyespots are seen with the electron microscope as dense osmiophilic granules (p. 433). They are considered to be associated with photoreception.

Microbodies and storage products

A variety of membrane bound vesicles associated with ER is found in a number of eukaryotic cells. Often crystalline inclusions appear in these vesicles. Their function is obscure but storage of proteinaceous material or of enzymes has been suggested.

Glycogen is stored in many mature fungal cells often in the form of dense stellate granules.

Lipid droplets are also found in many fungal cells, particularly mature spores. In osmium fixed specimens they are areas of dense material, while after potassium permanganate fixation they are seen as electron transparent areas bound by what appears to be a single dense membrane. Freeze-etched preparations of yeast cells have shown these membranes to consist of fine concentric lamellae.

Cytoplasmic microtubules and miscellaneous organelles

Microtubules are found in the cytoplasm of suitably fixed preparations of many eukaryotic protists.

They are associated with the cell wall in many algae, while in motile cells they often occur close to the kinetosome. In motile cells of fungi belonging to the order Blastocladiales, there is apparently a close link between the kinetosome fibre pattern and the form and arrangement of microtubules which lies close to the membrane surrounding the nuclear cap.

Trichocysts are found in many eukaryotic ciliates and flagellates (p. 437). When discharged they consist of a striated fibrous shaft with a spine at the tip. Some trichocysts have a fibre pattern in their tips, similar to that found in cilia and flagella, suggesting their origin from a basal body or kinetosome.

Contractile structures or myonemes are found in some protozoa. These are usually bands of fibrous material running around the cell beneath the pellicle. Bands of contractile fibres occur in those diatoms which extrude mucilage which is passed in streams along the outside of the cell. Active transport of this mucilage to one end of the cell by means of these contractile elements, provides the propulsive force necessary for cell movement.

From the above review it may be concluded that within recent years the electron microscope has provided information leading to a new and complex view of the cytoplasmic organization of cells. One of the most striking features of the eukaryote is the development of membrane systems which effectively compartmentalize the cell, provide sites for enzyme location and chemical reactions, and along which active transport can occur. It is not unusual that many of these membrane-bound organelles are similarly constructed and common to a large number of eukaryotes. It is a reflection on the efficiency of the system and the limited range of molecular components involved.

Finally it should be emphasized that although for convenience the various cell components have been considered in this chapter under separate headings as isolated structures, it must be realized that in any one living cell, these components are closely involved with each other, playing their part in contributing towards the metabolism and function of the cell as a whole.

THE FUNGI

INTRODUCTION

The relatively large size of fungus cells and their possession of discrete membrane-bounded nuclei and typical mitochondria distinguishes the fungi from the bacteria. Although the yeasts and some primitive aquatic species are normally unicellular, most fungi are filamentous. The filaments, which are known as hyphae, are usually much branched and collectively form the vegetative thallus or mycelium. The filamentous habit permits a greater diversity of form than that seen among bacteria. In the higher fungi the hyphae may aggregate together to form complex solid structures which in some species may reach a considerable size (e.g., the fruit-bodies of some of the wood-destroying fungi are often more than a foot in diameter). Not only do the fungi show a much wider range of form than do the bacteria, but also their life-cycles are more complex.

HABITAT

Fungi occupy a wide variety of habitats. Some are aquatic, some inhabiting fresh water and others the sea. The majority occupy moist situations on land although a few, notably species of *Aspergillus* and *Penicillium*, are able to withstand somewhat drier conditions.

Like the majority of the bacteria, fungi are heterotrophic (p. 145). No fungus is able to grow well in the absence of organic food substances. No fungi possess photosynthetic pigments and none is able to grow satisfactorily with carbon dioxide as the *sole* source of carbon. Hence they are dependent on preformed organic substances as a source of energy. Many are entirely saprophytic, feeding on non-living organic matter of diverse kinds. These include many species which are troublesome as destroyers and spoilers of man's foodstuffs and other valuable stored products (p. 664). The saprophytic abilities of others can be harnessed to our use in a variety of industrial processes (p. 680). In the presence of an external supply of sugars or other suitable organic food, many exhibit astonishing synthetic powers and produce a great variety of complex organic substances. These include not only cell proteins and reserve foods, such as fats and complex polysaccharides, but also vitamins, alcohols, organic acids, enzymes, pigments, and antibiotic substances, such as the well-known penicillin (p. 681). Many fungi obtain their organic nutrients from living host organisms. Many are active parasites of plants, including the major crop plants (p. 623), and some cause diseases of man and domestic animals (p. 547). A number of fungi parasitize insects (p. 605) and yet others, the predacious fungi, are able to trap and destroy eelworms and soil protozoa (p. 608). Fungi are present wherever organic material occurs, if other environmental factors are favourable (p. 159).

NUTRITION; THE EFFECTS OF OTHER ENVIRONMENTAL FACTORS

The general nutritional requirements of micro-organisms have been considered in Chapter 5. All fungi require a relatively large amount of a carbon source and most use carbohydrates more readily than other carbon compounds. Species and often different strains of a particular species differ in the range of carbohydrates and other carbon sources that they can utilize, but in general glucose and fructose are suitable sources of carbon; glucose is usually included in artificial media for the culture of fungi. In common with other living organisms fungi require a source of nitrogen, potassium, sulphur, magnesium, and a number of micronutrient minerals (trace elements, p. 149). No fungus has been proved to fix atmospheric nitrogen. Many grow and sporulate well on a glucose-salts medium from which they can synthesize all the complex organic substances they require. Others require external supplies of one or more vitamins or growth substances, which they are unable to synthesize themselves. The commonest requirement is for thiamin, but many fungi are deficient for biotin or other growth substances. Some can synthesize enough of such compounds to permit vegetative growth but need an external supply for the initiation of sporulation (p. 386).

Other environmental factors such as H ion concentration of the substrate, temperature, aeration, light, water content of the substrate, and humidity, may limit fungal growth and/or reproduction. The requirements of different species and of different phases in the life cycle of a given species differ widely, but in general fungi are favoured by a slightly acid reaction (in contrast to most bacteria), temperature around 20 to 25°C, good aeration, a moist substratum and a humid atmosphere.

FINE STRUCTURE OF THE FUNGAL CELL

The cell wall

The vegetative cells and spores of nearly all fungi are surrounded by a definite cell wall. The zoospores and gametes of some primitive species lack such a wall and the vegetative thallus is naked in a group of primitive plant parasites, the Plasmodiophorales (which includes *Plasmodiophora brassicae*, the cause of the serious club root disease of cabbage and other crucifers, p. 392), in a number of internal parasites of other fungi, and in species of *Coelomomyces*, an internal parasite of mosquitos.

The structure of the fungus cell wall is architecturally like that of the walls of algae and other green plants, and consists of a network of microfibrils (Fig. 10.2B) with matrix substances filling the spaces between the fibrils.

The cell walls of most of the fungi examined consist of 90–80 per cent polysaccharides together with proteins, lipids, and sometimes also pigments, polyphosphates, and inorganic ions. In at least one yeast the amount of protein is as much as 40 per cent but this is not characteristic of yeasts as a group.

The microfibrillar skeleton is frequently chitin, occasionally cellulose, or, in yeasts, non-cellulosic glucans. The cementing matrix usually contains protein, lipids, and various polysaccharides. Some of the protein may be enzymic, some may be due to cytoplasmic contamination, but some must be firmly bound to the microfibrils since it is removed only by drastic treatment. The structure of the cell wall lipids is not identical with that of cytoplasmic lipids. The nature and proportions of the polysaccharides vary.

Further study of the composition of the cell walls of fungi seems likely to give data of considerable taxonomic value. A tentative outline, based on admittedly incomplete data, showing correlation between cell wall composition and the main taxonomic subdivisions of the fungi has been drawn up by Bartnicki-Garcia (1968) and is given in Table 10.1. It will be recalled that cell wall composition is also proving a useful criterion in the classification of bacteria, particularly the actinomycetes (p. 333), and of the algae (p. 437).

Table 10.1   Cell wall composition and taxonomy of fungi (after Bartnicki-Garcia, 1968).

Chemical category	Taxonomic group
I Cellulose–glycogen*	Acrasiales
II Cellulose–glucan*	Oomycetes †
III Cellulose–chitin	Hyphochytridiomycetes
IV Chitosan–chitin	Zygomycetes ‡
V Chitin–glucan*	Chytridiomycetes
	Ascomycetes (except yeasts)
	Deuteromycetes (except asporogenous yeasts)
	Basidiomycetes (except mirror yeasts)
VI Mannan–glucan*	Saccharomycetaceae
	Cryptococcaceae
VII Mannan–chitin	Sporobolomycetaceae
	Rhodotorulaceae
VIII Polygalactosamine-galactan	Trichomycetes

* Incompletely characterized.
† Oomycetes are unique among filamentous fungi in possessing a cellulose wall.
‡ Zygomycetes differ from the filamentous higher fungi in the absence of glucan in the cell walls of vegetative hyphae; glucan has, however, been demonstrated in the spores of *Mucor rouxii* and the mannan content increases with the assumption of the yeast form by this fungus.

The protoplast or cytoplast

The protoplasts of fungi resemble, but in many ways are less complex than, those of animals or green plants. They are, however, definitely eukaryotic in structure and thus considerably more complex than those of the bacteria (p. 284).

The *cell membrane* or *plasmalemma* is similar to that of other eukaryotic cells (p. 354). It is pressed more or less closely to the cell wall, except for certain areas where it curves inwards enclosing a mass of minute tubules or vesicles (the lomasome, Fig. 10.4C) between it and the wall. Lomasomes were at first thought to be confined to fungi but have since been reported in algae and higher green plants. They are variable in form and have been interpreted in various ways (Bracker, 1967). They have not been identified in living cells and could thus be artefacts.

The *endoplasmic reticulum* (ER) in vegetative hyphae is usually sparse, compared with that of other eukaryotes (p. 363), and consists of scattered vesicles or cisternae, seldom showing a complex lamellar structure or well-defined Golgi bodies (p. 363). Local aggregations of ER occur in some fungi at sites of particularly intense metabolic activity. *Ribosomes* (p. 363) are usually scattered in the endoplasmic matrix but in some fungi they are associated with the ER as with the granulated or rough ER of higher plants and animals.

In the tips of growing hyphae the endoplasm contains no vacuoles, but further back from the tip small *stellate vacuoles* (p. 363) are present. These coalesce in older parts of the hyphae, the cytoplasm being reduced to a thin peripheral layer lining the cell wall. The coalescence of vacuoles has been suggested as a possible mechanism involved in the initiation and maintenance of cytoplasmic streaming. The presence of vacuoles and the occurrence of streaming are in contrast to their absence in bacterial cells.

Mitochondria (p. 365) have been seen in all actively growing fungal cells so far examined but are absent or deformed in some starved or anaerobically treated ones. They often have fewer cristae (Fig. 10.3A) than have the mitochondria of higher plants and animals, but their shape and complexity varies with the state of activity of the cell.

Interphase *nuclei* of fungi resemble those of other eukaryotes (p. 359). Light microscope investigations of stained preparations have demonstrated chromosomes in a large number of genera but although typical mitosis has been reported for many of these, in others the mechanics of division remain obscure. Some of the first electron microscope studies made on dividing fungal nuclei failed to show either organized chromosomes or a spindle apparatus. However, more recent work with improved fixation techniques has demonstrated at least one, and in many cases both, of these features in a number of genera representing most of the major groups of fungi: e.g., the Myxomycete *Clastoderma debaryanum*, the Phycomycetes *Catenaria anguillulae* and *Saprolegnia ferax*, the Ascomycetes *Saccharomyces cerevisiae* and *Pustularia cupularis* and the Basidiomycetes *Coprinus lagopus*, *Armillaria mellea*, and *Polystictus versicolor*.

Despite the increasing fine structural evidence for chromosomes and spindles no clear pattern has emerged as to exactly how the chromosomes separate

---

Fig. 10.5 (A) *Catenaria anguillulae*. Early anaphase nucleus. Centrioles, chromosomal fibres, continuous spindle fibres, and chromosomes are evident. The nuclear membrane is still intact. (× 36,000. From Ichida, A. A. and Fuller, M. S. (1968) *Mycologia*, **60**, 141.) (B) *Podospora anserina*. A transverse section through the centriolar plaque/ astral ray complex showing the electron-dense material of the plaque from which radiate the microtubules of the astral rays. (× 17,500. A. Beckett, Bristol University)

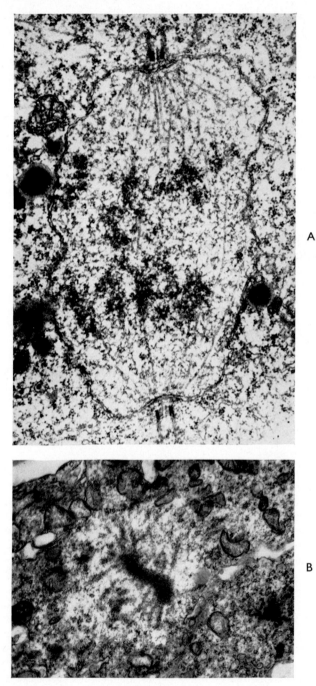

A

B

during anaphase, whether they are attached to the spindle by discrete kineto-chores and are pulled apart by contraction of polar fibres, or whether they are pushed apart by elongation of the interzonal microtubules of the central portion of the spindle. This latter mechanism has been suggested for *Catenaria anguillulae*. Neither is it yet possible to lay down any rules regarding the state of the nuclear membrane during division of fungal nuclei. In the majority of fungi so far examined the spindle is intranuclear, the persistent nuclear envelope only breaking down at the telophase separation of the daughter nuclei. In some, however (notably Basidiomycetes), the nuclear envelope breaks down at meta-phase leaving a spindle stretching between amorphous densely stained polar regions, e.g., *Polystictus versicolor*, *Coprinus lagopus*, and *Armillaria mellea*.

Centrioles have been readily demonstrated at the poles of dividing nuclei in fungi with motile stages in their life cycles, namely the aquatic Phycomycetes (Fig. 10.5A). Ascomycetes, which lack any form of motile stage, have been found to possess what are termed centriolar plaques. These are electron-dense thickened areas closely associated with the nuclear membrane, and lying at the poles of the spindle during division (Fig. 10.5B).

Similar but somewhat less regular structures have been demonstrated in the Basidiomycetes *Coprinus*, *Polystictus*, and *Armillaria*.

Differential staining shows that *glycogen* and *lipids* are the commonest *reserve foods* in the majority of fungi and these can also be recognized in electron micro-graphs of suitably fixed and stained material. At least some of the Oomycetes differ from other fungi in their lack of glycogen.

Some Lower Fungi are motile unicells or have motile phases in their life cycles. The *flagella* of these are of the complex type typical of motile eukaryotic cells (Fig. 10.3D) in contrast to the relatively simple ones of the bacteria.

THE FUNGAL THALLUS

Unicellular fungi

The simplest type of fungal thallus is that of the primitive unicellular fungi or chytrids and some allied forms. These show many affinities with the protozoa, but with only a few exceptions they possess a definite cell wall during the greater part of their life-history. Among them is a number of species causing diseases of economic plants (e.g., *Synchytrium endobioticum*, the cause of black wart disease of potato) and others may occur in epidemic form on planktonic algae, but the majority is of little or no economic importance.

In contrast, the yeasts, which also normally exist in the unicellular condition, are of the greatest importance to the industrial microbiologist, chiefly owing to their ability to ferment sugars with the production of carbon dioxide and alcohol (p. 661). These organisms are almost certainly not primitive but have probably degenerated into the unicellular condition from a previous filamentous state. Even those yeasts which are usually predominantly unicellular will, under certain cultural conditions, produce chains of cells resembling the hyphae of normal fungi but less stable than those of typical filamentous species. The component cells may become elongated and the filaments are then known as 'pseudo-

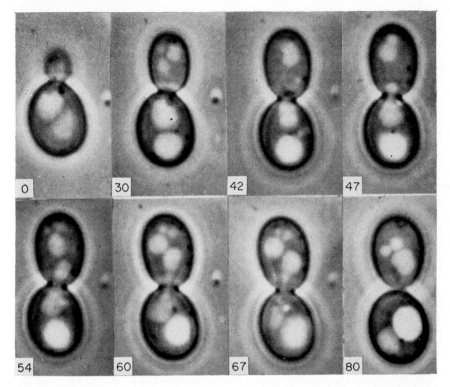

Fig. 10.6 Time-lapse phase-contrast photomicrographs of living cell of a budding yeast, *Saccharomyces cerevisiae*, showing budding of cell and division of nucleus. Figures represent minutes after first photograph was taken. In first, second, and third photographs nucleus is entirely within parent cell. As bud enlarges nucleus becomes dumbbell-shaped and extends into the bud (fourth, fifth, and sixth photographs), finally dividing, one daughter nucleus remaining in the parent cell and one in the bud. The last picture shows the bud rounded off and separated from parent. (× 3,000. From Robinow, C. F. (1966) *J. cell. Biol.*, **29**, 129)

mycelia'. The life-histories of the yeasts show a striking resemblance to those of certain normally filamentous species. Moreover, many species of filamentous fungi of various groups will assume a yeast-like condition under certain circumstances (p. 377).

Yeast cells are usually globose or ellipsoidal but are occasionally almost cylindrical. They are surrounded by a definite cell wall, which is thin and elastic in young cells but may become thickened and rigid in older individuals.

Under favourable conditions yeast cells divide rapidly. In most species, including those used in the fermentation industries, the cells multiply by budding. A small protuberance or bud grows out from the parent cell, enlarges to reach approximately the same size as the parent and is finally cut off and separates from it (Fig. 10.6). Buds develop from one or more sites on the parent cell according to the species. Where the process is taking place rapidly the cells

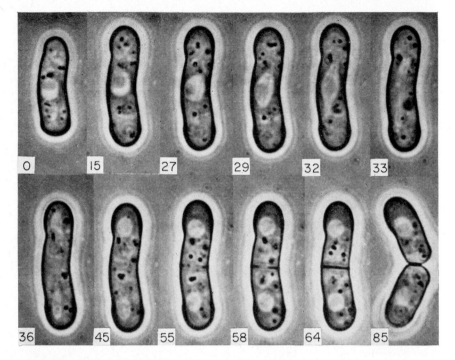

Fig. 10.7 Time-lapse phase-contrast photomicrographs of a fission yeast, *Schizo-saccharomyces versatilis*, showing nuclear division followed by septum formation and separation of resulting daughter cells. Figures represent minutes after first photograph was taken. First six photographs show stages in division of nucleus (light body); the next two show daughter nuclei migrated to poles of cell, the next three show septum growing inwards at middle of cell and the last shows the daughter cells separating. (× 1,520. From Robinow, C. F. and Bakerspigel, A. (1965) in Ainsworth, G. C. and Sussman, A. S. *The Fungi*, Vol. I. Academic Press, New York)

may remain partially attached to one another. Some groups of yeasts, known as the fission yeasts, show binary fission (Fig. 10.7), resembling that of bacterial cells (p. 311). Both methods of multiplication result in similar masses or colonies of cells which are known as sprout mycelia. Some yeasts produce spores (p. 395), but these are part of the sexual cycle and are formed in a manner different from that of bacterial endospores (p. 291).

Filamentous fungi

The majority of fungi are filamentous and consist of a mass of branched hyphae. These may be aseptate, the whole vegetative thallus then consisting of a single multinucleate (coenocytic) cell, or they may be regularly septate from an early stage in their development.

The aseptate forms, characteristic of the lower fungi or Phycomycetes ('algal fungi'), are considered to be the more primitive. Growth in length of an individual coenocytic hypha takes place at or just behind the tip. Branches may develop from points farther back along the hypha and these in their turn grow by increase in length of the apical portions. The wall of the growing tip is thin and extensible and the cytoplasm, as in the young yeast cell, is more or less homogeneous. Farther back from the tip the walls are relatively rigid, as a result of secondary deposition of material, and the cytoplasm is granular and vacuolated. Active streaming of the cytoplasm and its contained granules along the hypha can be readily seen and it has been shown that food materials are carried in this way towards the growing tip, branches, or developing reproductive organs. The nuclei are usually small, but they can be demonstrated by suitable staining.

Cross walls are formed in normally aseptate species in connexion with reproduction; dead or injured parts of the hypha are also cut off by septa. Old petri dish cultures of *Pythium* or *Mucor* often show young coenocytic hyphae at the edges of the plate and empty, regularly sepate ones in the older central part of

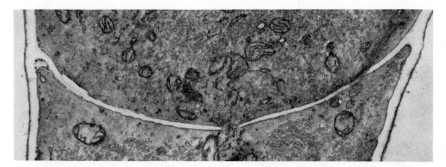

Fig. 10.8 Simple pore in septum between conidiophore (lower cell) and ampulla (spore-bearing cell) of *Oedocephalum roseum*. The septum grows inwards across the young conidiophore leaving a central pore, through which organelles (including nuclei and mitochondria) pass into the developing ampulla. The photograph shows vesicles and cisternae passing through the pore. (× 9,000. B. Dixon, University of Bristol)

the colony, indicating that as the older mycelium exhausts the food supply the hyphae die and walls are formed in succession cutting off the extending dead portion.

The higher fungi are septate from an early stage. Linear growth of these, too, is limited to the extreme tip of the hypha, and the septa form behind this extending zone at a distance from the tip characteristic for the species and to some extent influenced by environmental conditions. Fungal septa do not form as a plate across the spindle of a dividing nucleus, as in higher plants, but grow inwards from the wall as a ring or annulus more or less independently of nuclear division. The wall does not usually extend right across the hypha but a central pore is left, through which there is protoplasmic continuity from cell to cell.

In many fungi the pore is a simple hole in the septum (Fig. 10.8) but in some of the higher basidiomycetes (p. 399) it is a more complex structure. The rim of the septum bordering the pore is swollen and a complex system of membranes develops in the vicinity of the pore (Fig. 10.9). Such a complex system is termed

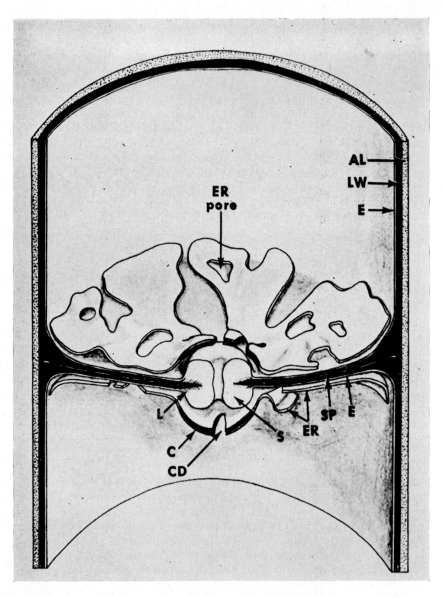

Fig. 10.9 Semi-diagrammatic sketch of a mature septum of *Rhizoctonia solani*.
AL = amorphous layer of lateral wall, LW = lateral wall, E = plasmalemma (ectoplast), ER = endoplasmic reticulum, SP = septal plate, composed of cross-wall
lamellae, S = septal swelling, L = lamellae within the septal plates, C = septal pore
cap, CD = discontinuity in the septal pore cap. (From Bracker, C. E. and Butler, E. E.
(1963) *Mycologia*, **55**, 35)

a dolipore. In spite of the complexity of the adjacent membrane system, nuclei and other organelles are able to pass through the dolipore. During passage from one cell to an adjacent one nuclei usually become vermiform, regaining their original shape on reaching the neighbouring cell. A septate hypha, therefore, is not completely divided into separate compartments; protoplasmic streaming and nuclear migration can go on. The partial septa may give added rigidity to the hypha. It is from this type of septate mycelium that the complex fruit-bodies of the higher fungi are formed. A further biological advantage is that 'repairs' to a damaged hypha can be more readily accomplished by the plugging of the pore in the septum next to the injured part than by the production of a complete new wall as is necessary when an aseptate hypha is damaged.

The cells of a septate mycelium become vacuolated as they age just as the coenocytic hyphae or yeast cells do. They may be regularly uninucleate or binucleate or may contain an indefinite number of small nuclei.

Anastomosis between hyphae of the same species takes place freely and nuclei are known to migrate across the bridging section from one hypha to another. This is of great significance in the life-cycles of many fungi since nuclei of different origin may thereby be present in a single hypha (heterokaryosis, p. 113).

Many species of both coenocytic and septate types tend to break down into a yeast-like mass of cells, either with ageing of the mycelium or in response to particular conditions, notably partial anaerobiosis or immersion in a strong sugar solution. The hyphae, which are normally cylindrical, become bead-like and the 'beads' round off and separate as *oidia* which may then start budding as in a yeast (Fig. 10.10B). Some fungal parasites of animals exhibit the phenomenon of dimorphism and are yeast-like when growing within the host and filamentous in artificial culture. Conversely some plant parasites, e.g., the grain smuts and the peach leaf curl fungus (*Taphrina deformans*), are filamentous in the host plant but produce only a yeast-like mass of cells on culture media.

Just as individual yeast cells may become full of reserve food material and develop a thick wall, in response to certain unfavourable conditions, so individual cells of filamentous hyphae or of normally thin-walled spores of some species, e.g., *Fusarium* spp., may round off and become surrounded by a thick wall. These *chlamydospores* (Fig. 10.10A) are probably resistant to adverse conditions and may play an important part in the survival of the fungi which form them.

Considerable differences in form may be seen between hyphae of the same thallus. Not only are the hyphae which bear the reproductive bodies often different from the vegetative ones in mode of branching, pigmentation and response to stimuli, such as light and gravity, but a mycelium may be made up of one or more types of purely vegetative hyphae. Those which penetrate the substratum are often of rhizoidal nature and quite different from the aerial hyphae (Fig. 10.17A). Special branches of distinct form may be produced, such as the haustoria, which are pushed into the host cells by many intercellular plant parasites (Figs. 19.5, A–D, 19.6) or the various rings and loops with which predacious fungi trap and entangle nematodes (Fig. 18.3). Many fungi produce long runner hyphae or stolons which creep over the substrate and give rise to rhizoidal penetrating branches and aerial hyphae at intervals, as with the saprophytic *Rhizopus* or the parasitic *Gaeumanniomyces graminis* syn *Ophiobolus graminis*.

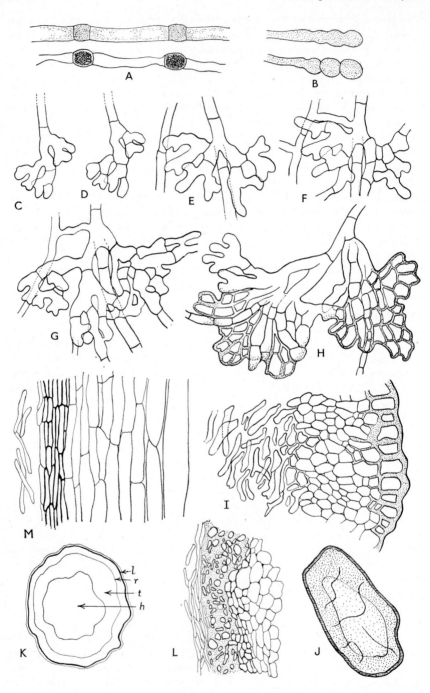

Large plectenchymatous structures

Further differentiation of fungal thalli is achieved through the aggregation of hyphae to form large structures of a more or less solid nature. The septate hyphae of certain higher fungi may become closely interwoven to form a mass, the individual cells of which are often inflated and globose. The resulting structure bears a superficial resemblance to the tissues of the higher plants, but is distinguished from these by its development from separate hyphae and the inability of the cells of the component filaments to divide in more than one plane. Such a pseudo-tissue or plectenchyma may, however, show differentiation into zones of different structure and function.

Stromata

Some fungi produce a mass of plectenchyma of regular or irregular shape on or in which spores or fruit-bodies are formed, e.g., the candle-snuff fungus, *Xylaria hypoxylon* or the ergot of rye, *Claviceps purpurea* (Fig. 10.11B, C). These are known as stromata (singular, stroma). Sometimes parts of the substratum or host tissue are embedded in the stroma as with the mummified fruits of Rosaceous plants attacked by the 'brown rot' fungi (*Sclerotinia* spp).

Sclerotia

Other fungi produce more or less globose bodies which are termed sclerotia (Fig. 10.10, C–J, 10.11A). These are usually differentiated into an outer protective layer of thick-walled cells and an inner mass of cells with thinner walls and containing reserve food substances, usually oily substances and glycogen (Fig. 10.10J). Sclerotia of species of *Rhizoctonia* and some related fungi are of looser construction and the component cells are all thick-walled and full of oily contents. Both types of sclerotium are better able to survive periods of unfavourable conditions, such as starvation or drought, than are the more delicate and relatively unprotected vegetative hyphae. Sclerotia are important in the survival of many plant parasites in the soil or on decaying plant remains.

Fig. 10.10 Chlamydospores, oidia, sclerotia and rhizomorphs. (**A**) *Mucor racemosus* (× 250). Upper hypha showing early stages in formation of chlamydospores; lower one with mature chlamydospores. (**B**) *Mucor racemosus* (× 250). Upper hypha showing irregular swellings which are the first stages in formation of oidia; lower one showing oidia rounding off.

(**C–J**) *Botrytis cinerea* (**C–I** × 500; **J** × 20). Stages in formation of a sclerotium. (**C–G**) drawn at approximately 4-hour intervals to show development of sclerotial initial by dichotomous branching and septation of the hyphae; (**H**) later stage of same initial showing coalescence of hyphae, and thickening of pigmentation of the walls; (**J**) section through mature sclerotium showing thick outer rind; (**I**) portion of this rind enlarged.

(**K–M**) *Armillaria mellea* (**K** × 10; **L** and **M** × 500). (**K**) T.S. of rhizomorph to show arrangement of zones; *l*, loose hyphae on outside of rhizomorph; *r*, thick-walled pigmented cells of rind; *t*, thin-walled cells of inner zone; *h*, hollow centre; (**L**) T.S. of part of the three zones; (**M**) L.S. of same.

(**A–B**, after Hawker, L. E. (1950). *Physiology of Fungi*. Univ. Lond. Press; **C–H, J,** after Willetts, H. J. (1949). *Thesis*. University of Bristol; **I**, after Townsend, B. B. and Willetts, H. J. (1954). *Trans. Br. mycol. Soc.*, **37**, 213; **K–M** after Townsend, B. B. (1954). *Trans. Br. mycol. Soc.*, **37**, 232)

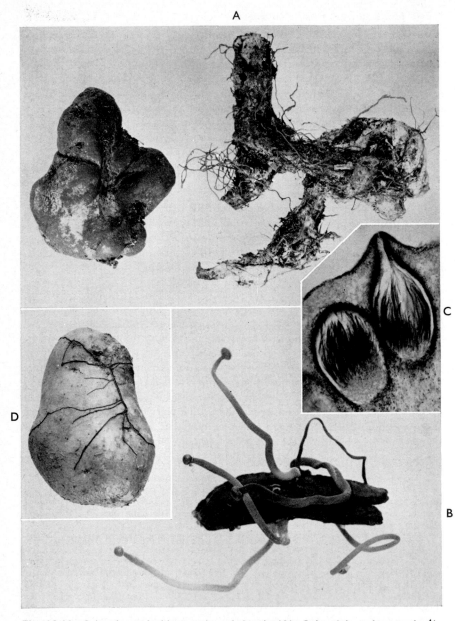

Fig. 10.11 Sclerotia and rhizomorphs of fungi. (**A**) *Sclerotinia tuberosa* ($\times \frac{4}{5}$). Sclerotium (left) removed from rotten rhizome, (right) of cultivated *Anemone nemorosa*. (Moore, W. C. (1946) *Trans. Br. mycol. Soc.*, **24**, 251.) (**B**) *Claviceps purpureum* ($\times 2\frac{1}{2}$). Germinating sclerotium showing stalked stromata with spherical fertile heads. (W. C. Moore.) (**C**) *C. purpureum* ($\times 100$). Photomicrograph of part of fertile head showing embedded perithecia. (L. E. Hawker.) (**D**) *Armillaria mellea* ($\times \frac{2}{3}$). Rhizomorphs on potato. (W. C. Moore)

Mycelial strands and rhizomorphs

Hyphae running parallel to each other may become aggregated to form strands of mycelium or may form more highly organized cylindrical or strap-shaped, often branched, structures, which are known as rhizomorphs. The latter may be relatively simple in structure or may show a high degree of differentiation into an outer protective layer of thick-walled cells and an inner layer of thin-walled elements, through which food materials may be conducted. Morphologically, complex rhizomorphs may be considered as much elongated sclerotia. They enable fungi to spread through the soil or over the surface of the host plant or other substratum while remaining in organic connection with the original mycelium. Food from the original supply or food base is carried along the strand to its growing point. The black rhizomorphs of the destructive honey agaric, *Armillaria mellea* (Fig. 10.10K, L, 10.11D), which destroys roots and other underground parts of many plants, including forest trees, fruit trees, and ornamental trees and shrubs, can spread through the soil for distances up to several metres provided that they remain attached at the base to a decaying root or tree stump from which they are able to draw food. They are thus able to reach, attack and colonize plants at some distance from the original victims. This ability to travel through the soil for long distances makes this a most difficult fungus to eradicate. Complex mycelial strands are also formed by the dry-rot fungus *Merulius lacrymans* (p. 401) and enable it to travel from one piece of timber to another over intervening areas of soil, brickwork, or metal.

Complex fruit-bodies

The characteristic spores of most higher fungi are borne on or in complex fruit-bodies (Fig. 10.18). Those of some of the giant puff-balls or the wood-destroying bracket fungi (p. 401) may reach a diameter of more than 30 cm. Such large structures must obviously possess a more or less rigid framework or 'skeleton' and must have some provision for the conduction of relatively large amounts of food materials to the growing tips and edges. Analysis of their structure shows that they are made up of several different types of hyphae, each of which plays a special part in the organization and maintenance of the fruit-body.

REPRODUCTION

Reproduction of fungi is usually by spores. These are most often single colourless cells, but in some species the spores may consist of two or more cells, may be pigmented, may have thick sculptured walls, or may bear appendages or be surrounded with a gelatinous layer (Fig. 10.12). The spores of a particular species are remarkably uniform in size and structure, and accordingly spore characters are largely used in the identification of fungi. Spores are usually readily detachable from the parent thallus. Some are shed by special and often complicated mechanisms. Many are small and easily carried by wind or other agency and are thus dispersed over considerable distances. This ease of dispersal is a major factor in the success of fungi in colonizing suitable habitats. Under

suitable conditions the spore germinates and one or more germ tubes (young hyphae) grow out (Figs. 10.13, 10.14).

Physiology of reproduction

If medium is inoculated with fungal spores or mycelium the resulting colony at first consists entirely of vegetative hyphae. Some colonies remain in the vegetative condition, either because their genetic constitution is such that spore production is impossible or because of unsuitable nutritional or other factors. Most, however, will sooner or later change to the reproductive phase and produce asexual or sexual spores or both. The duration of the initial vegetative phase depends upon both internal and external factors.

Internal factors

The most important internal factor controlling sporulation is the genetical constitution of the fungus. Thus a single strain of a heterothallic species will not produce spores under any conditions and a heterokaryon in which the majority of nuclei are derived from a non-sporulating strain will do so only sporadically (p. 113).

Even when the genetical constitution permits sporulation, other internal factors, such as the number and distribution of nuclei or the concentration of various complex substances in the hyphae, may be unsuitable. For example, it has been suggested that certain Pyrenomycetes do not form perithecia until the concentrations of thiamin and biotin in the hyphal tips have reached a certain level. This type of internal factor is, however, almost certainly the result of a particular combination of external factors.

External factors

The nature and concentration of the food supply is of great importance in determining the initiation, development, and maturation of reproductive structures. Usually, but not always, nutritional conditions favouring vegetative growth also favour the production of asexual spores, but those inducing sexual reproduction are often very different. In general the concentration of nutrients, and particularly of carbohydrates, optimal for sporulation, either asexual or sexual, is lower than that giving maximum dry weight. For example, the production of ascospores by the Pyrenomycete, *Sordaria fimicola* is possible only over a smaller range of concentration than that permitting vegetative growth. Although a high concentration of nutrients thus tends to prevent sporulation,

---

Fig. 10.12 Fungus spores as seen with the scanning electron microscope (Stereoscan). **(A)** *Zygorhynchus moelleri* mature zygospore showing sculptured wall (× 450). **(B)** part of same enlarged (× 2,500). **(C)** *Tuber puberulum* mature ascospore with reticulately thickened wall (× 1,600). **(D)** Group of spiny conidia of *Cunninghamella elegans* (× 1,000). **(E)** *Elaphomyces granulatus*, mature ascospore (× 2,200). **(F)** *Genea hispidula* mature ascospore (× 1,800). **(G)** Group of lemon-shaped basidiospores of *Hymenogaster luteus* (× 700). These appear smooth under light microscope but Stereoscan reveals that surface is rough; note basal cylinder which fits over sterigma. All taken by permission of Prof. H. E. Hinton, Bristol University, with instrument provided by Science Research Council, U.K. (E) taken from Hawker, L. E. (1968). *Trans Br. mycol. Soc.*, **51**, 493; the rest unpublished.

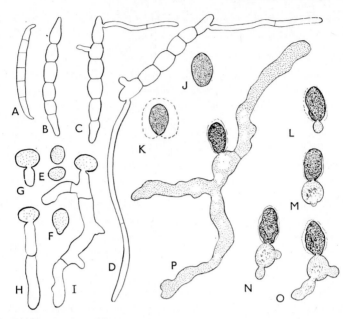

Fig. 10.13   Germination of fungus spores.

(A–D) *Fusarium lateritium* (×550). (A) sickle-shaped, septate conidium. (B) same after soaking in distilled water; note swelling of cells due to uptake of water. (C, D) stages in germination; germ tubes emerging from any cell of the spore, elongating and developing into septate hyphae; note constriction of hyphae at point of emergence from spore.

(E–I) *Botrytis cinerea* (×550). (E) ungerminated conidia. (F) conidium swollen through uptake of water and with bulge where germ tube is about to emerge. (G) early stage of germination with germ tube beginning to elongate. (H, I) later stages of germination, germ tube septate and branching.

(J–P) *Sordaria fimicola* (×550). (J) ascospore. (K) same showing gelatinous 'halo' as seen when mounted in nigrosin. (L) early stage in germination, contents extruding into small vesicle, spore shrinking. (M–P) later stages in germination, showing the original vesicle giving rise to several hyphae which then become branched and later septate.

(A–I after Hawker, L. E. (1950) *Physiology of Fungi*. Univ. Lond. Press; J–P after Hawker, L. E. (1951) *Trans. Brit. mycol. Soc.*, **34**, 181)

spores are usually produced only on a well-nourished mycelium. A commonly used method of inducing sporulation in fungi is to transfer an established mycelium from a rich to a dilute medium.

The composition of the nutrient substrate is as important as its concentration. Many Pyrenomycetes form perithecia earlier and in greater number when the source of carbon is a di- or polysaccharide than with a hexose sugar. Moreover, the concentration optimal for fruiting is often higher with a relatively complex carbohydrate. This has been shown to be due partly to the rate at which a particular fungus hydrolyses the various carbohydrates, the best results being obtained when the resulting concentration of hexoses remains optimal over a relatively long period. The ease of phosphorylation of various carbohydrates has also been shown to be important.

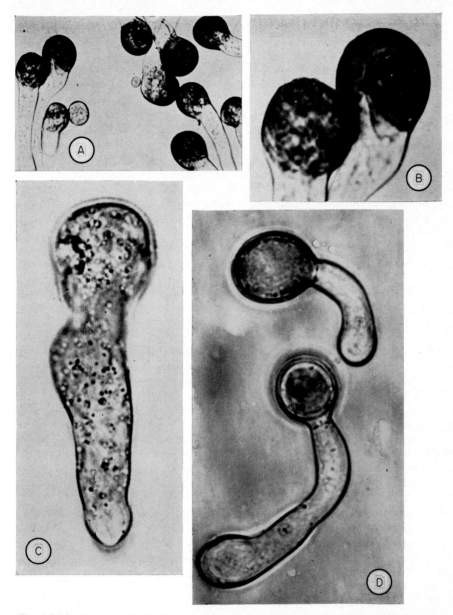

Fig. 10.14 (**A**) Germinating sporangiospores of *Rhizopus stolonifer* after 6 hr. in glucose-salts medium (×750). (**B**) Part of same enlarged. Note rupture of pigmented sporangiophore wall and emerging germ-tube enveloped by a thin wall. (J. A. Ekundayo, Bristol University.) (**C**) Germinating sporangiospore of *Rhizopus sexualis* after 5 hr. in 2 per cent malt extract, seen in optical section. (×2,500. R. J. Hendy, Bristol University.) (**D**) Germinating conidia of *Botrytis cineria* after 3 hr. in water. Note septum at base of germ-tube. (×1,000. R. J. Hendy, Bristol University.) From Hawker (1966) in 'The Fungus Spore', M. F. Madelin, ed. *Colston Papers*, No. 18, Butterworths, London.

The nature of the nitrogenous substance available and the carbon–nitrogen ratio in the medium also influence spore production. The mineral element requirements of most fungi are more exacting for spore production than for vegetative growth, and sporulation often does not begin until the concentration of minerals is significantly above that permitting minimum growth. Thus if certain species of *Fusarium* are grown without added magnesium no spores are produced, the sparse vegetative growth being supported by traces of magnesium present as an impurity in the other ingredients of the medium. Similarly the amount of copper necessary for measurable growth of *Aspergillus niger* is less than that for conidial formation. Moreover, colourless or light-coloured spores are formed sparsely at concentrations which do not permit the development of the characteristic black pigment. Finally while it has not been demonstrated that calcium is essential for vegetative growth of fungi it has been clearly established to be necessary for the formation of perithecia by *Chaetomium globosum*. Calcium is difficult to remove completely from the medium and it is likely that the requirements for vegetative growth are below, and those for fruiting are above, the amount remaining after the most stringent purification of the ingredients of the medium.

The amount of certain vitamins and other complex organic substances necessary for spore production is also greater than for vegetative growth. Many organisms, e.g., certain strains of *Sordaria fimicola*, are able to grow and even to sporulate feebly in the absence of an external supply of vitamin B_1 (thiamin), but show greatly increased fruiting when this vitamin is added to the medium. The addition of thiamin increases the rate of respiration during the early stages of colonial growth. A peak period of respiration has been shown to precede sporulation in *S. fimicola*, *Rhizopus sexualis* and some other fungi (Fig. 10.15). The effect of thiamin on spore production may thus be correlated with its effects on respiration.

Physical factors such as temperature, water content of the substrate, humidity, H-ion concentration of the substrate and aeration all influence sporulation, although the mechanisms by which they do so are largely obscure. In general the range of each of these over which spore production takes place is narrower than that permitting vegetative growth.

Some fungi are strikingly influenced by light; others grow and sporulate equally well in light or in darkness.

Gravity also influences the direction of growth of the spore-bearing structures of many fungi. Many sporangiophores and conidiophores are negatively geotropic and the fruit-bodies of many higher fungi are strongly influenced. The sporophores of most agarics (toadstools) show a complicated response to gravity. The stipe is usually negatively geotropic, the pileus is diageotropic and expands horizontally under the influence of gravity and the gills grow vertically downwards and are thus positively geotropic. This mechanism brings the various parts of the fruit-body into the most favourable position for spore discharge and dispersal.

It has become clear from recent studies that the conditions favouring the initiation of sporulation are not necessarily optimal during the later stages of development and maturation. It is important that when the effects of environ-

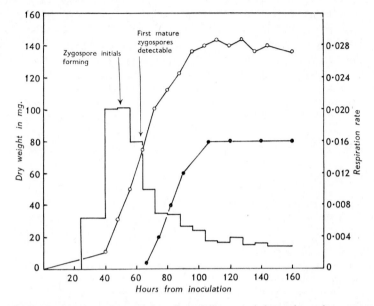

Fig. 10.15 Respiration, dry weight of mycelium, and formation of zygospores by *Rhizopus sexualis*. Respiration rate (mg. carbon, as CO_2, per hour per mg. dry wt.) is represented by the histogram. The dry weight of cultures in mg. is indicated by blank circles and the estimated quantity of zygospores (in arbitrary units) by solid circles. Arrows indicate the times at which zygospore initials and mature zygospores respectively could be detected. (P. M. Hepden)

ment on reproduction are being studied the process should be considered as a number of separate stages. For example, the production of zygospores by *Rhizopus sexualis* is inhibited by a temperature of *ca.* 10°C during the early stages, but after the gametangia have been delimited development continues at this temperature. The conditions permitting formation of perithecia of certain Pyrenomycetes do not always allow the formation of viable ascospores. Fruit-body initials of various species may be produced in quantity under conditions which permit the development of only abnormal fruit-bodies.

The differential effect of external conditions on the various stages of spore production supports the view that these developmental stages are under hormonal control and that the formation by the fungus of the various hormones concerned is influenced by the environment. Such control of a sequence of stages by distinct substances synthesized by the organism itself has been conclusively demonstrated for oospore formation by species of the water mould *Achlya* and less conclusively for more complex reproductive phases in some other fungi.

Types of reproduction

Most fungi produce more than one kind of spore. Most commonly, asexually produced spores (the so-called *imperfect stage*) are produced in large numbers when conditions are favourable to growth and these serve to spread the fungus rapidly. Later, often when conditions are no longer so favourable, either through exhaustion of food supply or from other causes such as seasonal fall in temperature, most fungi pass into the so-called *perfect stage*. In many species spores are then produced as the direct result of the fusion of sexual cells or branches, or after a period of secondary growth resulting from such a fusion. In a large proportion of the higher fungi, sexual fusion of gametes or of gametangia has been lost and the only trace of sex left is the fusion of nuclei at some stage during the production of the 'perfect' spores. Many fungi not only have no sex organs but are not known to produce the perfect stage under any circumstances (p. 401).

Asexual reproduction

Motile asexual spores (zoospores) are found only among the more primitive aquatic or semi-aquatic groups of fungi. The zoospore has no cell wall during its motile phase. It travels by means of one or more flagella, the number and arrangement of which is constant for the group. Zoospores are formed in special sac-like cells, or zoosporangia, from which they escape either by the rupture of the sporangial wall or through special pores developing in it. After swimming for some time in water, which may be only a film of dew on a host leaf, the zoospores settle down, withdraw their flagella and encyst, that is, they become surrounded by a cell wall. Under suitable conditions they germinate, usually by putting out a germ tube which develops into a typical hypha.

In one large group of the lower fungi (the Zygomycetes, p. 393) the spores are never motile and are surrounded by a definite cell wall. In most species of this group the spores are formed in multispored sporangia but there is a tendency for the number of spores in the sporangium to be reduced in some species and in others the asexual spores are borne singly and are then known as conidia.

In the higher fungi sporangia are not formed and in many groups the conidium is the typical asexual spore (Fig. 10.19). Conidia are often formed in enormous numbers and as they are usually light and readily detached from the parent hypha, or conidiophore, they are particularly efficient agents of spread of the fungus.

Sexual reproduction

The sexual spores of the lower fungi are the direct product of the fusion of gametes or gametangia. These may be isogamous (i.e., of equal size) and motile, as in most of the chytrids, motile but heterogamous (differing in size or pigmentation), as in a few chytrids and some other aquatic groups, or they may be non-motile, as in the important group of the Oomycetes (p. 392). In the Zygomycetes the characteristic sexual spore is the zygospore, produced by the conjugation of two sexual branches, which may be equal or unequal in size according to the species, and which are termed gametangia. Some species form zygospores freely in monospore cultures, and are then said to be *homothallic*. Others form them only by conjugation of hyphae derived from two different strains of opposite mating type, usually termed 'plus' and 'minus' strains. These are said to be *heterothallic.* Heterothallism was first demonstrated in this group but has since been found in all the major groups of fungi.

In the higher fungi the sexual spores are seldom the direct result of the fusion of sexual branches. Even in the yeasts and related fungi, in which fusion of isogamous or heterogamous cells or branches takes place in a number of species, the result is not a single spore but a sporangium-like cell, the ascus, in which the ascospores are formed (p. 393). In some higher Ascomycetes a fusion of sexual branches takes place and from them a secondary mycelium grows out and finally bears the asci. In most of the higher Ascomycetes and in all the Basidiomycetes definite sexual branches are no longer formed. In both groups, however, production of the perfect stage spores is preceded by a nuclear fusion and a reduction division. These spores are haploid, germinating to give a haploid mycelium. The exact point in the life-cycle at which this haploid mycelium becomes diploid differs in different species. In a few species the diploid condition arises in the spore itself.

The sexual spores of many fungi are resistant to adverse conditions, such as temporary drought or the cold of winter, either through the possession of a thick wall, as with the various resting spores, such as the oospores or zygospores of the Phycomycetes, or through being enclosed in protective fruit-bodies. Thus while the asexual spores are agents of multiplication and spread, the sexual or perfect stage spores tend to be agents of survival, although in many higher fungi they may also function as dispersal agents.

Germination

The final stages of the reproductive cycle, after the formation, discharge and dispersal of spores, is spore germination (Figs. 10.13, 10.14) and the initiation of a new mycelium. Germination cannot take place until the spore is mature and unless the spore wall is permeable to air and water. Some thick-walled spores, such as the oospores of Oomycetes or teleutospores of most rusts, remain dormant for long periods before germination. Most asexual spores and many ascospores and basidiospores are capable of germination as soon as they are shed from the parent hypha and, with some exceptions, retain their viability for only relatively short periods.

Even when a spore is in the right condition it germinates only if environmental factors are suitable. Many spores contain a relatively large amount of nutrients and are thus able to germinate in water or, as with the conidia of the powdery mildews, in humid air. When the reserves originally present in the spores are exhausted the germ tubes cease to grow unless they have reached an external supply of food. In the natural habitat pure water is seldom encountered, even raindrops on leaves have been shown to contain leaf exudates. Some spores with inadequate reserves of vitamins or other food substances require an external supply of these for germination.

With many external factors, such as temperature or H-ion concentration of the substrate, the range permitting germination is usually narrower than that permitting mycelial growth. Thus germination normally takes place only under conditions suitable for further growth of the germ tube and the establishment of a new mycelium.

OUTLINE OF CLASSIFICATION
(with examples from the major groups)

Owing to their number and wide range of form the fungi have not yet been classified satisfactorily and no two mycologists would produce identical arrangements. In this book only the main groups and some important genera will be described; for further details the reader is referred to the books listed at the end of this chapter.

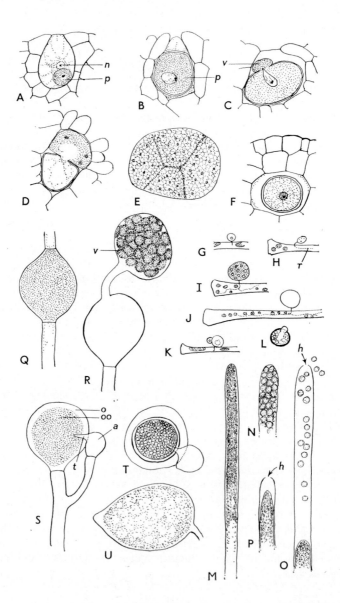

THE LOWER FUNGI (PHYCOMYCETES)

Thallus of simple organization; unicellular (but not reproducing by budding) or vegetative phase filamentous and aseptate, or occasionally septate. Classified according to presence or absence of motile (flagellated) stages and the morphology and number (i.e., one or two) of flagella where present.

Examples: (1) The chytrids (Chytridiales) are one of a number of groups the motile cells of which possess a single posterior whiplash type flagellum. In many species the whole of the unicellular thallus takes part in reproduction (i.e., they are holocarpic). Some species possess thread-like absorbing or attachment organs (rhizoids) which do not take part in reproduction and in others the reproductive bodies are connected by fine vegetative strands. Most species are aquatic, some are parasitic on higher plants.

The chytrids not only are simple in structure but have a simple life cycle. Parasitic species may, however, cause considerable damage to the host. Zoospores of *Synchytrium endobioticum*, the cause of black wart disease of potato (Fig. 19.3D), penetrate the surface cells of the potato tuber and cause these to enlarge and proliferate. More zoospores are then produced (Fig. 10.16, A–E) and released in large numbers into the soil. Successive generations of these may be formed but eventually similar motile cells fuse in pairs and after penetration produce resting spores (Fig. 10.16F), which may remain dormant for as long as ten years before germinating to produce active zoospores. In some chytrids sexual fusion is between unequal cells, e.g., in *Rhizophidium planktonicum*, which has been shown to cause fluctuation in numbers of the planktonic diatom *Asterionella* in Windermere and other lakes (p. 516), sexual fusion is between a smaller male cell, resembling an encysted zoospore, and a larger female thallus, resembling a young zoosporangium (Fig. 10.16, G–L).

Fig. 10.16 Chytrids and Oomycetes.
   (**A–F**) *Synchytrium endobioticum* (× 230). (**A**) young prosorus (*p*) in enlarged host cell. *n* = host nucleus; (**B**) prosorus (*p*) still in uninucleate condition, just prior to migration of contents; (**C**) prosorus, contents including nucleus migrating into vesicle (*v*); (**D**) vesicle after migration of most of contents of prosorus, and after nuclear division has commenced; (**E**) early stage in segmentation of prosorus to give sorus consisting of a few multinucleate zoosporangia; (**F**) resting spore (zygote), produced from fused gametes, division of host cells has occurred so that spore is becoming deep-seated. (After Curtis, K. M. (1921) *Phil. Trans. R. Soc., B*, **210**, 409).
   (**G–L**) *Rhizophidium planktonicum* (× 650). (**G** and **H**) early stages in development of sporangium from zoospore which has settled on cell of the diatom *Asterionella formosa*. Note thread-like rhizoid (*r*). (**I**) mature sporangium; (**J**) empty sporangium; (**K**) young female thallus with adherent male cell; (**L**) mature zygote (resting spore) with empty male cell still attached. (After Canter and Lund (1948) *New Phytol.*, **47**, 238).
   (**M–P**) *Saprolegnia* sp. (× 250). (**M**) young sporangium not yet cut off by basal wall but with denser granular contents than those of parent hypha from which it is not otherwise differentiated. (**N**) apex of mature sporangium containing zoospores. (**O**) sporangium dehiscing, showing some zoospores still inside and base already beginning to develop another sporangium. *h* = pore at apex. (**P**) apex of young sporangium which has nearly filled old empty one.
   (**Q–T**) *Pythium debaryanum* (× 750). (**Q**) young intercalary zoosporangium. (**R**) germinated terminal sporangium with nearly mature zoospores in vesicle (*v*). (**R**, after Matthews, V. D. (1931) *Studies on the Genus* Pythium. Univ. N. Carolina Press.) (**S**) oogonium (*o*) and antheridium (*a*) at fertilization stage, note single oosphere (*oo*) and fertilization tube (*t*). (**T**) fertilized oogonium containing thick-walled oospore antheridium empty and still attached to oogonium. (**U**) *Phytophthora cactorum* (× 750). Pyriform sporangium just before differentiation of zoospores.

Other related groups (Blastocladiales, Monoblepharidales) show differentiation of the thallus into vegetative and reproductive parts (i.e., are eucarpic).

(2) The Plasmodiophorales is a group of parasites, mainly of higher plants, in which the vegetative stage is represented by a naked mass of protoplasm, the *plasmodium*, which, however, differs in origin from that of the slime moulds, and in which motile zoospores and gametes possess two anterior flagella of whiplash type and very unequal length as in the slime moulds (p. 422).

The best-known example of this group is *Plasmodiophora brassicae*, the cause of club root of cruciferous plants (Fig. 19.3B).

(3) The group Oomycetes, fungi of aquatic or damp habitats, are characterized by motile stages (absent in a few genera) bearing two flagella of approximately equal length, one of whiplash and one of tinsel type, attached anteriorly or laterally; by the cellulose nature of the microfibrils of the cell wall; and by sexual reproduction involving the passage of the contents of a male branch (antheridium) into the female organ (oogonium), where fertilization of one or more female cells (oospheres) leads to the production of usually thick-walled resting spores (oospores).

The OOMYCETES include:

(A) **Saprolegniales** (water moulds). Aquatic or occurring in wet soil, mostly saprophytic.

The water moulds produce coarse, sparsely branched mycelium on various organic materials. *Saprolegnia* produces numerous zoospores in terminal club shaped or cylindrical sporangia (Fig. 10.16, M–P), which are little more than slightly swollen hyphal ends cut off by cross walls from the parent hyphae. The zoospores are pear-shaped and escape from the sporangium through an apical pore. After a time they settle down and encyst. Later the cysts germinate to give kidney-shaped zoospores which after a further motile phase again encyst and finally germinate by putting out a germ tube. This double motile phase is known as diplanetism. In other genera the first motile phase may be brief or lacking, only kidney-shaped zoospores being formed, or there may be no motile phase at all. Under suitable conditions *Saprolegnia* produces oogonia containing several female cells or oospheres, each of which is fertilized by a male cell or antheridium and is then known as an oospore.

Leptomitus, usually placed in a separate order (Leptomitales), differs from *Saprolegnia* in having a beaded, but still aseptate, mycelium and is found chiefly in polluted water.

(B) **Peronosporales.** Semi-aquatic or terrestrial; including many parasites of higher plants.

Pythium has simple, usually globose sporangia (Fig. 10.16Q, R) scattered over the mycelium. Zoospores develop in a vesicle extruded by the sporangium. The simpler species of *Phytophthora* resemble *Pythium* but their sporangia are pyriform (Fig. 10.16U) and vesicles are seldom formed. *Ph. infestans* (causing blight of potatoes) produces its sporangia on branched sporangiophores. The downy mildews (obligate parasites of higher plants) produce specialized branched sporangiophores bearing numerous deciduous sporangia which may function as conidia and germinate by germ tubes. Sexual reproduction throughout the group is relatively uniform and differs from that of *Saprolegnia* in the presence of only one oosphere in each oogonium and in the differentiation of the cytoplasm of the young oogonium into a central ooplasm, which produces the oosphere, and an outer zone or periplasm, which is used up in the formation of the thick, often sculptured wall of the fertilized oosphere or oospore (Fig. 10.16S, T).

(4) The group ZYGOMYCETES contains largely terrestrial species important as soil organisms and agents of decay of organic materials (Mucorales). Some are parasitic on insects (Entomophthorales), others on soil protozoa (Zoopagales).

Motile stages are lacking; asexual reproduction is by sporangiospores or conidia. Sexual reproduction is by the fusion of equal or unequal gametangia to form a zygospore.

The mycelium of the Mucorales consists of rather coarse, loosely growing hyphae (Fig. 10.17A). The colonies are at first white and become grey or brown with age or as a result of spore production. Most species (e.g., species of *Mucor, Rhizopus, Absidia, Zygorhynchus,* and *Phycomyces*) produce globose sporangia containing numerous spores and borne on special upright hyphae known as sporangiophores (Fig. 10.17, B–D). In some species the sporangiophores show characteristic branching, e.g., *Syzygites.* The large multi-spored sporangia are usually produced terminally on the sporangiophore, the tip of which in most species projects into the sporangium as a globose or hemispherical columella. As the sporangium grows, cytoplasm can be seen streaming into it until the contents are noticeably denser than those of the ordinary hyphae and finally the columella is laid down. Fissures then develop in the dense cytoplasm of the sporangium cutting it up into uninucleate polygonal blocks which round off, become surrounded by a cell wall and form a dark pigment. The mature sporangium is entirely filled with these spores. The sporangial wall is usually thin and ruptures to release the spores, but in some highly specialized genera, such as the coprophilous (i.e., growing on dung) *Pilobolus,* part of the wall is thickened. Some genera, e.g., *Thamnidium,* produce smaller sporangia (sporangioles, containing only a few spores) in addition to the multispored sporangium; other produce only sporangioles and a few produce true conidia, e.g., *Cunninghamella.*

Sexual reproduction is by the fusion of two cells, the gametangia, which may be of equal or unequal size according to the species, to give a thick-walled zygote (the zygospore). In the early stages of conjugation two branches approach each other and enlarge. These are termed the progametangia. Soon after the tips of these come into contact each progametangium cuts off a terminal gametangium from the remaining part of the swollen branch, which is termed the suspensor. Most species are isogamous, e.g., *Mucor, Rhizopus* (Fig. 10.17, E–I), the two gametangia being similar, but others show differences in size or behaviour, e.g., *Zygorhynchus* (Fig. 10.17J, K). In some species bristles are produced from one or both suspensor cells, e.g., *Absidia, Phycomyces*; in others the zygospore is formed in a vesicle growing out from one or both gametangia, e.g., *Piptocephalis*; in *Mortierella* the otherwise simple zygospore is protected by a sheath of sterile hyphae and in *Endogone* the conjugating branches are grouped together in a definite fruit-body in which the developing zygospores are surrounded by sterile hyphae.

THE HIGHER FUNGI

Thallus consisting of septate hyphae or of a sprout mycelium; asexual reproduction (imperfect stage) usually by conidia; sexual reproduction or perfect stage, when present, by characteristic spores (ascospores or badidiospores), often borne in or on complex fruit-bodies.

There are two groups; the Ascomycetes, the perfect stage spores (ascospores) of which are formed endogenously in a sac-like ascus, usually eight ascospores in each ascus; and the Basidiomycetes, the perfect spores (basidiospores) of which are borne exogenously on special cells (basidia). In addition, a third group is usually

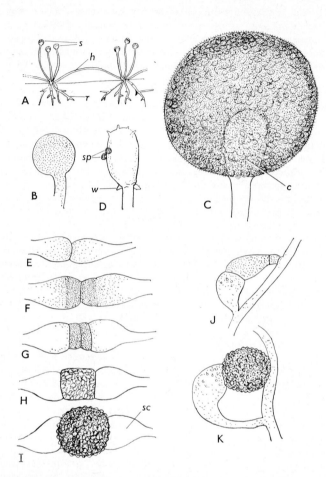

Fig. 10.17 Zygomycetes. (**A**) Growth habit of *Rhizopus* (diagrammatic, not to scale) showing sporangia (*s*), stolons or runner hyphae (*h*) and rhizoids (*r*).

(**B–D**) *Mucor plumbeus* (×500). (**B**) young, undifferentiated sporangium. (**C**) nearly mature sporangium showing columella (*c*) and partially differentiated spores. (**D**) columella of dehisced sporangium showing spines characteristic of this species, remains of sporangium wall (*w*) and two spores (*sp*) adhering to the columella.

(**E–I**) *Rhizopus sexualis* (×100). Stages in conjugation and formation of zygospore. (**E**) two equal-sized progametangia in contact and enlarged. (**F, G**) stages in differentiation of gametangia. (**H**) gametangia fused to form young zygospore which is rounding off and developing a thick sculptured wall. (**I**) nearly mature zygospore, spherical and with fully pigmented thick wall; suspensor cells (*sc*) empty.

(**J, K**) *Zygorhynchus moelleri* (×500). (**J**) young gametangia of unequal size. (**K**) nearly mature zygospore. Note large suspensor cell.

recognized, the Deuteromycetes or Fungi Imperfecti the members of which lack perfect stages. Since these are mostly forms closely resembling the asexual or imperfect stages of known Ascomycetes they are more properly treated with them. However, since many of them are of importance industrially or as parasites it is convenient to retain this 'form' group and to use the available artificial keys to identify not only those lacking a perfect stage but also others in which this stage is formed only occasionally or under unusual conditions.

I   *Ascomycetes*

This group was formerly subdivided according to the presence or absence of fruit-bodies (ascocarps) and to the gross morphology of these when present. It is now recognized that there is a fundamental difference between forms in which the asci have a single layered wall (unitunicate) and are formed in a layer (hymenium) within a fruit-body (Ascohymenomycetes) and those in which the asci have a two-layered wall (bitunicate) and are formed singly or occasionally in small groups in a stroma (Loculoascomycetidae). The breakdown of the matrix of hyphae surrounding the locules may produce a fruit-body resembling a hymenial form of entirely different development. Hence microscopic and developmental characters must be considered.

Examples:

(1) **Hemi-Ascomycetidae.** Asci formed singly, fruit-bodies not formed.

(A) **Endomycetales (Saccharomycetales).** Asci formed singly from conjugated or single cells, mostly saprophytic; a few parasitic on animals.

This group is limited to yeasts known to form ascospores and to some obviously related mycelial species, many of which assume a yeast-like habit in old cultures or under certain conditions, as when grown in media of high sugar content.

Saccharomyces is a large genus which includes nearly all the species employed in the production of alcohol and in baking. The cells vary in shape from globose to elongate and multiply by multipolar budding. Pseudomycelia are sometimes formed (p. 373). The formation of asci is preceded by conjugation of cells in some species, but this conjugation may occur many cell generations before ascus formation, so that the yeast may exist in either haploid or diploid forms, according to the stage of its life-cycle.

Schizosaccharomyces is a fission yeast (p. 374) dividing vegetatively by simple fission. True mycelium is not formed. *S. octosporus*, which was isolated from dried fruit from Greece and Turkey, normally forms 8-spored asci following conjugation of pairs of cells.

For convenience many organisms of yeast-like habit, but not closely related to the ascosporogenous yeasts, are considered with them. The so-called 'mirror yeasts', the family Sporobolomycetaceae, shoot off their spores by a drop mechanism similar to that of certain Basidiomycetes, to which they may be related. If a petri dish colony is inverted a mirror image of it is formed on the lid by the discharged spores. The mirror yeasts are epiphytic on leaves and are of no economic importance.

The asporogenous (i.e., non-sporing) yeasts or Cryptococcaceae are sometimes treated with the Fungi Imperfecti, but are commonly included with the ascosporogenous yeasts by the industrial mycologist. They include the common air-borne contaminant of laboratory cultures, the pink yeast *Rhodotorula*; the colourless *Torulopsis*; *Candida*, a pseudomycelial form, of which *C. albicans* is pathogenic to man and animals; *Cryptococcus* which contains species pathogenic to animals; and a few other genera of little or no economic importance.

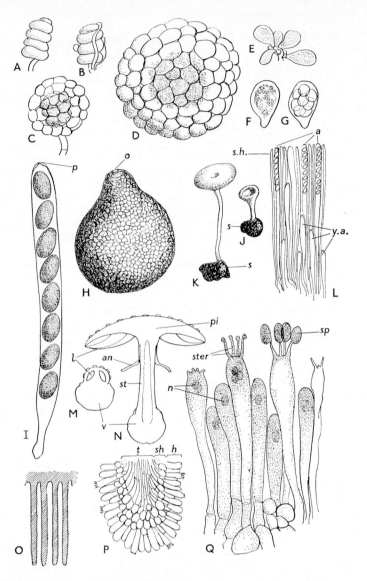

Fig. 10.18 Ascomycetes and Basidiomycetes.

(**A–G**) *Aspergillus* (*Eurotium*) *herbariorum* (×600). (**A**) coiled archicarp (female branch). (**B**) same at a slightly later stage with sterile hyphae growing up from base. (**C**) young fruit-body (ascocarp, cleistocarp) with archicarp still visible inside. (**D**) older fruit-body, dense mass of ascogenous hyphae visible in centre but details obscured by wall (peridium). (**E**) group of young asci. (**F**) older ascus, contents aggregating to form spores. (**G**) mature ascus containing eight ascospores.

(**H–I**) *Sordaria fimicola*. (**H** ×40) mature fruit-body (perithecium) showing short neck (beak) and ostiole (*o*). (**I** ×600) ascus containing eight black ascospores arranged in a single row (uniseriate). Thin place in wall at apex of ascus indicates pore (*p*) through which spores escape later.

(**J–L**) *Sclerotinia sclerotiorum*. (**J, K** natural size) stalked fruit-bodies (apothecia)

(B) **Taphrinales.** Parasites on vascular plants.

(2) *Eu-ascomycetidae.* Asci unitunicate, borne on secondary ascogenous-hyphae, usually in a fruit-body

Series α. *Plectomycetes,* asci globose, evanescent, arranged irregularly within a closed spherical fruit-body (cleistocarp) or a flask-shaped one opening in an ostiole at the apex (perithecium).

(A) **Plectascales (Aspergillales)** fruit-body a closed, spherical cleistocarp; most saprophytic, some parasitic on plants or animals.

Most members of this group produce numerous conidia by which they are able to spread rapidly under suitable conditions. The simplest forms, such as *Byssochlamys fulva* (a cause of spoilage of canned fruits), produce asci in groups arranged irregularly on ascogenous hyphae without any peridium. Others, such as *Gymnoascus* (common on dung and other decaying organic material), surround the asci with a thin weft of hyphae, thus producing a simple fruit-body of the cleisto-carp type. The cleistocarps of the more advanced genera, such as *Monascus, Aspergillus (Eurotium)* (Fig. 10.18, A–G) and *Penicillium* are surrounded by a definite wall or peridium consisting of one or more layers of cells. The asci of all genera are evanescent and the mature fruit-body contains a mass of ascospores. Several of the dermatophytes, formerly classed as Deuteromycetes, are now known to form ascocarps of a form resembling the simple cleistocarp of the Gymnoascaceae.

(B) **Microascales** in which the fruit-bodies are flask-shaped includes a number of serious parasites of forest trees (p. 674). The genus *Ceratocystis* is of importance in timber decay. The perithecia have unusually long necks. In some species the small detached asci travel up the neck and explode on reaching the ostiole. In other species the asci are evanescent and the spores are discharged in a stream of mucilage.

Series β. *Ascohymenomycetes,* asci usually clavate or cylindrical, seldom evanescent, arranged parallel to one another in a hymenium in a cleistocarp, perithecium or a cup-shaped fruit-body (apothecium) or a complex derivative of the latter.

(A) **Erysiphales.** Asci parallel in layer (hymenium) or single in a cleistocarp, asci persisting until maturity, ascospore discharge often explosive; plant parasites.

The powdery mildews, important obligate parasites of higher plants, are confined to the surface of the host and derive nourishment from the epidermal cells by means of haustoria (Fig. 19.5D). The mycelium produces numerous erect chains of conidia (the *Oidium* stage) which give the powdery appearance to the infected host plant. Later thick-walled fruit-bodies (cleistocarps) are formed.

---

developing from sclerotia (*s*). (**J**) cup-shaped fertile part of fruit-body still with incurved rims. (**K**) fertile part expanded and recurved. (**L** × 200) part of hymenium (fertile layer) which lines the cup, showing mature asci (*a*), young asci (*y.a.*) and sterile hairs (para-physes, *s.h.*).

(**M–Q**) *Amanita muscaria.* (**M**) unexpanded, (**N**) expanded fruit-body ($\times \frac{1}{5}$). *pi* = pileus or cap, *st* = stipe or stalk, *l* = lamellae or gills, *an* = ring or annulus, *v* = volva. (**O**) diagrammatic section through gills. (**P** × 50) L.S. gill showing central part of trama (*t*) with elements running vertically, subhymenial layer (*sh*) and hymen-ium (*h*) consisting of basidia. (**Q**) L.S. portion of hymenium (× 500) showing basidia at all stages from the beginning of elongation to final collapse after spore discharge. *n* = nucleus, *ster* = sterigma, *sp* = spore.

(B) **Pyrenomycetes.** Ascocarp a perithecium (some forms now transferred to the Loculoascomycetidae).

Many are plant parasites, some saprophytic species destroy cellulose or wood. The perithecia of *Chaetomium* are more or less globose, usually covered with characteristic coiled or branched bristles and the ostiole is rudimentary or lacking. The asci are globose and evanescent but are produced in a parallel layer. *Chaetomium* has no specialized method of spore discharge and the spores ooze out through the ostiole, if present, or through a tear in the peridium. This genus thus has some features reminiscent of the Plectascales, while the initial arrangement of the asci is characteristic of the Pyrenomycetes. In more typical genera the perithecium has a beak or neck, opening by a definite ostiole, the asci are clavate (club-shaped) or cylindrical, and dehisce explosively, discharging the spores to a considerable distance. Some such as *Sordaria*, found on dung, garbage, and in soil (Fig. 10.18, H–I) or *Neurospora*, the red bread mould, are of importance in industrial mycology. *Neurospora* is also of interest as the subject of much fundamental research on fungal genetics and nutrition.

(C) **Discomycetes.** Ascocarp an apothecium or a derivative of this, hymenium usually extensive. Many saprophytic, growing on soil or dung, some parasitic (including such important ones as the genus *Sclerotinia*, members of which cause brown rot of rosaceous fruits, many diseases of bulbous ornamentals, rotting of stems such as the potato haulm, etc.).

The large genus *Peziza* produces its asci in a large fleshy cup-shaped apothecium. The asci are numerous, cylindrical and tightly packed, parallel to one another in an extensive hymenium. The spores are shot off by the bursting of the asci. In some genera the apothecium may be stalked (e.g., *Sclerotinia*; Fig. 10.18, J–L) or the cup may become distorted by uneven growth to give the curious fruit-bodies of *Helvella* or of the edible morel (*Morchella*). The fungal partner (mycobiont) of most lichens is a member of this group (p. 617). The truffles (Tuberales), some species of which are edible, producing closed hypogeous ascocarps, are thought to have been derived from a form like *Peziza* by progressive infolding and fusing of the hymenium.

(3) *Loculoascomycetidae.* Asci bitunicate, formed singly or in groups in locules in a stroma, plant parasites, some of economic importance, e.g., *Venturia inaequalis*, the cause of apple scab.

(4) *Laboulbeniomycetidae* (p. 606). Insect parasites, not closely related to any other Ascomycetes.

II   *Basidiomycetes*

This group is divided into the *Heterobasidiomycetidae*, the basidia of which are septate or deeply cleft or, if single-celled, developed from a thick-walled cell (teleutospore) and the *Homobasidiomycetidae*, the basidia of which are septate and usually borne in an extensive hymenium on complex fruit-bodies (basidiocarps).

(1) *Heterobasidiomycetidae,* mostly specialized parasites of higher plants.

(A) **Ustilaginales** or smuts, number of basidiospores on each basidium variable.

The most economically important smuts are those which attack the grain of cultivated cereals. The bunt or covered smut of wheat, *Tilletia caries*, is seed-borne and the seedlings become infected when contaminated seed germinates. The smut mycelium is restricted to the growing points of the host and finally becomes concentrated in the young grain, where it breaks up to form numerous black spores, the brand spores or smut spores. These remain within the outer covering of the grain until the latter is broken during threshing. The spores are then released and healthy grain becomes contaminated with them. In some other species of grain smut (the loose smuts) the spores are released earlier, are deposited on the stigmas of healthy flowers and the germ tubes grow down into the ovary where the mycelium remains dormant within the grain until the latter germinates. Other smuts produce spores in the anthers or in the vegetative parts of the host plants. The smut spores germinate to form a short hypha, the promycelium or basidium, which usually bears a variable number of spores (sporidia). In many species the sporidia fuse in pairs.

(B) **Uredinales** or rusts; basidium (promycelium) transversely septate, usually bearing four sporidia or basidiospores; obligate parasites of vascular plants.

The rusts are a group of obligate plant parasites, many of which are of economic importance, e.g., the cereal rusts and certain species attacking conifers. *Puccinia graminis*, the black stem rust of wheat, is a heteroecious rust, that is, it alternates between two unrelated hosts (the barberry or certain related species and the wheat or some other grasses). The basidiospores or sporidia are released in spring and are able to infect the leaves of barberry. Small globose orange bodies (spermagonia or pycnidia) soon develop beneath the upper surface of the leaves, and later, clusters of large cup-like sori (cluster cups or aecidia) develop on the underside. The aecidiospores are unable to re-infect the barberry but instead their germ tubes are able to enter the leaves of wheat through the stomata and under favourable conditions establish a mycelium within the leaves of susceptible varieties. After a short incubation period numerous rusty red sori (uredosori) develop and large numbers of uredospores are carried away by the wind and spread the disease to other wheat plants. Later in the season these are replaced by black two-celled, thick-walled spores (the teleutospores) which do not germinate until the spring, when they produce short promycelia consisting of four cells, each of which bears a single sporidium. Not all rusts are heteroecious, some carrying out their life-cycle on a single host. Many species lack one or other of the spore forms and are known as short-cycle rusts.

(C) **Tremellales;** basidia transversely septate, longitudinally septate or deeply cleft, produced in hymenia on surface of gelatinous fruit-bodies, many parasitic on trees or growing on fallen timber.

(2) *Homobasidiomycetidae* Basidia 1-celled, usually in complex fruit-bodies.

(A) **Hymenomycetes.** Hymenium exposed before maturity, mushrooms, toadstools, bracket fungi, elf clubs, etc. (Fig. 10.18, M–P).

Agaricus campestris (the common field mushroom) is typical of the most advanced order of the Hymenomycetes, the Agaricales (agarics, commonly known as mushrooms and toadstools). The fruit-bodies begin as small globose masses of hyphae borne on branched mycelial strands. These develop into the familiar mushroom with a stalk or stipe which bears an expanded cap or pileus, from the under side of which thin plates or lamellae (the gills) hang down vertically. These are covered by the hymenium of basidia each bearing four basidiospores (most cultivated mushrooms are forms of *A. bisporus* and have 2-spored basidia). The gills are protected until they are nearly mature by a thin membrane consisting of several layers of hyphae, which later ruptures along the edge of the pileus as the latter expands, and remains attached to the stipe as a ring or annulus. The gills are at first white and become pink, purple and finally almost black as the spores ripen and become pigmented.

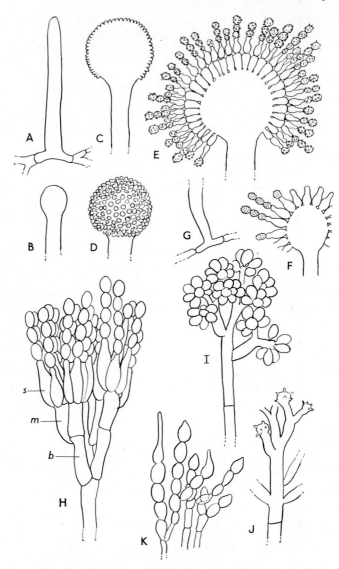

Fig. 10.19 Formation of conidia by moulds.
   (A–E) *Aspergillus niger* (A × 170, B–E × 260). (A) young conidiophore growing up vertically from foot cell. (B) tip of slightly older conidiophore beginning to swell up to form a vesicle. (C, D) young vesicle with developing sterigmata (C in optical section, D in surface view). (E) young fruiting head (in optical section) showing secondary sterigmata bearing chains of conidia.
   (F–G) *Aspergillus niveo-glaucus* (× 300). (F) typical fruiting head showing single series of sterigmata (only part of head drawn). (G) foot cell from base of conidiophore. (A–G, after Thom, G. and Raper, K. B. (1945) *A Manual of the Aspergilli*, Baillière, Tindall and Cox, London.)
   (H) *Penicillium expansum* (× 900). A typical asymmetric type of penicillus showing

Other agarics differ in spore colour, absence of a ring, and in a few species, including the death cap (*Amanita phalloides*), in the presence of an enveloping universal veil (volva) through which the growing fruit-body finally bursts leaving a cup-shaped structure at its base and often a few scraps of the torn volva adhering to the pileus.

Another order of the Hymenomycetes, the Polyporales, contains many economically important wood-destroying fungi. The fruit-bodies of members of the family Polyporaceae are more or less bracket-shaped and the hymenium lines numerous pores on the under side of the fructification. These pores may be merely shallow depressions, as in the dry-rot fungus (Fig. 21.1A), *Merulius lacrymans* (which is consequently sometimes placed in a separate family Meruliaceae), or, more commonly, they are long and tubular, as in species of *Polyporus* or the perennial woody fruit-bodies of *Fomes* and *Ganoderma*.

(B) **Gasteromycetes.** Hymenium enclosed until maturity in complex fruit-body; saprophytes or mycorrhizal fungi. (Puff-balls, earthstars, bird's nest fungi, stinkhorns, etc.)

III   *Deuteromycetes (Fungi Imperfecti)*

Perfect stage unknown or rare (Fig. 10.19).

(A) **Hyphomycetes (Moniliales).** Conidia borne directly on the mycelium or on specialized conidiophores which are usually entirely free or may be found in tufts or pulvinate masses; saprophytes and parasites on plants and animals some of which are of major economic importance.

This group is a heterogeneous collection of species, the only characters common to all being the production of conidia and the absence or extreme rarity of the perfect stage. In the absence of the latter the subdivision of the group is based on arbitrary conidial characters. Although modern systematists have attempted to devise a more natural arrangement, the classical one, devised mainly by Saccardo, is still the most generally used for identification of species.

In this system the Hyphomycetes are divided into four families as follows: (i) Moniliaceae, conidiophores, if present, distinct from one another, or spores borne scattered over undifferentiated hyphae, spores and conidiophores bright coloured or colourless, not dark; (ii) Dematiaceae, similar, but spores or conidiophores, or both, dark coloured; (iii) Stilbaceae, conidiophores interwoven to form a short cylindrical or spine-like *coremium* (synnema); (iv) Tuberculariaceae, spores produced in fascicles in dense patches, *sporodochia*. There is much overlapping between these sub-families. Thus *Aspergillus niger* which has black spores is obviously closely related to other species of *Aspergillus* with white or brightly coloured ones, but if the system is strictly followed should be placed in the Dematiaceae. Some species of *Penicillium*, which normally produce distinct conidiophores, produce coremia under certain environmental conditions. Despite these and other similar difficulties, this arrangement remains the most workable one

---

sterigmata (*s*), metulae (*m*) and branches (*b*). (After Raper, K. B. and Thom, G. (1949) *A Manual of the Penicillia*, Williams and Wilkins, Baltimore.)

(**I–J**) *Botrytis cinerea* ($\times$ 500). (**I**) conidiophore, bearing clusters of conidia. (**J**) conidiophore showing sterigmata from which spores have been shed.

(**K**) *Sclerotinia* (*Monilia*) *fructigena* ($\times$ 250). Chains of poorly differentiated conidia (drawn from a culture).

available for the rapid identification of these fungi. Current studies of these fungi may well produce a more natural classification. Further subdivision of these families is based on spore characters.

(B) **Sphaeropsidales.** Conidia borne in globose or flask-shaped fruit-bodies (*pycnidia*) or in cavities within host or stroma; plant parasites (Fig. 19.3A).

(C) **Melanconiales.** Conidia and conidiophores in cushion-like masses (*acervuli*) developing below surface of host and bursting through at maturity; plant parasites.

(D) Some species not known to produce spores of any type are classed as **Mycelia Sterilia** and include a number of plant parasites.

PROTOZOA

INTRODUCTION: FORM, LIFE CYCLES, AND NUTRITION

This category embraces those eukaryote Protista which have the combination of locomotion and heterotrophic nutrition that is characteristic of animals, together with some other organisms that clearly have close relations with them. By definition the Protozoa are simple animals, but, since the evolutionary divergence that led to 'higher' animals and plants probably involved flagellated eukaryotes, among which there is still abundant evidence of adaptive evolution, it is not surprising that some organisms which are indisputable plants show clear evidence of relationship with genuine animals. A complete account of the flagellate Protozoa would therefore necessarily include a number of autotrophic organisms that can also justifiably be classified as algae.

Since the different types of Protozoa show such diverse features a general description of a 'typical' protozoan cannot be devised. A general impression of the characteristics and diversity of Protozoa can more easily be gained from the account which follows.

It is often stated that the Protozoa are unicellular, but while there are many species within the group that have the characteristics of single cells, there are many protozoan species with two or more nuclei enclosed within a single membrane and many more species which have a colonial organization of interconnected individuals. The size of protozoan individuals ranges from a few μ to several mm. Protozoa frequently contain a range of specialized and complex organelles and may have complex life cycles in which the individual passes through a number of specialized stages. The organization and life of the protozoan individual is different in character from that of a cell of a multicellular organism, and many protozoologists prefer not to refer to an individual protozoan as a cell. Multicellular animals must be assumed to have evolved from protozoan ancestors in the distant past, but there is a lack of evidence to support speculations about the forms of Protozoa which may have been the starting points for these evolutionary developments.

The major sub-groups of Protozoa are separated on the basis of their organelles of locomotion and characteristics of their life cycles. All members of two of the groups are parasitic, with limited powers of locomotion and complex life cycles; they have spores of characteristic types. In the Cnidospora the complex spores contain one or several amoeboid sporoplasms (the infective individuals) and possess one or several polar filaments which are ejected before emergence of the sporoplasms from the spore. The spores of the Sporozoa contain infective sporozoites of characteristic shape; they are simpler and do not have polar filaments. Members of the Ciliophora (ciliate Protozoa) are characterized by the possession of numbers of cilia (organelles of unilateral beat), and (normally) two types of nucleus. The flagellate Protozoa (Mastigophora), which have only one type of nucleus and move by means of flagella (organelles with an undulatory beat), are now grouped with the amoeboid Protozoa (Sarcodina), whose organization characteristically depends on pseudopodia for locomotion and food capture. The occurrence in some organisms of both flagella and pseudopodia, either simultaneously or at different parts of the life cycle, led to the taxonomic linking of the flagellates and amoebae to form the Sarcomastigophora.

The nutrition of some of the pigmented flagellates may be exclusively photoautotrophic, but the majority of Protozoa are heterotrophs, dependent upon organic molecules as sources of carbon and nitrogen as well as for energy. Many protozoans (phagotrophs) eat other organisms or large fragments of these by phagocytosis, taking the organic material into food vacuoles for digestion. Such vacuoles may be formed at a permanent 'cell mouth' or cytostome in ciliates and some flagellates, or at many parts of the body surface in amoeboid organisms and some others. Numerous Protozoa (osmotrophs) feed saprozoically, absorbing soluble organic materials, with or without pinocytosis; in many species the relative importance of this method of feeding has not been evaluated, but in some forms, including many parasites, it is probably the only mode of nutrition. Since the Protozoa include both autotrophic and heterotrophic forms, it is not surprising that dietary requirements range from a simple selection of inorganic salts to a complex mixture containing many amino acids, vitamins and other growth factors.

The waste products of metabolic activities of Protozoa are believed to escape by diffusion, although in some organisms crystalline deposits may accumulate, internally or in skeletal structures. The contractile vacuole is believed to serve an osmoregulatory rather than an excretory function.

Studies of protozoan ultrastructure indicate that the basic form of a species can be related to a characteristic cyto-architecture which is primarily fibrous, but which also includes membrane elements and secreted structures. The structure and formation of this architecture has been fully studied in relatively few species, but the organization of some skeletal structures and of the ciliate and flagellate pellicles shows abundant evidence of the utilization of microtubular fibres, bundles of thin filaments and membranous vesicles, both as components of the complex structures and as morphogenetic organizing systems.

In most protozoan species the ability to reproduce is retained by all individuals although they may have a complex body organization. Usually reproduction involves binary fission of an organism to form two more or less equal daughters, so that no parent remains. Such reproduction involves not only the replication of the nucleus, but also the production of new organelles; in many cases existing organelles may be broken down and a new set produced in each of the daughters. Fission of an organism to produce a larger number of daughters is not uncommon; in the best-known example, called schizogony, the nucleus of an organism divides many times in succession, and then each nucleus separates from the main mass with a small amount of cytoplasm and many daughter organisms are formed.

Some sort of sexual process occurs in all major groups of Protozoa, although there are species in several groups which are not known to have any form of sexual activity, and there are also forms with modified sexual phases which have lost the full genetic potentialities of meiosis and cross-fertilization. Some flagellates and all sporozoans are haploid and show a meiotic division of the zygote nucleus, while other flagellates, the ciliates, and some sarcodines are diploid with meiosis in the formation of the gametic pronuclei. A life cycle involving intermediary meiosis and an alternation of a haploid generation with a diploid generation (comparable in development with that of most cryptogamic land plants) is shown by several species of foraminiferan sarcodines and may be normal throughout that group. The occurrence of sexual phases may be restricted to certain parts of the life cycle, as in foraminiferans and sporozoans, or may involve 'unspecialized' individuals. Examples are known in both flagellates and ciliates where the likelihood of fertilization between clones (rather than within clones) is increased by mating-type systems, in which only individuals of different mating-types show sexual fusion. These mating types are analogous to the cross-pollination devices of some higher plants. They bear no direct connection with sexuality, since, in ciliates at least, each individual is monoecious and produces gametes of both types.

In some Protozoa the life cycles are regular and include a consistent sequence involving divisions, sexual fusion, cyst formation, etc.; such regularity is characteristic of parasitic forms. The life cycles of other Protozoa are more dependent upon environmental changes of a physico-chemical or biological nature for the determination of cyst formation or characteristic types of reproduction, e.g., resistant cysts are commonly produced by soil Protozoa which are subjected to periods of drought, or predatory Protozoa may encyst when the supply of food organisms is exhausted. The alternation of a motile larva and sessile adult might be regarded as a simple sort of cycle, but such organisms may also show sexual and cyst phases of irregular frequency.

Protozoa may occupy any trophic level in the food chains of natural communities where free water is present. Autotrophic flagellates are abundant in the sea and in fresh waters as well as in symbiotic associations with animals at various levels of organization, including other Protozoa. Herbivorous members of the sarcodine and ciliate groups in particular form important items in the diet of a large number of animals including carnivorous ciliates and sarcodines. Of particular importance in the economy of many communities, in damp terrestrial habitats as well as in truly aquatic ones, are the saprozoic and bacteria-feeding Protozoa which make use of the substances and organisms involved in the final 'decomposition' level of food chains and themselves form the food of organisms at a higher trophic level, thereby recirculating organic matter.

THE BIOLOGY OF THE MAIN GROUPS OF PROTOZOA

Sarcomastigophora

This subphylum includes flagellates, sarcodines and the taxonomically enigmatic opalinids. They are forms moving by flagella or pseudopodia, or both, and have only one type of nucleus. The flagellates and opalinids can further be distinguished from ciliates by the form of binary fission, which in the flagellated organisms is basically longitudinal, producing two mirror-image daughters, while in ciliates anterior and posterior daughters are produced by a transverse division running across the kineties (p. 416).

Many of the simpler flagellates possess chlorophyll and other pigments, or are heterotrophic derivatives of the autotrophic forms. These flagellates are classed by protozoologists as Phytamastigophorea, many of which are described in the section on Algae (p. 439). The class Zoomastigophorea contains mainly parasitic forms with fairly complex structure as well as a small number of simple heterotrophic forms for which no relationship with any group of the Phytamastigophorea has yet been established.

Many pigmented flagellates feed osmotrophically or phagotrophically and at the same time use photosynthesis. Colourless, heterotrophic species occur in most of the major groups of algal flagellates; many of these are frequently classed as protozoa, and receive scant treatment in discussions on algae. Amongst the chrysomonads (= Chrysophyta) there are pigmented forms whose nutrition is supplemented with phagotrophy by means of pseudopodia, often aided by the collection of particles by flagellar activity. One of these flagellates, *Ochromonas malhamensis* (Fig. 10.20A), has a dietary requirement for vitamin B_{12}, and has been used in bioassay for this vitamin. Among the heterotrophic euglenoids are *Peranema* (Fig. 10.20C), whose diet includes algae and protozoa which are taken in through a cytostome which opens separately beside the apical invagination from which the flagella emerge, *Copromonas*, which feeds on bacteria, and osmotrophic species like *Astasia* and *Euglena*. One of the commonest of all fresh-water flagellates is the osmotrophic cryptomonad *Chilomonas* (Fig. 10.20B). The predominantly marine dinoflagellate group (part of Pyrrophyta) includes many species that are parasitic in animals or plants as well as free-living phototrophic forms and many phagotrophic forms with diverse body forms.

The majority of these plant flagellates have few flagella associated with simple root structures, and the body surface is often strengthened with cellulose or proteinaceous pellicular thickenings or scales. By contrast, members of the Zoomastigophorea frequently have large numbers of flagella whose bases are associated with complex fibrous organelles, and the body surface is seldom stiffened with anything more than sub-pellicular microtubules.

The collar-flagellates form a distinctive group of zooflagellates; the characteristic form of cell (Fig. 10.20D), with its single flagellum surrounded by a fine collar of microvilli, is also found in the multicellular sponges, and the common possession of this unique structure suggests a phylogenetic connection between these two groups. Many collar-flagellates are sessile, with or without stalks, and some of them secrete a lorica (a loose case within which the animal lives). These loricate forms particularly are liable to be confused with the bicosoecid flagellates

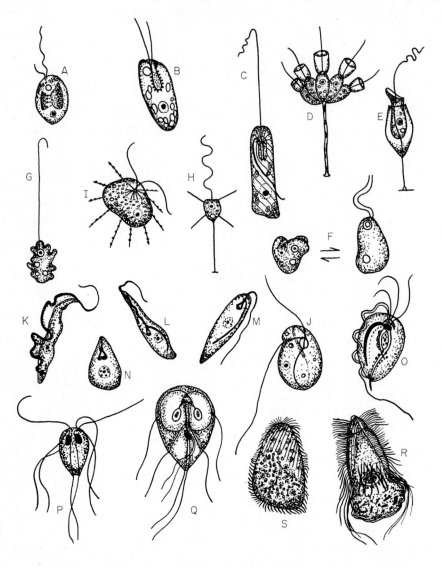

Fig. 10.20 Flagellate and opalinid Protozoa. (**A**) *Ochromonas* sp. 10 μ. (**B**) *Chilo-monas paramaecium* 25 μ. (**C**) *Peranema trichophorum* 50 μ. (**D**) *Codonosiga botrytis* 15 μ. (**E**) *Bicoeca* sp. 12 μ. (**F**) *Naegleria gruberi* 20 μ. (i) amoeboid and (ii) flagellate phases. (**G**) *Mastigamoeba* sp. 15 μ. (**H**) *Actinomonas* sp. 10 μ. (**I**) *Dimorpha mutans* 20 μ. (**J**) *Bodo saltans* 10 μ. (**K**) *Trypanosoma brucei* 25 μ. (**L**) crithidial form of tryp-anosome 25 μ. (**M**) leptomonad form of trypanosome 25 μ. (**N**) leishmanial form of trypanosome 15 μ. (**O**) *Trichomonas* sp. 20 μ. (**P**) *Hexamita intestinalis* 10 μ. (**Q**) *Giardia muris* 10 μ. (**R**) *Trichonympha campanula* 200 μ. (**S**) *Opalina ranarum* 500 μ. Drawn by Margaret Attwood from live material and a range of published illustrations; normally several sources of information have been used for each figure. The average length of each organism is given in the legend.

(Fig. 10.20E), which, however, lack the collar, and which have two flagella, the shorter of which attaches the flagellate to the base of the lorica. Organisms of both groups use the flagella to collect bacteria and organic particles from the water.

A number of forms showing features of both amoebae and flagellates may frequently be classed with the zooflagellates; these include species which are capable of transformation from an amoeboid form to a flagellate, like *Naegleria* (Fig. 10.20F), which is common in soil, and forms which possess flagella and pseudopodia simultaneously, in which case the pseudopodia may be lobed, e.g., *Mastigamoeba* (Fig. 10.20G) or fine axopodia, e.g., *Actinomonas* (Fig. 10.20H) and *Dimorpha* (Fig. 10.20I).

The most important group of flagellate parasites are *Trypanosoma* and its relatives. They share with *Bodo* and some related forms the possession of a DNA-rich mitochondrial organelle called the kinetoplast, which occurs in the region of the flagellar base. *Bodo* (Fig. 10.20J) has two flagella, one anterior and one trailing posteriorly, and is commonly found in water rich in organic matter, when it feeds on bacteria which are taken in through a tubular channel at the cytostome. Some related biflagellate forms are parasitic. The body form of *Trypanosoma* varies at different stages of the life cycle; in the familiar form from the blood plasma of vertebrates (Fig. 10.20K), the elongate flagellate has a single flagellum which runs forward from its basal body near the posterior end of the organism, and is connected to the body along much of its length by an extension of the flagellar membrane which forms an undulating membrane. In other stages of the life cycle the flagellum may originate near the middle or at the anterior end of an elongated body, or the body may be rounded and lack an emergent flagellum; these stages are known respectively as crithidial (Fig. 10.20L), leptomonad (Fig. 10.20M), and leishmanial (Fig. 10.20N) stages, since their characteristic forms are found in the related genera *Crithidia* and *Leptomonas*, parasites in insects and other invertebrates, and *Leishmania*, which is responsible for kala azar and other diseases of mammals in which white blood corpuscles are entered by the flagellate. All forms may not be present in the life cycle of all trypanosomes. There are many species of *Trypanosoma* which spend part of their life cycle in the circulatory systems of vertebrate hosts and part of the cycle in blood-sucking invertebrates, such as insects and leeches. *Trypanosoma gambiense* and *T. rhodesiense* which cause sleeping sickness in man in Africa, are transmitted by tsetse flies; parasites sucked by the fly from an infected man enter the stomach of the insect, where they multiply, undergoing changes in form, and then migrate to the salivary glands, whence they may be injected into the next man bitten by the insect. There are non-pathogenic species as well as the well-known pathogens which cause diseases of man and domestic animals; it seems that severe disease is caused only by trypanosomes which leave the blood plasma and enter other tissues.

The more complex zooflagellates are sometimes referred to collectively as metamonads, and it is convenient, if artificial, to subdivide these into poly-mastigote and hypermastigote flagellates. Several species of the polymastigote *Trichomonas* (Fig. 10.20O) occur as parasites in man and domestic animals, and this type illustrates several features of these metamonad flagellates, although it is one of the simplest. There is a complex of structures called the karyomastigont formed by an association of the nucleus, the flagella and internal organelles originating near the flagellar basal bodies. Many species of *Trichomonas* have three free flagella and one recurrent flagellum attached to the body by an undulating membrane; the basal bodies of all four flagella lie close together. Of the three large organelles which arise near the basal bodies, the axostyle, which runs down the middle of the body and may project posteriorly, is a hyaline rod enclosed in a sheath of microtubules, and the costa and parabasal fibre are striated. The thick fibre which curves round the side of the body beneath the undulating membrane of the recurrent flagellum is the costa, which is thought to be contractile and responsible for changes in body shape, and the slender parabasal fibre runs near the nucleus and is associated with parabasal (Golgi) bodies. *Trichomonas* is generally believed to be saprozoic, but there are some reports of phagotrophy, mostly involving bacteria.

In some other polymastigote flagellates there may be two simpler karyo-mastigont systems, e.g., in *Hexamita* (Fig. 10.20P) and *Giardia* (Fig. 10.20Q), which are parasitic in vertebrates, or one to several hundred karyomastigonts in some species symbiotic in wood-eating insects. The details of the mastigont system vary widely; in some the axostyle is an extensive array of microtubules and is responsible for active undulations of the body. The hypermastigote flagellates also inhabit the intestines of insects: they are often xylophagous, e.g., *Trichonympha* (Fig. 10.20R). Members of this group have a single nucleus, a large number of flagella arranged in tufts, spiral rows or longitudinal rows, and a variable pattern of associated fibres connected closely to the flagellar bases or more remote from them. The body is large and often extremely complex, but usually retains a naked area through which fragments of wood and other materials are taken in phagocytically.

The opalinid 'flagellates' are saprozoic parasites which are found in the intestine of cold-blooded vertebrates, principally anuran amphibians. In the best-known genus, *Opalina* (Fig. 10.20S), the body is extremely flat, carries a dense covering of ciliary organelles in oblique longitudinal rows and contains a large number of nuclei. Binary fission is typically longitudinal, along the rows of cilia, and sexual reproduction involves the fusion of two small multiflagellate gametes.

The opalinids differ from ciliates in these two reproductive features as well as in the absence of nuclear dimorphism; they clearly belong to a different phylogenetic line. The body surface between the ciliary rows is strengthened by a series of ribs which contain a single vertical row of microtubules running parallel to the body surface, and the ribs and ciliary rows are underlain by regular arrays of membranous inclusions. In two related genera there are only two similar nuclei and a less flattened body.

Sarcodines do not possess surface specializations of the type characteristic of the pellicle of ciliates and flagellates; their surface flexibility permits the production of cytoplasmic extensions called pseudopodia, which are used in locomotion and in phagocytic feeding. The form of the pseudopodia and the characteristics of any shell, test, or skeleton are features used in classification.

The plasticity of body form is most evident in such naked forms as *Amoeba*

(Fig. 10.21A), where the pseudopodia are broadly lobed. The structure is simple, perhaps because of the use of the whole body in movement, so that large permanent organelles could not be retained; smaller structures like the nucleus, contractile and other vacuoles, mitochondria, and granular inclusions move around with the more fluid interior cytoplasm. Contractile activity of the surface layer of gelled cytoplasm is held to cause the cytoplasmic streaming that is the most obvious feature of the movement of the organism. These lobose amoebae

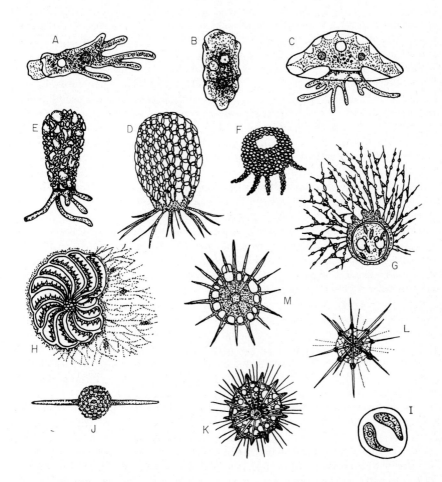

Fig. 10.21 Amoeboid Protozoa. (**A**) *Amoeba proteus* 500 μ. (**B**) *Entamoeba histolytica* 20 μ. (**C**) *Arcella vulgaris* 100 μ. (**D**) *Euglypha* sp. 150 μ. (**E**) *Difflugia* sp. 200 μ. (**F**) *Centropyxis aculeata* 130 μ. (**G**) *Gromia* sp. 150 μ. (**H**) *Elphidium crispum* 1 mm.. (**I**) *Babesia bigemina*, two trophozoites within an erythrocyte, 5 μ. (**J**) Internal siliceous skeleton of a radiolarian 750 μ. (**K**) *Aulancantha* sp. 1·5 mm. (**L**) *Acanthometron* sp. 500 μ. (**M**) *Actinophrys sol* 50 μ. Drawn by Margaret Attwood from live material and a range of published illustrations; normally several sources of information have been used for each figure. The average length of each organism is given in the legend.

range in size from a few microns to several millimeters in diameter, and may have one, two, or many nuclei. While the majority of them are free-living forms, there are also some important parasites, e.g., *Entamoeba histolytica* (Fig. 10.21B), which is responsible for amoebic dysentery in man, and which passes as a cyst from the faeces of one host to the mouth of another.

Some testaceous (shelled) forms have lobed pseudopodia which emerge from the aperture of the test to move the organism and collect food. The basis of the test is a chitinous membrane which may be thickened with further chitin (*Arcella*, Fig. 10.21C), secreted siliceous plates (*Euglypha*, Fig. 10.21D), or encrusted with sand grains (*Difflugia*, Fig. 10.21E) or diatom frustules (*Centropyxis*, Fig. 10.21F). These organisms are free-living inhabitants of moss, soil, and fresh water; they can retreat within the shell in adverse conditions.

Slender branching pseudopodia called filopodia are a characteristic of some forms, both naked and testate (e.g., *Gromia* (Fig. 10.21G)). These filopodia are distinguished from the fine, granular, branching and anastomosing pseudopodia of the foraminiferans (Fig. 10.21H), which are referred to as reticulopodia from the pseudopodial network that they form. Foraminiferans have a basically chitinous test of one or many chambers, the walls of which are usually perforated, and may be thickened with foreign material or a calcareous deposit. The pseudopodia emerge from the shell aperture, or the perforations, or both. The particles on the pseudopodia show a constant streaming motion which apparently indicates the flowing of strands of cytoplasm throughout the pseudopodial network, and shows how food particles and bacteria caught by the reticular fishing net may be carried to the main body of the organism. In some species an alternation of a diploid generation reproducing asexually and a haploid generation producing gametes has been described. The initial chamber of the shell is often smaller in the diploid generation than in the haploid one; the former, called the microspheric form, is often multinucleate, and may show nuclear dimorphism. The gametes in some species are biflagellate, and in others are amoeboid; autogamy is also known. The foraminiferans are almost exclusively marine. They have been abundant since the early palaeozoic, and because their shells leave good fossils they are of value to geologists in the identification of rock strata.

Many protozoologists believe that the slime moulds, particularly the cellular (acrasian) forms are derived from amoebae, and classify them with these pseudopodial forms. They are dealt with in detail in the next section (p. 421).

The piroplasms form an enigmatic group of small organisms that are amoeboid during part of their life cycle. *Babesia* (Fig. 10.21I), the cause of Texas red-water fever in cattle, is the best-known example, and is transmitted by ticks. Infective forms injected with tick saliva into the blood of a cow enter red blood corpuscles and divide to produce small pear-shaped bodies which break open the corpuscle and enter others; haemoglobin is released when the corpuscle bursts, and is excreted by the cow. If a tick sucks up parasitized corpuscles, the piroplasms mature to form gametes in the stomach of the tick, and fuse to form zygotes which migrate through the tissue to the ovary. Within the ova of the tick the zygotes multiply and migrate through all the tissues of the developing embryonic ticks, including the salivary glands, from which they may be injected into a cow by even a new-born tick. This life cycle shows remarkable adaptation of the parasite life cycle to the habits of the tick, and was the first protozoan life cycle which was shown to involve an arthropod vector.

The spherical floating amoebae (Actinopodea) form a distinct class of sarcodines. The pseudopodia, which are slender and radiate all round the body, are usually supported by an axial structure, when they are called axopodia; some forms have filopodia or reticulopodia. Many species produce some sort of skeletal material, either internally or externally.

The marine Radiolaria are best known from their skeletons of silica, e.g., Fig. 10.21J, which are abundant in the ooze of certain parts of the ocean floor, and are common fossils. The skeleton is deposited around a membranous central capsule with pores that allow communication between an inner region, containing one or more nuclei and other bodies, and an outer region in which food vacuoles occur (Fig. 10.21K). Many species have symbiotic algae in the outer region, but may also feed on small animals and plants caught by the filose or reticular pseudopodia. Some of the best-known species are colonial forms. Reproduction is not well known, but may include binary and multiple fission, and a sexual phase involving flagellate gametes is said to occur in certain radiolarians. In the Acantharia (Fig. 10.21L), which are also marine and which have often been grouped with Radiolaria, there is a skeleton of radial spines impregnated with strontium sulphate which support the pseudopodia. Their body organization appears broadly similar to that of radiolarians, with fine reticulate pseudopodia in addition to the axopodia; symbiotic algae may occur within the capsule. Some acantharians have been shown to reproduce by multiple fission which takes place in an encysted stage, and leads to the production of numerous flagellate spores.

A central capsule is not present in the Heliozoa (Fig. 10.21M), which are predominantly inhabitants of fresh water. They generally have axopodia supported by bundles of microtubular fibres around which the pseudopodial cytoplasm streams, and some forms have siliceous scales or spines, or a reticulate skeleton. Ciliates or other prey organisms are caught by the axopodia and surrounded by cytoplasm. One or many nuclei may be present in the central part of the body, and the surface cytoplasm between the basal parts of the axopodia is vacuolated. Heliozoans divide by binary or multiple fission, and may also reproduce sexually, in some cases autogamously (e.g., *Actinophrys*).

Sporozoa

All members of this subphylum are parasites, often with complex life cycles (Fig. 10.22). In the main class the Telosporea, the infective stage is a vermiform sporozoite which may be injected naked into a host by an insect vector in a few forms or, more usually, may hatch in the host intestine from spores which are eaten by the host. Since they do not have ciliary organelles, even the most active sporozoites show only weak motility, gliding or performing slow flexural movements. The sporozoites migrate to specific host cells and feed saprozoically or sometimes phagocytically within the cell. At this stage parasites of the sub-class Gregarinia leave the host cell and continue to grow in an organ cavity of the host body, feeding saprozoically, e.g., in the seminal vesicle of the earthworm in the case of *Monocystis* (Fig. 10.22A). When the feeding stage (trophozoite) of the

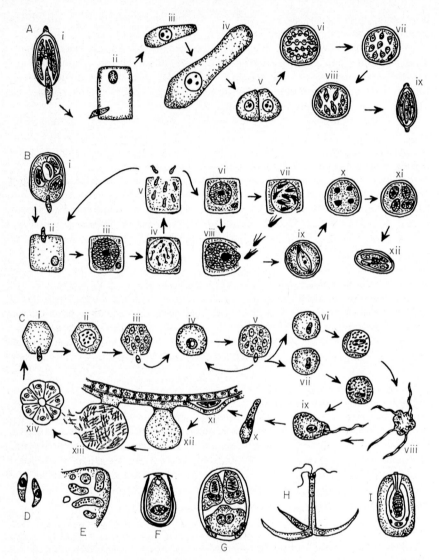

Fig. 10.22 Sporozoa and Cnidospora.

(**A**) Diagrams illustrating the life cycle of a monocystid gregarine. A sporozoite which hatches from the sporocyst (**i**) in the gut of an earthworm migrates and enters a cell in the seminal vesicle of the host (**ii**). Having outgrown the host cell the trophozoite lives free in the seminal fluid (**iii** and **iv**) and eventually pairs with another mature tropho-zoite (**v**). Within the gametocyst (**vi**) each gametocyte produces many gametes; fertilization occurs to form zygotes (**vii**), each of which develops into a thick-walled sporocyst containing several sporozoites (**viii**) which may later be released from the gametocyst (**ix**).

(**B**) Diagrams illustrating the life-cycle of an eimeriid coccidian. A sporozoite which has left the sporocyst emerges from the oocyst (**i**) and enters a cell of the host intestine (**ii**), grows (**iii**), and divides (**iv** and **v**). The merozoites released at the breakdown of the parasite is mature, it encysts with another mature trophozoite, and in both individuals nuclear divisions are followed by cytoplasmic separation to form numerous gametes. These fuse in pairs and each zygote becomes a spore within which a number of sporozoites develop. Reproduction in this case occurs only at spore formation (sporogony); in a few gregarines the trophozoite also undergoes multiple fission (schizogony) to produce a new generation of feeding individuals.

In the sub-class Coccidia the trophozoites do not grow very large since they remain intracellular (Figs. 10.22B, C). After growth they undergo schizogony to produce many new infective individuals (merozoites) which enter other cells, so that a very heavy infection can be built up by repeated schizogonies. Trophozites ultimately develop into gametocytes, some of which give rise to single macrogametes and others to several microgametes, which may be flagellate. Fusion of gametes in pairs produces zygotes, each of which develops into a few or many sporozoites, either naked or enclosed in a spore. Naked spores are characteristic of the Haemosporina (e.g., *Plasmodium*), and in the Eimeriina (e.g., *Eimeria*) the sporozoites are enclosed in a walled spore (or sporocyst).

Members of the genus *Eimeria* (Fig. 10.22B) cause disease (coccidiosis) in domestic mammals and birds as well as in other animals. Spores that are eaten hatch in the intestine and sporozoites enter the epithelial cells of the intestine or associated glands. In these cells the parasites grow and undergo repeated schizogonies, each one producing hundreds of minute merozoites; severe tissue damage or even death is caused. Certain trophozoites mature to form gametes, and the small biflagellate microgamete seeks out a macrogamete and fuses with it to form a zygote. Each zygote produces a resistant cyst wall (oocyst) within

---

host cell enter other intestine cells and may repeat the growth and schizogony stages (**iii, iv,** and **v**) or may become gametocytes (**vi**) from which flagellated microgametes (**vii**) or stationary macrogametes (**viii**) are formed. The zygote encysts (**ix**) and within this oocyst meiosis takes place (**x**); a number of sporocysts develops (**xi**), each of which may contain several sporozoites (**xii**).

(**C**) Diagrams illustrating the life cycle of *Plasmodium*. A sporozoite which has been injected into the blood of the vertebrate by the mosquito first enters an endothelial cell in the liver (**i**) where it grows (**ii**) and undergoes schizogony (**iii**) to release many merozoites which enter red blood corpuscles (**iv**) within which schizogony again occurs (**v**). Merozoites released by erythrocytes may undergo repeated schizogony (**iv** and **v**) or may develop into gametocytes (**vi** and **vii**). If these gametocytes enter the stomach of a mosquito they mature to form long microgametes (**viii**) or rounded macrogametes which are fertilized by microgametes (**ix**). The zygote (**x**) migrates between the stomach wall cells of the insect (**xi**) where it grows (**xii**) and divides to form many spindle-shaped sporozoites (**xiii**); these migrate within the insect to the salivary glands (**xiv**), whence they may be injected into a vertebrate and start a new cycle (**i**).

(**D**) *Toxoplasma gondii* 8 μ.
(**E**) *Sarcocystis tenella* 10 μ.
(**F**) Spore of *Haplosporidium* sp. 10 μ.
(**G**) Spore of *Myxobolus pfiefferi* 12 μ.
(**H**) Spore of *Triactinomyxon ignotum* 30 μ.
(**I**) Spore of *Nosema* sp. 5 μ.
Drawn by Margaret Attwood from live material and a range of published illustrations; normally several sources of information have been used for each figure. The average length of each organism is given in the legend.

which meiotic division of the nucleus takes place and a number of sporocysts develop, each containing a number of sporozoites. The number of sporocysts per oocyst and the number of sporozoites per sporocyst are generic features of diagnostic importance, e.g., in *Eimeria* these numbers are 4 and 2 respectively, while in *Isospora* (a species of which occurs occasionally in man) the oocyst contains 2 sporocysts, each with 4 sporozoites. The whole of the life cycle in these forms takes place in a single host individual.

Malaria, which is probably the most important human disease caused by a pathogenic protozoan, is caused by species of *Plasmodium* which are transmitted from man to man by *Anopheles* mosquitoes. Many other species of *Plasmodium* occur in other terrestrial vertebrates. Slender sporozoites some 10 μ long, are injected with the saliva of the insect into the blood of the vertebrate. The sporozoites first enter cells around the blood vessels in the liver (Fig. 10.22C), and after growing for several days they undergo schizogony to produce hundreds of very small merozoites which enter red blood corpuscles. Inside the erythrocytes the trophozoites grow as intracellular phagotrophs, maturing in about 2 days. The full-grown trophozoite undergoes schizogony, bursting the erythrocyte and liberating merozoites and toxic products into the blood. The merozoites may enter other erythrocytes and continue a sequence of schizogonies, releasing toxic products into the blood at more or less regular intervals and producing the recurrent fever characteristic of malaria. Sooner or later some of the trophozoites develop into gametocytes which remain dormant within corpuscles in the bloodstream until sucked by a mosquito. In the stomach of the insect the gametocytes develop into gametes, producing either four microgametes or one macrogamete. The motile zygotes formed settle between the stomach wall cells, where they grow and divide many times to produce large numbers of naked sporozoites. When these are mature they are released into the haemolymph of the insect, migrate to the salivary glands and enter the salivary ducts, whence they may be injected into another man and complete the cycle.

The majority of telosporeans are small, specialized parasites with highly developed reproductive abilities. The body organization is generally simple, but complex structures are found in gregarine trophozoites, where the surface region may contain arrays of microtubular fibres, and the interior has a full range of cell organelles including mitochondria, Golgi bodies, and other membranous organelles. In many forms two or three unit membrane layers are present around the body surface.

Toxoplasma and *Sarcocystis* are two genera of rather different parasites which show some similarities to coccidian forms, but appear to lack both sexual reproduction and spores. They are placed provisionally in a separate class, the Toxoplasmea, within the Sporozoa. *Toxoplasma* occurs very widely in various tissues of birds and mammals, including man, and is probably only weakly pathogenic. The trophic individuals (Fig. 10.22D) are 5 to 10 μ long and are usually found intracellularly, where they multiply by binary fission or perhaps schizogony. The means of transmission is unknown. There is no resistant spore. *Sarcocystis* is found in the muscles of terrestrial vertebrates, including man. Tubular bodies found in the muscles contain vast numbers of banana-shaped parasites, some 10 μ long (Fig. 10.22E). These may multiply by binary fission, but little is known of their biology or of their means of transmission. One report claims that the spores produce septate mycelia of a fungal type.

Another curious small group of parasites is the Haplosporea (e.g., *Haplosporidium*). These have simple resistant spores (Fig. 10.22F), but do not show sexual reproduction. They are normally parasites of invertebrates, but have been reported from fish. The amoeboid infective individual which emerges from the spore enters cells of the intestinal epithelium and grows to produce a multinucleate form which subdivides into uninucleate bodies, each of which may develop into a spore. Spore formation involves more stages in some forms. This group forms a link with the Cnidospora, since both have an amoeboid sporoplasm; and the spores are similar, except for the absence of a polar filament in spores of the Haplosporea.

Cnidospora

Parasites of this subphylum have complex spores (Figs. 10.22G, H, I) surrounded by a single membrane or a two- or three-valved structure, they contain one to six polar filaments, usually coiled in polar capsules, and one to many sporoplasms. The more complex spores found in the class Myxosporidea are formed from several cells during sporulation, and there is some doubt that these organisms are really protozoans. They occur in lower vertebrates and many types of invertebrates. In the host intestine, where the polar filaments are said to be extruded and anchor the spore, the amoeboid sporoplasms hatch from the spore and migrate through the gut epithelium to the organs characteristic of the species. Within the tissues the parasites grow, become multinucleate and split up into multinucleate masses, in each of which one or two spores develop. In *Myxobolus*, which causes the fatal boil disease of such fish as the carp, the parasite grows in the muscle or connective tissue producing large bulges at the surface of the fish which burst to release vast numbers of spores, each with two polar capsules and a single sporoplasm (Fig. 10.22G). The freshwater worm *Tubifex* is often parasitized by *Triactinomyxon*, whose spore has three valves, three polar capsules and eight sporoplasms (Fig. 10.22H). It is often stated that nuclear fusion occurs in the formation of the sporoplasms of myxosporidians, but the complete cytogenetic cycle remains controversial.

The simpler spores of members of the Microsporidea are of unicellular origin (Fig. 10.22I). The chitinous membrane of the spore has a single valve which encloses a single tubular polar filament, through which the single amoeboid sporoplasm emerges on hatching. The parasites occur in animals of most phyla, but are most commonly found in insects and fishes, where they multiply within host cells and cause hypertrophy of the host tissues. The spores hatch in the gut of the host and the sporoplasms migrate through the gut wall and around the body to enter specific cells within which they multiply. Later the trophozoites mature to form spores. The parasite remains uninucleate throughout its life cycle, except for the final stages of spore production in some species. The best-known microsporidian diseases are caused by species of *Nosema* which parasitize intestinal cells of honey bees and the cells of various tissues of the silkworm.

Ciliophora

The most distinctive features of typical members of the Ciliophora are the possession of two types of nucleus and the possession of many cilia which are used in characteristic ways in movement and in feeding.

The two types of nucleus found in ciliates are the smaller diploid micronucleus and the larger polyploid macronucleus. At least one of each is normally present in every individual, but in some species there may be several or many of either or both types of nucleus. In a few well-known species, e.g., *Tetrahymena*, *Didinium*, there are strains which lack micronuclei. It is believed that the macronucleus has a somatic function, providing for the routine synthetic activities of the cell, while the micronucleus is not concerned with these activities. Both types of nucleus divide during binary fission of a ciliate, and nuclei of both types pass to each daughter. However, during the sexual process, which in ciliates is of a type called conjugation, only the micronucleus provides genetic continuity.

Conjugation involves the exchange of genetic material between two ciliates which come together with a temporary, local, cytoplasmic fusion; a zygote nucleus is normally formed in both ciliates during the process. At conjugation the micronucleus divides meiotically and two of the products of this division form the haploid pronuclei. A pronucleus from each conjugant ciliate crosses the cytoplasmic bridge to the body of the other ciliate and fuses with the stationary pronucleus to form a syncaryon. The two conjugants separate and undergo a process of nuclear reorganization; only later does binary fission occur, so that increase in numbers is not an essential part of the sexual process. During conjugation the old macronucleus of each conjugant ciliate begins to disintegrate, and one of the products of the first mitotic division of the micronuclear syncaryon develops to form a new polypoid macronucleus by repeated replication of nuclear material. The details of conjugation show some variation in different species, e.g., in some forms one conjugant is much smaller than the other, the macroconjugant absorbs both the cytoplasm and pronucleus of the microconjugant, and mutual fertilization cannot occur. A form of self-fertilization may occur, in which the two pronuclei fuse with each other within the single individual; this process of autogamy does not require the pairing of ciliates, but may occur in ciliates that have paired, when it is called 'selfing'. Autogamy is more common in aged clones, while conjugation is characteristic of younger clones.

The ciliary organelles, together with their associated basal fibril systems, are a dominant feature of the pellicular organization of most ciliates. The basic arrangement of these cilia in characteristic, longitudinal, meridional rows called kineties (Fig. 10.23A) can be seen in most ciliates, except those in which the ciliature is restricted to certain body regions or to certain parts of the life cycle. The kinety is made up of a row of kinetids, which are regarded as the units of pellicular organization.

Each kinetid consists of a cilium arising from a basal body (kinetosome), to which are attached fibres which run off into the cytoplasm beneath the surface membrane. The ciliary base is surrounded by a pair of pellicular alveoli, so that over most of the body surface there are three unit membranes, the inner two being separated by a fluid-filled space. In some ciliates two cilia occur at the centre of each kinetid. The largest fibres of the kinetid in such species as *Paramecium* and *Tetrahymena* are the kinetodesmal fibres (striated aggregates of micro-filaments) which pass forward for a number of microns from their attachment to the basal body, and run with similar fibres from other kinetids of the same row to form a 'cable' of fibres called the kinetodesma. This kinetodesma can be seen in the light microscope, and is known to occur along the (animal's) right side of each kinety (Fig. 10.23B). Small groups of microtubular fibrils also arise near the basal body (Fig. 10.23C) and may terminate in close association with other fibril systems. Sub-pellicular fibrils, usually microtubular, are common in the surface layers of ciliate protozoa, and, while many are associated with ciliary bases, it is likely that others are not. In some ciliates (e.g., *Stentor*, *Spirostomum*) large arrays of microtubular fibrils seem to have replaced the striated microfilamentous kinetodesmata; they occur in almost the same position, but may have a different function in these highly contractile ciliates. These fibre systems of the pellicle are believed to function in maintaining the shape and rigidity of the body, and in the alignment of the kinetids; there is no evidence that they co-ordinate the activity of the cilia. Fibrous systems with an active contractile function occur in the stalk of *Vorticella* and in the sub-pellicular 'M' bands of *Stentor* and *Spirostomum*; they seem to form a specialized addition internal to the basic pellicular structures.

Associated with the pellicle are often arrays of organelles like mitochondria and specialized structures like trichocysts, which are capable of explosively ejecting a thread which may have toxic properties. There are also specialized regions of the pellicle, particularly at the cytostome, but also at the cytopyge (cell anus) and at the site of opening of the contractile vacuole.

Specialization of the ciliature for feeding or locomotion is a feature of most ciliate groups. Frequently this specialization takes the form of aggregation of cilia into bands or compound organelles, and is well illustrated by the mouth organelles of *Tetrahymena* (Fig. 10.23A, B). Along the (animal's) right side of the mouth is a single row of cilia which lie so close together that they appear to form a membrane, the undulating membrane (not to be confused with the quite different structure in flagellates). At the left side of the mouth are three compound cilia in the form of flat plates called membranelles, each one based on a rectangle of basal bodies. These four organelles together function to bring food particles to the cytostome which lies between them. In many ciliates both the undulating membrane and the membranelle row may show increased development, e.g., *Condylostoma* (Fig. 10.23L), while in many others only the long row of membranelles (the adoral zone of membranelles or A.Z.M.) persists. Such ciliates as *Euplotes* (Fig. 10.23R) possess compound cilia with a specialized locomotory function; these are called cirri, and they take the form of an elongate cone, being based on a more or less circular patch of basal bodies. The component cilia of compound organelles of the membranelle or cirrus type are not fused together, but normally move together as a single unit that is much more powerful than a single cilium.

The classification of ciliates reflects the distribution and specialization of their ciliary organelles. Among the holotrich ciliates, which characteristically have a

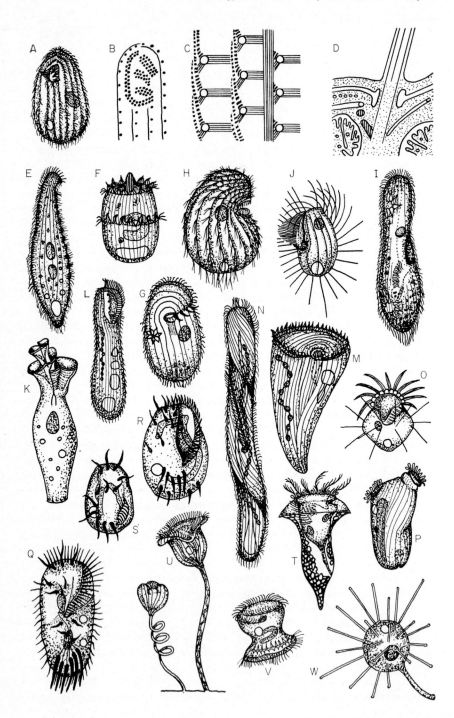

fairly uniform covering of cilia over the whole body, there is a progressive development of mouth ciliature. In the simplest holotrichs, the gymnostome ciliates, the feeding is usually macrophagous, through a mouth that may be capable of extreme dilation, e.g., as a slit (*Loxophyllum*, Fig. 10.23E) or as a proboscis reinforced by fibres (*Didinium*, Fig. 10.23F), or may be specialized to take in algal filaments or diatoms through a 'basket' of fibres (*Nassula*, Fig. 10.23G); in these forms cilia are not actively used in passing food to the mouth, although the fibrous organelles may originate from ciliary roots. Trichostome ciliates frequently have abnormally long cilia around the mouth, which may be used in feeding, e.g., collection of bacteria by *Colpoda* (Fig. 10.23H), but these cilia are not formed into compound mouth organelles, and they represent only a slightly modified region of body cilia in a depression (vestibulum) around the cytostome. In the hymenostome ciliates compound feeding cilia occur in a region around the mouth called the buccal cavity, as in *Tetrahymena* (Fig. 10.23A), where they collect food particles of the size of bacteria. The development of specialized buccal cilia in hymenostomes has taken diverse forms, including the complex arrangement in the tubular buccal cavity of *Paramecium* (Fig. 10.23I) and the large sail-like undulating membrane of *Cyclidium* (Fig. 10.23J). Among the holotrichs are classified the chonotrichs (Fig. 10.23K), an enigmatic group of ciliates which lack body cilia and live attached to crustaceans, and several specialized parasitic groups whose members have well-developed body ciliature but may have small mouths or no mouth at all.

A long adoral zone of membranelles is found throughout the spirotrich sub-class. This group shows specialization from such forms as *Blepharisma* or *Condylostoma* (Fig. 10.23L) with complete mouth ciliature and complete body ciliation, firstly by loss of the undulating membrane in, e.g., *Stentor* (Fig. 10.23M)

Fig. 10.23 Ciliated Protozoa. (**A**) *Tetrahymena pyriformis* 50 μ. (**B**) The arrangement of ciliary bases and kinetodesmata in the region of the buccal cavity of *Tetrahymena*. Down one side of the mouth is a double row of kinetosomes; only the outer of these rows has ciliary shafts which form the undulating membrane; at the other side of the mouth are three groups of basal bodies which mark the bases of the membranelles. Somatic kinetosomes lie alongside the kinetodesmal fibres. (**C**) The arrangement of fibres around the kinetosomes of *Tetrahymena*. Each basal body gives rise to a striated kinetodesmal fibre which runs forward, and to two groups of microtubular fibrils, one of which runs to one side and the other runs backwards; a third group of microtubular fibrils lies above the kinetodesmata, but does not make contact with the kinetosomes. (**D**) A longitudinal section through the base of a cilium of *Tetrahymena*. Beneath the cell membrane is an alveolar space in the pellicle and below this lie the fibres associated with the kinetosome, including the striated kinetodesmal fibres and the transverse and longitudinal microtubular fibrils shown in (**C**); two mitochondria lie nearby. (**E**) *Loxophyllum helus* 200 μ. (**F**) *Did nium nasutum* 150 μ. (**G**) *Nassula aurea* 200 μ. (**H**) *Colpoda cucullus* 75 μ. (**I**) *Paramecium caudatum* 250 μ. (**J**) *Cyclidium glaucoma* 25 μ. (**K**) *Spirochona gemmipara* 100 μ. (**L**) *Condylostoma arenarium* 500 μ. (**M**) *Stentor coeruleus* 1 mm. (**N**) *Spirostomum ambiguum* 2 mm. (**O**) *Halteria grandinella* 30 μ. (**P**) *Epidinium ecaudatum* 120 μ. (**Q**) *Oxytricha* sp. 120 μ. (**R**) *Euplotes patella* 100 μ. (**S**) *Aspidisca* sp. 30 μ. (**T**) *Tintinnopsis campanula* 100 μ. (**U**) *Vorticella* sp., expanded and partially contracted, body length 100 μ. (**V**) *Trichodina pediculus* 60 μ. (**W**) *Podophrya collini* 50 μ. Drawn by Margaret Attwood from live material and a range of published illustrations; normally several sources of information have been used for each figure. The average length of each organism is given in the legend.

and *Spirostomum* (Fig. 10.23N) and then by reduction of the body ciliature, e.g., in *Halteria* (Fig. 10.23O) which has a few long body cilia, in the entodiniomorph ciliates (symbiotic in the rumen of herbivorous mammals) which have only one or a few tufts of membranelles (Fig. 10.23P), in the hypotrich ciliates like *Oxytricha* (Fig. 10.23Q), *Euplotes* (Fig. 10.23R) and *Aspidisca* (Fig. 10.23S) with dorso-ventrally flattened bodies whose ventral cilia are formed into cirri which are involved in complex locomotory behaviour, and in the pelagic loricate tintinnids (Fig. 10.23T). Many of the spirotrichs are bacterial feeders, but among the large forms there are omnivores, quite a number of which may eat fragments of organic detritus as well as live food, and active carnivores, e.g., *Stentor*, which may eat multicellular rotifers or other stentors as well as a wide range of smaller ciliates and flagellates.

The origins and relationships of the peritrich ciliates are uncertain, but it seems likely that they may have been derived from fairly advanced holotrichs. In feeding individuals of sessile peritrichs like *Vorticella* (Fig. 10.23U) and *Carchesium* cilia are restricted to a band which leads to the mouth, although they develop an aboral ciliary ring in the migratory phase, and the stalk originates from a tuft of ciliary structures called the scopula. Mobile peritrichs like *Trichodina* (Fig. 10.23V) have aboral bands of cilia for locomotion as well as the oral bands for feeding. The majority of peritrichs probably feed on bacteria. Species of *Trichodina* and related genera are ectoparasitic on aquatic organisms, and may kill their hosts, vast numbers on fish gills may suffocate the host.

The suctorian ciliates lack cilia in the adult feeding stage (Fig. 10.23W), although the larval form is provided with kineties or bands of locomotory cilia. Suctorians are usually sessile organisms which feed on ciliate protozoa by means of tentacles through which the prey cytoplasm is sucked into the body of the predator. The tentacles have special adhesive bodies which enable them to keep hold of a ciliate which swims into them. There is no evidence of any homology between the tentacles and cilia, although the tentacle has an internal organization of microtubular fibres. Many suctoria have stalks that are developed from a scopula of ciliary organelles.

OUTLINE OF THE CLASSIFICATION OF THE PROTOZOA

Phylum PROTOZOA

Subphylum **Sarcomastigophora.** With flagella or pseudopodia or both, a single type of nucleus, and usually without spore production. Sexuality involves syngamy.

Superclass **Mastigophora.** The flagellates.

Class **Phytamastigophorea.** The pigmented flagellates and their relatives, e.g., *Ochromonas, Euglena, Peranema, Chilomonas, Noctiluca.*

Class **Zoomastigophorea.** The animal flagellates, e.g., *Codonosiga, Bicosoeca, Bodo, Trypanosoma, Trichomonas, Trichonympha.*

Superclass **Opalinata.** With numerous ciliary organelles and 2 to many nuclei, parasitic.

Class **Opalinidea.** e.g., *Opalina.*

Superclass **Sarcodina.** Pseudopodial forms, body at least partly naked, but shell or skeleton often present.

Class **Rhizopodea**. With lobed, filose or reticular pseudopodia, e.g., *Amoeba, Arcella, Gromia, Elphidium, Dictyostelium.*

Class **Piroplasmea**. Small piriform or amoeboid parasites in vertebrate erythrocytes, transmitted by ticks, e.g., *Babesia.*

Class **Actinopodea**. Spherical floating forms with axopodia or slender pseudopodia or both, often with skeleton or spicules, e.g., *Aulacantha, Sphaerozoum, Acanthometra, Actinophrys.*

Subphylum *Sporozoa.* Usually with simple spores containing sporozoites, parasitic.

Class **Telosporea**. With spores and characteristic sporozoites, e.g., *Monocystis, Eimeria, Plasmodium.*

Class **Toxoplasmea**. No spores, but cysts with many naked trophozoites, e.g., *Toxoplasma, Sarcocystis.*

Class **Haplosporea**. With simple spores and amoeboid sporoplasm, e.g., *Haplosporidium.*

Subphylum *Cnidospora.* Spores with polar filaments and amoeboid sporoplasms, parasitic.

Class **Myxosporidea**. Spore of multicellular origin, e.g., *Myxobolus, Triactinomyxon.*

Class **Microsporidea**. Spore of unicellular origin, single tubular polar filament, e.g., *Nosema.*

Subphylum *Ciliophora.* With cilia and two types of nucleus. Sexuality involves conjugation.

Class **Ciliatea**

Subclass HOLOTRICHIA. Body ciliature usually simple and uniform, mouth ciliature simple, e.g., *Loxophyllum, Didinium, Nassula, Colpoda, Paramecium, Tetrahymena, Pleuronema.*

Sublass PERITRICHIA. Body cilia absent in mature form, oral cilia in rows winding to mouth, often attached to substrate by stalk or basal disk, e.g., *Vorticella, Carchesium, Trichodina.*

Subclass SUCTORIA. No cilia in mature stage which feeds by tentacles, larva ciliated, e.g., *Podophrya, Dendrocometes.*

Subclass SPIROTRICHIA. Mouth cilia composed of many membranelles, forming a long adoral zone, body cilia usually reduced or compounded to form cirri, e.g., *Blepharisma, Stentor, Spirostomum, Halteria, Euplotes, Aspidisca, Epidinium.*

THE SLIME MOULDS (MYXOMYCOTA)

INTRODUCTION

The slime moulds are a heterogeneous assemblage of protists requiring organic nutrients (organotrophic). All at one stage in their life-cycles consist of motile single vegetative cells, which may be amoeboid or fusiform, with or without flagella, and show what generally are considered to be animal characteristics, notably no cell wall, a capacity usually for phagocytosis, and motility. One or more of three characteristic sorts of development thereafter occur in the life-cycle; (1) formation of a motile multinucleate mass of protoplasm known as the plasmodium, (2) aggregation of individual cells as motile masses of cells, known as pseudoplasmodia, which show a greater or lesser degree of organization and co-ordinated behaviour or (3) transformation of single cells, plasmodia, or pseudo-plasmodia into stationary spore-bearing fructifications which tend to be rather plant-like in that substances which give mechanical support and which possibly include cellulose are produced.

Some authorities classify the slime moulds as primitive fungi, others as protozoa, others as a separate group. Four main subdivisions have been recognized. The status of these subdivisions (and hence the endings of the names used) varies in different taxonomic treatments of the slime moulds.

MYXOMYCETES (TRUE SLIME MOULDS)

Myxomycetes, of which there are several orders, have a distinctive life-cycle. The spore liberates one or more uninucleate protoplasts which feed and multiply and eventually may fuse in pairs. The resulting zygotes develop into plasmodia (Fig. 10.24A) which are wall-less, multinucleate masses of cytoplasm that are capable of locomotion. Eventually the plasmodia transform into fruit-bodies in

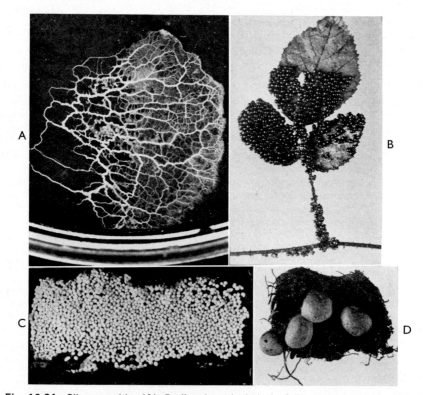

Fig. 10.24 Slime moulds. (**A**) *Badhamia utricularis* ($\times 2 \cdot 5$). Reticulate plasmodium growing on plain agar. Plasmodium has reached edge of petri dish. (**B**) *Leocarpus vernicopus* ($\times \frac{1}{2}$). Sporangia on leaf. (**C**) *Trichia varia* ($\times \frac{2}{3}$). Sporangia on dead wood. (**D**) *Lycogala miniatum* ($\times \frac{1}{2}$). Aethalia on dead wood.

which cellulose or chitin or both have been reported. The Myxomycetes are a homogeneous group and not likely to be related closely to the Acrasiales from which they differ in possessing flagellated stages in their life-cycle and a true plasmodium. They comprise two subclasses—Myxogasteromycetidae (or Endosporeae) and Ceratiomyxomycetidae (or Exosporeae)—of which the latter contains only a single species (p. 425).

Most of the four hundred-odd species are probably cosmopolitan. The plasmodia generally occur on damp decaying twigs and logs, but some can live indefinitely beneath water. Most appear to be independent of the substrate on which eventually they fruit, but nevertheless some prefer either hard or soft woods or leaves for this (Fig. 10.24, B–D).

The unicellular, usually uninucleate, haploid spores are enclosed in smooth or sculptured walls. Germination is influenced by factors similar to those that affect this process in true fungi. Some spores under favourable conditions germinate after only 20 minutes, others take up to 1–2 weeks, while in some species germination has proved very difficult to obtain. On germination, 1–4 naked cells escape through a crack or irregular pore in the spore coat. These cells may be myxamoebae or anteriorly flagellated swarm cells and in many species the one can change into the other. Such transformations are accompanied by changes in the exterior layers of the plasmalemma. Microelectrophoretic investigations on *Physarum nudum* suggest that the surfaces of swarm cells are predominantly protein, but those of myxamoebae are predominantly acid protein or mucoprotein, while those of the eventually formed plasmodia are predominantly fats. Though swarm cells often appear to be uniflagellate, they are usually unequally biflagellate, or at least are potentially so. Two basal bodies are consistently present. The shorter of the flagella may be visible only after special fixation. Both flagella are of the whiplash type and have the $9 + 2$ fibrillar sub-structure characteristic of eukaryotic cells (p. 361). Flagellum-like pseudopodia also have sometimes been reported. The cells feed by ingesting solid particles such as bacteria, and can multiply rapidly, but under unfavourable conditions may encyst until favourable conditions return. Contractile vacuoles are active in swarm cells.

In almost every species that has been critically examined, fusion between pairs of haploid cells is a prerequisite of plasmodium production. Nevertheless there are reports that plasmodia in some species have arisen without prior gametic fusion. Some species are homothallic, but in others plasmodia arise only when appropriate clones of cells are paired. In some heterothallic myxomycetes there appear to be multiple alleles which govern mating type so that either mating type in one race of a species may fuse with either type of another race. Even among heterothallic species plasmodia are known sometimes to have arisen from single spores. There is evidence for three causes of this: (1) that the spore in question was binucleate, (2) that at least one of the cells it produced had mutated to a genotype that allowed mating to occur, (3) that the myxamoebae sometimes can actually self-fertilize (i.e., be apogamous).

Cell fusion is closely followed by fusion of nuclei. Thereafter nuclear divisions take place without cell division so that a multinucleate diploid plasmodium forms; further nuclear divisions are synchronized. Sometimes the plasmodium coalesces with nearby zygotes and small plasmodia, while even large plasmodia may fuse together completely if there is genetical identity at a particular locus on their diploid genomes. However, fusions between particular plasmodia of *Physarum polycephalum* are followed by their death. Plasmodia sometimes cut off large numbers of myxamoebae and swarm cells of unknown ploidy. This probably represents a mean of reproduction.

There are several types of plasmodia. The simplest, the *protoplasmodium*, is a minute sheet of undifferentiated granular protoplasm with barely detectable cytoplasmic streaming. When eventually it fruits it forms a single sporangium. The *aphanoplasmodium* is a rather inconspicuous, very flat sheet of transparent, non-granular protoplasm which virtually lacks a sheath of gelled ectoplasm but exhibits streaming. The *phaneroplasmodium* is the most complex type. When young it resembles the protoplasmodium, but it grows to become a conspicuous, thick, fan-shaped mass of granular cytoplasm organized into veins of gelled ectoplasm within which flow currents of endoplasm (Fig. 10.24A). Studies of the fine structure of plasmodia have revealed mitochondria with tubular cristae, and in some plasmodia fibrils about 15 mμ diameter. The interest in this latter observation lies in the possibility that contractile proteins are involved in the mechanisms of cytoplasmic streaming and plasmodial locomotion.

Plasmodia from the field are often cluttered with relics of ingested food particles, but there is evidence that plasmodia may be nourished by soluble foods also. They are sometimes coloured, often in shades of yellow, orange, or buff, but the chemical nature of the pigments is unknown. Plasmodia which are pigmented need exposure to light before they will fruit. Studies on spectral sensitivity and light absorption suggest that a photoreceptor may be associated with fruiting which is common to a number of species. Other pigments may serve protective functions.

Protoplasmic streaming and locomotion occurring most conspicuously in phaneroplasmodia are intimately linked processes. Two contractile proteins, viz., myxomyosin and myosin B, and possibly others also are implicated in streaming, but streaming may not be primarily responsible for locomotion. Its function may be to circulate materials, and myxomyosin may serve less to motivate streaming than to control the pattern of flow. The flow follows a network of veins (Fig. 10.24A) at speeds as great as 1 mm./sec., and reverses its direction periodically.

A plasmodium under unfavourable conditions such as sudden drought may rapidly become sclerotic. Streaming ceases, the cytoplasm gels, and cell walls develop which divide it into a large number of multinucleate macrocysts. These are long-lived, and on rehydration rapidly regenerate the plasmodium.

Vegetative plasmodia eventually give rise to spore-bearing fructifications. The plasmodium usually differentiates into small clumps which become sporangia that may remain sessile or develop stalks. Sporangia are usually a millimetre or so high, and may be borne directly on the substrate or on a membranous hypo-thallus. Nuclear divisions occur within, and the formation of cell walls gives rise to large numbers of spores. In some species the plasmodia do not cleave into discrete primordia. Instead the poorly differentiated fructification resembles a prostrate strand of the plasmodium, the 'plasmodiocarp'. In others, the primordia coalesce as they develop and form an 'aethalium' which may be very large (e.g., up to 30 cm. across) and which may or may not manifest its composite nature at maturity. Within the sporangium there may be a system of threads, the capillitium, which is concerned with spore release. This consists of irregular tubes or filaments, with or without intermingled nodular masses, and sometimes contains calcium carbonate. In some genera the threads are beautifully and regularly ornamented. Most sporangia at maturity have a double wall, the outer one tough and sometimes impregnated with calcium carbonate, the inner thin and transparent.

The development of fruit-bodies has been critically studied in only a small number of genera. Generally a membrane (the *hypothallus*) is produced on the upper surface of the plasmodium and sporangial primordia enlarge by welling up beneath it. The upgrowth of the primordium produces a sporangial region borne on a narrower support within which a central core of material is laid down. The whole of the developing sporangium is ensheathed in a membrane which is continuous with the hypothallus and which forms the peridium and the outer wall of the stalk. The development of members of the order Stemonitales differs from this. Their hypothallus is deposited *beneath* the plasmodium, and within the young sporangial primordia stalks are produced which grow apically as the protoplasm of the primordia flows up their outsides and forms the sporangial heads. The eventual peridium is not continuous with the hypothallus and stalk, for it is a separate structure formed on the outer surface after other parts of the fruit-body have matured.

The capillitium usually arises by the deposition of material in or on a system of vacuoles which develops within the sporangial protoplasm. About the same time, the nuclei in the sporangia divide. Most evidence supports the view that reduction division precedes spore formation, but reports that meiosis is delayed even until the spores are maturing may indicate some variation in this respect. The spores, which arise by cleavage of the cytoplasm, are typically uninucleate but nuclear division within the spore has been reported.

Ceratiomyxa fruticulosa, the single species in the Exosporeae, differs from the foregoing account in a number of respects. The plasmodium when about to fruit emerges from the rotting wood in which it lives and produces a minute cushion-like body from which arise digitate processes that may branch and sometimes anastomose. On their surface the plasmodium takes on a network appearance which becomes a mosaic of cells, each of which produces a single spore on a slender stalk. When first delimited the spores are diploid, but as they mature a reduction division yields four haploid nuclei. The so-called spore is in fact widely interpreted as a much reduced sporangium. It germinates by solution of its wall and releases haploid swarm cells. These multiply and eventually fuse in pairs. Single zygotes appear to form plasmodia directly without coalescence with other zygotes.

Only a few myxomycetes have been induced to complete their entire life-cycle in artificial culture, and so far none has done so in the absence of other micro-organisms which have served as food. *Physarum polycephalum* has been cultured in the plasmodial state in a defined liquid medium, showing that at least vegeta-tive growth is possible with solely soluble nutrients.

The classification of myxomycetes is based principally on features of the fructifications, such as whether spores are borne externally or internally, the colour of spores in the mass, the presence of calcium carbonate in various parts of the fruit body, the structure of the stipe if present, and the form, ornamentation, and location of the capillitium.

ACRASIOMYCETES OR ACRASIALES (Cellular slime moulds)

There are three families: Dictyosteliaceae, Guttulinaceae, and Sappiniaceae, though the latter are somewhat doubtful members of this order and will receive no further consideration. The general life-cycle follows the sequence: germination of the spore to yield an amoeba, multiplication of amoebae, aggregation, migration of the entire aggregate, and finally transformation of the aggregate into a *sorocarp* containing spores. Throughout the sequence there is a steady increase in organization and in the precision with which shape is defined. There is no true plasmodial stage.

Differences between genera and species are most obvious at the stage of fruiting. Members of the Dictyosteliaceae form delicately proportioned stalked sorocarps; members of the Guttulinaceae form somewhat irregular and amorphous sorocarps with ill-defined stalks and spore-masses. About twenty species of Acrasiales are known and all are free-living within the soil. They appear to prefer forest to grassland soil, and occur chiefly in the surface humus and in mouldering leaf litter. For their isolation a diluted suspension of soil may be mixed with a suspension of bacteria, such as *Escherichia coli*, which serve as food, and be poured over a dilute hay infusion agar buffered with phosphate to pH6. Clones of amoebae should appear in 3–4 days.

The spores of Acrasiales are firmly walled. In an aqueous medium they split and each releases an amoeba. The amoebae are small, uninucleate, and nonflagellate and live in water films in which they feed by ingesting bacteria. If nourished properly they divide every few hours. They can be grown in twomembered cultures in association with a wide variety of bacteria, and recently myxamoebae of a *Polysphondylium* species have even been cultured in a defined liquid medium. Their cellular organization is typically eukaryotic. Food vacuoles are abundant and contractile vacuoles continually empty water to the exterior. When eventually the food supply is depleted, the amoebae stop multiplying, their food vacuoles progressively disappear, and after a pause the cells begin to aggregate.

Details of subsequent behaviour vary with different species but have been well studied in members of the Dictyosteliaceae. The amoebae elongate and orient themselves radially about a number of centres towards which they begin to migrate by amoeboid movements. This may readily be observed in plate cultures. The converging amoebae form streams in which the cells adhere end to end. Aggregation streams sometimes break up and reaggregate elsewhere, and centres may form only to melt away again, but eventually the pattern of aggregation becomes confirmed. Since feeding has ceased, all further development and movement are sustained only by endogenous reserves.

The way in which aggregation is controlled and accomplished is complex.

The process involves chemotaxis governed by a substance named acrasin that is produced by the amoebae themselves. Recent work indicates that for at least some of the Acrasiales, acrasin is cyclic 3′,5′-adenosine monophosphate (cyclic AMP). Chemotaxis requires that a cell in some way perceives a gradient in concentration of the chemotactic principle. There is, however, no single gradient of acrasin concentration associated with an aggregation pattern of amoebae. The system that operates appears to depend on acrasin having more than a single effect. Firstly an amoeba orients itself in the direction of a source of acrasin. Secondly acrasin induces responsive amoebae to produce acrasin themselves. Thirdly it induces amoebae to become adhesive one to another. The probable mechanism of aggregation is that a particular cell or group of cells produces acrasin briefly. The pulse diffuses outwards and induces nearby cells to orient themselves so that they are moving up the gradient and to produce a further pulse of acrasin. By repetition of this process a wave of orientation and acrasin production is propagated centrifugally. Successive pulses of acrasin production are initiated at the centre as the cells there recover from the preceding pulse. Acrasin is very labile as produced, for it is accompanied by an inactivator, which may be the enzyme phosphodiesterase that hydrolyses cyclic 3′5′-AMP to 5′-AMP. The simultaneous production of an inactivator may be biologically advantageous for it effectively increases the concentration gradient and also prevents the development of a high and uniform background concentration in confined environments.

The nature of the cells that initiate aggregation centres has aroused controversy. It has been suggested that there is a small fixed proportion of special initiator cells in a population, but current opinion favours the view that there are simply some cells or groups of cells that are slightly ahead of others in their development and so seize and maintain an initiative. The number and distribution of aggregation centres has been found to be affected by diffusible and volatile factors produced by the organisms themselves.

As long as cells that are aggregating remain separate they appear to be guided only by chemotaxis, but once they begin to meet, a process known as 'contact following' can operate. An amoeba contacting a moving chain of cells is indifferent to its direction of movement until the back end of the cell that it is touching moves past. It is to this that it then adheres. Short chains of cells thus lengthen and join up with others, and eventually continuous centripetal streams form. The behaviour of cells on joining a stream may be understood in the light of what is known of the mechanism of movement in amoebae of Acrasiales. This is thought to involve the active contraction of cell surface towards the posterior rim of the roughly cylindrical cell and its active enlargement around the anterior rim. The result is that cytoplasm moves forward within a *stationary* surface membrane which is continuously being extended in the forward direction and continuously being removed at the back. The anterior and posterior cell surfaces are perhaps stabilized through their contact with contiguous cells, and it is to a stable end that the newly arriving amoeba adheres.

The conical aggregate of cells eventually acquires a fairly definite shape, topples on to its side, and proceeds to migrate over the surface of its substrate as if it were a single multicellular organism. This allows the organisms in the natural environment to move from the wet conditions ideal for growth to the dry conditions best for sporulation. The phase of migration may however be eliminated in dry conditions or if the aggregate of cells is small. The migratory stage has been termed the 'grex'; the term pseudoplasmodium, though widely used, has little to commend it. In some species, for example *Dictyostelium mucoroides* and species of *Polysphondylium*, the grex as it migrates continually secretes a postrate cellular stalk which is left behind, but in other species the grex forms no stalk and may then be termed a 'slug'. The slug of *Dictyostelium discoideum* ranges from about a fifth to five millimeters long and consists of a few hundred to hundreds of thousands of cells. Its size depends on the number of amoebae entering the aggregation and the density of amoebae in the culture. A high density leads to a large grex. The slug has a distinct polarity and becomes increasingly differentiated inside as it migrates. Its anterior third comprises cells which differ histochemically, serologically, and in size from the rest. These anterior cells will eventually form the stalk of the sorocarp, while the posterior ones will form the spores. Those amoebae that arrived first at the aggregation centre become the anterior pre-stalk cells; those that arrive later become the posterior pre-spore cells. In *Dictyostelium discoideum* those that are the very last to arrive eventually form the basal disc that is characteristic of the sorocarp of this particular species. Up to a certain stage, the differentiation of cells within a slug is reversible, for slugs cut in half will give rise to normal sorocarps containing all three types of cell, i.e., stalk, spore, and basal disc. Further, cells removed from aggregates, slugs, or immature sorocarps will multiply and produce clones of cells able eventually to form normal sorocarps. The differentiation of these cells is therefore only phenotypic.

The grex migrates at about 0·25–2 mm./hour by a slow gliding movement, all the while leaving a collapsed slime sheath behind. The larger the slug the greater its velocity. Its direction of movement is influenced by illumination and extremely small gradients of temperature, but appears to be indifferent to acrasin concentration. There is much evidence that cells in the interior of the grex advance by their own amoeboid movement, so that they contribute to the movement of the grex as a whole, and are not simply carried along by the amoebae that are in contact with the substrate. The means by which the interior cells are able to obtain traction is an intriguing problem. It has been suggested that the slime sheath may extend throughout the grex and thereby provide traction for all the cells, but a more satisfactory explanation has emerged with the discovery that individual amoebae of Acrasiales move as a result of the forward transport of the granular cytoplasm within an apparently stationary cell surface which is being constantly augmented near the front of the cell and contracted at the back. A cell surrounded by others can thus advance independently because the surfaces of cells that adjoin it are stationary.

Eventually the grex undergoes the process known as culmination by which the erect sorocarp is produced. Culmination is hastened by a degree of desiccation, a rise in temperature, and possibly the removal of a volatile self-produced inhibitor. In the Dictyosteliaceae the sorocarp consists of a spore mass raised on a stalk built of vacuolated parenchymatous cells within a non-cellular cellulose sheath. The transformation of a grex into a sorocarp involves extensive and well-ordered movements of the constituent cells and their differentiation into special-ized types. In *Dictyostelium discoideum* culmination proceeds as follows (Fig. 10.25). The grex halts, becomes rounded, and its tip rises to become a papilla on top of the cell mass. A stalk sheath initial then appears beneath the papilla as a more or less cylindrical, hyaline membrane, and the cells that are within this sheath become large and vacuolated. More pre-stalk cells migrate into the open upper end of the stalk sheath which is growing in length by the deposition of sheath material by the pre-stalk cells as they ascend its outside prior to entering its open end. The lengthening of the stalk is accompanied by its downward

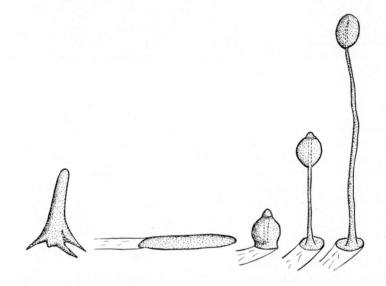

Fig. 10.25 Development of a cellular slime mould. *Dictyostelium discoideum*. The aggregate; migration of the grex which leaves behind it a collapsed slime sheath; and culmination stage. (Modified after Bonner, J. T. (1959). *The Cellular Slime Moulds.* Princeton University Press)

extension through the supporting mass of cells until its first-formed basal part contacts the substrate. Thereafter further extension elevates the growing end of the stalk. The pre-spore cells of the grex follow the pre-stalk cells in their ascent of the outside of the stalk, develop cellulose walls, and become a lemon-shaped or spherical mass of spores. In *D. discoideum*, but not other species, the very last cells of the grex remain around the base of the stalk where they develop into the basal disc. In *Polysphondylium* rings of cells are successively pinched off from the hind end of the grex as it ascends the erect portion of the stalk, and each forms a whorl of small sorocarps, each essentially like that of *Dictyostelium*.

The number of amoebae which enter an aggregate governs the size of the eventual sorocarp. Normally that of *D. discoideum* is 0·5–5 mm. long. The remarkable capacity for regulation of differentiation and form of the organized grex and sorocarp is revealed by the fact that though a normal sorocarp may consist of hundreds of thousands of cells, those of a mutant of the same species consist of as few as ten or twelve cells. Regardless of the number of cells, the proportion of pre-spore to pre-stalk cells remains constant.

The stalks of sorocarps always tend to be perpendicular to the substrate whatever its plane, so cannot be oriented by gravity. An unidentified gas formed by the fruit-bodies themselves and to which they are sensitive is instead involved in orientation.

Not all Acrasiales have sorocarps as precisely organized as those in the Dictyosteliaceae. In *Guttulinopsis vulgaris* the amoebae become grouped into membranous compartments in both sorocarps and aggregates. Almost all of the amoebae in the upper part of the sorocarp become spores but in the lower part and in the aggregate itself some become spores and some degenerate.

The life-cycle of members of the Acrasiales may be completed in as little as about three days. Sexuality has not been demonstrated unequivocally, though it may exist.

PROTOSTELIDA

This recently discovered group of organisms is classified as an order of slime moulds, among which it probably holds a primitive position. The organisms are widespread and common in soil, dung, and plant remains from which they may be isolated and grown in two-membered culture with an appropriate bacterium, yeast or mould to serve as food.

The spore liberates a single amoeboid protoplast which proceeds to divide. In one genus alone the protoplasts are anteriorly flagellate. Cell and nuclear divisions are not always closely associated, and protoplasts with several nuclei may appear. In some species multinucleate plasmodia arise, while coalescence of plasmodia is common.

Plasmodia at sporulation break into smaller protoplasts whose further development is essentially like that of single amoebae of non-plasmodial species. The individual protoplast becomes domed and then hat-shaped. A sheath develops over its upper surface and on withdrawal of the protoplast from beneath the rim the latter makes contact with the substrate. The protoplast then differentiates a basal core—the 'steliogen'—which proceeds to secrete a non-cellular, tubular stalk. This is progressively lengthened, enveloped closely by the original sheath, so that the protoplast and steliogen become raised above the substrate. The remnant of the steliogen is usually then walled-off as the apophysis, a slight swelling which lies immediately beneath the spore which forms from the elevated protoplast. Occasionally the latter forms two spores. The spores of species which have an apophysis are usually deciduous, and in one genus are discharged forcibly. Histochemical evidence suggests that the spore wall and stalk may contain cellulose. There is no evidence at present of sexuality in these organisms.

LABYRINTHULALES OR LABYRINTHULIDAE

This is a small group of aquatic organisms. The genus *Labyrinthula* contains nearly ten described species, most of which are parasites, chiefly of algae, but possibly some of the names will eventually prove to be synonyms. The best-known species is *L. macrocystis* which causes a destructive disease of *Zostera marina* (eel-grass) along Atlantic shore lines. Its predominant growth phase is the 'net plasmodium'. This is a lacy network of filaments on which spindle-shaped uninucleate cells (average size 18×4 microns) move by an obscure mechanism. These cells are neither flagellate nor amoeboid. They glide along the outside of the elastic and glutinous filaments at speeds up to 150 μ/minute, but in certain species, such as *L. roscoffensis*, it has been reported that the filaments are tubes and the cells move inside them. The filaments of *L. macrocystis* are produced by the cells themselves. At the advancing edge of an active colony there are clumps of cells, and from the tips of individual cells single, fine filaments suddenly shoot out. These are 8 to 10 times longer than the cells themselves, and are apparently endowed with a capacity for independent motion. They meet and fuse to yield the track on which the cells will continue their advance. The cells divide by fission. They are nourished by absorption of dissolved nutrients; in only one species has ingestion of particles been reported. Individual cells may encyst, and sometimes 5–100 cells mass together, acquire a pellicle, and form a sorus. In due course the pellicle ruptures and the many emergent cells become spindle-shaped and resume normal development. Heterokont zoospores have been reported in one species. Cells have also been seen to form large pseudo-plasmodial clumps that slowly move. There is no evidence of sexuality in these organisms. Within eel-grass, aggregates of cells lie on filaments in air spaces in the leaves, from which position they invade the cells whose walls they are able to penetrate.

Almost all the species are marine. It has been claimed that two fresh-water species have tracks that are probably composed of filamentous pseudopodia, and for these the genus *Labyrinthorhiza* has been proposed. Another genus, *Labyrinthomyxa*, includes *L. marina* (formerly *Dermocystidium marinum*) that causes heavy mortality of the commercial oyster (*Crassostrea virginica*) of North America, and another species that parasitizes clams.

THE ALGAE

INTRODUCTION

Together with the Cyanophyta (p. 346) the algae form the major carbon-assimilating micro-organisms and since seventy per cent of the global surface is water the importance of algal photosynthesis cannot be under-estimated. It is probable that at least as much carbon is fixed by algae as by the land flora, since although most algae are microscopic, the rate of turnover of the population is great and continuous throughout the year; in this sense they form 'evergreen' populations. The processes involved in this fixation of carbon are essentially similar to those of higher plants and the same pigment, chlorophyll *a*, is present in all algae. They do, however, possess a wider range of accessory pigments some of which are involved in photosynthesis and although starch is the reserve product of some algal groups (as with higher plants), various other polysaccharides, fats, etc., are the main reserve substances of others. Likewise, cellulose, often in a microfibrillar form is the wall component of some, but other carbohydrates, e.g., mannose, xylose, sulphated polysaccharides, alginic acid, fucoidin, laminarin, etc., are present in the walls of other species. Certain groups also form intricate organic or inorganic (silica or calcium carbonate) scales or deposit these in a pectin or mucilaginous base.

HABITAT

Algae require moisture for the movement of their reproductive cells but like most other plants many can exist for long periods in habitats which are subject to intense desiccation. Thus they are common as epiphytes on trees especially in moist tropical regions but also in temperate zones, on bare rock surfaces especially when moistened by sea-spray, and on soils, many of which are extensively colonized by diatoms, flagellates, and non-motile coccoid and filamentous algae. In all fresh-water and marine situations algae thrive down to depths to which photosynthetically usable light penetrates. Moving from the land into any body of water there is an intermediate zone where the soil (or rock) algal flora intermingles with that of the underwater soil. On the permanently submerged silt or rock surfaces characteristic algal floras occur; the former tend to be composed of motile and the latter of attached forms. Similar but not identical species coat the larger aquatic plants (algae, Bryophytes, or Angiosperms) and also occur on many animals, particularly molluscs and hydroids. The boundary between the land and the sea has its very characteristic algal flora, visually dominated by the large brown and red algae but these also are richly coated with microscopic species. In the open water a selected number of species of many algal groups are maintained in circulation by the turbulent motion of the water; this community extends over the whole ocean surface as a thin veneer of carbon-fixing species. A further marine algal community exists in and on the under surface of sea ice and is of considerable importance in polar waters. The ecology of algae is further considered in Chapters 12 and 13.

FINE STRUCTURE OF ALGAL CELLS

The main features of algal cells resemble those of other eukaryotic protists (p. 354), but some organelles show characteristic features and others, such as chloroplasts are found among micro-organisms only in algae.

Plastids

Both pigmented (chloroplasts, chromoplasts) and non-pigmented (amyloplasts) plastids occur in algal cells, though the latter are so far known only in certain siphonaceous Chlorophyta. Whatever the pigmentation all eukaryotic algal plastids are membrane bound and composed of a granular matrix through which is dispersed a varying number of elongated flattened vesicles; in disrupted dinoflagellate chromatophores they are seen to be circular. These vesicles are termed thylakoids (Figs. 10.1 and 10.26).

In addition, the plastids may contain pyrenoids, starch grains, osmiophilic granules and pigment-containing granules grouped to form the stigma (eyespot). The bounding membrane is usually a double structure corresponding to two closely appressed unit membranes (*sensu* Robertson) and in some algae it is continuous with elements of the endoplasmic reticulum. Indeed the endoplasmic reticulum may form additional layers around the plastid, e.g., in the Haptophyceae where there is a double endoplasmic reticulum membrane outside the double plastid one. However, in dinoflagellates, euglenoids, and Xanthophyta the plastid membrane is distinctly three layered. The thylakoids are usually stacked into lamellae which run along part of or the whole length of the plastids. Material of the matrix does not penetrate inside the flattened thylakoids. The arrangement of varying numbers of thylakoids into lamellae (bands or stacks) is characteristic of different algal groups. In the Red Algae the thylakoids run singly through the matrix (cf. the situation in the Cyanophyta, p. 347, which differ, however, in the absence of a bounding membrane) whilst in all other algal divisions they are grouped, e.g., two thylakoids in the Cryptophyta, two to six thylakoids in Euglenophyta and Chlorophyta, and three in dinoflagellates, diatoms, Chrysophyta, and Phaeophyta.

Highly characteristic 'girdle' lamellae encircle the inner lamellae in Phaeophyta, Chrysophyta, and Xanthophyta and some diatoms; in the latter they are distinctive features even of the male gametes (*Lithodesmium*). Most algae do not possess the characteristic stacks of disk-like vesicles, known as grana, found in the chloroplasts of Angiosperms but invagination and folding of the thylakoids result in 'pseudograna' in *Carteria* and perhaps even true grana in some *Acetabularia* species.

Colourless strains of some *Euglena* species can be induced by growth at high temperatures since chloroplast replication is then inhibited whilst cell division is not. Cell division in these circumstances produces some cells containing only a single chloroplast; each of these on further division yields one without any plastids. These cells *cannot* then be induced to form chloroplasts—a fact which is further evidence for the theory that chloroplasts cannot arise *de novo* but only from existing chloroplasts.

Fine arrays of microtubules have been found in the chloroplasts, vegetative cells, zoospores, and eggs of *Oedogonium*. They are obliquely striate in longitudinal section. Comparable microtubules have not been found in any other group of green algae.

Pyrenoids are either immersed in the plastid or appear as outgrowths from it, e.g., in some dinoflagellates and Phaeophyta. They are areas of fine-grained dense material without bounding membranes except in some diatoms where a distinct boundary layer has been seen. In all algal groups the thylakoids are displaced by the aggregation of the pyrenoid matrix, but one or two thylakoids of the lamellar system may continue into and sometimes through the matrix. In some Chlorophyta the normally flattened thylakoids reduce to tubules as they pass into the pyrenoid and in cross section these tubules contain peripheral rod-like structures. These rod structures may be the reduced inner thylakoids of the lamellae and the tube itself may thus be derived from the outer membrane of the group of thylakoids forming the lamella. Starch is formed usually as discrete

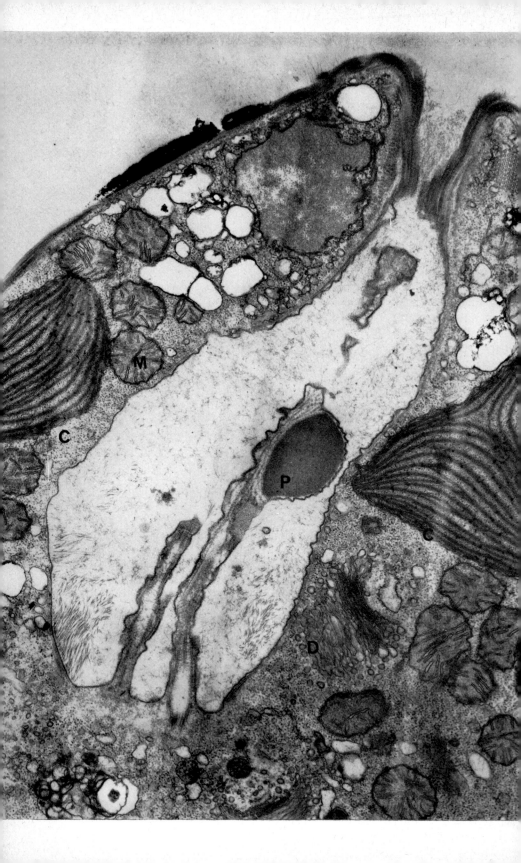

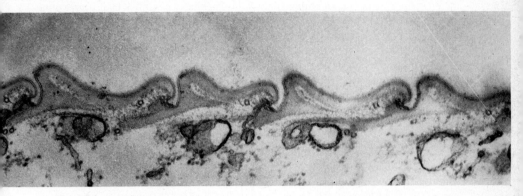

Fig. 10.27 Transverse section of the periphery of a *Euglena* cell showing several pellicular strips with muciferous bodies lying beneath each strip together with various microtubules. (Electron micrograph by G. F. Leedale, University of Leeds)

plates around the pyrenoid, though in the diatoms which do not form starch there is no apparent reserve product associated with this structure. Light microscopy suggests that the projecting pyrenoids are unconnected with starch formation, but the electron microscope reveals that some of them have a thin layer of some reserve substance surrounding the matrix, e.g., in Dinoflagellates where in the species with stalked pyrenoids the plastid membrane also covers the pyrenoid and the polysaccharide material is deposited *outside* this membrane. Similarly in the euglenoid algae the paramylon granules lie outside the chloroplasts. The starch grains in *Scenedesmus* are cut off from the pyrenoid by intrusion of lamellae. In this genus the pyrenoid disappears completely at cell division and arises *de novo* in the daughter chloroplasts.

In *Platymonas* and *Prasinocladus* of the Prasinophyceae the pyrenoid is unusual in that it is penetrated by branching canaliculi entering from a direction facing the nucleus and lined by the chloroplast membrane in the former and by this and also projections of the nuclear membrane in the latter.

Eyespots are composed of collections of osmiophilic granules in which carotenoid material is dissolved. When they occur within the plastid, they form either a single or multiple row(s) of granules located between thylakoids usually adjacent to the outer chloroplast membrane. Slight modifications of the chloroplast membrane and the plasmalemma in the region over the eyespot have been reported. The eyespots in the gametes of fucoid Phaeophyta appear as a series of chambers containing the pigment rather than as separate granules and it is

Fig. 10.26 *Lepocinclis ovum* v. *butschlii*. A euglenoid flagellate closely related to *Euglena*. Electron micrograph of cell apex showing the anterior invagination with the two flagella arising from the base. The larger flagellum which extends out of the invagination, though not seen to do so in this section, bears the swollen photoreceptor (p). The pellicular strips can be seen entering the opening of the invagination but ceasing there. Chloroplasts (C), mitochondria (M), and a dictyosome (D) are also visible (×20,000). (Electron micrograph kindly supplied by G. F. Leedale, University of Leeds)

noteworthy that in these organisms the hind flagellum is swollen as it passes over the eyespot and although not actually attached to the gamete in this position it is in some way pressed close against the cell membrane. The eyespot in euglenoid flagellates is situated free in the cytoplasm usually around the base of the anterior invagination but again consists of clusters of osmiophilic granules and some microtubular elements (also seen in some Chlorophyta). The eyespot granules are clustered in groups of 2–3 surrounded by a membrane of unit membrane dimensions, but the eyespot as a whole has no bounding membrane. Osmiophilic granules are common in the chloroplasts of many algae including euglenoid genera and there is some evidence that they can be channelled between the thylakoids towards the eyespot when this also is present.

Colourless amyloplasts occur in some siphonaceous Chlorophyta and resemble chloroplasts but contain a single large starch grain together with a small concentric mass of lamellae. It is interesting that this small concentric aggregation of lamellae also occurs in the chloroplasts of those species containing amyloplasts but not in those siphonaceous species without amyloplasts.

Mitochondria

Though basically similar to those of other plants, algal mitochondria show subtle variations between groups, particularly in the development of the cristae (tube-like, plate-like, extending to various degrees into the lumen, etc.). In *Micromonas pusilla* there is only a single mitochondrion but in most cells they are numerous but often precisely located within the cell, e.g., around the nucleus and between the lobes of the chloroplast (*Chlamydomonas eugamatos*), between the plasmalemma and the chloroplast (other *Chlamydomonas* spp. *Carteria*, and colonial species). They are often closely associated with flagella bases (kinetosomes) and in the Prasinophycean genus *Heteromastix* a single mitochondrion lies pressed against the inside of the 'C' shaped chloroplast.

Dictyosomes

Dictyosomes (Golgi bodies) are very distinctive components of algal cells and are often precisely located. They consist of stacks of elongate vesicles dilated at the periphery. In some colonial Chlorophyta they surround the nucleus and are contained within a series of 'amplexi' formed of endoplasmic reticulum which is continuous with the outer layer of the nuclear membrane. Associated with the dictyosomes are vesicles which are budded off from the cisternae and in the Haptophyceae and Prasinophyceae it is within these vesicles that the characteristic scales of the body and flagella are formed (p. 447). In the dinoflagellate *Gonyaulax* the dictyosomes form a 'spherical shell' closely associated with the concave side of the 'C' shaped nucleus and within this area vesicles, presumably formed from the dictyosomes, give rise to the trichocysts (p. 446). The dictyosomes in *Chlorella* pair prior to cell division to form corresponding 'mirror' images between which the new partition membrane is formed.

In the diatoms special membrane-limited vesicles appear during cell division and within these the new silica valves are deposited. The surrounding membrane has been termed the silicalemma and it is quite distinct from the plasmalemma.

Vacuoles

Contractile vacuoles appear as empty membrane-bounded spaces often with subsidiary cisternal elements around them and fusing to form the main vacuole. In dinoflagellates more complex vacuoles (pusules) are associated one with each of the flagellar canals. These are in clusters of approximately 40 vesicles, each with a double membrane and opening by means of a narrow pore, either into a central vesicle and then into the flagellar canal or individually into the canal. These pusules are permanent vacuoles unlike the temporary contractile vacuoles which enlarge and discharge and enlarge again.

Miscellaneous organelles

Structures comparable to the lomasomes of the fungi are found in the green algal genera *Chara* and *Nitella* of the Charophyta and in *Platymonas* of the Prasinophyceae but have not been seen in other groups.

Muciferous bodies are often located beneath the plasma membrane. They may discharge through it or into microtubules, e.g., in euglenoids.

Trichocysts of the dinoflagellates are complex organelles which eject fine, banded, probably proteinaceous threads from the cell. The threads are extruded through thin areas in the wall after the mature trichocyst has migrated from the central dictyosome area. The trichocyst has a single membrane around a central square core of amorphous material. The enclosing membrane is thickened by hoops or spirals of material. Along the upper end of the core small backwardly directed tubules occur and attached to the apex of the core is a group of twisted fibrils ending in even finer ones which just touch the closing membrane. Discharge is accompanied by a change in hydration of the core which takes place in a matter of milliseconds. Trichocysts occur also in some other algal groups but details are not yet available.

Cell walls

Wall structure in the algae is extremely variable, though usually based on polysaccharides within or on which other substances are deposited. Mucilages are common but these may have structural microfibrillar elements within them, e.g., in colonial volvocalean genera. In many Chlorophyta the microfibrillar material is cellulose and in the classic examples *Cladophora* and *Chaetomorpha* it is deposited in layers of microfibrils running at different angles to the long axis of the cells. Shorter microfibrillar units of xylose or mannose form the wall material in some of the siphonaceous green algae.

Two groups of algae which have received considerable attention in the last decade, the Haptophyceae and Prasinophyceae have a 'wall' layer composed of scales which are formed within the cells in the vesicles arising from the dictyosomes. In the former group they consist of organic scales, in which mannose has been detected, and which have a fibrous structure with an inner face with radiating fibres and an outer one of randomly arranged fibres. In fact the two faces are really two sides of a flattened disk and in one subgroup, comprising the Coccolithophorids, it is within such scales that the calcium carbonate is deposited (only two genera have so far been investigated and too wide a generalization should be avoided at the moment). Simple scales without the calcified part (coccolith) are also found on the body of the cell. Thus in the coccolithophorids at least two scale types can be demonstrated on any one cell and in the other section of the group containing the genus *Chrysochromulina* several precisely arranged purely organic scale types can be found on any cell (Fig. 10.29H). In the Prasinophyceae the situation is even more complex, for here it is common to find two scale types on the flagella and three types over the body of the cell (e.g. in *Pyramimonas*). In yet another genus (*Platymonas*) fine scale-like structures, also produced in the vesicles derived from the dictyosomes, are liberated and coalesce outside the cell to form the theca which surrounds the cell. The dino-flagellates have long been known to have an 'armoured' cell wall consisting of precisely arranged plates joined together by sutures. These plates at least in some species are sandwiched between outer and inner membranes and the inner membrane connects at the edge of the plates with the under-lying plasmalemma.

THE RANGE OF VEGETATIVE FORM

The tremendous range of vegetative morphology can be reduced to a series of types—(1) unicells which may be non-motile (*Chlorella*), motile by means of flagella (*Chlamydomonas, Euglena, Ceratium*), motile by extrusion of mucilage or other means (desmids and diatoms). (2) aggregations of cells into colonies—motile (*Volvox, Synura*), non-motile (*Pediastrum*), (3) filaments—unbranched (*Ulothrix*), branched (*Ectocarpus*), (4) complex thalli based on filamentous construction (Rhodophyta and some Phaeophyta), (5) 'parenchymatous', i.e., without distinct filamentous structure in the mature stages (*Ulva, Laminaria, Fucus*), (6) coenocytic—either simple unicellular (*Protosiphon*) or branched unicellular (*Bryopsis*) or divided into multinucleate segments (*Cladophora*).

LIFE-CYCLES OF ALGAE

All these algae have characteristic life-histories; the simplest is shown by *Fucus*, where the spermatozoids and eggs are the only haploid phase in the life-history. In others there are varying degrees of isomorphic/heteromorphic development of gametophytes and sporophytes culminating in the most complex of all plant life-histories which characterize the Rhodophyta. What they do all have in common is the formation of gametangia which lack a sterile wall layer around the fertile tissue; this distinguishes them from all higher plant groups. The sexual process may involve simple fusion of nucleated cytoplasmic masses or of flagellated gametes and non-flagellated egg cells. Accessory reproduction of either or both gametophyte and sporophyte may be achieved either by non-motile spores or by motile zoospores.

CLASSIFICATION OF ALGAE

The classical concept of algae as consisting of four basic groups distinguished on the chemical basis of pigmentation into the blue-green (Cyanophyta), green (Chlorophyta), brown (Phaeophyta), and red (Rhodophyta), was an over-simplification. The blue-green algae are now considered to be distinct from the rest and are treated as prokaryotes (p. 346). Detailed studies during the last decade and particularly the increasing use of electron microscope studies has led to a clearer segregation of eukaryotic algae into a series of divisions (phyla) each of which requires consideration to give a balanced view of the group. In this classification great stress has always been placed on the characteristics of the motile cells and of the reproductive processes which tend to be relatively constant throughout each division; indeed many anomalies which existed in the older schemes have now been shown to be due to unfortunate juxtaposition of unlike algal groups. Silva (in Lewin, 1962) recognized ten basic phyla and to these must now be added two classes (Haptophyceae and Prasinophyceae); applying the same criteria as those used by Silva these may one day warrant raising to divisions. These twelve groups fall distinctly into a predominantly green pig-mented series (Chlorophytes), a brown pigmented series (Chromophytes), and a red pigmented series (Rhodophytes).

Chlorophyte series

Four major groups are included here. Apart from chlorophyll α they also possess chlorophyll β and small amounts of carotenoid pigment. Starch or starch-like polymers are the common reserve product.

Euglenophyta (Fig. 10.28, A–D)

The main pigmented genera of this phylum are *Euglena, Lepocinclis, Trachelo-monas,* and *Phacus* (Fig. 10.28, A–C) all equally abundant, particularly in organic-ally polluted fresh waters. A marine, photosynthetic counterpart to *Euglena* is *Eutreptia* which, however, has two equal flagella. Although photosynthetic they all require organic molecules in the form of vitamins (that is, they are photoauxotrophic), e.g., *Euglena gracilis,* requires the vitamin B_{12} complex and has been used in the biological assay of this. Many green species are also facultative heterotrophs, that is, they can grow in the dark when supplied with organic compounds (acetate is one which is readily used) and they are often grouped with other algae as the acetate flagellates. Most genera can be cultured in a bi-phasic medium with soil at the base of the culture vessel and liquid above. The interior invagination (Fig. 10.26) is not a gullet and solid particles cannot be ingested through it; the complex contractile vacuole system discharges into this invagination. Within it the two very unequal flagella arise; the shorter does not extend to the outside and is smooth (i.e., does not have fine hairs or 'flimmer') but the larger may be as long as or longer than the cell, has a single row of hairs and near the base a swelling which is the photo-receptor. No other algal or protozoan group has hairs in a single row. The so-called 'pellicle' of euglenoids is pliable and stretchable allowing the cell to change its shape. It has recently (Leedale, 1967) been shown to consist of a series of strips of proteinaceous material wound around the cell like a bandage or rather a number of bandages which fuse in pairs at the base and apex (Fig. 10.27). The intricate means by which these 'pellicular strips' engage one another is shown in Fig. 10.27, together with the associated muciferous bodies, microtubules and external wart-like granules which are present in some species. This skeletal system is really an

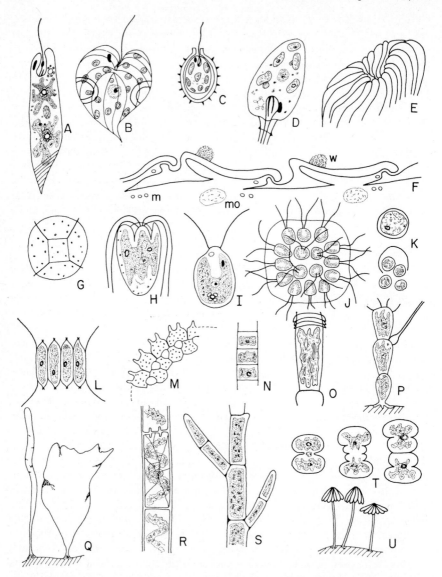

Fig. 10.28 Chlorophyte genera. (**A**) *Euglena.* (**B**) *Phacus.* (**C**) *Trachelomonas.* (**D**) *Colacium.* (**E**) The apex of a *Euglena* showing the pellicular strips curving into the anterior invagination. (**F**) Diagrammatic cross section of the pellicular strips as seen in electron microscope section. Showing the articulation of adjacent strips, mucilaginous warts (w), mucilage organs (mo), and microtubules (m). (**G**) *Pterosperma.* (**H**) *Pyramimonas.* (**I**) *Chlamydomonas.* (**J**) *Gonium.* (**K**) *Chlorella.* Cell and formation of four autospores. (**L**) *Scenedesmus.* (**M**) Fragment of *Pediastrum* colony. (**N**) *Ulothrix.* (**O**) *Oedogonium,* one cell of unbranched filament with cap cells. (**P**) Base of *Bulbochaete.* (**Q**) *Enteromorpha* (left) and *Ulva* (right). (**R**) *Spirogyra.* (**S**) *Cladophora.* (**T**) *Cosmarium,* cell and two stages in cell division. (**U**) *Acetabularia.*

endoskeleton since the cell membrane is external to it. The chromatophores vary greatly in shape and size but the paramylon granules (a β-1:3 linked glucan) are a most characteristic diagnostic feature as also is the large red eyespot which is *free* in the cytoplasm. Recently some species have been shown to contain endophytic bacteria which occur in the cytoplasm and in some cases within the nucleus. *Euglena* can be converted into a colourless, protozoan organism by heat or streptomycin treatment which destroys the chromatophores. Treated individuals are quite viable and continue growth heterotrophically. One green genus, *Colacium* (Fig. 10.28D) is colonial, attaching itself by secretion of a mucilage stalk at the apical end and yet another, *Euglenamorpha* has three emergent flagella each with basal flagella swellings. Other genera are found free in nature and are colourless, e.g., *Astasia* (which has no eyespot or flagellar swelling), *Peranema* which has two long flagella one directed forwards during swimming and the other thinner and trailing. There are no authentic records of sexual reproduction in this group; asexual reproduction involves longitudinal cleavage.

Chlorophyta (Fig. 10.28, I–U)

This is the largest and most complex group of green algae and has been split into a variable number of sub-groupings by different authors. The variously shaped chloroplasts possess pyrenoids around which starch (an α-1, 4 glucan) is deposited, though granules of starch may also be free in the cell. Wherever an eyespot is present it is *embedded* in the chloroplast. Contractile vacuoles open directly to the exterior and the two (sometimes four) flagella arise on either side of an apical papilla; they are equal in length and although devoid of 'flimmer' (p. 361) many do have a felt-like coating. The cell wall is frequently composed of cellulose and is often microfibrillar. Sexual reproduction is widespread and when motile gametes are involved these are usually biflagellate though all stages through to oogamy are found. Chlorophyta are most widespread in fresh waters; Bourelly (1966) describes 520 genera from fresh water. One or two groups, however, are predominantly marine, e.g., the Bryopsidophyceae.

The simplest cells are merely spheres of cellulosic wall enclosing the cytoplast which contains a single chloroplast and a nucleus, e.g., the ubiquitous *Chlorella* (Fig. 10.28K) which grows in laboratory glassware, condensers, transparent tubing, etc. Such a cell divides into four similar small daughter cells which then enlarge. This simple and rapid mode of division has encouraged considerable physiological study of this genus. In the same group as *Chlorella* are a large number of other single-celled algae and also genera which form colonies when the products of division remain together, e.g., *Scenedesmus* (Fig. 10.28L). Some of these algae also form motile zoospores but their major characteristic is that they are non-motile in the vegetative phase. Some, e.g., *Trebouxia*, are the phycobiont of many lichens (p. 617). Unicells or groups of cells embedded in mucilage and often provided with channels through the mucilage which can be mistaken for flagella, comprise another sub-group of which *Tetraspora* is one of the commoner genera. The vegetative cells often retain the contractile vacuoles and eyespots, derived from the motile stage. Of greater complexity but generally placed first in the classification of the Chlorophyta are the motile cells of the *Chlamydomonas* type (Fig. 10.28I) which have all the characteristics of a

Chlorella-like cell plus contractile vacuoles, an eyespot and two flagella (emerging from pores in the cellulose wall). Not all motile genera possess a true cell wall whilst yet others resemble *Trachelomonas* in having a distinct theca around the cell. Aggregation of chlamydomonad-like cells into mucilaginous colonies (e.g., *Gonium, Volvox*) and into non-mucilaginous aggregates (e.g., *Spondylo-morum*) form a parallel to the colonial habit of the coccoid forms.

The desmids are specialized coccoid cells with a highly sculptured, poroid wall usually consisting of two halves joined by a bevelled edge at an isthmus, e.g., *Cosmarium* (Fig. 10.28T) and *Staurastrum*. Mucilage is extruded through the pores and frequently forms a thick capsule around the cell. Bacteria are commonly found lodged in the channels between the blocks of mucilage which form the capsule. Some genera of desmids form long filaments owing to the possession of special spines or mucilage pads which prevent the separation of the daughter cells during division. In addition to forms with a wall composed of two halves there is a second smaller series in which the wall is composed of a single piece; here again some genera produce short loose filaments. The desmids are exclusively fresh-water species of world-wide distribution—they are most frequent in acid waters but a few well-defined species grow in alkaline situations. Desmids are often found in mucilaginous clumps on wet rock faces and on peaty soils. Many are cosmopolitan but a few genera have very restricted geographical distributions; arctic and tropical species can be distinguished.

The remaining groups of the Chlorophyta are all filamentous or siphonaceous, i.e., multinucleate thalli or in one group of the Bryopsidophyceae they are siphonaceous-like thalli but with a single nucleus, e.g., *Acetabularia* (Fig. 10.28U). Filaments may be unbranched, e.g., *Ulothrix, Spirogyra, Oedogonium* (Fig. 10.28N, R, O) or branched, e.g., *Chaetophora, Bulbochaete* (Fig. 10.28P). In some genera the initial filamentous morphology has been expanded into a pseudo-parenchymatous state, e.g., *Ulva* and *Enteromorpha* (Fig. 10.28Q). Many have a basal cell which is modified as an attaching organ whilst some have adopted a creeping habit and are firmly attached to the substrate by mucilage. They play a very important part in the microbiology of epiphytic associations and are themselves hosts to numerous other algae, protozoa, bacteria, and fungi. Some are particularly adapted to an aerial environment, e.g., *Apatococcus* and *Desmococcus* (formerly known as *Pleurococcus, Protococcus*, etc.) which grow on tree bark and develop a branched filamentous form only in moist situations or in culture. A further group, of which *Trentepohlia* is the commonest, are also terrestrial, growing on rock and tree surfaces and yet others, e.g., *Cephaleuros*, are serious leaf parasites of commercial crops including tea and coffee. This section is also exceptional amongst the Chlorophyta in forming oil rather than starch and also copious carotenoid pigments which give an orange coloration to the thallus.

The siphonaceous genera are mainly marine in distribution—some have cross walls delimiting multi-nucleate segments of the thallus, e.g., *Cladophora* (Fig. 10.28S) whilst others have branching siphons which aggregate to form complex leaf and rhizome-like structures of considerable size. Many are calcified and form sub-tidal turf-like growths. Amongst them is the interesting uninucleate plant *Acetabularia* (Fig. 10.28U) which has proved a valuable tool in biochemical and morphogenetic research. It consists of a cap, a stalk and a rhizoid by means of which it is anchored to the substratum. Although the organism is several centimetres long, it has only a single nucleus which is situated in the rhizoid. The cap is of complex structure, and its form is a species characteristic. If the cap is amputated, it regenerates. Such regeneration occurs even in the absence of the nucleus, and enucleate individuals survive for several months. It is possible by amputating the cap and at the same time replacing the nucleus by one transferred from another species, to produce a system in which the formation of a cap is jointly controlled by the nucleus of one species and the cytoplasm of another. The form of the regenerated cap at first resembles that of the species from which the cytoplasm was derived but finally becomes identical with that of the species from which the nucleus was obtained. It was suggested that factors controlling differentiation of the cap are produced by the nucleus and that these persist for some time in the cytoplasm after removal of the original nucleus, thus permitting regeneration of enucleate individuals.

Asexual reproduction is common in the Chlorophyta and achieved in a variety of ways. In the filamentous Conjugatophyceae convolutions of the cross walls result in fragmentation of the filaments. The desmids split apart at the isthmus and two new semi-cells are formed (Fig. 10.28T): thus every desmid cell has one old and one new semi-cell. Thick walled spores (akinetes) are formed in some filamentous species, e.g., *Cladophora*. Production of bi-quadri-flagellate zoospores is by far the commonest mode of asexual reproduction—these are formed usually by cleavage of the cytoplasm within the cell wall to form four or more cells each of which acquires two isokont flagella—they then escape either after dissolution of the parent cell, e.g., *Chlamydomonas*, or through special pores, e.g., *Cladophora*. In *Oedogonium* these zoospores are provided with a ring of paired flagella. Attachment is usually at the flagella pole in those which germinate to form an attached cell or filament. Sometimes the products of division are non-motile, e.g., in *Chlorella*. There is never cleavage of the vegetative flagellate cell such as occurs in *Euglena* and conversely *Euglena* never produces zoospores.

Sexual reproduction is achieved in many species by the fusion of motile gametes produced in a similar manner to the zoospores and often morpho-logically indistinguishable from these. In a few species the gametes are unequal in size and in some the process is oogamous. Many groups of the Chlorophyta exhibit a range from isogamy to oogamy whilst others, e.g., the Oedogoniales, are completely oogamous. The zygote usually forms a thick-walled resting cell. In the Conjugatophyceae the gametes are formed from all or part of the cytoplasm of the cell and they are non-flagellate masses which are transferred either into a conjugation tube between the two filaments or completely into the filaments of opposite strain. Homothallism and heterothallism both occur.

Although not as obvious as in the Phaeophyta (p. 452) both isomorphic and heteromorphic alternations of haploid and diploid phases occur in the Chlorophyta. Many of these life-histories require further study. Some of the heteromorphic genera have an alternation between a filamentous or thalloid phase and a unicellular (*Codiolum*) stage which is often endophytic in other algae.

Prasinophyceae

This group of algae has been clarified mainly during the last decade; before this many of the species were included as anomalous forms of the Chlorophyta. Many genera are motile and usually have four rather thick flagella arising from an apical pit around which the cell is extended into lobes (e.g., *Pyramimonas*, Fig. 10.28H); there are, however, several other arrangements. The cell surface and also that of the flagella is covered by layers of scales and hairs. These can be demonstrated only with the aid of the electron microscope. The scales are of organic material and in some species it has been shown quite clearly that they form within vesicles in the cells. The pyrenoid is often very large and in some genera penetrated by fingers of the chloroplast or nuclear membranes or both. Whilst some species are motile cells in the vegetative stage, others, e.g., *Halosphaera* and *Pterosperma* (Fig. 10.28G), are coccoid and yet others are dendroid (e.g., *Prasinocladus*), the typical motile cells being produced only during asexual reproduction. Sexual reproduction has not been recorded. There is still much to be learnt about the members of this group and of their ecology—they occur in both fresh-water and marine habitats and the non-motile spheres of *Halosphaera* and *Pterosperma* are frequent in marine plankton where they float to the surface.

Charophyta

This group is very distinctive, forming macroscopic radially organized plants which grow from apical cells which cut off cells at the base each one of which divides again to give an upper nodal cell which then gives rise to whorls of branches and a lower inter-nodal cell which enlarges greatly but is incapable of branching. They are anchored by a rhizoidal system and have complex oogonia and antheridia. Since the large internodal cells may grow several inches long they have proved suitable plants for many biochemical/biophysical studies. They are common fresh-water/brackish plants growing anchored in the sediment of ponds.

Chromophyte series (Fig. 10.29, A–S)

This is a heterogeneous collection of algal groups in which the ratio of chlorophylls to carotenoid pigments is such that in most groups the plastids are brown coloured. In addition, chlorophyll *b* is absent and carbohydrate reserve products tend to accumulate outside the chromatophore; in many instances oil globules are prominent in the cells. There are many different modes of flagellation but two apical flagella occur only in a relatively small number of genera and then in some they are accompanied by a further organelle, the haptonema. There are at least seven distinct algal groups in this series.

Cryptophyta

Only the flagellate genus *Cryptomonas* (Fig. 10.29A) is at all common but hardly a body of fresh water can be sampled without encountering this genus; it is found also in marine habitats. *Cryptomonas* is readily recognizable from the lateral invagination at the upper end out of which the two somewhat unequal

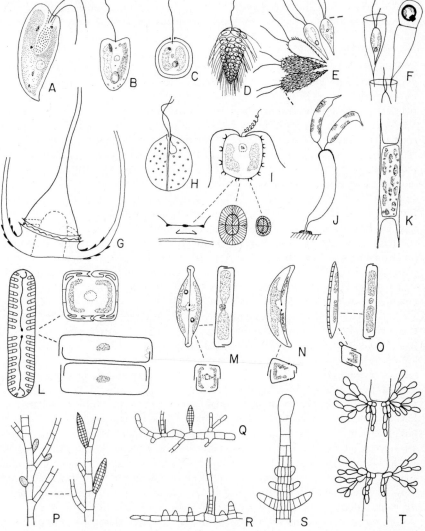

Fig. 10.29 Chromophyte genera. **(A)** *Cryptomonas.* **(B)** *Chromulina.* **(C)** *Chryso-coccus.* **(D)** *Mallomonas.* **(E)** Fragment of *Synura* colony, two lower cells showing scales. **(F)** Fragment of *Dinobryon* theca, on right with resting spore. **(G)** *Ceratium.* **(H)** *Exuvaiella.* **(I)** *Chrysochromulina* cell and scales of two sizes and in section on the cell (below). **(J)** *Ophiocytium.* **(K)** One cell of a *Tribonema* filament. **(L)** *Pinnularia,* valve view of a cell, cross section and cell division. **(M)** *Caloneis,* valve view, girdle view, and cross section. **(N)** *Cymbella,* valve view and cross section. **(O)** *Nitzschia,* valve view, girdle view, and cross section. **(P)** *Ectocarpus,* unilocular plant (left), plurilocular plant (right). **(Q)** *Streblonema.* **(R)** *Myrionema.* **(S)** Apex of *Sphacelaria.* Rhodophyte genus. **(T)** Fragment of branching thallus of *Batrachospermum* showing axial cells and branches.

flagella arise. The invagination is often lined with rows of trichocysts. Pigmentation is very varied but olive/green to brown is a common coloration although blue and red forms occur and are placed in related genera by some authorities.

Pyrrophyta (Dinophyta)

This phylum assumes considerable importance in the microbiology of the oceans (p. 516). The flagellates (Dinoflagellates) are often extremely abundant in the surface layers of the sea and must contribute a considerable amount to the primary production within the phytoplankton; many species occur in fresh water. They are striking flagellates many with a distinct 'armour' of interlocking plates (hence the prefix 'dino'). Equally abundant are the unpigmented genera but these are best considered as protozoa of the class Mastigophora (p. 405). Nonmotile stages of pigmented dinoflagellates are common symbionts (the zooxanthellae of earlier authors) in corals, clams, sea anemones, etc. Recent work suggests that even the apparently 'unarmoured' forms have thin plates embedded in the outer layer of the cell. Two basic morphological types are encountered. One has a theca in the form of two watchglass-like halves sealed together around the edge and with the two flagella arising from adjacent pores and extending out from the cell (e.g., *Exuviaella*, Fig. 10.29H). The other type has a series of plates fused together to form an epicone, a median furrow and a hypocone in which there is a longitudinal furrow (e.g., *Peridinium*). The two unlike flagella emerge through pores where the two furrows join—one runs around and within the median furrow and the other runs down the longitudinal furrow out into the surrounding medium. The plates are rather precisely orientated and joined together by sutures within which some slight growth of the cell occurs. The cells are often flattened and some genera have the angles developed into long spines (e.g., *Ceratium*, Fig. 10.29G). The chromatophores are usually brown disks with stalked pyrenoids (cf. similar pyrenoids in the Phaeophyta). The nucleus is often large and the chromosomes stainable at all times. Many dinoflagellate cells have trichocysts beneath the cell membrane. These trichocysts develop from the Golgi vesicles and eject banded proteinaceous threads. Highly characteristic sac-like structures are found associated with the flagella pores. These are the so-called 'pusules' and there are often smaller associated sacs opening by constricted tubes into the central sac. Their function is not yet known although they may be involved in either uptake or excretion. They are not, however, delicate structures such as the contractile vacuoles of Chlorophyta.

Cell division is achieved by a splitting of the cell along an oblique plane between the plates followed by the growth of the new halves—thus dinoflagellates are rather like desmids and diatoms in that part of the mature cell is derived from the past generation and part is younger and reformed by deposition of new plates. The exact details of asexual reproduction require further investigation. Sometimes the cells remain together after division to form colonies. Sexual reproduction has rarely been recorded and only recently has fusion between anisogametes been authenticated in *Ceratium*. This is followed by encystment of the zygote and germination after several months with the production of motile cells; meiosis occurs during formation of these motile cells.

Cysts which have been called 'hystrichospheres' by paleontologists are now known to be formed by dinoflagellates and can be found in marine plankton. On germination they form typical dinoflagellate cells. Some marine dino-flagellates produce a toxin which is poisonous to fish and man but not to shell-fish which feed on the dinoflagellates. Normally the concentration in the shell-fish is negligible but during times of intense growth of the dinoflagellates it becomes concentrated in lethal quantities and the sale of the shell-fish has to be prohibited. This is normally only a tropical or sub-tropical occurrence but recently cases of non-lethal poisoning in the north of England were attributed to it. The reasons for the massive growths of the dinoflagellates are not known. They are sometimes so intense that the sea is coloured by them—the so-called 'red tides'.

Another interesting property of some dinoflagellates is their ability to emit flashes of light, i.e., they are bioluminescent as are certain bacteria and fungi. This bioluminescence has been shown to be under an endogenous 'biological clock' control system. It is most evident during the night.

Chrysophyta

Many genera of this phylum are flagellate but unlike other groups there is a considerable range in type of flagellation. Coccoid, tetrasporal and filamentous genera also occur, but it is not known whether some of these are stages in the life history of flagellate forms. The simplest cells have a single flagellum and discoid brown chromatophores. Electron microscopic investigations of the Chrysophyta are essential to elucidate many points of structure. Several apparently uni-flagellate species, e.g., *Chromulina* (Fig. 10.29B), have been shown to have a second flagella base within the cell often associated with an eyespot. In some of these very fine *organic* scales have been found covering the cell membrane. In the larger flagellate, *Mallomonas* (Fig. 10.29D) the body is covered by relatively massive *siliceous* scales (Fig. 10.30F) each provided with a long siliceous bristle. Colonial flagellate forms also occur, e.g., *Synura*, which has two slightly unequal flagella and each cell is coated with numerous siliceous scales (Figs. 10.29E, 10.30E). In yet other genera the cell is enclosed in a theca, e.g., *Chrysococcus* (Fig. 10.29C), which is comparable to *Trachelomonas* in that a spherical theca encloses the cell leaving only a pore for the single flagellum. In *Dinobryon* the thecae are vase shaped, composed of microfibrillar material and joined together to form colonies (Fig. 10.29F).

Many Chrysophyte flagellates form characteristic siliceous cysts. These are endogenous structures formed within the cytoplasm of the cell, which slowly move into the cyst which is then 'plugged'.

Sexual reproduction by an isogamous fusion has been reported in some species.

Haptophyceae

This is the second group of algae which has been formulated over the last decade. It is probably of great importance in the microbiology of the seas. The whole group of flagellates formerly known as the Coccolithineae have been transferred to this group. These are biflagellate planktonic flagellates which secrete intricately formed calcareous scales on to the outside of the cell. In some

genera these have now been seen to form an outer layer superimposed on fine organic scales with characteristic fibrous markings on them. The calcareous scales are called coccoliths and they can be found abundantly in some chalk deposits indicating that this group is an ancient one and has been an important constituent of marine plankton over a very long period of geological time. Non-motile stages also occur as filamentous plants on rocks around the coasts. Yet another group of biflagellate cells, exemplified by *Chrysochromulina*, forms intricate organic scales in layers on the outer membrane (Fig. 10.29I). There is, in addition, a further organelle adjacent to the two equal flagella and this is the haptonema; it is a long, often coiled, tube of similar dimension to the flagella but constructed differently with three concentric membranes surrounding a group of seven or eight microtubules. The haptonema is sometimes swollen at the end and acts as an attachment organ. The cells have two or more brown chromatophores.

Xanthophyta

This is the apparent anomaly in the 'chromophyte' series, since there is not sufficient carotenoid pigment in the chromatophores to obscure the green colour given by the chlorophylls. In other features they are, however, a distinct group with many characters which ally them with the Chrysophyta—Bacillariophyta groupings. Motile vegetative cells and gametes are 'heterokont', i.e., they have one long and one short flagellum, the former having a coating of fine hairs. The old name for the group was Heterokontae. The chromatophores are almost always parietal and disk-like; oil, but never starch, is stored in the cells and the cell walls tend to have a greater proportion of pectin than cellulose. In some genera the cell wall is composed of striate H-pieces which fit together very closely (e.g., *Tribonema*) but their actual nature is only revealed by chemical treatment. Apart from the 'heterokont' flagellation, no single character can be used to distinguish a species of the Xanthophyta from one of the Chlorophyta. Cells with discoid plastids and/or without starch are rare in the Chlorophyta and such forms should always be regarded as possible heterokont algae.

Flagellate species are rare and of the coccoid species *Ophiocytium* (Fig. 10.29J) is perhaps the commonest. The unbranched filaments of *Tribonema* are abundant in fresh waters; the branching filmentous genera occur also on soils. The commonest genus is *Vaucheria*, which is a representative of the siphonaceous

---

Fig. 10.30 (**A, C**, and **E**) Transmission electron micrographs. (**B, D**, and **F**) Scanning electron micrographs. (**A**) A centric diatom *Stephanodiscus* in oblique girdle view (upper left) and valve view (lower right). (**B**) Another species of *Stephanodiscus*. Complete cell appearing like a petri-dish with a girdle band overlapping top and bottom valve. Note the concentric indentations on the valve face. (**C**) A *Navicula* species showing the raphe system and pores through the silica wall. (**D**) Another *Navicula* species showing the boat-like nature of the cell. (**E**) A field of scales from *Synura* spp. and *Mallomonas* showing the perforate base plates, various ornamentation, and spines on these. (**F**) A single scale of a *Mallomonas* species showing the raised V-shaped central ridge, the overturned rim and the protuberance from which the spine arises. (Magnifications: (**A**) ×4,000, (**B**) ×3,560, (**C**) ×4,000, (**D**) ×2,500, (**E**) ×4,500, (**F**) ×5,000. F. E. Round, Bristol University)

habit and is oogamous with male gametes which are of typical heterokont form. Asexual reproduction is achieved by means of a large swarmer which is multi-nucleate, and associated with each nucleus is a pair of smooth slightly unequal flagella.

The Xanthophyta are a somewhat neglected group containing a very large number of mainly freshwater genera of great interest to all soil and aquatic microbiologists.

Bacillariophyta

Few phyla of algae are as well circumscribed as the diatoms (Bacillariophyta). The cardinal feature of the group is the possession of a siliceous skeleton composed of discrete segments which fit together to form basically a 'pill-box' or 'date box' structure which can, however, be modified by curvature of one or more of the three planes (Fig. 10.29, L–O; 10.30, A–D). Within this hyaline silica case the diatom has the usual organelles, including a centrally suspended nucleus, together with oil globules and two or more brown chromatophores to which are attached deposits of the polysaccharide, *chrysose*. A large number of genera are motile but without visible means and the problem of motility has not yet been satisfactorily solved, though there seems little doubt that systems of fissures and pores in the silica skeleton known as the raphe system are involved. The cell membrane lies internal to the silica wall which can therefore in one sense be considered as an exoskeleton. However, in some genera there is conclusive evidence that there is also a fine organic 'skin' outside the silica. The siliceous parts consist of upper and lower lids (epitheca and hypotheca, which in 'centric' forms are shaped like the two halves of a Petri dish). These thecae have intricate pores, often radiating out from the central point but changing form at the rim which itself may be supplied with a series of spines and tubes (Fig. 10.30B). The two thecae do not overlap one another but are joined by two or more 'hoops', known as girdle bands, which overlap one another and also the edges of the two thecae. These 'hoops' may be plain bands of silica or they may be poroid. The exact mode of formation of these parts and their spatial relationship appeared clear from light microscope studies, but the electron microscope, and particularly the reflecting electron microscope, is revealing much finer detail (Fig. 10.30, A–D). At cell division the nucleus (all diatoms are diploid in the vegetative phase) divides and the daughter nuclei move to lie one adjacent to the hypotheca and one adjacent to the epitheca and then two new thecae are laid down in membrane-limited vesicles between the two nuclei. Since these are formed internally they are slightly smaller than the parent thecae. The exact fate of the old girdle bands and the mode of formation of new bands is not yet clear. When the new thecae are fully formed the two cells separate (Fig. 10.29L). As this process continues the diatom gets *smaller*—thus the very large centric forms can slowly reduce in size from around 400 μ in diameter down to 50 μ in diameter. It is not known with certainty whether all diatoms undergo this reduction in size; some may have thecae which are flexible enough to allow the new thecae to expand to the same size as the old. Nevertheless, diatoms are similar to desmids in that each cell has one old and one new half.

Reduction in cell size is reversed only by the occurrence of sexual reproduction which has, however, been adequately studied in only a very small number of species. In some 'centric' forms it is oogamous, cells being converted into single oogonia or into numerous spermatozoids each of which has a single apical flagellum. In *Lithodesmium* the flagellum is highly unusual in that the two central micro-tubules are missing. The zygote (auxospore) enlarges and becomes surrounded by a siliceous wall (perizonium). On germination new thecae and girdle bands are all formed endogenously within the auxospore. In some genera the auxospore is produced by enlargement of the cytoplasm and parting of the thecae without any sexual process and in others, particularly in 'pennate' diatoms (see below), it is formed after conjugation between pairs of diatoms and the fusion of amoeboid masses of cytoplasm (cf. desmids).

The box-like structure of the 'centric' diatoms is repeated in the 'pennate' forms but here the cells tend to be elongate so that viewed from above (the valve view) they are cigar shaped and from the side (girdle view) they are oblong (e.g., *Pinnularia*, Fig. 10.29L). Both these planes can, however, be curved or contorted in various ways (Fig. 10.29, N–O). In addition the 'pennate' diatoms tend to have the rows of pores arranged in lines on either side of the centre line leaving a clear central area which may be completely solid (a pseudoraphe) or it may have two raphe fissures running from central nodules to apical nodules. The fissures and associated nodules, although not obvious with the light microscope are in fact complicated systems of pores in the silica wall; the whole complex is termed the raphe system and it is this which is involved in cell movement. Various theories have been proposed involving extrusion of substances through the fissures/pores, flow of cytoplasm or water along the fissures, undulating membranes in the fissures and even cilia projecting through them. The latter is certainly not true and motility is probably achieved by a combination of processes involving extrusion of substances and adhesive/surface tension properties of the cell.

Colony formation occurs in both groups, the daughter cells formed at cell division remaining attached, assisted by the interlocking of spines and/or combined with extrusion of mucilage. Others secrete mucilage from special pores to form attachment disks or stalks.

Ecologically they are the most widespread of all algae, common from the surface of soils and wet rocks, through all freshwater, brackish and marine habitats to the undersurfaces and interstitial spaces of the sea ice around the poles. Since the silica incorporated in the skeletons is relatively insoluble, under natural conditions the dead cells which sediment out of lakes and over the ocean beds build up considerable fossil deposits. Some of these when raised up on to the land are used commercially, i.e., the so-called diatomite (kieselgühr) which is often composed of almost pure silica and is used in industrial filtration processes, etc. The deposition of the siliceous skeletons is sequential and hence examination of cores taken from lakes or ocean basins reveals details of earlier diatom floras and yields information on past changes in the environment. Finally since silica is absorbed in a soluble form from the water in which the diatoms live the concentration of this substance in the water often changes radically during the growth cycles and it does not accumulate in the water.

Phaeophyta

The 'brown algae' in the classical sense are usually considered as macroscopic structures outside the realm of microbiology. There is, however, a considerable number of small filamentous and encrusting forms, moreover microscopic stages in the life cycles of larger forms (e.g., the Laminariales) which require microbiological techniques for their study. The brown chromatophores have club-shaped pyrenoids. Around the nucleus there are often clusters of fucosan vesicles containing a tannin-like substance. The cell walls are characterized by the occurrence of the polysaccharides, laminarin and alginic acid in addition to cellulose and fucoidin. The sugar alcohol, mannitol, is frequently found in brown algal cells.

The simplest morphological types are branched filaments often growing by means of intercalary meristems, e.g., *Ectocarpus* (Fig. 10.29P). In species of *Sphacelaria* (Fig. 10.29S) there is a precise segmentation at the apex to give alternating 'nodal' and 'internodal' cells, only the former giving rise to branches. Some species, although filamentous, grow as creeping filaments, e.g., *Steblonema*, whilst others form upright branch systems from the basal creeping filaments, e.g., *Myrionema*. Some of these genera grow between the tissues of other macroscopic algae.

The motile cells are always pear-shaped with two unequal flagella inserted in a lateral position, one pointing forwards and bearing hairs, the other curving backwards and attached to the body in the region of the eyespot. The motile cells may act as zoospores, i.e., they germinate to form the haploid or diploid generation which produced them or they may act as gametes (isogametes or the male gametes of the oogamous genera).

The Phaeophyta are almost entirely marine in distribution and occur mainly in the coastal region, the microscopic forms being attached to other algae or growing on rock surfaces.

Rhodophyte series (Fig. 10.29T)

The final, and in many ways the most complex, group of the algae falls outside both the 'chlorophyte' and 'chromophyte' series but in some respects shows similarities to the Cyanophyta. These are the 'red algae' which are so abundant as macroscopic growths in coastal waters. Many, however, including almost all those which occur on soils or in fresh water, are microscopic.

The plastids contain the phycobilin pigments, phycoerythrin and phycocyanin, the former in sufficient quantity to impart a red coloration to the algae. A variety of starch (Floridean starch) is formed and the cell walls contain, in addition to pectin and cellulose, gel-like, sulphate polysacchrides (one of which is agar, a sulphated galactose polymer).

Only a few unicellular red algae are known, e.g., *Porphyridium*, which is a soil alga; the remainder are filamentous (e.g., *Batrachospermum*, Fig. 10.29T), though the filamentous nature tends to be obscured in the larger red seaweeds. As with Phaeophyta there is a large number of creeping and upright branching filamentous forms. These often occur as red crusts on rocks in the intertidal and subtidal zone, or as minute red filamentous pustules on rocks or plants. Some are also endophytic in other algae and a few genera have become completely parasitic.

The succession of phases in the life history and the changes associated with the post-fertilization developments of the female organ or carpogonium are more complex than in any other plant group. As in the Cyanophyta there are no known flagellate stages and the male cell is a minute single-celled structure (spermatium) budded off the filaments in the male gametangia. These spermatia are produced in large numbers on the male plants and transported by water currents to the female plants where they adhere to the long-drawn-out trichogyne of the carpogonium. After fusion of the nuclei in the carpogonium the diploid nucleus is transferred by various means to other cells in the thallus and eventually diploid carpospores are formed which on germination grow into a diploid plant usually identical with the sexual plants. It is only on the diploid (commonly known as the 'tetraspore plant') that reduction division takes place leading to the production of four spores (tetraspores), two of which germinate into male plants and two into female.

The microbiological activity of the numerous fine filamentous forms growing at solid/water interfaces in the inter-tidal and sub-tidal zone is as yet relatively unknown. Several are known to produce antibiotic compounds, many precipitate calcium/magnesium carbonate to form rock-like strata, and many play an important part in the economy of coral reefs.

SYNOPSIS OF THE CLASSIFICATION OF ALGAE*

* NB. This is not a complete classification but merely a guide to the genera mentioned in this book. Except in the Rhodophyta only orders are listed and other groupings have been excluded.

A *Prokaryotic*

a	Cyanophyta	Blue-green algae
	Chroococcales	*Aphanothece, Chroococcus, Holopedia, Merismopedia, Microcystis*
	Chamaesiphonales	*Chamaesiphon*
	Nostocales	*Aphanizomenon, Anabaena, Calothrix, Cylindrospermum, Lyngbya, Nostoc, Oscillatoria, Phormidium, Rivularia, Richelia, Spirulina, Symploca, Tolypothrix, Trichodesmium*
	Stigonemales	*Stigonema.*

B *Eukaryotic*

I **Chlorophyte series**

a	Euglenophyta	
	Euglenales	*Euglena, Lepocinclus, Phacus, Trachelomonas*
	Peranematales	*Peranema*
b	Chlorophyta	
	Volvocales	*Carteria, Chlamydomonas, Chlorogonium, Gonium, Pandorina, Spondylomorum, Volvox*
	Tetrasporales	*Tetraspora*
	Chlorococcales	*Characium, Chlorella, Chlorococcum, Pediastrum, Scenedesmus, Trebouxia*

Ulotrichales	*Ulothrix*
Ulvales	*Enteromorpha, Ulva*
Chaetophorales	*Apatococcus, Aphanochaete, Cephaleuros, Chaeto-phora, Desmococcus, Stigeoclonium, Trentepohlia*
Oedogoniales	*Bulbochaete, Oedogonium*
Cladophorales	*Cladophora*
Dasycladales	*Acetabularia*
Codiales	*Bryopsis*
Desmidiales	*Cosmarium, Staurastrum*
Zygnemales	*Spirogyra*

c Charophyta
Charales	*Chara, Nitella*

d Prasinophyceae
Pyramimonadales	*Platymonas, Pyramimonas*
Halosphaerales	*Halosphaera, Pterosperma*

II Chromophyte series

a Cryptophyta
Cryptomonadales	*Cryptomonas*

b Pyrrophyta
Prorocentrales	*Exuviaella*
Gymnodiniales	*Amphidinium*
Peridiniales	*Ceratium, Peridinium*

c Phaeophyta
Ectocarpales	*Ectocarpus, Myrionema, Streblonema*
Sphacelariales	*Sphacelaria*
Fucales	*Fucus*

d Xanthophyta
Heterococcales	*Ophiocytium*
Heterotrichales	*Tribonema*

e Bacillariophyta
Eupodiscales	*Coscinodiscus, Cyclotella, Melosira, Sceletonema*
Rhizosoleniales	*Rhizosolenia*
Biddulphiales	*Chaetoceros*
Fragilariales	*Asterionella, Fragilaria, Grammatophora, Licmorpha, Opephora, Rhaphoneis, Synedra*
Achnanthales	*Achnanthes, Cocconeis*
Naviculales	*Amphiprora, Amphora, Caloneis, Diploneis, Gyrosigma, Mastogloia, Navicula, Pinnularia*
Bacillariales	*Nitzschia*
Surirellales	*Campylodiscus, Cymatopleura, Surirella*

f Chrysophyta
Ochromonadales	*Dinobryon, Mallomonas, Synura*
Chromulinales	*Chromulina, Chrysococcus*

g Haptophyceae
Prymnesiales	*Chrysochromulina* and all genera producing coccoliths

III Rhodophyte series

a Bangiophycidae
Porphyridiales	*Porphyridium*

b Florideophycidae
Nemalionales	*Batrachospermum*

BOOKS AND ARTICLES FOR REFERENCE AND FURTHER STUDY

Fine Structure of the Eukaryotic Cell

BRACKER, C. E. (1967). Ultrastructure of fungi. *A. Rev. Phytopath.* **5**, 343–374.

FULLER, M. S. (1966). Structure of the uniflagellate zoospores of aquatic Phycomycetes. In *The Fungus Spore*: Proceedings of the 18th Symposium of the Colston Research Society (M. F. Madelin, ed.), pp. 67–84, Butterworths Scientific Publications.

GRIMSTONE, A. V. (1961). Fine structure and morphogenesis in protozoa. *Biol. Rev.* **36**, 97–150.

KENNEDY, D. (1965). *The Living Cell.* (Collected reprints from *Scient. Am.*) 296 pp.

LEEDALE, G. F. (1967). *Euglenoid Flagellates.* Prentice-Hall, Englewood Cliffs, New Jersey, 242 pp.

PITELKA, D. R. (1963). *Electron-microscopic structure of protozoa.* Pergamon Press, Oxford, London, New York, Paris, 269 pp.

—————— and CHILD, F. M. (1964). The locomotor apparatus of ciliates and flagellates. Relations between structure and function. In *Biochemistry and Physiology of Protozoa* (S. H. Hutner ed.). **III**, 131–198. Academic Press, New York and London.

VICKERMAN, K. and COX, F. E. G. (1967). *The Protozoa.* J. Murray, London, 58 pp.

The Fungi

AINSWORTH, G. C. and BISBY, G. R. (1961). *A Dictionary of the Fungi*, 5th Edn. Commonwealth Mycological Institute, Kew, Surrey, 475 pp.

—————— and SUSSMAN, A. S., eds. *The Fungi.* (1965). Vol. I *The Fungal Cell.* (1966). Vol. II *The Fungal Organism.* (1968). Vol. III *The Fungal Population.* Academic Press, New York.

ALEXOPOULOS, C. J. (1962). *Introductory Mycology.* 2nd Edn. Wiley, New York, 613 pp.

BARTNICKI-GARCIA, S. (1968). Cell Wall Chemistry 1. Morphogenesis and taxonomy of fungi. *A. Rev. Microbiol.*, **22**, 87–108.

BURNETT, J. H. (1968). *Fundamentals of Mycology.* Edward Arnold, London, 546 pp.

BULLER, A. H. R. (1909–1950). *Researches on Fungi.* Vol. I–VI, Longmans, Green, London. Vol. VII, University Press, Toronto.

HAWKER, L. E. (1974). *Fungi: An Introduction.* 2nd Edn. Hutchinson University Library, London, 216 pp.

INGOLD, C. T. (1965). *Spore Liberation.* Clarendon Press, Oxford, 210 pp.

MADELIN, M. F., ed. (1966). *The Fungus Spore*, Colston Papers No. 18. Butterworths, London, 338 pp.

LANGE, M. and HORA, F. B. (1963). *Collins Guide to Mushrooms and Toadstools.* Collins, London, 257 pp.

RAPER, J. R. (1968). *Genetics of Sexuality in Higher Fungi.* Ronald Press Co., New York, 283 pp.

RAPER, K. B. and THOM, G. (1949). *A Manual of the Penicillia.* Williams & Wilkins, Baltimore, 875 pp.

—————— and FENNELL, D. I. (1965). *The Genus Aspergillus.* Williams & Wilkins, Baltimore, 686 pp.

SMITH, G. (1969). *An Introduction to Industrial Mycology.* 6th Edn. Edward Arnold, London, 400 pp.

Protozoa

CHEN, T. T., ed. (1967–1969, Vol. 4 in the press). *Research in Protozoology* (in four volumes). Pergamon Press, Oxford.

CORLISS, J. O. (1961). *The Ciliated Protozoa*. Pergamon Press, Oxford, 310 pp.

GRASSE, P. P., ed. (1952–1953). *Traité de Zoologie*, Vol. 1. *Protozoaires*, Part I (1071 pp.) and Part II (1160 pp.). Masson, Paris.

GRELL, K. G. (1956). *Protozoologie*. Springer-Verlag, Berlin, 284 pp.

HONIGBERG, B. M. and Committee (1964). A revised classification of the phylum Protozoa. *J. Protozool.*, **11**, 7–20.

HUTNER, S. H., ed. (1964). *Biochemistry and Physiology of Protozoa*, Vol. 3. Academic Press, New York and London, 616 pp.

—— and LWOFF, A., ed. (1955). *Biochemistry and Physiology of Protozoa*, Vol. 2. Academic Press, New York and London, 388 pp.

JAHN, T. L. and JAHN, F. F. (1949). *How to Know the Protozoa*. Brown, Dubuque, Iowa, 234 pp. (For identification.)

KUDO, R. R. (1966). *Protozoology*. 5th Edn. Thomas, Springfield, Illinois, 1174 pp.

LWOFF, A., ed. (1951). *Biochemistry and Physiology of Protozoa*, Vol. 1. Academic Press, New York and London, 434 pp.

MACKINNON, D. L. and HAWES, R. S. J. (1961). *An Introduction to the Study of Protozoa*. Oxford University Press, London, 506 pp.

PITELKA, D. R. (1963). *Electron-microscopic Structure of Protozoa*. Pergamon Press, Oxford, 269 pp.

SLEIGH, M. A. (1962). *The Biology of Cilia and Flagella*. Pergamon Press, Oxford, 242 pp.

WARD, H. B. and WHIPPLE, G. C. (1959). *Fresh-Water Biology*. 2nd Edn., edited by W. T. Edmondson. Wiley, New York, 1248 pp. (Chapters 6, 8, 9 and 10 for identification of Protozoa.)

The Slime Moulds

ALEXOPOULOS, C. J. (1963). The myxomycetes II. *Bot. Rev.*, **29**, 1–78.

—— (1966). Morphogenesis in the myxomycetes. In Ainsworth, G. C. and Sussman, A. S., eds. *The Fungi*, 2, Ch. 8, 211–234. Academic Press, New York and London.

ALLEN, R. D. and KAMIYA, N., eds. (1964). *Primitive Motile Systems in Cell Biology*. Academic Press, New York and London, 642 pp.

BONNER, J. T. (1963). Epigenetic development in the cellular slime moulds. *Symp. Soc. exp. Biol.*, **17**, 341–358.

—— (1965). Physiology of development in cellular slime moulds (Acrasiales). In Ruhland, W., ed. *Encyclopedia of Plant Physiology*, **15**, Pt. 1, 612–640. Springer-Verlag, Berlin, Heidelberg and New York.

—— (1967). *The Cellular Slime Moulds*. 2nd Edn. Princeton Univ. Press, 205 pp.

CAVENDER, J. C. and RAPER, K. B. (1965). The Acrasieae in nature; I, II and III. *Amer. J. Bot.*, **52**, 294–296, 297–302 and 302–308.

GRAY, W. D. and ALEXOPOULOS, C. J. (1968). *Biology of the Myxomycetes*. Ronald Press Co., New York, 288 pp.

GREGG, J. H. (1966). Organization and synthesis in the cellular slime moulds. In Ainsworth, G. C. and Sussman, A. S., eds. *The Fungi*, 2, Ch. 9, 235–289. Academic Press, New York and London.

LISTER, A. A. (1925). *A Monograph of the Mycetozoa*. (3rd Edn., by G. Lister.) British Museum, London, 294 pp. and 220 plates.

MARTIN, G. W. and ALEXOPOULOS, C. J. (1969). *The Myxomycetes*. University of Iowa Press, Iowa City, 576 pp.

OLIVE, L. S. (1967). The Protostelida—a new order of Mycetozoa. *Mycologia*, **59**, 1–29.

SHAFFER, B. M. (1962). The Acrasina. In Abercrombie, M. and Brachet, J., eds. *Advances in Morphogenesis*, **2**, 109–182. Academic Press, New York and London.

—— (1963). The Acrasina. In Abercrombie, M. and Brachet, J., eds. *Advances in Morphogenesis*, **3**, 301–322. Academic Press, New York and London.

VON STOSCH, H.-A. (1965). Wachstums- und Entwicklungsphysiologie der Myxomyceten. In Ruhland, W., ed. *Encyclopedia of Plant Physiology*, **15**, Pt. 1, 641–679. Springer-Verlag, Berlin, Heidelberg and New York.

YOUNG, E. L. (1943). Studies on *Labyrinthula*: the etiologic agent of the wasting disease of eel-grass. *Am. J. Bot.*, **30**, 586–593.

The Algae

BONEY, A. D. (1966). *A Biology of Marine Algae*. Hutchinson Educational Press, London, 216 pp.

BOURELLY, P. (1966–1968). *Les algues d'eau douce. Initiation à la systematique.* Tome 1 *Les algues vertes*. 511 pp. Tome 2 *Les algues jaunes et brunes*. 438 pp. N. Boubée et Cie., Paris.

DAWSON, E. Y. (1966). *Marine Botany. An Introduction*. Holt, Rinehart & Winston, Inc., New York, 377 pp.

FRITSCH, F. E. (1935, 1945). *The Structure and Reproduction of the Algae*, Vol. I. 791 pp. Vol. II. 939 pp. Cambridge Univ. Press.

FOTT, B. (1959). *Algenkunde*. Fischer, Jena, 482 pp.

KYLIN, H. (1956). *Die Gattungen der Rhodophyceen*. Lund, Sweden, 673 pp.

LEEDALE, G. F. (1967). *Euglenoid Flagellates*. Prentice Hall, Englewood Cliffs, New Jersey.

ROUND, F. E. (1965). *The Biology of the Algae*. Edward Arnold, London, 269 pp.

SMITH, G. M. (1950). *The Fresh-water Algae of the United States*. McGraw-Hill Co., New York, 719 pp.

—— ed. (1951). *Manual of Phycology*. Chronica Botanica Company, Waltham, Mass., 375 pp.

11
Principles of Microbial Taxonomy

SOME DEFINITIONS

The terms of taxonomy have been subjected to almost as many variant defi-
nitions as those of mathematical statistics—a science with which it has many
points of contact.

In this chapter *classification* means 'the act of arranging a number of objects
(of any sort) into groups (or *taxa*) in relation to attributes possessed by those
objects'. The word *classification* is also applied to the result of any such arrangement. *Taxonomy* (Greek táxis = arrangement; némo = I manage) is concerned, *inter alia*, with definition of the aims of classification, the design of rules
by which arrangements may be achieved, and with the evaluation of the end
results. In brief, taxonomy may be described as the scientific study of classifications, and it is with this that the present chapter is concerned rather than with
the description of any particular classification of micro-organisms.

In biological classifications the primary objects (animals, plants, bacteria)
are usually arranged in groups which are themselves members of larger groups
(and so on) in such a way that any item or any group appears as a member of
only one larger grouping, i.e., the groups are non-overlapping. This method of
classification is the familiar *hierarchical* system which can be conveniently
(but not accurately—see later) represented by a 'family tree' or *dendrogram*.

For ease of reference, the units at each level (taxonomic rank) of a hierarchical
system are given distinctive names or labels—a branch of taxonomy known as
nomenclature. Every student of biology will be familiar with the system of
nomenclature normally used for living organisms, which derives from that used
by the great eighteenth-century taxonomist Linnaeus (Carl von Linné). In this

system the basic unit (the *species*) is given two names—one denoting its membership of a taxon at the rank that we label *genus* (generic name) followed by a second denoting the particular species (specific name). These names are written in a latinized form and constitute a so-called *latinized binomial* (e.g., *Staphylococcus aureus*, *Clostridium tetani*). Taxa of higher rank (families, orders, etc.) are given single latinized names with characteristic endings (e.g., Enterobacteriaceae, family; Eubacteriales, order). The naming of newly discovered organisms or of newly proposed taxa of higher ranks is governed by rigid rules standardized by international agreement.

So familiar is this system that it is perhaps worth emphasizing that it is by no means the only possible one. The simplest system would be merely to label the different types of organism with some sort of catalogue number which referred to a listed description. A much more useful approach might be similar to one proposed for the naming of viruses, viz., the virus is given a group name (probably latinized) which is followed by a descriptive formula akin to that used by botanists in floral diagrams (or to the antigenic formula of a *Salmonella* species, p. 577). This latter method is, in fact, reminiscent of that often used by Linnaeus, who sometimes followed his latinized generic name with up to a dozen descriptive 'specific' epithets.

Ideally, the coining of new names is contrived to convey as much information as possible about the organism or taxon. Unfortunately, both the restriction to latinized binomials and, often, the rules of precedence make this aim difficult to achieve.

Having very briefly stated the areas of taxonomy with which classification and nomenclature are concerned we must consider the meaning of *identification*. In essence, this simply involves the comparison of an 'unknown' object (e.g., a newly isolated bacterium, a collected plant or animal) with all similar objects that are already known. If the 'unknown' object matches up with a 'known' one we say that the former has been *identified*; if not, it may be considered to be a 'new' species, variety, or strain and, when adequately described, is added to the list of 'known' objects. In practice this act of comparison is normally carried out not between two actual objects but between the 'unknown' *isolate* and a recorded description of previously discovered bacteria, plants, animals, etc. The inadequacy of recorded descriptions of many microbial 'species' can sometimes make accurate identification very difficult, if not impossible.

It is not always appreciated that neither *identification* (as defined above) nor *nomenclature* need necessarily be connected with classification. In astronomy the identification of a particular star is made by matching the observed celestial co-ordinates with those recorded in star catalogues for stars previously described. Nomenclature is either a matter of a number in the star catalogue or of a name deriving from its membership of the constellation that suggested shapes to the eyes of early astronomers, e.g., α-centauri. In contrast, the classification of stars is based upon grouping together stars of similar physical nature (e.g., red dwarf, nova, blue giant). It is interesting to note that this sort of classification results in an arrangement in which stars of the same state of development are generally grouped together, i.e., it becomes an *evolutionary* classification—a point which will be taken up in the next section.

THE AIMS OF CLASSIFICATION

If the unsuspecting student of biology is asked 'Why do we classify living organisms?' the chances are high that his reply will contain a reference to 'showing evolutionary relations between the organisms'. So strong has been the influence of evolutionary criteria on taxonomy during the post-Darwinian period that production of such *phyletic* classifications is often thought to be the sole aim of the taxonomist. It is therefore necessary to consider whether other possible aims are valid and, indeed, whether any other approach might lead to classifications of greater value than the purely phyletic.

To do so we must first make the distinction between *special* (or *artificial*) classifications and *natural* classifications. A special classification is one made for a single, defined purpose: it assists in finding the answer to a specific question. A well-known example is the classification of enteric bacteria according to the biochemical differential tests, as used by the water bacteriologists (p. 585). The purpose of this classification is to group together those organisms which may indicate recent faecal pollution of a water supply and to separate these from other similar bacteria which do not have this significance. When a bacterial isolate is identified as falling within a particular group of this system an answer to the question of possible faecal pollution is obtained. A further example is the system of classification, used by medical bacteriologists, which places great weight on the pathogenicity of an organism in separating it from otherwise very similar bacteria, e.g., the anthrax bacillus from 'anthracoid' bacilli such as *Bacillus cereus*; the diphtheria bacillus from other 'diphtheroid' Corynebacteria. The question answered here is whether a fresh isolate is likely to cause disease—a question of paramount importance to the medical bacteriologist.

Such classifications are perfectly valid and perform an important function, but they make no pretence to be natural systems. In special classifications an organism may be separated from its fellows by differing in a single *key* attribute (e.g., toxigenicity) whereas the residue may be grouped under a common taxonomic title (e.g., species name) and yet differ between themselves in several attributes.

What then constitutes a *natural* classification? To answer this question we must first examine what guided the taxonomists during the pre-Darwinian period. We find that the taxonomic logic of this era can be traced back to the ideas of Aristotle; in particular to his Logical Division Theory, which governed the ideas of Linnaeus and held sway up to the beginning of the present century. The basic notion was that organisms (or any other items) should be classified according to their *essential nature*, i.e., according to 'what they really are'. This idea is linked to the Aristotelian notion of the *species infimae*—the ultimate unit of classification which became the basis of the Linnaean *species*. The *species infimae* was rather analogous to the atom of classical chemistry: it was the smallest unit into which more complex groupings could be broken down by repeated division into components. A classification based on such principles would be *sensu stricto* 'natural' but it is easily applied only to the classification of items which are clearly defined, e.g., geometrical shapes: one could construct a *genus* 'triangle' as a plane figure bounded by three straight lines and subdivide this genus into scalene, isosceles, and equilateral *species*. Here the 'essential natures' are known by definition.

When attempts were made to apply this logic to the classification of living organisms taxonomists were faced with two connected difficulties which were really impossible to overcome. The first and most fundamental of these is that Aristotle's principle is one of *deductive* logic and yet taxonomists tried to apply it to situations where only *induction* is possible. We cannot deduce that dogs are different from cats, we can only recognize that they differ on the basis of our observations (because we do not know the essential nature of 'dog' or 'cat'). The second difficulty is that of biological variation which makes the decision of which attributes are more 'essential' than others even more likely to be arbitrary.

Following the publication of Darwin's works on the origin of species, the earlier approach to classification was replaced by one that was thought to be at least equally 'natural', viz., the phyletic system. Once the doctrine of evolution had been accepted it seemed reasonable to argue that organisms of similar 'essential nature' would have shared common lines of descent. The great advantage to taxonomists of the phyletic approach was that speculation about which attributes reflected most accurately the essential natures of organisms was replaced by decisions based on more tangible evidence such as fossil records.

Even so, difficulties still remain. To mention only three: (1) fossil records are seldom adequate; (2) biological variation (both phenotypical and genetical) still poses the problem of the taxonomic level at which organisms are to be separated from each other; (3) the *homology* of various structures or other attributes is often in doubt. This point, which is a consequence of the lack of evolutionary evidence, refers to the identity of attributes which appear to be similar but which may in fact have evolved along quite different lines of descent (*convergent evolution*), e.g., the similar structural form of the truffles (Ascomycetes) and false truffles (Gasteromycetes), or may be due to quite different processes. If we consider micro-organisms in particular with regard to these three points we find that (1) fossil evidence is virtually non-existent; (2) biological variation is great—owing both to the ease with which mutations may be expressed (especially in laboratory environments) and to extreme phenotypic adaptability, e.g., the possession of inducible enzymes and other 'switchable' control systems; (3) the homology of attributes is frequently unknown; to mention one example, we can divide all bacteria into Gram-positive or Gram-negative organisms but there may be more than one mechanism by which cells avoid decolorization.

The problem of convergent evolution and homology raises a question of fundamental importance to the formulation of the aims of natural classifications. Darwin believed that evolutionary convergence was never so great as to obscure ancestral resemblances and that therefore natural taxa were always phyletic. With regard to higher plants and animals, this may well be so; the Biblical grouping of the bat with birds and the whale with fish is clearly a 'special' classification based on only superficial resemblances. The lack of fossil evidence makes it much more difficult, if not impossible, to decide whether apparently similar micro-organisms have evolved from a common ancestral organism or whether convergence, due perhaps to the selective pressures of sharing a similar habitat, has been responsible.

For the sake of illustration, let us suppose that we have two bacterial strains that share a large number of *what appear to be* similar attributes. Let us further suppose that we also know that the lines of evolution of these strains converged from very different origins. Would the objective of a natural classification be best achieved by grouping these strains together on the basis of their mutual overall similarity (*phenetic* classification) or by separating them so as to reflect their different origins (*phyletic* classification)? An argument for the phyletic approach might be that this best reflects the 'essential natures' of the two strains, to which the counter argument might be that, because of convergent evolution, their 'essential natures' have become similar.

Such arguments could be continued indefinitely, were it not possible to approach the question from a different angle, i.e., by asking 'what is the *practical* value of a natural classification?'. The answer to this question is that, to be of maximum use, a natural classification should have good predictive value (information content). This means that if we are told that an organism is classified in Group 'X' of a hypothetical taxonomic scheme then, from our knowledge of the general properties of Group 'X', we should be able to predict that the organism in question *certainly* has features a, b, c, d; *probably* has features p, q, r; and *may* have features x, y, z. When approached from this viewpoint it is possible to describe a natural classification as one which gives the greatest information to the 'general' user. In contrast, a special or artificial classification yields particular information to the specialized user. If we accept this distinction we may now resolve the hypothetical question posed above, since it is clear that the phenetic classification would allow the most *general* predictive properties, whereas the phyletic system would offer information that is primarily of use to the student of evolution, i.e., it is a special classification.

It is possible to see a resemblance between the grouping of organisms on the basis of phenetic classification and the use of statistical parameters in characterizing sets of data. For example, if we are told that a certain set of data is normally distributed about a mean value of 10·0 with Standard Deviation of 0·2 we may predict that in roughly 95 per cent of cases a single datum, randomly selected from the set, would have a value between 9·6 and 10·4, although we should be unable to predict the *exact* value of the datum. This is similar to saying that organisms placed in hypothetical Group 'X' of a phenetic classification would be expected to vary about an 'average' or modal form only between defined limits; although we should be unable to predict the *exact* complete description of a single organism selected randomly from Group 'X'. Indeed, in the systems of Numerical Taxonomy, to be outlined later, it is possible to describe these limits in numerical terms. Again, if the range of bacterial variation were so great that between each 'typical' or modal strain there was an almost continuous gradation of 'intermediate' strains (and some authors have suggested that this may indeed be so) a phenetic classification would still have practical use in much the same way that a histogram may allow us to group—and so handle— what is in fact a continuous spectrum of data.

So useful is this method of classification that we commonly apply it in everyday speech. For example, if we are told that Major Smith is a 'military' type and Dr Jones is a 'donnish' type of person we build up certain mental pictures of these individuals. When we actually meet Major Smith and Dr Jones we are not surprised that they differ in details from our preconceived mental pictures but we should be amazed if they do not share a large number of attributes with our mental 'images'.

MONOTHETIC AND POLYTHETIC CLASSIFICATIONS—THE CONCEPT OF WEIGHT

From what has been written above, it is obvious that the best phenetic classification is one built on comparisons based upon as many attributes as possible. Organisms which share a large number of attributes would *cluster* together to form a 'natural' group and such groups would separate from each other at 'points of rarity', i.e., at combinations of attributes which never, or very rarely, occur. If 'points of rarity' are absent it means that a continuous spectrum of 'intermediate' types of organism exists and the classification is then arbitrary (but could still be useful, as explained above). A phenetic classification based on overall similarity is termed *polythetic*.

The student of biology will certainly have encountered what is termed *monothetic* classification, since this principle is much used in the construction of artificial dichotomous keys for identification of both higher organisms and micro-organisms. The essence of such a system is that certain *key characters* are selected, the possession of which automatically places the organism to be identified into a group which is itself subdivided according to the presence (or absence) of other key characters.

We see immediately that once a key character is selected it assumes great *weight* (importance) in determining the classificatory position of an unknown organism and we should therefore inquire whether we are justified in giving some characters more weight than others. It is obvious that, in principle, the use of key characters could nullify the aims of 'natural' classification outlined above. For example, if a new strain of bacterium were discovered that differed in a single key character from bacteria already classified together in a group, and yet had a large number of characters in common with that group, we should be forced to place the strain in a separate group according to the monothetic system, whereas it would obviously join the existing group in a polythetic system.

It is easy to justify the use of certain key characters in artificial classifications, since they may reflect the very criteria that were used in setting up the classification. For example, a special classification based on the criterion of *pathogenicity* would justifiably separate *Corynebacterium diphtheriae* from closely related 'diphtheroid' bacilli on the sole key character, toxigenicity, which thus takes on over-riding weight. In the case of natural (phenetic) classifications the justification of weighting certain characters is less easy. One possible justification is in cases where we know that certain characters are homologous whereas we are unsure about others. Here we may logically argue that greater weight should be given to the homologous characters in deciding the classification. A second possibility is to argue that more weight should be given to those characters that are strongly correlated with others, a single one of these could then be used as a key character; e.g., a Gram-positive reaction in bacteria usually shows correlation with cell-wall structure, penicillin-sensitivity, sensitivity to basic dyes, etc. Two things follow from this example: First, that the same weighting would be obtained by giving *equal* weight to each of the individual correlated characters, which would then act in concert in influencing the classificatory position. Secondly, that if we eventually found that all of these correlated characters stemmed from a single genetical feature then their weight would disappear since all would be expressions of the same thing; a paradox that has been pointed out by Sneath (1957a). However, we are usually in doubt about the homology of apparently similar characters in micro-organisms, nor do we at present know the precise genetical reasons for observed correlations between characters. There is, therefore, an increasing trend in microbial taxonomy towards the idea that, in our position of ignorance, the best natural classification is one based upon comparison of micro-organisms with respect to as many characters as possible, *each character being given equal weight* in contributing to the grouping and separation of different organisms (i.e., a polythetic system). Once such a classification has been made it is *then* possible to search for key characters which may be of use in a method of *identification*. It is, however, still unlikely that single key characters could be used as in the familiar dichotomous system, rather a set of such characters would have to be examined together in order to narrow down the possible classificatory location of an unknown organism (see, for example, Cowan and Steel, 1965).

The idea of phenetic classification based on characters of equal weight is not new. Indeed this approach stems from one used by the eighteenth-century French botanist, Michel Adanson, and it is now usual to apply the term *Adansonian* to such classifications.

NUMERICAL TAXONOMY

Microbiologists, particularly bacteriologists, have long felt the state of microbial taxonomy to be unsatisfactory. The widely used classification of bacteria (embodied in *Bergey's Manual of Determinative Bacteriology*) is a mixture of phenetic classification (but based on very different numbers of character comparisons in the different groups) and a quasi-evolutionary approach (e.g., the type of flagellation is used in this way by analogy with the classification of protozoa). Moreover, the classification is arranged in the familiar Linnaean hierarchical system and yet it is obvious that the criteria applied to what constitutes a *species* are very different in the different 'genera' (e.g., the serotypes of *Salmonella* are given specific rank, whereas those of the pneumococcus are described as types of a single species (*Diplococcus pneumoniae*). Again, the weighting of certain features results in the classification of some organisms in groups with which they have very little overall similarity (e.g., *Corynebacterium pyogenes*).

These criticisms do not indicate that the present system is useless (which is certainly not true—indeed, it has been used in other chapters of this book) but rather that a more uniform approach based on Adansonian principles would almost certainly be more self-consistent and therefore a better natural classification. One disturbing aspect of the present system is that if a group of bacteria is re-examined across a set of criteria (characters) completely different from those already employed in making the existing classification, it is possible that the classification may have to be radically altered in order to accommodate the new information. This instability is unlikely to be a feature of the Adansonian approaches described below (see later).

During the past decade an increasing number of publications have appeared concerned with what is now known as *Numerical Taxonomy*. This branch of the subject owes much to the availability of high-speed digital computers, and interest in its application to bacterial classification was stimulated by the pioneer work of P. H. A. Sneath. Normally the term *Numerical Taxonomy* is applied to systems of classification which are basically Adansonian but in which the degree of similarity of organisms is assessed in quantitative, rather than merely qualitative, terms. There are many advantages in having some numerical estimate of the degree of phenetic similarity or difference between a pair of organisms, of which the most obvious is that it can provide a rational basis for deciding the levels of taxonomic rank. It has already been shown by Sneath's studies that there is at least as much difference between the 'species' of certain genera as there is between the 'genera' themselves.

The techniques of numerical taxonomy

A detailed discussion of these techniques will not be attempted here, and the interested reader is referred to the publications listed at the end of this chapter. What follows is an outline intended to illustrate the general principles.

There are several distinct approaches to Numerical Taxonomy, but all start by:
1. Collecting the organisms, or groups of organisms, to be compared, which are now known as *Operational Taxonomic Units* (OTUs).
2. Observing these OTUs for presence or absence (or quantity) of a large set of *characters*.
3. Drawing up of a table of OTUs *versus* characters.

A character is usually defined as an attribute about which a single statement can be made, e.g., 'present' or 'absent' or some quantitative measurement. It is important to give careful thought to what constitutes a single character before drawing up the OTU × character table. Some attributes are obviously not proper characters, e.g., the number of the OTU in the collection. Other apparent characters may not be permissible because they are redundant, i.e., are expressions of an already listed character. For instance, if an OTU ferments both glucose and sucrose with the formation of acid and gas this may generate three distinct characters, viz., Acid from glucose; gas from glucose; sucrose fermented. It is improper to score 'gas from sucrose' as a separate character if we know that the fermentation of sucrose involves an initial hydrolysis to glucose, which is subsequently fermented to acid and gas.

Furthermore, it is essential to the principle of Numerical Taxonomy that each of the OTUs should be examined across the complete set of characters, so that true comparisons may be made. Care must be taken, however, not to make comparisons that are illogical. Suppose that one OTU ferments glucose to acid and gas, whereas a second OTU does not ferment glucose at all. In the case of the first OTU we may score a positive character for each of the attributes; acid from glucose; gas production. However, with regard to the second OTU we may score a negative character for lack of production of acid from glucose, but it is now illogical to score a result for 'gas production' since this depends on the prior formation of formic acid which we have already noted as absent. We therefore score 'No Comparison' (NC) for gas production by OTU number 2, which means that this character cannot be used when comparing the similarity of OTU number 2 with any other OTU.

Further questions are prompted by practical considerations, such as:

1. Since observation of characters is necessarily carried out under the artificial conditions of the laboratory, can we make a true comparison of micro-organisms which might behave differently in their natural environment?
2. If we have among the OTUs some organisms that can carry out certain reactions at one temperature of incubation but not at a higher in comparison with organisms that can carry out the same reaction only at the higher temperature, should we then use different temperatures of incubation in order correctly to characterize the different OTUs?

The answer to the first question is that a comparison of micro-organisms under laboratory conditions (1) is the 'best we can do' and (2) according to our practical definition of a 'natural' classification, is satisfactory because other investigators will be observing the micro-organisms under similar conditions. The answer to the second question is more difficult. If there are many temperature-sensitive reactions, we may bias the comparison of OTUs, compared under standard conditions, towards an emphasis of dissimilarity when the temperature-sensitivity may be due to only a few underlying causes (which we do not know).

In our position of ignorance of the complete genetical and biochemical bases of observed characters it is generally considered best to compare OTUs over a *rigidly standardized* set of tests. Although it is almost inevitable that certain of these conditions will introduce bias when measuring the degree of similarity between pairs of OTUs, this course of action is adopted for two chief reasons: (1) practical expediency; (2) if sufficient characters are observed the bias should be 'diluted out' in much the same way that an arithmetic mean is not greatly affected by a few aberrant data, especially when they occur on both sides of the mean. Of course, tests should not be used to generate characters when it is *known* that bias is inherent in the test condition. For example, we may adjust the sensitivity of a test for urease production so that it is read as positive only with those *Enterobacteriaceae* that we call *Proteus* spp. To use this test in a phenetic comparison of *Enterobacteriaceae* would obviously introduce bias, since we have prejudged the issue by distinguishing certain species as urease-positive beforehand. Such a test, however, would be perfectly valid if applied to unrelated organisms.

The kinds of morphological, structural, and metabolic attributes commonly used as classificatory characters are typified by those which appear elsewhere in this book in descriptions of the various micro-organisms. More recently attention has turned to other potentially valuable sources of characters. These include (1) cell-wall chemistry, (2) electrophoretic studies on esterases and other soluble proteins, (3) infra-red adsorption spectra, (4) DNA base composition, and (5) gas chromatography of cell pyrolysis products (see references at end of chapter).

From this discussion it is clear that comparisons of OTUs based on a large number of characters are likely to be more accurate (free from bias) than comparisons based on only a few characters (we have already discussed the bias that can be caused by the use of a single key character). *How many* characters should we observe? Guide-lines to the answer may be obtained from elementary probability theory, which tells us that we are most likely to succeed in distinguishing different organisms when the number of characters is of the same order as the number of OTUs, and that we should have limited confidence in an *S*-type Similarity Coefficient (see below) calculated on the basis of less than 50 characters.

A special difficulty may exist when an attempt is made to compare organisms that have very different growth-rates under standardized conditions (e.g., pathogenic and saprophytic *Mycobacteria*). Here there is clearly the possibility of bias due to comparison of characters that depend on metabolic rate when similarities are calculated after an incubation period that is suboptimal for the slower-growing strains. We may either incubate all strains so that the reactions of the slowest grower are realized—when difficulties may arise due, for example, to alkaline reversion in carbohydrate fermentation tests with the fast-growing strains, or we may have recourse to special methods of calculation that attempt to separate effects due to *Vigour* (growth-rate) from that due to *Pattern* (Sneath, 1968).

Methods of comparing similarities

After an OTU × character table has been compiled, all possible pairs of OTUs are compared and their similarities computed. There are three basic methods by which measures of similarity may be computed, only one of which has been much applied to micro-organisms. These are:

1. Correlation coefficients.
2. Measures of taxonomic distance.
3. Similarity coefficients (S).

The first two methods have the advantage that characters which are expressed as quantitative data may be more or less directly incorporated into the calculations of similarity.

The correlation coefficients are closely related to the commonly used statistic r, which expresses the degree of correlation between two sets of bivariate data and can vary from $+1$ (absolute correlation), through 0 (no correlation at all), to -1 (absolute negative correlation). Thus two organisms that were absolutely identical in all characters studied would generate a coefficient of $+1$, two organisms that were absolutely opposite in every character (if this were possible) would generate a coefficient of -1, whereas a coefficient of 0 would indicate no correlation of the characters of the first organism with those of the second.

Measures of taxonomic distance attempt to plot the relative positions of the OTUs in multi-dimensional space (one dimension for each character studied) in such a way that if two OTUs were identical their *mean taxonomic distance* would be 0 whereas if they were absolutely dissimilar their mean taxonomic distance would be $+1$.

However, it is the similarity coefficients (S) that have found most application in recent studies of microbial classification mainly owing to the ease with which they can be computed and the results handled in subsequent stages of the classification. These S coefficients require that the character data must be coded in binary form, i.e., 1 $(+)$ for the possession of a character, 0 $(-)$ for the absence of a character, and NC for 'No Comparison'. It follows that quantitative data must be broken down into a set of single characters, and there are two chief methods of doing so, viz., the additive and the non-additive methods. Suppose that we have three OTUs one of which produces no penicillinase, a second produces a small quantity of the enzyme, and the third a large amount under comparable conditions, i.e.,

OTU	Penicillinase
A	−
B	+
C	+ +

In the additive method of coding we may decide as follows:

OTU	Character a	b	c
A	−	NC*	NC*
B	+	+	−
C	+	+	+

Here character a codes for presence or absence of the enzyme, b codes for production of a small amount, and c for an *additional* amount. However, because we cannot distinguish $a+b+$ from merely $a+$ we should probably delete character b altogether since it contributes no additional information.

The same data coded by the non-additive method gives:

OTU	Character a	b	c
A	−	NC	NC
B	+	+	−
C	+	NC	+

Here character a codes for 'production of penicillinase', b codes for 'production of + penicillinase', and c codes for 'production of + + penicillinase' in a non-additive fashion. OTU C must therefore be scored NC for b since production of a + + quantity would mask production of a lesser amount. The reader

* Sneath (1962) scores these entries as negative in order to preserve the quantitative gradation but admits to logical redundancy.

may have noticed that here again character *b* does not give any additional information to that provided by character *a*; accordingly, character *b* would be deleted with the result that, in this simple example, the results of codings are identical by the two methods, viz.,

| OTU | Character | |
	a	*c*
A	−	NC
B	+	−
C	+	+

However, if we consider a fourth OTU (D) that produces an even larger amount of enzyme (+ + +) we should obtain:

OTU (D)	*a*	*c*	*d*
Additive method	+	+	+
Non-additive method	+	NC	+

In general, the difference between the two methods increases as the number of characters allotted to the quantitative data increases. Since the additive method generates a greater number of comparisons it tends to over-emphasize differences which could be due to differences in growth-rate, etc. (i.e., vigour), and so tends to bias the *S*-value in the direction of dissimilarity. For this reason the non-additive approach is generally preferred.

Once the OTU × character table has been drawn up it is possible to represent the comparison of a pair of OTUs thus:

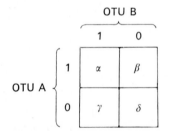

where α represents the total number of characters for which both A and B are scored +, β represents the total number of characters for which A is scored + but B is scored −, and so on. Thus α and δ represent the number of characters on which A and B are scored similarly, whereas β and γ represent the number of un-matched characters. 'No Comparisons' are ignored in making these entries. Such tables can be drawn up for all possible pairs of OTUs.

There are two chief ways in which similarity coefficients have been calculated for application to microbial classification. One, known as S_{SM}, includes both positive and negative *matches* in calculating the degree of similarity, thus:

$$S_{SM} = (\alpha + \delta)/(\alpha + \beta + \gamma + \delta)$$

The other, known as S_J, bases the comparison only on the positive matches, thus:

$$S_J = \alpha/(\alpha + \beta + \gamma)$$

The point at issue in the choice between the two methods is whether two 'absences' is a valid criterion of similarity. In general, S_{SM} is currently favoured on the grounds that for many qualitative characters the coding as '+' or '−' is arbitrary. For example, penicillin-sensitivity may be scored as either '+' or '−' according to whether one thinks of resistance as an active or passive phenomenon. The danger in including negative matches is that it is possible to bias values of S towards excess similarity by choosing a large number of features which the organisms do not possess. However, this applies also to some positive characters and here again it is hoped that introduced biases are 'smoothed out' by observing a sufficiently large number of characters. It is usual to delete as redundant any character which is *uniformly* positive or negative (apart from NC entries) for all OTUs under study, otherwise bias towards excess similarity would certainly occur.

It is obvious that both forms of S may vary from 0·000 (absolutely no matches) to 1·000 (100 per cent matches). Moreover, the dependence of S on the number of matches is absolutely linear, e.g., if on the basis of 100 characters two OTUs were 100 per cent similar ($S = 1·000$) a third OTU which had a single mismatch with either of the former would drop its value by 1 per cent ($S = 0·990$). This feature constitutes one of the large advantages of the similarity coefficient (particularly S_{SM}) over the other methods of comparison outlined above: it is possible to grasp the meaning of differences between S-values very easily.

When S-values have been calculated for all possible pairs of OTUs (and here the contribution of the high-speed computer is evident) they are tabulated in a *similarity matrix*. This is a table of OTUs × OTUs, which is symmetrical about its principal diagonal, since the S-value between OTUs A and B is obviously the same as that between B and A. The values on the principal diagonal are all 1·000, since these consist only of self-comparisons. The similarity matrix is therefore usually recorded in a triangular form, omitting these redundant entries.

At this point it may be helpful to introduce a very simple hypothetical example where five OTUs are compared over only ten characters.

OTU × Character Table

| | OTU | | | | |
Character	A	B	C	D	E
1	1	1	0	0	0
2	1	1	0	0	0
3	1	0	0	0	0
4	0	0	1	0	1
5	1	1	1	1	1
6	0	0	1	1	1
7	1	1	0	0	1
8	1	1	0	1	1
9	0	0	1	0	1
10	0	0	0	1	1

Similarity Matrix

Values of S_{SM}

OTU	A	B	C	D	E
A					
B	0·9				
C	0·2	0·3			
D	0·4	0·5	0·6		
E	0·3	0·4	0·7	0·7	

Cluster analysis

After numerical estimates of the degrees of similarity between all possible pairs of OTUs have been generated, the next step is to form the groups (or clusters) which are the basis of the final classification. When using S coefficients there are three main ways in which this operation, known as cluster analysis, may be tackled:

1. Single linkage
2. Average linkage
3. Total linkage.

The method that has been most applied to microbial classification is that of *single linkage*. Although it has certain disadvantages (see below) its ease of computation and manipulation makes the method eminently suitable, at least for preliminary studies. Its use may be illustrated by reference to our simple example.

First, the similarity matrix is scanned at a high level of S and the pairs of OTUs that have mutual S-values *at least as great* as the scan level are listed. Suppose we begin by scanning at a level of $S = 1.000$ (absolute similarity), no such values appear in our example above. We next decrease the scan level by an arbitrarily selected amount that has to be chosen by reference to the scatter of S-values actually obtained (or to some other criterion). In our example a decrement of 0·2 (20 per cent) would seem suitable. Thus the next scan level becomes $S = 0.8$ and we obtain a single pair of OTUs.

Level	OTU-pairs
$S = 0.8$	A,B ;

Decreasing by a further amount of 0·2 we list further entries:

Level	OTU-pairs
$S = 0.6$	A,B ; C,D ; C,E ; D,E ;

At this level of scan the principle of clustering by single linkage can be applied; i.e., OTU-pairs are fused to form a single cluster if any one OTU of one pair has an S-value at least as great as the scan level with any one OTU of a second pair (or of an already existent cluster). To return to the example, we see that the last three OTU-pairs satisfy this criterion and fuse into a single cluster, whereas the pair A,B remains isolated:

Level	Clusters
$S = 0.6$	A,B ; C,D,E ;

Proceeding, we obtain:

Level	Clusters already formed	New OTU-pairs
$S = 0.4$	A,B ; C,D,E ;	A,D ; B,D ; B,E ;

The new OTU-pairs fuse into a single cluster (A, B, D, E;) by the criterion of single linkage, but this cluster has elements in common (at $S = 0.4$) with the two existing clusters. Therefore the five OTUs form into a single group at $S = 0.4$ and the clustering process ends.

It is now possible to represent the results of clustering by means of a *dendrogram*, or 'family tree', resembling that of the usual hierarchical classifications.

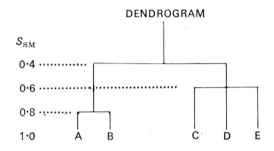

DENDROGRAM

Although this form of representing the results of a cluster analysis is exceedingly useful, it is relevant to point out two distortions inherent in it. One is the fact that the points of fusion of branches of the dendrogram are shown as occurring at single levels of S, whereas the actual S-values causing the fusion occur anywhere between the limits set by the arbitrarily chosen decrement. The second is that a true spatial representation of the relations between the various OTUs and Clusters would require multi-dimensional space; distortion is therefore inevitable in a two-dimensional dendrogram.

Nevertheless, the method allows a tentative classification of the OTUs having the great advantage of being based on numerical estimates of the levels at which differences and similarities appear. At what level we decide to label members of a cluster 'strains', 'species', 'genera', and so on (or to abandon these terms) is still a matter of choice and agreement, but we now have a numerical 'yardstick' to guide us in this decision.

The method of clustering by single linkage has an inbuilt disadvantage which could make for unsatisfactory grouping. Suppose the cluster A, B, C, D formed because A linked with B, B with C, and C with D. It is evident that A might be quite dissimilar from D and yet would still be clustered with it. In fact, it is easy to show that if we know $S_{A,B}$ and $S_{A,C}$ (where these are S_{SM}-values) then $S_{B,C}$ may have a *minimum* value equal to

$$|1 - (S_{A,B} + S_{A,C})|$$

When $S_{A,B} = S_{A,C} = 0.5$, $S_{B,C}$ can be as low as zero, as is obvious from the following example:

		OTU	
Character	A	B	C
1	1	1	0
2	1	1	0
3	1	1	0
4	1	0	1
5	1	0	1
6	1	0	1

Fortunately, in practice good results are commonly obtained in spite of this potential snag and a method is available that allows a check on the occurrence of serious distortion due to single linkage. In order to understand the nature of this check it is necessary to consider what is meant by *mean similarity*. Mean similarity may be computed either between the members of a single cluster (i.e., within-cluster mean) or between the members of two separate clusters (between-cluster mean).

The within-cluster mean represents the *average* similarity shown between all possible pairs of OTUs within the cluster. Thus, in our example, the cluster C, D, E was formed at $S = 0.6$. The S-values to be utilized in calculating the within-cluster mean for this example are:

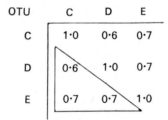

OTU	C	D	E
C	1.0	0.6	0.7
D	0.6	1.0	0.7
E	0.7	0.7	1.0

Two forms of the within-cluster mean may be obtained. The 'square' mean (Γ mean) is the average of all 9 values in the square matrix shown above, i.e., $\Gamma = \frac{7}{9} = 0.\dot{7}$. The 'triangle' mean ($\Delta$ mean) ignores the redundant comparisons and the self-comparisons, and is therefore the mean of the 3 values in the triangle, i.e., $\Delta = 2.0/3 = 0.\dot{6}$. The two sorts of within-cluster mean bear a simple relation to each other: Δ is less than Γ, but the two become similar as the number of OTUs in the cluster increases.

If we compare the mean values obtained above with the level of S at which the cluster was formed ($S = 0.6$) we see that the means are greater than the clustering level. This indicates that the cluster is homogeneous with respect to the mutual similarities between the individual members. If OTUs had been included by single linkage that showed low levels of S with some existing members of the cluster, i.e., if the cluster had become heterogeneous, then the within-cluster mean would have been depressed below the clustering level by an amount dependent upon the degree of heterogeneity. It is this feature that provides a check on the validity of single linkage methods of analysis.

The between-cluster mean has only one form of computation. Here each

OTU in the first cluster must be compared with each OTU in the second cluster. In the example two clusters exist at $S = 0.6$:

1. A, B;
2. C,D,E;

The between-cluster mean is obtained from the rectangular matrix of S-values:

		Cluster (1)	
		A	B
	C	0·2	0·3
Cluster (2)	D	0·4	0·5
	E	0·3	0·4

Here there are no redundancies and the between-cluster mean is $2.1/6 = 0.35$: an average measure of the degree of similarity between the two clusters.

Between-cluster means may themselves be used as a basis for clustering: the so-called method of *average linkage* referred to above. The essence of this approach is that, at each level of clustering, individual OTUs join existing clusters, and existing clusters fuse together, only if the *mean* similarity between the OTU and its potential cluster, or the *mean* similarity between two clusters, is at least as great as the chosen level of S. This approach largely removes the danger, inherent in the single-linkage method, of creating clusters which appear to be more homogeneous than they really are; the check on the within-cluster mean may be incorporated as an additional safeguard.

There are a number of different techniques that have been used to apply the method of average linkage to classification studies but all of them require more labour, and more skilful computer programming, than does the method of single linkage—often without producing a very different result.

The method of *total linkage* represents a further extension of the attempt to ensure homogeneous clusters. In this approach the criterion of linkage is that an OTU is allowed to join a cluster only if it has the required level of S with *each* existing member, and two clusters fuse only if *each* member of the first cluster has the required level with *each* member of the second. This approach has been little used in microbial classification.

The matches hypothesis

The advantage of having a *numerical* estimate of similarity for use as a guide in making decisions on classification has already been stressed. Numerical (Adansonian) Taxonomy offers a second substantial advantage over methods that rely on qualitative, or on arbitrarily weighted, judgments. This is embodied in the *matches hypothesis*, which supposes that there is some true measure of similarity which could be computed if every possible character could be taken into account, and that the deviation from it of an actual calculated S-value (based on a 'sample' of all the possible characters) will be accounted for by sampling error. Thus a second estimate of S made between the same pair of OTUs, but based on an independent set of characters, should tend to give a value similar to that first obtained, i.e., estimates should be self-consistent. This notion is similar to that used in mathematical statistics where estimates of the true mean (μ)

of a Normally Distributed population, obtained from the observed means ($\bar{x}$) of randomly selected samples, cluster around μ in a manner that is predicted by the sampling error (variance).

With regard to S-values the matches hypothesis seems to be borne out in practice, and the sampling error is approximated by the prediction of the Binomial Distribution:

$$\text{Standard deviation of } S \approx \sqrt{[S(1 - S)/N]}$$

Here S is taken as the probability of occurrence of a 'match' and N is the member of comparisons (characters) observed.

The advantage of self-consistency is that further studies carried out on groups of organisms already classified according to the principles outlined above are unlikely to necessitate radical changes in classification—a property that is not true for a number of existing classificatory schemes, where a new study may dictate substantial re-arrangement of taxa.

Applications of numerical taxonomy to micro-organisms

During the past decade various investigators have applied Numerical Taxonomic methods to different groups of micro-organisms. These include: *Chromobacterium, Bacillus, Micrococci, Streptococci, Corynebacteria, Mycobacteria, Basidiomycetes*, and root-nodule bacteria—to mention but a few.

The results of these studies tend, in general, to confirm the prediction of the matches hypothesis, i.e., where the existing classification has been largely phenetic and based on many characters it is confirmed, with minor deviations, by the numerical study. However, even in these cases the great advantage of having some sort of quantitative criterion on which to base points of separation and combination is evident. In examples where the existing classification has been biased by reliance on a few weighted characters the numerical studies have shown up discrepancies. For instance, in a study of pigmented bacteria made by Sneath (1957b) it was found that the S-value between the species *Chromobacterium violaceum* and *Chromobacterium lividum* was similar to that obtained between those *species* and the genera *Serratia* and *Pseudomonas*. This finding would suggest that perhaps these genera should be 'demoted' to specific rank in a rational classification.

It will be obvious from the outline of Numerical Taxonomy given above that an overall classificatory study on micro-organisms in general can be carried out only by actually comparing representative organisms over a wide range of characters. The problems of data-collection and of computation make this a formidable task and the studies so far have been largely confined to more or less well-defined groups of micro-organisms. It is not entirely satisfactory to use the usual recorded descriptions as a source of data for Numerical Taxonomic studies. Often the characters recorded for the different organisms—even within a classificatory group—either do not belong to the same set, or are incomplete for any one organism, or have been obtained under different conditions. Moreover, the descriptions often record a result as 'variable' or show a range when it is the actual responses of representative organisms that are important.

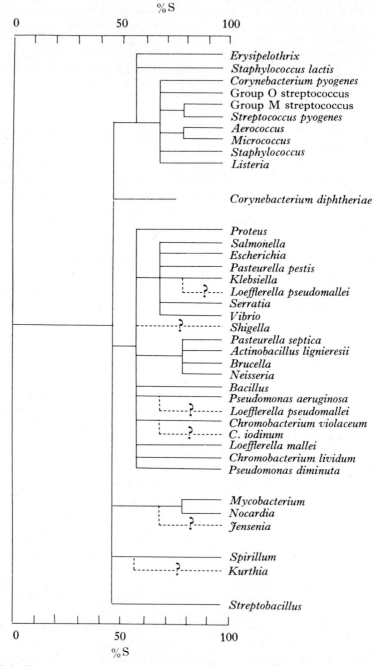

Fig. 11.1 Taxonomic dendrogram of a wide range of bacteria. (From Sneath and Cowan (1958). *J. gen. Microbiol.*, **19**, 551–565 ; also cited in Symposium (1962) p. 34, cited in references)

Nevertheless, some attempts have been made to gain an idea of how Numerical Taxonomy compares with existing wide classifications by using published data. An example for bacteria is shown, in the form of a dendrogram, in Fig. 11.1. Here, three main groups can be distinguished; the Gram-positive cocci, the Gram-negative rods, and the 'Actinomycetales'. When examined in detail, however, various examples of divergence from accepted classification become evident, e.g., *Corynebacterium pyogenes* clusters with the Gram-positive cocci more closely than with *Corynebacterium diphtheriae*; *Proteus* is as different from the *Salmonella-Escherichia* group as it is from *Bacillus*.

Although at present microbiologists will continue to use existing classifications in order to make possible communication of information, nevertheless the increasing interest that is being shown in Numerical Taxonomic studies gives promise of a more consistent and more rational (and, therefore, more generally useful) scheme of microbial classification. Indeed, as has been remarked by A. J. Cain (Symposium, 1962), just *because* it is unencumbered with fossil evidence and a sexual (interbreeding) definition of species, and because many characters can be easily studied, there is no reason why bacterial classification should not become the most firmly based of all taxonomies.

BOOKS AND ARTICLES FOR REFERENCE AND FURTHER STUDY

General

AARONSON, S. and HUTNER, S. H. (1966). Biochemical markers and microbial phylogeny. *Q. Rev. Biol.*, **41**, 13.

BURGES, A. (1955). Problems associated with the species concept in mycology. In J. E. Lonsley, ed. *Species Studies in the British Flora*. Buncle, Arbroath.

COWAN, S. T. (1965). Principles and practice of bacterial taxonomy—a forward look. *J. gen. Microbiol.*, **39**, 143–153.

—— (1970). Heretical taxonomy for bacteriologists. *J. gen. Microbiol.*, **61**, 145–154.

—— and STEEL, K. J. (1965). *Manual for the Identification of Medical Bacteria*. Cambridge Univ. Press, 217 pp.

SAVILE, D. B. O. (1955). A phylogeny of the Bascidiomycetes. *Can. J. Bot.*, **33**, 60–104.

SNEATH, P. H. A. (1957a). Some thoughts on bacterial classification. *J. gen. Microbiol.*, **17**, 184–200.

—— (1957b). The application of computers to taxonomy. *J. gen. Microbiol.*, **17**, 201–226.

—— (1968). Vigour and pattern in taxonomy. *J. gen. Microbiol.*, **54**, 1–11.

SOKAL, R. R. (1966). Numerical taxonomy. *Scient. Am.* **215**, No. 6, 106–116.

—— and SNEATH, P. H. A. (1963). *Principles of Numerical Taxonomy*. Freeman, San Francisco, 360 pp.

12th Symposium of the Society for General Microbiology (1962). *Microbial Classification*. G. C. Ainsworth and P. H. A. Sneath, eds. Cambridge Univ. Press, 483 pp.

Special taxonomic criteria

BARTNICKI-GARCIA, S. (1968). Morphogenesis and taxonomy of fungi. *A. Rev. Microbiol.*, **22**, 87–108.

BOOTH, G. H., MILLER, J. D. A., PAISLEY, H. M. and SALEH, A. M. (1966). Infrared spectra of some sulphate-reducing bacteria. *J. gen. Microbiol.*, **44**, 83–87.

COLMAN, G. and WILLIAMS, R. E. O. (1965). The cell walls of streptococci. *J. gen. Microbiol.*, **41**, 375–387.

GOTTLIEB, D. and HEPDEN, P. M. (1966). The electrophoretic movement of proteins from various *Streptomyces* species as a taxonomic criterion. *J. gen. Microbiol.*, **44**, 95–104.

HAWKES, J. G., ed. (1968). *Chemotaxonomy and Serotaxonomy*. Academic Press, New York and London, 299 pp.

LUND, B. M. (1965). A comparison by the use of gel electrophoresis of soluble protein components and esterase enzymes of some group D streptococci. *J. gen. Microbiol.*, **40**, 413–419.

MANDELL, M. (1966). Deoxyribonucleic acid base composition in the genus *Pseudomonas*. *J. gen. Microbiol.*, **43**, 273–292.

PUNNETT, T. and DERRENBACKER, E. C. (1966). The amino acid composition of algal cell walls. *J. gen. Microbiol.*, **44**, 105–114.

REINER, E. and EWING, W. H. (1968). Chemotaxonomic studies of some Gram-negative bacteria by means of pyrolysis—gas–liquid chromatography. *Nature, Lond.*, **217**, 191–194.

ROSYPAL, S., ROSYPALOVÁ, A. and HOŘEJŠ, J. (1966). The classification of micrococci and staphylococci based on their DNA base composition and Adansonian analysis. *J. gen. Microbiol.*, **44**, 281–292.

Applications of numerical taxonomy

KENDRICK, W. BRYCE and WERESUB, L. K. (1966). Attempting neo-Adansonian computer taxonomy at the ordinal level in the Basidiomycetes. *Systematic Zoology*, **15** (4), 307.

References to the application of Numerical Taxonomy to various groups of micro-organisms may be found in *J. gen. Microbiol.* from 1957 onwards.

12
Microbiology of Soil

INTRODUCTION

Early work in soil microbiology was almost entirely confined to the study of bacteria. The pioneer work of Beijerinck and Winogradsky in the late nineteenth century on bacteria involved in the transformation of sulphur and nitrogen compounds provided explanations for some of the chemical changes which were known to take place in the soil and to be related to soil fertility. These studies were the forerunners of many concerned with the soil bacteria and the processes they mediated, but it was not until about thirty years later that comparable interest began to be shown in the fungal flora and its activities. The microfauna was comparatively neglected for even longer and only recently has interest in the soil algae increased.

Much work which has been done on soil bacteria involves isolation, characterization, and establishment of the *in vitro* capacity of the organisms to mediate processes occurring in the soil. Similarly fungi, algae, protozoa and some small invertebrates from the soil have been isolated and identified but their biochemical capacity has been studied to a lesser extent. Following on and to some extent simultaneously with this work, the numbers and, later, the distribution of the soil population were established. Another approach was that of physiological studies on soil as a whole; processes, such as nitrification, taking place within it were followed.

The accumulated knowledge gained from these studies strongly indicates that particular organisms are associated with particular processes but there is no direct evidence of this, and now there is a resurgence of interest in the establishment of the roles of various soil organisms. The technical problems in achieving this could be immense, since an ideal method should be capable of giving a complete description of an environment, which measures only a few microns in each direction, without disturbing it in any way.

Meanwhile, the problem of world food supply is pressing and, therefore, factors affecting fertility of the soil, on which all agricultural processes are based, are of immediate practical importance. Carbon and nitrogen transformations have been studied intensively with respect to this as, to a lesser extent, have those involving compounds of other elements such as phosphorus, manganese, and iron. The interactions of environmental and biotic factors which control the rate and direction of these processes, and hence crop yields, are also intensively studied. Another approach to increased agricultural production is the use of various synthetic organic chemicals for the control of pests and this introduces certain problems. In particular, the problem which faces soil microbiologists is that of persistence in the soil of some of these compounds which are toxic.

THE SOIL ENVIRONMENT

Soil consists of mineral particles and organic residues which are, to a greater or lesser extent, wet. The interstices of these materials are usually occupied by a gaseous mixture but when the soil is waterlogged they become full of water.

The mineral particles are derived from the parent rock by weathering, which includes physical processes such as the action of heat, rain, running water, ice, and blown sand. According to the nature of the parent rock, chemical processes such as oxidation and solution by weak acids, arising from carbon dioxide dissolved in water, also play a part. The size of the resultant particles ranges from that of clay to gravel (see Table 12.1) and the whole range of sizes may be found

Table 12.1   Size of soil particles (mm).

Clay	< 0.002
Silt	$0.002–0.02$
Fine sand	$0.02–0.20$
Coarse sand	$0.20–2.00$
Gravel	> 2.00

in one soil, or at the other extreme, perhaps as a result of sorting by water, particles of only one size may be found.

Organic residues from plant and animal sources are added to these mineral particles. The size of these additions varies greatly from one situation to another as does the state of decay of the residue when it reaches the soil, but this diverse organic matter is incorporated into the soil by the action of soil organisms and, if it is not first oxidized, is converted to humus, the dark, amorphous colloidal material (p. 495) which plays such an important part in soil processes. The result of these physical and chemical processes is not a homogeneous mixture of mineral and organic particles, considerable variation in the soil constitution occurring at points separated by only very short distances. The mineral particles are bound together by organic material and form soil crumbs. These aggregates are extremely important for soil fertility since they increase the pore space essential for the good drainage and aeration necessary for high levels of microbial activity. The volume occupied by the various soil constituents is illustrated for one type of soil in Table 12.2.

Table 12.2 Approximate percentage of the total volume occupied by various components of a fertile soil.

Mineral material	51%
Pore space	40%
Organic material	9%
Organisms	<1%

The main source of soil water is precipitation, and after heavy rain the pores will be full, but in a well-structured soil much of the water will drain away under the influence of gravity leaving water only in the capillary spaces and bound to inanimate soil constituents. The gravitational and capillary water is available to soil organisms and plays an important part in solute transport, whereas the bound water is not available and does not function in solute transport.

The soil atmosphere differs from the air above the surface in that there is usually substantially more carbon dioxide and less oxygen in the soil gaseous phase. When gaseous exchange is hindered by permanent waterlogging, anaerobic conditions develop and not only carbon dioxide but also methane, hydrogen, and hydrogen sulphide may accumulate.

In addition to the point-to-point variation in soil constitution at a particular level it can readily be seen with the naked eye that the soil overlying the parent rock exists in distinct layers or horizons. A pit dug in the soil will reveal a profile of these horizons (see Fig. 12.1).

The upper layers, containing most of the organic matter and in which the larger part of the soil microflora and fauna is found, are referred to as the A horizons. The B horizons contain little organic matter and organisms are sparse. The C horizon consists of the parent mineral material of the soil with only minute amounts of organic matter. This description is much simplified and in fact the horizons are often subdivided, for example, into the A_0, A_1, and A_2 horizons and in some soil types the organic matter may be distributed in such a way that the B_1 horizon contains more organic matter than the A_2.

The chemical nature of a soil affects the type and size of its contained population. The soil also acts as a buffer reducing the range of physical conditions as compared with those found on and above its surface and its effectiveness as a

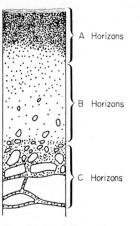

A Horizons

B Horizons

C Horizons

Fig. 12.1   Soil profile.

buffer will also depend on its chemical composition and physical make-up. The more fertile soils on the whole ensure a more constant supply of water and nutrients than do infertile ones. However, even in fertile soils adjacent habitats separated by only a few microns may be extremely favourable for growth and reproduction of micro-organisms on the one hand and completely inhospitable on the other.

THE SOIL INHABITANTS

It is impossible to assess accurately the soil population partly because of sampling problems and partly because of inadequacies of technique (see p. 487). Nevertheless, in Table 12.3 some estimate of the numbers in the various main categories of micro-organism is given together with an estimate of the biomass of each of the groups. The biomass value is possibly more meaningful than

Table 12.3   Numbers and biomass of some organisms in the top 15 cm. of agricultural soil.

	Number of organisms per gram	Biomass g/m^2
Bacteria	9.8×10^7	160
Actinomycetes	2.0×10^6	160
Fungi	1.2×10^5	200
Algae	2.5×10^4	32
Protozoa	3.0×10^4	38
Nematodes	1.5	12
Earthworms	0.001	80

numbers alone because it takes into account the mass of organisms and hence possibly gives a more realistic indication of the activity of a particular group of organisms in the soil. The data in Table 12.3 have been assembled from a number of sources and hence are not strictly comparable but they nevertheless give a useful indication of the relative abundance of the main groups of soil organisms.

It can be seen that although bacteria are more numerous than any other group of organisms the biomass of the fungi is larger than that of the bacteria. Also the combined biomass of protozoa, nematodes, and earthworms is not much different from that of the more important groups of the soil microflora. If the biomass of all soil invertebrates is taken into account the total would be about a quarter or a half of that of the microflora. Often in the past the importance of animals in the soil processes has been underestimated and this biomass value probably gives a truer indication of their significance. Animals do not exhibit the biochemical versatility of bacteria but they are mobile and also possess a gut flora which is transported and provided with a controlled environment much less liable to the sudden fluctuations that occur in soil.

Bacteria

Most bacterial types could probably be isolated from the soil if the search were diligent enough and if the right techniques were used; however, this does not mean that all the bacteria isolated are soil inhabitants. This description can be applied only to relatively few species which are fairly well categorized. These show considerable diversity in their morphology, most morphological types being represented. Physiologically, they range from aerobes to obligate anaerobes, from heterotrophs to autotrophs, and from saprophytes to symbionts and parasites.

Members of the genera *Pseudomonas*, *Achromobacter*, and *Bacillus* will be found in most aerobic soils; where conditions are anaerobic *Clostridium* will occur. The numbers of such organisms will often increase dramatically when suitable substrates are added to the soil. The Actinomycetes too may show a similar quantitative increase under such conditions and favourable alterations to the soil will cause a multiplication of autotrophic organisms such as the nitrifying bacteria *Nitrosomonas* and *Nitrobacter* and the sulphur oxidizers of the genus *Thiobacillus*. Organisms whose numbers increase in this way as a result of special soil conditions may be regarded as being part of the *zymogenous* or fermentative flora of the soil.

The bacteria generally found to be the most numerous, and often comprising a majority of the population of the soil, are those belonging to the genus *Arthrobacter*. These organisms do not form short regularly curved or flexuous rods or produce endospores but may occur as any of the other known morphological types. Indeed some authorities think that soil bacteria identified as *Micrococcus* by morphological examination will usually prove to be *Arthrobacter* if cultured. Bacteria of this genus are variable not only in their morphology but also their Gram-staining and are the most important members of the *autochthonous* or indigenous population. In any one situation their numbers remain relatively constant and are not affected by soil amendment. *Agrobacterium* species are also members of this group of organisms. The distinction between the zymogenous and autochthonous flora is often not as clear cut as these remarks would indicate, and instead of describing an organism as a member of the zymogenous flora it may be said to show a zymogenous response to particular soil conditions.

The size of the bacterial population depends not only on the nutrients available but on other environmental factors too. Temperature, moisture, and indirectly gas content of the soil atmosphere and pH will vary with the prevailing climatic conditions and with depth in the soil. These variations in conditions will introduce temporal and spatial variations in the soil population which can be followed by sampling at intervals of time and at different depths within the soil. Temperature fluctuations that occur above the surface are buffered to some extent by the soil but there are still rapid fluctuations in the soil population correlated with these changes. Seasonal increases and decreases of temperature result in a raising and lowering of the average level of microbial activity as well as in qualitative changes in the composition of the population. Too much moisture results in waterlogging of the soil and too little in a shortage of water for metabolism, both resulting in a reduction of soil activity. Bacteria tolerate a soil reaction between pH 4 and pH 10 but the most favourable pH for the majority is just on the alkaline side of neutrality.

The spatial distribution of bacteria in the soil is complex. Generally there is a decrease in numbers with depth (Table 12.4) which is a reflection of decreasing

Table 12.4 Numbers of micro-organisms of the major groups present in various horizons determined by the dilution plate method. From Starc, A. (1941). *Arch. Mikrobiol.,* **12,** 329.

			Organisms/gram of soil $\times$ 10^3				
Horizon	Humus %	Depth cm.	Aerobic bacteria	Actinomycetes	Anaerobic bacteria	Fungi	Algae
A_1	3·00	3–8	7,800	2,080	1,950	119	25
A_2	1·28	20–25	1,804	245	379	50	5
A_2–B_1	0·91	35–40	472	49	98	14	0·5
B_1	0·37	65–75	10	0·5	1	6	0·1
B_2	0·41	135–145	1	–	0·4	3	–

organic matter content of the soil. There are, however, instances where it is known that the maximum bacterial population does not correspond with the level containing most organic matter, possibly because the conditions favour the development of one of the other groups of soil organisms rather than bacteria. Superimposed on the general distribution are local variations due to a number of factors. For example, plant roots cause an increase in the population in the soil adjacent to them (p. 508); large populations develop in association with big pieces of organic matter such as lengths of fungal mycelium and plant remains. Bacteria not associated with residues of this kind exist in the soil in small colonies, 60 per cent of which contain 2–6 cells per colony, 20 per cent contain 7 or more cells per colony, and others occur as individuals. These colonies are often associated with the film of colloidal material which coats the soil mineral particles. Within a soil crumb it is found that the population is largest near the surface and falls to its lowest level at or near the centre.

Fungi

As with bacteria the problem exists of distinguishing true soil inhabitants from organisms present by chance in the sample examined, but experience has shown that certain organisms occur constantly and others frequently. Isolation experiments suggested that in general *Mucor, Penicillium, Trichoderma*, and *Aspergillus* predominate and that *Rhizopus, Zygorhynchus, Fusarium, Cephalosporium, Cladosporium*, and *Verticillium* occur commonly but all of these grow quickly and sporulate copiously and are therefore favoured by the dilution plate method (p. 488). Direct examination of the soil by special methods (p. 489) shows that Basidiomycetes (p. 398) are numerous and that dark sterile hyphae probably of members of the Dematiaceae (p. 40) are common. Representatives of these last two types are seldom isolated from soils by the dilution plate technique. The biomass of fungi in cultivated soils often exceeds that of any other group of organism and in acidic environments these organisms are usually numerically dominant too. Addition of organic matter to a soil stimulates the fungal flora in the same way as it does the zymogenous bacterial population. Hence the soil fungi are regarded mainly as zymogenous, spores forming the major part of the autochthonous fungal flora.

Water also affects fungi in the same way as it does bacteria. A minimum level is required for activity, although since more fungi than bacteria have structures resistant to desiccation, the fungal population of arid environments is greater than that of bacteria. Waterlogging and the consequent reduction of oxygen concentration inhibits most filamentous fungi as the majority are strict aerobes, hence this group is virtually absent from permanently saturated soils.

Some soil fungi are predators (p. 512); those attacking nematodes may play a part in the control of these animals, some of which are troublesome pests; others prey on amoebae and other protozoa. The soil is also an important reservoir of plant pathogenic fungi (p. 510) although the number of organisms of this type is small as compared with the total population.

Fungal mycelium is interwoven among the soil particles and plays an important part in binding these together, thus improving the texture of clay soils. Mycelium is also intimately associated with larger organic particles which it frequently penetrates. Hyphae may become associated together in strands containing only a few threads or, particularly in Basidiomycetes, into complex rhizomorphs which are commonly found in the surface layers of forest soils. Generally, as can be seen in Table 12.4, the major part of the fungal flora occurs in the upper soil horizons where there is most organic matter, although the relative decrease in numbers with depth is not as marked as with other groups of organisms not excluding the anaerobic bacteria. When the qualitative differences in the fungal flora occurring at different depths are examined it emerges that this distribution reflects the occurrence of carbon dioxide tolerant species at the greater depths.

Algae

Recent investigations suggest that soil algae are more widespread and abundant than was formerly supposed since, even in regions as dry as the Negev desert in Israel, rich algal floras have been recorded both on the surface and also underneath stones and even within cracks in the stones. If the sand is sufficiently translucent a layer of algae may occur in the slightly moister region under the upper millimeter of sand. In cold deserts (even in Antarctica) similar floras exist, particularly on the under surface of stones. This flora consists mainly of cyanophytes (e.g., *Schizothrix* and *Microcoleus*) and coccoid Chlorophyta. The fact that they can live on hot and cold desert soils emphasizes the fact that even other apparently inhospitable soils have a soil flora. Indeed these floras have been shown to be extensive wherever studies have been made. The algal biomass on surface soil must be great since it is frequently visible to the naked eye. Certain algae have been found at considerable depths in undisturbed soil but it is unlikely that they are present in an active state since although some species are capable of heterotrophic growth in the dark, utilizing carbohydrate substrates, it is impossible that they could compete effectively with heterotrophic soil bacteria and fungi under these conditions. Thus relatively large algal populations far below the surface of the soil must be the result of wash-down from the surface, the individuals persisting as resting spores, etc. Some algae which do not obviously form resting stages have been reported from dried, stored soil, suggesting extreme longevity for many species.

The soil algae include flagellate, coccoid, or filamentous species. The coccoid species reproduce via motile stages (e.g., *Chlorococcum*) whilst almost all the diatoms occurring are motile. Motility or multiplication via motile cells is an obvious advantage in sites which are periodically wet. Attached species, however, also occur in suitable situations.

Algal species living in soils are notoriously difficult to identify; many are smaller celled versions of aquatic species and others are ecophenes (i.e., ecological growth forms). As a result insufficient is known of soil algae to list the commonest soil forms but *Nostoc*, *Cylindrospermum*, and *Anabaena* of the Cyanophyceae, *Chlorella* and *Chlorococcum* of the Chlorophycae, together with certain diatoms, are frequently isolated from soil samples. However, in one study of soil samples taken from a small area, representatives of 16 families, and at least 30 genera, were recorded amongst which *Chlorococcum*, *Chlamydomonas*, and *Hormidium* were the most widely distributed. In addition, 16 new chlorophycean algae were described thus illustrating the paucity of knowledge of these rich floras.

Cyanophyta are common constituents of neutral to alkaline soils and also of saline soils. The growth of soil algae undoubtedly affects the surface soil—depletion of some nutrients has been demonstrated and the addition of organic matter is sometimes considerable, e.g., in algal crusts on desert soils and on the surface of peats. Several of the Cyanophyta occurring on soils fix atmospheric nitrogen. Although most soil algae are phototrophs some are capable of heterotrophic growth although the extent of this in natural soils is not known. Antibiotic activity is also reported in isolates of some soil algae and there is no doubt that a combination of factors, of which some may be antibiotic, result in an intricate mosaic of growth on the surface of natural soils.

The soil fauna

The soil fauna contains numerous protozoa and representatives of most groups of metazoa, most of the latter falling outside the scope of this book. However, their activities in some soil processes are of such moment that they cannot be excluded on grounds of size alone. The smallest species, protozoa and nematodes, are most widely distributed in the soil. The former generally occur in the water film surrounding soil particles and the latter usually in the soil of the upper horizons among plant roots. The larger animals are on the whole much more mobile yet particular types are usually found in characteristic environments. Some are found under stones or large pieces of debris on the surface, some in the surface soil and others in deeper layers in the soil, but most occur at or near the surface. Some animals such as earthworms are exceptional in that they frequently move from horizon to horizon and hence play a very important part in the transport of materials not only within the soil but from the surface to the soil proper. Prior to this mixture of organic residues with the soil a large number of animals have taken part in the primary attack of the debris. Several arthropods may be involved, such as the springtails, woodlice, millipedes, earwigs, and so on, but molluscs and various worms also play a part.

On very wet or very dry soils in which few animals exist surface litter tends to accumulate. Under these conditions fungi will be the principal agents of decomposition. Where animals are absent from a soil it is significant that little of the humus so important in producing a well-structured fertile soil occurs. Yet, paradoxically, cultivation tends to reduce the animal population and the efforts of man in incorporating organic materials into the soil must be substituted for those of other animals.

METHODS FOR STUDYING SOIL ORGANISMS

The complexity of the soil environment, the diversity of the population, and the frequently minute size of the habitats make it extremely difficult or impossible to study the *activity* of particular organisms in the soil.

Methods for studying the microflora

Knowledge of the types of organism in particular soils has accumulated as a result of the application of isolation and identification methods. The work of Beijerinck and Winogradsky pointed the way to obtaining particular types or functional groups from the soil by the use of selective media, elective media, and enrichment culture techniques. All media are, to some extent, *selective* but some are specially designed to discourage organisms other than those required. For example, a combination of rose bengal and streptomycin in an agar medium virtually eliminates the growth of bacteria and also restricts the rate of growth and the spreading of some fast-growing fungi, thus allowing the development of slower growing types. *Elective media*, while encouraging the required organism, do not positively deter the growth of others. *Enrichment cultures* are used to obtain a large population of organisms carrying out a particular transformation. For example, if cellulose is added to a soil sample which is then incubated, cellulose-decomposing organisms will usually be present in large numbers after a suitable period and can then be isolated by selective or elective methods more easily than from non-enriched soil. However, large numbers of bacteria in soil may be overlooked by these methods. Other techniques, such as the schemes whereby bacteria are classified according to their nutritional requirements, have therefore been devised to include these organisms. Bacteria are first isolated on a soil extract agar medium. Isolates are then inoculated into fluid media of varying complexity (i.e., in nitrogen source and, in some, the addition of various growth factors). By this means it is possible to group each isolate according to the level of complexity of its nutritional requirements. Estimates of the relative incidence of these groups in soil may thus be obtained and these have proved useful, for example, in studies comparing rhizosphere with non-rhizosphere soils (Table 12.6, p. 509).

A common method of enumerating microflora is the *dilution plate method*. A soil sample is taken and care exercised to minimize fluctuations in the population before it is counted. It is thus desirable to prepare dilution plates as soon as possible after taking the sample and hence a preliminary estimate of the numbers of organisms in the sample cannot be made before making the count proper. A wider range of dilutions is thus required than would be made when counting a better-known population. Arable soils are known to contain normally about one million bacteria per gram and this gives some guide as to the range which should be covered by the dilutions. In preliminary preparation of the sample the soil must be thoroughly broken down so that the bacteria which occur associated with soil particles and in aggregates are completely dispersed. In view of the heterogeneity of the soil, dilutions from a large number of samples should be made rather than making many replicates from dilutions of a single sample. The medium chosen for dilution plate counts is often one such as nutrient agar, potato-dextrose agar, or soil extract agar and it should be borne in mind that these media will favour the fast-growing heterotrophs and that certain groups will be completely missed. With suitable media and incubation conditions, however, organisms such as autotrophs and anaerobes may be counted. When assessing the population of a soil by this method, account should be taken of the probable bias introduced by the germination of spores, etc., which may be numerous in a soil but not active and which may germinate rapidly and be numerically important or dominant on a dilution plate. This method has many difficulties but is nevertheless useful, particularly in comparative studies of soils or soil treatment. It also permits the isolation of certain groups of the soil microflora. Copious mixtures of cyanophyta, chlorophyta, and diatoms frequently result from plating soil samples on to suitable agar media. By the use of selective media and incubation conditions many groups of bacteria can be enumerated and representatives isolated, although it has recently been found that one numerically important group (p. 490) is missed by conventional techniques.

The dilution plate method is less satisfactory for fungi, since slow growing, non-sporing ones may be overlooked or become overgrown and difficult to isolate. Several modifications have been devised for the isolation of fungi. Waksman merely placed soil crumbs on the surface of an agar medium suitable for fungal growth. Warcup buried the crumbs in the medium by pouring cooled melted agar on to them. This method prevents bacteria spreading on the surface of the agar from soil particles on the surface and sometimes allows the isolation of a greater number of species than does the soil dilution plate method, although it does tend to favour faster-growing species and give a similar picture of the flora to that given by the dilution method. Chesters designed an immersion tube which consists of a glass tube about six inches long, the sides of which have four to six invaginated narrow tubes and which is filled with agar. A core is removed from the soil, the tube put in its place and, after a suitable period, the agar is removed from the tube and cut into lengths which are plated out. This method favours fast growing mycelial fungi which can compete effectively in growing through the narrow tubes and which can grow in the depths of the agar.

A less commonly used method of counting soil bacteria is the *membrane filter method* which involves suspending the soil in sterile water, treating the suspension with a Waring blender to disperse the particles and clumps of bacteria and then diluting it as for plate counts. The dilutions are filtered through a membrane which is then transferred to plates of a suitable nutrient medium. The *most probable number count method* which is routinely used in the bacteriological examination of water (p. 585) has also been applied to counting soil organisms. It is valuable since it is easily modified to count organisms responsible for particular processes in the soil.

The three methods described are all indirect means of enumerating soil microorganisms. Other methods involve the direct examination of soil samples. In the simplest of these methods aliquots of the quantitatively made soil dilutions are spread on slides and after the film has dried and been stained the organisms in a large number of random fields are counted. A disadvantage of this method is that as the soil suspensions dry on the slides the organisms in suspension may become unrandomly distributed. A method introduced by Thornton and Gray overcomes this difficulty by mixing a counted suspension of indigo particles with the soil suspension and placing small drops of the mixture on a slide. These are allowed to dry and are then stained with erythrosin. During the drying process the distribution of particles will alter but it is assumed that since the indigo particles and bacteria are approximately the same size their distribution will be affected similarly. The ratio of numbers of bacteria to indigo particles in a number of microscopic fields is then counted and from this it is possible to estimate the number of bacteria per gram of soil. A third direct method of counting the soil microflora (that of Jones and Mollison) involves grinding up soil samples and washing the heavier sand fraction out and adding a molten 1·5 per cent agar solution. The suspension is then shaken vigorously and allowed to stand for a very short period so that any remaining sand grains can sediment. It is then immediately pipetted on to a haemocytometer slide and the well covered with a coverslip. When the agar is set the slide is immersed in water and the agar film recovered on to an ordinary microscope slide. The film is allowed to dry, stained, and twenty random fields are examined for micro-organisms. From the area of the microscope field, the thickness of the agar film and the dilution of the soil suspension, the number of bacteria, pieces of mycelium, and spores per gram of soil can be determined. This method overcomes the problem of irregular distribution of micro-organisms on a slide and it is also possible to distinguish at least some of the organisms which are viable by allowing micro-organisms in the agar to develop into colonies under suitable incubation conditions. By this method it was possible to count 95·7 to 98·4 per cent of bacteria added to sterile soil.

Direct methods of counting invariably yield a higher value for the soil population than do indirect methods. This is probably not due to any great extent to inability to distinguish between viable and non-viable organisms since in one study of Australian soils it was found that the carbon dioxide production corresponded closely with what would be expected from the activities of the microbial population as estimated by direct counts. Thus the discrepancy between estimates obtained by using direct and indirect methods which is often a factor of tens or even hundreds (Table 12.5) must be attributed to a failure to culture all viable organisms from the soil.

Table 12.5   Estimates of numbers of bacteria in three soils by direct and indirect methods.

Soil	Direct count	Dilution plate count
1	$3·7 \times 10^8$	29×10^6
2	$1·0 \times 10^8$	8×10^6
3	$2·4 \times 10^8$	1×10^6

Recently, evidence has been provided of numerous micro-aerophilic, actino-mycete-like soil organisms which do not grow under conventional cultural conditions. This may be the group, or one of them, responsible for the large differences.

There have been many attempts to devise methods which reveal the form and distribution of the microflora in the soil. One of the earliest attempts at this, which is now generally known as the Rossi-Cholodny slide method, involves burying a clean glass slide in the soil for an appropriate period. It is then removed, one side is cleaned, and the other suitably stained and examined. The relationship of organisms to soil particles can often be seen by this method, which has been used in studying changes in the soil population due to soil treatment such as addition of fertilizers and also in examining the relationship between micro-organisms and plant roots.

Attempts to examine soil directly in the field with a microscope were not very successful, except with fungi which are more readily observed than are bacteria. Techniques for removing soil from the field whilst preserving the relationship of the soil structure and the soil organisms for laboratory examination have proved more useful. The soil is impregnated with various substances such as

Fig. 12.2   (A) Bacteria on the surface of a quartz sand grain in soil to which peptone was added ($\times 1,275$). (B) Bacteria associated with fungal mycelium on a root, in soil to which peptone was added ($\times 2,620$). (Both photomicrographs supplied by Dr T. R. G. Gray)

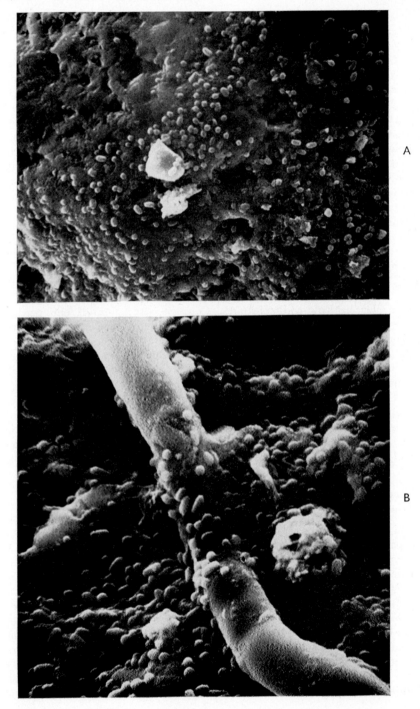

A

B

gelatine and resins which harden permanently. The embedded material is then either sectioned or ground in the same way as are geological specimens and examined microscopically.

The scanning electron microscope offers a new approach to direct observation of organisms in the soil. In undisturbed soil samples some relationships between organisms and with inanimate soil constituents are revealed (Fig. 12.2A and B).

One of the major problems in interpreting results from *in vitro* experiments is knowing to what extent the conclusions drawn are applicable to the micro-organism in the soil. There is no certain way of telling whether the organism is in an active state or not, nor of identifying the organisms observed by the direct methods. The use of fluorescent antibodies specific for particular species may provide one approach to the latter of these two problems. This technique has been used to identify *Aspergillus flavus* hyphae and spores and to study the nematode-trapping fungus *Arthrobotrys conoides* in artificially inoculated soil. In addition, *B. subtilis* cells and spores and *B. circulans* spores have been identified in soil and some information obtained which suggest that these two species have different roles in the soil.

The application of these newer methods should lead to greater knowledge of the distribution of particular organisms in the soil and thus association with particular habitats and substrates. Constant association of one particular organism with a particular micro-environment and a knowledge of its physiological capacity *in vitro* will, however, still provide only presumptive evidence of its role in that situation. The provision of definitive evidence will still depend on the ability to examine the activity in the soil itself of only a very few microbial cells.

The scope of physiological studies on the soil has been enlarged recently by the increased interest shown in cell-free enzymes in the soil. It is possible that these enzymes, whether they are truly extracellular or liberated by cell lysis, may persist in the soil for long periods, stabilized by adsorption on to soil particles, and thus play a significant part in soil processes.

Methods for studying the soil fauna

The larger animals can be isolated and counted relatively easily by sorting and sifting them out of soil samples. The soil protozoa and nematodes are not separable by these methods and other approaches have to be adopted. The nematodes may be separated by flotation methods which are applicable to larger animals too. Soil protozoa may be examined for directly in much the same way as is the microflora but this approach has not been very successful and, although some workers have isolated testacean types by bubbling carbon dioxide or air through the soil and ciliates by applying an electric current, cultural techniques are most widely used. The soil sample is normally plated on to an agar medium containing cells of a suitable edible bacterium such as *Aerobacter aerogenes*. Protozoa developing on the plates after a suitable incubation period at 20–22°C can then be isolated for further study or, by a suitable modification, quantitative estimates of the population can be obtained by a most probable number count.

Overall estimates of activity

Estimates of the activity of the soil populations with respect to a particular activity such as nitrification can easily be made with a soil perfusion apparatus. Essentially this is a device for continuously circulating liquid through a soil sample. In this particular instance the solution would originally contain ammonium ions and the rate of the transformation to nitrate (p. 499) would be followed by measuring the level of this ion in samples withdrawn periodically.

The total activity of the soil population can be measured by determining oxygen uptake and/or carbon dioxide evolution. Some workers perform determinations on small soil samples in a Warburg manometer over short periods, others use large samples in specially designed apparatus and follow uptake and evolution of the gases over longer periods of time. Both methods have their disadvantages but are useful for comparative studies.

TRANSFORMATIONS IN THE SOIL

The soil population is responsible for transformations in the soil which are important for its continued fertility and which ensure the removal of natural litter from the surface of the earth. Many of the transformations are cyclic systems comprising sequential reactions of compounds containing a common element such as carbon, nitrogen, sulphur, etc. However, it is rare to find even the majority of the steps in a cycle occurring in a particular situation at any one time. Some stages in these systems result in an amelioration of the soil, whereas others decrease its fertility, thus both the encouragement and control of these stages is agriculturally desirable. In some instances stages in a cycle are mediated only by organisms which are extremely sensitive to environmental conditions, others take place as a result of the action of a wide variety of micro-organisms and proceed under a wide range of conditions.

The carbon cycle

Quantitatively the most important series of transformations in which soil organisms are concerned is that involving carbon compounds. Organic residues added to the soil contain about 50 per cent carbon which is eventually converted to carbon dioxide as a result of the action of the soil population. The microflora is responsible for the evolution of 95 per cent of the gas produced and animals for only 5 per cent. Carbon dioxide is fixed mainly by green plants but all heterotrophic organisms also use small quantities of this gas and the rate of process is such that it has been estimated that the atmospheric supply would be exhausted in twenty-five to thirty years if it were not replenished in this way. A much simplified representation of the cycle is given in Fig. 12.3.

Decomposition of organic residues

The main types of carbon compounds added to the soil are plant and animal remains in which the carbon is chiefly included in high molecular weight compounds. Soluble molecules account for only a very small proportion of the total. In plants the main types of high molecular weight compounds which occur are

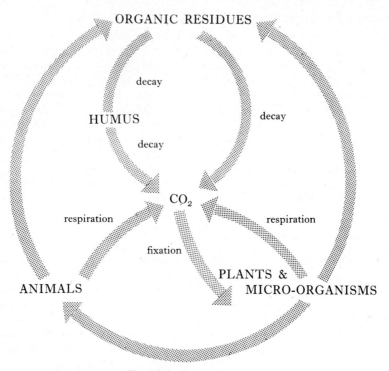

Fig. 12.3 The carbon cycle.

cellulose, hemicellulose, and lignin, together with smaller quantities of fats, waxes and oils, proteins, and nucleic acids. In micro-organisms and animals a similar range of high molecular weight compounds exists. However, the main structural components are different. In invertebrates and fungi the polymer chitin occurs widely and in bacteria a chemically very similar aminopolysaccharide is found, although other polymeric compounds account for a substantial part of the total carbon in these organisms (p. 282).

On the surface of the soil, residues are attacked by bacteria, fungi, and animals, the latter usually eating remains that have been partly digested by the microflora. The predominant microbial types involved and the rate of the attack depend on the chemical nature of the residue, the environmental conditions and the nature of the underlying soil. For example, nitrogenous materials are generally degraded more rapidly than residues with high $C:N$ ratios. Of the environmental conditions, moisture is particularly important and in dry conditions the attack will be slow. Temperature also affects the rate of degradation and the composition of the attacking flora. This will also be affected by the nature of the underlying soil which determines to a large extent the population available for the degradation.

The animal population in the soil also plays a role in the decay process. The activities of springtails, bark lice, and earwigs, for example, increase the surface

area of the residue available to the microflora and various worms eat the partially decomposed material and transport it into the soil proper. Further decay
occurs in the gut as a result of the action of the microflora. Within the soil the
conditions are more favourable for the microflora and the rate of degradation
is increased. In particular, the role of bacteria becomes more important in most
types of soil and a zymogenous response occurs involving an increase in the
numbers of most of the heterotrophic soil bacteria.

At every point in this process, the initial attack on the high molecular weight
polymers is by enzymes liberated into the soil or acting at the outside surface of
the producing cell. Environmental factors may alter the rate of decay by their
effect both on the microflora and on the enzymes outside the cell.

Most of the carbon contained in organic residues is oxidized to carbon dioxide
but some is incorporated in microbial tissues which themselves will eventually
be decomposed. The remainder is incorporated into humus. This is the dark-
coloured amorphous organic material which is so important for soil fertility as is
evidenced by the infertility of soils from which it is absent. The processes
whereby it is formed are not understood, although it may be that soil animals
are indispensable since in soils where there is no soil fauna humus is not found.
Similarly its chemistry is obscure. It is not a single chemical substance and its
composition varies but it is essentially polyphenolic in nature and contains some
amino acids and amino sugars.

Degradation of pure substrates

In view of the complexity of the processes involved in the breakdown of
natural residues much of our understanding of the microbiology of the degradation of carbon-containing polymers has come from the study of the decay of
pure substrates such as cellulose and chitin.

Cellulose decomposition has been studied by burying strips of cellophane
supported on glass coverslips. In general the course of the decomposition in
approximately neutral soils is similar. Fungal hyphae colonize the surface
rapidly; these include *Botryotrichum*, *Chaetomium*, *Humicola*, *Stachybotrys*,
and *Stysanus*. They produce in the cellulose dense 'root' systems which
hydrolyse the substrate. After a few weeks the mycelium dies off and attack by
bacteria, sometimes simultaneously with animals, occurs. The animals evidently
are important in this stage of the degradation since in their absence cellophane
persists for long periods. In acid soils the fungal flora is qualitatively different
and the bacterial development less. In addition to the pH of the soil the degree
of cultivation is probably important in determining the composition of the
flora degrading cellulose. Cultivation causes an overall increase in the number of
cellulose decomposing micro-organisms but the increase in bacterial numbers is
much greater than that of the fungi. In particular *Cytophaga* and *Sporocytophaga*
may be increased three- or four-fold.

The availability of nitrogen is a factor limiting the role of breakdown of any
organic residue but is particularly important with respect to pure cellulose which
contains none. The average C : N ratio of microbial cells is about 10:1, thus when
material with a higher proportion of nitrogen than this is added to a soil, nitrogen is not limiting, indeed there will be an excess over demand and release

of inorganic nitrogen will occur. Conversely residues with higher C : N ratios than this are decayed at a rate related to the exogenous nitrogen supply, other things being equal. Cellulose decomposition stops when the nitrogen content of the soil falls below 1·2 per cent.

In vitro an association between *Sporocytophaga* and *Azotobacter* on cellulose media has been demonstrated in which the latter fixes nitrogen after exhaustion of the supply in the medium. It has also been shown that if straw, which is principally cellulose, is incorporated into the lower layers of an agricultural soil, the crop yield in the first year is often low as compared with control soils. However, if instead the straw is incorporated into the top few centimeters, crop yields do not diminish and it has been established that in this instance conditions favour nitrogen fixation and straw decomposition. Thus it may be that *in vivo*, non-symbiotic nitrogen fixation (p. 501) provides nitrogen for the microflora degrading cellulose.

In contrast to the decomposition of cellophane the degradation of chitin in neutral and slightly alkaline soils is mainly achieved by actinomycetes and eubacteria. In well-aerated soils actinomycetes are predominant. However, in waterlogged anaerobic soils both cellulose and chitin are attacked by *Clostridium* species. In the alkaline C horizon of one forest soil, bacteria and actinomycetes were dominant only in the early stages of the colonization. Later fungi became relatively more important. Conversely, in an upper horizon of this soil in which the pH was 3·5, fungi were dominant at first but were later replaced by actinomycetes. This is surprising although 97 per cent of the chitin-digesting types in a Dutch soil of pH 4·7 were actinomycetes.

Cellulose and chitin account for much of the carbon added to soil yet it is impossible to predict fully the composition of the microflora involved in the various stages of their degradation even when added to the soil as purified substances. Thus it is clear that much remains to be learnt about the decomposition of natural residues.

Decomposition of humus

The amount of humus in a mature soil varies little from year to year provided it is not depleted as a result of cultural practices. Humus is relatively resistant to degradation either because it is intrinsically stable or because it is adsorbed on to clay particles, but whatever the reason, only a small fraction of the total is broken down in any one year. As with the degradation of plant and animal residues added to the soil, the rate of breakdown is influenced by environmental factors such as aeration, temperature, pH, and in addition the amount of organic matter in the soil; when the amount is large the rate of humus decomposition is high. This is probably because the zymogenous flora attacks the humus fraction in addition to the added organic residues. Normally, however, humus degradation is achieved mainly by the autochthonous flora.

Methane formation and oxidation

In permanently waterlogged anaerobic soils, *Clostridium* species may convert organic matter to organic acids, carbon dioxide, and hydrogen which are in turn converted by certain strictly anaerobic bacteria to methane. Only a few bacteria are capable of methane production which, although they are morphologically dissimilar, are included in the family Methanobacteriaceae. This process is of no significance in agricultural soils which are temporarily waterlogged but a combination of poor drainage and high organic matter content may very rarely produce conditions such that methane is produced and liberated at the surface as gas. In waterlogged anaerobic paddy field soils where methane generation occurs, a methane oxidizing flora exists at the surface and converts most of the gas to carbon dioxide. Methane oxidizing bacteria also occur in several other soil types including well-drained ones in which they are more numerous in the lower layers of the soil.

Nitrogen cycle

The activity of some of the organisms involved in the series of processes which constitute the nitrogen cycle (p. 498) results in nitrogen being made available to plants in an assimilable form. These organisms therefore act in an analogous way to those involved in the carbon cycle and which make carbon dioxide available as a result of oxidation of carbon-containing compounds. An important difference between the two cyclic systems is that nitrogen turnover determines productivity in most agricultural situations, hence the activity of those soil organisms metabolizing nitrogen compounds in such a way as to remove nitrogen in a form assimilable by higher plants from the soil, may be very important with respect to loss of soil fertility.

There are two other conspicuous differences between the carbon and nitrogen cycles in terrestrial environments. Firstly, carbon is made available to plants in gaseous form as carbon dioxide, whereas nitrogen is utilized by plants mainly in the form of nitrate ions. Secondly, most of the micro-organisms involved in the carbon cycle are heterotrophic, although in aquatic environments autotrophs are important whereas many of those taking part in the nitrogen cycle are autotrophic.

Living organisms contain about 10–15 per cent nitrogen and the annual turnover of this element has been estimated to be between 10^9 and 10^{10} tons per year. The cycling of nitrogen is mainly achieved by biological processes but certain chemical reactions mediated by lightning and ultra-violet light can cause fixation of atmospheric nitrogen. In addition, nitrogen compounds in solution can be converted by chemical processes under certain soil conditions.

The various transformations involving nitrogen can conveniently be summarized in diagrammatic form (Fig. 12.4). This illustration is incomplete in that it includes only the main biological processes involved in nitrogen conversion discussed below.

Decomposition of organic nitrogen-containing compounds

Until recently, it was thought that almost all nitrogen in soil was in the form of organic compounds but it has been found that both available ammonium and nitrate ions may under certain circumstances be present, although their levels fluctuate rapidly and they do not usually account for more than 2 per cent of the total soil nitrogen. Three main types of organic molecule contain much of the soil nitrogen. Between 20 per cent and 50 per cent is combined in amino

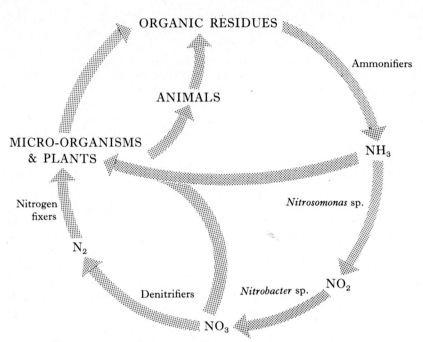

Fig. 12.4   The nitrogen cycle.

acids, 5–10 per cent in amino sugars and only about 1 per cent in purine and pyrimidine bases. Thus about half or more of the total bound nitrogen in the soil is not accounted for. Some of this may be in the form of non-exchangeable ammonium ions bound to clay particles, but the majority is combined in compounds whose chemistry is not understood.

The three types of compound mentioned above occur in organisms mainly as polymers but may be added to the soil as monomers as a consequence of autolytic processes. More usually the residue incorporated into the soil will be of high molecular weight and will initially be degraded by extracellular enzymes or enzymes liberated into the soil by cell lysis. A wide variety of soil bacteria are capable of hydrolysing protein. *Bacillus* and *Pseudomonas* species particularly are responsible for the breakdown of pure protein added to the soil. These and other bacteria such as *Arthrobacter* are probably responsible for the degradation in alkaline soils of proteinaceous material from natural sources too. But in acidic soils fungi are predominantly responsible. The fate of the amino acids produced depends on their type and the soil conditions. Some, such as glutamic acid, may be utilized directly, most will be deaminated with the formation of ammonia and the corresponding acid, but there are a few such as threonine, lysine, and methionine which are attacked only very slowly. Aminopolysaccharides are broken down by a much more restricted flora than are proteins but in the average agricultural soil, these macromolecules do not occur very commonly. They are far more important in marine sediments where there are anaerobic organisms;

it is probable that their breakdown is performed mostly by species of *Clostridium*. These polymers are usually hydrolysed to the constituent amino sugar and then deaminated. Nucleic acids are broken down by a relatively complex series of processes. Firstly, the polymer is degraded to form the constituent nucleotides. Several of the common fungi, such as *Aspergillus* and *Penicillium*, play a part in this as do the bacteria, *Clostridium*, *Bacillus*, and *Achromobacter*. The nucleotides are then dephosphorylated and then the residual base-sugar complex is broken down. The nitrogen in the base is finally liberated in the form of urea, ammonia, or β-alanine according to the particular base and the soil conditions as a result mainly of bacterial action. Thus the main product of the breakdown of organic residues is ammonia. The fate of this depends on the C:N ratio of the soil. When carbohydrate is abundant the ammonia will be used immediately, but when the C:N ratio has fallen to 12:1, it will be oxidized by nitrifying bacteria or utilized by other organisms.

Nitrification

The oxidation of ammonia to nitrite and nitrate is achieved chiefly by organisms of only two autotrophic genera, *Nitrosomonas* and *Nitrobacter*. There are numerous examples of *in vitro* demonstrations of nitrification by heterotrophic bacteria and fungi but the significance of the activity of these organisms in the soil is small when compared with the autotrophic process.

Nitrosomonas is a small, Gram-negative rod with polar flagella. It is strictly aerobic and mediates the overall reaction:

$$2NH_4^+ + 3O_2 \longrightarrow 2NO_2^- + 4H^+ + 2H_2O$$

The intermediate steps in this oxidation are not yet confirmed.

Nitrobacter has the same characteristics as *Nitrosomonas* but is slightly smaller and converts nitrite to nitrate. The process can be represented by the equation:

$$2NO_2^- + O_2 \longrightarrow 2NO_3^-$$

Compared with the transformations involved in the production of ammonia these processes are very sensitive to environmental conditions and are carried out effectively only in neutral or slightly alkaline soils. At soil pH values below 6, ammonium ions usually accumulate and in very alkaline soils high concentrations of nitrite may be found. This is due to the sensitivity of *Nitrobacter* to free ammonia and it is generally true that this organism is more sensitive to adverse conditions than is *Nitrosomonas*. Extremes of temperature and drought may also cause accumulation of nitrite. Nitrification is also suppressed by addition of carbohydrate to the soil. This is not due to the inability of these autotrophic organisms to tolerate organic matter, indeed they thrive in manure where there are large quantities, but to competition for ammonium ions with the large heterotrophic population involved in carbohydrate degradation. Both *Nitrosomonas* and *Nitrobacter* are strict aerobes, consequently waterlogging of the soil and concomitant reduction in gaseous exchange renders them inactive.

The supposed importance of nitrification with relation to soil fertility rests upon the fact that nitrate ions are preferable to ammonium ions for plant nutrition. Nitrification takes place in most cultivated soils but in many forest, orchard, and grassland soils the nitrifying bacteria are inactive because of soil or climatic conditions. The plants in these situations must therefore assimilate ammonium nitrogen. Ammonium ions have the advantage that they are chemically stable in acidic conditions and as they may be bound to negatively charged clay particles, are not leached out of the soil. Nitrate and nitrite, in contrast, are easily leached out and nitrite is also converted in acid soils to gaseous nitrogen and nitrous oxide thus causing a net loss of nitrogen from the soil. There are, however, some disadvantages associated with the nitrogen in the soil remaining as ammonium ions. Only a relatively small proportion of the ammonium ions bound to clay particles is exchangeable and available for assimilation. Above certain levels they may be toxic to plants and in very alkaline soils they may be volatilized. Notwithstanding these problems, it has been suggested that nitrification may be an agriculturally undesirable process and a pyridine derivative suppressing *Nitrosomonas* has been marketed. The value of such control is not yet certain but it is clear that attempts to control nitrification must be directed at *Nitrosomonas* and not *Nitrobacter*, since it is high levels of nitrite which are phytotoxic.

Denitrification

The process of denitrification in the strict sense involves the reduction of nitrate or nitrite to molecular nitrogen or to nitrous oxide. It does not include assimilation of these ions by plants or micro-organisms or loss by leaching. Thus this term refers only to those biological processes which lead to a loss of nitrogen from the soil as gaseous products. This distinction must be made clear since the initial stages of the denitrification and assimilation pathways are probably the same.

$$HNO_3 \longrightarrow HNO_2 \text{------} \begin{cases} \rightarrow NH_3 & \text{Assimilation} \\ \rightarrow N_2O \\ \quad N_2 \end{cases} \text{Denitrification}$$

The products of denitrification are not assimilable by higher plants and most micro-organisms, and thus this process is deleterious to soil fertility. However, since a wide range of ubiquitous forms are capable of causing denitrification it is possible to effect control only by alteration of soil conditions.

Nitrate is used as an alternative electron acceptor to oxygen. Thus aeration of a soil will reduce denitrification although even in a very well aerated soil anaerobic micro-environments providing suitable conditions for denitrification will still occur. The level of organic matter and the pH of the soil are also important factors. The former provides the energy source for the heterotrophic denitrifying bacteria and it is found that in acidic soil with a reaction lower than pH $5 \cdot 0$–$5 \cdot 5$ little or no nitrate loss occurs. However, in acid soils nitrite is chemically unstable and is converted to nitrogen and nitrous oxide.

In spite of the enormous population potentially capable of causing denitrification this activity is apparently confined to bacteria, particularly the genera *Pseudomonas*, *Achromobacter*, and *Bacillus*. Fungi and actinomycetes are probably not involved in soil.

Non-symbiotic nitrogen fixation

A wide range of bacteria is capable of non-symbiotic nitrogen fixation in the laboratory. This includes the aerobes *Azotobacter, Beijerinckia, Azotomonas, Pseudomonas,* and *Nocardia,* the facultative anaerobes, *Klebsiella, Bacillus polymyxa* and *Rhodospirillum,* and the anaerobic *Clostridium pasteurianum, Desulphovibrio, Methanobacterium, Chromatium,* and *Chlorobium.* Although *Azotobacter* is an extremely active nitrogen fixer *in vitro,* the efficiency of its nitrogen fixation is low, only 5–20 milligrams of nitrogen being fixed per gram of sugar oxidized. Thus to fix 5–20 pounds of nitrogen per acre these organisms would require and entirely utilize 1,000 pounds of carbohydrate. It is therefore improbable that *Azotobacter* makes any great contribution to soil fertility by nitrogen fixation in temperate agricultural soils. Other free-living organisms capable of fixing nitrogen are probably unimportant in most soils, although there is evidence that in some specialized environments some do, almost certainly, contribute to soil fertility. In some very fertile Egyptian soils *Azotobacter* cells are very numerous and may have a significant role. In tropical paddy fields an extensive population of blue-green algae develops and it is thought that the nitrogen fixed by these organisms accounts for the ability to take successive crops of rice from these fields without any nitrogenous supplementation of the soil. Also in one sub-arctic soil there is some indication that nitrogen fixation by a variety of non-symbiotic organisms may be significant.

Symbiotic nitrogen fixation

Awareness of the advantageous effects of leguminous crops on soil fertility can be traced back to biblical times. However, it was not until much later that Atwater, in 1885, and Hellriegel and Wilfarth, in 1888, demonstrated that such plants as peas and clover can utilize nitrogen from the air when they are nodulated; plants without nodules requiring nitrate for growth. In 1888 Beijerinck isolated a bacterium, now included in the genus *Rhizobium,* from nodule material and from soil. It was concluded that as a result of this association of microorganisms in the nodules nitrogen was fixed. However, rigorous demonstration of this was obtained only when ^{15}N labelled nitrogen was used.

Neither of the partners in this symbiotic association is capable of nitrogen fixation in the absence of the other but they are not dependent on each other for existence. The legume is capable of using combined nitrogen in the form of nitrate and the bacteria exist independently even in soils which have not recently supported a legume crop. The population in soil is, however, larger when the macrosymbiont has recently been grown there.

Rhizobia are Gram-negative forms which when grown on normal media are rods. When freshly isolated or when grown on media containing alkaloids or glucosides and when within the nodule they adopt the bacteroid form. These are irregular, sometimes branched, star or club-shaped cells which stain unevenly. The bacteria in this genus are classified into species according to their host range (cross-inoculation group) although an Adansonian approach (p. 464) to the classification of this genus has now been made. Different strains of *Rhizobium*

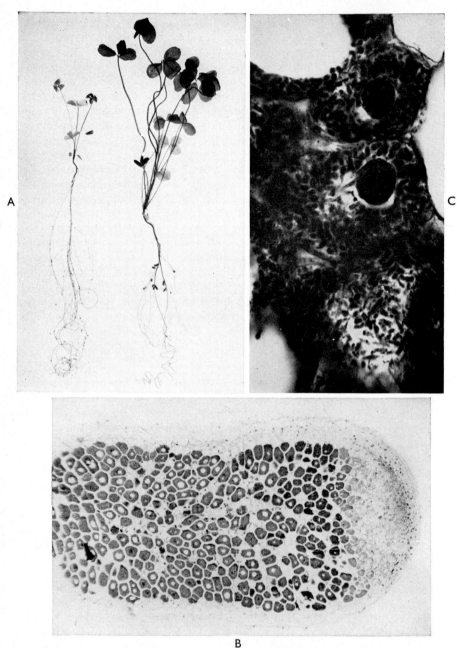

Fig. 12.5 Root nodules of leguminous plants. (**A**) Plants of red clover ($\times \frac{1}{2}$), grown in nitrogen-free sand; plant on left inoculated with ineffective strain of *Rhizobium leguminosarum* (note stunted aerial parts and numerous small nodules on roots), plant on right inoculated with effective strain of the bacterium (note well-developed aerial

are different in their infectiveness and effectiveness in nitrogen fixation. These two properties are controlled by DNA and non-infective strains can be transformed (p. 118) to infectiveness with DNA from an infective strain. Similarly it is possible to transform rhizobia from one cross-inoculation group to another. Mutations to ineffectiveness are common in the soil and to minimize the chance of infection by an ineffective strain legume seeds may be treated with an effective one before sowing. Effective nodules are large and few in number compared to ineffective ones on the same host (Fig. 12.5A).

The process of infection and nodule formation and the effects of the various factors controlling these processes are complex and not fully understood, but, provided that the legume and the *Rhizobium* strain are compatible and the nitrogen level of the soil surrounding the root is not too high, nodule formation will usually occur. As far as other environmental factors are concerned conditions best for root growth are normally also most favourable for nodulation.

Within the rhizosphere (p. 508) of the legume the growth of rhizobia is stimulated by root exudates which also promote the growth of competitors. However, it is possible that substances specifically encouraging *Rhizobium* may be secreted too. The bacterium stimulates root hair elongation by indole acetic acid (IAA) and then root hair deformation results from another *Rhizobium* product possibly in conjunction with IAA which alone does not cause deformation. The root hair is next induced to form polygalacturonase (PG) by bacterial polysaccharide material. The bacteria at this stage are attached to the root hair and become trapped in folds formed as a result of the deformation. This enables high IAA and PG concentrations to develop and invagination of the root hair wall will then occur. This invagination becomes the infection thread. It contains the bacterial cells and grows down the root hair into the root cortex where cells are stimulated to enlarge and divide, a process paralleled by division of the rhizobia which are now mainly bacteroids (Fig. 12.5B and C). Concomitantly the pigment leghaemo-globin which is involved in the fixation of nitrogen and which gives effective nodules their pink colour, is developed.

The fixation of nitrogen by this symbiotic process is of tremendous agricultural importance. Average amounts of nitrogen added to the soil by a leguminous crop are frequently in excess of 100 lb. per acre per year and in one instance under a grass-clover crop in New Zealand, more than 500 lb. of nitrogen were fixed per acre in one year.

Several examples of micro-organisms inducing nodule formation by roots of non-leguminous plants are known. Among the plants indigenous to Britain which show this relationship are *Alnus*, *Hippophae*, *Elaeagnus*, and *Myrica*. The natures of the microsymbionts associated with these plants are uncertain and have been described by various authors at different times as being fungi, bacteria, and actinomycetes but none has been isolated in culture and their identity must remain in doubt, although recent work indicates that the endophytes may be

---

parts and few but large root nodules). **(B)** L.S. effective nodule of red clover (× 80)⋅ Note cortical cells filled with bacteroids (deeply stained) and absence of infection of growing tip. **(C)** Cortical cells of similar nodule (× 1,200) to show presence of bacteroids. (Rothamsted Experimental Station, Copyright)

actinomycetes. There is strong evidence that these associations result in nitrogen fixation.

Mechanism of nitrogen fixation

The problem of the mechanism of nitrogen fixation, that is, the conversion of chemically inert nitrogen gas at ordinary temperature and pressure, is common to both the symbiotic and non-symbiotic process. Owing to the availability of cell-free N-fixing systems from free-living nitrogen-fixers our understanding of the non-symbiotic process is very much better than that of the symbiotic process, where the list of outstanding problems is formidable. A cell-free preparation from *Clostridium pasteurianum* can be split into two fractions one of which is involved in metabolizing pyruvic acid:

$$CH_3.CO.COOH + Pi \longrightarrow CH_3.CO \sim Pi + CO_2 + 2[H]$$

$$CH_3.CO \sim Pi + ADP \longrightarrow CH_3.COOH + ATP$$

The other fraction, which is repressible by ammonium ions, is involved in actual nitrogen fixation and includes the molybdenum-containing enzyme nitrogenase and requires ferredoxin (Fd) which is first reduced:

$$Fd_{ox} + n[H] \longrightarrow Fd_{red}$$

and then, in return, reduces nitrogen:

$$Fd_{red} + N\text{-ase} + N_2 \longrightarrow Fd + N_2.N\text{-ase}_{red}$$

$$N_2.N\text{-ase}_{red} + nATP \longrightarrow NH_3 + nADP + nPi + N\text{-ase}$$

Cell-free systems derived from *Azotobacter* fix nitrogen when supplied with reducing power and energy from the *Clostridium* fraction. Thus although ferredoxin is absent from *Azotobacter* and an alternative hydrogen carrier must be involved; the mechanism is reductive and not dissimilar from that of *Clostridium*.

Sulphur transformations

Many of the organisms taking part in and playing a quantitatively significant part in the sulphur cycle (Fig. 12.6) are not active in the soil, e.g., the photosynthetic bacteria involved in the oxidation of sulphides which are important in the control of pollution in natural waters.

In most agricultural soils sulphur is present in sufficient concentrations to meet the requirements of the soil population and crop plants. This supply derives from the parent rock, from organic residues and also, in agricultural areas adjacent to industrial regions, in substantial amounts from volatile compounds. In some soils sulphur is limiting and may, with benefit, be applied as a fertilizer.

Most soil sulphur is combined in organic compounds. In some Australian soils 93 per cent is in this form, 41 per cent as amino acid sulphur and 52 per cent as sulphate esters present in sulphated polysaccharides. The remaining 7 per cent is 6 per cent sulphate and 1 per cent sulphide and other reduced compounds.

Higher plants normally use sulphate ions as their sulphur supply although they can also assimilate sulphur-containing amino acids. Micro-organisms are,

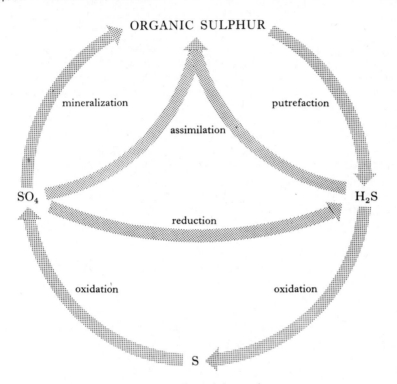

Fig. 12.6 The sulphur cycle.

as a group, more versatile and can utilize most of the sulphur compounds and are instrumental in making sulphur available to higher plants in a utilizable form.

Decomposition of organic sulphur-containing compounds

If cystine or cysteine are added to a well-aerated soil, sulphate is rapidly formed by a route not involving hydrogen sulphide. However, when these amino acids are incorporated in a protein, at least some hydrogen sulphide will be produced even in an aerobic soil. In waterlogged anaerobic soil hydrogen sulphide will be the major product. Methionine metabolism is less well understood but it is possible for sulphate, sulphide, or a volatile mercaptan to be produced.

Environmental factors affect the mineralization of sulphur in the same way as they affect that of nitrogen and carbon, but they are less well understood. However, the C : S ratio of a residue apparently has to be less than about 200 before sulphur is made available to higher plants.

Oxidation of hydrogen sulphide

In aerated agricultural soils high levels of hydrogen sulphide do not occur. *Thiobacillus* species, some of which can oxidize this gas, are present in cultivated soils in small numbers but it is uncertain whether these autotrophic organisms play a part in the oxidation since it occurs spontaneously in the presence of oxygen.

In waterlogged soils high levels of hydrogen sulphide may occur and characteristic black sulphide deposits may be found where ferrous or other metallic ions are also present. Hydrogen sulphide may be oxidized by filamentous colourless sulphur bacteria such as *Beggiatoa* where there is sufficient oxygen, possibly at the surface of muds.

Oxidation of sulphur

The addition of sulphur to soil induces the development of a large population of *Thiobacillus* species and the lowering of the soil pH due to the production of sulphuric acid by these organisms, one of which, *Thiobacillus thiooxidans*, can tolerate a pH of below 0·5. The pathway of sulphur oxidation by these bacteria is not clear although it seems certain that several sulphur compounds of intermediate oxidation states between sulphur and sulphate are involved.

Reduction of sulphate

The organisms chiefly responsible for reducing sulphate in the soil are members of the genus *Desulphovibrio*. These organisms are widespread in nature and occur in a wide range of environmental conditions. They can be found growing both heterotrophically, using such compounds as carbohydrates, organic acids, and alcohols as electron donors, and also autotrophically, using molecular hydrogen as the electron donor.

In agricultural soils these organisms will not play a very important part, since the aeration conditions will be such that they will not be significantly active. However, in waterlogged paddy field soils where anaerobic conditions prevail, they will be responsible for the generation of hydrogen sulphide which may damage the roots of rice plants growing there although evidence of such a toxic effect is not conclusive. Apart from this, sulphate-reducing organisms are economically important in a number of ways. They are responsible for causing extensive corrosion to underground iron pipes by the removal, and use as an electron donor, of hydrogen which normally forms a protective layer around these pipes in the absence of air. Further corrosion then occurs. It is also thought that they may play a part in liberating oil from oil-bearing shales. In the past they probably took part in the formation of sulphur deposits by producing hydrogen sulphide which was then oxidized by micro-organisms or by purely chemical processes.

Assimilation

Hydrogen sulphide in addition to being oxidized may be assimilated as such by some micro-organisms, but sulphate is the form in which sulphur is more usually taken up by these organisms and by higher plants. Within the cell sulphur is mainly in a reduced form and hence a reduction process must be involved in its incorporation.

Transformations involving phosphorus

Phosphorus is needed in substantial quantities by living organisms for nucleic acid synthesis and in lesser amounts for phospholipids and other organic phosphate compounds. Soils are frequently deficient in this element and the supply from plant and animal residues is often supplemented with a phosphate fertilizer.

The transformations of phosphorus do not, like those of sulphur and nitrogen, involve autotrophic bacteria. There are other differences too, although there are also many similarities between these processes.

Organic residues added to the soil are degraded by the heterotrophic microflora, the rate of process depending on the environmental conditions as do the rates of sulphur and nitrogen mineralization. In fact, the conditions for rapid ammonification are the same as those for speedy phosphate release. Substantial amounts remain bound in organic material and in Russia attempts have been made to increase crop yield by introducing into the soil *Bacillus megaterium* var. *phosphatium*, a bacterium which releases inorganic phosphate from the combined form, but increases in yield claimed were often within the limits of experimental error. The phosphorus content of the added residue determines whether phosphate is made available to crop plants or whether it is incorporated into microbial tissue. The level above which phosphate is made available is about 0·2 per cent.

Phosphate is also made available to plants as a result of the liberation by micro-organisms of organic acids which dissolve insoluble inorganic phosphate compounds in the soil.

Other transformations

In addition to the major elements dealt with above, organisms require smaller amounts of several others. Crop plants and the soil microflora obtain their supply from the soil and for a long time considerable interest has been shown in the occurrence and availability of some of the other elements, since an inadequate supply causes deficiency diseases in plants. The effects of these elements on the soil microflora has, however, until recently been more or less completely neglected and even now very little is known. The fungal flora appears to be less sensitive to iron and manganese availability than are the bacteria. *Bacillus* and *Arthrobacter* have a particularly high requirement for iron. A much greater, although incomplete, knowledge exists of the transformation of the various compounds of these two elements and their effect on the availability of the ions to plants.

In soils of normal pH, transformation of manganese is the result of biological activity. Only at pH values in excess of 8 is this element converted to the insoluble manganic form, usually manganic oxide (MnO_2), by chemical means. At a pH of below 5·5 manganese exists predominantly as the divalent manganous form which it is often stated is the form utilized by plants. There is, however, now some evidence that certain manganese oxides also can act as a source of assimilable manganese. But as a result of the activity of the manganese-oxidizing flora soils may become deficient in available forms of this element and a reduction in the crop yield may occur. Bacteria oxidizing manganese are common only in the upper layers of heavy soil but are more widely distributed in sandy soil—a reflection of the aeration of the two soil-types presumably. Certain heterotrophic bacteria, such as *Pseudomonas*, *Corynebacterium*, and *Flavobacterium*, can oxidize manganese as can two recently described budding bacteria. One, *Metallogenium*, is widely distributed in soil but apparently only develops in association with a fungus. The other, *Pedomicrobium*, which resembles closely *Hypomicrobium* (p. 339), may, according to species, oxidize not only manganese but also iron.

Reduction of manganese compounds to a form readily utilizable by plants can result from the production of acid and/or by a reduction of the redox potential of the soil as a result of the activity of the microflora. However, the importance of bacteria in making manganese available to plants is uncertain since the return of aerobic and neutral pH conditions results in its reoxidation.

In contrast to the role of the soil microflora in manganese transformation the part played by bacteria in the conversion of iron compounds in neutral and alkaline soils is unimportant. Some heterotrophic bacteria may cause deposition of iron from organic complexes by using the organic compound as an energy source and it is possible, but not certain, that iron deposited in this way may play a part in the formation of the iron pans which occur in some soils. Certain other heterotrophic bacteria can also reduce ferric to ferrous iron under anaerobic conditions. Where sulphide ions occur this will be precipitated as ferrous sulphide. Most of the iron bacteria including the filamentous forms such as *Sphaerotilus* and the unicellular forms such as *Gallionella* are active in water as are several of the manganese bacteria.

The availability of several other elements including calcium, magnesium, potassium, zinc, copper, molybdenum, cobalt, and boron is affected by biological activity and arsenic and selenium can be converted into volatile compounds by the action of certain soil fungi.

INTERACTIONS AND RELATIONS OF SOIL ORGANISMS

The numerous and varied soil organisms interact with one another in a number of ways. Because of the density of the population, interaction between soil organisms is often competitive. The outcome of competition between two organisms may be determined because one is better adapted than the other to the prevailing nutritional and other environmental conditions. Alternatively or additionally, one may produce an antagonistic substance which inhibits or kills the other. All members of the soil population compete to some extent with other organisms whereas only a part of the population is involved in relationships such as symbiosis, pathogenesis, and predation.

The rhizosphere

In the region of the soil under the influence of plant roots, i.e., the *rhizosphere*, many of the types of interaction mentioned above occur. It is this region of the soil which provides the plant with its soluble nutrients and through which any soil organism influencing the plant via its root must pass or transmit its effector. It has therefore been extensively studied.

The extent of this zone depends on the activity of the plant and the type and state of the soil, but generally the influence of the root extends for only a few centimetres. Its effect on the soil population is often expressed as the R/S ratio, that is, the number of organisms in rhizosphere soil as compared with the number in the same soil beyond the influence of the root. R/S values of over 100 are obtained for bacteria under certain conditions. Bacterial response to the rhizosphere conditions is greater than that of other groups. The nematode population in the rhizosphere of some plants is substantial and sometimes where a high bacterial response is recorded the R/S ratio for protozoa is also large. Fungal numbers, estimated by the dilution plate count, increase only slightly in the rhizosphere although more than do those of algae which may be unchanged or even decreased.

Differences between the nutritional requirements of the rhizosphere flora and that of the soil flora in general are shown in Table 12.6. There is a substantial

Table 12.6   Nutritional groups of bacteria in rhizosphere and control soil. From Lockhead, A. G. and Thexton, R. H., *Can. J. Res.*, C. **25**, 20.

Group	Nutritional requirements for maximum growth	Control soil %	Control soil Number per g.	Rhizosphere soil %	Rhizosphere soil Number per g.	Times increase in rhizosphere
	Plate count		37,500,000		532,000,000	14·2
I	Grow in basal medium	12·0	4,500,000	22·5	119,700,000	26·6
II	Require one or more amino acids	6·8	2,550,000	25·0	133,000,000	52·2
III	Require growth factors	23·1	8,660,000	15·0	79,800,000	9·2
IV	Require amino acids and growth factors	16·2	6,080,000	15·0	79,800,000	13·1
V	Require yeast extract	16·2	6,080,000	11·7	62,200,000	10·2
VI	Require soil extract	6·8	2,550,000	5·8	30,800,000	12·1
VII	Require yeast extract and soil extract	11·1	4,160,000	2·5	13,300,000	3·2

increase in the number of bacteria requiring one or more amino acids and bacteria which require a complex nutrient supply are relatively less abundant although the total number of bacteria in each group is markedly increased.

The types favoured are Gram-negative rods belonging to the genera *Pseudomonas*, *Achromobacter*, and *Agrobacterium* and probably other members of several functional groups (e.g., the ammonifiers, denitrifiers, aerobic cellulose decomposers and the nitrifying bacteria, *Nitrosomonas* and *Nitrobacter*, although it has been recorded that the latter group is inhibited as are the anaerobic cellulose decomposers and nitrogen fixers). *Bacillus* species decline somewhat in number although there is a qualitative change in the population of this group. The fungal population also changes qualitatively. As with the bacteria the species recorded depend on the isolation technique used.

The plant influences the rhizosphere flora mainly by sloughing off dead cells from the growing root tip and by exudation of soluble organic compounds also from the young part of the root. These compounds are released in substantial quantities and wide variety; for example, 21 amino acids have been found. In addition the root affects the microbial environment by raising the carbon dioxide level, decreasing the oxygen concentration and taking up nutrient ions and water.

The microflora influences the plant root in a number of ways. The possibility of a specific stimulation of *Rhizobium* by legume roots has been mentioned earlier (p. 501) as has the effect of *Rhizobium* on root hairs. The rhizosphere microflora also affects the root by the secretion of polypeptide membrane-active antibiotics which stimulate further leakage of plant cell contents into the soil. The micro-organisms also affect the availability of certain inorganic nutrients such as phosphate and manganese (p. 507), compete with the plant for nutrients and water, and affect the root by producing carbon dioxide and taking up oxygen.

The increase in size of the microbial population demonstrates effectively that as a whole the microflora benefits from the presence of plant roots in the soil. It is more difficult to assess the significance of the rhizosphere organisms for the plant in most instances. However, there is evidence that plants grown in sterile soil grow less well than those in soil inoculated with micro-organisms. Also, such plants are more susceptible to infection by reintroduced soil-borne plant pathogens.

One instance where there is no doubt of the beneficial effect of a root-soil micro-organism relationship is the symbiotic association between legumes and *Rhizobium* (p. 501). Another is the mycorrhizal association between fungi and higher plant roots (p. 620). The ectotrophic mycorrhizae can be regarded as a special case of the rhizosphere effect in which one fungus is dominant on the root surface.

Plant pathogens

The soil is a major reservoir of plant pathogenic species in particular root-infecting fungi. Wherever these pathogens exist in the soil they will be in competition with other soil organisms and it has been suggested that the species most highly adapted for the parasitic existence are ones least likely to persist for long in the soil and that less specialized parasites will have a greater competitive saprophytic ability. The phase of a parasite's existence in the soil is, moreover, the period when it is most susceptible to environmental influences and control measures are more likely to succeed at this time than at any other.

Two examples of the relative sensitivity of parasites to environmental influences during their saprophytic phase may be cited. The causative agent of potato scab (*Streptomyces scabies*) can be suppressed by green manuring, that is, by the ploughing-in of vegetation. It is thought that this is due to the antagonism of an increased population of free-living actinomycetes brought about by the green manuring. A strain of flax resistant to *Fusarium* wilt has been shown to excrete hydrogen cyanide into the rhizosphere. The cyanide allows good growth of *Trichoderma viride* but depresses that of the *Fusarium*, which is further affected by the antagonism of the *Trichoderma*.

The partial sterilization of the soil, either by means of steam or dry heat or by treatment with chemicals such as carbon disulphide, formalin, and chloropicrin, is now frequently practised with valuable glasshouse crops. The disinfection process destroys plant pathogens but also ultimately brings about other changes such as an increase in the numbers of bacteria and of certain resistant species of fungi. The multiplication of bacteria is probably mainly due to free organic nitrogen and carbon released from dead fungi and protozoa. The destruction of predators (particularly protozoa), antagonists, and competitors may be a subsidiary factor.

Succession

The metabolic activities of the dominant micro-organisms in a particular situation usually alter the environment in such a way as to favour development of another organism which then becomes dominant. Such a change is referred to as a succession and is exemplified, for instance, by the replacement of the dominant fungal flora initially developing on buried cellulose films (p. 495) by bacteria. The effect in the micro-environment is to exhaust the colonized material but the net result is the addition of humus to the soil and the development or maintenance of soil fertility which results in higher over-all microbial activity.

In a few instances the activities of the dominant microflora changes the environment in such a way that succession is impossible. The formation of peat bogs is one example of this. The digestion of the plant material under anaerobic conditions results in the production of considerable quantities of acid which so lowers the pH that biological processes virtually cease.

Antibiosis

In laboratory experiments it is frequently found that antagonism is exhibited between organisms isolated from the soil. This is commonly seen on soil dilution plates and a clear zone round a colony into which adjacent colonies have not spread usually indicates antibiotic production by the central one. The frequency with which antibiotic-producing micro-organisms can be isolated from soil indicates that they are ubiquitous and numerous. Indeed, the soil is the chief source of organisms producing antibiotics which are subsequently used for therapeutic purposes (p. 331). Nevertheless, the significance of this type of antagonism in nature is not yet satisfactorily established. Antibiotics produced by a particular organism *in vitro* are not necessarily produced in the soil and, if they are produced, the part they play in the ecological relationships in the soil is difficult to establish. Insufficient note has been taken of the importance of micro-habitats where local suitably high concentrations of organic matter for antibiotic production could well occur and where levels of antibiotic too low to be detected easily, but high enough to be effective in a restricted situation, might be found. There is now evidence that some organisms, at least, produce antibiotics in such habitats but it remains to be demonstrated that their production by an organism is important in its relationships with other soil inhabitants. The antibiotic may be rapidly degraded by chemical or biological agents in the soil, or may be adsorbed on to clay particles, but the possibility of its effectiveness at the micro-ecological level exists.

The phenomenon of fungistasis, widespread among soil fungi, is likely to be due to the sum of the effects of various antibiotics produced by the active soil microflora.

So far the low molecular weight compounds conventionally regarded as antibiotics have been considered. But in the soil, high molecular substances such as bacteriocins may also be important antagonists. Indeed, since micro-organisms in the soil usually inhabit ecological niches which are relatively restricted it would seem probable that antibiotic action will be more common between organisms of a particular type than between organisms of different types. Thus it seems possible that bacteriocins (p. 266), which have a very narrow spectrum of activity directed at organisms of the same type as the producer organisms, may be more important than the wider spectrum antibiotics which act against organisms of different types.

Lytic enzymes may also be a factor in antagonistic relationships within the soil but although it is known that certain soil bacteria produce extracellular enzymes which lyse fungal mycelium, the importance of this in the soil is in some doubt.

Predation

It is sometimes said that organisms with cell walls such as algae, bacteria, and fungi cannot be predators because they cannot engulf their prey. If this view is adopted the micro-fauna comprise the only microbial predators. Protozoa prey on one another and also on algae, bacteria, yeasts, and some even on nematodes. The latter and other small animals also feed on bacteria.

Taking a wider view of predation, several other soil micro-organisms exhibit this behaviour. Several fungi trap and kill nematodes and others prey on amoebae and rhizopods using the dead animals for food. Some myxobacteria (p. 339) kill eubacteria with a secreted antibiotic, lyse the dead cells with an extra-cellular enzyme mixture and utilize the soluble products.

To what extent predation affects the growth of higher plants is uncertain. It has been shown in one soil at Rothamsted that there was an inverse relationship, which fluctuated daily, between the numbers of bacteria, assessed by the dilution plate method, and the numbers of the two most prominent amoebae. This was confirmed later but it was also established that the total numbers of bacteria in the soil did not vary in this way, and it is now generally agreed that the protozoal population of a soil does not, as was once suggested, adversely affect soil fertility by reducing the beneficial bacterial population.

The activities of the predacious fungi have been much investigated in the hope that they could provide an effective means of controlling root parasitic eelworms. Early efforts in this direction were discouraging although it was found that in some instances, if the soil organic content was high enough, predacious fungi developed to such an extent that crop yields in infested soils were increased.

PESTICIDES AND MICRO-ORGANISMS

It has become obvious relatively recently that many substances added to the soil remain in it for long periods without being degraded. Pesticides of one sort or another, herbicides, fungicides, insecticides, and nematocides, comprise a particularly large and important part of this class of compound. While persistence may be advantageous from the point of view of pest control, the effects of these compounds, which if very resistant may accumulate significantly, on other organisms and ultimately on man give cause for serious concern.

One approach to the problem of resistance to degradation (i.e., recalcitrance) is to investigate the susceptibility of different compounds to decay and to study the effect that modification of their chemical structure has on the rate of the process. With variously substituted phenols and benzoates, it has been established that the type, position and number of the substituents affect the rate of degradation of the compound. For example, chlorophenols which are fungicidal and chloro-benzoates which are herbicidal are both more resistant than the corresponding unsubstituted phenols and benzoates.

Another approach is to study the degradation of non-persistent pesticides, to discover the pathways and to characterize the enzymes involved in their break-down. As a result of this approach a considerable amount is known about the decay of herbicides. Regrettably less is known about insecticides such as DDT, aldrin, and dieldrin, which are more toxic for mammals than many other pesticides and which are known to persist for up to ten and between six and more than nine years respectively. It is fortunate that pesticides appear to have little if any lasting effect on the soil microflora. Even soil where populations have been severely depleted as a result of soil fungicide and fumigant treatment are eventually recolonized. It is also fortunate that some of the more recalcitrant insecticides such as aldrin and dieldrin are removed from the soil by volatili-zation, but this process is slow and the gravity of the problem should not be underestimated.

BOOKS AND ARTICLES FOR REFERENCE AND FURTHER STUDY

ALEXANDER, M. (1961). *Introduction to Soil Microbiology*. John Wiley, New York and London, 472 pp.
——— (1965). Biodegradation: problems of molecular recalcitrance and microbial fallibility. *Adv. appl. Microbiol.*, **7**, 35.
BROCK, T. D. (1966). *Principles of Microbial Ecology*. Prentice-Hall, Englewood Cliffs, New Jersey, 306 pp.
BURGES, A. (1963). Some problems in soil microbiology. *Trans. Br. mycol. Soc.*, **46**, 1.
——— and RAW, F., eds. (1967). *Soil Biology*. Academic Press, London and New York, 532 pp.
CHESTERS, C. G. C. (1949). Concerning fungi inhabiting soil. *Trans. Br. mycol. Soc.*, **32**, 197.
DUDDINGTON, C. L. (1957). *The Friendly Fungi: a new approach to the eelworm problem*. Faber and Faber, 188 pp.
GARRETT, S. D. (1956). *Biology of the Root-Infecting Fungi*. Cambridge Univ. Press, 293 pp.
——— (1963). *Soil Fungi and Soil Fertility*. Pergamon Press, Oxford and London.
GIBSON, T. (1964). Progress in agricultural microbiology. *J. appl. Bact.*, **27**, 1–6.
GRAY, T. R. G. and PARKINSON, D. eds. (1968). *The Ecology of Soil Bacteria*. Liverpool University Press, Liverpool, 681 pp.
MCLAREN, A. D. and PETERSON, G. H. (1967). *Soil Biochemistry*. Edward Arnold, London, and Marcel Dekker, New York, 509 pp.
NICHOLAS, D. J. D. (1963). Biochemistry of nitrogen fixation. In P. S. Nutman and B. Mosse, eds., *Symbiotic Associations. 13th Symp. Soc. gen. Microbiol.*, Cambridge Univ. Press, pp. 92–124.

PARK, D. (1968). The ecology of terrestrial fungi. In G. C. Ainsworth and A. S. Sussman, eds. *The Fungi*, vol. 3, pp. 5–39. Academic Press, New York and London.

RUSSELL, E. J. (1957). *The World of the Soil*. New Naturalist Series, Collins, London, 285 pp.

—————— (1963). *Soil Conditions and Plant Growth*. 9th Edn. Longmans, London.

SCHALLER, F. (1968). *Soil Animals*. University of Michigan Press, Ann Arbor, 144 pp.

THORNTON, H. G. (1956). The ecology of micro-organisms in soil. *Proc. R. Soc. B.*, **145**, 364.

WILLIAMS, R. E. O. and SPICER, C. C., eds. (1957). *Microbial Ecology. 7th Symp. Soc. gen. Microbiol.* Cambridge Univ. Press, 388 pp.

13
Microbiology of Water

THE AQUATIC ENVIRONMENT

Seventy per cent of the global surface is covered by water in which micro-organisms occupy almost every niche and are intimately linked with the biological processes of decay and production. Rain, snow, or hail remove large numbers of micro-organisms from the air. Hence water reaching the earth by precipitation is not sterile. Over most of the earth's surface rain water collects micro-organisms from the surfaces of plants, buildings, or soil on which it falls. Large numbers are acquired from the soil and pass out with the drainage water into streams and rivers and thence into freshwater lakes and eventually into the ocean or inland seas. Water from natural springs or artesian wells is usually relatively free of organisms owing to the filtering effect of natural percolation through the surface rocks. The number and type of micro-organisms in surface water varies according to the source of the water, its organic and inorganic content and with geographical, biological, and climatic factors.

The organisms inhabiting large volumes of water are insulated from the extremes of climate found on land and so are rarely frozen or exposed to harmful high temperature, desiccation, or sudden changes in concentration or chemical composition of the water. Organisms inhabiting the soil water or small volumes of water such as pools, artificial ponds or ditches, however, are exposed to greater extremes and are liable to periods of desiccation when the water dries up. Only those species able to form resting stages are likely to survive in such habitats. In running water the effect of the current is important.

Most naturally occurring water contains nutrients and other substances in solution, together with colloidal and particulate matter. Pure water probably never occurs in nature but the concentration of dissolved substances varies from negligible amounts in some upland waters to progressively higher concentrations in fresh water which has collected by drainage of agricultural and industrial areas, brackish waters of estuaries, the saline waters of the oceans and the extremely saline waters of some inland lakes in regions of excessive evaporation, e.g., the Dead Sea. The most concentrated of these is still a relatively dilute solution, but the range is sufficient to exert a selective effect on the organisms. Some organisms inhabiting tidal estuaries are able to tolerate a wide range of salinity but few, if any, extend over the whole range. In fresh water the concentration of nutrients does not determine the species present but influences the 'productivity' or number of individuals developing. Productive waters are termed eutrophic, unproductive ones, oligotrophic.

AQUATIC MICRO-ORGANISMS

The algae: primary producers of organic matter

The algae are the main producers of organic matter in aquatic habitats through photosynthesis although they occupy the water column only down to the point where photosynthetically available light penetrates. This may be to the bottom of a shallow lake or to the seabed near the coast but is only a relatively thin 'veneer' of water in the open ocean.

The factors affecting the growth of algae and the means of study are similar in both marine and freshwater habitats. The organisms in the two environments are, however, almost entirely distinct, few if any species being common to both. The same genus may occur, but not the same species, e.g., *Asterionella formosa* (Fig. 13.1, E1) is a common freshwater form and *A. japonica* (Fig. 13.1, F1) is the marine counterpart. There are two major spheres in the aquatic environment; the benthic, which encompasses a vast range of habitats all associated with solid/liquid interfaces and therefore usually with the bottom, e.g., the surfaces of rocks, silt, plants, animals, etc. (Fig. 13.2 summarizes this), and the planktonic, which encompasses all the niches within the water mass itself and within which the organisms must either float or be capable of swimming. Study of the biology of the organisms in these aquatic spheres is the province of biological limnology and oceanography, though these are intimately interconnected with physical and chemical limnology and oceanography which studies the framework within which the organisms grow.

The algal flora tends to be quite discrete in each habitat. In the benthos two distinct life forms are found, one attached and non-motile, the other unattached and capable of horizontal and vertical movements. The attached species (Fig. 13.1A, B) grow on rock or stone surfaces (epilithic) either as single cells or in groups, and include many diatoms (e.g., *Synedra*, *Meridion*, *Licmophora*), the blue-green *Chamaesiphon*, creeping filaments or semicircular mucilage colonies (e.g., *Hildenbrandia*, *Rivularia*, *Chaetophora*), or filamentous outgrowths attached by basal holdfast cells (e.g., *Cladophora*, *Sphacelaria* and many small Rhodophytes). Many of these tend to be very firmly 'glued' to the substrate, some even penetrating it (e.g., the lime boring species of Cyanophyta). Some produce considerable amounts of mucilage in which they are embedded whilst others precipitate calcium carbonate around themselves and build up nodular growths some of which become fused with the rock. 'Beach rock' is formed in a similar manner in many tropical regions by the cementing of sand grains and calcium carbonate amongst the algal cells. Very much smaller algal cells are attached amongst the coating of bacteria on both freshwater and marine sand grains (epipsammic flora, Fig. 13.1A, B); the majority belong to the Bacillariophyta, though some Cyanophyta and Chlorophyta are also reported. A very dense flora develops in some situations; the cells are usually attached by mucilage pads or short mucilage stalks and they tend to be so small that they are protected by the slight concavities of the grains. A similar attached flora occurs on the undersurface of sea ice and even penetrates into the interstitial water.

The second large group of attached species occurs on larger plants (epiphytic, Fig. 13.1A, B) and on animals (epizoic). The exact relationship between these associations and those on inanimate matter is not clear, though it is certain that many species occur in only one or the other habitat. The same morphological types are present as in the lithophilic associations but there is a greater degree of host specificity, e.g., on some large marine algae a single diatom species such as *Isthmia* or *Licmophora*, is dominant, whilst on others a large number occurs. The former is often the situation also when small Crustacea are found with epizooic algal populations. On many freshwater plants a seasonal succession of epiphytes occurs. The relationship between the growth of the host, the seasonally changing environments, and the epiphytes is not yet clear. It is also becoming apparent that the metabolic activities of the host affect the epiphyte populations by the excretion of nutrients and metabolic products. Owing to the intensive concentration of the epiphytic biomass at a convenient surface it forms one of the most important 'grazing grounds' for protozoa, etc., and this flora is grazed often almost exclusively by the vegetarian fish, particularly of tropical waters. Amongst the primarily attached forms of the true epiphyton occur numerous other unattached organisms; algae such as desmids, coccoid and flagellate green algae, motile diatoms, etc., collectively termed metaphyton. It is also extremely rich in bacteria, fungi, and protozoa.

Wherever sediments are deposited in water, shallow enough to receive adequate radiation, a rich motile algal flora grows (epipelic, Fig. 13.1C, D). The prime necessity for life in this habitat is that the species are able to maintain themselves on the surface, either because they are motile (e.g., flagellates, diatoms, and Cyanophyta), or by forming flocculent masses which sediment more slowly than do the inorganic particles. These algae frequently form growths obvious to the naked eye and some mat together and may often be seen rising to the surface buoyed up by the bubbles of oxygen formed during photosynthesis. This habitat abounds in species of diatoms which have raphe systems on the two valves and many of these display an innate diurnal rhythm of motility. The motility rhythm is even more pronounced where this association occurs on inter-tidal sediments. In this latter habitat the algae (diatoms and *Euglena* species) move beneath the silt at sunset and move up on to the surface during the early morning. Tidal cover during the daylight period induces a downward movement *before* the tide actually reaches the algae; this rhythm is one which

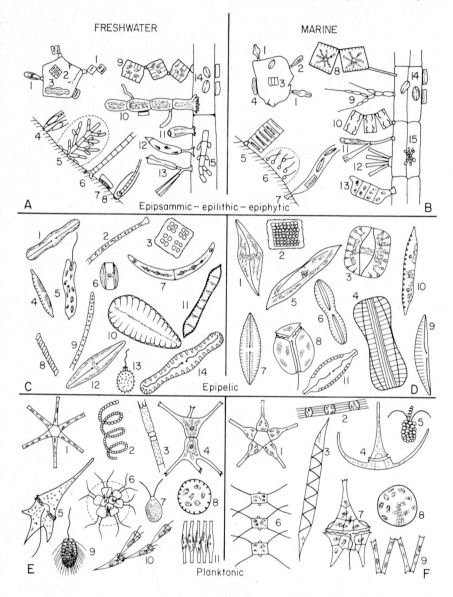

Fig. 13.1 Examples of the algae occurring in the various freshwater and marine associations. (**A**, **B**) Freshwater and marine epipsammic, epilithic, and epiphytic.

(**A**) 1. *Opephora*, 2. *Fragilaria*, 3. *Achnanthes* on sand grains, 4. *Chamaesiphon*, 5. *Chaetophora*, 6. *Ulothrix*, 7. *Cocconeis*, 8. *Calothrix* on rock, 9. *Tabellaria*, 10. *Oedogonium* (also with *Achnanthes, Cocconeis*, and *Eunotia*), 11. *Characium*, 12. *Ophiocytium*, 13. *Gomphonema*, 14. *Cocconeis*, 15. *Aphanochaete* on plant material.

(**B**) 1. *Raphoneis*, 2. *Opephora*, 3. *Fragilaria*, 4. *Amphora* on sand grains, 5. *Fragilaria*, 6. *Rivularia*, 7. *Navicula* (in mucilage tube) on rock, 8. *Striatella*, 9. *Acrochaetium*,

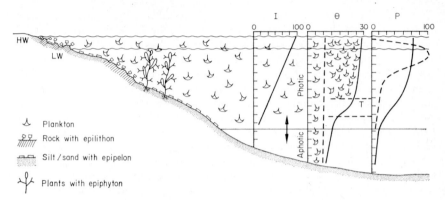

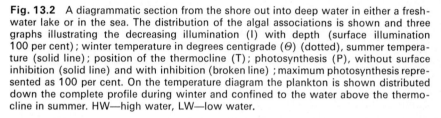

Fig. 13.2 A diagrammatic section from the shore out into deep water in either a fresh-water lake or in the sea. The distribution of the algal associations is shown and three graphs illustrating the decreasing illumination (I) with depth (surface illumination 100 per cent); winter temperature in degrees centigrade (Θ) (dotted), summer temperature (solid line); position of the thermocline (T); photosynthesis (P), without surface inhibition (solid line) and with inhibition (broken line) ; maximum photosynthesis represented as 100 per cent. On the temperature diagram the plankton is shown distributed down the complete profile during winter and confined to the water above the thermocline in summer. HW—high water, LW—low water.

can often be demonstrated under constant laboratory conditions, showing that it is under cellular control and is not impressed by the environment.

The same diatom genera are often present on both freshwater and marine sediments but the species are different, e.g., *Caloneis*, *Diploneis*, *Navicula*, *Pinnularia*, *Amphora*, *Cymatopleura*, *Campylodiscus*, *Nitzschia*, and *Surrirella* are all common, but whereas *Diploneis* and *Amphora* are represented in the freshwater epipelon by four of five species there are up to 100 or more in the marine epipelon. Cyanophyta (e.g., *Aphanothece*, *Merismopedia*, *Anabaena*, *Oscillatoria*) and desmids (e.g., *Closterium*), are more common in fresh water than in the sea, where the other components tend to be dinoflagellates and occasional Euglenoids.

---

10. *Grammatophora*, 11. *Licmophora*, 12. *Synedra*, 13. *Isthmea*, 14. *Cocconeis*, 15. Filamentous Rhodophyte on plant material.

(C) Freshwater—epipelic. 1. *Caloneis*, 2. *Oscillatoria*, 3. *Merismopedia*, 4. *Nitzschia*, 5. *Euglena*, 6. *Amphora*, 7. *Closterium*, 8. *Spirulina*, 9. *Phormidium*, 10. *Surirella*, 11. *Cymatopleura*, 12. *Navicula*, 13. *Trachelomonas*, 14. *Pinnularia*.

(D) Marine—epipelic. 1. *Gyrosigma*, 2. *Holopedia*, 3. *Campylodiscus*, 4. *Amphiprora*, 5. *Tropidoneis*, 6. *Diploneis*, 7. *Navicula*, 8. *Amphidinium*, 9. *Amphora*, 10. *Nitzschia*, 11. *Mastogloia*.

(E) Plankton—freshwater. 1. *Asterionella*, 2. *Anabaena*, 3. *Melosira*, 4. *Staurastrum*, 5. *Ceratium*, 6. *Pandorina*, 7. *Chlamydomonas*, 8. *Cyclotella*, 9. *Mallomonas*, 10. *Dinobryon*, 11. *Fragilaria*.

(F) Plankton—marine. 1. *Asterionella*, 2. *Sceletonema*, 3. *Rhizosolenia*, 4. *Ceratium*, 5. A coccolithophorid. 6. *Chaetoceros*, 7. *Peridinium*, 8. *Coscinodiscus*, 9. *Thalassionema*.

N.B. The various genera are drawn diagrammatically and are *not* to scale.

The plankton does not support such a variety of life forms since the habitat is isotropic. One small variant is formed by the algae, bacteria, and protozoa which live at the water/air interface (neustonic); they have been little studied. The only common motile forms are flagellates some of which are capable of vertical migration especially under relatively calm conditions. The other components tend to sink slowly and are kept in suspension only by water movement, e.g., the bulk of the diatoms (*Asterionella, Fragillaria, Cyclotella, Stephanodiscus,* and *Melosira* in fresh water and *Asterionella, Chaetoceros, Rhizosolenia, Bacteriastrium, Guinardia, Leptocylindrus,* and colonial *Nitzschia* species in the marine habitat). A few species have buoyancy mechanisms, e.g., gas vacuoles in Cyanophyta (*Microcystis, Anabaena, Aphanizomenon*), oil globules (*Botryococcus*) and unknown mechanisms in forms such as *Halosphaera.* Even the forms without apparent buoyancy have some subtle mechanism which aids their maintenance in the plankton since death is always followed by rapid sedimentation. Some species appear to be able to maintain themselves at given depths, e.g., some dinoflagellates are 'shade plants' only found in the sub-surface ocean water, others are 'sun plants' and predominate at the surface. The mechanism for maintaining this is not known. All these organisms are large forms easily collected by drawing fine silk nets through the water. However, between these large particles and passing through most collecting nets are a whole assemblage of minute organisms (less than 5 μ in diameter) often collectively called nannoplankton with the smaller species sometimes separated off as μ-plankton. These are algal and bacterial components and they may be very numerous although in fresh water their total biomass is probably less than that of the larger species. In the oceans many of these forms are flagellates (e.g., *Chrysochromulina*) and many are extremely delicate and very difficult to collect and preserve. Even less is known of their distribution and biology than for the larger forms, since techniques involving direct observations are not possible and although plating techniques reveal some, this is a selective procedure. In nature the species comprising the plankton are probably never randomly dispersed either horizontally or vertically (see below). Swarms of species occur in a complex interaction between growth rates, grazing, sexual reproduction, nutrient depletion, etc. In large lakes and the oceans even more extensive patches occur in the various current systems. Grazing by animals is not as general as many books lead one to believe whilst epiphytism of one alga on another, of bacteria and protozoa on algae, of parasitic fungi and protozoa on and in algae are all very common.

Three interrelated major parameters affect the whole aquatic microbiological system, viz., the radiation flux reaching the sea or lake surface, the thermal properties of the water, and the nutrient flux within the water. Primary fixation of carbon by algae is dependent on the radiation flux and this varies from a very high figure at 0° latitude, where two minima and two maxima occur every twelve months, to the single maximum towards the poles where, for a short summer period, the daily flux exceeds that of the tropics but where at all other times light is absent or severely limiting to algal growth. Most algae grow intermittently as do annual land plants and the commencement of growth may be related to the light climate. Additionally radiation is absorbed and its spectral composition modified as it penetrates water, thus even in pure water (which does not exist in nature) some 2 per cent of blue light and 4 per cent of the yellow is absorbed in 1 m. Since particulate matter, including the suspended algae, bacteria, etc., reduces the amount transmitted these figures are approached only in the purest oceanic water whilst in coastal water 40 per cent or more may be absorbed in the first metre. The effect of this light absorption can be readily shown if the algae living on the sediments from the shore are studied down a depth profile. These can be estimated either as cell numbers, cell volume, or by estimation of some cell component (e.g., chlorophyll α). In a relatively clear lake (e.g., L. Windermere) this population has virtually disappeared between 6–8 m. but in a very small lake with turbid water (e.g., Abbots Pond near Bristol) a comparable population exists only down to 2 m. Measurement of a planktonic population in this way is not possible since water currents continuously mix the population through all depths (p. 519) and hence light limitation of growth is better determined by inoculating plankton into stoppered bottles which can be hung at each depth. 'Growth' may be measured over short intervals by measuring photosynthesis (Fig. 13.2) either by the change in oxygen content in the bottles or by uptake of ^{14}C supplied in a soluble form. However, as Fig. 13.2 shows, the highly illuminated surface water often has a lower rate of carbon fixation than has the sub-surface; this is related to inhibition of carbon fixation by high light intensity. In general the red end of the spectrum is absorbed rapidly and the blue penetrates to the greatest depths. Thus algae at various depths are exposed to light of different wave-lengths as well as to different amounts.

The second major effect of radiation entering the water is its heating effect, which if the water was pure and undisturbed would show an exponential fall with depth. However, owing to cooling of the surface by evaporation and movement via wind-induced currents, a very characteristic pattern develops for all except very shallow waters. The radiation heats the surface waters and this is transmitted to lower depths by wind-induced currents. When heating becomes excessive in early spring the resultant reduction in density of the surface heated water makes it difficult to mix this water into the denser cold layer below (rather like pouring water on to syrup and blowing on the surface to mix the two liquids). The layer of warm water then rests all summer on the cold lower layer and the region of rapid temperature drop between the two is known as the thermocline (Fig. 13.2); the water mass is said to be thermally stratified and active circulation is then confined to the upper layer. In the autumn, surface cooling and increased wind stress eventually break down the stratification, sometimes within 24 hours, and the water then circulates to a greater depth, which may be to the bottom in many lakes or through 100 m. or more in the open ocean. Thus during the most unfavourable time of the year the algae are circulating throughout a greater depth and therefore spend a longer period in the region where photosynthesis is impossible. In the summer, however, they are trapped in the warm surface water which is also well illuminated (Fig. 13.2).

This stratification has two major effects on the nutrient supply to the micro-organisms, firstly absorption of nutrients is excessive in the surface water above the thermocline and nutrient concentrations may thus be reduced to limiting amounts. Replenishment is slow, since transport across the thermocline is hampered by relative lack of water movements. Secondly, in the lower water, oxygen is being depleted by respiration of all the inhabitants, both micro- and macro-organisms, and is neither being replaced by downward movement, since the thermocline is a barrier, nor by photosynthesis, since it is below the limit of light penetration. Various degrees of de-oxygenation result and these are accompanied by a steady build-up of nutrients in this region. In lakes these tend to diffuse out of the sediments when the surface changes from an oxidized to a reduced state. Thus the surface few metres of freshwater lakes and the upper 100 m. or so of the ocean may be rich in organisms but depleted of nutrients whilst below this depth is a rich reservoir of nutrients unavailable to the surface crop until mixed by some agency which breaks down the thermocline. In the sea this nutrient-rich water upwells in some coastal areas (e.g., off the Peruvian coast) and this results in tremendous algal productivity. The breakdown of thermal stratification in fresh water allows the deep nutrient-rich water to mix with the surface water but this usually occurs in autumn by which time light intensity is often too low to allow much growth before the spring.

The factors affecting the growth of micro-organisms in nature have been studied for only a few species. Nutrient depletion has been shown to be responsible for the decline of the growth rate of such diatoms as *Asterionella* and *Stephanodiscus*. Since diatoms have an absolute requirement for silica and this substance is only supplied slowly, via inflows or by re-cycling from the lower waters, it can be easily exhausted. This effect is pronounced on organisms in the plankton but as might be expected less obvious amongst the benthic forms which are living close to a supply of silica in the sediments or possibly diffusing from some plant stems. Other algae which require carbon and not silica for their cell walls are unlikely to be limited by lack of wall material since carbon is continuously supplied via solution of carbon dioxide and the interaction of pH, carbonates, and bicarbonates. Instead nutrients such as nitrogen and phosphorus become limiting to many algae especially in mid-summer when supply from the lower waters is limited. There is other evidence, however, to show that some algae, such as *Dinobryon*, grow in lakes when the concentration of the major ions has been reduced and this is coupled with further experimental evidence that high phosphate concentrations are deleterious.

The onset of growth of vernal species is not related to a build-up of nutrients, since these have been in high concentration throughout the winter, but is triggered by increase in radiation flux. Temperature is still very low and although it does affect the growth rate is not the prime factor. Other species (e.g., *Oscillatoria rubescens* and *Melosira italica*) behave as winter or cold-water shade plants. Both these species tend to disappear at least from the surface waters of lakes in spring. *M. italica* has been shown to remain alive on the sediments for several years and its sudden appearance in the plankton in the autumn is due to mixing of this material upwards by autumn gales. It is a heavy species and if ice cover occurs it sinks owing to the reduction of wind-induced turbulence. In spring, as the temperature increases and stratification begins, it sinks out of the surface water.

Since all species in the aquatic environment have patterns of seasonal growth it is obvious that the problems are complex, and, in addition, there are well-authenticated examples of biological interactions which influence the primary fluctuations correlated with the cycle of physico-chemical variables. Indirect effects of the production of antibiotics have been demonstrated in culture, but although these are undoubtedly important, it is extremely difficult in the natural habitat to separate effects due to antibiotics from those of other factors. Much more drastic, however, are the effects of parasitism, e.g., of the chytrid *Rhizophidium* (p. 391) on planktonic diatoms. Detailed investigations over the last two decades have revealed a large number of such parasites on almost all freshwater planktonic species and recently a whole series of minute animal parasites have also been found in or on planktonic algae. So far few such cases have been detected amongst benthic organisms but this may be merely due to lack of detailed study. Recently algal viruses have been reported (p. 268) and these too may have a profound effect on algal populations.

An imperfectly understood aspect of the chemical composition of the environment is its effect on the species composition of all the algal associations. No two water masses are alike in their species complement but certain broad features can be discerned particularly where there is a single over-riding factor (e.g., saline lakes tend to be dominated by certain Cyanophyta, such as *Spirulina*; very acid, low-calcium lakes by desmids and Chrysophyceae; organically rich waters by Euglenoids). However, almost all algae can be cultured in defined media containing the common elements plus the vitamins thiamin, cyanocobalamin and biotin, all of which are generally present in natural waters. Although an obvious simplification, it is clear that certain species are abundant in low-calcium/acidic waters (e.g., the diatoms, *Frustulia*, *Stenopterobia*, *Actinella*) whilst others occur only in higher calcium/alkaline waters (e.g., *Gyrosigma*, *Cymatopleura*, *Epithemia*). Similar correlations can be made for almost every group of algae. In the marine environment, however, there is no such easy distinction, since the waters are all alkaline and salinity varies only slightly, at least in the open oceans which appear to have a similar flora throughout the world, though ecological races do occur related to temperature and salinity. Brackish waters of varying degrees are an exception and have characteristic floras. Further study may also reveal differences in the benthic flora.

Heterotrophic micro-organisms

Heterotrophic micro-organisms are usually present in natural waters in direct proportion to the amount of organic matter available. Bacteria, aquatic fungi, and protozoa are associated with the various communities of algae described above. By their activities they break down organic compounds and thus return plant nutrients to the water.

Bacteria are found wherever sufficient organic matter is present and vary in number with the amount of such food material available. They are numerous around submerged vegetation and in and just above the mud layer at the bottom of both fresh water and the sea. In the absence of algae, competition for oxygen is intense on the bottom sediments and oxygen becomes limiting for aerobic species. Anaerobic bacteria and a few fungi with an unusually low oxygen requirement then predominate.

The number of bacteria present in freshwater plankton is directly correlated with the amount of organic matter present and hence is greatest in eutrophic waters. In waters relatively unpolluted by intestinal organisms, both pigmented and non-pigmented bacteria occur, including species of *Pseudomonas*, *Chromobacterium*, *Achromobacter*, *Flavobacterium*, and *Micrococcus*. Organisms washed into water from soil are mainly spore-bearing bacilli, including species of *Bacillus* (aerobic) and *Clostridium* (anaerobic), but also include members of the coli-aerogenes group (Eschericheae) normally found on plants and decaying vegetation. Nitrifying bacteria and species of *Streptomyces* are frequently present. In water polluted with animal or human excreta many organisms characteristic of the intestinal canal survive for considerable lengths of time and include *Escherichia coli*, *Streptococcus faecalis*, *Proteus vulgaris*, and *Clostridium perfringens*. Occasionally intestinal pathogens, such as *Salmonella typhi*, and, in certain parts of the world, *Vibrio cholerae* and other water-borne pathogens may be found.

Bacteria, including species of *Pseudomonas*, *Vibrio*, *Flavobacterium*, and *Achromobacter*, are present in the surface layers of the sea but their numbers differ widely with conditions. They are most numerous where the organic content of the water is high, that is near coasts, particularly where land drainage occurs, in harbours and in areas of highly productive plankton. Various counts show decreases from more than 1,000,000 per ml. in harbours and around 400,000 per ml. near the shore to as low as 10–250 per ml. a few miles offshore. Among the seaweeds of the Sargasso Sea the count rises to more than 1,000 per ml. A large proportion of marine bacteria are pigmented and thus protected from the sun's rays. Some, such as the purple sulphur bacteria, may be sufficiently numerous to impart a red tinge to the water. Many are phosphorescent. Species of actinomycetes and yeasts have also been recorded. Marine bacteria tend to be smaller than freshwater ones, but sulphur bacteria and the spirochaetes are exceptions. All are salt-tolerant (halophilic) and most are relatively intolerant of high temperatures. Marine bacteria play similar parts in the nitrogen, sulphur, phosphorus, and carbon cycles in the sea to those played by other species in the soil (p. 493). Some bacteria may be troublesome by eroding metal e.g., ships' plates) immersed in sea water.

Where oxygen is available, as in moving water, or in epiphytic algal communities, a wide range of aquatic fungi is found. Reference has already been made (p. 391) to the chytrids parasitic on algae. Other species are parasites on other aquatic fungi or on protozoa. Many are saprophytic on organic matter. Stems of such freshwater aquatic plants as *Phragmites* and species of *Scirpus* are parasitized by aquatic Ascomycetes, including many Discomycetes. A few brown seaweeds are parasitized by species of *Mycosphaerella*, some Pyrenomycetes and some Fungi Imperfecti. Pyrenomycetes (e.g., species of *Ceriosporopsis*) also attack and erode wood submerged in the sea. Decaying leaves of deciduous trees in well-aerated streams and lakes support a characteristic flora of aquatic Hyphomycetes. These are taxonomically unrelated forms and although most of them resemble one another in the production of conidia with projecting arms or spines or consisting of a curved or branched row of cells, these superficially similar spores arise by quite different processes. Their shape, however, is well adapted to their habitat and they readily become entangled in submerged leaves which they colonize after germination. Many of the aquatic Ascomycetes have ascospores of a similar shape to the conidia of the aquatic Hyphomycetes. *Saprolegnia* (Fig. 10.16, M–P) and related species and *Monoblepharis* (found on submerged sticks) are also limited to well-aerated water. These produce motile spores (zoospores) and resting spores (oospores) which are non-motile and relatively thick walled.

Where oxygen is scarce, as in deposits of leaves or other organic matter at the bottom of stagnant ponds or where excessive growth of bacteria has depleted the supply, the fungus flora is quite different from that of similar substrata in well-aerated waters. Certain Hyphomycetes, such as species of *Clathrosphaerina* and *Helicodendron*, which have distinctive coiled spores may be present as hyphae but usually produce their conidia only when removed from the water and placed in humid air. They have been described as 'aero-aquatic' fungi. Members of the Blastocladiales may also be present. It has been shown that *Blastocladia pringsheimiana* is almost anaerobic and that it produces its resting spores only under conditions of high carbon dioxide concentration.

Most water moulds of the family Saprolegniaceae are less common in strongly eutrophic waters than in rather less productive ones, but the related *Leptomitus lacteus* (the so-called sewage fungus), increases with the organic content of the water and is particularly common in water polluted by sewage or organic industrial effluents where it may become a serious nuisance by actual mechanical blocking of channels and by exhaustion of the available oxygen supply.

Aquatic protozoa have been considered in a previous chapter (p. 402). They are common in both fresh and salt waters.

Planktonic protozoa (ciliates, flagellates, and the Heliozoa or 'sun animalcules' which possess radiating strands of protoplasm) and other animals, particularly the grazing species, also fluctuate more or less with the phytoplankton. They are numerous in marine plankton, and include flagellates, foraminifera (members of which sink and build up thick calcareous deposits on the ocean floor, e.g., the globigerina ooze), radiolaria (which similarly contribute to siliceous deposits, e.g., radiolarian ooze) and ciliates. Other small animals and larval forms are also present. Grazing of the phytoplankton by these microscopic animals and by the young fishes may be a factor in the fluctuation of the algae and even be a contributory cause of the frequent disappearance of phytoplankton under conditions of good nutrition.

MICRO-ORGANISMS AND THE PROBLEMS OF WATER POLLUTION, SEWAGE DISPOSAL, AND WATER SUPPLY

Pollution of natural waters

The increasing use of fertilizers and the disposal of wastes in streams is resulting in rapidly increasing nutrient content (eutrophication) in many natural water bodies (e.g., Lake Erie), and enormous algal crops result, followed by their decay giving massive pollution of shores. In spite of considerable study few algicides have proved effective and non-toxic to other organisms. Sodium pentachlorophenate has been used but it is toxic to fish. Quinone derivatives, e.g., 2,3-dichloronaphthoquinone (dichlone), are more effective and apparently less harmful. Similar problems of excessive algal growth arise in industrial plants, e.g., in power-station cooling towers, and ponds. Here the above chemicals and quaternary ammonium compounds (p. 210) and tetrachlorobenzoquinone (chloronil) have been used successfully. The potent fish-killing alga *Pyrmnesium parvum* has been controlled in Israeli fish ponds by treatment with ammonium sulphate. The degree of pollution of both fresh and marine waters can be determined by studying the algal growth. Benthic algae are very sensitive indicators and in the most extreme polluted zones only algae such as *Oscillatoria chlorina* and *Spirulina jenneri* occur and in the slightly less polluted zones the diatoms *Nitzschia palea* and *Gamphonema parvulum* appear. Pollution often reduces the diversity of the algal flora but within limits increases enormously the growth of the resistant species, owing to the removal of competitors and predators. This is particularly noticeable in marine habitats where removal of molluscs, etc., results in intense growth of diatoms and allows the sporelings of green algae such as *Enteromorpha* to become established. A further deleterious effect of algae is their growth on all objects placed in water or which conduct water for industrial processes. Ships are particularly prone to colonization by algae and in spite of the use of algicidal paints there is still no completely satisfactory solution to this expensive problem.

The recent growth in number of nuclear power stations, from which a large amount of excess heat has to be dissipated, is giving cause for concern over 'heat pollution' of rivers and estuaries. If the waters of these are used for cooling not only is there direct danger to fish, but interference with the numbers and types of micro-organisms may disrupt the food chain or favour harmful species.

Sewage disposal

The problem of the disposal of domestic and industrial wastes is considered in Chapter 22. It is appropriate, however, to consider some aspects of the problem here.

In open sewage oxidation ponds green flagellates and coccoid algae grow in distinct sequence and are essential to the purification process since they supply the oxygen for the bacterial growth. Unfortunately they have to be removed, by the ton, which itself provides problems. They can be processed into cattle food by dilution with other feeding stuffs; undiluted algae from these plants are *too* concentrated a protein source for direct consumption. It has been calculated that in the United States a total of six million acres of such ponds will be needed to deal with the water-borne wastes of that country by 1990 and that recovery of the algae from such systems would feed a quarter of the country's livestock. The initial engineering problems of pond design, recovery of algae (by centrifugation or flocculation), dewatering (centrifugation, drum drying or sand bed drying), and processing have already been overcome. The incidental or intended effect of the algal growth in these processes is the removal of vast quantities of inorganic salts from the waters. Experiments are now proceeding in an attempt to utilize this property for the extraction of these salts from polluted waters draining either from industrial plants or from irrigation schemes.

The disposal of trade wastes presents many problems. Some industrial liquors, such as the effluents from flax retting (p. 692), from factories dealing with sugar beet or milk, or from slaughterhouses or breweries, can be mixed with sewage or treated in similar disposal plants. Other trade wastes may contain poisonous organic material (e.g., effluents from factories processing leather, cellulose, textiles, or glue) or poisonous inorganic substances (e.g., effluents from coke ovens, which contain inorganic cyanides, from metal works, where acids and metallic compounds will be present, or from lead or copper mines). These cannot be dealt with by ordinary methods of sewage disposal without preliminary treatment to remove the toxic substances. A different type of pollution is that from china-clay mines where a fine deposit of clay covers the bottom of streams is continually replenished, thus preventing growth of most micro-organisms. This is only a local effect and is not a danger to public health.

Purification of water supplies

Storage

When water is stored in a reservoir the amount of suspended food material normally tends to decrease, both by sedimentation and as a result of the activities of micro-organisms. This in turn leads to a reduction in the number of micro-organisms and in particular of pathogenic bacteria which are unable to survive in competition with saprophytic species.

Populations of planktonic and benthic algae similar to those in natural water occur in water supply reservoirs. These algae are controlled by similar factors to those in nature and the same principles apply with the added complication of control by man, e.g., thermal stratification may be prevented by inflow design, pumping, etc. The algae are here normally beneficial since they absorb certain nutrients, supply oxygen to the water, have some anti-bacterial effect, assist filtration, etc., but these activities can readily become deleterious when the algal growth becomes too great. Water reservoirs receiving acid drainage water with low nutrient concentrations rarely give trouble, but the same water impounded in lowland reservoirs or river water containing large amounts of dissolved chemicals can produce enormous algal crops (the so-called 'water blooms'). These may decay and produce very unpleasant by-products, causing taints or odours. If the water is being drawn off during such a growth the beds of sand through which the water is filtered in the treatment plants become clogged within a few hours and have to be drained and the surface layer of sand removed. Algae, such as *Cladophora*, may grow actually on the filters and then must be removed. Control of the algae in the reservoirs may be achieved by adding small amounts of copper sulphate to the water, but this does not always have the desired effect since the removal of one alga is often followed by the growth of another and as is often the case after copper sulphate treatment this may be a smaller and even more troublesome species. Reservoir management, including the installation of jets at the inlets to cause the water to circulate, switching to alternative reservoirs, drawing off the water at the most suitable level from a stratified reservoir, and the regulation of the storage period to give the maximum reduction of bacteria with the minimum increase of algae, is important in overcoming the problem of excessive algal growth.

Filtration

Even if no additional pollution occurs during the process, storage of water in open lakes and reservoirs is insufficient to render it safe for drinking, hence artificial methods of purification must also be employed. The water is first piped to a filter plant which may be a *slow sand filter* suitable for reasonably pure waters and for use where adequate space is available for the large filters needed, or a *rapid sand filter*, suitable for turbid waters or where insufficient land is available.

SLOW SAND FILTERS. The natural process of filtration through rocks is imitated by allowing the waters to percolate slowly through clean sand. Purification does not, however, depend only on mechanical straining of the water by the sand but is achieved mainly by the activity of the micro-organisms present in the system. After filtration has proceeded for several days a slimy gelatinous mass composed of bacteria, protozoa, and algae (the 'Schmutzdecke') accumulates, particularly in the upper layers of the sand. This layer brings about biological oxidations and reductions, slowly closes up the pores of the filter thus rendering it more effective by slowing down the rate of filtration, and provides predatory protozoa which feed on bacteria. The net result is that the water emerging from the filter is chemically and biologically purer than before filtration. Eventually the slime layer becomes too thick and slows filtration too much. The filter bed is then cleaned, relaid with fresh sand and filtration is renewed. A newly prepared filter bed is of low efficiency and the water from it cannot be used until a new slime layer has developed.

RAPID SAND FILTERS differ from the slow filters in having a smaller filter area and in the rapidity and entirely mechanical nature of the process. The organic matter in the water is first precipitated by chemicals such as ammonium aluminium sulphate, and the sticky, bulky precipitate is allowed to sediment in large settling basins. Bacteria as well as any colouring matter in the water are readily adsorbed on to the precipitate and are thus carried down with it. The clean supernatant fluid is finally run over a sand filter to remove any precipitate still in suspension and the purified water is collected at the bottom of the filter.

BOOKS AND ARTICLES FOR REFERENCE AND FURTHER STUDY

HUTCHINSON, G. E. (1957, 1967). *A Treatise on Limnology*. Vol. I, *Geography, Physics and Chemistry*, 1015 pp. Vol. II, *Introduction to Lake Biology and the Limnoplankton*, 1115 pp. John Wiley and Sons, Inc., New York.

JACKSON, D. F., ed. (1968). *Algae, Man and the Environment*. Syracuse University Press, 554 pp.

KRISS, A. E., MISHUSTINA, I. E., MITSKEVICH, N. and ZEMTSOVA, E. V. trans Syers, E. (1967). *Microbial Populations of Oceans and Seas*. Edward Arnold, London, 287 pp.

LEWIN, R. A., ed. (1962). *Physiology and Biochemistry of Algae*. Academic Press, New York and London, 929 pp.

ROUND, F. E. (1965). *The Biology of the Algae*. Edward Arnold, London, 269 pp.

RUTTNER, F. trans. Frey, D. G. and Fry, F. E. J. (1953). *Fundamentals of Limnology*. Univ. of Toronto Press, 295 pp.

SOUTHGATE, B. A. (1950). *Treatment and Disposal of Industrial Waste Waters*. H.M.S.O., London.

14
Microbiology of Air

INTRODUCTION

The atmosphere of the earth contains many minute particles of solid matter, a large proportion of which are of biological origin, some being viable. Most viable airborne particles are spores which are to some extent suited for survival in such an environment for a limited period, but most eventually succumb to the rigours of airborne travel. Not only spores of fungi, myxomycetes, bryophytes, and pteridophytes, but also pollen grains, moss gemmae, propagules of lichens, cells of algae, vegetative cells and spores of bacteria, cysts of protozoa, and virus particles may occur in the air and constitute the 'air spora'. The largest of these particles are the pollens which may range up to 200 μ in diameter, though most are 20–50 μ. Most airborne fungal spores are between 3 and 30 μ in diameter, most bacterial cells about 1 μ in diameter, and viruses 0·1 μ. Methods for studying the air spora and some results of their application are described below. The air spora comprises cells which have entered the atmosphere at various places and by various means. Thus the interpretation of the spora in a given air mass requires an understanding of the probable movements that brought it there. It is therefore necessary to consider first some physical aspects of the atmosphere.

Air is in constant motion, the kinetic energy of this circulation deriving primarily from radiation from the sun. The amount of radiant energy received by the earth is balanced over a period of time by the amount radiated back into space, but in the course of the year the region between 40°N and 40°S receives more energy than it loses by radiation, while the regions nearer to the poles lose more than they receive. Consequently there must be a net transport of heat from the equatorial belt towards the poles. This is effected chiefly by winds generated by the differential heating of the atmosphere, warm air flowing towards the poles, and cold air towards the equator. The vertical transport of energy eventually to be radiated into space is augmented locally by convection currents, especially over land in intense sunshine. Because of the drag of the ground the various major currents of air generate a layer of turbulence which extends upwards for perhaps 500–1,000 metres. However, in the immediate vicinity of the ground and other surfaces the air is virtually still or flows in an orderly fashion. This constitutes the laminar boundary layer which may be several metres thick on clear nights when the wind is light, but around a blade of grass on a windy day may be only about a millimetre in thickness.

ESTIMATION OF MICRO-ORGANISMS IN THE AIR

Qualitative and quantitative study of the particles in the air requires first that they be trapped. The air spora may be trapped by filtration, centrifugation, electrostatic deposition, or by accelerating the air to a high speed so that airborne particles collide with suitably placed sticky surfaces or fluids. No single method of trapping is universally satisfactory because of the diversity of particle sizes. Small particles will tend to follow flow-lines of air as it travels around intercepting surfaces. Only if particles are large enough or travelling fast enough will they strike the surface. If the surface itself is large then only extremely large or fast-moving particles will be caught. For reasons such as these, different methods have been used to trap different sorts of airborne micro-organisms. Most techniques used nowadays sample the spores by drawing air at known rates through a suitable trap, and give results faster than do passive sampling methods in which sticky surfaces are freely exposed to the atmosphere. Slides and petri dishes have been exposed in the slipstreams of aircraft, but their relatively large size discriminates against small particles of the size of most fungal spores or less. Two widely used aspirator traps are the cascade impactor, in which sampled air passes through successive jets at different velocities so that particles of different size ranges are trapped at different stages, and the Hirst spore trap in which the particles impinge on a slowly moving target so that the numbers of particles captured from the air at different times can be recorded.

Methods used to collect samples from air in buildings are similar in principle but the more cumbrous apparatuses, such as the Hirst spore trap, are unsuitable. It is possible to use animals directly as a means of assessing airborne contamination with animal pathogens but quantitative studies are only possible if a single organism can give rise to an initial lesion. The level of contamination of the air in a confined cubicle from a patient suffering from pulmonary tuberculosis has been determined by this means. Controlled volumes of air evacuated from the cubicle were led to cages housing guinea pigs. Inhalation produced lesions (tubercles) in their lungs and the number of tubercles found at post mortem were approximately proportional to the number of bacilli present in the air that they breathed.

After the constituents of the air spora have been trapped by mechanical means it is generally necessary to identify them. Trapped particles may be identified visually under the microscope if they are sufficiently large and distinctive. Pollen, various spores, lichen propagules, algae, and protozoa may be identified in this way with more or less precision. A total count of bacteria may be obtained by direct inspection but these micro-organisms can seldom be identified unless they possess specific staining reactions (as, for example, does *Mycobacterium tuberculosis*).

If components of the trapped spora are too small or too uncharacteristic to be identified or countable by visual means it is generally necessary first to culture them on artificial media or appropriate host organisms or tissues. Trapped fungi and bacteria may sometimes be identified after cultivation on artificial media, though because culture media are selective and organisms sometimes antagonistic the method has limitations. The use of fluorescent antibody techniques or specific bacteriophages can allow colonies to be identified at very early stages. Serological tests can also be applied. Viable counts may be obtained by inoculating known volumes or dilutions of a fluid used to scrub the organisms from the air on to culture media. More generally, bacterial counts are made on 'settle plates' of solid culture media exposed directly to air, time being allowed for particles to settle. This technique can be made more sensitive and quantitative by the use of the 'slit-sampler' which draws a known volume of air through a narrow slit immediately above a rotating plate of nutrient medium, the plate making one complete rotation during the exposure.

A number of factors can greatly influence the counts. These include the choice of nutrient media and the temperature of incubation. Often single organisms, particularly pathogens, fail to multiply into colonies unless media of a highly nutrient quality, more complex than is normally required, are used. The use of inhibitory selective media that inhibit the majority of airborne bacteria and select for a limited group (e.g., crystal violet blood agar for *Streptococcus pyogenes* in fever hospital wards) frequently gives a lower count than expected. On the other hand, higher counts may be realized if excessive turbulence of the sample fluid, as in bubbler sampling, disperses clumps of organisms. In any method involving the cultivation of trapped cells it is necessary to recognize that the very process of trapping may kill some of them.

Airborne viruses and bacteriophages have been recovered from air in suitable sampling fluids such as 10 per cent skim milk. Animal viruses are inoculated into living animal tissues including embryonated eggs and tissue cultures (p. 237). Bacteriophages are mixed with a suitable host culture in soft agar and poured on to the surface of a nutrient agar plate (p. 256).

ORIGIN, DISTRIBUTION, AND MOVEMENT OF THE AIR SPORA

Spores remain suspended in the air for as long as their fall speeds are less than the speeds of frequently recurring upward air currents. The terminal velocity of a falling particle in air is proportional to the square of its radius, and for a body 20 μ in diameter it is about 1 cm./sec. Thus for most fungal spores it is 0·05–2·0 cm./sec. and for most pollens 1–10 cm./sec. Convection and turbulence can generate upward air speeds considerably in excess of these velocities, but spores must first of all enter regions where this sort of movement is frequently encountered before they begin to rise. Most of the air spora derives from the surface of vegetation or vegetable debris above ground level. The laminar boundary layer which adjoins these sources is too tranquil for shed spores to remain long in suspension. However, many organisms have structures or mechanisms which introduce spores more or less directly into the turbulent zone. Under many circumstances it is necessary that the spore be raised only a few millimetres above the parent surface in order to do this. Numerous organisms expel their spores forcibly for distances ranging from a few to many millimetres by active mechanisms in which the propulsive force is generated within the organism. Modifications towards this end are found among ferns, fungi, and bryophytes. Among the fungi, forcible spore release dependent on turgid cells is accomplished by most Ascomycetes, most Entomophthorales, and a few conidial fungi. Mechanisms are also known which involve hygroscopic twitching or the breakdown of cohesion in water within gradually dehydrating cells. Basidiospores and the ballistospores of mirror yeasts (Sporobolomycetaceae, p. 395) are forcibly discharged by an obscure mechanism for short distances from the parent cells which themselves are frequently raised on fruit-bodies or on substrates so that free gravitational fall introduces the spores into turbulent air. Other fungi launch their spores by passive means, the environment contributing the requisite energy. Thus slimy spores may be launched in droplets sprayed by rain-drop impact. Rain drops may also expel clouds of spores from puffballs on which they fall, and may dislodge spores from conidiophores by vibration and air tremors. Gusts of wind sometimes disrupt the laminar boundary layer and pick up spores. The elevation of spores of many species on erect conidiophores places them in situations where their chances of being swept up by such gusts are greater.

Viruses, bacteria, algae, and protozoa have no special mechanisms of take-off. Animal hosts by sneezing and coughing may assist the launching of some bacteria and viruses, but this is biologically significant only in restricted spaces (p. 535). Adventitious physical disturbances are probably the agents mainly responsible for the take-off of these various organisms. Seasonal variations in numbers of airborne bacteria may reflect seasonal peaks of agricultural activities which raise dust to which organisms may be attached.

Another sort of periodicity, this time diurnal, is shown by populations of fungal spores near ground level. Spores of mirror yeasts are most abundant in the air spora before dawn, spores of *Phytophthora infestans* late in the morning, and spores of *Cladosporium*, *Alternaria*, and *Ustilago* in the early afternoon. Although this periodicity could reflect diurnal patterns of air turbulence or movement, the major cause is probably diurnal periodicity of spore release. Release of the spores of many fungi is influenced by environmental factors among which light is often important. Also some release mechanisms depend on the activity of turgid cells, so it is likely that only when the environment is moist will release occur. It is for this reason that after rain there is often the appearance near ground level of a distinctive 'wet-air spora', rich in basidiospores, ballistospores of mirror yeasts, and ascospores. This replaces rather than augments the 'dry-air spora', which consists of pollen grains and the dry spores of *Cladosporium*, powdery mildews, *Alternaria*, smuts, and rusts, and which is largely washed out of the air by the same fall of rain that generates the wet-air spora.

Once a spore has taken off it will rise or fall in the air according to the relative influences of turbulence and gravity. Spores liberated in stable air encounter little uplift and soon settle to earth, but upward movement may be swift in unstable conditions. Upward air speeds of 30 cm./sec. are common near the ground, while well above ground these may exceed 600 cm./sec. The vertical profile of spore concentration in many parts of the world generally shows a decrease in concentration with height. Spores may even be undetectable at heights of 2–6 kilometres. Nevertheless, sampling of very large volumes of air with balloon-borne equipment at much greater heights has yielded small numbers of viable cells of moulds, yeasts, and bacteria. Theory predicts that concentration will decrease logarithmically with height and actual observations sometimes show this relationship, but various factors may modify the profile. For example, over a source of spores there is constant replenishment at the base of the profile, but away from the source a high concentration near ground level may no longer be maintained. Also rain which evaporates before reaching the ground may sweep spores from a higher to a lower altitude. Bacteria and fungal spores are the most numerous members of the air spora near ground level where an average concentration of spores in country air in summer is 10,000/m^3, but over short periods the concentration may greatly exceed this.

Two aspects need to be distinguished in considering the horizontal distribution of spores in the air: firstly the behaviour of clouds of spores arising from a single restricted source and secondly the background concentration of spores from a myriad distant sources. The concentration of spores in the atmosphere decreases rapidly with distance from a source and with time from the moment of release. More than 90 per cent of spores from near-ground sources are often deposited within 100 metres. The progressive dilution of the spore cloud by upward migration of the remainder leads to spores from the local source becoming undetectable at near-ground level within a short distance. The spore cloud may be envisaged spreading downwind as a widening, gently ascending plume. The new technique of plotting the course and shape of clouds of particles by lidar, an optical analogue of radar in which pulses of light from a laser reflect from atmospheric particles, may add much to what is known of the movement of such clouds. Despite the numbers of spores entering the atmosphere in the ways described, very few are ubiquitous and catches made at sea show rapidly decreasing numbers of micro-organisms in the air as distance from land increases. Few organisms other than marine bacteria are caught in mid-ocean.

Spores which continue their ascent may eventually rise into regions of high wind where they may remain suspended for several weeks. Under these conditions they are exposed to low temperatures, desiccation, and intense and harmful solar radiation before they eventually descend. Survival of long journeys of this sort is unlikely, though it should be noted that very low concentrations of viable fungal and bacterial cells have sometimes been detected at heights in the range 18–27 km.

The airborne flight of a spore draws to a close when various factors tend to accelerate the spore's downward velocity. Rain is an important factor in this process. Inside rain clouds the majority of particles in excess of 0·2 μ diameter become the nuclei of cloud droplets which aggregate and fall as rain. Falling rain drops range up to about 5 mm. in diameter. Whether spores are captured by falling drops depends on whether the spore has sufficient inertia to resist displacement by air lying ahead of the falling drops. Spores which are larger than 2μ diameter, as are most eukaryotic spores, are readily intercepted by raindrops. Raindrops 2 mm. in diameter are reported to have the greatest collecting efficiency and the efficiency increases with the size of the spore. Another mechanism of deposition is sedimentation in association with boundary layer exchange, which is the process by which spores from a cloud of particles overhead diffuse into the boundary layer of air in which settling is mainly gravitational. The spores thus constantly replenish the population in the ground layer of still air in which even minute spores can sediment. Deposition of spores is also achieved by their impaction against solid objects. Impaction can be efficient for relatively large spores ($> 10 \mu$ diameter) encountering small sticky objects at high speeds but is not efficient for small particles or for large obstacles. The relative importance of these various deposition processes varies with the circumstances. Close to the source of a spore cloud impaction and boundary layer exchange are important while rain-wash has less effect because the height and breadth of the spore cloud are small. However, rain assumes more importance as a cause of deposition as the spore cloud rises and broadens.

SOME CONSEQUENCES OF THE EXISTENCE OF AN AIR SPORA

Airborne dispersal of spores out of doors has considerable biological and economic importance. Numerous plant diseases are caused by air-borne fungi (p. 399). The rapid spread of some such diseases may be the result of many successive acts of dispersal and infection rather than long individual journeys, but long-distance dispersal is very important in the epidemiology of some. For example, uredospores of the wheat rust *Puccinia graminis* do not survive the cold winters in the northern parts of the north American wheat belt, nor the hot summers in the southern part. Spring-sown wheat in the northern U.S.A. and Canada is showered by rust uredospores blown northwards from rusted autumn-sown wheat in Mexico and Texas. Sometimes the spores travel large distances in a single hop, while at other times a succession of shorter hops is punctuated by phases of infection and multiplication. A reverse air movement later in the season carries spores south to infect the next autumn-sown wheat crop in the south. Migrations of uredospores of cereal rusts have been reported also in India, Russia, and elsewhere.

Because most human and animal pathogens spread from host to host directly, being unable to multiply in the non-living environment, outdoor air is not a serious source of infection. It is noteworthy that the principal exceptions to this generalization are fungus diseases such as histoplasmosis and coccidioidomycosis, both of which are caused by organisms able to multiply in the soil. Many sorts of airborne biological particles, particularly pollen grains and spores of certain fungi, are, however, important causes of allergies in man (p. 578).

THE MICROBIOLOGY OF AIR INSIDE BUILDINGS

Many organisms normally found in the outside atmosphere will also be present in the air inside buildings where they are introduced by air currents but the main sources of contamination arise from within the buildings. Since most airborne micro-organisms have no special structures to facilitate their dispersal into the atmosphere, they are dependent on physical disturbance for their 'take-off' (p. 532). The microbial flora of the air within any single building at any one time therefore depends on the number and kind of organisms present and the mechanical movements within the enclosed space. These are usually the result of ventilation and animal or human activity. An air flow in a room varies widely according to the shape, size and furnishings of the room, the movement of its occupants and the design and manner of operation of heating and ventilation systems. Air flow is particularly pronounced near a concentrated heat source. Micro-organisms present in dust are from animal or vegetable sources. Many are ejected from the respiratory tract of animals and man in droplets of moisture. Once airborne, organisms are not reduced in number or dispersed as in the air outside buildings, but numbers continue to increase unless the room is adequately ventilated or otherwise treated (p. 593).

Dust as a vehicle of airborne contamination may arise from textiles (especially bedding, handkerchiefs, and clothing contaminated from contact with man) and desquamated skin scales and hairs which are being continuously shed. In buildings housing animals, the air pollution is particularly high owing to the presence of hay, straw or other fodder, bedding and dried excreta, and contamination from the animals' coats. The denser particles settle rapidly but particles of 1 μ or less in diameter remain suspended permanently. The dry-sweeping of floors, the dusting of objects, shaking of cloths, making of beds, movement of people, and draughts can break up the original substrates into finely divided particles or disturb settled dust and cause it to become airborne.

The other important source of micro-organisms is the respiratory tract. During talking, coughing and sneezing air is forced, under considerable pressure, through the nose and the constricted apertures between the teeth. Organisms from the upper respiratory tract are thus ejected into the air. Fluid picked up by this expelled air is broken down into droplets. Sneezing is the most vigorous of these mechanisms and a single sneeze may generate as many as one million droplets less than 0·1 μ in diameter and thousands of larger drops, mainly from the saliva at the front of the mouth. Not all expelled droplets become airborne; the larger ones exceeding 100-200 μ in diameter fall to the ground before evaporation but the smaller ones rapidly evaporate down to their non-volatile residuum and remain suspended as 'droplet-nuclei'.

The survival of airborne micro-organisms will depend on many factors. Spores which can tolerate dehydration will survive longer than vegetative cells and the length of survival of all micro-organisms is increased in humid atmospheres away from the bactericidal rays of sunlight which penetrate glass windows to only a limited extent (p. 594). The presence of organic material, as from saliva, skin, etc., gives added protection to many organisms.

The microbial content of the air inside buildings may include viruses, pathogenic and non-pathogenic bacteria, and fungi. Endospores of the genera *Bacillus* and *Clostridium*, especially *Cl. perfringens*, are commonly found in inhabited rooms, hospital wards, and even operating theatres. *Rhodotorula* and other yeasts, and spores of species of *Aspergillus*, *Penicillium*, *Mucor* and other moulds are commonly present. These may contaminate food and moist, perishable organic materials, such as leather, and by inhalation may cause respiratory infections of man and animals and allergic reactions, such as asthma.

Many viruses, including those of influenza, the common cold, virus pneumonia, measles, German measles, and smallpox are dispersed by air and may produce infection when inhaled. Airborne bacterial pathogens include the organisms causing scarlet fever and tonsillitis (*Streptococcus pyogenes*), tuberculosis (*Mycobacterium tuberculosis*), diphtheria (*Corynebacterium diphtheriae*), and whooping cough (*Bordetella pertussis*). All but the last of these have been demonstrated in the dust of fever hospital wards. Various occupational diseases result from the inhalation of contaminated dust associated with particular occupations. Thus 'woolsorter's disease' was caused by the inhalation of dust particles contaminated with spores of the anthrax bacillus from wool imported from parts of the world where anthrax is common; this is now controlled by the prior treatment of the wool with formalin. The development of intensive methods of incubating, hatching, and rearing chicks in large numbers in cabinet incubators is responsible for an increase in the severity and extent of Pullorum Disease caused by *Salmonella pullorum*. Not only is the disease transmitted in eggs from diseased hens (p. 595) but also fluff from infected chicks, which is heavily contaminated with the bacterium, dries and becomes dispersed throughout the atmosphere of the hatching chamber. Inhalation of this results in extensive lung infection, as many as 90 per cent of the chicks being infected in the first week of life.

Certain fungi with airborne spores are responsible for pulmonary disease. These may be derived from sources outside buildings, such as mouldy grain, straw, damp vegetation, and compost heaps, some of which may be brought inside farm buildings as bedding or fodder. Outbreaks of aspergillosis in pigs, for instance, have been traced to consignments of bedding straw contaminated with *Aspergillus fumigatus*, and penguins, when exposed to the stress of transport, frequently become infected from contaminated bedding and develop aspergillosis when kept in captivity. The spores of *A. fumigatus* have been found in 80 per cent or more of samples of dust examined from city houses. This fungus can produce one or other of several forms of disease in man. It is able to grow in the mucoid secretions of bronchi, without invading the tissues, producing a hypersensitive state (p. 578), the patient demonstrating an allergic response either to the fungus already in his respiratory system or when later spores of the same species are inhaled. Aspergilloma is an X-ray detectable solid lesion resulting from a saprophytic colonization by *A. fumigatus* of an old lung cavity such as may be caused by a healed tuberculous cavity; the lesion can be surgically removed. In the invasive type of aspergillosis the fungus invades lung or other tissues, often secondary to any grave systemic disease, the fungus probably contributing to death of the patient.

Spores of other fungi, including *Cladosporium herbarum*, cause allergies in man. The serious disease of agricultural workers known as 'farmer's lung' is, however, caused by the development of hypersensitivity to spores of thermophilic actinomycetes (p. 328) derived from mouldy hay and not to spores of *Cladosporium* and other fungi even though these are also present in the same hay. Ultimately the inhalation of only a limited number of actinomycete spores may cause severe pulmonary symptoms.

Until recently many microbial agents responsible for ripening of cheese gained entry into the curd from the air of the dairy. Hence certain localities became famous for particular types of cheeses owing to the presence of specific organisms (e.g., species of *Penicillium* giving their characteristic flavour to gorgonzola, stilton, roquefort or camembert cheese). Nowadays cheese is largely produced on a factory scale and the ripening processes are initiated by the introduction of pure cultures as 'starters' (p. 655). Bacteriophages, acting against bacterial starters, may be also present in the air of cheese factories where they present an important problem.

BOOKS AND ARTICLES FOR REFERENCE AND FURTHER STUDY

GREGORY, P. H. (1973). *The Microbiology of the Atmosphere*. 2nd Edn. Leonard Hill Limited, London, 257 pp.
—— and MONTEITH, J. L., eds. (1967). *Airborne Microbes. 17th Symp. Soc. gen. Microbiol*. Cambridge Univ. Press, 385 pp.
WILLIAMS, R. E. O., BLOWERS, R., GARROD, L. P. and SHOOTER, R. A. (1966). *Hospita Infection*. 2nd Edn. Lloyd-Luke Limited, London, 386 pp.

15

Micro-organisms and Vertebrates: their normal and pathological relationships

INTRODUCTION

The skin and mucous membranes of animals are continuously exposed to contamination with micro-organisms. These contaminants include saprophytes, pathogens, and many intermediate forms. Most organisms fail to establish themselves in any particular site on the body, partly as a result of the self-sterilizing action of many tissues and partly owing to competition between micro-organisms for the same site. Certain species, not normally pathogenic, are particularly suited to colonization of various sites; some are present in a particular site under almost all conditions, others may be there only temporarily. Thus in man *Escherichia coli* is a normal intestinal parasite whereas *Staphylococcus aureus* is present on the skin of only some individuals and at certain times.

Organisms regularly present constitute the 'normal flora' of these sites. Little is known of the nature of the association between micro-organisms and the healthy host tissue, but it is thought that there exists a dynamic equilibrium rather than mutual indifference between the two. Some organisms of the normal flora may assume a pathogenic role under certain circumstances, e.g., when the host resistance is lowered by injury to tissues, or by diseases that diminish resistance. Pathogenic organisms may or may not be present in the normal flora but most have a special capacity to spread from host to host, establish themselves, and later invade deeper tissues from the site initially colonized.

NORMAL FLORA

The term 'normal flora' may be misleading since the environment can greatly influence the types of micro-organisms present. Nevertheless, there are certain bacterial species which are invariably found in particular sites in the healthy animal. Thus it is possible to give general descriptions of normal flora for a given animal species. The normal flora of corresponding sites in different animal hosts varies widely and is profoundly influenced by such factors as body temperature and diet. Most work on this subject has been done on man.

Normal flora of the human skin

The normal flora of the skin and sebacious glands is restricted because of the natural disinfecting powers of the skin. It consists of a few species of bacteria including Gram-positive cocci (mainly *Staphylococcus albus* and *Sarcina* spp.) and diphtheroids. These are all non-pathogenic but sometimes the pathogenic species, *Staphylococcus aureus*, is found on the face and hands, particularly in individuals who are nasal carriers. These organisms persist even after intensive washing. Many other organisms, not generally considered to be part of the normal flora, may be present temporarily in numbers largely determined by the standard of personal hygiene, especially in moist crevices of the skin and surrounding the body orifices, such as the anus. In the fatty and waxy secretions of skin, lipophilic yeasts are common; *Pityrosporum ovalis* occurring most commonly on the scalp and *P. orbiculare* on the glabrous skin. In the secretions of ears and genitalia, saprophytic acid-fast bacilli (e.g., *Mycobacterium smegmatis*) and diphtheroids are common. The tears secreted on the conjunctiva of the eyes contain the enzyme lysozyme, which dissolves bacteria on the eye, and this keeps the number of organisms in this site at a low level.

Normal flora of the human respiratory tract

Diphtheroids and *Staph. albus* are found in limited numbers in the nose. *Staphylococcus aureus* is often present in large numbers in the nose, this being the main carrier site of this important pathogen. In addition to these organisms, larger numbers of many more species of bacteria are to be found in the nasopharynx; predominantly mixed viridans streptococci (e.g., *Streptococcus salivarius*) and *Neisseria* spp. (e.g., *N. pharyngis*). The nasal sinuses are normally sterile in a healthy person. Occasionally pathogens, such as pneumococci, *Haemophilus influenzae*, *Neisseria meningitidis*, *Klebsiella pneumoniae*, and allied organisms, may be harboured in the nasopharynx of apparently healthy individuals.

The trachea, bronchi, and pulmonary tissues are, in the healthy state, almost free of micro-organisms. Air is inhaled through the tortuous nasal passages which act as effective filters. Bacteria adhering to particulate matter are removed at this site. Most organisms that pass this filter are trapped on the mucous surfaces of the nasopharynx and trachea, being swept upwards by ciliated epithelial linings of the tract, and are subsequently removed by expectoration or swallowing. There is thus no normal flora of the mucosa of trachea and bronchi.

Normal flora of the genital tract of the human female

In the sexually mature human female the vagina is acidic, owing to the production of lactic acid from glycogen in the epithelial linings, a defence against other bacteria. Consequently the main flora consists of mixed acid-tolerant organisms, including the predominant *Lactobacillus acidophilus* (Döderlein's bacillus), together with corynebacteria and anaerobic streptococci. Mycoplasmas (p. 344) have been isolated on a number of occasions. Before puberty and after menopause the vaginal secretions are alkaline and contain normal skin organisms, e.g., staphylococci, streptococci, coliforms, and diphtheroids. The organs of the urinary tract are generally free of bacteria, possibly as a result of the flushing action of the urine.

Normal flora of the human alimentary tract

The presence of particles of food, epithelial debris, and secretions makes the mouth a nutritionally favourable habitat for a great variety of bacteria. The flora is subject to great variation but always includes very large numbers of α- and non-haemolytic streptococci and, in addition, non-pathogenic species of *Neisseria*, anaerobic spirochaetes, *Fusobacterium* (particularly between gums and teeth) and lactobacilli (in the acidic cavities of teeth, especially in the presence of dental caries). The yeast-like fungus *Candida albicans*, the cause of 'oral thrush' and other diseases, is also normally present (Fig. 15.1) but causes disease only in persons, particularly infants, suffering from malnutrition or in adults with debilitating diseases, e.g., diabetes, or in whom the normal bacterial flora is disturbed by the use of antibiotics, steroids and cytotoxic drugs.

The majority of vegetative organisms, other than acid-fast bacilli, originating in food or from the mouth are destroyed by the hydrochloric acid of the gastric juices so that the stomach, when empty, is relatively free of living micro-organisms. In the small intestine the influence of bile secretions becomes gradually apparent; the highly buffered foods revert to an alkaline pH and the numbers of bacteria increase, reaching a maximum in the caecum and colon. The flora includes coliform bacilli (particularly *Escherichia coli*), entero-cocci, clostridia (mainly *Clostridium perfringens*), species of *Bacillus*, Gram-negative pleomorphic non-sporing anaerobes of the genus *Bacteriodes* (this genus constitutes the predominant group of organisms), saprophytic acid-fast bacilli and many Gram-negative rods of varied types. Large numbers of organisms die during their passage down the gut and approximately one-third of the dry weight of faeces consists of bacteria.

Diet greatly influences the predominant flora. This is illustrated by the changes in flora in a child as a result of variations in diet during early life. In the breast-fed infant the flora consists predominantly of *Lactobacillus bifidus* which, during the early days of life, may form 90 per cent of the intestinal flora. With change in diet other organisms, characteristic of the adult alimentary canal,

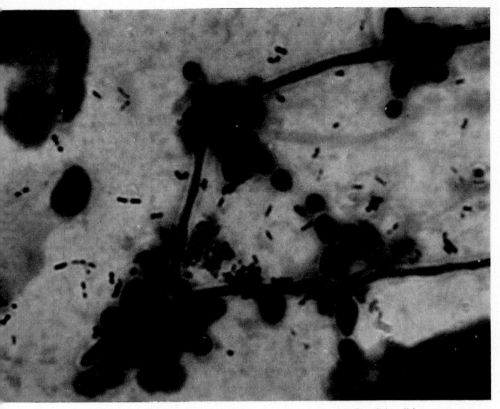

Fig. 15.1 Smear of human sputum showing pneumococci and *Candida albicans* as budding forms and pseudomycelium. (Gram stain; ×1,500; A. H. Linton)

gradually become predominant. A high meat diet results in an increase in pro-teolytic organisms.

Origin of the alimentary flora

The alimentary tract of the foetus is sterile but at birth becomes rapidly and progressively infected by way of the mouth and anus. Immediately after birth, organisms commonly found in the mouth of the new-born are of the same type as those in the vagina of the mother. A few days later other organisms, charac-teristic of the normal mouth flora of adults, develop and it is thought that these originate from the naso-pharynx of the mother. The intestinal flora of infants gradually changes with time as other organisms gain access until a complex mixture of organisms resembling that in the adult is attained.

The importance of the mammalian intestinal flora

Considerable variation in intestinal flora is found among mammalian species. *Escherichia coli*, for example, is far less frequently found in herbivores than in carnivores. In guinea pigs lactobacilli comprise approximately 80 per cent of the

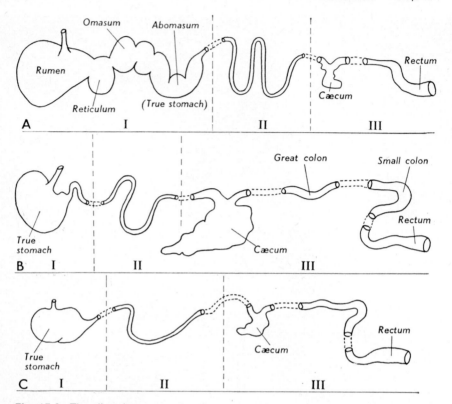

Fig. 15.2 The digestive tracts of different animals showing how ruminants differ from other animals. (**A**) Herbivorous ruminant, e.g., cow. (**B**) Herbivorous non-ruminant, e.g., horse. (**C**) Omnivorous animal, e.g., rat. (**i**) pre-absorptive part of gut. (**ii**) small intestine. (**iii**) large intestine. (After Kon, S. K. (1953). *Voeding*, **14**, 137)

flora, the remainder consisting of organisms from soil and air. Consequently the importance of the intestinal flora must be considered in relation to the host species. Unlike the flora of herbivores, discussed below, the flora of carnivores is probably not essential to life. Animals, including guinea pigs and chicks, have been successfully reared on a balanced and sufficient diet under completely sterile conditions without showing stunted growth. With chicks any lack of normal ability to digest their food completely was made good by eating more. In these animals, therefore, life has been shown to be possible in the absence of a gut flora, although there is an indication that the enzymic activities of the bacteria assist in the digestion of the food. However, the administration of broad spectrum antibiotics over a long period may result in vitamin deficiency, presumably because these vitamins are normally synthesized by intestinal organisms.

In contrast to carnivorous or omnivorous animals, which depend wholly upon the secretion of enzymes into their alimentary canal to digest their food, herbivorous animals rely upon the enzymic activities of micro-organisms to bring about the initial digestion. In these animals the gut flora is essential to life and a symbiotic relationship exists between the host and the microbial flora of the gut. The breakdown of vegetable matter requires a large chamber where the food can be retained and slow transit through the digestive tract to allow sufficient time for microbial degradation to take place. The alimentary tract of herbivorous animals is modified to include an enlarged portion or portions of the tract, such as the caecum and colon in the horse, or an enlarged portion of the oesophagus, including rumen and reticulum in addition to the true stomach, in ruminants (Fig. 15.2). In the latter the diet is subjected to two fermentations, one preceding and the other following the conventional mammalian digestive process.

Microbiology of the rumen

The food of ruminants is complex, containing carbohydrates (mainly cellulose), proteins, fats, numerous organic compounds, and minerals. Cellulose, derived from plant cell walls, is broken down by the enzyme cellulase secreted by bacteria. This energy-yielding carbohydrate is converted into microbial cells and metabolic by-products including carbon dioxide, methane, ammonia, and acetic, propionic, lactic, and butyric acids. The bacterial proteins and their products of metabolism, including vitamins, are food for the host.

A complex mixture of micro-organisms is involved in the breakdown of food. An extremely dense population of many species of bacteria and protozoa results from the favourable conditions provided in the rumen. The rumen consists of a system not greatly different from that of a laboratory steady-state continuous culture (p. 180) with a regular supply of ingested food and a continuous removal of fermentation products and food residues. The rumen bacteria are adapted to live at acidities between pH 5·5 and 7·0, under anaerobic conditions and at a temperature of 39–40°C.

Techniques for the study of the microbiology of the rumen include the isolation of pure cultures on artificial media, containing sterilized rumen fluids, to which inorganic salts and specific carbohydrate, e.g., cellulose, is added. This is essential for the study of precise metabolism of particular organisms. On the other hand, the biochemical activities of micro-organisms in the rumen are so complex and interrelated, one organism often utilizing the breakdown products of another, that it is often necessary to study the system as a whole rather than the limited activities of pure isolates. Direct microscopy and the study of the biochemical activity of washed mixed cell suspensions, obtained through artificially produced fistulae or openings in the functioning rumen, have yielded much information.

Only insignificant numbers of intestinal organisms common to man and other omnivorous animals (p. 540) have been demonstrated. The specialized flora for the breakdown of cellulose and its degradation products include micro-organisms unique to the rumen.

Cellulose decomposition is associated with iodophilic bacteria (i.e., organisms which reveal starch-like substances when stained with iodine) found in large numbers attached to plant debris in the rumen contents. 'Chewing the cud' doubtless facilitates the disintegration of the cellulose fibres, thus allowing bacterial attack to commence at the broken ends. In these, cavities of decomposition are etched out by bacterial enzymes. The same morphological forms are consistently found in these cavities.

Yeasts of the genera *Candida* and *Trichosporon* have been found in small numbers in the rumen of the cow, sheep, and goat. Many protozoa of varied kinds are found in the rumen, the majority being ciliates. The part they play in rumen digestion is not clear. It has been shown that certain species (e.g., *Entodinitum neglectum*) fail to grow in artificial culture if cellulose is omitted from the medium. It does not follow, however, that cellulose is digested by the protozoa since in all cultures there is active fermentation by cellulolytic bacteria. The necessity for cellulose in the medium could therefore be due to a requirement by the protozoa for the metabolic products of the bacteria digesting the cellulose. It is generally thought that protozoa play no vital role in rumen metabolism.

The diversity of bacteria in the rumen is striking. Some types are found regularly in the rumen flora; others are highly host specific. The flora varies according to the chemical composition; some bacteria have a broad and others a narrow range of biochemical activity. Some of the typed species of rumen bacteria are considered according to the substrate attacked.

Cellulose digestion

Three kinds of rod-shaped cellulolytic bacteria present in large numbers in the rumen are *Bacteriodes succinogenes*, *Butyrivibrio fibrisolvens*, and *Clostridium lochheadii*. *Bacteroides succinogenes* is a Gram-negative, crescent-shaped organism with migratory and biochemical activities similar to the myxobacterium *Cytophaga succinicans* (p. 339). It ferments glucose and cellobiose in addition to cellulose; from cellulose, succinic acid is the most important product. Cellulolytic cocci include *Ruminococcus flavefaciens*, which grows in chains, is Gram-positive and produces yellow colonies, and *Ruminococcus albus*, a Gram-negative coccus which characteristically produces white colonies.

Starch digestion

Amylolytic, but non-cellulolytic, rumen bacteria include *Streptococcus bovis*, *Bacteroides amylophilus*, and *Selenomonas ruminantium*, a crescent-shaped flagellate cell. The fermentation products vary among strains but include formic, acetic, propionic, butyric, and lactic acids.

Hemicellulose digestion

Hemicelluloses are plant carbohydrates insoluble in water, which constitute a large part of the forage consumed by ruminants. Most of the cellulolytic and amylolytic bacteria attack this carbohydrate.

Other important rumen bacteria include *Veillonella alcalescens*, which is concerned with fat metabolism and lactate utilization, and *Peptostreptococcus elsdenni*, also an important lactate fermenter which follows the acrylic pathway of propionate formation.

Synthesis of vitamins of the B complex

The amount of B vitamins in rumen contents is found to be several times greater than that expected on the basis of the vitamin content of the food ingested.

This is the result of synthesis by rumen bacteria but not protozoa. The vitamin concentration in the rumen is not influenced by the amount present in the diet; metabolic control of bacterial synthesis maintains a steady level. This explains the fact that the concentration of these vitamins in ruminant's milk remains steady irrespective of fluctuations in the diet. Although the synthesis of vitamins by bacteria in the alimentary canal is not limited to ruminants, it is probably of particular importance to these animals since for the larger part of the year they graze on dried pastures of poor quality which are deficient in vitamins.

Protein synthesis

The most common nitrogen source for ruminants is provided as protein in forage. The proteins are broken down by ruminant micro-organisms into peptides, amino acids, and ammonia, these products being used in different ways for the synthesis of microbial protein. This in turn becomes available to the host when bacteria die and are digested by the ruminant. Protein is therefore used by the host in the form of rumen bacteria and this constitutes the major part of the ruminant's nitrogenous food.

This knowledge has opened up a new approach to nutritional problems by adding urea or ammonium salts to diets deficient in protein, with the effect of allowing cheaper high-fibre feedingstuffs (e.g., straw) to be economically utilized. The microbial degradation of protein in the rumen inevitably means loss through ammonia and urea. Efforts are therefore being made to protect food proteins (e.g., by formalin) so that they may enter the abomasum before undergoing digestion.

The influence of antibiotics on alimentary flora

Considerable changes in faecal flora result from the administration of sulphonamides and broad spectrum antibiotics. Antibiotics, often in combination, are sometimes prescribed prior to alimentary surgery in order to render the bowel virtually free of living bacteria, with a view to avoiding septic complications. This has met with mixed success. In a proportion of patients, the normal bowel flora is superseded by antibiotic-resistant staphylococci, resulting in a staphylococcal enteritis. Overgrowth of other antibiotic-resistant micro-organisms, such as *Candida* spp., may also produce complications.

The addition of small amounts of antibiotics to the feed of poultry and swine has sometimes been shown to increase food conversion and rate of weight-gain. Antibiotics have also been tried with ruminants but only young animals have shown an advantage in food conversion. When later the ruminants' nutrition depends on rumen micro-organisms the addition of antibiotics at low levels to the food has negligible effect, but in higher amounts may interfere with digestion. The mechanism of antibiotic growth stimulation is not fully understood but it is thought that the antibiotics check the growth of toxic organisms during the early days of life when the gut of the new-born animal, which is thinner and absorbs nutrients more efficiently than the adult gut, is first colonized by micro-organisms. There are, however, reasons why this practice of feeding antibiotic-supplemented foods should be discontinued. Even sub-inhibitory levels of antibiotics are selective for antibiotic-resistant organisms, and if these organisms carry the genetic determinants of infective drug resistance (p. 582) transfer to enteric pathogens may occur; as a result the pathogens would become resistant to the antibiotics being used and would not subsequently yield to antibiotic therapy.

Coprophilous fungi

The dung of various herbivorous animals supports a large fungus flora. Many of these are chance contaminants from air or soil but a few species, including representatives of the Zygomycetes (*Pilobolus* spp.), Ascomycetes (*Sordaria* and related genera, e.g., *Coprobia*, *Ascobolus*), and Basidiomycetes (e.g., some species of *Coprinus*) are always associated with particular species of animals. These coprophilous species may be taxonomically unrelated but are biologically similar in their adaptation to the habitat. Their spores are ingested with the grass or other animal food and pass through the alimentary canal of the animal. Many of them germinate during passage through the gut, thus gaining a start which enables them to compete with faster growing species, such as *Mucor* spp., which reach the dung only after excretion. It is not known whether these germinated spores play any part in the breakdown of cellulose in the gut, but their part would certainly be a minor one compared with that of the more numerous rumen and intestinal cellulolytic bacteria. The spores of many coprophilous species are thick-walled or enveloped in mucilage, some contain black or purple pigments and many germinate only after exposure to relatively high temperatures or treatment with digestive juices, i.e., under conditions similar to those obtaining in the animal's gut.

A succession of fungi develops on the excreted dung, including true coprophiles and chance contaminants. Coarse mycelium and sporangia of species of *Mucor* and some other Zygomycetes are the first to develop and are accompanied, or closely followed, by the truly coprophilous *Pilobolus* (Fig. 15.3C). Species of the Zygomycetous genera *Chaetocladium* and *Piptocephalis* may occur parasitically on the mycelium of other Zygomycetes. When the first wave of Zygomycetes has partially subsided apothecia of discomycetous genera develop, closely followed by perithecia of *Sordaria* and related genera. Finally fruit bodies of a number of species of *Coprinus* and some other agarics are formed.

Most coprophilous species are phototropic (Fig. 15.3) and many of them discharge their spores to considerable distances by various explosive mechanisms. The orientation of the spore-bearing parts towards the direction of greatest light, combined with violent spore discharge, gives the spores the maximum chance of falling away from the substratum on to the surrounding grass. The zone of response to light varies with the group of fungus; thus the sporangiophore of *Pilobolus*, the neck of the perithecium of *Sordaria* and other Pyrenomycetes, and the apices of the asci of the Discomycetes are phototropic.

A somewhat similar relationship occurs between species of *Basidiobolus* (a Zygomycete of the family Entomophthorales, p. 393) and frog and lizard dung. The conidiophores of these fungi are phototropic and the conidium is shot off by a mechanism involving the bursting of a sub-conidial vesicle. The conidia are eaten by beetles which may later be eaten by the frog or lizard, in the gut of which the spores germinate.

Fig. 15.3 Reaction to light of some coprophilous fungi. **(A)** *Sordaria tetraspora* (seen in optical section ×45). **(B)** *Dasyobolus immersus* (×38). **(C)** *Pilobolus klenii* (×9). **(D)** *Coprinus fimetarius*; left, young sporophore primordium orientated by positive phototropism; right, mature expanded fruit-body liberating spores. (approx. nat. size). (After Ingold, C. T. (1953). *Dispersal of Fungi*. Clarendon Press, Oxford)

ANIMAL DISEASES CAUSED BY MICRO-ORGANISMS

Introduction

In contrast to the equilibrium established between the normal flora and the host, in disease the host–parasite relationship favours the parasite at least initially. In this section emphasis will be placed on the mechanisms by which the parasite produces disease; in the next chapter the resistance of the host to infection will be considered. Disease-producing parasites either directly damage the animal tissues or disturb the bodily function. Such disturbance produces physical changes in the host recognized as signs of disease and these range from minor tissue damage to death.

Infectious diseases of animals may be caused by bacteria (Fig. 15.4 and 15.5), fungi, protozoa, or viruses. A few parasitic bacteria and fungi are able to propagate in nature outside the animal host and assume a pathogenic role when introduced into the body. For example, *Aspergillus fumigatus* is commonly present in soil and on plants or plant products such as grain or straw, but when the spores, in sufficient quantity, are inhaled by birds, or occasionally by other animals including man, the respiratory disease aspergillosis may result. In contrast most pathogens, whilst able to survive for varied periods of time outside the animal

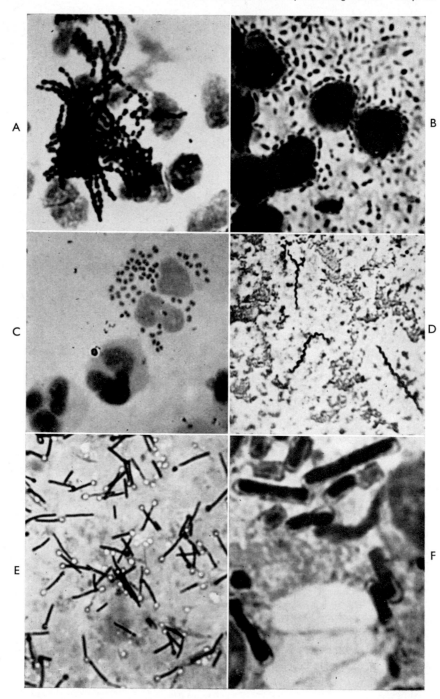

host, normally do not multiply away from the animal. The majority of pathogenic bacteria and fungi have been grown under laboratory conditions devised to resemble the host environment. Many grow relatively easily; others only with great difficulty because they require highly nutritious media and prolonged incubation. Thus *Mycobacterium johnei*, the causal pathogen of chronic enteritis in cattle and sheep, will grow only on egg media containing extracts of other mycobacteria and on primary isolation takes upwards of two months to produce visible growth. A few have not so far been grown in artificial culture or, if so, have not retained full virulence, e.g., *Mycobacterium leprae*, the causal pathogen of leprosy, and *Treponema pallidum*, the causal pathogen of syphilis.

The survival of a pathogenic micro-organism depends on many interdependent factors. Under natural conditions, if an obligate parasite is to survive, it must be able to infect a susceptible host in which it can propagate. A high degree of infectivity is essential to the survival of those organisms which do not have a resistant phase. Other factors, such as the numbers of susceptible hosts available, also limit the survival of the pathogen. In island populations and other isolated communities, where the number of susceptible hosts is limited, certain diseases, such as measles, may flourish for a time then rapidly die out owing to this limiting factor. An organism which seriously damages its host may not survive, since if the host is killed the organism perishes unless it can first reach another susceptible individual. On the other hand, an organism of low virulence may fail to survive owing to inability to overcome the body defences of the host. Among the most successful of the common pathogens are those which have a high degree of infectivity but relatively low virulence under natural conditions, e.g., the virus of the common cold.

Dispersal of pathogenic micro-organisms

Escape of pathogens from infected host

The successful survival of a parasite depends on its ability to escape from one host and be transmitted to and infect another susceptible animal. Pathogenic micro-organisms usually escape from the living animal in excreta from the

---

Fig. 15.4 Various bacteria in preparations of pathological material. **(A)** *Streptococcus agalactiae* in pus from mastitis of bovine udder (× 2,200). Gram-stained. (K. E. Cooper). **(B)** *Diplococcus pneumoniae* in deposit of cerebro-spinal fluid from a human case of meningitis (× 2,000). Gram-stained. (A. H. Linton). **(C)** *Neisseria gonorrhoeae* in smear of pus from the cervix of a woman with gonorrhoea (× 2,000). Note that the diplococci are phagocytosed by a polymorph leucocyte. Gram-stained. (A. H. Linton). **(D)** *Treponema pallidum* in a section of liver from a baby with congenital syphilis (× 2,500). The spirochaetes have been impregnated with silver (Levaditi's stain) to make them visible (Department of Bacteriology, University of Bristol). **(E)** *Clostridium tetani* in a smear from an infected wound (× 2,200). It is very rare to find so many organisms in an infected site—usually they cannot be detected in direct smears. Gram-stained (K. E. Cooper). **(F)** *Bacillus anthracis* in a smear from spleen of an infected mouse (× 2,200). Note the capsulated, non-sporing bacilli arranged singly or in pairs. This is characteristic of the appearance of this organism in body tissues and contrasts with the long chains of non-capsulated, usually sporing bacilli seen in smears from artificial cultures. Methylene blue-stained. (K. E. Cooper)

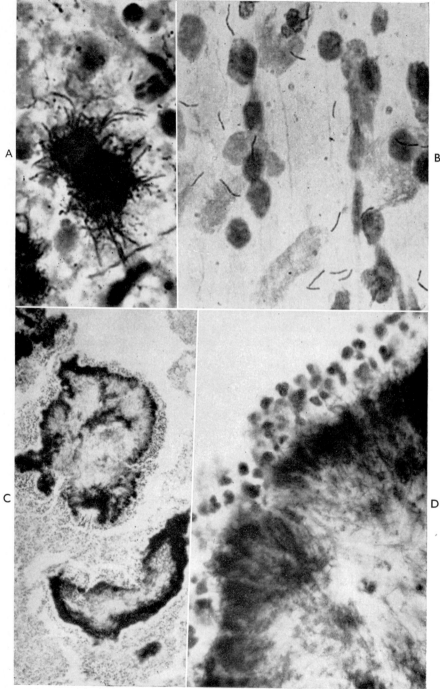

alimentary tract, in secretions from mucous membranes of the respiratory tract, in organic debris such as desquamated skin and hair, and in pus from infected wounds. Coughing and sneezing assists the escape of organisms from the respiratory tract.

Transmission of pathogens between hosts

This usually occurs between hosts of a single animal species but sometimes infection arises between hosts of different species. Examples of the latter are the infection of man by dermatophytic fungi from infected domestic animals, by *Pasteurella multocida* from a cat bite or scratch, or infection of a dog with the tubercle bacillus derived from its master.

Certain diseases are contracted by direct association with an infected host. This may occur by inhalation of droplets from an individual with a respiratory disease (e.g., tuberculosis, common cold virus), by direct contact with an infected host either venereally (e.g., syphilis and gonorrhoea in man or vibriosis in cattle) or generally (e.g., ringworm infection).

Contamination of specific sites may result in infection. For instance, infections of the human external ear may be caused by *Aspergillus niger* and *A. fumigatus*; following mastoid operation the resultant cavity, in which debris collects due to impaired drainage, frequently becomes infected by any of the following: *Candida* spp., *Aspergillus niger*, *A. fumigatus*, *A. flavus*, and *A. terreus*. Bacteria also often infect these sites.

Many diseases are contracted indirectly by ingestion of food contaminated at source from an infected animal or contaminated subsequently by polluted water or by excreta from an infected patient or healthy carrier (e.g., food poisoning, typhoid, tuberculosis, brucellosis, etc., p. 586). Horses may be indirectly infected with anthrax by feeding on hay grown the previous summer on anthrax-contaminated soil, since the spores survive throughout the winter months. Another method of indirect infection is by penetrating the skin with a contaminated instrument, as often occurs with tetanus, or by contamination of an abrasion, as in anthrax in man.

Some pathogens are transmitted to a new host by arthropod vectors. This is the mode of transmission of the plague bacillus and is very common with many virus, protozoan, and rickettsial infections (p. 612). The rickettsiae of typhus fever, transmitted by lice, infect the susceptible host by insect excreta entering either by the bite of the vector or through an abrasion caused by the patient's scratching.

The length of time during which pathogenic organisms survive outside their hosts varies enormously from species to species. Some, like the organisms

---

Fig. 15.5 Various actinomycetes in preparations of pathological material. (**A**) *Nocardia* sp. in a smear of pus from an infected mesenteric lymph node of a cat (× 2,000). Gram-stained (A. H. Linton). (**B**) *Mycobacterium tuberculosis* type *human* in human sputum (× 2,000). Stained by Ziehl Neelsen's method (K. E. Cooper). (**C**) *Actinomyces bovis* in a section of an actinomycotic lesion of a bovine jaw (× 140). Gram-stained (A. H. Linton). (**D**) A portion of the edge of the 'ray fungus' structure shown in (**C**) (× 2,000). The organism is seen growing out into the surrounding granulation tissue from the central mass of mycelial growth. Gram-stained (A. H. Linton)

responsible for human venereal diseases (*Neisseria gonorrhoeae* and *Treponema pallidum*) which are mainly transmitted during coitus, rapidly die outside the body. Many pathogens of the respiratory tract are more hardy but are favoured by fairly rapid transmission in cough spray, although infection may persist in contaminated dust and may be inhaled later; this has often been shown to occur in cross-infection in fever hospital wards (p. 593). In general, pathogens of the intestinal tract have much greater resistance and can survive for many days, or even months, outside the body. Spore-bearing pathogens, such as the causal organisms of tetanus, gas-gangrene, and anthrax, survive as spores in a fully virulent condition in manured soils for years.

Routes of infection

Many fully virulent micro-organisms are able to cause disease in the animal body only if they enter by specific routes. *Clostridium tetani*, the cause of tetanus, is frequently found without harmful effects in the intestinal tract, especially of domesticated animals (horse, pigs, etc.), to which it gains access via the mouth. It assumes a pathogenic role usually if introduced into a deep wound, either following surgical manipulation or accidental penetration, in which anaerobic conditions essential for its growth are found. Many alimentary infections, on the other hand, are unable to establish themselves when introduced into abrasions but readily infect via the mouth. The same organism may show varying degrees of virulence when introduced into the same host by different routes; thus anthrax bacilli and staphylococci are more virulent when they enter through skin abrasions than when ingested; the tubercle bacillus is more virulent when it enters by the respiratory than by the alimentary route, probably owing to the differences in the defence mechanisms.

Closely linked with the portal of entry is the affinity of many organisms for particular tissues, e.g., the pneumococcus for lung tissue, the leprosy bacillus for skin tissue, the meningococcus for the meninges of the brain, leptospirae for the liver and kidney, and *Vibrio cholerae* for the intestine.

Pathogenicity of micro-organisms

Since Koch's observation in 1878 on the characteristic symptoms and anatomical lesions in animals caused by six morphologically distinct micro-organisms, it has been well established that each pathogenic organism produces a specific range of disease. Certain pathogens produce the same symptoms in a variety of animals hosts, e.g., paralysis resulting from the toxin of the tetanus bacillus. Others are able to produce different forms of disease in the same host. For instance, *Streptococcus pyogenes* (Fig. 15.6) in man can cause tonsillitis, scarlet fever, erysipelas, or wound infections. Some organisms are highly pathogenic to certain hosts and non-pathogenic to others, as, for instance, *Corynebacterium diphtheriae* which is pathogenic naturally to man and experimentally to the guinea pig but to which the rat is completely resistant. Other pathogens produce disease of varying severity in different animal hosts, as illustrated by *Bacillus anthracis* which produces an acute, usually fatal septicaemic disease in cattle, sheep, and horses, inflammatory oedema of the pharynx in pigs and dogs, and a malignant pustule at the site of entry through the skin in man.

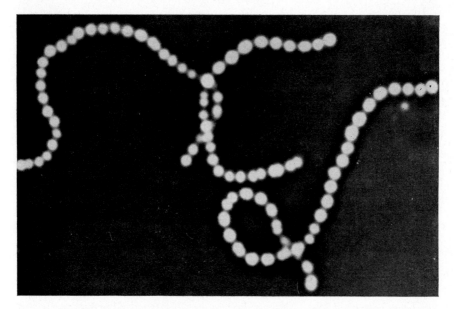

Fig. 15.6   Chains of Lancefield Group A Streptococci isolated from a septic sore throat. Negative staining (×3,600). (C. F. Robinow)

Only those strains of known pathogenic species able to overcome the resistance of the host and propagate in the animal body are virulent; those which cannot do this are termed avirulent. There is, however, no clear distinction between the two, since virulence is a relative term dependent upon many factors which can be altered under a variety of environmental influences. The virulence of many bacterial and viral pathogens may be diminished by such procedures as artificial culture, culture at temperatures above the optimum, or passage through animals in which the disease is not normally found. These methods are used to produce attenuated vaccines for artificial immunization (p. 565). Frequently loss of virulence is associated with a change from the 'smooth' colonial form to the 'rough' (p. 313). Selection of smooth virulent colonies from a rough avirulent culture is not readily accomplished, but in certain instances can be achieved simply by animal passage when the few virulent organisms present in the culture will assume a pathogenic role and later be isolated from pathological material in pure culture. Rough avirulent pneumococci can be transformed into a virulent form by growing them in the presence of killed suspensions of whole, smooth pneumococci, or simply in the presence of the genetically important transforming substance extracted from smooth pneumococci (p. 118). Virulent toxin-producing strains of the diphtheria bacillus have been isolated from cultures of avirulent non-toxin-producing strains after exposure to particular bacteriophages, indicating that in this species there is an association between lysogeny (p. 264) and virulence; however, not all lysogenic strains of the diphtheria bacillus are virulent.

The properties which distinguish a virulent from an avirulent bacterium are twofold, namely, 'invasiveness', i.e., the ability of the organism to establish itself and propagate in the host tissues, and 'toxicity', i.e., the ability of the organisms to form toxins which destroy or damage tissues or impair their physiological functions.

Invasiveness

Apart from those organisms which are introduced into the body through wounds, most micro-organisms initially lodge on the skin or mucous membranes through which they pass in order to invade the underlying tissues. Most mucous membranes, particularly those of the nasopharynx and intestine, are liberally supplied with lymphatic channels near to the surface. Organisms impinging on the membranes of the nasopharynx are constantly being removed by ciliated epithelial cells or engulfed by wandering phagocytic cells (p. 561) which carry them away to the lymph channels. If the organisms escape these defences and multiply, infection may be initiated. Local multiplication of organisms such as *Staphylococcus aureus* (Fig. 15.7) may occur in crypts of the skin, in hair follicles, or sebaceous glands, thereby setting up local damage followed by penetration into the sub-epithelial tissues. Skin secretes substances which are bactericidal and these must be overcome by micro-organisms before they can multiply or reach a penetrable surface. Many pathogens are thus destroyed before penetration

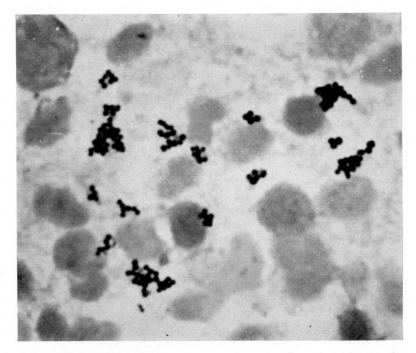

Fig. 15.7 Clusters of *Staphylococcus aureus* in smear of pus from abscess. Gram-stain (× 1,500). (A. H. Linton)

is achieved. After penetration the subsequent course of the infection varies with the type of organism and the nature of the internal defences of the host. The infection may be confined to local abscess formation, as is common with staphylococci, or the organism may penetrate through subcutaneous tissues to give rise to spreading skin lesions, as in streptococcal skin infections, e.g., erysipelas in man caused by *Streptococcus pyogenes*, or a clinically similar condition in swine caused by *Erysipelothrix rhusiopathiae*. Some pathogens after local multiplication spread via the lymph channels to the lymph nodes, which rapidly become enlarged owing to the stimulation of defence mechanisms in these glands. These may check the spread of infection; otherwise organisms continue to spread either in the lymph or the bloodstream. Organisms entering the bloodstream are rapidly circulated to all parts of the body (bacteraemia) and they may set up local foci of infection in various organs from which pus may escape into the bloodstream (pyaemic spread). Progressive multiplication of the organism in the blood stream is termed septicaemia.

Toxicity

The ability of different pathogens to invade the body varies considerably. Many with very little power to invade are, however, extremely harmful through their ability to produce highly potent toxins. The tetanus bacillus, for instance, has no invasive powers but, when introduced into a deep wound, multiplies at the local site and generates a powerful toxin which has a specific affinity for nerve cells and brings about destruction of those parts of the central nervous system which it reaches, resulting in symptoms of paralysis. The diphtheria bacillus does not invade beyond the tissues of the throat, nasopharynx, and bronchi but its toxin, produced *in situ*, is absorbed and circulated around the body in the bloodstream to produce the various generalized symptoms of diphtheria, in particular, damage to heart muscle tissue. Many strongly invasive organisms, such as streptococci and the gas-gangrene anaerobic sporing bacilli, are producers of potent toxins which result in gross tissue destruction. The various bacterial toxins are divided into two classes, namely, endotoxins and exotoxins.

ENDOTOXINS are structural components of bacterial cells released after death and disintegration of the cell, and are then found in bacteria-free filtrates of cultures. In the body endotoxin release occurs when the bacteria are phagocytosed or otherwise lysed by natural components of the body fluids (e.g., complement in the presence of antibody, p. 570). Endotoxins are produced mainly by Gram-negative pathogens and non-pathogens (e.g., species of *Salmonella*, *Escherichia*, *Proteus*, *Shigella*, *Brucella*, and *Neisseria*). Irrespective of the organism from which the endotoxin came the symptoms produced in the host are similar and include fever, diarrhoea, local haemorrhage, especially in the visceral tissues, fall of blood pressure and shock. The severity of the effect depends on the dose of endotoxin and the sensitivity of the animal to it; sensitivity is variable. Endotoxins have been chemically extracted from a number of bacterial species and have been found to consist of complex molecules of protein, lipid, and polysaccharide, corresponding to the O-somatic antigens of the bacterial cell walls

(p. 576). Serum antibodies are produced following immunization with bacterial endotoxins; these combine with the O-somatic antigens in the whole cell producing antibacterial effects (e.g., bacteriolysis) but the antibodies do not completely neutralize the toxic action of the endotoxins when these are released from the cell. Endotoxins are more heat stable than exotoxins but do not act at such high dilution. They produce their effects on the host by damaging small blood vessels by a hypersensitive type of reaction, the sensitivity of the patient arising from past exposure to the same antigens which induced the state of hypersensitivity (p. 578).

EXOTOXINS, in contrast to endotoxins, are readily separable from the living producer cells and may be obtained by growing the bacteria in a suitable liquid medium and separating the toxin from the cells by centrifugation or filtration. Those which have been isolated as pure chemical substances (e.g., tetanus and diphtheria toxins) are simple proteins. They are frequently highly toxic for the host, the lethal dose of pure crystalline tetanus and botulinus toxins for mice being of the order of 0·0000001 mg. They act as enzymes and as such are generally unstable to chemicals (e.g., formalin slowly converts them to toxoid, p. 565), to oxygen (e.g., oxygen-sensitive haemolysins), and to heat. Sensitivity to heat varies widely; diphtheria toxin is sensitive to 65°C, *Clostridium perfringens* α-toxin to 100°C, and staphylococcal enterotoxin and the toxin of botulinus to higher temperatures. Some exotoxins exhibit specific affinity for tissues; for instance, tetanus bacilli and other bacteria produce neurotoxins which destroy cells of the central nervous system, the symptoms of paralysis being produced several days after the toxin is released into the body. Other toxins are rapid in action and are named according to the effects they produce; haemolysins, as produced by streptococci, lyse red blood cells; leucocidins, e.g., the leucotoxin of *Staphylococcus aureus* (Fig. 15.7), destroy leucocytes, the wandering phagocytes of the blood stream. The necrotoxins exhibit more general tissue destruction, such as the general destruction of tissue by the gas-gangrene group of Clostridia. The same toxin may produce a number of effects. For instance, the α-toxin of *Staphylococcus aureus* acts as a haemolysin, a leucocidin, a necrotoxin, and a lethal toxin. In most instances the mechanism of action is unknown. The α-toxin of *Cl. perfringens* acts as a lecithinase, a lecithin-splitting enzyme which can attack lecithin-containing cell membranes such as those of red blood cells. The toxin of diphtheria appears to be the protein moiety of the bacterial cytochrome *b* enzyme and may act by interfering with synthesis of cytochrome *b* in the host's tissues. Some organisms are able to produce toxins resulting in a tissue rash (e.g., *Streptococcus pyogenes*, which produces an erythrogenic toxin causing the rash of scarlet fever). Specific neutralizing antibodies (antitoxins) can be demonstrated in the bloodstream of an animal following infection with an exotoxin-producing bacterium, or these may be stimulated by artificially immunizing the animal with a toxoid (p. 565).

OTHER CELL PRODUCTS. Some bacterial cells produce substances which influence the course of an infection in ways other than by toxic action. For instance, pathogenic staphylococci produce the enzyme coagulase which induces the coagulation of blood plasma, thereby setting up a barrier to the circulating body defences and assisting the organism to establish itself in the host body. In contrast, streptococci produce the enzymes streptokinase, which dissolves fibrin clots, and hyaluronidase, which breaks down hyaluronic acid (a muco-polysaccharide which holds together the mesodermal structures of subcutaneous tissues), thereby facilitating spread of the organism from the primary focus.

Size of the infecting dose

It is very rare for a single bacterial cell to give rise to disease. More usually a critical minimum number of cells must be present in the infecting dose before invasion of the host can occur. Very large numbers of spores of certain pathogenic fungi, such as *Aspergillus fumigatus*, are necessary to establish an infection. Even when micro-organisms are able to set up an infection, the rate of onset of symptoms, the severity of the disease, and the mortality rate, are often directly proportional to the size of the initial infecting dose. The more virulent the pathogen, the smaller the minimum dose required to produce infection.

BOOKS AND ARTICLES FOR REFERENCE AND FURTHER STUDY

AINSWORTH, A. S. and AUSTWICK, P. K. C. (1959). *Fungal Diseases of Animals*. Commonwealth Agricultural Bureau. (Review Series No. 6), 148 pp.

BRANDE, A. I. (1964). Bacterial endotoxins. *Scient. Am.*, **210**, No. 3, 36.

BRUNER, D. W. and GILLESPIE, J. W. (1966). *Hagan's Infectious diseases of Domestic Animals*. 5th Edn. Baillière, Tindall and Cox, London, 1105 pp.

BURNET, F. M. (1953). *The Natural History of Infectious Diseases*. Cambridge Univ. Press, 356 pp.

HOWIE, J. W. and O'HEA, A. J., eds. (1955). *Mechanisms of Microbial Pathogenicity*, *5th Symp. Soc. gen. Microbiol*. Cambridge Univ. Press, 333 pp.

HUNGATE, R. E. (1966). *The Rumen and its Microbes*. Academic Press, New York and London, 533 pp.

LOVELL, R. (1958). *The Aetiology of Infective Diseases*. Angus and Robertson, Sydney, 136 pp.

MARPLES, MARY J. (1969). Life on the human skin. *Scient. Am.*, **220**, No. 1, 108.

NUTMAN, P. S. and MOSSE, B., eds. (1963). *Symbiotic Associations. 13th Symp. Soc. gen. Microbiol*. Cambridge Univ. Press, 356 pp.

SMITH, H. (1968). Biochemical challenge of microbial pathogenicity. *Bact. Rev.*, **32**, 164–184.

SMITH, I. M. (1968). Death from staphylococci. *Scient. Am.*, **218**, No. 2, 84.

STABLEFORTH, A. W. and GALLOWAY, T. A., eds. (1959). *Infectious Diseases of Animals due to Bacteria*, Vols. I and II. Butterworth's Scientific Publications, London, 870 pp.

VAN HEYNINGEN, W. E. (1968). Tetanus. *Scient. Am.*, **218**, No. 4, 69.

WEBSTER, J. (1970). Coprophilous fungi. *Trans. Br. mycol. Soc*, **54**, 161–180.

WILLIAMS, R. E. O. and SPICER, C. C., eds. (1957). *Microbial Ecology. 7th Symp. Soc. gen. Microbiol*. Cambridge Univ. Press, 388 pp.

WILSON, G. S. and MILES, A. A. (1955). *Topley and Wilson: Principles of Bacteriology and Immunity*. 4th Edn. Edward Arnold, London, 2563 pp.

16

Micro-organisms and Vertebrates: host resistance and immunity

INTRODUCTION

The success or otherwise of a virulent parasite in establishing itself in a host is not solely a property of the parasite, but is greatly influenced by the resistance of the host. This may be characteristic of the host species or may be peculiar to the individual.

Innate resistance

It is well known that certain diseases are limited to particular species of animals, as with gonorrhoea and diphtheria found naturally only in man. Reasons for host specificity are not usually known but the following are examples where some information is available. Anthrax, a pathogen of most animals, is not found in birds, presumably owing to their higher body temperature; it is well known that this organism when grown at a temperature of 42°C becomes avirulent. Differences in susceptibility of various animal species to brucellosis may in part be due to the presence of a factor (the smooth selecting factor) in their serum which maintains the organism in the smooth virulent phase. In non-susceptible animals, such as the rat, this factor is absent and the organisms rapidly degenerate into the rough, avirulent phase, as in artificial culture. More recently it has been shown that the carbohydrate erythritol, a growth-promoting substance for *Brucella* spp., was consistently found in significant concentrations in the tissues where the organisms normally localize in susceptible animals, but not in the corresponding tissues of non-susceptible ones.

Not only does susceptibility vary between species, but different breeds or races within a species may show variation in resistance to the same pathogen. East African hairy sheep are reputed to be more resistant to anthrax than wooled breeds, and negroes are often more susceptible than the white man to tuberculosis.

Individual resistance

The individual animal, in addition to its innate resistance, has many forms of defence against the pathogenic organisms; these may be considered under two headings, general host resistance and specific immune responses or acquired immunity.

GENERAL HOST RESISTANCE

Health

It is widely recognized that animals in good health are better able to resist infections than those in poor health. The health of an individual can be influenced by many environmental factors, a few of which will be considered.

Factors affecting the primary defences of the body

Before an organism can reach the body tissues, the epithelial coverings, e.g., skin and mucous membranes, must be penetrated (p. 554). In a healthy animal, when these natural barriers are intact, many pathogenic micro-organisms are removed before they can invade the body. Any factor able to damage these natural barriers (e.g., minor abrasions of the skin, or burns by acids or heat) deprives the animal of this protection.

Diet

Adequate food is essential for good health, and there is no doubt that variations in diet are associated with alterations in resistance to infection. Epidemic diseases frequently follow times of food deficiency such as in war or famine.

The influence of diet on resistance is a very complex problem interconnected with many other factors, and little experimental data is available. Many workers claim that deficiency of vitamins A and C, and of proteins, particularly during the period of rapid body growth, decreases resistance to infection. It is thought that this is due to a lowering of efficiency in the production of phagocytes and antibodies (see below, p. 562). Indirectly a deficiency of one nutrient in the normal diet may lead to unusual habits which predispose to disease. In certain areas of the cattle ranges of South Africa, a deficiency of phosphorus in the soil leads to a deficiency of this element in the grass on which the cattle feed. The latter then chew bones of animals that have died on the open veldt. The carcasses from which these bones come are frequently contaminated with *Clostridium botulinum* type C which produces a potent neurotoxin in the decomposing carcass. Cattle chewing the bones ingest the preformed toxin and contract the disease 'Lamziekte' which is a form of toxic food poisoning analogous to botulism in man (p. 588).

Fatigue

Fatigued animals are more readily infected than others. This appears to be the main predisposing factor in the so-called 'shipping fever' of animals. This clinical condition possibly involves a number of infectious agents which frequently include a myxovirus of parainfluenza group 3 and *Pasteurella multocida*. When animals are transported long distances without proper facilities for rest and under badly ventilated conditions, the resistance of the animal is lowered and organisms normally carried in their respiratory tracts assume a pathogenic role, infect the lungs, and produce pneumonia.

Climatic variations

Sudden changes of weather, particularly increased humidity and chilling after a warm, dry spell, are known to influence the resistance of animals, particularly to respiratory infections (e.g., aspergillosis of young poultry is most severe in cold wet weather in spring). Prolonged exposure to cold has been shown to depress antibody formation (p. 562). On the other hand the incidence of infection by the common cold virus under carefully controlled experimental conditions has been shown not to be associated with chilling.

Occupational hazards

Exposure of individuals to unusual hazards, such as close contact with infectious agents or damage to normal defences, increases the risk of infection. Particles of silica or other material regularly inhaled by stone-masons, miners, and others cause irritation of the lungs which predisposes to pneumonic diseases, tuberculosis in particular. The effects of increased contact with infectious agents may be illustrated by a number of examples; brucellosis in veterinary surgeons due to contact with infected cattle; woolsorter's disease due to inhalation of anthrax spores by workers sorting untreated imported wool; erysipeloid (in contrast to erysipelas caused by streptococci) in butchers and abattoir workers as a result of infection with *Erysipelas rhusiopathiae* through abrasions caused by contaminated pig bones; Q-fever by inhalation of dust (contaminated with rickettsias) by workers in close contact with infected cattle.

Irradiation

The effect of radiation on the resistance of the animal is dependent on the degree of exposure. Small doses of X-rays have been shown by some workers to increase resistance, probably owing to a stimulation of the immunity-producing centres of the body. On the other hand, there is no doubt that large doses of ionizing radiations, whether X-rays or atomic radiations, decrease the resistance of the host by damaging the lymphoid tissue so that it can no longer produce antibody in response to infection (p. 563). Radiation may also damage bone marrow tissue, thereby impairing the production of red blood cells and leucocytes in the adult mammal.

The presence of other parasites

Damage caused to the host by one organism may increase the risk of infection to a second, as in black disease of sheep, in which infection of the liver with *Clostridium oedematiens*, a blood-borne infection from the gut, occurs if liver flukes have previously caused necrosis of the liver, thereby producing suitable anaerobic conditions for the rapid propagation of the *Clostridium*.

Age and sex

Susceptibility to infection varies considerably with the age of the animal. New-born animals are frequently very susceptible to organisms, such as those of the normal gut flora (p. 540), which are not usually pathogenic to older animals. Infection often occurs during the critical period whilst the animal is acquiring its gut flora and before it has built up a resistance against these organisms. Thus young children succumb to infection by specific strains of *Escherichia coli*; young lambs frequently become infected with *Clostridium perfringens* from the soil and develop lamb dysentery. New-born animals may, however, be more resistant to certain infections than are slightly older ones if they have received a temporary immunity conferred on them from the mother (p. 563). Once this temporary immunity is lost, young animals are frequently more susceptible to certain infections than are mature ones. For instance, babies in their first year of life are usually immune to measles because of passive immunity transferred from the mother but in the next few years are highly susceptible. With some diseases, mature animals are actually more susceptible than the young. *Brucella abortus* is unable to establish itself in calves but readily invades the uterus and udder in the cow and the testicles in the bull. It has been suggested that sex hormones or other growth promoting substances (p. 558), present only in the adult animal, are essential for the establishment of this pathogen.

General cellular resistance

When bacterial cells pass through the natural barriers of the primary defences of the body they are treated in the same way as inert particles by the cellular defences. Specialized cells, known as phagocytes, rapidly engulf the organisms and unless killed by the toxins of the bacteria, begin to digest and destroy them. These phagocytic cells may be fixed in the liver, spleen, and lymph glands and include macrophages or histiocytes; other phagocytes may be circulating in the blood and consist mainly of large mononuclear leucocytes and polymorphonuclear leucocytes, the so-called white cells of the blood.

Many virulent organisms kill phagocytes, resulting in the formation of pus, and continue to multiply. The subsequent course of the infection then depends on whether the animal is stimulated to produce sufficient phagocytes to overcome the rapidly multiplying bacteria. In this the cellular response is aided by the other forms of immunity and any artificial assistance that may be provided by chemotherapeutic treatment (p. 580). Frequently virulent organisms are able to resist phagocytosis by means of special properties of their capsule which give them some degree of protection.

SPECIFIC IMMUNE RESPONSES

It is widely known that many diseases of childhood rarely occur in adult life, the early occurrence of the disease establishing a lasting immunity in the individual. This acquired immunity is the result of either or both of two mechanisms of defence against microbial infection, namely, antibody-dependent or humoral immunity and cell-mediated immunity (p. 580). These are interrelated but will be considered separately.

Humoral immunity depends on the formation of substances termed antibodies which react with and neutralize the infective properties of the invading micro-organism. In addition to its importance as a mechanism of immunity, the specificity of antibody reactions has provided a means of demonstrating differences and similarities between substances capable of stimulating antibody production and has led to the emergence of the science of immunology. This is widely used in the identification and classification of micro-organisms, the diagnosis of disease, and the immunochemical analysis of components of micro-organisms and other organic substances.

Antigens

Substances able to induce antibody production are termed antigens. They are organic substances which, when introduced naturally or artificially into the tissues of a living vertebrate host, elicit the formation of antibodies with which they will combine often in a visible manner. Antigens are usually protein in nature but certain polysaccharides and polypeptides may be antigenic. Other substances, such as lipids and nucleic acids, are antigenic when conjugated with proteins, the latter possibly operating by retaining the antigenic determinants in the body for a sufficiently long time to elicit an antibody response. By virtue of their specific chemical groupings at the molecule surface, lipids and nucleic acids determine the specificity of the antibody produced against the protein complex. Antigens have a large molecular weight of at least 5,000, i.e., they are too large to be disposed of by the host's normal physiological functions; they can be soluble or particulate and are essentially foreign to the host. Micro-organisms and many of their products are therefore strongly antigenic.

Antibodies

An antigen may enter the animal host and gain access to the tissues during the course of a natural microbial infection or it may be introduced artificially. Under this stimulus antibodies are formed principally by cells of the lymphoid tissue, especially plasma cells, which are distributed throughout the reticulo-endothelial system of the animal body. Plasma cells are not normal body cells but develop from small lymphocytes under an antigenic stimulus.

Antibodies are serum globulins which have a low solubility in water and are completely precipitated by half-saturated ammonium sulphate. By the technique of electrophoresis (i.e., under the influence of an electric field), the negatively charged serum globulins migrate to the positive cathode. The slowest diffusing fraction of an immune serum, the gamma globulin, contains most antibody. These immunoglobulins can be separated by chemical and physical means into at least four major components, IgG, IgA, IgM and IgE (Table 16.1).

Table 16.1   Comparison of immunoglobulins.

Class of immunoglobulin	IgG	IgA	IgM	IgE
% of total in normal serum	75	21	7	<1
Molec. Weight	150,000	150,000+	900,000	196,000
Following primary antigenic stimulus	0	0	+	0
Major component after secondary stimulus	+	0	0	0
Complement binding	+	0	+	0
Transport across placenta	+	0	0	0
Fix to homologous tissue	0	0	0	+

Each is antigenically distinct and the dominant component in the serum at any time depends upon a number of factors. For instance, IgM is chiefly formed in young animals and some lower animals, such as Elasmobranchs. (More primitive vertebrates and invertebrates do not possess the capacity to produce antibody.) A primary antigenic stimulus results in IgM formation whilst subsequent stimulation by the same antigens produces IgG. Only IgG can pass from mother to foetus by the placental circulation and thus only IgG is found in the serum of the new-born. Particulate antigens tend to stimulate IgM production.

Natural active immunity

The production of antibody by the animal host is known as active immunity. The titre of antibody increases during the course of an infection and, by combining with the infecting micro-organisms or their products, aids the cellular defences of the body to function with impunity. After the disease has been controlled, the titre remains high for months or years, depending on the identity of the original infection, but eventually begins to fall. When, however, the animal, which is specifically immune, is exposed to another attack by the same pathogen, antibody production occurs rapidly and only rarely does the infection establish itself a second or subsequent time. This sequence of events may be compared with antibody production following a second or subsequent 'booster' injection as practised in immunization (p. 566). Frequently a subclinical infection, which does not produce symptoms of disease, stimulates antibody production and may confer a lasting immunity on the individual. This may explain why some animals never succumb to a particular disease to which others of the same host species fall ready victims.

Natural passive immunity

During early foetal life antibodies are not formed under an antigenic stimulus but later the foetus is able to produce antibodies and full immunological competence is reached soon after birth. Antibody to specific pathogens can, however, often be demonstrated in the blood serum of newly born animals. These antibodies have not been produced by the young animal itself but by its mother in response to antigenic stimuli to which she has been exposed. The antibodies are passed to the young either via the placenta whilst the animal is *in utero*, as in man, rabbits, guinea pigs and rats, or in the first milk (colostrum) which is ingested by the animal immediately after birth, as in cattle, horses, sheep, goats and pigs. Passive immunity persists in the young for up to several months but is nevertheless of great value in giving protection against infections that are likely to be met prior to the animal becoming fully immunologically competent. If calves are deprived of their mother's colostrum, they frequently succumb to pathogenic *Escherichia coli* present in their environment which produces a lethal septicaemia, associated with profuse scouring and emaciation; calves receiving colostrum usually exhibit scouring only and survive. The colostrum must contain specific antibodies to the types of *E. coli* encountered if protection is to be achieved. By natural exposure to these certain serotypes before the calf is born, the cow is stimulated to produce specific antibodies which are later passed on to the calf in the colostrum.

Artificial active immunity

The methods by which artificial active immunity may be achieved simulates that resulting from natural exposure to infection. Antibody production can be induced artificially by administering relevant antigens to the patient in the form of vaccines of dead or living micro-organisms or their products (Table 16.2).

Table 16.2   Some agents used in artificial immunization.

Active Immunity	
Disease	*Agent*
Living agents	
1. *Unmodified* Rubella (German measles)	controlled exposure to natural infection
2. *Attenuated* Tuberculosis	B.C.G. (Bacillus of Calmette and Guérin)
Poliomyelitis	poliovirus (Sabin)
Brucellosis	strain 19 in cattle
Measles	measles virus
Smallpox	vaccinia virus
Anthrax	spore vaccine of *B. anthracis*
Rabies	rabies virus (goat, rabbit or egg-adapted)
Yellow fever	yellow fever virus
Killed or inactivated agents	
1. *Micro-organisms* Enteric fever (e.g., typhoid)	T.A.B. vaccine
Whooping cough	*Bordetella pertussis*
Poliomyelitis	poliovirus (Salk)
Cholera	*Vibrio cholerae*
Influenza	influenza virus
Measles	measles virus
Leptospirosis	*Leptospira* sp.
Foot and mouth disease	foot and mouth virus
Canine distemper	distemper virus
2. *Toxoids* Diphtheria	diphtheria toxoid
Tetanus	tetanus toxoid
3. *Bacterins* Infections by anaerobic spore-bearing bacilli in animals	anaerobic bacilli and toxoid

Table 16.2 continued

Passive Immunity	
Disease	*Agent*
Diphtheria	antitoxin
Gas gangrene	,,
*Tetanus	,,
Canine distemper	,,
Animal infections with sporing anaerobes	,,
†Rubella (German measles)	human gamma globulin (pooled)

* The risk of serum reactions (e.g., anaphylaxis p. 578) is so high that antiserum therapy is now replaced by antibiotic therapy when indicated.

† This mild disease is highly dangerous to the human foetus during the first three months of pregnancy and protection is given to women at risk using pooled human gamma globulin.

It is extremely important that vaccines be prepared from organisms antigenically similar to prevalent members of the same species in order that antibodies corresponding to the virulent pathogens may be formed. Only occasionally are fully virulent organisms used as vaccines because of the danger of initiating the disease in susceptible individuals. Immunization against avian infectious laryngotracheitis utilizes the fully virulent virus, but this is introduced at an unusual site on the mucous membrane of the bursa of Fabricius, a pouch of the cloaca, and care must be taken to ensure that the virus does not gain access to the respiratory system. A harmless inflammatory reaction results but immunity against the virulent respiratory disease is achieved.

Living vaccines are usually prepared with organisms of lowered virulence. These are known as attenuated strains and may be obtained by growing organisms under artificial conditions in the laboratory (e.g., the Bacillus of Calmette and Guérin—B.C.G. vaccine for tuberculosis), or attenuated mutants may be selected such as the S19 strain of *Brucella abortus* and the non-paralytic strain of poliovirus oral vaccine (Sabin) used for protection against poliomyelitis. Occasionally, related strains, such as vaccinia virus for vaccination against smallpox, or strains adapted by passage through an animal host (e.g., goat-adapted rabies virus) or egg-adapted viral and rickettsial vaccines grown in chick embryos or their membranes, may be used.

Resistance to certain infections can be achieved by the use of killed bacterial and inactivated viral vaccines. These include bacterial vaccines against whooping cough, the enteric fevers (T.A.B., a vaccine against typhoid and paratyphoid A and B fevers), cholera, plague, and leptospirosis and viral vaccines against poliomyelitis (Salk vaccine, in contrast to the Sabin vaccine already mentioned), foot-and-mouth, swine fever, influenza, and measles. In many diseases in which major tissue damage is the result of bacterial exotoxins, protection can be achieved by stimulating the production of antitoxins by the host. Toxins are not used for this purpose on account of their toxicity but this can be reduced either by ageing or by treatment with formalin which converts them to *toxoids*. Toxoids are antigenic but not toxic and can be safely used to produce antitoxin against the toxin. These include tetanus, diphtheria, and staphylococcal toxoids.

Animals may be protected against many anaerobic infections caused by clostridia (e.g., lamb dysentery, black disease of sheep, blackleg of cattle) by the use of bacterins which are formalin-killed, whole cultures and include both organism and toxoid.

When a non-living vaccine is used there is no continuous antigenic stimulus as in an infection and usually two or more injections are required to produce an adequate level of immunity. The first injection sensitizes the animal and gives a low titre of antibody which can be detected in the bloodstream from 8–12 days, dependent upon the site of the injection (Fig. 16.1). This is termed the 'primary' response. A second injection given several weeks later then results in a more rapid and marked response, as indicated by a rising antibody titre which can be demonstrated from about 48 hours onwards. This is termed the 'secondary' response. The titre once established, persists for varying periods of time but usually begins to fall after several months. A 'booster' dose given at infrequent intervals stimulates a rapid response, comparable to the secondary response

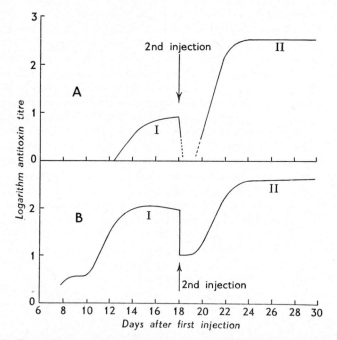

Fig. 16.1 The primary and secondary antitoxic responses of two rabbits to staphylococcal toxoid. Curves I (A and B) show the slowly appearing primary response following injection at time 0, by A, intravenous and B, subcutaneous routes respectively. Curves II (A and B) show the rapid secondary response in both animals to intravenous injection made eighteen days after the primary injection. The fall in antibody titre immediately following a second injection of a large amount of the corresponding antigen is known as the *negative phase*. This is thought to be due to removal of circulating antibody by combination with the antigen. (Modified from Burnet, F. M. and Fenner, F. (1949). *The Production of Antibodies*, Macmillan, London)

of the initial immunization, and restores a significant titre and corresponding immunity.

Artificial passive immunity

Several weeks are needed to establish a satisfactory immunity by active immunization and, to be fully effective, inoculation must be carried out before the animal is exposed to infection. Circumstances arise when it is necessary to give immediate protection, as when a susceptible animal is known to have been in contact with infection or is showing symptoms of the disease. It is then possible to protect by injecting antiserum containing the specific antibody against the infection. This is prepared by immunizing a suitable animal (usually the horse) and separating the serum from blood aseptically taken. Thus a dog which is likely to be exposed to the distemper virus could be given a prophylactic dose (i.e., treatment to ward off infection) of antiserum containing neutralizing antibodies against the virus. Similarly when diphtheria was prevalent in Great Britain infected children were treated with diphtheria antitoxin to neutralize the lethal effects produced by the toxin of the diphtheria bacillus. Just as passive immunity naturally transferred from mother to offspring is of a relatively short duration, so artificial passive immunity gives protection for only about three weeks. Administration of this foreign serum to the patient presents certain risks. The serum proteins from a foreign host are themselves antigenic and produce antibodies upon injection. Patients who have previously received injections of the same foreign serum may develop a hypersensitivity to it and produce an anaphylactic reaction upon receiving a later injection (p. 578). For this reason antiseral therapy is not practised as frequently as it used to be, more reliance being placed on combined antiseral and antibiotic therapy as soon as symptoms arise.

In sheep, a combination of active artificial and passive natural immunity is used to protect young lambs against lamb dysentery within the first few days of life. The pregnant ewe is actively immunized with a formalin-killed whole culture of *Cl. perfringens*. The resultant high titre of antibody is transferred to the new-born lamb in the colostrum of the ewe's milk thereby providing adequate protection. Typical values of antitoxin titres in the sera of the ewe and lamb, together with the titre in the colostrum over a period of time, are shown in Fig. 16.2.

Theories of antibody production

Several theories have been postulated to explain the mechanism of antibody production.

Instructive theory

The older 'instructive' theory, suggested that the antigen entered the antibody-producing cells and acted as a template for the synthesis of serum proteins from amino acids, thereby producing molecules whose chemical configuration was the mirror image of the antigen. For the continued production of specific antibody this theory required the constant presence of the antigen, which is most unlikely. A modified form of the 'instructive' theory has been postulated

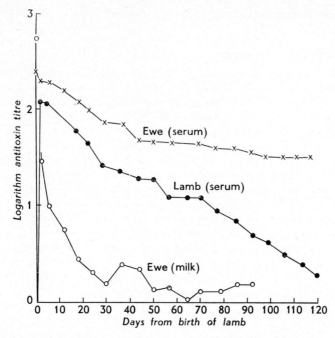

Fig. 16.2 Passive transfer of antitoxin in colostrum. The concentration of antitoxin in the blood of a ewe, in her colostrum and milk, and in the blood of her lamb from birth onwards. The ewe was immunized during pregnancy. (After Mason, J. H., Dalling, T. A., and Gordon, W. S. (1930). *J. Path. Bact.*, **33**, 783)

more recently. This proposes that the antigen, upon entering the body, is first phagocytosed (engulfed) by non-antibody producing macrophages in the tissues of the host. It is considered that in these cells the antigen is processed and information transferred to cells of the lymphoid series with the ultimate synthesis of specific antibody by plasma cells. This theory requires the replication of the information-carrying substance in order that daughter cells will be capable of producing similar antibody.

Selection theories

The clonal selection theory of Burnet envisages that within the host there exist many different clones of serum protein-producing cells that between them are genetically capable of synthesizing specific antibody against every possible antigen. Each clone of cells, capable of producing antibody against a limited number of antigens, would arise from a single cell, itself derived by mutation of ancestral cells in early embryonic life. Each cell clone remains dormant until the antigen enters the body when the complementary cell group is selected and stimulated to proliferate and produce quantities of specific antibody. Subsequent to the initial stimulation, the cell line persists indefinitely in the host, continuing to produce antibody at a diminishing rate but always capable of producing increased quantities of antibody when stimulated by re-exposure to the same kind of antigen. This theory sought to explain why the host does not normally produce antibodies to self-proteins. It was suggested that cell clones corresponding to self-proteins are destroyed late in embryo life and are not present in adult life to produce antibodies to these.

More recently modifications of this theory are favoured. Following experiments on tolerance to self-tissues in mice it was thought that Burnet's concept of a complete destruction of antibody-producing cells against self-proteins at the end of foetal life, is not absolute. The cell groups, genetically able to produce such antibody, are not destroyed but suppressed by the overwhelming and persisting concentration of self-protein in the body and hence no antibodies are produced against them. Similar but temporary tolerance or the suppression of antibody formation against foreign antigens can also be achieved in the adult by artificially injecting massive doses of the antigen.

In auto-immune diseases certain tissues are no longer recognized as self-tissues and antibody is produced which combines with these tissues and damages them. In allergic ophthalmitis, for example, mechanical damage to one eye may expose antigens of the interior of the eye which do not normally come into contact with the vascular system. These antigens are therefore not recognized as self-tissues and antibodies are formed against them. Several weeks later the other eye is destroyed by interaction between its tissues and newly formed antibody, accompanied by cellular infiltration.

Antigen–antibody reactions

When an antigen is mixed with antiserum containing its homologous antibody, physico-chemical forces induce the antigen and antibody to combine together. Each antibody molecule possesses two combining sites of equal specificity, thus one molecule can bridge two molecules of antigen. Since antigens are multivalent the result is the formation of a lattice structure which is often macroscopically visible as in the agglutination of red blood cells or bacteria.

The detection of an antigen–antibody reaction is only possible if demonstrable reactions take place. These include agglutination, precipitation, complement fixation, killing and lysis of whole cells, capsule swelling, phagocytosis, and neutralization of the physiological activity of the antigen. Antibodies used to be named according to the type of reaction they produced, such as agglutinin, precipitin, lysin, and opsonin. It is now appreciated that the same antibody may produce a number of effects depending on the environmental circumstances. An agglutinating antibody, for instance, may render cells susceptible to phagocytosis and killing or lysis in the presence of complement (p. 572), in addition to agglutinating the particulate antigen. On the other hand, some antibodies may be deficient in certain properties, as when an agglutinating antibody fixes complement but poorly.

Agglutination reaction

This reaction occurs between homologous antibody and particulate antigen in the presence of electrolyte. When a suspension of cells (bacteria or red blood cells) is mixed with homologous antiserum a visible clumping or agglutination occurs. Agitation of the mixture by shaking or stirring speeds up the reaction, which also occurs more rapidly at higher temperatures. By use of a standard suspension of cells and a series of dilutions of antiserum the antibody can be titrated. This test is widely used for blood grouping and the identification and diagnosis of bacteria and their infections. Agglutinins, in addition to agglutinating bacteria, may also immobilize motile ones, an observation used in the Treponema Immobilization Test for the diagnosis of syphilis.

Soluble antigens may be adsorbed on to carrier particles, e.g., bacterial polysaccharide on to red blood cells (sometimes treated with tanning agents, e.g., bis-diazo-benzidine, to aid the adsorption) and these cells agglutinate when mixed with antiserum against the bacterial polysaccharide.

Precipitin reactions

Soluble antigens, such as serum and bacterial polysaccharides, when mixed with homologous antibody in suitable proportions, precipitate in the presence of electrolyte. Excess of either reagent results in soluble non-precipitating complexes being formed. For good precipitation reactions it is therefore necessary to determine the optimum conditions for a visible reaction to take place. The reaction may be used as a ring test, where a ring of precipitate forms at the fluid interphase when antigen is layered on to an antiserum; as a flocculation test when floccules of antigen–antibody complex are formed in a mixture of both reagents; as a gel diffusion test (often called the Ouchterlony test after the first worker to describe it), in which both reagents diffuse towards each other in an agar gel (Fig. 16.3A).

In gel diffusion, lines of precipitate, corresponding to each antigen–antibody reaction, are formed in the gel at points where optimal levels of reactants meet. This technique is sometimes superimposed on a primary separation of immuno-globulins by immuno-electrophoresis. Antigens are caused to move through an agar gel under the influence of an electric current by which separation is accomplished. Antiserum, containing antibody to the various antigens, is later diffused from a source at right angles to the prior movement of the antigens and lines of precipitate form at the points reached by the various antigens (Fig. 16.3B).

Lysis by complement

Antibody may combine with living bacteria and usually bring about their agglutination, but the organisms still remain viable and fully pathogenic. However, a component of normal serum, known as complement, will combine with bacterial cells treated with homologous antibody bringing about their death and, in some species, their lysis. Both death and lysis frequently occur with the Gram-negative bacteria, as with *Vibrio cholerae* originally described by Pfeiffer, but no lysis occurs with Gram-positive bacteria. This is thought to be due to the greater thickness of cell walls in Gram-positive bacteria, the antibody not being able to make close contact with the cell membrane. Complement combines with most antigen–antibody systems with certain exceptions, e.g., toxin–antitoxin complexes.

Red blood cells are also lysed in the presence of both homologous antibody and complement; this has been used as a model system for the elucidation of

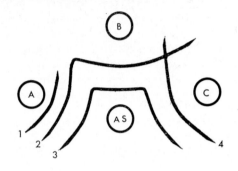

A

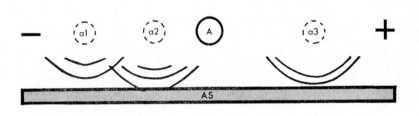

B

Fig. 16.3 Precipitation reactions in agar gels. **(A)** Lines of precipitation laid down in an agar gel when three antigen complexes (A, B, and C) are set up in an *immuno-diffusion* test against an antiserum (AS). 1. Antigen common to complex A only. 2. Antigen common to complexes A and B (hence joined). 3. Antigen common to complexes A, B, and C. 4. Antigen common to complex C only (hence crossing line 2). **(B)** Immunoelectrophoresis. Antigen A, placed in an agar cup, has been subjected to an electric field B which separates fractions a1; a2; and a3. Each of these fractions includes a number of antigens which migrate in the electric field at the same rate. Subsequent normal diffusion in the agar gel against a composite antiserum (AS) reveals the constituent antigens.

the mechanisms of complement lysis. Indeed, red cell lysis, first described by Bordet and Gengou in 1901, is now extensively used in the laboratory for detecting the presence of free complement *in vitro* and especially in the complement fixation test.

Recent electron microscope studies have shown that complement lysis of red cells is the result of a large number of holes being produced in the cell membrane (80–100 mμ diameter) through which the haemoglobin leaks out. The initial addition of homologous antibody to the red cells did not give this lysis; only the final addition of complement produced the holes. Further qualitative studies showed that one molecule of IgM (p. 563) was sufficient for the formation of one hole, whereas 3,000–6,000 molecules of IgG were required per cell in order to form one hole. The greater efficiency of IgM is thought to be due to its larger molecular size covering a greater number of denaturation sites on the cell membrane.

Complement is a non-specific component of all animal sera and is not increased by any immunization procedures. The serum level of complement varies from animal to animal, but guinea pigs fed on green food produce a sufficiently high level for their serum to be employed as the laboratory source of complement.

Complement is not a single substance but a mixture of at least nine substances (C_1', C_2', C_3', etc.) of globulin or mucoprotein composition. Certain components are heat labile and heating an immune serum at 56°C for 30 minutes 'inactivates' complement; activity can be restored by the addition of fresh serum.

When complement is added to an antigen–antibody mixture the components are adsorbed in a defined order (i.e., C_1', C_4', C_2', $C'_{3,5,6,7,8,9}$) at a site adjacent to the antigen–antibody junction.

The complement lysis of red cells or bacterial cells, which can be observed to take place *in vitro* (see below), will also take place in the tissues of the immune animal. For instance, complement participates in haemolytic anaemia of the new-born, bringing about the haemolysis of red blood cells in the presence of specific antibody. A further possible role of complement is in assisting polymorphs to lyse ingested bacterial cells. It should be noted that the neutralizing antibody, which constitutes the basis of viral immunity, can neutralize the infectivity of virus particles in the absence of complement.

Complement fixation

The lysis of red blood cells by complement is the basis of a widely used serological test known as the complement fixation test. Many antigen–antibody reactions, involving viral or soluble antigens, are not visible, but it is possible to demonstrate the reaction by superimposing the visible lysis of red blood cells in the presence of complement.

To detect specific antibody in a patient's serum, a standard volume of the neat or diluted serum (heated at 56°C for 30 minutes to destroy its natural complement) is mixed with a standard amount of the specific antigen. To this is added a minimal quantity of titrated fresh normal serum (usually guinea pig) as a source of complement. This combines with the antigen–antibody complex if the patient's serum contains antibody. To demonstrate that this reaction has occurred, sensitized sheep red blood cells (i.e., red blood cells treated with homologous antiserum) are added. These cells will only lyse in the presence of free complement and hence act as an indicator of the experimental system. If antibody is present in the patient's serum, it will combine with the antigen and fix complement—the consequence will be no lysis of the red blood cells. On the contrary, if no antibody is present, the complement will be free and available to lyse the sensitized red blood cells. These reactions are illustrated in Fig. 16.4.

Phagocytosis

Foreign particles found in the body are phagocytosed by macrophages and monocytes in the tissues and polymorphonuclear leucocytes in the peripheral

Positive test

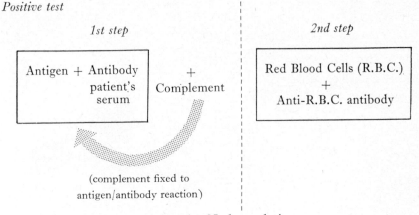

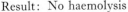

Result: No haemolysis

Negative test

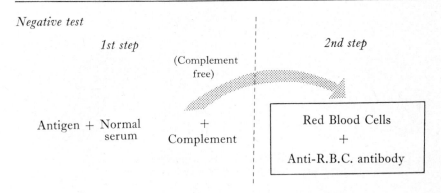

Result: Haemolysis

Fig. 16.4 The complement fixation test. (See text for explanation)

blood of the host. In contrast to inert particles, virulent bacteria are removed only after prior modification usually by antibody. In the presence of complement, the antibody treated bacteria are killed, after which they are removed as inert particles (Fig. 16.5). Components of normal serum function to a lesser degree. For instance, opsonins, both IgM and IgG of normal gamma globulin, probably act by neutralizing the negative charge on the bacterial cell so that they are repulsed to a lesser degree by the negatively charged phagocytes. A measure of the opsonic index, i.e., the ability of a patient's serum to render bacterial cells susceptible to phagocytosis, was used at one time to measure the patient's natural resistance to infection.

Neutralization

Antibody is usually able to neutralize the toxic or infective power of a micro-organism.

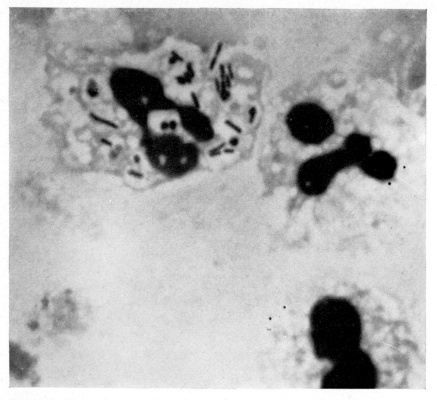

Fig. 16.5 Phagocytosis of *Escherichia coli* in polymorphonuclear leucocytes in a centrifuged specimen of urine from a case of cystitis.

Neutralizing antibodies constitute the basis of animal immunity against virus diseases. These antibodies are stimulated by the protein components of the virus capsid (p. 223), and the neutralization of virus infectivity entails the firm adsorption of the homologous antibody molecules to the capsid proteins. This reaction occurs *in vivo* and *in vitro*. Subsequent penetration of an animal cell by the neutralized virus particle results in its digestion by phagolysosomal enzymes rather than its re-emergence within the cytoplasm as an infectious particle.

The assay of the virus neutralizing antibodies contained in a serum is carried out in the laboratory by means of a neutralization test. Standard amounts of the virus are mixed with a range of dilutions of the serum, allowing neutralization to proceed to finality. Aliquots of each mixture are inoculated into either animal tissues or tissue cultures to determine the greatest serum dilution which is able to neutralize the virus infectivity.

Virus neutralizing antibodies persist for a long time in the animal, thereby providing prolonged immunity. For example, the immunity against yellow fever and German measles tends to be life-long. The apparently short-lived immunity to certain viruses (e.g., influenza, common cold viruses) is not due to a failure to produce a sufficient immune response but to the large number of serotypes, or to their genetic variation, infection with one not providing immunity against another.

Bacterial exotoxins are antigens and the antibody (antitoxin) they stimulate combines with the toxin to neutralize its toxic effects on the animal host. Endotoxins are poor antigens and their toxicity is usually incompletely neutralized by antibody stimulated by them. Exotoxins can be assayed by determining the quantity of the more stable homologous antibody required to neutralize their lethal effects or local skin reactions in animals. Other tests, such as the specific neutralization of haemolytic and lecithinase activity of particular exotoxins can be determined *in vitro*. Exotoxins are not destroyed by antitoxins but may be recovered from toxin–antitoxin complexes, and, because of this, complexes of these reagents are unstable as immunizing agents.

Haemagglutination inhibition

Certain viruses or products of their intracellular replication (haemagglutinins) are able to agglutinate red blood cells of various animal species *in vitro*. This reaction is termed haemagglutination (HA) and, since under controlled conditions the degree of agglutination is proportional to the quantity of virus, the HA test can be used as an assay method for the virus. The different degrees of agglutination are readily detected by allowing the agglutinated red cells to settle out in round-bottomed tubes or in the cups of perspex plates, giving different settling patterns. The sensitivity of the HA test is not great; with influenza virus one haemagglutinating unit (i.e., the quantity of virus which will agglutinate 50 per cent of the red cells in the tube) may be 10^6 virus particles. Further, since both infectious and non-infectious particles haemagglutinate, the test is a measure of the total virus in a preparation, not of infectivity.

Following infection with a haemagglutinating virus, antibodies are formed which inhibit the HA reaction, and these are termed haemagglutination-inhibiting antibodies. These may be assayed in the laboratory by means of the haemagglutination-inhibition (HAI) test. Standard amounts of virus are mixed with a range of dilutions of the serum, and, after a suitable incubation period, standard amounts of a 1 per cent red cell suspension are added to each mixture. The settling patterns of the red cells in the tubes or cups determine the HAI endpoint of the serum. Serum from a convalescent patient gives a four-fold or greater rise in titre compared with serum taken from the same patient during the acute stage of the illness.

Specificity of antigen–antibody reactions

Antigen–antibody reactions are highly specific, making it possible to recognize differences between proteins and other antigens of closely related plant or animal origin and even between tissues within the same host. This specificity depends primarily on the nature of the chemical groupings of opposite charge on molecules attracting each other. In addition, the spatial arrangement of the molecules, dependent upon the *ortho-*, *meta-*, or *para-* positions of chemical groups and stereoisometric spatial arrangements (*laevo-*, *dextro-*, or *meso-*),

imprint their influence upon the shape of the molecules and enable the antigen
and antibody to combine together in close proximity, thereby allowing short-
range inter-molecular forces to operate. These include van der Waal's forces,
electrostatic forces, and hydrogen bonding. The union of an antigen with its
antibody is usually very firm but the avidity (i.e., the strength of the attracting
forces between antigen and antibody) varies from system to system and in
some cases the union is reversible. No profound changes occur in either
molecule as a result of combination, as is seen in antitoxin–toxin reactions
from which the unchanged toxin can be fairly readily released.

The intact bacterial cell possesses multiple antigens on the cell wall surface
in addition to those of flagella, fimbriae, and capsules. Many antigens are common to different species of the same genus and occasionally different genera,
as for instance, antigens common to the rickettsiae (p. 345) of typhus and typhus-
like fevers and certain strains of *Proteus*, where the sharing of antigens forms the
basis of the Weil–Felix diagnostic test for these diseases. Antigens which are
specific for individual species or strains are those most useful in distinguishing
micro-organisms.

Our present knowledge of the chemical structure is most complete for polysaccharide antigens. In some organisms the nature of the chemical group
conferring specificity has been determined; the determinant group is often a
relatively small part of the whole molecule. The antigenic specificity of the
polysaccharide capsule of Lancefield Group A Streptococci is dependent on the
presence of N-acetyl glucosamine protruding as a side chain from a backbone of
rhamnose in the polymer; in Group C, specificity is attributed to a side chain of
N-acetyl galactosamine.

Similar work has been done on the chemical basis of serotyping in *Salmonella*. This genus consists of a large group of related Gram-negative bacilli responsible for a range of diseases of man and animals. The majority are flagellate, flagellar antigens being called H-antigens (from the German, Hauch, i.e.,
with emanation). The somatic antigens of the cell wall are termed O-antigens
(from the German, ohne Hauch, i.e., without emanation). The O-antigens are
common to a limited number of species enabling *Salmonella* to be subdivided
into groups (designated by capital letters). The H-antigens (when in their
specific phase) are usually limited to one or a few species and these antigens
enable the species to be serotyped.

As with the streptococci the specificity of each serotype is a function of the
chemical structure of particular determinant groupings on the antigen molecule,
the nature of which has been worked out for some of the O-antigens. In these
the specificity has been shown to be due to the presence of specific carbohydrates
on the polysaccharide molecules of the cell wall structure. By gentle hydrolysis,
the *Salmonella* polysaccharide can be split into hexamines, heptoses, hexoses,
pentoses, and deoxy-sugars. Comparisons between the carbohydrates and
the serological analyses show that the terminal dideoxyhexoses (of which only
five have been identified) determine the specificity of the antigenic determinants.
Each *Salmonella* species examined has been shown to possess only one of these
sugars which is characteristic of the serological group based on O-antigens
(Table 16.3). For instance, paratose is characteristic of O-antigen 2 in species

Table 16.3 Somatic antigens and carbohydrate units of specific O-antigens of selected salmonella groups A, B, D and O.

Salmonella species	Group	O-antigens	Hexoses			6-deoxy hexose	Terminal sugar 3·6 dideoxyhexoses			
			Gal.	Glu.	Man.	Rham.	Abequose	Colitose	Paratose	Tyvelose
S. paratyphi A	A	1, **2**, 12	+	+	+	+	−	−	+	−
S. paratyphi B	B	1, **4**, 5, 12	+	+	+	+	+	−	−	−
S. typhimurium	B	1, **4**, 5, 12	+	+	+	+	+	−	−	−
S. abortus equi	B	**4**, 12	+	+	+	+	+	−	−	−
S. budapest	B	1, **4**, 12	+	+	+	+	+	−	−	−
S. stanley	B	**4**, 5, 12	+	+	+	+	+	−	−	−
S. salinatus	B	**4**, 12	+	+	+	+	+	−	−	−
S. typhi	D	**9**, 12, vi	+	+	+	+	−	−	−	+
S. enteritidis	D	1, **9**, 12	+	+	+	+	−	−	−	+
S. gallinarum	D	1, **9**, 12	+	+	+	+	−	−	−	+
S. adelaide	O	**35**	+	+	−	−	−	+	−	−
S. monschaui	O	**35**	+	+	−	−	−	+	−	−

Group specific O-antigens are in heavy type.

falling in Group A, abequose of antigen 4 in Group B, tyvelose of antigen 9 in Group D and colitose of antigen 35 in Group O. The specific sugar always occupies the terminal position but the antigen specificity may be modified by the way the sugar is linked to the molecule or by the sugar inmediately next to it on the side chain.

Hypersensitivity

Contact with an infectious disease usually results in an increased resistance by the host to subsequent exposure to the same infectious micro-organism. Occasionally the converse occurs, the animal becoming abnormally sensitive to the infectious agent or other environmental antigens. Many types of hyper-sensitivity are encountered which may be divided according to whether the host response to contact with the sensitizing agent is immediate or delayed. In the former the host reaction is the result of antigen-antibody reactions; in the latter the sensitivity is mediated by host cells specifically sensitized to the antigen.

Immediate hypersensitivity

Immediate hypersensitivity has been classified into three types diagram-matically represented in Fig. 16.6. Type 1 occurs in patients previously sensi-tized to a particular foreign antigen. The antibody produced is thought to be absorbed to the surface of tissues cells. When a second injection of the same antigen is given, the interaction of antigen–antibody at cell surfaces results in tissue damage and the release of pharmacologically active substances such as histamine. These substances cause the smooth muscle of particular organs to contract which results in profound shock. The classical example is *anaphy-laxis* (meaning 'against protection') which may follow a second injection of a foreign serum, as in serum therapy (p. 567) or, as more recently experienced, by exposure to antibiotics. In small animals anaphylactic shock may result in death; in man and large domesticated animals symptoms of shock include bronchial spasm, pulmonary oedema (outpouring of fluids into the tissues), and urticaria (skin rashes). When the antigen is introduced locally, as the inhalation of pol-len or dust, etc., in hay fever and asthma, respectively, the symptoms of exces-sive secretion of the mucous membranes or bronchial spasm occur locally although the patient may be generally sensitive.

Type 2 occurs when circulating antibody reacts with an antigenic component of tissue cells bringing about their damage. This may occur in acute glomerular nephritis, a disease of the kidneys, and rheumatic fever. Both these conditions follow infection with specific types of streptococci. Antibodies produced against the streptococci cross-react with the basal membrane of kidney substance and heart muscle respectively. Since antibodies begin to appear in the serum after about ten days, the symptoms of these diseases do not arise until sometime after the streptococcal infection has cleared up.

Type 3 is demonstrated by the toxic symptoms which follow an antigen–antibody reaction in the blood in the presence of excess antigen as in serum sick-ness which frequently follows injection of large volumes of serum into certain hosts. Antibodies produced against the foreign serum are demonstrated in the patient's serum after the normal period of about 10 days. These react with

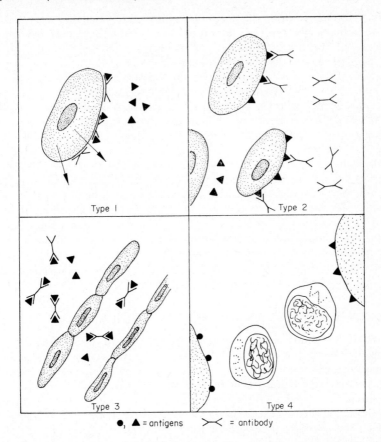

Fig. 16.6 Highly diagrammatic illustration of the four types of allergic reaction which may be deleterious to the tissues and harmful to the host. Type 1. Free antigen reacting with antibody absorbed to the cell surface results in the release of histamine and other pharmacologically active substances from the cell. Type 2. Antibody reacting either with cell surface or with antigens or haptens which becomes attached to cell surfaces: complement plays a major destructive role. Type 3. Antigen and antibody reacting in antigen excess forming complexes which are toxic to cells. Type 4. Specifically modified mononuclear cells (indicated by intracellular shadow of antigens) reacting with antigen deposited at a local site. (After Gell, P. G. H. and Coombs, R. R. A., 1963)

excess antigen (the foreign serum) still circulating in the blood and form toxic complexes which produce symptoms of serum sickness.

Delayed hypersensitivity

Delayed hypersensitivity (Fig. 16.6, Type 4) occurs in patients with chronic infections such as tuberculosis. In these diseases cell-mediated immunity plays a greater role than antibody immunity. As a result of long exposure to the bacterial antigens, lymphocytes and other mononuclear cells become sensitized to products of the bacterial cells (e.g., tuberculin), with which they react specific-ally when these substances are introduced locally either by natural infection or by injection. The area becomes red and swollen owing to the infiltration of cells into the local site and the reaction reaches a peak after 48–72 hours; hence the term delayed hypersensitivity. This reaction is used as a clinical test to determine if a patient has been previously, or is at present, infected. No tissue reaction occurs in a patient who has had no previous exposure to the tubercle bacillus. In animals the test is known as the Tuberculin test and in man, the Heaf test. A Heaf-negative patient can be converted to a Heaf-positive patient by BCG vaccination (p. 565).

Cell-mediated immunity, stimulated by chronic infection, provides another mechanism of host resistance. It can be passively transferred by living lymphoid cells but not by serum and, as with passive antibody immunity, passive cell-mediated immunity is of relatively short duration. Not all cell-mediated immune responses are advantageous to the host. Exposure to foreign tissues induces this type of immune response and homografts (i.e., tissue transplants from an-other host of the same species not genetically identical) are rejected after about two weeks by infiltration of specifically sensitized lymphocyte-like cells.

CHEMOTHERAPY AS AN AID TO RESISTANCE

Antibody immunity aids an animal either by neutralizing bacterial toxins so that they cannot damage the host or by inactivating the parasites in the presence of complement so that they can be removed by the phagocytic cells in the same manner as inert particles. Chemotherapeutic agents, by their bacteriostatic or bactericidal actions, similarly aid the body defences to deal with the invading organisms. Satisfactory chemotherapeutic agents are selectively active against specific parasites without being toxic to the host and are extremely valuable in controlling infections already established in the body. In particular, the anti-biotics are highly selective inhibitors, each one showing a defined spectrum of activity against micro-organisms owing to their specialized modes of action (p. 74). The clinically useful antibiotics can be grouped according to their range of antibacterial activity (Table 16.4). Some are active against Gram-positive bacteria and pathogenic Gram-negative cocci but relatively inactive against Gram-negative bacilli; a second group is active to a greater or lesser degree against most species of bacteria, the larger viruses and rickettsiae, and are referred to as broad spectrum antibiotics; a third group is particularly active against the Gram-negative bacilli. A very few, e.g., streptomycin, are active against the tubercle bacillus.

The use of chemotherapy during the early stages of an infection will prevent the production of antibody immunity since the stimulating micro-organisms will be killed and removed from the body. Chemotherapy may not therefore always be immediately desirable since it may prevent development of lasting immunity.

Certain bacteria, such as *Streptococcus pyogenes*, pneumococci, and meningo-cocci, are consistently sensitive to penicillin. Other bacterial species, normally susceptible to a particular drug, frequently produce resistant variants; with these

Table 16.4   Antibacterial agents used in clinical practice. (Modified from M. Barber)

Group 1 Active against: Gram-positive bacteria and Gram-negative cocci	Group 2 Active against: Gram-positive and Gram-negative bacteria, Bedsoniae and rickettsias	Group 3 Active against: Gram-negative bacilli
Penicillins	Semi-synthetic penicillins	←— Streptomycin
Benzy penicillin (G)	Ampicillin	←— Kanamycin
Phenoxymethyl penicillin (V)	Carbenicillin	←— Neomycin
Cephalosporin —→	Tetracyclines	Polymixin (colistin)
Erythromycin	Chloramphenicol	
Lincomycin	Sulphonamides	
Novobiocin Fucidin Vancomycin Ristocetin Bacitracin		

The arrows indicate a broader spectrum of activity by particular antibacterials than those organisms listed in the group.

organisms the choice of antibiotic for treatment of an infection must be based on prior tests of their sensitivity *in vitro.*

Resistance to antibiotics may arise by a number of possible routes.

1. Mutations within the bacterial population may result in resistant variants being formed from the sensitive organisms. When this occurs during treatment the sensitive population may be killed by antibiotic therapy leaving the few resistant mutants to proliferate and cause a recurrence of the disease. This type of complication can often be avoided by the use of high dosage of the antibiotic or combined therapy, a second drug dealing with mutants resistant to the first.
2. Within a bacterial species a number of strains are frequently found to be naturally resistant. This is characteristic of *Staphylococcus aureus,* many strains of which produce varying quantities of the enzyme penicillinase which destroys penicillin. Exogenous infection with such a strain means that the pathogen will not respond to penicillin and consequently another suitable drug, such as one of the penicillinase-resistant semi-synthetic penicillins, e.g., methicillin, must be used. Within a closed community (e.g., hospitals), sensitive strains of staphylococci are readily killed by widespread use of antibiotic therapy but this results also in a build-up by selection of resistant

strains ('hospital strains') within an isolated community. Both staff and patients may become carriers of resistant staphylococci in their noses and on their skin and cross-infection among patients entering hospital rapidly occurs.

3. Infective drug resistance differs from other forms of resistance in that the resistance can be transferred to sensitive cells by conjugation (p. 114). The resistance factor which is transferred is a self-replicating extrachromosomal particle carrying genetic information. Infective drug resistance is transferred between Gram-negative bacteria of both the same and different genera and species. This fact is of great epidemiological importance since sensitive pathogens may receive the resistance factor from non-pathogens in the alimentary canal. The resistance factor may carry resistance to only a single antibiotic but resistance of a multiple type may occur. Any environmental influence, such as the feeding of sub-inhibitory antibiotic food supplements to animals, which will select for resistant members of the normal non-pathogenic flora, is potentially dangerous since this resistance may be subsequently transferred to pathogens during an infection and these, in turn, will not respond to antibiotic therapy. Should these resistant pathogens be transferred to man the antibiotics against which the organisms are resistant will no longer be useful in treatment. For instance, many *Salmonella typhimurium* infections of man derived from cattle (especially calves) carry multiple resistance to antibiotics

BOOKS AND ARTICLES FOR REFERENCE AND FURTHER STUDY

ANDERSON, E. S. (1968). The ecology of transferable drug resistance in the Enterobacteriaceae. *A Rev. Microbiol.*, **22**, 131–180.

BERNHARD, R. (1967). Where do antibodies originate? *Sci. Res.*, September.

BLAIR, M. D. (1965). Hay fever and allergy. *Sci. J.*, **1**, No. 5, 66.

BURNET, F. M. (1954). How antibodies are made. *Scient. Am.*, **191**, No. 5.

——— (1961). The mechanism of immunity. *Scient. Am.*, **205**, No. 1.

DAVIS, B. D., DULBECCO, R., EISEN, H. N., GINSBERG, H. S., and WOOD, W. B. (1968). *Microbiology*. Harper and Row, Publishers. New York and London, 1464 pp.

DOURMASHKIN, R. (1964). How antibody attacks cells. *New Sci.*, No. 399, 81–82.

GELL, P. G. H. and COOMBS, R. R. A. (1963). *Clinical Aspects of Immunology*. Blackwell Sci. Pub., Oxford, 883 pp.

GOLD, E. R. and PEACOCK, D. B. (1970). *Basic Immunology*. John Wright & Son, Bristol, 416 pp.

HUMPHREY, J. H. and WHITE, R. G. (1964). *Immunology for Students of Medicine*. 2nd Edn. Blackwell Sci. Pub., Oxford, 498 pp.

LOVELL, R. (1958). *The Aetiology of Infective Diseases*. Angus and Robertson, Sydney, 136 pp.

LUDERITZ, O., STAUB, A. M., and WESTPHAL, O. (1966). Immunochemistry of O and R antigens of salmonella and related Enterobacteriaceae. *Bact. Rev.*, **30**, 192–255.

NOSSAL, A. J. V. (1964). How cells make antibodies. *Scient. Am.*, **211**, No. 6, 106.

PORTER, R. R. (1967). The structure of antibodies. *Scient. Am.*, **217**, No. 4, 81.

SELA, M. (1967). Synthetic antigens. *Sci. J.*, **3**, No. 8, 61.

SPIERS, R. S. (1964). How cells attack antigens. *Scient. Am.*, **210**, No. 2, 58.

STEWART, F. S. (1968). *Bacteriology and Immunology for Students of Medicine*. 9th Edn. Baillière, Tindall and Cassell, London, 603 pp.

WILSON, G. S. and MILES, A. A. (1964). *Topley and Wilson's Principles of Bacteriology and Immunity*. 5th Edn. Edward Arnold, London, 2630 pp.

17
Micro-organisms and Animals: prevention and control of disease

Most microbial diseases of animals and man arise from an exogenous source. One important method of disease control is achieved, therefore, by treating or removing the source of infection to prevent pathogenic micro-organisms from reaching susceptible hosts. Some of the more important sources of infection and their control will be considered.

DRINKING WATER AS A SOURCE OF INFECTION

Some pathogenic organisms are water-borne and when present in drinking water may infect susceptible hosts. Outbreaks of diseases, such as typhoid fever and cholera in man, were at one time widespread but the control of water supplies and improved hygiene in food handling have virtually eradicated these diseases from the technically more advanced countries. Water-borne pathogenic organisms usually gain access to the water from animal and human excreta and sewage. Water may also be polluted by poisons of non-microbial origin from factory wastes. Constant control is therefore necessary to ensure that drinking water for human consumption conforms to both chemical and bacteriological standards of purity. In addition, water for drinking purposes may be artificially purified (p. 527) and chlorinated.

Chlorination of drinking water

Neither natural purification nor filtration renders water safe for drinking, and disinfection is finally achieved by the addition of chlorine. Chlorination destroys all water-borne pathogenic bacteria and also pathogenic intestinal protozoa, such as *Entamoeba histolytica*, the cause of amoebic dysentery. When chlorine is added to water a portion combines with reduced organic material in the water and is no longer effective. This 'chlorine demand' varies considerably with different waters according to the amount of organic matter present and consequently the amount of chlorine that must be added to give a safe residual amount of free chlorine (0·2 to 0·4 part per million) also varies. The small amount of free chlorine necessary to render water safe is harmless and usually cannot be detected, but waters with an abnormally large 'chlorine demand' may have an unpleasant flavour after chlorination, due to the presence of such chlorine compounds as chlorophenols.

Bacteriological standards of water supplies

Bacterial pollution of water may originate either from individuals with clinical symptoms of disease or from symptomless carriers of enteric (i.e., intestinal) pathogens such as typhoid bacilli.

Pathogens are difficult to detect in water for a number of reasons. They may be present only sporadically owing to intermittent excretion and dilution and, unless the water is being continuously polluted from an infected source, the pathogens usually disappear before the disease is recognized in someone infected by drinking the contaminated water. The isolation of pathogens necessitates the concentration of large volumes of the suspected water and the use of selective media, techniques which are complex and unsuitable for routine examinations. In contrast, harmless members of the normal faecal flora are constantly being excreted and may persist much longer in polluted water. Bacteriological tests to determine the suitability of water for drinking are designed therefore to detect the presence in the water of organisms of the normal flora of the gut. The detection of these organisms is relatively simple but their presence is indicative of faecal contamination which renders the water potentially dangerous. Simple routine tests for faecal pollution, including viable counts and the demonstration of organisms of known faecal origin, are therefore employed at regular intervals to screen supplies of water used for drinking purposes.

Viable counts

Counts are made from dilutions of the water sample on a standard medium after incubation at 22°C and 37°C. Most saprophytic bacteria grow at 22°C and all organisms able to grow on the standard medium at this temperature will be included in this count. Since the number of saprophytes present is likely to be proportional to the amount of organic matter available for their nutrition, this count gives a rough indication of the relative amount of organic material present in the water.

The majority of organisms growing at 37°C will be parasites or potential parasites of man and animals, derived from soil, excreta, or sewage. A high count at 37°C, relative to the count at 22°C almost always indicates pollution with animal or human excreta. In unpolluted water the ratio of the count at 22°C to that at 37°C is usually greater than 10:1; in polluted water the ratio may be 1:1.

Demonstration of organisms of faecal origin

The water is further examined for the presence of specific bacteria of known intestinal origin. Certain species, as *Escherichia coli*, *Streptococcus faecalis*, and *Clostridium perfringens*, usually present in the human intestine and frequently in those of other animals, are useful indicators of recent pollution with excreta. In Great Britain, *E. coli* (*Bacterium coli* faecal type I) is used as the indicator strain in official tests.

Many coliform bacteria are found on vegetation and in soil but *E. coli* is found only transiently outside the alimentary canal and its presence in water is thus almost certain evidence of recent faecal pollution. The demonstration of even a few *E. coli* in water is sufficient to condemn it as unfit for human consumption even though no pathogens have been found in it.

Two tests are used to detect *E. coli*: (1) the presumptive coliform count and (2) the identification of isolated strains by differential biochemical tests.

PRESUMPTIVE COLIFORM COUNT. This count is performed in a selective nutrient fluid medium to which bile salts, lactose, and a pH indicator are added. The bile salts inhibit the growth of most non-intestinal organisms but not that of *E. coli* and allied organisms. The presence of coliform bacilli under these conditions is demonstrated by the production of acid and gas by fermentation of lactose. For statistical reasons a range of volumes and dilutions of the water under test is added to appropriate volumes of the bile-lactose medium. The greater the number of *E. coli* present and the larger the sample of water examined, the higher will be the chance that *E. coli* will be contained in any particular sample and vice versa. The distribution of coliform positive reactions in the range of volumes tested from each water sample is referred to probability tables and the probable number of coliform bacilli in 100 ml. of the original water sample is deduced. The standards required vary for different types of water but, in general, an unchlorinated water with more than 5 coliform bacilli per 100 ml would be regarded as unfit for drinking. Chlorinated water would obviously be expected to have no viable coliforms present.

DIFFERENTIAL COLIFORM TESTS. Fermentation of lactose in the presumptive coliform count may be due to any one of a number of coliform bacilli and not necessarily to *E. coli*. It is essential, therefore, to determine whether the organism detected in the count is the type known to be parasitic in the gut. The organism is first isolated on solid medium from one of the tubes showing fermentation and differential biochemical tests are performed. If biochemical reactions typical for *E. coli* are obtained and if the presumptive coliform count is in excess of the maximum number permitted, the water is regarded as potentially dangerous and condemned as unfit for human consumption.

FOOD AS A SOURCE OF INFECTION AND DISEASE

Many diseases may be contracted by the ingestion of contaminated food. The food may be infected at source, as in meat, milk or eggs from infected animals or birds, many of the natural infections of these animals being common to man; vegetables grown in contaminated soil or irrigated by polluted water (e.g., watercress beds) may also be contaminated. Foods may also become contaminated with pathogenic organisms during processing for preservation or preparation for eating. Pathogens may be introduced by flies or, most important of all, by the handling of foods by human carriers of food poisoning organisms (see below). Most risk is entailed with foods prepared for consumption the following day, the organisms multiplying if the food is kept at a temperature conducive to growth.

Food poisoning

A special group of diseases contracted by the ingestion of food are designated 'food poisoning'. Outbreaks of these are usually explosive in character and may be related to ingestion of a specific meal or common food source.

Food poisoning may be due to poisons derived from plant or animal sources, or to chemicals added inadvertently or as preservatives at too high concentrations, or to the presence of harmful micro-organisms or toxins produced by them. The symptoms produced by different poisons may be similar and the final diagnosis thus depends upon laboratory tests. The time from the consumption of the poisonous food to the onset of symptoms varies from 10 minutes to 2 hours with some forms of chemical poisoning, to around 6 hours with bacterial toxins, 12 to 72 hours when poisoning is due to living bacteria and 2–3 days when poisonous toadstools are the cause.

Bacteria causing food poisoning

In Great Britain the term 'bacterial food poisoning' is legally restricted to poisoning resulting from the ingestion of food containing certain bacteria (or their toxins) which are not included among other specific infectious diseases notifiable to the Minister of Health. Thus, although the dysentery bacillus may be ingested in food and cause gastro-enteritis, a common symptom of bacterial food poisoning, it is not officially classed as a food poisoning organism since it causes a notifiable disease in its own right.

The types of foods involved in food poisoning are those favouring growth of the causative organisms, so that large numbers of bacteria or their products are present in the food at the time it is ingested. Outbreaks of food poisoning frequently involve a large proportion of the people who have eaten a particular food.

Two groups of agents are recognized: pathogenic bacteria which can infect the alimentary canal and produce symptoms of gastro-enteritis, and bacteria (or their toxins) which do not infect the gut but produce various toxic symptoms when ingested with food.

INFECTIVE FOOD POISONING. In this form of food poisoning the infective agent is able to establish itself in the gut, multiply rapidly and produce symptoms of disease over a number of days, often for about a week or longer. This form of poisoning is mainly limited to species of *Salmonella*. When infection is established, endotoxins (p. 555), resulting from the breakdown of dead bacterial cells, cause irritation to the gut linings resulting in symptoms of acute gastroenteritis.

The Salmonella group includes many hundreds of types of serologically related organisms which are found in almost every species of animal, including birds and reptiles. In Great Britain meat, milk, synthetic cream, and eggs (particularly duck eggs) are the foods most commonly contaminated with salmonellae. Meats from cattle and pigs have been found to be a frequent source of infection, particularly when made up into processed foods such as brawns, sausages or pies, which may be incompletely cooked or subsequently contaminated during handling. Ideally salmonella should be absent from food but if adequately cooked, these pathogens are destroyed and there is no risk of food poisoning, since there are no residual toxins to produce food poisoning symptoms.

Symptoms of salmonella food poisoning vary in severity but include acute gastro-enteritis accompanied by severe headaches followed by nausea, vomiting, abdominal pain, and diarrhoea. Fever frequently occurs. These symptoms usually arise between 12 and 72 hours after the meal has been consumed, by which time the organisms are established in the gut. Symptoms of toxic food poisoning may arise earlier if sufficiently large numbers of organisms are ingested (p. 589). In favourable cases the symptoms subside within a week, but in a small percentage of patients organisms invade the tissues from the gut and may cause death. Even when symptoms disappear, the patient may continue to excrete the pathogens and thereby remain a potential source of infection for susceptible individuals and of contamination of food. If this continues for a long period after recovery, the patients are termed carriers. It is most important to ensure that such people are not employed in food handling.

BACTERIAL CELLS OR THEIR PRODUCTS AS CAUSES OF FOOD POISONING. In this form of food poisoning no infection of the gut occurs but bacterial cells or exotoxins of bacterial origin produce symptoms of food poisoning when ingested in foods. Since these toxins are absorbed by the gut wall they are called enterotoxins. They are preformed in the food by growth of the pathogenic organism and although the vegetative cells may be killed by heating the food prior to eating, many of the toxins, being moderately resistant to heat, remain and cause symptoms of food poisoning when ingested. The most important bacterial toxic food poisonings are caused by the toxins of specific strains of *Staphylococcus aureus* and *Clostridium botulinum*, the latter frequently causing a fatal form of food poisoning known as botulism. Strains of *Clostridium* perfringens also cause toxic food poisoning and certain non-pathogenic bacteria, such as *Proteus* sp., not generally considered to be toxigenic, produce acute gastro-enteritis when consumed in sufficiently large numbers.

Staphylococcal food poisoning. Most outbreaks of this type of food poisoning are associated with milk and its products (e.g., cream fillings in cakes and milk deserts such as custards) and prepared meat (e.g., pies or brawns). These foods favour the rapid multiplication of staphylococci (and consequent toxin production) in warm weather if they are not refrigerated. In non-pasteurized milk the organisms may come from an infection of the cow's udder, but since staphylococci are frequently found on human skin and in the naso-pharynx, there are many opportunities for contamination of the food during preparation. Fortunately only a few types of staphylococci (generally phage types of group III) are able to produce the enterotoxin which is formed in significant amounts only when the bacterial count reaches 10^5 to 10^6 per gram of food.

Symptoms of staphylococcal food poisoning may arise after only 1 hour but more usually after about 12 hours from ingestion of the toxin, and include violent vomiting, diarrhoea, and prostration but there is no fever and recovery is rapid, often within 24 hours. This type of poisoning is only occasionally fatal.

Botulism. Clostridium botulinum is a spore-bearing anaerobe occurring in soil. It may be ingested with food but does not multiply in the body. Inadequate sterilization of foods originally contaminated from the soil and stored under anaerobic conditions conducive to the growth of this organism are responsible for the majority of outbreaks. In Europe most outbreaks of botulism follow the eating of preserved meats such as hams or sausages. In the U.S.A. home-canned but inadequately sterilized vegetables and fruits are most usually responsible, and in Russia some cases have been attributed to salted fish. No case of botulism is known to have arisen as a result of eating fresh foods.

The preformed toxin, when ingested, is absorbed through the mucosa of the stomach and upper part of the intestine. The symptoms differ from all other forms of food poisoning in that the toxin acts directly upon the central nervous system. Nausea and vomiting usually first occur within 24 hours and are followed several days later by paralysis of specific muscles due to damage of the nerve centres controlling them. Thus double vision results from paralysis of the eye muscles, difficulty in swallowing follows paralysis of the throat muscles. If paralysis of the muscles of the respiratory tract occurs, death from respiratory failure results.

Several types of *Cl. botulinum* are known, each of which produces a serologically distinct toxin. In man, Type A is most prevalent in the U.S.A., whilst in Europe Type B is commonest. Other human outbreaks have been attributed to Type E, which is mainly of marine origin and, though tolerant of lower temperatures, has spores which are more easily killed by heat. No human cases of botulism caused by types C and D were known before 1958, but some were reported in 1958 and 1961.

Clostridium poisoning. Particular strains of *Cl. perfringens* when ingested in large numbers produce a characteristic form of food poisoning. In Great Britain a variant of *Cl. perfringens* type A is the responsible strain; this strain is only feebly toxigenic but its spores exhibit a greater heat tolerance than other strains. Cooked meat contaminated with this strain makes an ideal medium for germination of the spores which have survived the cooking process and is frequently incriminated in this form of poisoning. It is not known whether the symptoms are due to the organisms themselves, the toxins they produce, or breakdown products of the meat resulting from bacterial activity.

Non-specific bacterial causes of food poisoning. Occasionally outbreaks of food poisoning have been reported in which the foods involved were heavily contaminated with species of *Proteus, Escherichia, Streptococcus,* or *Bacillus.* It is generally thought that large numbers of these organisms have an irritating effect on the gastro-intestinal mucosa and set up symptoms of toxic food poisoning. This emphasizes the fact that gross contamination of food with organisms generally considered to be non-pathogenic, followed by storage at temperatures conducive to bacterial propagation, presents a potential hazard to the consumer.

LABORATORY DIAGNOSIS OF BACTERIAL FOOD POISONING. Confirmation of clinical diagnosis rests either on laboratory isolation of the causal organism, principally from faeces, or demonstration of the toxin either in the patient or in the incriminated food, if some is still available.

Since in staphylococcal food poisoning the causal organism is rarely present in the faeces of patients, isolation and subsequent bacteriophage typing is attempted from food, vomit, etc. Often the staphylococci have been killed by subsequent heating of the food and then confirmation of this type of food poisoning is only presumptive. Whilst many animal tests to detect staphylococcal enterotoxin in food and culture filtrates have been explored, the only reliable method is to feed small amounts of culture filtrates to human volunteers in an attempt to reproduce food poisoning symptoms.

The organism causing botulism is spore-bearing and its spores are highly heat-resistant. It may survive in cooked food, from which it may be isolated. As a further test, the toxin, obtained as a sterile filtrate from a suspension of the food or gut contents, is injected intraperitoneally into mice. If mice, protected against botulism by the injection of specific antiserum survive whilst non-protected ones die, proof of the presence of botulism toxin is established.

Poisonous fungi

Even edible species of fungi are liable to prove indigestible and to cause minor illness if eaten in excess and individuals may be allergic to particular species. A few agarics, however, are very poisonous. The most deadly species in Britain is the death cap, *Amanita phalloides.* The consumption of even a small piece of a fruit-body of this species may be fatal and at best will cause a long and serious illness. Attempts have been made to prepare a serum against this species but this is effective only if it is known at once that the death cap has been eaten. More often the victim is unaware that he has eaten a poisonous variety and the symptoms develop after only one or two days by which time treatment is useless. The best safeguard is the ability to recognize this and a few other poisonous species. The toxins of *A. phalloides,* which cause degeneration of liver and kidneys, and progressive paralysis of the central nervous system, are not destroyed by cooking, but those of some other species are inactivated by heat, so that fruit-bodies which are dangerous when eaten raw are safe after adequate cooking. For instance, the Ascomycete, *Gyromitra esculenta* produces a heat-labile toxin, helvellic acid, which acts on red blood cells.

The sclerotia of the Pyrenomycete *Claviceps purpurea* the ergot of rye, are of medicinal value, owing to the presence of an alkaloid which is extracted and used to stop excessive bleeding, particularly in childbirth. They are also poisonous; rye grass infected with ergot can cause abortion in sheep and cattle, and flour prepared from grain admixed with the 'ergots' (sclerotia p. 379) causes serious disease (formerly known as St. Anthony's Fire) in persons consuming it. Another Pyrenomycete, *Gibberella zeae*, infects the ears of cereals and renders the grain poisonous to man, pigs, and dogs but not to sheep or cattle. The darnel grass fungus, a non-sporing systemic endophyte of *Lolium temulentum*, is said to render the seeds of this grass poisonous to stock, but this has not been proved and it has been suggested that in the instances reported contamination with ergot had occurred.

Control of food supplies

The increase in urban populations during the present century and improvements in methods of food preservation have led to large-scale transport of basic foods from the producer to the consumer areas. This has inevitably increased the risk of infection of many people from a common food source. This risk can be considerably reduced by suitable precautions.

Methods employed for improving the keeping quality of a food (i.e., avoiding spoilage) are often adequate to render a food safe for eating. The handling of vegetables, eggs (p. 658), meat (p. 651) are considered in Chapter 20. Milk and its products are also considered in Chapter 20 but the control of milk supplies by which a safe supply is guaranteed and the statutory requirements will be described below.

Milk supplies

Consumer-milk is usually transported by bulk-collection services; it is therefore most essential that adequate measures of control are observed since contaminated milk from one source may be mixed with a large volume of clean milk. The first requirement is good animal husbandry and dairy technique to produce a clean product of high quality. As an additional safeguard most milks are heat-treated to kill pathogenic bacteria which may be present and at the same time to reduce the number of contaminants thereby improving keeping quality.

METHODS OF HEAT TREATMENT OF MILK. Heat treatment is the most widely used method of preservation of milk. Three methods are currently used.

Pasteurization: This method of partial destruction of the microbial population by heat was first introduced by Pasteur to kill contaminating organisms which interfered with the fermentation processes in the manufacture of wine. Its application to milk was first used in Denmark to safeguard pigs against infection from bovine pathogens, but its widest industrial application today is in treating milk for human consumption. By holding the milk for a defined time at a standard temperature, as, for instance, 15 seconds at 161°F (71·7°C) in the 'High Temperature Short Time' process, most non-sporing organisms, including all non-sporing pathogens, are killed. This renders the milk safe for drinking and extends its keeping quality.

Boiling: Greater numbers of micro-organisms are killed when milk is held at a temperature not less than 100°C for a period, which ensures that it will comply with the turbidity test (see below), and the bottles are sealed immediately afterwards. This process imparts a caramelized flavour to the milk and homogenization occludes a visible cream line. Milk subjected to this process is often called 'Sterilized' but total sterility is not achieved and it will not keep indefinitely at normal temperatures.

Ultra heat treated: By this method, introduced in Great Britain in 1965, the milk is exposed to a temperature of not less than 270°F (132·2°C) for at least one second. Usually this treatment renders it sterile and therefore gives it excellent keeping qualities. There is no cream line, the cream being dispersed throughout the milk as in homogenized milk. From 10 to 20 per cent of vitamins are destroyed and a slight flavour is imparted, but this becomes less upon storage. Because of the excellent keeping quality of the milk it is likely to become very popular since less frequent deliveries to the consumer will be possible and export to other countries is facilitated.

STATUTORY STANDARDS FOR MILK. In Great Britain milk supplies are regularly tested to ensure that they conform to a reasonable degree of cleanliness or have been exposed to the correct heat treatment for the particular designation specified. The one official test to which an 'untreated' milk must comply is the *Methylene Blue Test*. A standard solution of methylene blue is added to a sample of the milk on the day after it left the farm and the sample is then incubated at 37°C. A satisfactory milk will not reduce the dye to the colourless leuco-form within 30 minutes. Methylene blue is a redox potential indicator and is reduced by the microbial activity in badly contaminated samples.

Pasteurized milk is required to pass two official tests prescribed by the Ministry of Health. *A Methylene Blue Test* similar to the one described for untreated milk is performed on the milk 24 hours after it has been pasteurized. This determines the bacterial cleanliness of the milk at the time when it would normally reach the consumer, and measures contamination which may arise from improperly cleaned bottles subsequent to processing. The other test, the *Phosphatase Test*, is a quantitative method of detecting the amount of the natural enzyme, phosphatase, normally present in milk. In pasteurized milk, most of this enzyme is destroyed and only a minimal amount is detectable. This test checks that the milk has been adequately pasteurized.

'Sterilized' milk must conform to the *Turbidity Test*. Proteins are denatured by boiling, but if the milk has received inadequate heating, soluble proteins remain and may be detected by the turbidity test. The milk is first saturated with ammonium sulphate and filtered; a tube of the clear filtrate is then plunged into boiling water and kept in it for 5 minutes. The formation of a turbidity, due to the heat denaturation of proteins, indicates that the milk had been insufficiently heated. Ultra Heated Milk is tested by a *Colony Count*. A standard loopful of the milk is cultured in 5 ml. of yeastral milk agar at 37°C for 48 hours. Not more than 10 colonies are permitted if the milk is to pass the test.

Neither the phosphatase nor the turbidity test can be applied to Ultra Heated Milk since phosphatase and undenatured proteins remain after the ultra heat treatment.

ANIMAL AND HUMAN RESERVOIRS OF INFECTION

The spread of infection from man to man, from animals to man, and to a lesser extent from man to animals, constitutes some of the more important pathways of cross-infection. The term zoonoses has been coined to include those diseases (protozoan, fungal, bacterial, and viral) which are naturally transmitted between vertebrate animals and man; more than 100 zoonoses are recognized. Because of their close association with man, domesticated animals are the sources of greatest danger to man. Prevention, control and eradication of these sources of infection are the concern of many countries.

Eradication of infected animals

Eradication programmes obviously cannot be practised in the control of human diseases although under certain circumstances examination and treatment is compulsory. Eradication is, however, proving a most successful approach to the control of a number of diseases of domesticated animals. Success is even greater in countries geographically separated from others. Great Britain, separated from the Continent by a sea barrier, has already achieved eradication of a number of serious diseases including tuberculosis and foot-and-mouth disease in domestic animals, and plans are already in hand to eliminate brucellosis. Initially a defined area is cleared of infection and designated disease-free, and this area is enlarged year by year until the whole country is free of infection. Reliable tests to detect infected animals are essential to success. For instance, in order to maintain the national herd free of tuberculosis, farm animals are regularly tested by the tuberculin test (p. 580) and positive reactors immediately destroyed.

Anthrax, an acute, fatal disease of cattle, is controlled in Britain by the enforcement of special methods of disposal of anthrax-infected carcasses. *Bacillus anthracis* readily produces highly resistant spores when exposed to air. Hence, if death from anthrax is suspected, the carcass must not be opened for post-mortem examination; this avoids exposing the organism to the air and contamination of the surrounding ground or pasture by spores which would thereby be produced. Instead, the cause of death can be confirmed by microscopical examination of blood smears taken from a peripheral vein and, if anthrax is diagnosed, the carcass is either incinerated on the spot or buried in quicklime well below ground level. In foot-and-mouth disease stringent measures of eradication are adopted in every outbreak both in Britain and N. America. The cost of compensating farmers for loss of livestock, even in such a serious outbreak as that of 1967-8, is considerably less than the widespread economic loss in milk yields and value of carcass meat which would result if the disease were allowed to become endemic in the animal population.

Isolation of infected animals

The spread of infection may be prevented by the segregation of infected animals from susceptible ones. This is often referred to as quarantine measures, the term being derived from the practice of segregating infected individuals for forty days (Latin—quadriginta) which dates from the Mosaic period in Jewish history. Quarantine measures are widely practised today in the control of animal and human diseases but are now based on knowledge of the length of time the patient remains infectious. Importation into Britain of many animal pets is controlled by quarantine measures at the ports to avoid introducing infections not normally present in the country, such as rabies in dogs and psittacosis in birds. Many countries require veterinary certification that dogs are free of leptospiral infection before they are allowed in. Immigrants are medically examined at ports of entry to ensure that they are free of infection. Patients infected with smallpox or similar highly infectious diseases are treated in isolation hospitals to restrict spread of infection.

Interference with the normal cycle of infection

Many viral and rickettsial infections are vector-transmitted (p. 610). The most successful control of these infections requires destruction of the transmitting vector. For instance, the general application of mosquito control measures has virtually stamped out epidemic yellow fever in man, the disease now existing only as an enzootic infection of jungle monkeys with man as an incidental host.

CONTROL OF AIRBORNE CROSS-INFECTION

The majority of pathogens are obligate parasites and consequently most are spread from infected animals and man to susceptible hosts. In contrast to plant pathogens, the number of infective agents dispersed are relatively few and these are rapidly diluted in the atmosphere to levels below which infection does not occur. With few exceptions airborne animal infections are therefore limited to indoor atmospheres.

As stated earlier (p. 535), the air inside buildings is being continuously contaminated with micro-organisms from human or animal occupants. These gain access to the air on dust from skin, hair, and clothing or in droplets from the mucous membranes of the respiratory tract, propelled by talking, coughing and sneezing. Respiratory infections arise by inhalation of pathogenic organisms and control measures aim at restricting these to a minimum.

Cross-infection by 'droplet' infection is restricted by avoiding overcrowding. This may be achieved by the adequate spacing of beds in hospital wards, of desks in schoolrooms, and of stalls in stables. Recent work has shown that subdividing large hospital wards into smaller units is more effective in reducing cross-infection than wide spacing of beds. It has long been held that the common cold virus is transmitted from patient to patient by droplet infection but here, too, recent work suggests that prolonged personal contact, such as occurs in the home or school, is more important than the casual contact. These studies reveal the difficulty of separating a purely airborne route from that of personal contact and invariably both pathways of infection have to be controlled at the same time.

The provision of adequate natural or mechanical ventilation with filtered air, reduces air contamination. The fact that more ventilation is practised in summer may account, in part at least, for less respiratory disease occurring at this time of the year. In many-storied buildings with mechanical ventilation requiring sealed windows, movement of contaminated air can occur from one floor to another by way of lift shafts or laundry chutes and these present a real hazard, especially in hospitals where frequently a build-up of infection in one ward may be passed to another on a different floor. In operating theatres positive air-pressure within the theatres excludes air contamination from adjacent rooms and corridors in addition to removing air contaminated from movement of personnel.

Both droplet and dust-borne air contamination are partly controlled in special environments, such as operating theatres and wards of infected patients, by the wearing of sterile protective clothing and face masks by the staff. Face masks must be frequently changed and although they filter only the larger salivary droplets this is considered adequate to prevent direct contamination of wounds.

Dust derived from animal and human sources is invariably laden with micro-organisms. This comes from hair, desquamated skin and fragments of textiles resulting from friction between layers of clothing or activities such as bed making. This is of particular importance when the bedding has been used by an infected patient. Contaminated dust settles on to horizontal surfaces, but is readily re-dispersed into the air by air currents, dry sweeping, dusting, and similar activities. By the use of water or, more effectively, oils and other mixtures which cause individual particles to adhere together in large aggregates, the raising of dust can be greatly reduced. Textiles, such as bed clothes, can also be treated with paraffin oil at concentrations which do not affect their appearance or comfort, thereby reducing dispersal of micro-organisms during bedmaking. Vacuum cleaning is a far more efficient way of removing dust without contamination of the air than other means, but precautions must be taken to prevent fine particles of dust from being blown out of the machine. These preventive measures alone, however, seldom provide complete control.

Organisms may be destroyed in air by the use of ultra-violet radiation, and aerosols or sprays of disinfectants. Ultra-violet radiation of wavelength 2,537 Å, whilst not being the most bactericidal part of the spectrum, is readily available from mercury vapour type lamps and is used to reduce the count of airborne bacteria and viruses within buildings. Since ultra-violet radiations produce an irritant effect on the conjunctiva of the eye the radiation source must be shielded from occupants in the room. This correspondingly reduces its field of effectiveness. Other factors must also be taken into account. Ultra-violet radiations have a limited range of action which varies as the square root of the distance from source; it is far less effective against bacterial and fungal spores than vegetative cells; organisms must be directly exposed to the rays and screening them from direct radiation or protection by thin layers of protein materials reduces the 'kill'. Ultra-violet radiations are more effective against organisms suspended in droplets than against those attached to dust particles. In contrast to chemical disinfectants, they are most effective at low humidity.

Chemical disinfectants have the advantage of penetrating to all parts of a room. To be efficient they must make contact with the airborne organisms and, for this reason, those in a gaseous or vapour state are more effective than others but less volatile substances are sometimes effective if sufficiently finely dispersed. They are used either as sprays or in a volatile form dispersed as aerosols. Disinfectant sprays consist of droplets larger than 150 microns in diameter and result in wet mists, the drops rapidly settle out, lay the dust and disinfect horizontal surfaces. Aerosols, of droplet size 5 to 150 microns in diameter, produce dry mists or fogs, which remain suspended in the atmosphere for long periods of time, travel for considerable distances from the source and, by virtue of their large surface area/volume ratio, rapidly build up high local vapour concentrations of volatile substances. All effective air disinfectants act at comparatively low levels of concentration when dispersed in the atmosphere, a feature of great practical importance since these levels are well below those toxic for man and animals. They can therefore be used in occupied buildings. For instance, 1 gram of sodium hypochlorite dispersed in 40,000 litres of air has been found to be bactericidal in a few minutes. It was thought that air disinfection with aerosols was brought about by aerosol particles themselves coming into contact with the airborne bacteria. This view does not take into account the rapidity with which aerosols act and it is now generally agreed that the lethal effects are due to the vapour derived from the aerosol particle producing a high lethal concentration localized around the micro-organism. The particle therefore acts merely as a mobile source of vapour, aiding its persistence and distribution. Humidity usually plays an important role, providing residual moisture on the organismal surface in which the germicide dissolves.

Not all disinfectants are suitable for air disinfection. Some, such as phenols, cresols, and quaternary ammonium compounds, although successful, have been rejected on practical criteria including odour, irritation, and unpleasant side effects. The most useful include chlorine, sodium hypochlorite, hypochlorous and lactic acid, propylene, and triethylene glycols and phenols of low volatility such as hexyl resorcinol and resorcinol. Many of these have been used successfully for the sterilization of air in public buildings and for the safeguarding of young poultry against pullorum disease in brooder houses. Formalin vapour, due to its toxicity, cannot be used in occupied buildings but is a most effective means of sterilizing the air of unoccupied animal and poultry houses. Reductions greater than 99 per cent in the coliform counts have been achieved by use of this reagent.

PREVENTION OF CROSS-INFECTION IN HOSPITALS

Cross-infection by any route is an important problem in hospitals where seriously ill and highly susceptible patients are often at risk to infection by particularly virulent antibiotic-resistant organisms from patients already infected. In addition to the preventive measures already considered, numerous precautions are now taken in hospital to prevent infection by direct and indirect contact. Indirect contact, the transfer of pathogens from an infected to an uninfected patient via hospital equipment, surgical instruments, or the hands of patients or staff, constitutes the greatest problem in hospital cross-infection. Many of these infections can be avoided by very careful aseptic and antiseptic techniques in surgical and nursing procedures but in recent years many hospitals have introduced improvements such as central sterile supply departments to ensure that instruments and dressings are correctly disinfected or sterilized. In addition, improvements are constantly being introduced in disinfectant measures such as the pre-operative disinfection of skin, disinfectant soaps and creams for doctors' and nurses' hands, disinfectant creams and sprays to prevent nasal colonization by staphylococci, and the treatment of carriers.

Many of these improvements have been made in the past two decades but still more are needed. With the increase in highly specialized surgery involving long and intricate operations with intensive post-operative care, the risks of infection are greater than ever before and consequently the study of methods of preventing infection will continue to be of the greatest importance.

BOOKS AND ARTICLES FOR REFERENCE AND FURTHER STUDY

ANDERSON, E. S. (1968). Transferable drug resistance. *Sci. J.*, **4**, No. 4, 71.

DACK, G. M. (1956). *Food Poisoning*. 3rd Edn. Univ. Press, Chicago, Ill., 251 pp.

DAVIS, J. G. (1955). *Milk Testing*. Dairy Institutes Ltd., London, 260 pp.

HARVEY, W. C. and HILL, H. (1967). *Milk: Production and Control*. 4th Edn. Lewis and Co. Ltd., London, 711 pp.

HINMAN, E. H. (1966). *World Eradication of Infectious Diseases*. C. C. Thomas, Springfield, Illinois, U.S.A., 223 pp.

HOBBS, B. C. (1968). *Food Poisoning and Food Hygiene*. 2nd Edn. Edward Arnold, London, 252 pp.

RAMSBOTTOM, J. (1945). *Poisonous Fungi*. King Penguin Books, London.

Report (1957). *The Bacteriological Examination of Water Supplies*, No. 71. H.M.S.O.

Report (1965). Statutory instruments. *The Milk (Special Designation) Regulations*, No. 1555. H.M.S.O.

VAN DER HOEDON, J. (1964). *Zoonoses*. Elsevier Pub. Co., Amsterdam, 774 pp.

WILLIAMS, R. E. O., BLOWERS, R., GARROD, L. P. and SHORTER, R. A. (1966). *Hospital Infection: Causes and Prevention*. 2nd Edn. Lloyd Luke Ltd., London, 386 pp.

WILSON, G. S. and MILES, A. A. (1964). *Topley and Wilson's Principles of Bacteriology and Immunity*. 5th Edn. Edward Arnold, London, 2630 pp.

18
Micro-organisms and Invertebrate Animals

INTRODUCTION

Everyone is aware of the importance of micro-organisms causing diseases of higher animals but this is only one aspect of the effects of micro-organisms on animals. Invertebrates too are associated with micro-organisms in a number of ways, including loose associations in which both organisms compete for food, the use of micro-organisms as food, diseases caused by them, and their dispersal by insect and other vectors. There are specialized associations of other sorts too. It is believed that the symbiotic bacteria which are present in the luminous organs of certain insects and cephalopods are wholly or partly responsible for the light that they emit. Most instances of luminescence in animals are, however, purely animal phenomena.

MICRO-ORGANISMS AND THE NUTRITION OF INSECTS AND SOME OTHER INVERTEBRATES

Competition for food and oxygen is often severe in mixed communities, such as those in soil or in water (Chaps. 12 and 13). The activities of micro-organisms may sometimes deprive animals of foodstuffs, but very many invertebrates are dependent, directly or indirectly, on certain micro-organisms for the digestion or other preparation of their food.

Micro-organisms as food for invertebrates

Bacteria, fungi, Protozoa, and even small algae, are used as food by some insects and other invertebrates. For example, aquatic micro-organisms serve as
food for mosquito larvae. The part played by grazing invertebrates in determining the number of algal cells in fresh water and marine habitats has been considered in Chapter 13 where it is pointed out that algae, as 'primary producers',
directly or indirectly provide food for aquatic animals.

The larvae of many flies and some beetles feed upon the fruit-bodies of the
higher fungi. Their tunnels are commonly seen in mushrooms, other agarics,
woody fruit-bodies, and truffles. A particular species of insect usually attacks
only a single species of fungus.

Other insects have a closer relationship with specific fungi which live in their
nests and on which they feed. Ambrosia beetles are wood-boring insects that
do not as a rule eat wood but instead feed on the 'ambrosia' which lines their
tunnels. This a velvety layer of fungus and in it are concentrated nutrients
drawn from the surrounding wood. Before newly formed adults leave the tunnels they feed on the fungus and also gather spores into special pockets so that
on arriving at new timber they seed their borings with spores and assure their
progeny of a food supply. The growth of other species of fungi in the tunnels
is suppressed by an unknown means. Once the galleries are deserted by the
beetles the ambrosia fungi are overrun by more vigorous species. Each species
of beetle is associated with a single species of fungus and most species of ambrosia fungi are each cultivated by only one species of beetle. A somewhat similar relationship exists between certain species of wood-boring and bark-boring
beetles and the fungus *Ceratocystis ulmi* (the cause of Dutch elm disease) and
the related blue-stain fungi, *Ceratocystis* spp. These fungi are carried from diseased to uninfected trees by the beetles.

Ants of the tribe Attii of the family Myrmicinae cultivate fungi in the so-
called 'fungus gardens' made from pieces cut from leaves. The queen ant carries
the specific fungus in a special pouch (the intrabuccal pocket) and deposits it
on her faeces, where it grows to start a new garden. The fungi, which have not
been fully identified, are induced to produce swollen hyphal tips—bromatia—
in spherical clusters on which the larvae are fed.

Algae live with invertebrates in various intimate associations that almost
certainly result in nutrients synthesized by the algae becoming available to the
animal, but it is seldom known whether nutrient transfer depends more on
digestion of the whole alga or on the escape of nutrients from the intact algal
cell. Green or yellow algal cells have been observed living in invertebrate animals
that include hydras and many marine hydropolyps, sponges, corals, turbellarians,
Echinodera (minute marine worm-like animals), rotifers, annelids, and molluscs.
The presence of symbiotic green algae in the hydra *Chlorohydra viridissima*
seems also to confer resistance to attack by a species of parasitic amoeba.

Fig. 18.1   Fungi associated with insects. (**A**) Fungus gardens in subterranean nest of
termites in West Africa ($\times 1/3$. M. F. Madelin). (**B**) The surface of a termite fungus
garden, showing surface growth of mycelium and spherical white masses of bromatia.
($\times 8$. M. F. Madelin)

A

B

The part played by micro-organisms in the preparation and digestion of the food of invertebrates

Micro-organisms may break down complex materials thus rendering them suitable as food for many invertebrates, such as the nematodes and insect larvae inhabiting the soil. It is significant that the adults of many sorts of insect are attracted by odours of fermentation or putrefaction and feed upon the decomposing substances or lay their eggs therein. Termites and the death-watch beetle (*Xestobium*) tend to attack timber already partially rotted by fungi. The life-cycle of this beetle may take as long as five years on sound timber and as little as eighteen months on partially decayed wood.

A more specific association exists between certain wood wasps and species of the basidiomycete genera *Stereum* and *Daedalea*. Fungal oidia contained in tiny pouches at the base of the ovipositor are introduced into wood when the female lays her eggs. The larvae that hatch feed on wood that is partly digested by the advancing mycelium of the implanted fungus. A similar relationship exists between certain flies of the genus *Hylemyia* and bacteria. The larvae of one species on hatching near to seed potatoes spend their first day abrading the surface of the potatoes. This allows bacteria to invade the underlying tissue. As the rot spreads so the larvae penetrate into the tissues and feed. Other *Hylemyia* species that attack onion and cabbage tissues similarly depend on a preceding attack by bacteria.

Many species of termites cannot themselves digest the wood that they eat, relying instead on intestinal Protozoa to do this for them. However, there are some termites that lack these Protozoa and instead cultivate fungus gardens. It is likely that the fungi in these gardens decompose cellulose and also provide vitamins for the termites. The fungus gardens lie in subterranean chambers within the termite nests. They are usually rounded, several to many inches across, firm, but sponge-like in appearance (Fig. 18.1). They are made of partly digested faecal matter from the termites. Mycelium of a particular fungus, usually a species of *Termitomyces*, permeates the mass and forms spherical groups of bromatia at the surface. The termites feed directly on the material that they have built into the fungus garden. Sometimes the masses of bromatia are eaten too, as they are by the fungus-growing ants (p. 598). *Termitomyces* species are commonly prevented from fruiting while the garden is actually being tended, but occasionally they fruit (Fig. 18.2). Though tended gardens remain effectively pure cultures, deserted ones soon become overrun by weed fungi. It is possibly by means of antibiotic material in their saliva that the termites curb weeds and prevent fruiting in their fungus gardens.

Certain midges cause galls in plants, and within these grow particular species of fungi which form a thick lining. The spores of the fungi may be implanted when the female midge lays her eggs. It is not known whether the fungus does benefit the insect, but it could do so by altering the gall tissue so that the sap-sucking insects can feed more easily.

The gut of many invertebrates has a characteristic flora including bacteria and flagellates, which by breaking down complex substances and synthesizing others is essential to the nutrition of the animal, as with the breakdown of cellulose by the intestinal bacteria of many wood-eating insects. Some of these insects

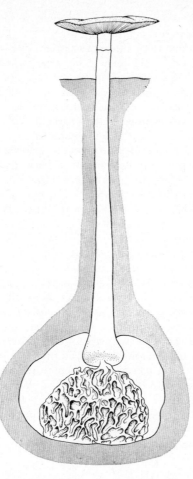

Fig. 18.2   Fungus–insect association. *Termitomyces* sp. (×1/3). Occasionally the fungus in a termite fungus garden (see Fig. 18.1) forms a fruit body which grows through the soil above the chamber containing the garden until it reaches the surface. The base of its stipe connects with the fungus garden.

have large evaginations of the hindgut termed 'fermentation chambers' within which micro-organisms decompose cellulose. The role of flagellates in the break-down of cellulose in the hindgut of the termite has already been described (p. 600). Termites deprived of their gut micro-organisms cannot live on their normal diet of wood. Some of the gut Protozoa themselves are dependent for cellulose digestion on bacteria located in vacuoles where particles of wood are engulfed. The shipworm (*Teredo*), a mollusc which bores holes in submerged wood, is similarly able to utilize wood with the aid of bacteria normally present in its gut. The blood leech formerly kept for medicinal purposes contains in its gut a pseudomonad bacterium that probably makes the digestion of blood possible. Similarly many ticks that feed on vertebrate blood contain symbiotic micro-organisms in their Malpighian tubules.

Many insects and some ticks and mites possess intracellular symbionts of a more specialized nature. These include bacteria, rickettsiae (p. 345), yeasts, and yeast-like fungi and are carried in special cells (bacteriocytes or myceto-cytes) or in special organs (mycetomes) often of considerable complexity. The eggs of these insects become infected, in some species by means of elaborate mechanisms. Almost all arthropods that harbour such symbionts share the characteristic that they have a diet that is incomplete, e.g., keratinic materials or vertebrate blood, which are deficient in vitamins; or plant sap, which is deficient in nitrogen; or cellulose, which is deficient in both. Experimental elimination of the symbionts may lead to derangements of nutrition or metabolism that may indicate, though not prove, the role of the symbiont. The microbial symbionts undoubtedly receive shelter and food from the insect host, but the exact part they play in the life of the host is not always clearly understood. The synthesis of vitamins by micro-organisms may provide invertebrates with their sole source of supply of these substances in habitats deficient or lacking in them. The facts that all individuals of a host species are infected under natural conditions and that the micro-organisms are limited to certain cells or organs and are controlled in number suggest a balanced symbiotic relationship. The removal of the myce-tomes from the females of the body louse (*Pediculus humanus corporis*) shortened their lives and impaired their fertility. This was not due to harmful effects of the surgical treatment, since treated females which became reinfected with the appropriate organism lived normally. Certain insects infesting grain and grain products and normally carrying symbionts were reared aseptically in the absence of the symbiont. They were then unable to live and grow on purified white flour but could do so if the flour were supplemented with vitamins of the B complex. These experiments suggest that the symbionts are able to synthesize vitamins which the insect host is unable to synthesize and that the reproduction of the host, in particular, is dependent on the synthetic abilities of the symbiont.

HARMFUL EFFECTS OF MICRO-ORGANISMS ON INVERTEBRATES

Micro-organisms causing disease in invertebrates

The organisms inhabiting the mycetomes of some species of insects are pre-vented, by some mechanism as yet unexplained, from seriously damaging the host. This relationship is thus one of controlled parasitism and may be compared with the relationships between legume and nodule bacteria, higher plant and mycorrhizal fungi, or algal and fungal partners in the lichens (p. 617). There are, however, many examples of invasion of invertebrates by micro-organisms in which the latter are not controlled and disease results.

A number of bacteria, fungi and Protozoa cause diseases of various inverte-brates. Owing to the economic importance of the hosts, diseases of insects have been studied, but little is known of diseases of other invertebrates. Virus dis-eases of insects have already been considered (p. 251).

Bacterial infections of invertebrates

Some of the bacteria that cause disease in man and higher animals, and for which arthropods are the vectors, harm their vectors also. For example *Pasteurella pestis* (the cause of plague), *Pasteurella tularensis* (the cause of tularemia), and certain species of *Borrelia* (that cause relapsing fevers) and *Salmonella* all cause disease in their arthropod vectors. The same is true of many rickettsiae. For example, the rickettsiae that some arthropods transmit multiply at the expense of the arthropod. While some multiply on the surfaces of host cells (e.g., of the gut epithelium) and cause no disease, others do so in the cytoplasm of infected cells or even enter the nuclei as well. Their relationships with arthropods range from obligate commensalism to parasitism accompanied by fatal disease. For example *Rickettsia prowazekii* (the cause of classical typhus in man) and *R. typhi* will invade cells of the epithelium of the gut of lice and multiply in their cytoplasm. The cells enlarge and are discharged from the gut epithelium which is left irreparably harmed. The lice soon die. Organs other than the gut may become infected if the rickettsiae are experimentally introduced into the body cavity.

Besides these organisms there are bacteria and rickettsiae that are specific pathogens of invertebrates.

Varieties of *Bacillus thuringiensis* (e.g., vars. *thuringiensis, entomocidus, sotto*) comprise an important group of microbial insect pathogens that form endospores accompanied by parasporal protein crystals. These bacteria cause disease chiefly in lepidopterous larvae. The effects of the pathogens are various, but silkworms and other lepidoptera in which the pH of the gut is high become paralysed very rapidly, e.g., within 80 minutes of ingesting the pathogen. This effect appears to be the result of the parasporal protein changing the permeability of the epithelium of the midgut so that the alkaline, strongly buffered midgut contents cause the pH of the relatively poorly buffered blood to rise. In other insects only the gut is paralysed at first, but this is followed by rapid growth of the bacteria, and septicemia. The endospores are able to germinate in the gut where the action of the parasporal crystals may have made conditions favourable for this event. The multiplying vegetative cells provide various water soluble harmful materials which augment the disease. These include the enzyme phospholipase which attacks the substances that cement the cells of the host together.

B. thuringiensis var. *thuringiensis* is grown in artificial media in large fermentation tanks (e.g., 40,000 litre capacity) to produce mixtures of spores and protein crystals that are marketed as microbial insecticides which can be applied to crops. This bacillus is not known to harm honey bees or birds, fish, and mammals, including man. However, it has been reported to kill earthworms.

A second important group of bacterial pathogens contains *Bacillus popilliae* and *B. lentimorbus*, the causes in white grubs of types A and B milky diseases respectively. The large-scale use in North America of Type A milky disease for the control of the Japanese Beetle, *Popillia japonica*, represents one of the major successes for artificial biological control of insect pests by microbial pathogens. The pathogens affect only certain closely related beetles of the family Scarabeidae. The bacterium gains access to the host's blood in which it multiplies and produces a marked turbidity, whence the name of the disease. In an area in which the disease becomes established, spores of the pathogen come to be present in the soil in enormous numbers. The pathogen cannot continue its development in dead insects, and requires special media and conditions to be cultured artificially. Commercial preparations of these bacilli for use as microbial insecticides are prepared from the bodies of specially infected Japanese beetle larvae.

Bacillus cereus, a widely distributed common soil saprophyte to which *B. thuringiensis* is closely related, is known sometimes to cause disease in divers insects. It multiplies in the gut and produces much phospholipase which either kills the host directly or facilitates the invasion of the body cavity by bacterial cells. Other insect-invading bacilli include *Bacillus larvae*, the cause of American foulbrood of bees. The etiology of European foulbrood of bees is, by contrast, confused. Several bacteria are common in bees showing this condition, and include *Bacillus alvei*, *Streptococcus faecalis*, *S. pluton*, and *Achromobacter eurydice*. The latter two species in association have been reported to cause the disease, but recently a free virus has also been implicated.

Obligate anaerobes are also known to cause diseases in insects and include *Clostridium brevifaciens* and *Clost. malacosomae*. Clostridia are probably commoner insect pathogens than was formerly thought. Their detection requires the use of special culture media.

There are a number of Pseudomonads and members of the Enterobacteriaceae that cause lethal septicaemias in a range of insect hosts when small doses (e.g., 10^4 cells) of them are injected into the body cavity, but do not readily infect when ingested unless injury helps them penetrate the gut wall. Such 'potential pathogens' are common causes of low and sporadic mortality in laboratory populations of a wide range of insects but rarely cause epizootics in either the laboratory or the field. Nevertheless *Serratia marcescens* is probably responsible for a considerable cumulative mortality among grasshoppers in North America throughout the season. This and similar bacteria have little if any potential for use as biological control agents by man.

There are many bacteria which cause disease if injected directly into the body cavity in massive numbers (e.g., 10^7 cells) but these are hardly to be regarded as pathogens. Small doses of non-pathogenic bacteria are rapidly eliminated from insect blood by cellular or humoral factors.

One of the few known naturally occurring bacterial diseases of marine invertebrates is gaffkemia, a fatal disease of the American lobster *Homarus americanus* caused by *Gaffkya homari*. The micrococcus multiplies in the host's blood.

Among rickettsiae with a distinct affinity solely for arthropods there is *Rickettsiella popilliae* which causes the fatal 'blue disease' of the Japanese beetle, *Popillia japonica*. The name of the disease refers to the colour that results from the scattering and reflection of light by the rickettsiae themselves in the fat-body cells of the host. *Rickettsiella melolonthae* attacks cockchafer grubs and has been introduced into other insects. It usually proves fatal to the grubs.

Protozoan infections of invertebrates

Many Protozoa live in association with invertebrates in relationships that range from commensalism to parasitism. About 1,200 of the approximately 15,000

known Protozoa are associated with insects. Some of these are vertebrate pathogens and the insect serves as a vector, but there are Protozoa that multiply within the vector and harm it. There are trypanosomes that spend part of their
life-cycle in the gut or salivary glands of insects and the rest in higher plants or
vertebrates. Infection of mosquitoes with species of *Plasmodium*, including the
malarial pathogen, may shorten their lives and prejudice their survival of unfavourable conditions. However, other protozoa are serious and more or less
specific pathogens of insects, helminths, molluscs, and other invertebrates. Such
parasites are to be found in all four subphyla of the Protozoa.

Among the Sarcomastigophora there are both flagellates and sarcodines that
cause insect diseases of which one of the best-known is amoebic or 'spring'
disease of the honey bee. The amoeba (*Malpighamoeba mellifica*) develops as an
extracellular parasite in the Malpighian tubes of the bee and causes a malfunctioning of the tubes that lowers the resistance of the bees to other infections
and is fatal for actively flying bees. A similar amoebic disease affects grasshoppers. Among ciliates pathogenic for insects there are species of *Tetrahymena*
that besides parasitizing and killing insects such as mosquitoes and chirono-
mids, can also live freely. Many of the Sporozoa are associated with insect
vectors to which they do little harm but others are strict insect pathogens.
Among these are many gregarines. Because schizogony (p. 404) leads to the
formation of a new generation of feeding individuals that need much host tissue
for their nutrition it is only among what are termed schizogregarines that serious
insect pathogens are found. For the same reason, entomophilous eugregarines,
being incapable of schizogony, are nearly all more or less harmless commensals
that live in the lumen of the gut. Little is known of the host specificity of schizogregarines. Sometimes they cause severe outbreaks of disease among insects
in the field. Coccidians, another group of Sporozoa, are mainly parasites of
invertebrates and some have insect vectors, but a few species are true parasites
of insects that include stored products pests, aquatic insects, and fleas. Another
group, the haplosporidians, invade such freshwater invertebrates as Copepoda
and Cladocera, but also some insects, among them mosquitoes, black flies, ants,
and the bark beetle *Ips typographus*. Infection ranges from chronic to fatal.
The haplosporidian, *Minchinia nelsonii*, is the probable cause of extensive mortality among oysters in the Delaware and Chesapeake Bay regions of the United
States, while another, *Haplosporidium tumefacientis*, causes a disease in the
Californian sea mussel. Among the Cnidospora most microsporidians are parasites of insects and some are well known as causes of loss of silkworms and honey
bees.

Fungal infections of invertebrates

Many aquatic Phycomycetes have been recorded as growing in or on aquatic
invertebrates which include flatworms, eelworms, rotifers, water fleas, copepods,
crustacea, mites, and molluscs. Many of these records undoubtedly relate to
parasitic infections, but in few cases has the pathology been investigated. Most
of what is known about fungal infection of invertebrates derives from studies
of entomogenous fungi. These are fungi that grow upon insects, living or dead.
Many infect live insects and cause disease. This capacity has been well developed

in some groups of fungi in which considerable numbers of closely related parasitic species have arisen. On the other hand there is a scattering of unrelated fungal species in which the habit has arisen presumably quite independently. The diseases that fungi cause range from fatal ones to mild cutaneous infections. Other entomogenous fungi characteristically grow on or in living insects in relationships of commensalism or even mutualism (p. 602). Finally there are some entomogenous fungi which are relatively unspecialized wound parasites, secondary invaders, or saprophytes.

Most of the fungi that produce fatal infections spread vegetatively through much or all of the host's body as hyphae or free cells or both. Such endoparasitic fungi probably infect by way of the integument in most instances, but infection proceeding from ingested spores has been demonstrated in some. Three sizeable specialized groups of endoparasites merit special mention. *Coelomomyces* is a genus of uniflagellate aquatic fungi (order Blastocladiales, p. 392) of which there are around thirty known species, almost all of which are highly specialized parasites of mosquitoes whose aquatic larvae they most commonly infect. The parasite eventually produces large numbers of sporangia throughout the body cavity of the still living host. In the Entomophthorales (class Zygomycetes, p. 393) there are many species that are specialized for parasitism of insects in subaerial environments. In most of these the parasite spreads to new hosts by means of specially modified sporangia that are forcibly discharged into the air from masses of hyphae arising within the living or recently dead host insects. Thick-walled spores of sexual or asexual origin are also formed and presumably provide for survival of the parasites from one season to another. The third important group are the numerous species of the ascomycetous genus *Cordyceps*. Many of these parasitize insects that eventually die in litter or in the soil. A somewhat club-shaped fructification that rises into the air after the death of the insect provides for the dispersal of the parasite's ascospores. Similar endoparasitism is displayed by a variety of other species widely distributed among the major groups of fungi. For example, among the parasitic Deuteromycetes are included species of *Beauveria* and *Metarrhizium*, both of which are particularly common and widespread, and a number of species of *Aspergillus*, including *A. flavus*.

Insect parasitism is uncommon in the Basidiomycetes except in the highly specialized genus *Septobasidium*. Scale insects feeding on a branch may become infected by species of this genus, but though they are stunted and sterilized by the internal infection they generally are not killed. Hyphae eventually emerge which connect with a mat of mycelium that more or less covers the colony of scales not all of which are parasitized. Consequently the fungus is nourished by some of the scales, but in turn it physically protects others from various natural hazards.

Ectoparasitism that generates mild non-fatal cutaneous disease is particularly well developed in the order Laboulbeniales (class Ascomycetes). About 1,500 species of these fungi have been described already but undoubtedly many others await discovery. Each fungal thallus comprises a minute cellular axis that is attached to the surface of the host's integument and bears antheridia or perithecia or both. Probably these fungi are nourished by way of haustorial penetra-

A

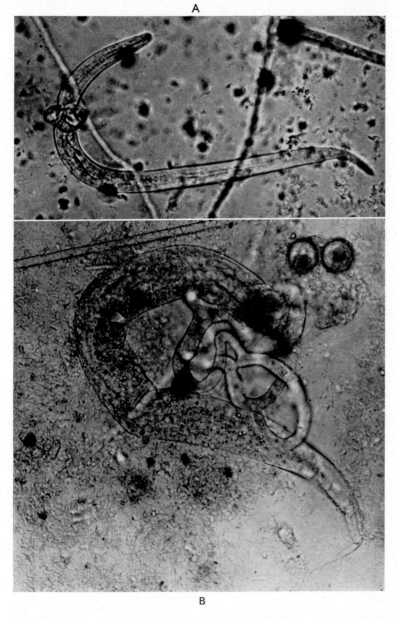

B

Fig. 18.3 Predacious fungi. (**A**) *Dactylaria gracilis* (×500). An eelworm is shown caught in a constricting hyphal ring (C. L. Duddington, by permission, *New Scientist*). (**B**) *Arthrobotrys robusta* (×500). An eelworm captured in a network of adhesive hyphae. (C. L. Duddington)

tions of the host's integument, though only in relatively few species have these actually been observed. Recently a member of this group has for the first time been cultured to a limited extent in the absence of a living host. Detached fly wings in a complex biphasic culture medium were used as a substrate.

There are some fungi that appear to be strictly superficial commensals. These live attached to the integument of insects and certain other arthropods, either on the outer surface or where the integument is invaginated as in the lining of the hind gut. This habit is characteristic of the class of Trichomycetes, the members of which produce anchored filaments of limited growth. A number have been cultured axenically. In nature they probably are nourished by materials dissolved in the aqueous phase that surrounds them.

Attempts to use entomogenous fungi to control insect pests of crop plants have not, on the whole, been successful. The introduction of the parasite into an area is seldom alone sufficient to achieve control, since it is usually present already but does not reach epidemic proportions unless conditions are just right. Modification of the environment to favour the parasite is a better line of approach (cf. the use of predacious fungi to control eelworms, below) but is seldom practicable. Nevertheless, in many localities there are naturally occurring fungi which exercise a powerful control on the size of insect populations. This control may be regarded as beneficial or deleterious depending on whether the insects concerned are harmful or useful.

Predacious fungi

A number of fungi, including some members of the Hyphomycetes (p. 401) and Zygomycetes (Zoopagales, p. 393) actually trap nematodes, soil amoebae, and sometimes rotifers and subsequently invade and devour them. The hyphomycetous predacious fungi (species of *Arthrobotrys*, *Dactylella*, etc.) trap their prey by complex loops of hyphae (Fig. 18.3); by hyphal rings the component cells of which inflate, thus contracting the ring, when an eelworm touches it while attempting to pass through (Fig. 18.3); or by hyphal branches with terminal sticky knobs to which the eelworm adheres. Once the prey is trapped the hyphae penetrate the skin and ramify through the body. Most nematode trapping fungi do not form their traps in the absence of eelworms. A substance named nemin that has been isolated from eelworms stimulates their formation. The members of the Zoopagales shed sticky conidia which adhere to the prey, usually an amoeba or other small soil animal, and put out germ tubes which penetrate into the body where a small thallus is formed. Since their spores fail to germinate on sterile media they are probably obligate predators. Most appear to be highly host specific. Some success in the control of parasitic eelworms has followed the modification of soil conditions to favour the predacious fungi (p. 485).

INVERTEBRATES AS AGENTS OF DISPERSAL OF MICRO-ORGANISMS

Viruses, bacteria, and the spores of many fungi are readily carried by insects, mites or other small invertebrates. While invertebrates, and particularly insects, play a large part in dispersal of micro-organisms picked up by chance, there are also many examples of specific relationships between micro-organism and dispersal agent. The most important aspect of dispersal of micro-organisms by animals (which are referred to as vectors) is the transmission of plant and animal diseases.

Arthropod vectors of virus diseases

Plant viruses

Many plant virus diseases can be transmitted only by an insect vector, and many more can be transmitted by insects in addition to mechanical and other methods (p. 248). The vector does not usually show any ill effects from the presence of the virus although some plant viruses are known to multiply within an insect vector. A few arthropods other than insects may act as vectors of particular diseases, e.g., the eriophyid mite which transmits wheat streak mosaic disease.

Most vectors of plant viruses are sucking insects (aphids, leafhoppers and, to a lesser extent, thrips, white flies, and mealy bugs). A few plant viruses are transmissible by biting insects. Tobacco mosaic and potato X viruses are not transmitted by sucking insects but can be transmitted by grasshoppers. Transmission by biting insects is probably largely mechanical and voracious feeders are likely to consume the portions infected in the early stages of feeding. Several beetles seem to transmit viruses by regurgitation.

Transmission by sucking insects is more complex. There is often considerable specificity between virus and vector, some viruses being transmitted by many species of the same group of insects, others by only a few.

The length of time necessary for an insect to acquire a virus by feeding on an infected plant varies greatly with different viruses and is determined by the nature of the virus, since the same insect may behave differently in the transmission of different viruses.

Some viruses may be transmitted by aphids which have fed for less than one minute on an infected plant before transfer to a healthy one. Such insects rapidly cease to be infective. These viruses are often more readily transmitted by insects which have fasted before feeding on an infected plant. Such insects may become less able to infect healthy plants with increase in the period of feeding on an infected one. One suggested explanation is that viruses of this type are more concentrated in the epidermis of the plant than in deeper tissues and that during a long period of feeding on an infected plant the insect tends to draw sap from the deeper tissues which may contain less virus. Differences in the ability of insects to transmit these viruses are possibly correlated with the relative ability of their stylets to absorb and release the virus particles.

A few viruses are transmitted only when the insect has fed for an hour or so on an infected host but the insect rapidly loses its power of infecting plants after removal from the diseased one. These viruses are usually either not transmitted mechanically or are so transmitted only with difficulty and it is thought that they may be present in lower concentration in the host and in deeper tissues than the first group.

A number of viruses transmitted by aphids and many or all of those transmitted by the leafhopper, white fly, or thrips remain in the insect throughout its life and the power of infecting healthy plants may even be transmitted to the progeny via the egg but not the sperm, which suggests a cytoplasmic location of the virus. Insects are not able to transmit such a virus immediately after feeding on diseased plants but do so only after an interval of hours or even days. This latent period is correlated with the time taken for the virus to pass through the gut wall into the bloodstream and thence into the saliva. Viruses that are transported along such an internal route are termed circulative. Puncturing the gut wall in non-vector races of a leafhopper enabled the insects to act as vectors. Multiplication of the virus within the insect vector has been demonstrated in, for example, the aphid *Myzus persicae* transmitting potato leaf roll virus. Plant viruses that multiply in their vector are termed propagative. At least some of the agents causing yellows-type diseases, such as aster yellows, are known to be circulative and propagative, but recent work indicates that the causal organisms are not in fact viruses, as formerly thought, but *Mycoplasma* or *Bedsonia* (p. 221).

The ability of insects to transmit viruses which cannot be transmitted mechanically may be due to instability of the virus when exposed to the air or to the deposition of the virus by the sucking insect directly in the phloem with a minimum amount of damage to the host cells. It is more difficult to explain why certain viruses, such as tobacco mosaic, which are readily transmitted mechanically are not transmitted by sucking insects. It is possible that such viruses are inactivated by some inhibitor within the insect's body.

Experimental transmission of those insect-carried viruses which cannot be successfully established by mechanical inoculation, involves the breeding of virus-free insects by a special technique and the use of cages to confine the insects on the experimental plants.

Animal viruses

Arthropods may be both reservoirs and vectors for animal viruses. In the former the virus is passed from one generation of arthropod to the next so that all the descendants of an infected progenitor will harbour the virus. For example, Colorado Tick Fever virus is passed from generation to generation via the egg. Vectors of animal viruses are primarily arthropods, and they may be further classified as mechanical (or casual) vectors and biologic (or essential) vectors.

MECHANICAL VECTORS provide a means of transport for the virus from host to host. The biting and sucking mouth parts of the vector become contaminated with virus, new hosts becoming infected when subsequently bitten or sucked.

The epidemiological paradox created by mechanical vectors in the epizootic transmission of rabbit myxomatosis in Great Britain and Australia, in attempts to control the rabbit populations in the two countries from 1950 to 1954, is of interest. In Great Britain rabbits infected with the virulent myxomatosis virus took to their burrows and died there. After death, the rabbit flea (*Spilopsyllus cuniculi*) left the carcass to find new hosts, thereby facilitating transmission of the virulent virus. On the South Coast, and in France also, *Anopheles atroparvus* played a relatively minor role in the transmission. In Australia the same initial pattern of the disease was seen, the fatally infected animals dying within their burrows. However, the Australian rabbit flea (*Echidnophaga* spp.) is of the 'stick fast' or 'stick tight' variety, which does not leave its host. Flea transmission was thereby largely eliminated. Only those few animals which recovered from the infection or those infected with an avirulent strain of myxomatosis virus, came out of the burrows to feed, and mosquitoes (e.g., *Anopheles annulipes*) there facilitated the transmission of avirulent myxomatosis on an epizootic scale. Since the avirulent virus provided full immunity, subsequent attempts to reintroduce the virulent virus met with complete failure. Carnivorous birds, feeding upon the carcasses of myxomatosis-infected rabbits, may also carry the rabbit flea to new localities and thus assist in the spread of the disease.

BIOLOGIC VECTORS are those in which the virus undergoes a separate replication cycle but does not produce disease in the vector. There is no transmission to subsequent generations by way of the egg. The arthropod-borne (Arbo) viruses are characterized by their dependence upon biologic arthropod vectors for infection of their vertebrate hosts. Yellow Fever virus, for example, is ingested by the mosquito (*Aëdes africanus* and *Aëdes simpsoni* in Africa; *Haemagogus* spp. in South America) whilst feeding on the infected animal host; the virus replicates in the cells of the gut of the mosquito, passes to the salivary glands via the haemolymph and there replicates a second time, and is discharged into the saliva from which infection of the next host occurs. The period required for virus replication within the vector varies with the environmental temperature, being prolonged at reduced temperatures.

The viruses of Dengue Fever, Rift Valley Fever and the various Encephalitides are similarly mosquito-transmitted. Sandfly Fever is transmitted by *Phlebotomus papatasi*, a 'sand-fly' or 'owl-midge' which is a blood-sucker also. Louping-ill and Colorado Tick Fever viruses are transmitted by ticks, which are also blood-sucking. All of these viruses, excepting Dengue Fever and Sandfly Fever viruses, infect other vertebrate animals in addition to man.

Dispersal of bacteria by arthropods

Insects and mites are usually externally contaminated with large numbers of bacteria and other micro-organisms picked up from the surroundings. The house fly in particular, owing to the nature of its skin, the hairy pads of the feet and its habit of crawling over and feeding on putrefying organic matter, including excreta, invariably carries a varied and numerous bacterial flora, which may include species, particularly enteric pathogens, pathogenic to man and domestic animals. An even larger number of organisms may be present in the gut of the house fly and many of these may be excreted in a viable condition. The number and nature of the organisms carried depend upon the surroundings of the fly. Thus the number of bacteria per fly in a particular city varied from $3\frac{1}{2}$ million in slum areas to less than 2 million in cleaner districts. The number in the intestine was 8–10 times as large as the number borne externally. Other insects, with smoother skins or with different feeding habits, carry a much smaller load of organisms. Moreover, the number in the gut may be reduced by the secretion of bactericidal substances by some species of insect. A notable example is the blow-fly larva which, despite its habits, has few viable bacteria in its gut.

One of the best-known examples of pathogens carried by specific insect or other arthropod vectors is that of the rickettsiae (p. 345). Many rickettsiae propagate in the lumen and epithelial linings of the alimentary canal of certain arthropods, such as lice, fleas, and mites, and sometimes invade the salivary glands. Infection of vertebrates follows biting by the arthropod vector or, more usually, the rickettsiae from a crushed arthropod or its faeces gain access through an abrasion of the skin often caused by scratching. Some are non-pathogenic to the arthropod host (as in murine typhus); others are pathogenic both to the vector and the vertebrate host (as in classical typhus).

Rickettsial infections of man may be divided into those which are 'demic', that is, transmitted from man to man, and those which are 'zootic', that is, transmitted from animals to man. The first group include, *Rickettsia prowazekii*, the cause of classical typhus, which is transmitted by the body louse *Pediculus humanus corporis*, and *R. quintana*, the cause of trench fever, a lice-transmitted disease which assumed importance during the trench warfare of 1914–18. The second group may be further divided according to the arthropod vector by which they are transmitted to man. Flea-borne rickettsiae include *R. mooseri*, the cause of murine typhus or shop typhus of Malaya. This organism produces a mild endemic infection in rats and is transmitted from rat to rat by the rat louse *Polyplex speculosus* and from rat to man by the rat flea *Xenopsylla cheopis*. Those transmitted to man by ticks include *R. rickettsii*, the cause of Rocky Mountain Spotted Fever, which is transmitted from rodents, sheep, and dogs to man by ticks such as the wood tick, *Dermacentor andersoni*. Q-fever, an influenza-like disease of man, caused by *R. burneti*, is tick transmitted between animal hosts such as opossums, bandicoots, dogs, cattle, sheep, and goats but usually infects man by inhalation of contaminated dust in abattoirs, laundries, etc., where infected animals or contaminated clothing are found. Mite-borne rickettsiae include *R. tsutsugamushi*, the cause of scrub typhus which is transmitted from rats to man by mites such as *Trombicula akamuchi* var. *deliensis*. Some are pathogenic only to animals, e.g., *R. ruminatium*, the cause of heartwater disease of sheep, goats and cattle, which is transmitted by the tick *Amblyoma hebraeum*. In addition, at least forty apparently non-pathogenic species of rickettsiae have been found in various insects and other arthropods.

In contrast to the rickettsiae very few bacterial pathogens are vector transmitted. A most important example is that of *Pasteurella pestis*, which, when flea-borne, causes bubonic plague in man. The animal reservoir of infection is the large grey rat (*Rattus norvegicus*) and the smaller black rat (*Rattus rattus*). *P. pestis* is pathogenic to the rat and at the terminal stage of the illness the ovoid bacteria are present in the blood in enormous numbers. Fleas common to both rat and man (*Xenopsylla cheopis* and *Ceratophyllus fasciatus*) engorge the organism-laden blood and when the rat dies, leave the corpse and in the absence of the normal host will bite man thereby transmitting the plague bacilli.

Certain spirochaetal diseases are transmitted by insects or ticks. *Borrelia recurrentis* which causes the European type of relapsing fever is louse-borne, man usually becoming infected by contamination of scratches with the body fluids of lice in which the spirochaetes occur. Those transmitted by ticks usually enter the vertebrate host when the latter is bitten by the vector. Cattle spirochaetosis resembles relapsing fever in being tick-transmitted and fowl spirochaetosis, a disease of fowls, geese, and other domestic birds, is usually tick-borne but may be mite-borne.

Insects are responsible for the spread of many bacterial parasites of plants including those causing such serious diseases as fire blight of pears, soft rots of many crop plants, gummosis of sugar cane and black rot of crucifers. Insects not only disseminate these pathogens but often inflict wounds which allow entry of the bacteria.

Arthropod vectors of Protozoa

Protozoa have been found associated with most insects and many other arthropods. Many of these Protozoa are parasites of other higher animals including man. The life-cycles of many are intimately linked with the insect vector and the vertebrate host. These have already been considered (p. 605).

Dispersal of fungi by invertebrates

Spores of many fungi may adhere to insects and other small animals and be transported by them, but a few species are specially adapted to insect dispersal.

The Gasteromycetous family of the Phallaceae (p. 401), represented in Britain by the common stinkhorn (*Phallus impudicus*), the lesser stinkhorn (*Mutinus caninus*) and rarely by other species, such as *Clathrus cancellatus*, produce masses of slimy basidiospores which have an unpleasant odour attractive to flies. These insects remove the mass of spores within a few hours of its emergence from the fruit body.

A number of fungi parasitizing the floral parts of plants are also normally dispersed by insects, which are often attracted by sugary secretions and characteristic odours. The ergot, *Claviceps purpurea* (Fig. 10.11) which parasitizes the ovary of rye and some other grasses, produces masses of minute conidia on the surface of the invaded ovary together with a sugary secretion or 'honey-dew' with an odour which attracts insects. The visitors readily carry the spores to healthy ears. Other fungi, such as *Ustilago violacea* (the anther-smut of campions and other members of the Caryophyllaceae), anther smuts of some other plants, and *Botrytis anthophila* (causing a systemic disease of red clover), produce their spores in the anthers of the host. The spores are dispersed by the same insects that disperse the pollen of normal flowers.

The part played by bark beetles in carrying the spores of *Ceratocystis ulmi*, the cause of Dutch elm disease, from diseased to healthy trees, has already been cited (p. 598). Many other examples could be given.

Insects also bring about the mixing of strains of certain species of fungi, notably those of rusts (Uredinales, p. 399) which produce spermatia in spermogonia. Craigie in 1933 showed that many rusts, including *Puccinia graminis*, the black stem rust of wheat, are heterothallic and unable to produce aecidia in the presence of only one strain. The spermatia of these fungi exude from the spermogonium in a sticky fluid of nectar, which may be sweetly scented (as with the thistle rust *Puccinia punctiformis*) and thus attractive to insects. When a spermatium is deposited on a hypha of complementary strain, projecting above the leaf surface, a bridging hypha is usually formed through which the nucleus of the spermatium migrates, and, after rapid division, brings about diploidization of the invaded mycelium.

Similarly mites may bring about the diploidization of species of *Coprinus* by transporting oidia from one strain of mycelium to another. It is likely that many such relationships occur.

Other invertebrates, such as slugs and nematodes, which feed on fungus fruit-bodies may assist in spore dispersal. Spores of some species are known to pass uninjured through the alimentary canals of slugs.

BOOKS AND ARTICLES FOR REFERENCE AND FURTHER STUDY

ANDREWS, C. H. (1967). *The Natural History of Viruses*, pp. 149–161. Weidenfeld and Nicolson, London.

BATRA, S. W. T. and BATRA, L. R. (1967). The fungus gardens of insects. *Scient. Am.*, **217**, 112–120.

BUCHNER, P. (1965). *Endosymbiosis of Animals with Plant Microorganisms*. Revised English Edn. (trans. B. Mueller). Interscience Publishers, New York, London, and Sydney.

BURNET, F. M. (1960). *Principles of Animal Virology*. 2nd Edn. pp. 390–392. Academic Press, New York and London.

COHEN, D. (1966). Epidemiology of virus diseases. In Prior, J. E., ed., *Basic Medical Virology*, 202, The Williams and Wilkins Co., Baltimore.

DUDDINGTON, C. L. (1968). Predacious fungi. In Ainsworth G. C. and Sussman, A. S., eds. *The Fungi*, vol. **3**, 329, Academic Press, New York and London.

EKLUND, E. M. (1966). Arbovirus. In Prior, J. E., ed., *Basic Medical Virology*. The Williams and Wilkins Co., Baltimore.

FENNER, F. (1968). *The Biology of Animal Viruses*, **2**, 765–769, Academic Press, New York and London.

GRAHAM, K. (1967). Fungal-insect mutualism in trees and timber. *A. Rev. Ent.*, **12**, 105–126.

HEIMPEL, A. M. (1967). A critical view of *Bacillus thuringiensis* var. *thuringiensis* Berliner and other crystalliferous bacteria. *A. Rev. Ent.*, **12**, 287–322.

HORTON-SMITH, C., ed. (1957). *Biological Aspects of the Transmission of Disease*. Oliver and Boyd, Edinburgh.

INGOLD, C. T. (1953). *Dispersal in Fungi*. Clarendon Press, Oxford.

LEACH, J. G. (1940). *Insect Transmission of Plant Diseases*. McGraw-Hill, New York.

MADELIN, M. F. (1966). Fungal parasites of insects. *A. Rev. Ent.*, **11**, 423–448.

——— (1968). Entomogenous Fungi. In Ainsworth, G. C. and Sussman, A. S., eds. *The Fungi*, vol. **3**, 227, Academic Press, New York and London.

MARAMOROSCH, K. (1964). Virus-vector relationships: Vectors of circulative and propagative viruses. In Corbett, M. K. and Sisler, H. D., eds., *Plant Virology*, 175, University of Florida Press, Gainesville.

PLOAIE, P. and MARAMOROSCH, K. (1969). Electron microscopic demonstrations of particles resembling *Mycoplasmo* or Psittacosis-Trachoma group in plants infected with European yellows-type diseases. *Phytopathology*, **29**, 536–544.

STEINHAUS, E. A., ed. (1963). *Insect Pathology*, vols. 1 and 2. Academic Press, New York.

SWENSON, K. G. (1968). Role of aphids in the ecology of plant viruses. *A. Rev. Phytopath.* **6**, 351–374.

19
Micro-organisms and Plants

INTRODUCTION

Interaction between micro-organisms and higher plants is of varied type, ranging from the usually beneficial effects of soil micro-organisms on soil fertility, and hence on plant growth, to destructive parasitism, resulting in damage or death of the plant.

De Bary in 1879 introduced the term *symbiosis* to cover all examples of the living together of different species, including the destruction of one organism by another. Later the term was more often used in a limited sense to denote a state of *controlled or reciprocal parasitism* in which each partner benefits to some extent from the presence of the other. The terms *commensalism* or *mutualism* are sometimes preferred in describing this condition but are not commonly used by plant pathologists. Animal pathologists use the term commensalism to denote a relationship in which one organism is benefited and the other is not obviously either benefited or harmed (e.g., the normal flora of the mouths of mammals, p. 540). A *parasite* derives its food from another living organism (the *host*), thereby injuring the latter to a greater or lesser extent, although some may benefit the host in other ways. A parasite causing serious disease of the host is termed a *pathogen*. Some parasites are unable to grow outside the living host and are termed *obligate parasites*. Others, although capable of living parasitically, are also able to grow and to complete their life histories in the absence of the host. These are *facultative parasites*.

CASUAL INTERACTIONS BETWEEN MICRO-ORGANISMS AND HIGHER PLANTS

The activities of micro-organisms in the soil, already described in Chapter 12, necessarily affect plants. Micro-organisms break down organic remains, thus releasing plant food into the soil; bring about nitrification, fixation of nitrogen, and many other processes; break down harmful substances such as fungicides and weedkillers and compete with plants for inorganic nutrients. A closer relationship is that between plant roots and the microflora of the rhizosphere or layer of soil surrounding them. Here exudations from the roots encourage the growth of an increased population of bacteria and other soil micro-organisms, which by their activities alter the composition of the soil and thus in their turn influence growth of the roots.

In swampy soils the presence of photosynthetic algae may increase the oxygen content of the soil water and thus benefit plant roots. A disease of rice caused by poor aeration of the roots is alleviated in this way.

Plants, by providing shade and by reducing the effect of wind and thus providing pockets of still air of high humidity, influence the 'micro-climate' in which micro-organisms grow and consequently influence the number and type of the latter growing among them.

MICRO-ORGANISMS EPIPHYTIC ON PLANTS

Numerous micro-organisms occur on the aerial parts of plants, subsisting on traces of nutrients exuding from the plants or present in dust falling upon them. These epiphytic species do not penetrate the plant and except in wet weather, when moulds may grow to an extent sufficient to reduce the light reaching the leaves, do not harm it. They are able to survive under severe climatic conditions, many of the bacteria being protected by a mucous sheath. Comparatively few species are represented, but the number of individual organisms may be very large. Estimates of the numbers of bacteria (mainly non-sporing rods) on newly harvested grain vary from 500,000 to 1,330,000 per gram. This is probably higher than the numbers on the actively growing parts of the plants, since spread on to the floral parts is encouraged by sugary liquids produced there. The fungi most frequently found as epiphytes on leaves are the mirror yeast, *Sporobolomyces*, the related *Tilletiopsis* and *Bullera*, and the common Hyphomycete, *Cladosporium herbarum*. The number of *Sporobolomyces* cells on wheat leaves is low until the leaves have lived about half their lives, when it increases rapidly, reaching a maximum on the dead leaves. The increase is probably correlated with the higher humidity obtaining at the base of crowded plants and with increased exudation of nutrients from ageing and dead leaves. *Cladosporium* is readily isolated from dead leaves of many species and from twigs and the bark of trees. The numbers of individuals and of species increase rapidly in wet weather and also on leaves covered with the sticky 'honey dew' resulting from aphis attack. Leaves so attacked may become completely covered with a sooty layer of spores and mycelium. The flora is then known as 'sooty mould'. The constituents of sooty moulds are external to the plant but may seriously harm it by blocking stomata, thus interrupting transpiration and aeration, and by screening the leaves from light, thereby reducing photosynthesis.

CONTROLLED PARASITISM ('SYMBIOSIS')

In the relationships described in the preceding sections the micro-organisms concerned are not in organic relationship with the plant. Plant parasites include some organisms mainly external to the host but deriving food from it by haustoria which penetrate the epidermal cells (e.g., the powdery mildews, Erysiphaceae), and others which spend all or part of their life-cycles within the tissues or even inside the cells. The effects of such parasites vary from death of the host to minor damage which may be negligible, or, as with the so-called symbiotic relationships, may actually be balanced by beneficial effects.

Early investigators considered that in such classic examples of symbiosis as the lichens, the nodules of leguminous plants and the mycorrhizas of orchids, both partners benefited and neither was seriously harmed. Modern work has shown, however, that in all these the balance between the partners is a delicate one and that if it is upset by a change in environmental conditions favouring one partner this may then damage or destroy the other. The relationship is thus one of controlled parasitism in which normally the mutual benefits are greater than the mutual disadvantages.

Lichens

Lichens are composite organisms consisting of a fungus (mycobiont) and an alga (phycobiont) in intimate association (Fig. 19.1). Both components of a

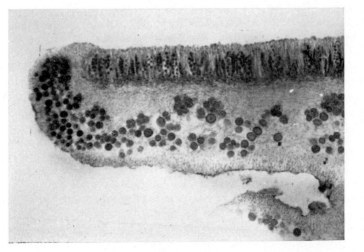

Fig. 19.1 *Physcia* sp. (×100). Section through part of an apothecium of a disco-lichen. Note hymenium of asci on upper surface and spherical algal cells among the fungal hyphae of the hypothecium (basal part) of the fruit-body. (L. E. Hawker)

lichen thallus show distinctive morphological and physiological characteristics not possessed by their free-living counterparts. The fungus is usually the dominant partner, but in certain lichens (Collemataceae) the thallus is a gelatinous mass closely resembling a free-living colony of the blue-green algal phycobiont. Current taxonomic practice bases primary classification on the fungal component. As a result, the same lichen 'genus' may include species with phycobionts from different algal genera. The same algal genus may be represented in several lichen genera differing markedly in the thallus form, e.g., the blue-green alga *Nostoc* is arranged in a definite layer (heteromerous condition) in *Peltigera* but is unstratified (homoiomerous) and predominant in *Collema*. The mycobionts of most lichens are ascomycetes; a few, mainly tropical, genera are basidiomycetes. Ascomycetous lichens are classified according to current views on taxonomy of fungi (p. 395). Most are ascohymenial, producing typical apothecia (p. 398) or, rarely, perithecia (p. 398); a few are ascolocular, producing bitunicate asci (p. 395). The mycobionts of some lichens are infertile, and although probably of ascomycetous affinity are classed in the Deuteromycetes. Agar cultures of the isolated mycobiont result in colonies whose morphology is unlike that of the parent lichen and which are sterile, making equation with free-living species difficult.

Cultures of the phycobiont are readily identified with known free-living algae. Out of 26 algal genera found in lichens, 17 are attributed to Chlorophyta (p. 441), 8 to Cyanophyta (p. 346) and one to the yellow-green Xanthophyta (p. 449). The green algal *Trebouxia* (rarely found in the free-living state) is the commonest phycobiont genus in temperate regions; *Trentepohlia* predominates in the tropics.

About 8–10 per cent of lichens have a blue-green phycobiont. The filamentous nature of many of these is lost in the lichen thallus (e.g., Graphidaceae, Peltigeraceae); it is retained in the gelatinous Collemataceae. Reproduction of the phycobiont within the thallus is generally limited to binary fission (which may be less rapid than in the free-living state) or the formation of akinetes or aplanospores (p. 443), the formation of zoospores or motile isogametes being rare. Isolation and culture of these algae are therefore necessary for identification.

Attempts to resynthesize lichens from their isolated components have met with only limited success, probably owing to the need for exact cultural conditions to suit both organisms without unduly favouring one of them.

Lichens may be arranged in three categories according to their growth form. (i) CRUSTOSE species are closely adpressed to the substratum from which they cannot be removed without damage to the thallus (e.g., *Lecanora*, *Verrucaria*). They may be embedded in the substratum or superficial with a stratified thallus (upper 'cortex' of closely woven hyphae, an algal layer and a medulla of loosely intertwined hyphae). (ii) FOLIOSE species have a leafy dorsiventral stratified thallus often attached to the substratum by rhizoid-like outgrowths (e.g., *Parmelia*, *Peltigera*). (iii) FRUTICOSE species are shrubby, finger-like, cord-like or strap-like, with a radial symmetry (e.g., *Cladonia*, *Usnea*). (iv) The SQUAMULOSE condition (*Cladonia*) is intermediate between crustose and foliose and lacks a lower cortical layer. *Cladonia* also produces upright simple or branched structures of radial symmetry (podetia) which may bear the reproductive bodies of the mycobiont.

Vegetative reproduction is effected in some species by soredia (loose masses of fungal hyphae and algal cells) or isidia (corticate outgrowths of the thallus).

Many lichens accumulate extracellularly a wide variety of unique chemical substances. These are useful in delimiting species but do not include substances of more than marginal economic value.

Most species grow very slowly. Many occur in exposed situations on soil, rocks, walls, tree trunks, and branches, under conditions where neither partner could survive alone. Others grow luxuriantly in damp habitats such as tropical forests or where the atmosphere is humid from sea spray and mists. The 'reindeer moss' (*Cladonia rangiferina*) and some other species form an important constituent of the vegetation of the Arctic tundra and others are the chief form of vegetation on rocky mountain tops and on sea shores above the tide line. Although many lichens grow best in warm humid conditions, they are more often found in less favourable habitats where competition from other plants is at a minimum. The generally held view that lichens are primary colonists and soil-forming organisms probably arose from their tolerance of dry conditions and favourable response to high light intensity. Most lichens, however, require at least a small amount of organic material in the substratum and do not necessarily hasten degradation of rock surfaces. Some lichen communities, such as those on a rocky shore affected by tides and spray, show pronounced zonation of species. Colonization of such habitats as twigs, sand dunes and abandoned arable fields may involve a succession of lichen species.

Lichens behave as hydrophilic gels, having no mechanism for controlling water absorption or loss. In the dry state (2–15 per cent water content) they are tolerant of extreme temperatures. Survival of particular species at $-216°C$ and of others at $+101°C$ have been reported. Saturation after a period of drought leads to a rapid rise in the respiration rate but the rate of photosynthesis rises more slowly.

Owing to their generally slow growth, lichens have not been favoured material for physiological study. The status and role of the two components is of particular interest. The rate of photosynthesis of the lichen thallus is much lower than that of a leaf of similar surface area under comparable conditions, but if calculated on a unit chlorophyll basis the rate is similar to that of higher plants. The small proportion of the thallus occupied by photosynthetic cells limits the net assimilation rate and may be one cause of low growth rate. Recent work has shown that the mycobiont obtains mobile carbohydrate from the phycobiont and metabolizes it, usually to a sugar alcohol. The nature of this carbohydrate depends upon the type of alga but its excretion is stimulated by the fungus. In the lichen *Xanthoria aureola* the alga *Trebouxia* excretes ribitol which is converted predominantly to mannitol by the fungus, but the ability to excrete ribitol is rapidly lost on isolation of the phycobiont. The corresponding mobile carbohydrates excreted by blue-green phycobionts are glucose or glucan. A number of lichens (e.g., *Collema*, *Peltigera*) with blue-green phycobionts have been shown to fix atmospheric nitrogen. Reports of the isolation of the N-fixing bacterium *Azotobacter* (p. 501) from a number of lichens suggested a state of 'polysymbiosis', for which, however, there is no positive evidence. The phycobiont clearly aids the nutrition of the fungus. In return it probably obtains minerals from the latter.

Little is known about mineral requirements of lichens but specific needs may be the reason why some species are confined to particular habitats such as calcareous rocks, specific trees, or bird perching-stones rich in excreta. High levels of inorganic ions are frequently reported in lichens. Eskimos were found to have increased levels of strontium 90 and caesium 137 originating from radio-active fall-out and acquired through caribou meat from the lichens eaten by the animals. A combination of high mineral absorptive capacity and slow growth rate may be one cause of the intolerance of lichens to atmospheric pollution, particularly with sulphur dioxide, and thus account in part for the loss of many species in S. England during the past century.

Root nodules

Another example of a micro-organism and a green plant benefiting each other is that of the root nodules of leguminous plants already described (p. 501). Here the bacterium *Rhizobium leguminosarum* is parasitic on the plant and causes the formation of small galls or nodules on the roots (Fig. 12.5). The bacterium undoubtedly derives most of its food supply from the host but in turn, by its powers of nitrogen fixation, it enables the leguminous host to grow well in soils so poor in nitrogen that most other plants cannot flourish. A fuller account of this important association is given on pages 501–503. Organisms inhabiting the nodules on the roots of certain non-leguminous plants (p. 504) and on leaves of certain members of the Rubiaceae also fix atmospheric nitrogen.

Mycorrhizas and mycotrophic plants

Roots of many plants normally contain fungal mycelium with results to the plant which are not obviously harmful and are often beneficial. Such complexes of root and fungus are termed mycorrhizas (literally, fungus roots). With many forest trees penetration of the roots takes place to only a slight extent and the mycelium is mainly external (ectotrophic mycorrhizas). In many other mycor-rhizas the fungus penetrates into the cortical cells or intercellular spaces of the host root (endotrophic mycorrhizas). In some plants the endophytic fungus is not confined to the roots but permeates the aerial parts also. Such plants are said to be mycotrophic.

Endotrophic mycorrhizas

The best-known example of an endotrophic mycorrhiza is that of certain orchids. It is a well-known horticultural fact that the seeds of many orchids, which are very small and relatively poorly differentiated, fail to germinate in sterilized soil but will do so in the presence of soil in which the same species has been previously grown. This is due to the presence of the mycorrhizal fungus in such soil. The germinating seeds become infected at an early stage. In the roots of older plants coils (pelotons) of fungal hyphae are present in certain cortical cells, the exact layer occupied varying with the host species (Fig. 19.2B). The meristematic tissue of the root tip is normally not invaded and the infected zone does not usually extend throughout the root, the part just below the junction with the aerial parts of the plant being free of infection. The tubers of most tuberous species and the aerial shoots of both green and chlorophyll-less

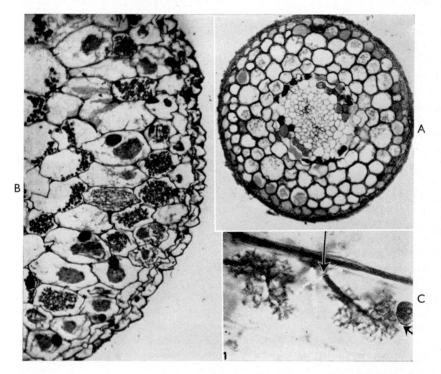

Fig. 19.2 **(A)** T.S. mycorrhizal root of pine (×100). Note external mantle of fungal hyphae, enlargement and separation of cortical cells with fungal hyphae in intercellular spaces (Hartig net) and tannin cells (showing as dark cells) in cortex. (L. E. Hawker.) **(B)** T.S. outer tissues of mycorrhizal root of *Neottia nidus-avis* (×100) showing coiled peletons of hyphae in some outer cortical cells and partially digested amorphous masses in others. (L. E. Hawker.) **(C)** A cortical cell of a root of *Allium ursinum* containing a much-branched arbuscule of a vesicular-arbuscular endophyte (×500). Note origin of arbuscule as a branch from an intercellular aseptate hypha. The dark-stained body on the right is the nucleus of a host cell. (Hawker, L. E., Nicholls, V. O., Harrison, R. W., and Ham, A. M. (1957). *Trans. Brit. mycol. Soc.*, **40**, 375)

ones are not infected and it has been shown that certain tubers produce an antibiotic substance which inhibits the growth of the mycorrhizal fungus. The root cells containing healthy hyphae are usually adjacent to others in which the hyphae are obviously disintegrating (digestive cells). It is assumed that the fungus obtains nutrients from the orchid and that the latter benefits by the synthesis of organic substances by the fungus which become available by the digestion of the hyphae. No critical proof of this hypothesis exists but it is known that some orchid seeds will germinate in the absence of the mycorrhizal fungus if sucrose is added to the culture medium.

The chlorophyll-less orchids, e.g., *Neottia nidus-avis*, are a special case since they are unable to carry out photosynthesis. These were formerly described as saprophytes, but it is probably more accurate to describe them as parasites on their mycorrhizal partners, although the exact nature of the relationship is unknown. The Japanese orchid *Gastrodia elata*, which largely consists of a colourless tuberous structure, is unable to flower unless it is invaded by *Armillaria mellea*, normally a destructive parasite of higher plants. The mycorrhizal fungus associated with most orchids is *Rhizoctonia repens* or some allied species, again closely related to actively parasitic species.

The roots of a wide range of higher plants, including many ferns, frequently contain endophytic fungi of a different type from those of the orchids. These are the 'arbuscular-vesicular' or 'phycomycetous' mycorrhizas. The endophytes are of a phycomycetous type with few or no septa and commonly produce large, globose, or irregular vesicles in the intercellular spaces or less commonly within the cells of the root cortex, and much branched, haustorium-like, intracellular structures, termed arbuscules (Fig. 19.2C). The arbuscules finally disintegrate and are probably digested by the host. There is evidence that the aerial parts of some species of host plant, such as the wild garlic, *Allium ursinum*, produce an antibiotic substance which prevents the invasion of bulbs and aerial shoots by the root fungus.

The identity of the endophyte is uncertain. There is good evidence that most commonly the endophyte is one of a number of a species of the curious zygomycetous genus *Endogone* (p. 393). These produce hypogeous fruit-bodies containing large chlamydospores, zygospores, or a mixture of both types of spore. Such fruit-bodies have been shown to be in direct association with the hyphae in the roots and, moreover, the endophytic association has been induced by inoculation of soil with the appropriate fruit-bodies. Isolated spores of size and structure similar to that of the chlamydospores of *Endogone* have been demonstrated in soil and have also induced the development of endophytes in the roots of seedlings under experimental conditions. Other fungi have been claimed as the endophyte in arbuscular-vesicular mycorrhizas, but of these only a species of *Pythium* isolated from such roots has induced the condition in seedlings, fern prothalli and liverworts grown under strictly aseptic conditions. It could well be that *Endogone*, *Pythium* and possibly other fungi are the fungal partners of the same plant species according to the environmental, and particularly soil, conditions. It is not certain that these endophytes are of general benefit to the host. It has, however, been shown that *Pythium* is pathogenic, endophytic or is unable to colonize the roots of seedlings of onion or lettuce according to whether conditions favour growth of the fungus or of the host plant. It has also been suggested, but not rigorously proved, that the establishment of the vesicular-arbuscular endophyte depends upon the presence of particular bacteria.

Ectotrophic mycorrhizas

The roots of many forest trees, including both conifers and hard woods, are frequently infected by fungi, mostly higher Basidiomycetes (agarics, *Boletus* spp., and some Gasteromycetes) or occasionally Ascomycetes (species of Tuberales, *Elaphomyces*, etc.). Infected roots branch more freely than uninfected ones and the laterals are short and swollen. In some species of conifers they dichotomize repeatedly to give large coralloid masses. A section through an infected lateral or 'short root' (Fig. 19.2A) shows that the hyphae occupy the intercellular spaces of the cortex (forming the so-called Hartig net) but are chiefly aggregated in a dense covering or 'mantle' outside the root. Penetration of the cortical cells may occur in early stages of infection and again as the roots die, but in normal mycorrhizal roots the hyphae are entirely external to the cells. The cortical cells are much enlarged and become separated by the destruction of the pectic substances of the middle lamella (i.e., the central membrane of the wall between two cells, p. 631). Many of them are filled with tannin, a substance known to be weakly inhibitory to fungi.

There is no doubt that the fungus enables the tree to grow on soils too low in nitrogen or in mineral content to permit satisfactory development in the absence of the fungus. In rich soils mycorrhizal roots are rare or absent. The most generally accepted explanation of the beneficial effect of the mycorrhizal fungi is that they increase the absorbing area of the root and that, since the hyphae of the mantle are continuous with a widespread mycelium in the surrounding soil, they may draw nutrients from a wider area than that reached by the tree roots. Experiments with labelled phosphorus and nitrogen have proved conclusively that these elements can reach the leaves of tree seedlings from soil by way of the fungal hyphae. Claims that mycorrhizal fungi actually fix nitrogen have not been substantiated.

In those examples which have been investigated the fungi forming ectotrophic mycorrhizas are invariably forms unable to synthesize thiamin and one must suppose that they obtain this vitamin and probably other nutrients from the roots. The relationship here is again one of controlled parasitism; the formation of tannin by the host cells and the failure of the fungus to establish itself within the cells of the living root are evidence of host reaction; the ultimate destruction of infected roots by the fungus is evidence of successful parasitic activity when host resistance is weakened. Nevertheless the net result is beneficial to both partners.

Systemic mycotrophic infections

In some plants, including a number of liverworts and many members of the Ericaceae and related plants, such as *Pyrola* spp., the endophyte is more uniformly distributed throughout the host. The identity of the endophyte is not known with certainty.

DESTRUCTIVE PLANT PARASITES (PLANT DISEASES)

Introduction

Symptoms of disease in plants can arise from a variety of causes and in some instances are due to external factors in the plant's environment. These include an excess or deficiency of mineral nutrients in the soil and adverse climatic conditions such as drought, frost or excessive solar radiation. Again, damage to plants sometimes follows the application of toxic chemicals, or results from 'drift' of such sprays applied to neighbouring plants. Most commonly, however, plant disease symptoms are due to attack by pathogenic organisms including fungi, bacteria, and parasitic nematodes. Damage is also caused by insect pests (which are beyond the scope of this book) and by viruses (Chap. 8).

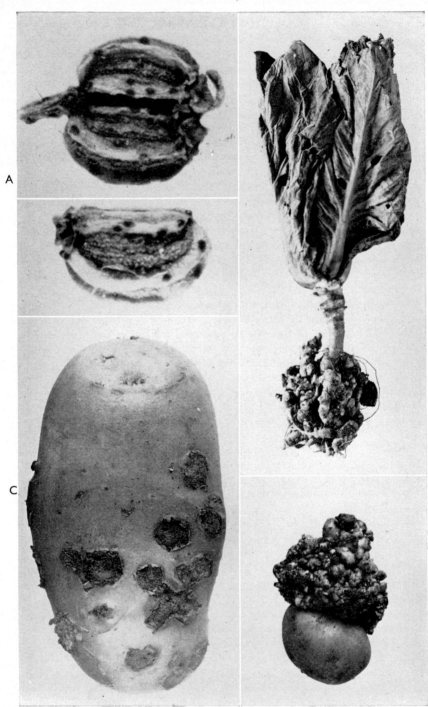

The symptoms of disease can take many forms (Figs. 19.3, 19.4) and are the reflection of changes at a cellular level, themselves arising from physico-chemical reactions between host and parasite. Some pathogens cause the rapid death of the entire plant, e.g., the destruction of potato haulm and tubers by the late blight fungus *Phytophthora infestans*. Others show marked specificity for certain plant organs, for example, damage by *Sclerotinia fructigena*, the cause of the brown rot diseases of fruit trees, is almost entirely restricted to fruit tissues. The intensity of attack varies greatly, from the complete destruction of tissue, to only minor interference with the host's physiology. Thus, the cereal smut fungi, although present in the host throughout its development, induce no visible symptoms until the host reaches the reproductive stage, when they attack the floral parts. Another type of parasite causes abnormal host stimulation, rather than destroying the tissue, e.g., the 'crown gall' disease caused by *Bacterium tumifaciens*.

Micro-organisms parasitic on plants include a wide range of taxonomic forms. The parasitic habit is widely distributed among fungi. Some groups such as the rust and smut fungi, are entirely composed of pathogens. Relatively few serious diseases of economic plants are due to bacteria. Almost all bacteria parasitic on plants are non-sporing Gram-negative aerobes.

The relation between host and parasite

Types of parasitism

The degree of dependence of the parasite on the host varies enormously. The rust fungi, the downy and powdery mildew fungi and a few species from other groups (e.g., *Synchytrium endobioticum*, the cause of black wart disease of potato) have been considered to be strictly obligate parasites, i.e., organisms which have not yet been grown in axenic culture (free from any other organism). Recently some have been grown on tissue cultures of their host and a few rust fungi have even been grown in axenic culture. Other plant parasites, such as many of the smut fungi, are grown on artificial media only with difficulty. Some (e.g., *Ophiobolus graminis*, the cause of take-all disease of wheat) although readily grown on artificial media in pure culture, are unable to survive the competition of other micro-organisms under natural conditions in the absence of the host. Many successful saprophytes which are part of the normal flora of the soil or of other habitats are able to attack certain plants under suitable conditions. These facultative parasites, which include such fungi as *Botrytis cinerea* and *Pythium* spp. are usually non-specialized parasites on a relatively wide range of host plants. They are in marked contrast to the highly specialized obligate parasites which have a strictly limited host range.

Fig. 19.3 Diseases of plants. (**A**) Leaf blight of celery (caused by *Septoria apiicola*) ($\times 30$). Infected seeds showing black pycnidia of the pathogen. (**B**) Club root of cabbage (caused by *Plasmodiophora brassicae*) ($\times \frac{1}{2}$). Cabbage plant with large gall on main root. (**C**) Common scab of potato (caused by *Streptomyces scabies*. Nat. size). Tubers of variety 'Crusader' showing scab lesions. (**D**) Wart disease of potato (caused by *Synchytrium endobioticum*. Nat. size). Tuber bearing cauliflower-like gall. (**A–D**, W. C. Moore)

The obligate parasite is entirely dependent upon the host for its food supply and its physiology must be closely adapted to that of the host. Since true obligates cannot be cultured on even highly complex artificial media it seems likely that they have lost the ability to carry out certain essential steps in their metabolism and that they are thus forced to rely on the enzyme systems of the host plant to make good this loss.

Obligate parasites may induce considerable hypertrophy of the host (e.g., the galls characteristic of black wart of potato or club root of crucifers or produced by certain rusts, such as *Gymnosporangium clavariaeforme* on juniper) but do not usually kill it rapidly. Many rusts are perennial within the host and cause a minimum of damage. Where death of the host does result from attack by an obligate parasite it is usually delayed until after the latter has sporulated. The hyphae of the majority of specialized parasites, whether obligate or not, are confined to the intercellular spaces of the host. Special branches, or haustoria (Figs. 19.5, A–D, 19.6), are formed by many of them and penetrate the walls but not the protoplasts of the host cells. It is assumed, but without conclusive proof, that these haustoria absorb food from the host and pass it back to the intercellular hyphae.

Facultative parasites, such as species of *Botrytis* or *Pythium* are usually less specialized, and their hyphae penetrate and rapidly kill the host cells. Such a parasite then lives saprophytically on the dead tissues of the host.

The attack by a parasite may be divided into three phases: the *pre-penetration phase*, covering the dispersal of the parasite, its arrival on a suitable new host and such further growth and development of the dispersal unit as may be essential before penetration of the host can take place; *the penetration phase*, that is, the actual entry into the host, and *the post-penetration phase*, which is the phase of the establishment of the parasite within the host and the subsequent exploitation of the tissues after penetration has been accomplished.

Phases in infection

PRE-PENETRATION PHASE. The methods by which a parasitic organism reaches a new host plant or a new part of the original host are many and varied.

It seems likely that many soil-inhabiting fungi parasitic on plants are capable of growing at least short distances through the soil and thus reach the host roots in the form of active mycelium. Some, such as *Armillaria mellea*, the honey agaric, which is a serious parasite of many woody plants, produce rhizomorphs (p. 381) which can grow through the soil and attack nearby roots with which they come into contact. Others produce resistant sclerotia (p. 379) on the surface of perennating parts of plants. On germination the sclerotia produce hyphae which, as with *Rhizoctonia solani* which produces flat black sclerotia on potato tubers, spread to and attack the susceptible young shoots of the host.

Fig. 19.4 Diseases of plants. (**A**) Apple canker (caused by *Nectria galligena*) (× ½). A canker which has started around the base of a side shoot; note peeling of bark. (**B**) Peach leaf curl (caused by *Taphrina deformans*) (× ½). Curling and wrinkling of leaves of nectarine. (**C**) Brown rot of apples (caused by *Sclerotinia fructigena*) (× ½). A cluster of apples showing spread of the disease from the originally infected fruit (1) via fruit (2) to fruit (3). (**A–C**, W. C. Moore)

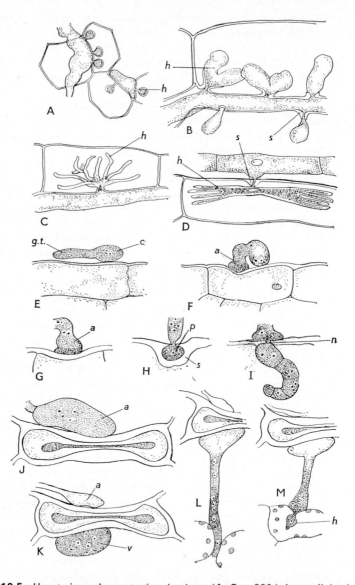

Fig. 19.5 Haustoria and penetration hyphae. (**A–C** ×280.) Intercellular hyphae of three downy mildews bearing intracellular haustoria. (After Fraymouth, J. (1956). *Trans. Brit. mycol. Soc.*, **39**, 79.) (**A**) *Albugo candida* in leaf of *Alyssum saxatile*, showing small globose haustoria (*h*); (**B**) *Peronospora parasitica* in leaf of *Capsella bursapastoris*, showing large sac-like haustoria. Note swelling (*s*) of host wall at point of penetration; (**C**) *Peronospora chlorae* in leaf of *Blackstonia perfoliata*, showing branched filamentous haustorium. (**D**) *Erysiphe graminis* (×280), powdery mildew of grass, haustorium in epidermal cell of oat. Note swelling (*s*) of host cell wall and narrow penetration hypha. (Composite drawing, partly after Smith, G. (1900). *Bot. Gaz.*, **29**, 369, and partly from fresh material)

Many plant diseases are seed-borne. The fungus may be present as dormant mycelium within the seed (as with the loose smuts of wheat and other cereals), as spores adhering to the surface (as with bunt or covered smut of wheat, or the species of *Pyrenophora* which cause foot rot of cereals) or as spore-bearing structures (as with celery leaf blight, caused by *Septoria apiicola*, which produces spores in pycnidia Fig. 19.3A), or as sclerotia. Seed-borne fungi are in a position to attack on germination of the seed and cause pre-emergence damping-off (in which the seedling is destroyed before it reaches the surface of the soil), damping-off of seedlings after emergence, and a variety of foot-rots or anthracnose diseases in which the plant is weakened even if it survives.

The aerial parts of plants become infected by infection units (*propagules*) carried to them by the agency of wind, water (including rain splashes), insects or other animals or by contact with infected plants. The infection unit of parasitic bacteria is the bacterial cell itself and raindrops or insects are the commonest dispersal agents. With fungi the infection unit is usually a spore and here the problem is threefold; namely the release of the spore from the parent mycelium or fruit-body (spore discharge), its transport to the host plant (spore dispersal) and the alighting of the spore on the host surface (Chap. 14).

It is obvious that weather is of importance in determining the production of spores or other propagules, their successful transport to the new host and their development up to the moment of initial penetration of the host tissues. Many fungus spores will germinate only when submerged in water. Those of the powdery mildews are an exception in germinating best in humid air. For the majority a thin film of rain or dew on the host surface will suffice, but if this dries up before penetration takes place the germ tubes may die. Temperature also influences the rate of germination. The relation between weather and the incidence of plant disease is being increasingly studied and it may soon be possible to issue warnings of the likelihood of epidemics of many diseases, as is already done for potato blight and, to a limited extent, for apple scab.

The germination of spores or other propagules may also be influenced by secretions from the host plant. Thus germination of conidia of *Botrytis cinerea* is stimulated or inhibited by volatile or diffusible substances given off by various plants. Most fungal spores contain sufficient essential nutrients to permit germination, but growth of germ tubes soon ceases in the absence of an external

---

(**E–I**) *Botrytis cinerea*, stages in penetration of leaf of *Vicia faba* (× 700). (After Blackman, V. H. and Welsford, E. J. (1916). *Ann. Bot.*, **30**, 389.) (**E**) conidium (*c*) germinating on surface of host leaf to form germ-tube (*g.t.*) ; (**F, G**) tip of germ-tube swelling to form an appressorium (*a*) where it is in contact with leaf; (**H**) penetration of cuticle and wall of epidermis, note swelling of penetration hypha (*p*) and reaction of host cell wall (*s*); (**I**) penetration completed and short hypha developed inside host. Note narrow neck (*n*) of hypha at point of penetration.

(**J–M**) *Puccinia triticina*, stages in penetration of leaf of wheat by germ-tube from a uredospore (× 700). (After Allen, R. F. (1926). *J. agric. Res.*, **33**, 201.) (**J**) Appressorium (*a*) formed at tip of germ-tube in contact with guard cell (seen in longitudinal section) of stoma of host; (**K**) penetration has taken place between guard cells and a sub-stomatal vesicle (*v*) has been formed; (**L**) hypha has grown from substomatal vesicle and is in contact with mesophyll cells, appressorium now shrivelled and empty; (**M**) haustorium (*h*) formed inside mesophyll cell.

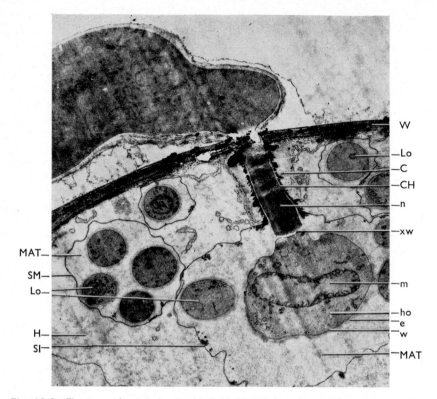

Fig. 19.6 Electron micrograph of a cell of *Hordeum vulgare* infected by *Erysiphe graminis* ($KMnO_4$ fixation). White spots are tears in the section. The haustorial neck connects the external hypha to the body of the haustorium through the epidermal cell wall. Since this is not a median section through the haustorial septum, the septal pore is not evident. c.f. Fig. 19.5D. ho = haustorial body, C = collar, CH = channel between haustorial neck and collar, e = fungal ectoplast, H = host cytoplasm, lo = haustorial lobe, m = mitochondrion, MAT = sheath matrix, n = haustorial neck, SI = sheath invaginations, SM = sheath membrane, w = fungal wall, W = host wall, xw = cross wall. (From Bracker (1968). *Phytopathology*, **58**, 12–30, Plate 8)

supply of food. Some spores, however, either fail to germinate or germinate only slowly in the absence of an external food supply. The amount of the infective material ('inoculum potential') is also important and many fungi are able to attack only when this is sufficiently large.

PENETRATION PHASE. Bacteria and some fungi enter the host only through natural openings, such as stomata or lenticels, or through wounds. Other fungi are able to penetrate the intact cuticle.

There is no conclusive evidence that chemotropism, i.e., growth of the hyphae towards an attractive substance or away from the products of their own metabolism, plays any part in directing them towards stomata. Hyphae seem to reach the stomata by chance but, once there, reactions take place, such as the production of an appressorium over the stoma by a rust hypha (Fig. 19.5, J–M), which suggest that some influence emanates from the stoma and that entry may then be guided by such a stimulus. Chemotaxis may be involved in the swimming of zoospores towards plant rootlets.

When fungal spores or bacteria alight on the newly cut surface of a vascular bundle they may actually be sucked into the vessel and there is evidence that some disease organisms, such as the apple canker fungus (*Nectria galligena*), enter mechanically in this way.

Penetration of the intact cuticle by certain fungi is almost certainly mechanical and is independent of chemical stimuli or enzyme action. Such fungi attach themselves firmly to any hard surface and may form an appressorium at the point of contact (Fig. 19.5, E–G). From this, fine penetration hyphae grow out and bore through the cuticle and epidermal cell wall (Fig. 19.5H, I). Artificial membranes of the right degree of hardness are similarly penetrated. This shows that penetration is mechanical, since if enzymes played a part one would expect hardness of a membrane to delay penetration, but not prevent it. The mechanism by which such mechanical penetration takes place is unknown.

There is now evidence that some fungi can secrete enzymes capable of digesting cutin, but the role of such enzymes in pathogenicity has not been clearly established.

POST-PENETRATION PHASE. After the fungus has gained access to the host by one means or another, the hyphae advance through the host tissues. In some diseases, e.g., leaf spots, their advance is restricted; in others, progressive lesions result. The development of symptoms, as already mentioned, is a reflection of changes at the cellular and subcellular levels, due in turn to physico-chemical changes brought about by the fungus. The nature of the host-parasite interaction is considered more fully below.

The physiological and biochemical nature of pathogenicity

TOXIC PRODUCTS OF FUNGI. The toxic compounds secreted by fungi which are involved in pathogenicity are chemically diverse but for convenience may be classified as follows.

(i) *Extracellular enzymes*, which attack specific plant structures, e.g., cellulose in the cell wall, pectin in the middle lamella region. There are various biochemical types of enzyme, but a full description is beyond the scope of this chapter.

(ii) *Non-enzymic toxins*, which may be of high or low molecular weight. Examples of compounds of high molecular weight are polypeptides and bacterial polysaccharides. Compounds of comparatively low molecular weight include fumaric and fusaric acids. The term '*vivotoxin*' has been used for toxins produced by a pathogen, the host, or both, as distinct from those secreted *in vitro* for which no role in pathogenicity has been proved.

Toxins often act at a considerable distance from their site of secretion—thus, the 'silver leaf' toxin secreted by *Stereum purpureum* is produced by the fungus in the *trunk* of the affected tree, but the marked silvering symptoms appear in the leaves (which are not invaded by the fungus) as a result of the separation of the epidermal cells by the toxin.

(iii) *Growth substances*, of which gibberellins secreted by *Gibberella fujikuroi* afford a good example. These compounds are responsible for the excessive elongation of the internodes of rice seedlings infected by the fungus, giving rise to the so-called 'bakanae' disease.

The action of groups (ii) and (iii) is characterized by modification of host metabolism rather than structural degradation.

As already mentioned, obligate parasites do not secrete highly toxic substances, and the biochemical nature of their action is more subtle and yet to be elucidated.

EFFECTS ON THE HOST AT THE CELLULAR AND SUB-CELLULAR LEVELS. In the deranged metabolism which may follow infection, it is often difficult to resolve primary from secondary effects.

(i) *Structural effects*, resulting from enzyme action, are frequently observed in the cell wall area due to the activity of pectolytic (Fig. 19.7A) or cellulolytic enzymes, or in the cell membranes (Fig. 19.7B) due to enzymes not yet characterized.

(ii) *Physiological effects*, due to toxins or the secondary effects of enzyme attack, also occur. The *respiration* rate generally rises, and there is also a shift to the pentose phosphate pathway, while the rate of *photosynthesis* generally falls. Cell death may occur. The criteria of death have not been clearly defined, but loss of semi-permeability of the cell membranes is generally taken as an indication of death. In the case of viral and at least some fungal infections, there are drastic changes in the proteins synthesized by the host cells; levels of sugars, amino acids and nucleic acids are frequently altered.

The nature of host resistance

Plants resist the entry and spread of pathogenic micro-organisms in a number of ways.

Disease escape

Plants may escape infection if they pass through the susceptible stage at a time or under conditions when the pathogen is absent. Thus early potatoes in England usually escape late blight (caused by *Phytophthora infestans*) since their growth is completed and they are harvested before the period in late July and August when the inoculum has appeared and the nights are warm and humid enough to permit development of the disease. Another form of escape depends on the presence of a highly water-repellent leaf surface, which prevents the spores of water-dispersed pathogens from remaining, and germinating, on the surface.

Structural factors

STATIC. The presence of a thick cuticle has often been thought to confer resistance to attack by parasitic fungi. Reappraisal of the significance of the cuticle as a

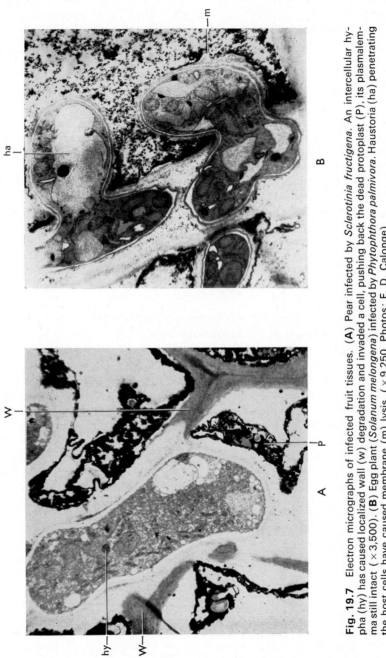

Fig. 19.7 Electron micrographs of infected fruit tissues. (**A**) Pear infected by *Sclerotinia fructigena*. An intercellular hypha (hy) has caused localized wall (w) degradation and invaded a cell, pushing back the dead protoplast (P), its plasmalemma still intact (×3,500). (**B**) Egg plant (*Solanum melongena*) infected by *Phytophthora palmivora*. Haustoria (ha) penetrating the host cells have caused membrane (m) lysis. (×9,250. Photos: F. D. Calonge)

barrier to infection now indicates that it is probably of little direct importance. Indirectly, however, it may be involved, e.g., resistance of plum varieties to brown rot (*Sclerotinia* spp.) has been correlated with a tough skin, but the effect of this is probably to prevent minor injuries which wound parasites need for entry.

Some varieties of strawberry appear to be relatively resistant to *Botrytis* grey mould because the fruits are borne clear of the foliage and thus dry more quickly after rain or dew.

DYNAMIC, the formation of structural barriers. Attempted invasion by a pathogen is often countered by a structural change in the host in response to the presence of the invader, or as a natural reaction to wounding. Thus common scab (due to *Streptomyces scabies*) and corky scab (due to *Spongospora solani*) of potato are usually limited to the surface layers of the host by the active formation of a cork barrier in response to attack, but *Spongospora* may cause severe cankering of tubers in wet soils where cork formation is checked. The speed of formation of cork or callus over a wound surface is often an important factor in disease resistance. Varietal resistance to *Fusarium oxysporum* f. *narcissi*, the cause of basal rot of *Narcissus*, is correlated with the degree to which the natural wounds made by the emerging roots are sealed by cork formation.

Chemical factors

There is now good reason to believe that, even more important in the resistance of potential host plants, are various chemicals. These may either be already present in healthy plants, or formed in response to attempted invasion.

STATIC. Many plants contain compounds toxic to micro-organisms and apparently in concentration sufficient to inhibit a potential invader. It is not easy to prove conclusively that these compounds are the primary cause of disease resistance, but there is some evidence to suggest such a role in certain host–parasite relationships. One of the earliest reports concerned the presence of protocatechuic acid in the scales of onion bulbs resistant to *Colletotrichum circinans*, and its absence in other susceptible varieties. The compounds for which such a role has subsequently been postulated in other diseases include polyphenols and their glycosides, coumarin derivatives, oxazolinones, and alkaloids. In addition, the pH of the cell sap may act as an important deterrent to the invasion of organisms with a different pH optimum for growth. Nutritional factors have also been shown to play an important role in resistance. A high nitrogen level often leads to susceptibility. The sugar level of plants has been suggested as an important determinant of their susceptibility to certain disease organisms.

DYNAMIC. Great interest has recently been focused on the significance of chemicals which may or may not normally be present in healthy plant tissue, but substantial amounts of which are formed as a result of attempted invasion. They may arise through the metabolic activity of either the pathogen or the host.

An example of the production of toxic metabolites by the parasite is the formation from apple juice by the brown rot fungus *Sclerotinia fructigena* of substituted benzoic acids which are themselves toxic to the organism.

A great advance in the study of host-parasite relations followed Müller's suggestion that substances not normally present in the host tissue might be formed by the plant in response to the presence of a potential parasite. He termed such compounds *phytoalexins*. The hypothesis gained great impetus when Cruickshank demonstrated a phytoalexin, which he named *pisatin*, in pea pods subjected to attempted invasion by conidia of *Sclerotinia fructicola*, which is not a parasite of peas. Later, he and his colleagues identified pisatin chemically. Other phytoalexins have since been identified, and certain resistance factors described earlier by other workers may perhaps be regarded as phytoalexins though there are divergent views on terminology.

A phytoalexin may now be defined as a *fungus–induced host metabolite, causally controlling fungal growth in a host–parasite interaction*. The phytoalexins appear to be specific to the host which produces them, but non-specific qualitatively to the invader, and in this respect differ from antibodies produced by higher animals in immunological responses: also, they are generally aromatic in nature and not proteins. In addition to being formed in response to invasion by a living micro-organism, they may also be formed in response to a chemical or physical stimulus. They are the products of biosynthetic pathways, and not of catabolism.

The success of a pathogen in its attempt to infect a host which is capable of producing a phytoalexin is envisaged as being determined by two factors: the quantity of phytoalexin which the particular pathogen induces the plant to produce (successful pathogens generally being less stimulating) and the toxicity of the phytoalexin to the given invader (frequently being greater to the non-pathogens) (Fig. 19.8).

Another 'dynamic' effect involving host metabolism, but not of intact living cells, is the phenomenon of hypersensitivity, which has been studied with particular reference to obligate parasites. If the response of the host cell to invasion is rapid death, with accompanying loss of the semi-permeability of the cell membranes, the obligate parasite can no longer live on the tissue, and as a result its attack is limited.

With facultative parasites, somewhat similar events may occur where the semi-permeability of host cell membranes is destroyed, either by physical wounding or by the pathogenic activities of the invader. As a result, the barrier which normally isolates host polyphenolases from appropriate substrates within the host cell is broken. Some of the resulting oxidized polyphenols are active inhibitors of extracellular enzymes, probably by virtue of their protein precipitating properties. In consequence, wounds may be protected from invasion, or the course of tissue invasion modified.

Control of plant diseases

Introduction

In recent years much attention has been directed to the protection of plants against disease with a view to reducing economic losses. In particular great progress has been made in control by chemical treatment.

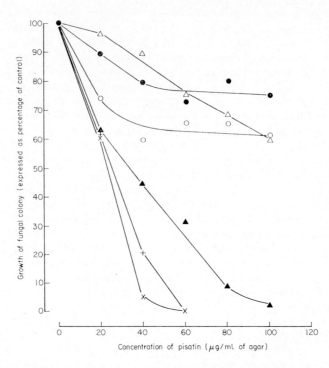

Fig. 19.8 Dosage-response curves of three pathogens and three non-pathogens of peas to pisatin (after Cruickshank (1962). *Aust. J. biol. Sci.*, **15**, 147–159, Fig. 2). Radii (mm) of control colonies are as follows:

	Pathogens of peas			Non-pathogens of peas	
●	*Ascochyta pisi*	20	▲	*Botrytis allii*	66
○	*Mycosphaerella pinodes*	21	+	*Colletotrichum lindemuthianum*	16
△	*Fusarium solani* var.		×	*Leptosphaeria maculans*	12
	martii f. *pisi*	32			

Intelligent control of a plant disease depends upon a detailed knowledge not only of the life-cycle of the parasite, particularly those stages at which it is most vulnerable to interference, but also of its mode of attack and the environmental factors which influence its growth and reproduction.

Cultural control

Much can be achieved by *general plant hygiene*, including the destruction of diseased crops and the remains of harvested crops, removal of diseased and dead parts of a plant, and the removal of species capable of acting as alternate hosts. For example, the eradication of all diseased material has been used in the U.K. in an attempt to check the spread of the recently introduced 'fireblight' disease of apple and pears (caused by the bacterium *Erwinia amylovora*).

The use of *clean planting stock* is an obvious precaution against the spread of disease. Virus diseases of many crops, including potatoes, strawberries, and apples, are kept in check by such means: potatoes for 'seed' are commonly obtained from Scotland and Ireland, where the comparative absence of aphis vectors, together with careful 'roguing' of the crop, enables tubers to be produced free of virus. Strawberry runners are generally obtained from plantations of special 'mother' plants from which aphids are eliminated by spraying, and apple rootstocks and varieties are now available from stock which has been freed of virus by carefully controlled heat therapy. *Legislation and import controls* are designed to prevent the spread of disease from one country to another. An example of successful legislation is the greatly reduced incidence of the black wart disease of potato (caused by *Synchytrium endobioticum*) following the introduction of laws to prevent the planting of susceptible varieties in infected soil.

Adjustment of cultural conditions may also prevent or lessen the effects of disease. For example, many parasitic fungi are favoured by a micro-climate of high humidity, and wider crop spacing results in more rapid drying of the plants, with a consequent decrease in disease incidence. Similarly, many diseases, such as the canker disease of apples, are favoured by a high level of nitrogen nutrition of the host, which leads to an excess of 'soft' growth: a routine measure which helps to check the disease is to reduce the nitrogen level by modifying the manurial programme. Water content, H-ion concentration of the soil, and light intensity are other factors which influence the susceptibility of plants to various diseases.

Chemical control

Chemical control is often necessary for adequate disease control, since changes in cultural practice are frequently insufficient by themselves to keep diseases in check. Agricultural practice results in many individuals growing together in close proximity, thus greatly facilitating the spread of pathogens, and increasing the need for adequate control measures.

The chemical control of plant disease rests primarily on *selective toxic action*, and the discovery of chemicals which although highly toxic to the pathogen are non-toxic, at least at the applied dosage level, to the plant. This requirement is not easily met, and many compounds of high toxicity to bacteria or to fungi have not found general application because they are *phytotoxic*, i.e., toxic to the plant. Apart from a lack of phytotoxicity, a low mammalian toxicity is also desirable, to obviate hazards to the worker and also to the consumer of food crops.

Fortunately, largely through the efforts of the chemical industry, suitable compounds are available. Initially, these were mainly inorganic, based on copper and sulphur, but the past thirty years have seen the increasing use of organic fungicides for plant disease control. Among the most successful of these have been a number of dithiocarbamates, certain phenyl mercury derivatives, and a substituted phthalimide known as 'captan'. Many of these organic fungicides are much more selective in their action than are copper and sulphur, and may also be used at lower concentration. However, their very specificity may be a disadvantage, and they often fail to control certain groups of fungi, such as the powdery mildews (p. 397). Substituted dinitrophenols are often used to deal separately with pathogens tolerant to the more general fungicides.

Biochemical studies on mode of action suggest that, in general, conventional fungicides inhibit a fairly wide range of enzymes in their 'target' organism, and do not owe their selective action to the fact that they inhibit enzymes present only in the pathogen. Their selectivity is due rather to the fact that they are taken up in much greater quantities by the parasite than by the host, which may be protected physically (for example, by its cuticle) at the site of application.

A chemical may be used in one of several ways to control plant disease.

AS A PROTECTANT, applied to a *healthy* plant to prevent its becoming infected by fungus or bacterium. Ideally a continuous film of chemical is applied by means of a spray or dust. In practice, complete coverage is rarely obtained, and the deposit after the spray dries is further eroded by rain and, to a lesser extent, other factors. Nevertheless, the residue of a good protectant fungicide will give a degree of protection for some days, after which a further application is often made to protect new growth. The toxic principle present in the residue may reach the fungal spore by diffusion through a layer of water bridging the deposit and the spore. There is also good evidence that the toxicant may act in the vapour phase, particularly against spores which do not require free water to germinate (Fig. 19.9). Fungicides for this type of application require a comparatively low water solubility or the ability to adsorb strongly to the host surface.

AS AN ERADICANT, applied to destroy a fungus already established, or becoming established on or in the host plant (or in its immediate environment). For this purpose, the chemical must be capable of killing the fungus or at least preventing it from sporulating ('anti-sporulant action'). Few compounds possess these attributes, but organo-mercurials and substituted phenols have often proved useful, while many volatile hydrocarbon soil fungicides and nematocides operate in a similar way, but as they are too phytotoxic to be used in the presence of plant roots are generally used only in fallow soil.

AS A SYSTEMIC, capable of entering the living plant and acting against the fungus from within. In order that the plant is not adversely affected, great specificity is essential, together with the ability to be translocated (generally in the xylem stream). The substituted benzimidazole (Benomyl) has a wide activity spectrum, and other systemic compounds are in commercial use against powdery mildew diseases. In some instances it seems that the applied compound is not fungicidal *per se*, but alters the host–parasite balance in favour of the former ('*conferred resistance*').

RESISTANCE TO FUNGICIDES may be a factor of increasing importance in the future, although at present the development of resistant strains does not occur frequently in the field. In general, since formation of a given enzyme is controlled by at least one gene, fungicides which act against only one or two enzymes are most likely to be those for which resistant strains may develop. In fact, most fungicides in current use act against a wide range of enzymes, as already mentioned,

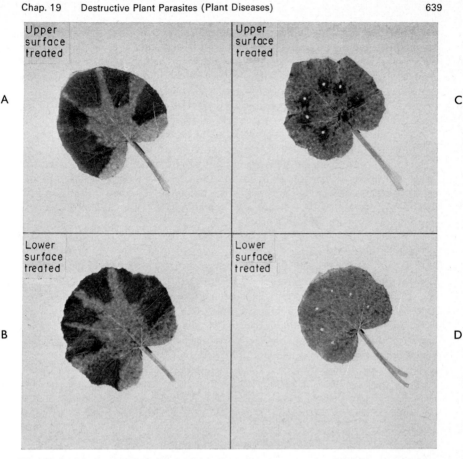

Fig. 19.9 Mechanisms of fungicidal action. Six drops, each of 0·001 ml. and containing 5×10^{-9} mole of fungicide, were applied to either upper or lower surfaces of leaves of vegetable marrow. The next day the *upper* surface was inoculated with spores of *Oidium* sp., and then incubated in a ventilated greenhouse for a further 8 days, when the *upper* surface was photographed. Infected areas show light grey in the photograph, healthy areas are darker.

(**A** and **B**) 5-Butyl-2-dimethylamino-4-hydroxy-6-methylpyrimidine. Activity through the leaf and to the margin indicates systemic activity.

(**C** and **D**) 4,6-Dinitro-2-(1'-propylpentyl)phenol. Absence of trans-laminar effect and a circular zone of inhibition around the deposit zone indicate vapour action. Note white spots indicating tissue necrosis at the site of application. (Photographs: Long Ashton Research Station)

and this may explain why resistance is less commonly evolved than in insects, where strains resistant to specific insecticides are frequently reported.

Biological control

The desirability of avoiding the use of synthetic chemicals in the plant environment is a general target which has long been recognized, but efforts to achieve disease control by the encouragement of organisms antagonistic or parasitic to the pathogen have seldom been successful. Examples of micro-organisms antagonistic to pathogens and which can alter the populations have been described. Common scab of potato, caused by *Streptomyces scabies*, can be reduced by green manuring, with its consequent increases in growth of soil saprophytes, and it has been shown that populations of *Pseudomonas mors-prunorum*, the bacterium responsible for bacterial canker of cherries and other stone fruits, are susceptible to the levels of saprophytic bacteria present with them on the leaf surface. The control of plant parasitic eelworms by predacious soil fungi (p. 608) has had some success and is worth further investigation.

Many viruses and some bacteria and fungi are carried by insects or by certain other arthropods (Chap. 18) and control of the vector, where possible, will also control the disease. This is not easy, since a single aphis may transmit a virus disease to several healthy plants, and it is not always possible to achieve a hundred per cent kill of insect pests.

Breeding for host resistance

In theory, the ideal way of controlling plant diseases is by the use, or breeding, of resistant varieties. The cereal rusts, certain wilt diseases, black wart of potato, and a number of virus diseases have been successfully controlled in this way. Nevertheless, the success of breeding has often been short-lived, particularly with cereal rusts, and late blight disease of potatoes. The breakdown of resistance is due to the remarkably flexible genetic system of the pathogen, which results in the continual production of new races of modified pathogenicity which are rapidly 'selected' by confrontation with the host. Much recent study of the phenomenon has been based on the '*gene-for-gene*' hypothesis of Flor, which states that for each gene which conditions reaction of the host to infection there is a corresponding gene in the parasite which controls pathogenicity.

Tolerant or carrier varieties are also useful as a means of producing crops in the presence of a disease, although their use has the serious disadvantage that they are a danger to susceptible varieties of the same or other species growing near. Thus some varieties of strawberry (e.g., 'Huxley') show tolerance to virus diseases which cripple others (e.g., Royal Sovereign) and the same type of phenomenon accounts for the apparent resistance of some hop varieties to *Verticillium* wilt.

The formulation of suitable control measures for a given disease clearly requires detailed knowledge of many factors, relative to the fungus, the host plant, the environment, and the choice of crop-protection chemicals. These factors are so varied that each disease needs to be considered on its own. Nevertheless, progress is being made to lessen the losses caused by plant disease which man's 'exploding' population can ill afford.

BOOKS AND ARTICLES FOR REFERENCE AND FURTHER STUDY

AGRIOS, G. N. (1969). *Plant Pathology*. Academic Press, New York, 629 pp.

BOND, G. (1963). The root nodules of non-leguminous Angiosperms in Nutman, P. S. and Mosse, B., eds. *Symbiotic Association 13th Symp. Soc. gen. Microbiol.*, 72–91.

BROOKS, F. T. (1953). *Plant Diseases*. 2nd Edn. Oxford Univ. Press, 457 pp.

BROWN, W. (1965). Toxins and cell-wall dissolving enzymes in relation to plant disease. *A. Rev. Phytopath.*, **3**, 1–18.

CRUICKSHANK, I. A. M. (1966). Defence mechanisms in plants. *World Rev. Pest Control*, **5**, 161.

GARRETT, S. D. (1970). *Pathogenic Root-infecting Fungi*. Cambridge Univ. Press, 294 pp.

GOODMAN, R. N., KIRÁLY, Z., and ZAITLIN, M. (1967). *The Biochemistry and Physiology of Infectious Plant Disease*. van Nostrand, London, 354 pp.

HALE, M. E. (1967). *The Biology of Lichens*. Edward Arnold, London, 176 pp.

HARLEY, J. L. (1959). *The Biology of Mycorrhiza*. Hill, London, 233 pp.

HORSFALL, J. G. (1956). *Principles of fungicidal action*. Chronica Bot. Waltham, Mass., U.S.A., 279 pp.

LARGE, E. C. (1940). *The Advance of the Fungi*. Cape, London, 488 pp.

MELIN, E. (1963). Some effects of forest tree roots on mycorrhizal Basidiomycetes In Nutman, P. S. and Mosse, B., eds. *Symbiotic Association 13th Symp. Soc. gen. Microbiol.*, 125–145.

MOSSE, BARBARA (1963). Vesicular-arbuscular Mycorrhiza: an extreme form of fungal adaptation. In Nutman, P. S. and Mosse, B., eds. *Symbiotic Association 13th Symp. Soc. gen. Microbiol.*, 146–170.

NUTMAN, P. S. (1963). Factors influencing the balance of mutual advantage in legume symbiosis. In Nutman, P. S. and Mosse, B., eds. *Symbiotic Association 13th Symp. Soc. gen. Microbiol.*, 51–71.

SMITH, D. C. (1963). Experimental studies of lichen physiology. In Nutman, P. S. and Mosse, B., eds. *Symbiotic Association 13th Symp. Soc. gen. Microbiol.*, 31–50.

TORGESON, D. C., ed. (1968). *Fungicides*. (Vols. I and II.) Academic Press, New York and London.

WHEELER, B. E. J. (1969). *Plant Diseases*. John Wiley and Sons Ltd., London, 374 pp.

WOOD, R. K. S. (1967). *Physiological Plant Pathology*. Blackwell, Oxford, 570 pp.

20
Microbiology of Food and Beverages

INTRODUCTION

The production of safe food, largely free from pathogenic and spoilage organisms, is vitally important with the rapid changes in the pattern of food distribution now taking place in highly industrialized countries. Cities are so large that the inhabitants are becoming increasingly dependent on regular supplies of reasonably priced food, pre-packaged and available all the year round. Such foodstuffs, marketed over a wide area, must be microbiologically safe. Only companies with large capital resources can afford the necessary amount of research and quality control to produce food to these high standards. Further, in order to keep the large food factories running economically, constant supplies of animals, vegetables, and fruit are needed, of standard size and quality. This in turn has led to intensive farming practices where extremely high standards of hygiene and scientific principles of nutrition are practised. Hence, food and beverage production, processing, and retail will become concentrated increasingly into very large units, with rigid microbiological control at all stages. The same pattern will be repeated with fish, crustaceans, and molluscs. If the population of the world continues to increase, recourse will have to be made to new foods, either entirely microbiological in origin, or bland starch and protein, suitably modified in flavour and nutrient status by the action of micro-organisms.

Food microbiologists are concerned with the level of spoilage organisms in the raw materials, often of animal and plant origin, the standard of hygiene during processing, the efficiency of preservation and storage methods, as well as the incidence of organisms responsible for food poisoning. Particular attention will be paid in this chapter to the changes induced in the natural microflora by methods of preparation and preservation. The public health aspects are discussed elsewhere (p. 586) and will be considered here only in relation to incorrect processing methods.

The natural microflora

The natural microflora of a food or beverage can consist of three main components, i.e., those associated with the raw material, those acquired during processing, and those surviving preservation and storage. They can be further subdivided into harmless organisms, producing either desirable or undesirable flavour changes in the food, and pathogens forming dangerous enterotoxins (p. 587).

Micro-organisms spoiling the aroma, taste, colour, or texture of a product may be moulds, yeasts, or bacteria. The precise flora that is present at any stage will depend on the nutrient status of a food, its temperature, pH, water content, etc., as well as on the nature of the organisms themselves. Thus, many moulds flourish below pH 3·8 in the presence of air, even at low levels of available water. Aerobic yeasts are similarly acid-tolerant and can spoil pickles, dry wines, and ciders. Some yeasts have specialized properties, thus *Debaryomyces* spp. are aerobic, salt-tolerant, and can utilize nitrite as a sole source of nitrogen; they are commonly found on the cut surfaces of cured meats. Fermenting yeasts can grow anaerobically at low pH and are thus associated with spoiled fruit juices; osmophilic yeasts are contaminants of sugar and highly sugared or salted products. Similarly, bacteria can be tolerant of heat (thermophilic, p. 160), cold (psychrophilic, p. 649), low pH (acid-tolerant), salt (halophilic), etc. Even foods without very marked characteristics carry an association of bacteria, e.g., fresh meat and fish have *Pseudomonas* and *Achromobacter* species that give way to a micrococci/lactobacilli association when the flesh is cured. Often the natural flora is succeeded by a factory flora, as in cheese and cider making. Many foods and beverages are stable only because of the early development of particular bacteria which reduce the pH by producing lactic acid and thus inhibit the development of food pathogens. Again, foods are often rendered unpalatable by spoilage organisms before pathogens have had time to develop in sufficient numbers. Sequential changes in the microflora of a number of foods are detailed later in this chapter.

Food pathogens include salmonellae (p. 587), *Clostridium botulinum* (p. 588), *Cl. perfringens* (p. 588), and some of the staphylococci (p. 587) and streptococci. All non-sporing pathogens are killed by pasteurization of the food but preformed enterotoxins are not destroyed by the temperatures used in pasteurization. Where the raw material contains pathogenic organisms, there is a danger of re-infection of the pasteurized product. Staphylococci are more salt-tolerant than many spoilage organisms and this can give them an advantage in some preserved foods. Faecal streptococci (*Strep. faecalis* and other group D streptococci) and coliforms, including *Escherichia coli*, are common in foodstuffs; but in this situation are not reliable indicators of faecal contamination, though they may be so in water supplies (Chap. 17). These organisms grow readily on food residues in badly maintained processing lines. Thus their numbers in foods are generally considered to give an indication of the degree of factory hygiene. They are normally absent from liquids pasteurized and processed in a closed system. Massive numbers (10^6 to 10^7/g) in other foods would be regarded as a potential hazard to health, since in such circumstances they have been suspected of causing food poisoning. The general food poisoning bacteria grow optimally at 37°C and little or no multiplication occurs under refrigeration, a technique widely used to restrict bacterial multiplication.

METHODS OF FOOD PRESERVATION

Any product, whether solid or liquid, can be sterilized if heated long enough and/or treated with a suitable concentration of a germicide. Uusually the treatment with substrates of neutral pH is so drastic that accompanying changes in the flavour, texture, and colour render the product unacceptable. Hence, for this type of product, techniques are designed to inhibit multiplication of the spoilage flora prior to consumption of the food. The treatments available are numerous, the choice depending not only on the nature of the microflora, but also on the chemical and enzymic constitution of the product. The effect of these treatments on the microflora will be discussed first in terms of the treatment itself and then in relation to specific foods and beverages.

Chemical inhibition

The use of herbs and spices in the Middle Ages to moderate or mask the effects of food spoilage is well known. Essential oils, e.g., clove, have been shown to have only mild anti-bacterial properties, so that they are now used only as flavouring agents.

Lowering the pH is an obvious method of controlling micro-organisms by chemical means. Lactic acid is preferred in many products, not only for its lack of distinctive flavour, but also because of its very flat neutralization curve, the pH range 3·5–5·5 being of particular interest in food preservation. This allows considerable pH change without unpleasant increase in acid taste. Both additions of the acid and encouragement of growth of lactic acid bacteria are used in practice. Growth of food pathogens is rare below pH 5·0; lactic acid bacteria grow readily down to pH 3·8 but only very slowly down to ca pH 3·0. Many moulds and yeasts will continue to grow at pH 2·3. Lowering the pH also improves the anti-microbial action of preservatives which are effective in the form of the undissociated molecule. Acetic acid has a similar function where its flavour is acceptable or traditional.

Organisms can also be inhibited specifically by the addition of small quantities of a chemical preservative. The ideal properties of an antimicrobial food preservative were formulated at the Fourth International Symposium on Food Microbiology (Molin, 1964) and are summarized below:

1. It is preferable that the preservative should kill rather than inhibit the micro-organisms. Having killed them it should itself decompose into innocuous products.
2. A bacteriostatic preservative would be equally satisfactory if it is destroyed only during final cooking, but if it is to be used in conjunction with thermal control processes it would need to have adequate heat resistance.

3. The range of specificity should correspond with the range of micro-organisms able to develop in the food, i.e., it must inhibit both food poisoning and spoilage organisms.

4. Any preservative used as a supplement to thermal processing should give a similar protection against *Clostridium botulinum* as that given by the standard thermal processing alone, i.e., a reduction in viable spores by a factor of 10^{12}.

5. A preservative should not be inactivated or removed by chemical reaction with the food, by some specific inhibitor in the food, or by products of microbial metabolism.

6. The preservative should not stimulate the development of resistant strains, and should be avoided totally if the same substance is also used therapeutically or as an additive to animal feeds.

7. There should be a chemical method for analysing the effective portion of the preservative.

No known antimicrobial food preservative has all these properties and there is a dearth of suitable compounds active in the pH range 5·0 to 7·0, a range which is important especially for wet foodstuffs of high nutritive value such as fish, meat, and milk.

Food preservatives can be divided into two groups according to their mode of action. In the first, which includes acids, esters, and phenols, the compound is absorbed by the solid components of the bacterial cell and, if of high lipoid solubility, is concentrated on the cell membrane and on various cell structures. The metabolic effects of salicylic acid, for example, are related to its influence on ATP formation in the mitochondria (p. 365). The action of such preservatives becomes increasingly less effective with an increase in the lipid and solid content of the food to which it is added. The antimicrobial effect of the second group, quinones and nitrofurans, depends primarily on their ability to penetrate cell membranes. Quinones are able to penetrate any type of cell, whereas nitrofurans have a selective action on bacteria which they can penetrate, but are unable to penetrate yeasts. Once inside the cell, excessive amounts of coenzyme are required to produce the corresponding hydroquinones and aminofuran compounds, ultimately disrupting electron transport in the cell. The role of preservatives in the prevention of spore germination is also important. No known preservative will prevent germination of *Bacillus cereus* spores, but at minimum inhibitory concentrations nisin, subtilin, diethyl pyrocarbonate, and sodium nitrite prevent growth immediately after germination. Spores that shed their spore wall are prevented from elongating into vegetative cells by sodium benzoate, whereas tylosin, sodium sorbate, sodium metabisulphite, and sodium chloride allow some increase in length but prevent cross wall formation. At greater concentrations all preservatives prevent any development after germination. Nisin is used to prevent gas formation by clostridia in cheese, but attempts to use it in other foodstuffs have not been entirely successful; part of its activity is lost during heating or curing and on storage it is eventually lost completely so that any bacterial spores present can then develop.

Knowledge of the range of organisms sensitive to a preservative is most important. Thus, growth of yeasts and moulds follows suppression of bacteria by the use of tetracyclines. Attempts to use tetracyclines for delaying spoilage in poultry carcasses led to the selective growth of salmonellae, which may be relatively resistant to antibiotics.

Acid foodstuffs and beverages are more easily conserved chemically. Sulphurous acid (sulphur dioxide) controls the acid-tolerant bacteria of ciders and wines, but moulds and fermenting yeasts, common to such environments, are highly resistant. Sorbic acid is used in cheese wrapping to inhibit surface mould growth. It has some activity against yeasts and in France it is permitted in sweet white wines, together with sulphur dioxide, which is antibacterial and an antioxidant. Benzoic acid, used as a preservative in soft drinks, is effective against many yeasts, but *Saccharomyces acidifaciens* can metabolize it to some extent in the presence of sugar. Diethyl pyrocarbonate is active against yeasts, particularly in the presence of alcohol, so that it would be more effective in wines than in freshly pressed fruit juices. Only a few countries permit its use at present, while the polyene macrolide type of compound, pimaricin, effective against moulds and yeasts, is not yet a permitted additive. All substances suggested for use as antimicrobial food preservatives must be submitted to a statutory two-year evaluation programme, that includes feeding to rats and dogs, before being permitted in foods and beverages.

Chemical inhibition of microbial growth can also be obtained with gases. Thus, with the original Boehi process non-sterile apple juice was impregnated with eight atmospheres pressure of carbon dioxide and stored at ambient temperature. The process had to be modified, since contaminating lactic acid bacteria (acid-tolerant and micro-aerophilic) grew and spoiled the product. The juice is now concentrated to a third or fourth of its original volume, saturated with the same percentage of carbon dioxide (0·8 per cent wt) and held at 2°C. There is a cheaper process in France for the sterile storage of flash-pasteurized juice at ambient temperature under a blanket of nitrogen gas. Stringent inspection is required since traces of oxygen would encourage a growth of any mould and yeast contaminants; gas production by the latter would burst the tanks.

Smoking meat, fish, cheese, etc. (normally in conjunction with salting), is an ancient method of food preservation brought up to date. Essentially it is a vapour absorption process, the solid particles playing only a minor role. Recent work has shown that the active components of the vapour include volatile fatty acids which have both a bactericidal and a residual mild bacteriostatic effect. Usually micrococci and staphylococci are inhibited, leaving a flora consisting mainly of lactic acid bacteria.

Dehydration and the use of concentrated solutions

A wide range of foods is preserved by dehydration or by the use of concentrated solutions of salt or sugar. The lower the moisture content of a product, the less liable it is to support microbial growth. A similar effect is obtained the higher the osmotic pressure of a food or beverage, whether this is due to added salt or sugar. Hence the terms xerophile, halophile, and osmophile are used to describe organisms likely to be found in extremes of such environments.

Moisture requirements of a micro-organism for survival or growth are expressed as water activity (a_w), i.e., equal to one-hundredth part of the corresponding relative humidity. Water activity can also be related to osmotic pressure and absolute temperature (for details see Scott, 1957).

Each species of micro-organism has its own characteristic optimum and range of a_w values. Bacteria normally exist only between 0·995 to 0·990 a_w, although staphylococci can exist down to 0·86 a_w and a few halophiles down to as low as ca 0·75 (saturated sodium chloride solution). Yeasts can withstand drier conditions than bacteria; 'osmophilic' species such as *Zygosaccharomyces barkeri* (*S. rouxii* var. *polymorphus*) can exist at a_w values as low as those tolerated by most moulds. However, some moulds can exist at lower a_w values than any other micro-organisms, some species tolerating an a_w as low as 0·62. Reducing the available water below the optimum serves merely to increase the lag phase and decrease the rate of growth; at a_w values 0·65 to 0·62 mould spoilage would be unlikely to become serious in less than one-and-a-half to two years.

It is possible to compare the susceptibility of all types of foods and beverages to microbial growth by using the concept of available water. Thus the a_w in frozen foods is given by the ratio of the vapour pressure of ice to that of water at the temperature under consideration. At $-5°C$, $-10°C$, and $-15°C$ the corresponding a_w values are 0·9526, 0·9074, and 0·8642, so that the inability of bacteria to grow on frozen food below ca $-5°C$ could be due to the limitations of unsuitable a_w. The ability of specific moulds to grow below this temperature is due to their tolerance both of low temperature and low a_w values. Some specific effects are also due to the toxic effect of ions; for an osmophilic yeast these are in the order of toxicity $K^+ < Na^+ < Mg^{++} < Ca^{++}Li^+; < Cl^- < SO_4^{--}$. It must not be forgotten that enzyme reactions can still continue even in a dry substrate at 0·35 a_w, albeit very slowly indeed and irrespective of whether water is a reactant (hydrolytic enzymes) or not (oxidizing enzymes).

Temperature

High temperature

In practice the terms 'sterilization' and 'pasteurization' are used to differentiate different levels of heat treatment. The second usually implies that only some of the spoilage organisms or food pathogens are destroyed. With both processes it is first necessary to know the amount of heat required to inactivate an organism or, if spore-forming, its spores. This will vary with the chemical nature of the food, e.g., its pH, water activity (a_w), and presence of inhibitors. Secondly, heat penetration studies must be carried out on the food in its actual container during the heating process. The rate of penetration is modified by the volume and shape of the container and the viscosity of the contents, presence of solid particles, etc. Thirdly, since the death/time curve approaches zero only at infinity, it is never possible to kill every organism in every container in every batch produced. Hence, some statistical limit must be agreed initially, e.g., a reduction in the number of viable spores of *Cl. botulinum* by a factor of 10^{12} in the canning of certain foods, such as wet foods at neutral pH.

Cases of botulism have been reported in the past, mainly in the U.S.A., following the consumption of vegetables, mushrooms, etc., canned or bottled in the home. The heat treatment given was sufficient to destroy the spoilage organisms that would normally keep *Cl. botulinum* in check, but left spores of the latter undamaged. If the subsequent storage temperatures were sufficiently high the spores grew and produced toxin, without altering the physical appearance of the food. These infections have now disappeared where governmentally sponsored and tested recipes are used in conjunction with domestic pressure cookers. With acid foods and beverages a relatively mild heat treatment will give satisfactory results, since *Cl. botulinum* will not develop at low pH.

The amount of heat required to inactivate bacterial spores (units of lethal heat) is usually calculated in terms of F values, or the number of minutes at 121°C. This assumes instantaneous heating and cooling but, as this is impossible in practice, the equivalent heating effects occurring during the heating and cooling processes must also be calculated from standard sets of tables or graphs The values would be inconveniently small if used for calculating heat requirements in pasteurization, which is carried out below 100°C. Instead, Pasteurization Units (P.U.) are used; for pickles the reference temperature is usually 82°C, while 60°C has been proposed for brewing. It is often difficult to determine the effectiveness of sterilization or pasteurization, except by incubating large numbers of samples. With clear liquids, membrane filtration of the contents of a specified number of containers is a routine measure, while chemical determination of the remaining amount of the enzyme phosphatase is a suitable test for milk and beer. For milk there is also the Short Methylene Blue Test carried out 24 hours after heating, to determine the amount of post-pasteurization infection (p. 591).

Normally canned goods other than acid or cured products are processed to at least the *Cl. botulinum* spore standard but difficulties are sometimes experienced with other heat-resistant organisms, e.g., the spores of thermophilic bacteria. The problem is greatest with very viscous products of neutral pH, like canned rice pudding, whose flavour and appearance would be spoiled by the excessive heating needed to kill extreme thermophiles. Whether the contents of the can remain sound will depend on the minimum growth temperature of the organisms remaining after heating and the temperature at which cans are stored. Canners now have rigid standards for the numbers of thermophiles they will accept in sugar, starch, and other additives.

Staphylococcal food poisoning from consumption of commercially canned vegetables has now been eliminated following recognition that the source of the infection was the operatives handling the newly processed cans. The cans are sterile when leaving the cooker, but 2 to 3 per cent of cans with good commercial seams will leak temporarily during cooling. Hence, infected liquid on the outside of the same could be drawn into the can while it was cooling. The amount of water entering is *ca* 300×10^{-4} ml for large leaks and 8 to 30×10^{-4} ml for small. The following measures have now been adopted in modern canneries to overcome this problem: (1) rigid control of cooling water chlorination, (2) no mechanical damage to the can seam while it is in motion, (3) keeping the wet runways between cooler and dryer to a minimum (surface count on the runway must be <500 organisms/4 sq. in.), (4) keeping any subsequent runways dry, (5) rigid sterilization programme for the equipment, (6) no manual handling of wet cans. Typhoid outbreaks, such as the one in Aberdeen in 1964, associated with imported canned meats have usually originated from the use of non-chlorinated river water, contaminated with sewage, for cooling the cans after heating.

Considerable developments are now taking place in sterile canning. The product is flash-heated in a heat exchanger to the required temperature, then filled into sterilized cans that are sealed and held for the required time before cooling. Many viscous products suffer much less damage with the improved heat penetration. Basically the same process is used for the U.H.T. process for milk (138°C for a few seconds) (p. 654) and for the flash-pasteurization of carbonated beer or cider under pressure prior to bottling.

Low temperature

Food and beverages may be chilled to between 0 and 5°C to delay spoilage while awaiting sale or consumption, or to $-18°C$ for extended storage. Only deep frozen foods will be considered in this section. Three main groups of organisms are important in this respect—food poisoning bacteria, psychrophiles, and moulds. Psychrophiles are defined as organisms that grow well on solid media at 0°C, forming colonies visible to the naked eye in one week. It is important to note that this is not their optimum temperature. Most psychrophilic bacteria are strains of the genus *Pseudomonas*, with some from the genera *Flavobacterium, Achromobacter, Alcaligenes, Escherichia*, and *Aerobacter*. They are very rare amongst Gram-positive bacteria. Psychrophilic yeasts usually belong to the genera *Candida* and *Rhodotorula*. Vegetative forms of moulds are more sensitive to cold than are spores. Spores of certain species of the genera *Monilia, Chaetostylum, Cladosporium, Aspergillus, Fusarium, Mucor, Thamnidium*, and *Botrytis*, are particularly resistant. Some of these are able to grow slowly at low levels of water activity and prevention of growth therefore requires the addition of CO_2 or exclusion of air from the package.

It is difficult to determine accurately the minimum temperatures at which organisms will grow. It is only in recent years that refrigerators capable of maintaining low temperatures consistently for weeks at a time have become available. The freezing of the medium alters the ability of the organism to grow, but the addition of antifreezes of low osmotic pressure (e.g., glycerol) or super-cooling without crystallization are beginning to show that some organisms can be induced to grow at $-5°$ to $-7°C$. Without these techniques, minimum growth temperatures are normally determined by extrapolation of the growth rate or the reciprocal of the lag period. Of course, the lag phase before growth occurs may be so long that the results of the experiment depend more on its duration than on the experimental conditions. With these provisos it is generally accepted that (1) the lowest temperature for bacterial growth is $-10°C$, (2) growth of *Staph. aureus* and *Salmonella* spp. has not been reported at $+6·7°C$ (Entero-toxin production by staphylococci is unknown below $+18°C$), (3) *Cl. botulinum* (types A, B, C, and D) do not grow or produce toxin below $+10°C$, and (4) *Cl. botulinum* type E can grow and produce toxin at $+3·3°C$ after prolonged incubation, but there is no recorded case of botulism arising from this type of activity at low temperatures.

Frozen foods are not sterile and most spoilage is due to improper handling prior to freezing. Freezing is not generally bactericidal unless it is carried out slowly and the produce subsequently stored at a comparatively high temperature (0 to $-10°C$). Fluctuating storage temperatures within this range also reduce the bacterial load; food pathogens are found to die more rapidly between these limits (maximal at $-2°C$) than they do below $-17°C$. Hence the commercial freezing temperature cannot be guaranteed to destroy pathogens, particularly with modern freezing techniques using *rapid* freezing at $-35°C$ followed by storage at $-18°C$. The new cryogenic technique of extremely rapid freezing, by spraying with liquid nitrogen, has hardly any effect on bacterial numbers at all. Hence prevention of food poisoning from frozen foods depends on the use of good quality foodstuffs, on maintaining impeccable hygienic standards during processing and on the avoidance of contamination and multiplication of micro-organisms in the thawed product. All vegetables and some meat and fish products are heat treated before freezing. The cooling period between the two processes should be as short as possible to avoid bacterial growth. Products containing uncooked meat, especially imported frozen boned meat (readily contaminated during the boning procedure) are particularly troublesome. Salmonellae, usually from infected animals, are liable to be present and even with some heat pre-treatment there is a danger of cross-infection between raw and cooked meat. Similarly egg products are likely to contain salmonellae and must be pasteurized at $65°C$ for $2·5$ minutes and not re-infected afterwards.

After a frozen food is thawed it does not then spoil any more rapidly than it would have done had it not been frozen. But microbial control must be exercised during thawing; this is done either in cold rooms or by using dielectric heating, the latter being too rapid for bacterial growth to occur. Normally even if unheated frozen vegetables are thawed and stored above the minimum growth temperature, the food is damaged by obvious taints from the harmless spoilage flora rather than by clostridia. The latter organisms appear to be inhibited by these psychrophiles. Similarly, food being de-frosted reaches the minimum temperature for growth of the psychrophilic microflora before that of the toxigenic staphylococci. Danger would arise in products containing fairly high concentrations of salt, since staphylococci may then grow faster than less salt-tolerant spoilage bacteria. The ability of *Salmonella typhimurium* to grow in the presence of a competing microflora appears to depend on the nature of the food. No frozen food should contain salmonellae, irrespective of whether it will be cooked before being eaten or not. This is perhaps an ideal situation, practicable only in pasteurized products and frozen vegetables. Faecal indicators and enterococci rarely grow below $5°C$ but their presence in substantial numbers in a food thawed above this temperature could falsely indicate insanitary processing conditions.

Two industrial processes based on both freezing and heating have been developed. The first is the dielectric defrosting of blocks of frozen meat, fish, and crustacea, which takes 20 to 30 minutes compared with 25 to 30 hours needed for cold room defrosting. The average temperature after dielectric defrosting is still just below $0°C$, consequently there is very little change in the bacterial count. The second is freeze-drying, when a frozen food is dried by microwave heating under vacuum at a temperature just below the incipient melting point. These products, however, need the same care in their pre-freezing and re-constitution phases as do other frozen products.

Miscellaneous methods

Clear liquids can be clarified and then filtered free of all micro-organisms (p. 208). Many millions of gallons of wines, beers, ciders, clear fruit juices, soft drinks, sugar syrups, etc., are sterilized annually in this manner.

In contrast to this well-tried method, efforts have been made over the past twenty years to use atomic or ionizing radiation for food preservation. Unfortunately off-odours and taints tend to form at the dose levels needed to sterilize foods.

SPECIFIC FOODS AND BEVERAGES

Only very brief mention can be made of some of the microbiological problems involved in the preparation of particular foods and beverages.

Meat

The microflora of freshly dressed meat from healthy animals is derived only slightly from the flesh (mainly the lymph nodes) but mainly from dirt on the animal, its faeces, the personnel and instruments in the abattoir, and the air flora of the chill rooms.

Animals must not be fatigued or distressed before slaughter, otherwise *rigor mortis* sets in early and the meat putrefies rapidly. The high muscle glycogen content of a rested animal ensures the continued presence of some lactic acid, which is unfavourable to spoilage bacteria in the meat. Efficient stunning suspends the heart action. Otherwise contaminants from the knife or from the cut area circulate around the body during the bleeding process. Strains of *Serratia*, *Achromobacter* and *Pseudomonas*, which are resistant to the bactericidal action of fresh mammalian blood, are particularly important in this respect. Immobilization of pigs with carbon dioxide prior to bleeding is helpful for both these reasons; its use could well be extended to cattle and sheep. During evisceration it is essential to remove the intestines from the carcass cutting area immediately; this reduces the amount of contamination by bacteria from the gut.

When the carcasses are hanging in the chill room (2° to 3°C for 2 to 3 days) before cutting up, the microflora can include bacteria, moulds, and yeasts (Ayres, 1955).

Most of the bacteria are mesophiles and tend to die out during chilled storage; only a few are associated with any specific defect or spoilage of stored meat. Moulds and yeasts may then form between one and ten per cent of the flora. Growth on the uncut surfaces which are covered with a layer of fat and connective tissue is very limited.

During chilled storage any of the following changes may take place, depending on the original microbial 'load':

1. Psychrophilic bacteria grow readily on moist surfaces of the meat, thus producing individual colonies that coalesce, especially if there is condensation, forming a slime and off-odours. *Pseudomonas* sp. are virtually the sole slime formers when the meat is stored at 10°C, while at 15°C approximately equal amounts of *Micrococcus* and *Pseudomonas* occur.

2. Putrefactive spoilage can occur in the deep tissue of large thick pieces. *Cl. novyi* and other putrefactive anaerobes have been implicated with bone taint, while salt-tolerant bacteria, able to grow between 0 to 3°C in bone marrow, are responsible for ham souring.
3. The mould *Cladosporium herbarum* can grow as black spots on chilled meat during transhipment. Carcasses of home-killed meat, held for long periods in chilled rooms formerly became infected with *Cladosporium, Rhizopus, Mucor*, and *Thamnidium* species but these moulds rarely develop with the shorter storage periods and damper conditions now in vogue.
4. Rancidity can be caused by lipolytic bacteria and yeasts.

After a chilling period the carcasses are transported, cut and re-chilled. Each cut re-distributes the organisms and adds to the total number of bacteria, so that minced meat, which is cut most of all, has the highest count. The microfloras of chilled carcasses and cut meat are mainly aerobic surface contaminants with very few anaerobic spore-formers. Good quality meat has an initial load of *ca* 10^3–10^4 bacteria/cm^2 and should not become slimy before 14 days at 0°C. When slime does occur the count is usually 3 to 10×10^7 bacteria/cm^2, including mainly *Pseudomonas* and *Achromobacter* species.

When meat has lost its freshness it becomes offensive and is not likely to be eaten. The consumer, however, may be unable to judge whether apparently fresh meat has come from a diseased animal and he is protected against this risk by inspection of animals slaughtered for food. Qualified inspectors examine the animal both before and after death and only satisfactory meat is released for human consumption. Premises where meat is sold or prepared are regularly inspected. The prevention of cross-infection from meat is the complete eradication of certain diseases from the domesticated animal population within a country (p. 592), as in Great Britain where tuberculosis has been eradicated from domesticated animals and plans are in hand to eradicate brucellosis.

Bacon

Lack of space permits mention of only one cured meat product, bacon. Originally bacon was prepared by packing split pig carcasses in dry salt for a period, giving a brown tough product. It is now customary to inject the meat with brine containing sodium chloride, potassium nitrate, sodium nitrite, polyphosphate, etc., followed by steeping in a similar solution for several days. The brine tanks contain large numbers of bacteria (10^9/ml), mainly salt-tolerant micrococci and some lactobacilli, part of whose function is the reduction of nitrate to nitrite. The sliced bacon is vacuum-packed in oxygen-impermeable plastic wrapping; the exclusion of air prevents the attractive pink colour of nitrosomyoglobin from fading. The bacterial flora of micrococci (10^5 to 10^6/g) is fairly stable at low temperatures, but it changes rapidly at higher storage temperatures. Slices with a normal salt content (5 to 7 per cent w/v in the aqueous phase) spoil after *ca* 15 days at 20°C with the scented sour smell derived mainly from lactic acid bacteria. At 30°C the slices putrefy from the proteolytic action of the mesophilic coagulase-negative staphylococci. Increasing the salt content in the bacon to 8 to 12 per cent reduces the number of lactic acid and proteolytic bacteria, thereby increasing the storage life at 30°C. Sliced bacon can support the growth of *Staphylococcus aureus* if the storage temperature is 30°C and the numbers of competing organisms are low. Packs of sliced bacon should therefore be kept cool during sale and in the home. The same is also true of cooked or smoked hams.

Fish

The flesh of healthy fish is largely free of organisms, but their external surfaces carry an appreciable bacterial flora (skin 10^2–10^7/cm^2, gill tissue and intestines 10^3–10^8/g). The aerobic bacteria of marine fish are largely psychrophiles, species of *Pseudomonas*, *Achromobacter*, coryneforms, *Flavobacterium*, micrococci, *Vibrio*, and possibly *Alcaligenes*. The percentage composition of the flora differs in fish from different parts of the world, with larger numbers of mesophiles on fish from warmer seas. The intestines usually contain some strictly anaerobic bacteria, similar to those found in deposits on the sea bottom. In certain parts of the world there has been a progressive increase in the incidence of botulism caused by fish contaminated with *Cl. botulinum* Type E, an organism able to grow at low temperatures. Intestines of fish caught in sewage polluted areas commonly contain typical coliform and food poisoning bacteria.

Fish caught by deep-sea trawlers is eviscerated and packed in ice, that caught inshore may be so treated after landing. Evisceration contaminates the flesh with intestinal organisms and a low storage temperature is essential to prevent these from multiplying with consequent spoilage of the fish. As on meat, *Pseudomonas* species are the most active spoilage organisms at 0°C.

Once the fish reaches the market cooling usually ceases, leading to an increase in the flora already present. Further contamination is likely from fish boxes, filleting knives and boards, etc. The nature and amount of the microflora developing depends upon the temperature during transit and sale.

Many fish are brined and/or smoked, with numerous permutations of treatment on both unsplit and split fish. The overall effect of brining is to reduce the Gram-negative bacteria and to increase Gram-positive types such as micrococci and coryneforms. Cold smoking, while reducing the total bacterial load considerably, has little effect on the composition of the flora. Moulds never occur on fresh fish but they are a problem on smoked fish, the spores being derived from the sawdust used in smoke production. In contrast, hot-smoked fish (i.e., processed at temperatures of 65–75°C for 30 minutes or longer) are usually sterile unless *Cl. botulinum* was present originally and the salt concentration is below 3 per cent. Probably one of the safest fish products, as far as food poisoning organisms are concerned, is fish sticks. For these, frozen rectangular blocks of cod and haddock fillets are cut or sawn to size, dipped in batter, fried at 205 to 260°C for 2 minutes, packed and re-frozen. The bacterial counts remain at a very low level since the interior of the sticks remain frozen throughout.

Milk

Any consideration of the microflora of milk should take cognizance of the marked changes that have taken place in British dairying over the past 20 years. Thus in 1940–42 only 6 per cent of 4,500 Welsh farms sterilized their dairy utensils with steam or hypochlorite after washing with detergent (Thomas *et al.*, 1967). In 1963–6 95 per cent of 3,000 farms used detergent-sterilizers containing hypochlorite, organic chlorine compounds or quaternary ammonium compounds. The more recent development of bulk tank collection has further reduced the bacterial content of raw milk.

Milk drawn aseptically from the udder shows a predominance of staphylococci (the great majority coagulase-negative) and diphtheroids (mainly heat sensitive corynebacteria), both groups being part of the cow's normal skin flora. Some udders harbour mastitis streptococci but, from 1962 onwards, coagulase-positive staphylococci have become the commonest cause of mastitis. However, thermoduric organisms are uniformly absent from milk collected aseptically. Under usual milking conditions further contamination occurs from milking utensils, the dust of the dairy, the udder of the animals, and from the milking operatives. Of these the most important source is the milking utensils which, if unsterilized, become coated with large numbers of bacteria and contribute enormous numbers of organisms to the milk.

Not unnaturally, single farm samples show wide variations both qualitatively and quantitatively, ranging from those in which udder organisms are predominant to those with a denser and more complex flora. Counts of thermoduric bacteria (micrococci, corynebacteria, aerobic spore-forming bacilli, and thermophilic strains of *Strep. faecalis*) above 10^2–10^3/ml are indicative of heavy infection from contaminated equipment. Bulk samples, particularly from non-refrigerated tanks, carry heavy contamination and give evidence of bacterial growth, particularly of *Strep. lactis* and *Strep. kefir*. Holding the raw milk unchilled (10 to 20°C) allows an active growth of the two last-named species, coagulase-negative staphylococci and Gram-negative rods (*Alcaligenes viscolactis* and fluorescent and non-fluorescent pseudomonads). Finally, the flora is likely to be dominated by *Strep. lactis*, resulting in souring owing to fermentation of lactose. Lower temperatures favour the growth of Gram-negative rods which turn the milk alkaline and eventually putrid by their proteolytic activity.

Under laboratory conditions, milk pasteurized by being held at 63°C for three minutes succumbs to deterioration by *B. cereus* (including *B. mycoides*). In commercial practice, whether flash (High Temperature Short Time—H.T.S.T.) or in-bottle pasteurization methods are used, *B. cereus* again overgrows the slower growing coryneforms, but is itself overtaken by Gram-negative rods derived from recontamination. These in turn are outgrown by the milk-souring organisms, *Strep. lactis* and *Strep. cremoris*. *B. cereus* causes not only rapid sweet curdling but also 'bitty' cream and 'ropiness'. 'Sterile' milk has long been produced commercially by heating the bottled milk above 100°C; the product, with a deep colour and caramelized taste, is still popular in the Midlands. It is likely to be replaced by Ultra-High Temperature treated (U.H.T.) milk, filled aseptically into sterile bottles and Tetrapaks (p. 591).

Butter

Butter is made from separated cream, which is usually soured, since butter made from soured cream keeps better than that made from sweet. The souring of the cream is brought about by organisms naturally present in the milk, or the milk may first be pasteurized and specific starters then added. Organisms concerned in the making of butter include *Streptococcus lactis* and *Strep. cremoris*, which ferment lactose and sour the cream, and two other capsulated strepto-coccal-like organisms, *Leuconostoc citrovorum* and *L. dextranicum*. The latter organisms attack citric acid, a by-product of lactose fermentation, to produce diacetyl which imparts a buttery aroma to the cream. Many other identified strains are also present and add their particular flavours to the butter.

The temperature at which the cream is soured is very important. Tempera-tures around 20°C are low enough to prevent growth of thermophilic spoilage organisms, which have survived pasteurization, and yet are sufficiently high to allow growth of the desirable streptococci.

In butter-making the soured cream is churned. This causes the fat globules to aggregate to form butter, leaving the buttermilk which can be drained off.

Bacteria are restricted to the moisture droplets throughout the butter. Wash-ing of butter to replace these droplets of buttermilk by water deprives bacteria of nutrients and limits their growth, thus improving the keeping quality of the butter. 'Working' of butter to break the moisture into smaller droplets also improves the keeping quality by limiting the available nutrients within a drop-let. This does not apply to moulds, which can penetrate the surrounding wall of fat and move to other droplets.

Cheese

Cheese is one of the most important milk products. The widely used Cheddar cheese has a close texture, firm mellow body, and a mild nutty flavour. The raw milk is heated to 68°C for 15 seconds to destroy most vegetative organisms, including coagulase-positive staphylococci (which have increased with the wider use of milking machines) as well as the udder-derived coagulase-negative strains. It is important that the milk be pasteurized without delay to ensure that any strains of *Staph. aureus* which may be present would not have had time to pro-duce enterotoxins. The pasteurized milk is not sterile and will contain thermo-duric organisms (p. 590). A starter is required to produce lactic acid *in situ* in the curd; this can be either *Strep. lactis*, *Strep. cremoris*, or a mixture of these, with or without *Leuconostoc cremoris*, etc. Great care must be taken not to build up specific bacteriophages in cheese rooms, particularly where a single strain of bacterium is always employed. Single strains are used when waxy, bland-flavoured cheese is favoured, whereas multi-strains commonly produce the more mellow, nutty flavour. Antibiotics used to control mastitis cause problems in cheese making and penicillinase-producing strains of coagulase-negative staphylo-cocci have been used both to remove the antibiotic and because their lipolytic action can contribute to the cheese flavour.

The starter rapidly gives a low redox potential, produces protein degradation products and a pH of 5·0, all conditions essential to the development of the lactobacilli and unfavourable to other bacteria; clostridia are inhibited by the antibiotic nisin produced by some strains of streptococci. If the streptococci fail, so also do the lactobacilli and the cheese develops such faults as taints and gas produced by coliforms and clostridia, and faulty colour by a variety of organ-isms. Toxigenic staphylococci would also find the conditions favourable.

Renneting with calf rennet follows immediately after the starter which also produces the optimum pH for casein decomposition and clotting. The coagulum
is cut into small pieces to release the whey, the degree of syneresis being dependent on the temperature and the rate of acid production by the bacteria. Finally
the curd is consolidated into blocks under light pressure until they become plastic, when they are broken into pieces, sprinkled with salt and pressed into
moulds. The shaped blocks are first dipped in molten paraffin wax to exclude
air and thus to prevent external mould growth and weight loss, and then ripened
at 13°C for quick process cheese or 5·5°C for the old slow-ripening type. The
optimum period of maturing varies from 2 to 14 months, according to the previous treatment and type of cheese required. During cheese ripening the numbers of lipolytic organisms are low; the lactic streptococci used as starters are
soon inhibited while the numbers of lactobacilli increase progressively during
the later stages of ripening of all cheeses, especially the hard varieties. The main
flora consists of homofermentative types (*L. casei* and *L. plantarum*); *L. brevis*
and *L. lactis* occur much less frequently. The total number of lactobacilli and
streptococci, both alive and dead, is extremely large and probably accounts for
as much as 0·1 per cent of the weight of the cheese. In Camembert cheese,
Strep. lactis is dominant during the first phase of growth, *Penicillium* and *Oidium*
in the second and *L. casei* in the third phase. Blue cheeses are made from open
textured cheese through the action of varieties of the mould *Penicillium roqueforti*.
The drying cheese coat supports the growth of moulds, yeasts (*Torulopsis* spp.),
streptococci, and lactobacilli. The microflora of the coat of Stilton cheese is
characteristic of the factory producing it. Clammy rind on Dutch cheese is
caused by the growth of corynebacteria that develop only at pH 5·3 and is favoured by the prior development of yeasts (such as *Debaryomyces, Candida,
Trichosporon* spp.) that decompose lactic acid and form ammonia from proteins,
thus altering the pH in favour of the corynebacteria. On the other hand, surface ripened cheeses such as Liederkrantz, Trappist, Tilsiter, etc., that have a
brown surface smear of aerobic bacteria, have antibiotic activity due to *Brevibacterium linens*.

Fermented milks

Many different strains of lactobacilli, either alone or in combination with
lactose fermenting yeasts and streptococci, are used to produce characteristic
fermented beverages. In Europe and the Middle East yoghurt is made from
milk thickened by heating or the addition of dried milk solids and fermented
with *Lactobacillus bulgaricus* and *Streptococcus thermophilus*. It has a custard-
like consistency and is often flavoured with ground fruit or nuts. Kefir is a
fizzy beverage made from fermented mares' milk using 'grains' that contain
yeasts, streptococci, lactobacilli, and micrococci. The grains are strained off
after their action is complete and added to the next batch of milk. Koumiss,
also made from mares' milk, is a greyish-white liquid with uncurdled casein,
often bottled and allowed to carbonate by the action of the micro-organisms. It
normally contains 1–2·5 per cent v/v alcohol. Acidophilus milk has been a
fashionable drink for several decades since it was thought that *Lactobacillus
acidophilus* would survive in the gut and replace so-called harmful bacteria.

However, it is doubtful whether this organism would survive without a high carbohydrate diet rich in growth factors.

Fermented Eastern foods and beverages

Fermentations of bland starch (cereal products) and protein (particularly fish) with moulds, yeasts, and bacteria have been practised for centuries in the Orient to add flavour, improve nutrient quality, or prevent spoilage. Some of these fermented foods and beverages are an important source of minerals and vitamins in the diet of large populations. Table 20.1 lists a few of the many

Table 20.1 Some fermented foods. (After Hesseltine and Wang, 1967)

Product	Raw material	Micro-organism used	Country of origin
Katsuobushi	Bonito (fish)	*Aspergillus glaucus*	Japan
Nam-pla (fish sauce)	Fish	Halophilic bacteria	Thailand, S.E. Asia
Shoyu	Wheat and soybean	*Aspergillus oryzae*, yeast and *Lactobacillus* spp.	Japan
Natto	Soybean	*Bacillus subtilis*	Japan
Tempeh	Soybean	*Rhizopus oligosporus*	Indonesia
Sufu	Soybean curd	*Actinomucor elegans*	China
Hamanatto	Soybean	*Aspergillus oryzae*	Japan. Called Tao-Si in Phillipines and Tu-Su in China
Ontjom	Peanut press cake	*Neurospora sitophila*	East Indies
Sake	Rice	*Lactobacillus* spp. and *Saccharomyces sake* (*S. cerevisiae*)	Japan

preparations known. Space does not permit detailed descriptions of these or of methods of preparation.

Fermented drinks, prepared from millet or other local cereals, are also an important source of vitamins in the diet of many African populations.

Lactic pickles

While vinegar is commonly used in the Western world in the preparation of pickles, many vegetables are preserved by fermentation with lactic acid bacteria. Thus in the preparation of Sauerkraut, the solid, round-headed cabbages are washed to remove the undesirable Gram-negative bacteria on the outer leaves (*Pseudomonas, Flavobacterium*, and *Achromobacter* spp.) and then sliced and mixed with brine. Sugar, mainly sucrose, nutrients, and mineral salts are leached into the liquid which is then fermented by a succession of three bacterial species, derived from the innermost leaves. The heterofermentative *Leuconostoc mesenteroides*, and *Lactobacillus brevis* produce lactic and acetic acids, carbon dioxide, alcohol, mannitol, and dextran while the homofermentative *Lactobacillus plantarum* produces lactic acid only. The optimum temperatures and salt concentrations are 13–18°C and 1·8–2·25 per cent salt. At the upper ends of these limits, the homofermentative species *Streptococcus faecalis* and *Pediococcus cerevisiae* tend to replace the less salt-tolerant *Leuconostoc mesenteroides*. At 3 per cent salt the product is tough and can develop a pink coloration with the growth of carotenoid-producing yeasts. At too low a salt concentration the product is unpleasantly soft due to the growth of bacteria capable of digesting pectin and cellulose. Pure culture inoculations have never produced the succession of species that gives the best flavoured product. Olives and small cucumbers are also brined and preserved by the action of lactic acid bacteria.

Cocoa and coffee

Fermentation is also used to prepare certain products for processing, as with cocoa and coffee beans, where removal of unwanted outer layers is facilitated by the action of micro-organisms.

Eggs

The contents of some 90 per cent of freshly laid hens' eggs are sterile; their shells and the interior of the remainder contain Gram-positive bacteria of ovarian origin. However, the predominant species inside rotten eggs are Gram-negative rods and extra-genital in origin, derived from the nesting material and not the faeces. The predominant species are *Alcaligenes faecalis*, *Aeromonas liquefaciens*, *Proteus vulgaris*, non-proteolytic strains of *Cloaca* sp., *Citrobacter* sp., and *Pseudomonas fluorescens*. These organisms possess combinations of proteolytic and lecithinase activity, pigment and hydrogen sulphide production. The progress of bacterial rotting follows a definite pattern. First there is contamination and penetration of the shell, followed by growth in the cell membranes. The extent of growth is controlled by the antimicrobial action of the conalbumin in the egg albumen. Eventually the yolk makes contact with the shell membrane and the organisms are able to utilize the glucose of the albumen and infect the complete contents of the egg.

Salmonellae can proliferate in eggs without producing any visible symptoms other than a faint turbidity in the albumen. *Salmonella typhimurium*, *S. anatum*, and *S. enteritidis* are frequently encountered even in eggs from clinically normal birds and the importance of duck eggs as a source of human salmonellae is well established. Humans can become infected by salmonellae not only from fresh whole eggs but also from frozen, liquid, or spray-dried products made from lower grade eggs. Because of the lack of obvious spoilage symptoms, salmonella infections are not apparent and the contents of such eggs are added to the main bulk for processing. This is in addition to any contamination from the hands of the process workers or from the containers into which the eggs are poured.

The control of infection from egg sources relies on reduction of the incidence of infection in the flock, reduction of contamination of the shells by faeces, special attention to the sites where eggs are laid, the sale of only clean, unwashed eggs, and the education of kitchen staff to encourage adequate cooking of food containing eggs or its products.

Bread

In the conventional breadmaking process, a dough of flour, water, salt, etc., is allowed to ferment with a culture of the yeast *S. cerevisiae*, originally overnight, now for 3 hours. The dough is worked, allowed to rise, then reworked, placed into tins, allowed to rise for a further 50 minutes (final proof) and baked. This is now being replaced by the high-energy system (Chorleywood Bread Process) in which the dough is worked mechanically. The amount of yeast is increased by 50 per cent and the addition of a fast-acting oxidizing agent and fat are essential parts of the process. Production time is decreased by some 60 per cent, much less space at controlled humidity is required in the bakery, the loaf has a lower staling rate, and the yield of bread is increased 4–5 per cent. Yeasts produced by controlled hybridization are now available and are particularly suitable for modern methods of breadmaking.

Both compressed yeast and bakery products are subject to contamination. Bacteria contaminating compressed yeasts are mainly lactobacilli, normally considered harmless; only heavy infections of the slime-forming *Leuconostoc mesenteroides* are liable to affect the baking properties of the yeast. Nowadays contamination with wild yeasts is rare, but before rigid sanitation programmes were instituted in yeast factories the presence of such aerobic yeasts as *Candida krusei*, *C. mycoderma*, *C. tropicalis*, *Trichosporon cutaneum*, *T. candida*, and *Rhodotorula mucilaginosa* was not uncommon. The widespread use of moisture-impervious wrappings has made control of moulds such as *Aspergillus* and *Penicillium* species and *Oidium* (*Oospora*) *lactis* of special importance, since they can form unsightly blotches on the surface of the yeast block beneath the wrapping. Acid calcium phosphate and acetic acid are commonly used as preservatives in bakery products to prevent the development of bacterial ropiness. Regulations are now being considered to allow the addition of propionates in bread, sorbic acid and its salts in flour, and both these in baked products for the prevention of mould growth. Reducing the available water content (a_w) to as low a level as possible without spoiling texture and appearance helps to delay mould growth but, in the absence of fungicides, cannot prevent it. Strict cleanliness within the bakery is absolutely essential to ensure that any waste flour is quickly removed, otherwise it becomes mouldy and spores are discharged into the atmosphere.

Sour dough bread is made with a mixture of *Lactobacillus brevis* (provides aroma), *L. plantarum* (provides crumb elasticity), and yeast (for raising power). Increasing the holding temperature decreases the percentage of yeast in the mix, until at 35°C it is practically yeast-free.

Beer and lager

Beer is made by fermenting a hop-flavoured extract of barley malt with a top-fermentation strain of *S. cerevisiae*. The bottom-fermenting yeast, *S. carlsbergensis*, is required for lager. Most beers in England are made by the first process, whereas lager predominates in the rest of the world.

Barley is converted into malt by allowing the soaked grain to germinate under controlled conditions of temperature and humidity, leading to the formation of α- and β-amylases, and the breakdown of the protein, hordein, to amino acids. The process is stopped by raising the temperature and removing the rootlets and plumules. The exact treatment depends on the type of malt required. Whereas in England the protein is fully converted into amino acids, lager malts still contain appreciable quantities of protein. Again, stout malts are dried at 105°C to give a dark-coloured extract; malts from distilleries and vinegar breweries receive no final kilning and still contain limit-dextrinase.

An extract or wort is now prepared from the ground malt, the exact conditions of extraction again being varied according to the type of beer. In England the starch is hydrolysed by the action of the amylases to glucose, maltose, malto-triose and -tetrose, and oligosaccharides; glucose and fructose are also present in the extract. Worts from lager also need extensive proteolytic action during mashing. Sometimes flaked barley or maize, or gelatinized unmalted cereal grits are added to supplement the starch supply. When extraction is complete the wort is boiled under pressure with hops, the tannins from which co-precipitate with any remaining proteins. A proportion of the humulone and lupulone complexes are converted into their corresponding isomers during boiling, ultimately giving bitterness to the beer and conferring some resistance to Gram-positive bacteria. The cooled and filtered wort is 'pitched' with yeast and allowed to ferment. Distillery and vinegar brewers' worts are not boiled (so that further saccharification due to the enzyme limit-dextrinase can take place during fermentation), neither are they hopped. Bacterial infection is often restricted by inoculating part of the wort with *Lactobacillus delbrueckii* and, when acidified sufficiently, boiling this before returning it to the remainder of the wort ready for yeasting; sulphuric acid may be used as an alternative method of reducing the pH. Again, some power-alcohol distilleries sterilize the wort chemically with 1,000 ppm ammonium fluoride and use a fluoride-adapted yeast.

The two types of brewing yeasts differ not only as to species but also in other properties. *S. cerevisiae* has round cells, most of the crop rises to the surface at the end of the fermentation and, characteristically, when supplied with raffinose ferments only one third of the molecule, leaving melibiose. *S. carlsbergensis* is cylindrical, forms a deposit and can ferment the whole of raffinose. There are strains of both yeasts with intermediate morphology and brewing properties; the latter can be modified of course by the brewing system employed.

Beer wort is yeasted with between 2 and 6 g moist yeast/litre, depending on its original specific gravity. After 12–18 hours at 15°C a rapid fermentation ensues, the temperature is raised by 3 to 7°C and the pH falls from 5·2 to 4·1. The thick yeast head that collects after $2\frac{1}{2}$ days is skimmed off, fermentation slackens and the beer, still containing some sugar, is ready for clarification and natural conditioning in tanks before sale. Some modern breweries use a non-flocculent yeast that ferments very rapidly; the fermentation is terminated at the desired sugar content by centrifugation. Traditionally, part of the yeast crop (often a balanced collection of strains) was collected and used for the next fermentation. Nowadays pure cultures are being used increasingly and are replaced after 12 to 20 brews, depending on the standard of hygiene. Lager has always been made with a pure culture, the crop being collected from the bottom of the tank after the main fermentation, subsidiary crops in the pre-fermentation and conditioning tanks being discarded. The pitching temperature, 5 to 9°C, is lower than with beer; fermentation at 12°C is more protracted (7 to 14 days), and storage at 2°C may take 6–40 weeks before the lager is considered sufficiently conditioned and stabilized for sale. Distillery worts of high specific gravity are pitched at 21°C with a heavy inoculum of a yeast specially bred for alcohol tolerance; fermenta-tion is rapid, the temperature rising to 25–30°C during the process. Either a fresh culture is used for each fermentation or the previous crop is acid-washed and re-used.

Cider and wine

Successful brewing is dependent not only on the biochemical properties of a yeast strain but also on the physical properties of the mass culture. In contrast, cider and wine were made until recently by controlling a natural flora of yeasts, moulds, and bacteria, so that only organisms with desirable properties were allowed to grow. The microbiology of cider will be discussed here, that of wine is similar in many respects.

In the traditional cider-making countries, the raw material consists of special varieties of apples, high in tannin and low in acidity. Nowadays these are supplemented with dessert and culinary apples graded as unsuitable for the fresh fruit market; other countries use these latter sorts as their sole source of raw material. The fruit as it arrives at the factory carries both an internal and an external microflora. In England this is mainly *Candida pulcherrima*, *Kloeckera apiculata*, and *Torulopsis famata*, together with the 'black yeast' *Aureobasidium pullulans* and occasionally rhodotorulae, *Candida*, and *Torulopsis* spp. *Saccharo-myces* spp. and bacteria are rare in sound fruit but mould-damaged specimens carry large populations, not only of mould spores but also of acetic acid bacteria and fermenting yeasts. Both the moulds and acetic bacteria produce sulphite-binding compounds that are of great significance at later stages of processing. Hence sound, preferably washed fruit, is essential. The apples are milled to a pulp and pressed hydraulically in cloth envelopes between wooden drainage racks. Formerly the racks and cloths were washed very infrequently, consequently they soon supported a dense population of *Saccharomyces* species; these were the yeasts mainly responsible for the subsequent fermentation. At first *Candida* spp. and other poorly fermenting yeasts are most numerous but they are rapidly overgrown by *Kloeckera* spp. after a few degrees drop in specific gravity. The latter then co-exist and in turn are overgrown by *Saccharomyces* spp. while in the fully fermented dry ciders film yeasts and acetic bacteria become more im-portant. The whole process is improved by treatment of the freshly pressed juice with sulphur dioxide, the actual amount being varied according to the pH. Below pH 3·3 100 ppm suffices, while 150 ppm is desirable between 3·3 and 3·8; to be effective more SO_2 is required above 3·8 than is allowed by law (200 ppm), but most factories process a mixture of apples whose juice pH is generally 3·5–3·6. It is vitally important that the concentration of sulphite-binding compounds be kept low in the juice, otherwise no free sulphur dioxide remains in the solution and this is the only effective moiety (sometimes referred to as undissociated sulphurous acid). After treatment only *Saccharomyces* spp. remain, so that in effect sulphiting selects a pure culture from the original microflora. However, with higher standards of cleanliness in the press rooms, the juices have a flora similar to that of the fruit, i.e., virtually free of fermenting yeasts. Hence, the addition of pure cultures of bottom-fermenting wine yeasts, following sulphiting, is becoming almost universal in English cider factories. It also means that both the time needed for fermentation and the flavour of the cider become more predictable, since there is no longer any need to rely on what were virtually factory contaminants.

A second fermentation of malic acid to lactic acid and carbon dioxide, due to lactic acid bacteria, nearly always occurs during either the yeast fermentation or storage of the cider. The organisms that can bring about this change include both homo- and heterofermentative lactic rods and cocci, including the organisms responsible for producing polysaccharide slime or ropiness. The bacterium found most frequently is *Lactobacillus pastorianus* var. *quinicus*. If citric acid is present, as in perry but not cider, diacetyl, lactic, and acetic acids are also formed and the flavour is thereby spoiled. The source of these bacteria is still in some doubt since they appear to be absent from many sulphited juices. Ciders stored in contact with air, especially those held in small wooden containers, soon show surface growths of *Acetobacter xylinum* and the film yeasts *Pichia membranaefaciens* and *Candida mycoderma*. The problem is now rare in large factories where juice sulphiting and very large storage tanks are normal practice.

Basically, wines are made by a similar process but many other fermentation treatments are possible, due to the much wider range of climatic conditions in which grapes are grown. Thus grapes can be 'raisinified' by being left on the vine or by being dried in the sun; wines made from their high gravity juices are fermented with osmophilic yeasts such as *Saccharomyces rouxii*. Sauterne is made from grapes naturally infected with the mould *Botrytis cinerea* that dehydrates the grapes, metabolizes some of the acid and produces glycerin and the mild antibiotic *botryticin*. Consequently the subsequent fermentation of the very sugary juice ceases prematurely, leaving a soft, sweet, luscious-flavoured wine. Climate also has an effect on the type of yeasts found on the grapes. The greater the ambient temperature, the greater the concentration of alcohol-tolerant *Saccharomyces* yeasts and sporing apiculate yeasts (*Hanseniaspora valbyensis*) and the fewer there are of their non-sporing counterparts (e.g., *Kloeckera magna* and *K. apiculata*). Finally a number of *Saccharomyces* species have been isolated from both cider and wine. These are *S. acidifaciens*, *S. carlsbergensis*, *S. cerevisiae*, *S. delbrueckii*, *S. elegans*, *S. florentinus*, *S. fructuum*, *S. oviformis*, *S. rosei*, *S. rouxii*, and *S. uvarum*. In addition, there are other *Saccharomyces* spp. apparently characteristic of each beverage. Again *Brettanomyces* spp. have been isolated twice from cider and it was only in the 1960s that they have been found in wine, in the Bordeaux area and in dry wines made within a 40-mile radius of Cape Town. There is one wine for which no parallel exists in cider-making, namely, sherry. The yeasts responsible for the characteristic fermentation of alcohols and organic acids, sometimes called *S. beticus*, *S. cheresiensis*, etc., develop as a veil on the surface of the dry wine when it is exposed to air. Unlike *C. mycoderma* and other spoilage film yeasts, the sherry yeasts form acetaldehyde and other characteristic flavours from the alcohol; the remaining process is a complicated blending system (the solera) also used for the sun-baked wine, Madeira.

BOOKS AND ARTICLES FOR REFERENCE AND FURTHER STUDY

ARIMA, K. (1957). Sake and similar yeasts. In Roman, W., ed. *Yeasts*. W. Junk, New York, Chapter 4, pp. 143–153.

AYRES, J. C. (1955). Microbial implications in the handling, slaughtering and dressing of meat animals. *Adv. Fd Res.*, **6**, 109–161.

BEECH, F. W. and DAVENPORT, R. R. (1970). The role of yeasts in cidermaking. In Rose, A. H. and Harrison, J. S., eds. *The Yeasts*. Academic Press, New York and London. Chapter 3, Volume 3 (in press).

CARR, J. G. (1968). *Biological principles in fermentation*. Heinemann Educational Books Ltd., London, 97 pp.

DAVIS, J. G. (1963). The lactobacilli. II. Applied aspects. *Progress in Industrial Microbiology*, **4**, 95–136.

FOWELL, R. R. (1967). Infection control in yeast factories and breweries. *Process Biochem.*, **2** (12), 11–15.

FRAZER, W. C. (1967). *Food microbiology*. 2nd Edn. McGraw-Hill, London, 512 pp.

GIBSON, T. and ABD-EL-MALEK, Y. (1957). The development of bacterial populations in milk. *Can. J. Microbiol.*, **3**, 203–213.

HERSON, A. C. and HULLAND, E. D. (1964). *Canned Foods: an introduction to their microbiology*. Chemical Publishing Co., New York, 291 pp.

HESSELTINE, C. W. and WANG, H. L. (1967). Traditional fermented foods. *Biotechnol. Bioengng.*, **9**, 275–288.

INGRAM, M. (1962). Microbiological principles in prepacking meats. *J. appl. Bact.*, **25**, 259–281.

LEITCH, J. M., ed. (1965). *Food Science and Technology*. Gordon and Breach Science Publishers, New York, Volumes 2 and 4.

LÜTHI, H. (1959). Micro-organisms in non-citrus juices. *Adv. Fd Res.*, **9**, 221–284.

MICHENER, H. D. and ELLIOTT, R. P. (1964). Minimum growth temperatures for food poisoning, faecal-indicator and psychrophilic micro-organisms. *Adv. Fd Res.*, **13**, 349–395.

MOLIN, N., ed. (1964). *Microbial inhibitors in food*. Almqvist and Wiksell, Stockholm, 402 pp.

MOSSEL, D. A. A. (1964). Essentials of the assessment of the hygienic condition of food factories and their products. *J. Sci. Fd Agric.*, **15**, 349–362.

PEDERSON, C. S. (1960). Sauerkraut. *Adv. Fd Res.*, **10**, 233–291.

PIRIE, N. W. (1969). *Food Resources Conventional and Novel*. Penguin Books Ltd., Harmondsworth, Middlesex.

POLLARD, A., BEECH, F. W. and BURROUGHS, L. F. (in press). Cider and perry. In *Encyclopedia on Food and Food Science*. John Wiley and Sons, Interscience Publishers, Inc., New York, Volume 2.

SCOTT, W. J. (1957). Water relationships of food spoilage micro-organisms. *Adv. Fd Res.*, **7**, 84–128.

SHEWAN, J. M. and HOBBS, G. (1967). The bacteriology of fish spoilage and preservation. *Progress in Industrial Microbiology*, **6**, 169–208.

TANNER, F. W., ed. (1944). *Microbiology of Food*. 2nd Edn. Garrard Press, Illinois, 1196 pp.

THATCHER, F. S. (1963). The microbiology of specific frozen foods in relation to Public Health: Report of an International Committee. *J. appl. Bact.*, **26**, 266–286.

THOMAS, S. B., DRUCE, R. G., PETERS, G. J., and GRIFFITHS, D. G. (1967). Incidences and significance of thermoduric bacteria in farm milk supplies: a reappraisal and review. *J. appl. Bact.*, **30**, 265–298.

21

Spoilage of Industrial Materials by Micro-organisms

INTRODUCTION

The most important industrial materials, other than foodstuffs, affected by
micro-organisms are cellulose and wood products, including wood itself, wood
pulp and paper, and textiles made from natural fibres such as cotton, flax, and
jute. These materials are attacked by fungi, and to a lesser extent by bacteria,
causing loss in strength and discoloration. Rots of growing timber, petroleum
microbiology, and damage to plastics and other products of lesser importance
will also be considered. Spoilage of food during industrial processing and uses
of micro-organisms in industry are treated elsewhere (Chaps. 20 and 22).

DEGRADATION OF CELLULOSE PRODUCTS

Importance and method of attack

Cellulose for industrial use is obtained from cell walls of higher plants either
directly (for example, cotton fibres are approximately 95 per cent cellulose) or by
removing other cell wall components, such as lignin, from wood as in the
production of paper pulp. Cellulose decay may cause great economic loss where
large quantities are handled, as in the military use of tents, clothing, and other
equipment, particularly in tropical climates. Losses may also be important in

stores of such expensive items as high grade paper, or valuable books and documents.

Cellulose is a polymer of β 1-4 D-glucopyranose with a molecular weight of approximately $1-1\cdot3 \times 10^6$, normally forming unbranched chains up to 4,000 Å long. In cell walls the chains are grouped into crystalline structures termed micelles. One molecule may be incorporated, along its length, into several micelles forming an interlocking system. The micelles are themselves grouped into micro-fibrils.

The enzyme cellulase (correctly called β 1-4 glucan, 4 glucanohydrolase), which enables micro-organisms to attack this complex system is not fully understood. However, two stages seem to be involved; first the crystalline micelle is broken down, and then the amorphous cellulose is hydrolysed (either endwise or in a random manner, depending apparently on the organism), to cellobiose. It is at present debatable whether these stages are carried out by two different enzymes or by one enzyme which changes during the overall reaction by combination with some product of the first stage. Cellulase production is induced when the organism grows on a cellulose substrate and it is extra-cellular, though in bacteria it may be closely bound to the cell walls or to the substrate. The cellobiose is further hydrolysed to glucose by cellobiase (β-glucosidase).

Fungal hyphae penetrate the fibre cell wall and grow in the lumen, eroding from within, while bacteria generally attack only the outer surface of the fibre to which they may become closely attached. Attack on cellulose may at first cause little loss in strength, though the decay is visible microscopically as eroded areas on the fibres. It may also be detected at this stage by an increase, as compared with unaffected fibres, in the swelling in alkali solution. This stage probably corresponds to the breakdown of the crystalline structure. As decay proceeds the cellulose chains are broken up and the strength of the material is rapidly lost.

Organisms associated with decay

Two fungi, *Chaetomium globosum* and *Stachybotrys atra*, are of outstanding importance in cellulose decay; both have a world-wide distribution, though the latter is more common in north temperate regions. Both occur on rotting vegetation in the soil. *Myrothecium verrucaria* and *Memnoniella echinata* are less common but have great cellulolytic powers and are often used in laboratory tests for decay resistance.

Other fungi, including *Aspergillus*, *Penicillium*, *Cladosporium*, *Fusarium*, *Alternaria*, and *Trichoderma*, having lesser cellulolytic powers are often isolated from cellulose substrates.

Many other fungi, not themselves capable of utilizing cellulose, may degrade dyes, sizes, and other finishers in the fabric. Though they cause no damage to cellulose they may produce stains or otherwise affect the quality of the product; such fungi as *Mucor*, *Aspergillus*, *Penicillium*, and *Fusarium* are often isolated from such damaged material.

Bacteria are less important than fungi in cellulose textile breakdown. They may, however, become of particular interest under anaerobic conditions, where fungi are less active, and may cause oxidation instead of hydrolysis of cellulose. *Cellvibrio*, *Cellulomonas*, *Cytophaga*, and a large number of Actinomycetes are aerobes, and *Clostridium* spp. such as *Cl. cellulosolvens* are anaerobes commonly found on cellulose substrates.

A specialized micro-flora is associated with wood pulp and paper production where large quantities of cellulose suspensions are handled. Many fungi are present in bulk storage vats; over one hundred species were isolated within a year on one survey. The most important genera are *Alternaria*, *Aspergillus*, *Cladosporium*, *Penicillium*, and a basidiomycete genus *Polyporus*. They presumably cause a loss of cellulose but more important they produce in association with bacteria such as *Flavobacterium*, *Pseudomonas*, *Desulfovibrio*, and *Clostridium*, a slime which is often coloured and foul-smelling. The slime affects the quality of the paper and increases the cost of machine maintenance. Bacteria may also cause corrosion of machines and pulp storage tanks.

Conditions for decay: decay prevention

The factors affecting the rate of decay are those governing the growth of the organisms, namely pH, temperature, nutrient availability, oxygen supply, and moisture. Apart from the latter these are generally not limiting, and anyway often cannot be controlled under conditions of use. The balance of these factors determines the particular species concerned, for example, bacteria are more common than fungi under anaerobic conditions, and vice versa. The cellulose substrate must be moist or in conditions of high humidity for decay to occur.

The commonest way of preventing decay is to keep the material dry. When this is not possible (for example, under tropical conditions, when use involves contact with moist soil or water, or in pulp mills where the cellulose is in suspension) then chemical preservation methods must be used. There is a wide variety of commercial products available, and these are routinely used to impregnate cellulose materials likely to be used under moist conditions. Preservatives are ideally toxic to micro-organisms but not to animals and should be virtually insoluble in water so that they are not leached out. Heavy metal salts, particularly those of copper, are commonly used.

Uses of micro-organisms in the manufacture of textiles

Micro-organisms are used in the retting of flax, hemp, and jute. This is a process whereby the useful fibres, either of cellulose alone or containing both cellulose and lignin, are separated from the rest of the plant stem. The stems are soaked in water and then micro-organisms, either naturally present or added as pure cultures, are used to remove the middle lamellae and cytoplasm, thus releasing the fibres which are then cleaned and dried. It was once a natural process carried out in rivers or bogs, where the main organism was *Clostridium*, or under aerobic conditions by *Cladosporium*, *Mucor*, and *Rhizopus*. Present methods involve the use of pure cultures of suitable organisms under controlled conditions of anaerobiosis.

Synthetic fibres and wool

Synthetic fibres based on cellulose, e.g., rayon, have now been superseded by nylon and polyester fibres such as 'Terylene'. Rayon (cellulose acetate) is resistant to attack, the molecular configuration having been changed sufficiently to prevent cellulases from operating. Though modern synthetics are likewise resistant the dyes and finishers may still be attacked.

Wool is largely protein and is quite resistant to decay. In a raw state when contaminated with dirt and oil there may be a fairly high microbial population. However, once the wool is cleaned and manufactured and provided dyes and finishers are suitably chosen, only some Actinomycetes cause problems.

Diseases associated with textile manufacture

Textile fibres often originate in tropical or sub-tropical countries and occasional cases of unusual diseases are reported amongst workers handling the raw cotton and jute fibres before processing. Cotton may be sterilized if necessary but wool, being protein, will not usually withstand this, and various diseases such as anthrax (caused by *Bacillus anthracis*) sometimes occur when foreign fleeces are handled. Some dermatophytic fungi, such as *Trichophyton* sp. may also be carried, or even live saprophytically, on wool.

DEGRADATION OF WOOD

Wood and the importance of its decay

Wood is the xylem (conducting tissue) of trees and contains fibres and vessels or tracheids which are usually dead, and in growing trees a small amount of live parenchyma also. The cell walls contain lignins and cellulose together with a selection of polysaccharides including mannans, xylans, and hemi-celluloses. Lignins are a group of complex polymers of syringaldehyde, vanillin, and coniferyl alcohol whose structure is not fully known and which are very resistant to decay. The cellulose in wood is not usually attacked in the ways described above for the pure product owing to the formation of complexes with the other cell wall constituents. Wood from Angiosperms is generally referred to as 'hardwood' and that from conifers as 'softwood' though these terms are not always true indications of relative hardness. Hardwood and softwood differ in their chemical and cellular composition and decay fungi are frequently restricted to one or other of them. The biochemistry of the decay and the enzymes involved are not well understood.

Timber is one of the world's main natural resources and is in great demand for constructional use, furniture, fuel, packaging, and increasingly for pulp production for paper, fibreboard, and fibrous packaging materials. In view of the very large quantities involved even a small percentage of decay during growth, conversion, storage or use represents a large economic loss. Estimates of decay loss in Britain vary greatly but in standing trees it is not less than 15 per cent of the annual timber yield. Decay of sawn timber by one fungus alone, 'dry rot' (*Merulius lacrymans*, Fig. 21.1), causes a loss of several million pounds sterling per year, and there are numerous less important rots also attacking timber in use.

A

B

C

Types of degradation, their occurrence, and appearance

'Hard' rots

This group of wood decay types is caused by Homobasidiomycetes and in-cludes all the important rots of standing and converted timber. There are two main types, 'brown' and 'white' rot. Brown rot occurs when the fungus utilizes the cellulose component of the wood and leaves the brown lignin, causing a rapid loss in strength. White rots are produced by fungi which attack both the lignin and the cellulose and by fungi using only cellulose, as in brown rots, but also bleaching the lignin. Most white rots produce a less rapid loss in strength than do the brown rots.

Further terms may be used to describe the appearance of the wood or the distribution of the fungus. Thus *Merulius lacrymans* produces a 'brown cubical rot', the wood is brown, due to the removal of cellulose, and is split into cubical blocks (Fig. 21.1B). Similarly *Fomes annosus*, a very important parasite of stand-ing timber, produces a 'white pocket rot', the white rot is confined to small spindle-shaped cavities in the wood (Fig. 21.2A). Numerous descriptive terms are also used referring to the position of the rot in a standing tree, *F. annosus* causes a root and 'butt rot' (a rot at the base of the stem) for example.

The gross appearance of the wood may also be altered by 'zone lines'. These are clearly defined black lines within the timber caused by masses of dark mycelium or by deposits of resins and gums produced by the tree to contain the rot (Fig. 21.2B).

The hyphae of brown and white rot fungi are found within the lumena of the wood cells (Fig. 21.2C). They occasionally pass through the cell walls, usually directly across the wall, by enzyme secretion. A hyphal tip comes into contact with the wall and dissolves a narrow hole through which it passes into the next lumen. This bore hole is often enlarged by subsequent enzyme action. In white rots the walls become noticeably thinner, but this does not occur with brown rots until an advanced stage of decay is reached.

Decay has a great effect on the properties of wood, but for many purposes, such as rough packaging and low grade constructional use, a small amount is tolerable. The effects include: (1) The density is reduced. The unusual lightness of a large piece of timber may indicate a pocket of decay in its centre which is not visible from the outside. (2) The shrinkage of rotten wood on drying is greater than that of sound timber. This is important during conversion and seasoning when a slight amount of decay, which may not otherwise be important, causes excessive distortion of planks. (3) The chemical constitution of the wood is altered, making it unsuitable for pulping. (4) As indicated above the colour of the wood may be changed, though when this becomes obvious the decay is usually

---

Fig. 21.1   (A) *Merulius lacrymans* (about two-thirds natural size); a large fruit-body of the dry-rot fungus on a door lintel. **(B)** *Merulius lacrymans* (about natural size); a board showing the typical brown cubical rot of and mycelial strands of the fungus. **(C)** *Coniophora cerebella* (about natural size); a board showing typical longitudinal cracking and black rhizomorphs of the 'cellar fungus'. (From Cartwright and Findlay (1958), by courtesy H.M.S.O.)

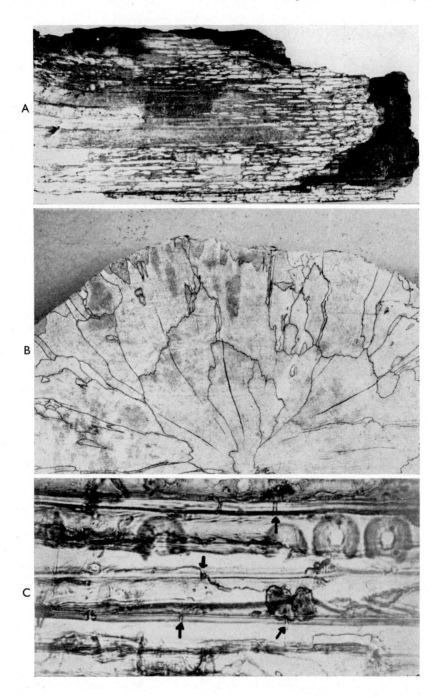

so far advanced that the wood is useless anyway. There are rare exceptions, however, when discoloration makes the timber more valuable; for instance, *Fistulina hepatica* (the beef-steak fungus) produces a brown pocket rot of oak which results in an attractive pattern on veneers. (5) The most important effect is a decrease in strength. There are different types of strength, such as tensile, bending, impact, or crush strength, and the amount of each type lost will depend on the type and degree of rotting. The grading of timber takes into account the type and position of the rot in the piece; the wood may then be put to a use where loss of strength in this position is not important.

In standing trees the causal fungi of brown and white rots may be parasites, such as *Fomes annosus* and the honey fungus (*Armillaria mellea*), or virtually saprophytic, such as *Fistulina hepatica*, causing serious rot only after the tree has been killed or severely weakened by some other agent. The two most important fungi in Britain causing rots are *F. annosus* and *A. mellea*. The former eventually kills the tree and may then live saprophytically using the dead stump and root system as an infection centre, spreading to neighbouring healthy trees by root grafts. It is world-wide in distribution and unfortunately very common, causing losses of up to 30 per cent of utilizable timber in badly affected plantations. *A. mellea* causes a fibrous white rot of a great variety of trees and shrubs. Conspicuous black rhizomorphs grow under the bark of infected trees and also out into the soil thus allowing extensive spread of the fungus through the soil (similar, though smaller, rhizomorphs are shown in Fig. 21.1C). Amongst the many other species causing damage in standing timber the following are sometimes important in Britain: *Polyporus schweinitzii, P. betulinus, P. sulphureus, Fomes ulmarius,* and *Ganoderma applanatum.*

Decay of converted timber in use may be caused by a very wide variety of fungi depending on the type of wood and the conditions of use. The decay of timber in houses is particularly important for economic reasons, and in Britain *Merulius lacrymans*, the dry-rot fungus (Fig. 21.1A and B), is of outstanding importance. It produces a brown cubical rot and is common in all types of buildings using wood in their construction. It requires damp conditions to become established, but thereafter can spread for long distances by mycelial strands which can pass through brickwork and other apparently solid barriers. In the course of hydrolysis of cellulose sufficient water is produced to enable the fungus to attack relatively dry timber when humidity is high due to poor ventilation. A less important, but still quite common, fungus causing rot in building timber is the cellar fungus (*Coniophora cerebella*). It is restricted to damp wood, in which it causes a brown rot with longitudinal cracks (Fig. 21.1C) and occasionally with cracks at right angles to the grain (which are not as prominent as in *M. lacrymans*). Characteristic thin dark rhizomorphs are formed over the surface of the wood. Other fungi occasionally causing damage include *Phellinus megaloporus* and *Poria xantha*

Fig. 21.2   (**A**) (about natural size). White pocket rot in oak (caused by *Hymenochaete rubiginosa*); note the well-defined areas of white rotten wood. (**B**) (about one-quarter natural size). Zone lines on the transverse face of a log. (**C**) (× about 400). Microscopic appearance of wood attacked by a brown rot fungus; note the hyphae in the cell lumens and the bore holes (arrowed).

which is common in greenhouses. *Poria incrassata* is particularly common in North America where it is more important than *M. lacrymans*.

Numerous other fungi are found as saprophytes on rotting wood under natural conditions where they are important in the carbon and nitrogen cycles of forests.

'Soft' rots

Though the existence of rot of this type has been known for more than 100 years it has only comparatively recently been investigated in detail. The causal organisms of soft rot are a small group of Ascomycetes, mainly Pyrenomycetes, and Fungi Imperfecti (Deuteromycetes). Soft rot is of interest for two reasons: firstly because it occurs under very wet conditions where rots by Basidiomycetes are not common. Secondly soft rot fungi are not controlled by many preservatives routinely used on timber, though copper chromate and copper chromium arsenate are effective. The rot is usually confined to specialized environments such as timbers in constant contact with fresh or salt water and has received particular attention because of the damage caused in softwood slats used in cooling towers of power stations. It is possible that soft rot is quite widespread but usually masked by the faster rotting and more conspicious Basidiomycetes.

Soft rot is so called because of the soft spongy texture of the outer layers of the attacked wood, usually without any discoloration. The rot is characterized, however, by its microscopic appearance (Fig. 21.3). The hyphae penetrate the

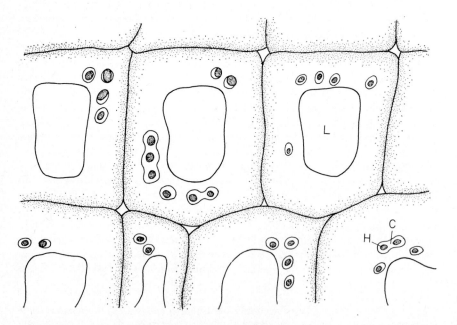

Fig. 21.3 (× about 2,000). Diagram of a transverse section of wood attacked by soft rot fungi; note the cavities (C) containing hyphae (H) in the secondary cell wall, and the absence of hyphae in the lumen (L) of the tracheids.

wood through the cell lumena and through rays but then enter the secondary cell walls of the wood cells and branch at right angles, forming cylindrical cavities with sharply pointed ends running the length of the cell and frequently following the orientation of the cellulose microfibrils (c.f. hard rots where the hyphae occur mainly in the cell lumen). The cylindrical cavities, enlarged by enzyme action, may fuse with each other in advanced stages of decay forming irregular shaped cavities in the secondary cell wall. This may also be eroded by enzyme action on its inner or outer surfaces.

The soft rot causes a loss of all types of strength in the surface layers of the wood. In large bulks of timber, for example marine piling, this may not be particularly important unless it allows excessive abrasion; in smaller sizes of timber, such as plywood hulls of small boats and the slats of water-cooling towers, the destruction may be virtually complete.

Many genera have been implicated in soft rot, but the most common include: *Chaetomium, Xylaria, Leptosphaeria,* and *Ophiostoma* among the Ascomycetes, and *Phoma, Bispora,* and *Phialophora* in the Fungi Imperfecti. There is no notable difference between the rots produced by the different species; nothing equivalent to the distinction between brown and white rots of the Basidiomycetes described above.

Bacterial rots

Little information is available on the rots caused by bacteria and they are thought to be relatively unimportant. Decay attributed to bacteria usually occurs under anaerobic conditions such as are found in wood deeply submerged in water or in waterlogged soil. Decay is slow and appears microscopically as minute eroded areas in the cell walls of the wood.

There are two situations in which it may be important: firstly in log storage ponds of pulp mills where large quantities of timber may be kept submerged for a considerable time. The bark and outer layers of wood become soft and slimy causing some handling difficulties but little loss of timber. Secondly, as mentioned earlier, bacteria may be important in slime production in the pulping process, particularly when mechanical rather than chemical methods of disintegration of the wood are used.

Chemical decay

This again is of minor importance but is mentioned here as it may be confused with fungal decay. It occurs in such obvious cases as when acid or alkali has been accidentally spilt but is known to be caused by the acidity of bird droppings and by acids produced when metals are corroded by salt waters. This type of decay can be distinguished from fungal attack by lack of hyphae in the wood and by its characteristic loose fibrous texture as the wood cells fall apart.

Staining of timber by micro-organisms

As with paper and textiles various fungi may cause stains without seriously affecting strength. The economic loss may, however, be considerable because the wood is unsuitable for decorative work where highest prices are paid. Common fungi such as *Penicillium, Aspergillus,* and *Trichoderma* are frequently isolated, but the most serious staining is caused by *Ceratocystis* spp. in sawn softwoods. These fungi live on the contents of the ray parenchyma cells, not attacking cellulose or lignin to any great extent, and produce blue or greenish discolorations.

Staining may be particularly associated with damage by wood- and cambium-boring beetles. The holes may furnish entry sites for naturally occurring fungi, the beetles may passively carry the spores on their bodies or even actively inoculate the brood galleries with fungi (carried in special structures on their bodies) on which the larvae feed.

Prevention of timber decay

The main concern is the prevention of the very important hard rots.

Prevention in standing timbers

Prevention of decay in standing timbers depends on: (1) The maintenance of trees in a healthy condition so that any infection that arises may be resisted by the production of resin which may wall off the fungus and by biochemical defence mechanisms. The health of the trees depends on the choosing of the correct species for a particular plantation, maintaining drainage, and correcting density by thinning. (2) Sources of infection must be kept to a minimum by removing old diseased trees and rotting logs. (3) Damage to the trees, which might allow access by fungi to the wood, is avoided as far as possible during felling and subsequent extraction of the timber.

The degree to which these measures are applied will depend on the nature of the crop. It may be better, economically, to practise none of the measures, their cost being greater than the anticipated loss by decay. All the above measures are, however, used in British forests and, in addition, the damage caused by *F. annosus* is severe enough to warrant coating stumps with creosote or borax to prevent spore germination, or with urea to supply nitrogen which encourages antagonistic organisms. Biological control by inoculating stumps with the antagonistic *Peniophora gigantea* is now commonly used. Ornamental trees, of aesthetic rather than economic value, may have wounds or rotten portions treated individually by preservatives or tree 'surgery'.

The prevention of decay during conversion and use

It should first be pointed out that timbers vary greatly in their natural resistance to decay, depending on their structure (close-grained being generally more durable than less dense types) and their natural content of fungicides such as phenols and tannins. Timbers in common use that are resistant to microbial decay include oak (*Quercus* spp.), red cedar (*Thuja plicata*), walnut (*Juglans* spp.) and many imported tropical hardwoods such as teak (*Tectona grandis*) and greenheart (*Ocotea rodiaei*). Wood from near the centre of the stem, the heartwood, is often more durable than the sapwood.

During the conversion of logs into sawn timber the most practical way of preventing decay is to saw as soon as possible and store the planks in well-ventilated stacks to dry (i.e., to 'season'). Modern techniques of drying timber under carefully controlled conditions of heat and humidity are safer but more expensive. At this stage, and in subsequent use, the timber is best preserved by removing water essential for fungal growth, and subsequently keeping it dry. Fungi will not normally grow when the moisture content is less than 20 per cent, which is barely obtainable under covered outside storage in this country.

If the wood cannot be kept dry in use then, as with cellulose products described above, chemical preservatives must be used. The commonest one is creosote, a product of the coal-tar industry, which is effective against most fungi and is cheap. It has the disadvantages that it is exceedingly unpleasant to work with, wood treated with it cannot be painted, it smells, and is leached comparatively rapidly by water. A wide range of copper salts are now frequently used where the defects of creosote cannot be tolerated. Impregnation with preservatives is usually done under pressure in large tanks to achieve adequate penetration of the wood.

MICROBIOLOGY OF PETROLEUM AND HYDROCARBONS

Degradation of petroleum, its products, and hydrocarbons

Bacteria are more active than fungi in the degradation of petroleum and its products, mainly because anaerobic environments are often involved. Decay depends on the presence of water and minerals but these are not usually limiting since crude oil is contaminated with brine from the oil beds and water is invariably present in storage tanks from condensation or for technical reasons concerned with its transport by sea.

Crude oil in transport and storage may be attacked by *Actinomyces* sp., *Mycobacterium* sp., *Pseudomonas aeruginosa*, and *Desulfovibrio* sp. These are originally introduced from soil contamination, but become established as a resident population in tanks and pipelines. It has been claimed that bacteria, particularly sulphate-reducing ones, exist in the oil deposits underground and indeed the absence of oil in potentially oil-bearing geological formations has been attributed to bacterial decay. However, the damage to crude oil is not usually severe, but includes a decrease in viscosity, loss of oil by oxidation and changes in the relative amounts of the aliphatic and aromatic fractions of the oil.

Petroleum products such as gasolene, kerosene, paraffin, and lubricating oil are all subject to attack in the presence of water. This again does not cause appreciable loss of the substance except that the octane content of gasolene may be reduced. The by-products of the attack may, however, produce deleterious effects such as the hydrogen sulphide content rising to unacceptable levels because of the action of sulphate-reducers. Methane may also be produced and combinations of gases have been blamed for bursting and spontaneous explosions in storage tanks. Attack on kerosene used for aircraft fuel has become important and is particularly associated with the fungus *Cladosporium resinae*. This may cause blocking of fuel lines, with fatal consequences, and the organic acids which are produced damage fuel tanks and pipes. The fungus is exceedingly difficult to eradicate as it can live in empty tanks on condensation water and with kerosene vapour as its sole carbon source. The problem is so serious that all jet fuel is now filter-sterilized on loading and frequent inspection of tanks is carried out, though this is very difficult with the intricate systems of modern aircraft.

Hydraulic fluids, both oil and glycol based, are frequently attacked, causing hydraulic failures as the pipes are blocked by microbial growth. Such fluids now usually contain additives to prevent growth of fungi and bacteria.

Problems in drilling for oil

Sulphate-reducing organisms such as *Desulfovibrio* often occur under anerobic conditions in bore holes and pipelines where water and minerals from the wall are also present. Hydrogen sulphide is produced which with the water and the cathodic effect of the iron pipes and minerals results in sulphuric acid which causes corrosion of the pipes. The organic acids produced by this and other organisms may also be important. Corrosion may be quite rapid ($\frac{3}{8}$ in. of steel dissolved in four years) and replacement of pipe sections at the bottom of oil wells may be impossible; at best exceedingly expensive remedies such as filling the pipe with cement and then reboring may be used.

Problems may also arise in the microbial attack on drilling fluids which are used to lubricate the bit and seal the bore when passing through permeable strata. These fluids are usually based on various clays, supplemented with colloidal starch and cellulose, both of which may be attacked causing severe damage to the well.

Micro-organisms may actually block injection wells. These are used to pump water into the oil-bearing strata so forcing the oil out. Fungi, bacteria, and algae may accumulate in the pipes or block the pores in oil-bearing rock preventing use of the injection well and reducing the oil yield. This may be combated by complex filtration processes or by using artesian water without exposure to sources of contamination.

Beneficial use of micro-organisms in the petroleum industry

Sub-soil micro-organisms are affected by gas, mainly methane, leaking from underlying oil beds, and it has been claimed that such oil deposits may be detected by studying the variation in numbers of methane-utilizing organisms. This method of prospecting has not yet been used commercially as far as is known, but much work has been done in the U.S.S.R. on the problem.

Various patents have been issued in the U.S.A. and elsewhere for the use of bacteria, such as *Desulfovibrio*, to increase the yield from oil wells by introducing them into the oil-bearing strata. The viscosity of the oil is decreased by the organisms and it was hoped that calcareous rocks would be dissolved by metabolic acids increasing the pore size in the rock and the gas pressure. Laboratory experiments have not, however, yielded consistent results.

Despite the discouraging results obtained from the above methods, further research, the increased demand for oil, and the inevitable necessity of utilizing sources at present uneconomic, could render them of importance in the future.

The fermentation of petroleum might be mentioned here, though major industrial fermentations are considered in Chapter 22. Various yeasts are used to ferment paraffin fractions of petroleum with added minerals, and protein powder is recovered from the organisms. This system is in commercial pilot operation and cost, efficiency of conversion, and nutritional value of the protein, though not its palatability, compares very favourably with conventional production from animals.

The degradation of crude oil by yeasts and bacteria has recently become an important beneficial effect in connection with marine pollution. The more volatile fractions evaporate and the destruction of the residual tar-like mass is slow. It may take place under alternating aerobic or anaerobic conditions as the oil sinks with the accumulated weight of micro-organisms and then rises again as such gases as methane accumulate in the mass. The final result is a hard pebble-like piece of unutilizable tar residues. The spreading of detergents is sometimes thought to be necessary to break up large patches of oil but is ultimately harmful as it may kill the bacteria and yeasts, as well as much other marine life, thus actually slowing the destruction of the oil.

MICROBIOLOGY OF MISCELLANEOUS INDUSTRIAL PRODUCTS

A large number of other industrial products are degraded by micro-organisms including such unlikely substances as glass (and associated lens coatings and adhesives), paints, stone, plastics, rubber, and electrical insulation.

Damage to glass is important only in optical instruments where precision ground surfaces are necessary. It occurs particularly in the tropics where high temperatures and humidity made it a serious problem in the Second World War. Fungi, of which *Aspergillus* is the most common, grow on lens coatings, adhesives, and dust particles. Their metabolic products etch the glass, or in extreme cases the mass of fungal hyphae obscures the lens. Prevention of damage depends on keeping the equipment as dry and dust-free as possible, and having tight fitting seals around the lenses. This will not, however, prevent damage under the worst conditions. The use of fungicidal vapour introduced into the air spaces of the equipment and even radio-active foil around lens and prism mountings may then be effective.

Paints may be discoloured by fungi growing in the wood beneath, or fungi may become established in dust on the paint surface and subsequently attack the paint itself causing it to peel from the surface of the wood. High humidity and moisture on the surface are usually required. Species of *Pullularia, Cladosporium*, and *Phoma* have been most commonly reported and are frequently to be seen on the paint of greenhouses in Britain. Bacteria, such as *Pseudomonas* and *Flavobacterium* may also cause degradation of paints in use. In addition they may attack the paint in storage, before application, under the anaerobic conditions in the tins. Water-based emulsions are more liable to this attack than oil-based paints, and deterioration may include fermentation and changes in viscosity and colour. A wide variety of antimicrobial additives are routinely used by paint manufacturers to combat this damage.

The deterioration of stone surfaces in cities with polluted air is well known and is attributable to the chemical action of such substances as sulphuric acid derived from sulphur impurities in fuels. The problem may be increased by growth of bacteria such as *Thiobacillus thio-oxidans* which oxidize sulphur compounds, thus increasing the amount of acid.

Rubber, both natural and artificial, is subject to attack by bacteria. The hydrocarbon content may be lowered by many genera including *Actinomyces*, *Serratia*, and *Pseudomonas* and even the fungi *Penicillium* and *Aspergillus* have been implicated in the spoilage of raw latex, though it is possible that they were living on impurities. Sulphur oxidizers such as *Thiobacillus thio-oxidans* may attack sulphur in vulcanized rubber, causing the production of sulphuric acid which destroys textile reinforcing and metal joints in hoses.

The damage done to plastics depends on their exact chemical nature, but attack on additives and adhesives is common. Some types of polyvinylchloride (P.V.C.) are particularly liable to attack by *Penicillium*, *Aspergillus*, and Actino-mycetes. This, combined with attack on rubber can cause serious effects on electrical insulations both in underground cables and complex circuits of electronic equipment in tropical regions. In the latter case spores may be carried in by mites and by their growth, initially on dust and grease but later on the insulation, may cause short-circuits. This again became particularly important during the Second World War. Modern miniaturized electronics and printed circuits often make it impossible to get at the failure and the whole piece of equipment must be discarded.

It is desirable, in the long run, that micro-organisms should develop enzyme systems capable of degrading modern synthetic substances such as plastics and detergents so that man has some convenient way of disposing of his increasing quantities of rubbish. The reverse approach of designing bio-degradable substances has been adopted in the short term and such detergents are at present in use. Micro-organisms are, however, very adaptable and there are signs that even such resistant substances as polythene can eventually be destroyed by a combination of chemical weathering and microbial attack.

BOOKS AND ARTICLES FOR REFERENCE AND FURTHER STUDY

CARTWRIGHT, K. ST. G. and FINDLAY, W. P. K. (1958). *Decay of Timber and its Prevention*. 2nd Edn. H.M.S.O., London.

CHESTERS, C. G. C. (1951). Mildews of textiles and related materials. *Research, Lond.*, **4**, 102–107.

DAVIS, J. B. (1967). *Petroleum Microbiology*. Elsevier Publishing Co., New York, 604 pp.

—— and UPDEGRAFF, D. M. (1954). Micro-biology in the petroleum industry. *Bact. Rev.*, **18**, 215–238.

HENDY, N. I. (1966). How fungi attack materials. *Sci. J.*, **2** (1), 43–49.

LEVY, J. (1965). The soft rot fungi: their mode of action and significance in the degradation of wood. *Adv. Bot. Res.*, **2**, 323–357.

NORKRANS, B. (1963). Degradation of cellulose. *A. Rev. Phytopath.*, **2**, 325–350.

PILPEL, N. (1968). The natural fate of oil on the sea. *Endeavour*, **27** (100), 11–13.

ROSS, R. T. (1964). Microbial deterioration of paint films. In *Developments in Industrial Mycology*, **6**, 149–163. American Institute of Biological Sciences, Washington, D.C.

SCHEFFER, T. C. and COWLING, E. B. (1966). Natural resistance of wood to microbial deterioration. *A. Rev. Phytopath.*, **4**, 147–170.

TURNER, J. N. (1967). *The Microbiology of Fabricated Materials*. J. & A. Churchill Ltd., London.

WALTERS, J. S. and ELPHICK, J. S., eds. (1968). *Biodeterioration of Materials: microbiological and allied aspects*. Elsevier Publishing Co., Amsterdam, London and New York.

WANG, C. J. K. (1965). *Fungi of Pulp and Paper in New York*. Technical Publication 87, State University College of Forestry, Syracuse, New York, 115 pp.

WOLFSON, L. L., MICHALSKI, R. J. and SAPIENZA, M. S. (1966). The effect of shipment on the reliability of microbiological analyses from paper mills. In *Developments in Industrial Mycology*, **7**, 173–178. American Institute of Biological Sciences, Washington, D.C.

ZOBELL, C. E. (1946). Action of micro-organisms on hydrocarbons. *Bact. Rev.*, **10**, 1–49.

22

The Industrial Uses of Micro-organisms

INTRODUCTION

Micro-organisms were used in industrial processes even before their existence was known. The production of fermented beverages and vinegar, and the leavening of bread are all traditional processes which have come down to us from time immemorial (Chap. 20). The discovery of micro-organisms with their multiplicity of highly specific biochemical activities, has stimulated a steady growth of industrial fermentation processes. Perhaps the most famous of all industrial fermentations is that of acetone-butanol production by *Clostridium acetobutylicum*. During the First World War, acetone, sorely needed for the manufacture of cordite, was produced by this fermentation.

Between the wars, the industrial use of micro-organisms was extended, but it was not until after the Second World War that the fermentation industries developed on the massive scale that we know today. The major stimulus was the need to produce antibiotics. Just prior to the war, some of the larger American companies had developed microbial fermentations to produce fumaric acid and this was one of the first examples of the use of neutral conditions. Hitherto, most industrial fermentations had been conducted under acid conditions where sterility was not highly critical. With the advent of neutral fermentations, there were new bio-engineering problems to be solved in order to maintain pure cultures on a large scale. The neutral conditions and the special sterile precautions required for the production of fumaric acid were exactly those required for the mass cultivation of antibiotic-producing micro-organisms in the post-war period.

Although the industrial processes using micro-organisms are too numerous to mention, many have declined and some have disappeared altogether. With the enormous amounts of crude oil being distilled for petroleum, there are many useful hydrocarbons in other fractions unsuitable for fuels. These form the starting materials for such organic solvents as acetone and butanol which can be produced more efficiently and cheaply from these sources. Thus, the need to use such organisms as *Clostridium acetobutylicum* has disappeared. In spite of the expansion of the petrochemical industries, there are some substances that are still most easily produced microbiologically and these may be conveniently discussed under the following headings: (1) Substances produced as a major end-product of microbial metabolism, (2) The use of micro-organisms to reshape molecules that do not participate in their normal metabolism, (3) Mixed culture fermentations, and (4) Fermentations for the production of micro-organisms.

The development of fermentation processes to their present degree of efficiency was a lengthy and costly procedure. It is, therefore, not surprising that commerical firms are reluctant to disclose their manufacturing methods. Indeed, some of the methods are such closely guarded secrets that they are not even patented. Thus, while the broad outlines of industrial fermentations are well known, the details often remain undisclosed.

SUBSTANCES PRODUCED AS A MAJOR END-PRODUCT OF MICROBIAL METABOLISM

Antibiotics

The penicillins

One of the most dramatic fermentation stories of the century started in the late twenties when Fleming recorded the effect of a stray mould, *Penicillium notatum*, upon some of his pathogenic bacteria. This observation was not pursued further until the pressure of wartime needs stimulated its fuller investigation in the 1940s and led to the founding of a new industry and the saving of many lives; a remarkable example of the application of microbiology on the industrial scale.

When *Penicillium notatum* was originally grown as a surface organism on liquid media, levels of only 10 units/ml. of penicillin were produced (an international unit of benzyl penicillin is 0·5988 μgm or 1,670 i.u./mg). With improved media this was increased to 200 units/ml. The introduction of the more slowly metabolized lactose, instead of glucose, together with corn steep liquor as a growth medium increased yields even further. This was again improved by the use of *Penicillium chrysogenum* followed by the selection of many mutants produced artificially with the aid of such mutagens as ultra-violet light.

Nowadays, this antibiotic is produced under submerged aerated conditions in stainless steel vats of 20,000 gallons capacity with the control of such variables as pH and glucose concentration. Glucose has supplanted lactose because it is now added in controlled doses giving the same slow carbohydrate metabolism. Amounts in excess of 7,000 units/ml of penicillin are attained under modern fermentation conditions.

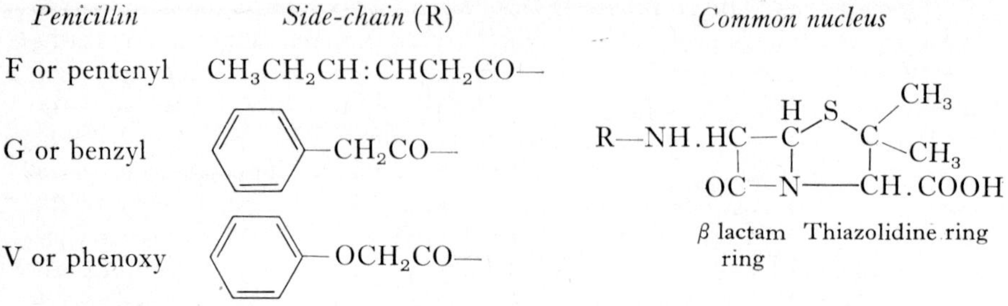

| *Penicillin* | *Side-chain* (R) | *Common nucleus* |

F or pentenyl $CH_3CH_2CH:CHCH_2CO—$

G or benzyl   〈benzene ring〉$—CH_2CO—$

$$R—NH.HC\underset{\underset{OC—N}{|}}{\overset{\overset{H\ \ S}{|}}{C}}\underset{CH.COOH}{\overset{C}{\diagdown}}\overset{CH_3}{\underset{CH_3}{\diagup}}$$

V or phenoxy   〈benzene ring〉$—OCH_2CO—$

β lactam   Thiazolidine ring
ring

Fig. 22.1   The structure of several penicillins showing the common nucleus.

Many other organisms are known to produce penicillins, including *Aspergillus*, *Trichophyton*, *Epidermophyton*, and *Cephalosporium* species. It was also discovered that penicillin made in Britain differed from the American product. Investigations showed that these two penicillins had the same basic molecular structure with different substituent groups. The American preparation is benzyl penicillin or penicillin G, while the British equivalent at that time was pentenyl penicillin or penicillin F. The various molecular structures are illustrated in Fig. 22.1.

Fig. 22.1 also shows penicillin V which with penicillin G are now the only two members of this group of compounds manufactured by large-scale fermentation.

During the course of chemical analyses on fermentation liquors, it was found that there was often more penicillin present than could be accounted for by the measurable antimicrobial activity. This led to the discovery of 6-amino-penicillanic acid (6APA) which has an NH_2 group on the β lactam ring (see Fig. 22.1) and hence no side chain. Although 6APA is biologically inactive, side chains can be substituted by acylation either chemically or enzymically. A wide range of organisms that contain penicillin acylases including *Escherichia coli* and *Alcaligenes faecalis* is now known. These will link specific side chains to 6APA producing additions to the penicillin family with novel properties. One such semi-synthetic penicillin, Cloxacillin, is illustrated in Fig. 22.2. It is stable to the action of β lactamase (penicillinase) and to acid, the latter property allowing administration by mouth.

The tetracyclines

These form another family of antibiotics with the general formula shown in Fig. 22.3. Oxytetracycline was first obtained from *Streptomyces rimosus* and

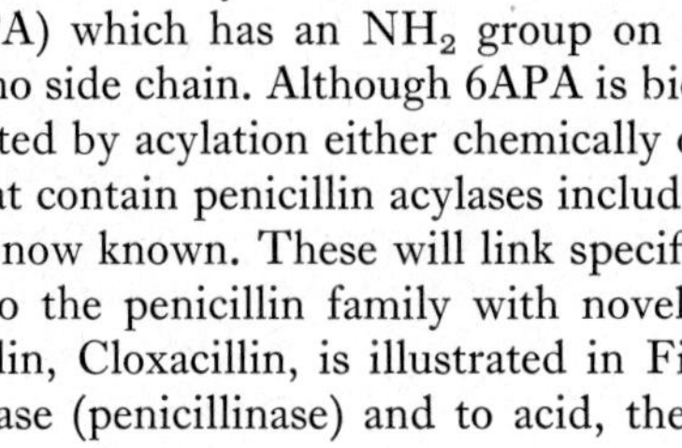

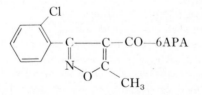

Fig. 22.2   Cloxacillin: sodium-3-ortho-chlorophenyl 5-methyl-4-isoxazolyl-penicillin monohydrate.

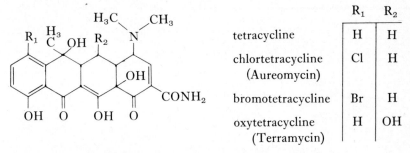

	R₁	R₂
tetracycline	H	H
chlortetracycline (Aureomycin)	Cl	H
bromotetracycline	Br	H
oxytetracycline (Terramycin)	H	OH

Fig. 22.3   The general structure of tetracyclines and variations.

chlortetracycline from *Streptomyces aureofaciens*. The latter is normally used for commercial production and may be grown on a variety of media including corn steep liquor, ground nut or soy bean meal. For the production of chlortetracycline, lard oil is also added while arachis oil is used in the manufacture of oxytetracycline. Sucrose or starch are the most suitable carbohydrate sources for this organism. Like many antibiotics, the tetracyclines are toxic to the organisms that produce them, and since the toxicity is thought to be due to sequestration of trace metals a plentiful supply of calcium or magnesium ions is added to overcome this effect. These organisms are favoured by a neutral pH and a temperature of 28–33°C. Aeration is essential.

Other antibiotics

Space does not allow the description of other antibiotics in detail, but Fig. 22.4 shows the sources and structures of three of the more important ones.

It should be remembered that although all the antibiotics discussed here are derived from fungi or actinomycetes, others such as subtilin and bacitracin are produced by bacteria. The usefulness of the latter is limited by their greater toxicity.

Dextran

Dextran is a polysaccharide composed of glucose molecules and may be found in slimy masses wherever sucrose is used. Sugar refineries and ham curing factories are prone to trouble from dextran-forming bacteria. Dextran can also block pipelines, make floors slippery, render food uneatable and beverages undrinkable. It has, however, an important use in blood transfusion where it is employed as a blood volume expander. Its application in this way is possible because it is relatively inert, neither causing pyrogenic (i.e., raised body temperatures) nor allergic reactions (p. 578) and will remain sufficiently long in the circulation to allow time for protein renewal in the blood plasma. It is stable to autoclaving and can be stored for long periods at room temperature without deterioration. One particular advantage is that it can be prepared with a known mean molecular weight. In Britain, for instance, a mean molecular weight of 110,000 is favoured, whereas in the U.S.A. 75,000 is the preferred molecular size.

Name	Source	Structure
Griseofulvin	*Penicillium patulum*	
Streptomycin	*Streptomyces griseus*	
Chloramphenicol	*Streptomyces venezualae* Now made synthetically; formerly known as chloromycetin	

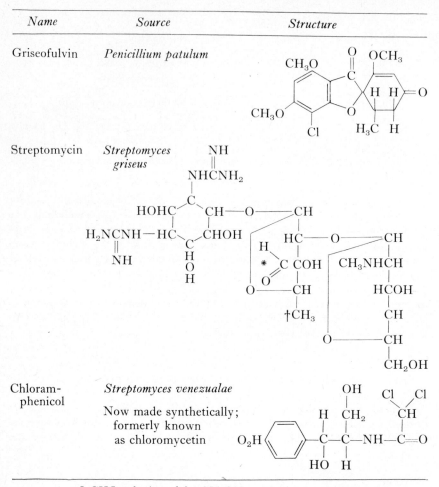

* CHO substituted for CH_2OH = dihydrostreptomycin.

† CH_3 substituted for CH_2OH = hydroxystreptomycin.

Fig. 22.4 The source and structure of some antibiotics.

There is only one bacterial strain used in Britain for dextran production, namely, the heterofermentative lactic coccus *Leuconostoc mesenteroides*. This is normally maintained in sucrose agar and is gradually worked up through intermediate volumes to a 1,000 gallon scale for commercial production. The medium used is sucrose plus yeast extract supplemented with ammonium and potassium phosphates. Incubation temperature is 23°C: aeration is unnecessary since the organism is micro-aerophilic. Additions of alkali are made to keep the pH above 5·0 because below this point dextran production is reduced.

The overall reaction for dextran production is as follows:

$$n \text{ sucrose} \xrightarrow{\text{dextran/sucrase}} (\text{glucose}) \, n + n \text{ fructose.}$$

The fructose is subsequently used as an energy source by the bacteria, while the glucose units are joined together into chains mainly with 1-6 linkages but sometimes 1-4 and occasionally 1-3. Branched chains vary in amount according to the particular strain of bacterium used. Hence *L. mesenteroides* is suitable for commercial production because it gives 95 per cent straight-chain molecules, these being the most acceptable for clinical use.

Dextran is removed from the culture fluid by acetone precipitation, and this so-called 'native dextran' is then ready for the final stage of degrading to the correct molecular size. The methods available for this molecular sizing operation are as follows:

1. *Acid hydrolysis.* Dextran is heated with acid, and the average molecular size is checked by viscosity measurement. This method has the advantage that it can be performed without separating the dextran from the fermentation liquor.
2. *Heating under pressure.* Degradation may be brought about by heating the dextran under pressure at 160°C in the presence of sodium sulphite to prevent oxidation, and calcium carbonate, to keep the pH at 7.
3. *Enzymic degradation.* Various micro-organisms contain dextranases, the one most frequently used being *Cellvibrio fulva.*
4. *Mechanical breakdown.* High frequency waves emitted by a piezo-electric quartz crystal when stimulated electrically will break down high molecular chains and give a product of considerable homogeneity. As with method (1), viscosity is the main criterion used for estimating molecular size.

Lactic acid

This is one of several acids produced microbiologically on a commercial scale. Others include citric, fumaric, itaconic, and gluconic acids, some of these being produced from moulds which require high aeration rates in the submerged conditions employed. Lactic acid, in contrast, is produced by microaerophilic bacteria belonging to the genus *Lactobacillus*. This acid is an odourless, colourless liquid having an acid flavour. It is used in a variety of ways finding applications in the drinks industry as a flavouring, in the preservation of food as an adjunct or substitute for vinegar, and in the form of calcium lactate as a convenient means of getting calcium into the body. Other uses are in the plastics and leather industries. A number of grades are produced varying in purity and concentration (22–85 per cent) according to the purpose for which they are required. Broadly, the grades may be classified as pharmaceutical, edible, and technical. Irrespective of grades, however, all types are made by the same process, and it is the rigour of the purification processes that decides the final grade.

The most common organisms used in lactic acid production are the homofermentative lactobacilli. In these, as with yeasts, the Embden–Meyerhof–Parnas pathways is a main energy-gaining mechanism. The bacterial pathway, however, differs from that of the yeasts because there is no pyruvic decarboxylase present, therefore no loss of CO_2 and no subsequent acceptance of hydrogen by acetaldehyde. Instead, pyruvic acid performs this role in these bacteria and, by accepting hydrogen, is reduced to lactic acid as shown below.

$$CH_3.CO.COOH + 2NADH \longrightarrow CH_3.CH.OH.COOH + 2NAD^+$$
$$\text{Pyruvic acid} \qquad\qquad\qquad \text{Lactic acid}$$

This type of lactobacillus produces a single end-product, namely, racemic lactic acid. Recent work has shown, however, that by varying temperature and type of medium, the relative proportions of the two optical isomers can be varied.

In the commercial production of lactic acid, a variety of substrates may be used, such as hydrolysed starch with barley, whey or molasses supplemented with nitrogenous material. The purer the starting material the less post-fermentation processes are required to produce pure lactic acid. The starting concentration of sugar is about 15 per cent, and this is reduced to 0·1 per cent in 4–6 days. Calcium carbonate is added to the mix to prevent a fall-off in productivity caused by a drop in pH. The organisms most favoured for lactic acid production are the high temperature group of homofermenters that includes *Lactobacillus delbrueckii*. This allows a working temperature of 50°C which has the advantage of suppressing possible contaminants with lower optimum growth temperatures. To maintain this unusually high temperature in vats that may be of 30,000 gallons capacity, heating coils are fitted and the medium is stirred to ensure uniform temperature and composition. Aeration is not applied, since this is an anaerobic process. As with all large scale batch fermentations, it is necessary to build up the inoculum by passing it through a series of tanks of increasing volume until finally the size is sufficient to initiate a rapid fermentation in the large vats used for commercial production.

When fermentation is complete, the yield is about 85 per cent of the fermentable hexose. The liquor is then heated to 80°C to kill the organisms and coagulate the protein. The end-product, present as the calcium salt, is acidified with sulphuric acid which frees the lactic acid and precipitates the calcium as insoluble sulphate. Heavy metals, at one time removed by treatment with sodium sulphide, are now taken out with ion exchange resins. Lactic acid is decolorized with powdered charcoal and is extracted from other constituents by the use of iso-propyl ether. The acid is then extracted from the ether by using water. Both of these processes can be operated continuously in counter-current towers. For the very pure grades, such as those required in the manufacture of plastics, a lactic acid ester is prepared, the impurities distilled off, and the lactic acid and alcohol recovered.

Amino acids

These substances are used for various clinical purposes, such as post-operative therapy, as supplements to low protein diets, and as food flavourings. A common ingredient of certain dehydrated foods such as soups is mono-sodium glutamate. Amino acids can be prepared by chemical synthesis, but this has the disadvantage of producing both optical isomers. In contrast, biosynthesis produces the L-isomer which is the form found in most biological systems. Sometimes, a combination of chemical synthesis and biosynthesis is employed as in the production of lysine from diamino pimelic acid. The production of the diaminopimelic acid is purely chemical, while the final step to lysine is biological.

The main amino acids prepared by fermentation are glutamic acid and lysine (Table 22.1). In the preparation of the former, the carbohydrate (usually glucose or sucrose) is present at a concentration of 10–20 per cent. Urea or NH_4^+ salts provide the nitrogen; magnesium, manganese, and iron, the metallic constituents, and such substances as pelargonic acid are added as stimulators. The pH is maintained between 7 and 8 by the addition of urea or ammonia gas. The concentration of biotin is most important because it controls the yield.

Table 22.1 Organisms and methods for amino acid production.

Amino acid	Method	Organisms
Aspartic acid	Produced by transamination from oxalacetate or fumaric acid and ammonia.	*E. coli* *S. marcescens* *Bact. succinum* *Bacillus* spp.
L-homoserine	This organism produces L-lysine with L-homoserine. Varying the concentration of biotin changes the ratio of the yields of glutamate, lysine, and homoserine.	*Micrococcus-glutamicus*
L-threonine	Made by a strain which has a growth requirement for diaminopimelic acid or one which also has a requirement for methionine. Mannitol or sorbitol instead of glucose increases yield.	*E. coli*
L-methionine	This organism is cultured in a medium containing methyl mercapto-α-hydroxy butyric acid.	*Pseudomonas* sp.
L-valine	Strains (a), (b), (c), and (d) all accumulate L-valine in the medium Strain (e) is an auxotrophic mutant requiring L-leucine or L-isoleucine. Biotin favours L-valine production.	(a) *Aerobacter cloacae* (b) *Paracolabactrum coliforme* (c) *E. coli* (d) *Aerobacter aerogenes* (e) *Micrococcus glutamicus*
L-ornithine	This organism accumulates ornithine and requires either L-arginine or L-citrulline as a precursor.	*Micrococcus glutamicus*
DL-alanine	This is a transamination reaction involving pyruvic and glutamic acids and a racemization.	*Corynebacterium gelatinosum*
L-phenylalanine	Strains (a) and (b) require tyrosine as a precursor. Yield is affected by biotin and may be stimulated by the presence of shikimic, phenyl pyruvic, and D-quinic acids.	(a) *Micrococcus glutamicus* (b) *E. coli*

continued

Table 22.1 continued

Amino acid	Method	Organisms
L-tyrosine	Phenylalanine is required to accumulate L-tyrosine. Shikimic, *p*-hydroxy-phenyl pyruvic, and D-quinic acids or tryptophan stimulate L-tyrosine production	*Micrococcus glutamicus*
L-tryptophan	Strains (a), (b), and (c) will convert indole in submerged culture to L-tryptophan. Strains (d) and (e) will convert 3 indole pyruvic acid to L-tryptophan by reductive deamination. Strain (f) produces L-tryptophan from anthranilic acid.	(a) *Claviceps purpurea* (b) *Rhizopus oryzae* (c) *Ustilago avenae* (d) *Micrococcus* sp. (e) *Serratia* sp. (f) *Hansenula anomala*

The organism commonly used in the manufacture of glutamic acid is a *Coryne-bacterium* sp. but strains of the genera *Brevibacterium, Micrococcus, Micro-bacterium*, have also been used. Removal of glutamic acid from the growth medium is easy because it is insoluble in acid conditions. High yielding media are treated as follows: the organisms are removed, the filtrate concentrated and acidified to pH 3·2, and the glutamic acid, thus precipitated, is recovered by filtration. Lower yielding media are treated with a suitable ion exchange resin, and this method not only separates but also purifies the glutamic acid.

Lysine preparation is similar to that of glutamic acid, the usual organism being a mutant of *Micrococcus glutamicus*. Biotin is less critical than for the production of glutamic acid, whereas the concentrations of homoserine, threo-nine, and methionine have a profound effect upon yield. Lysine is more difficult to separate from the growth medium than glutamic acid, but ion exchange resins offer the best means of extraction. In the final stages, it is usually obtained as lysine monohydrochloride and the pure amino acid produced by recrystallization.

Many other amino acids can be made in smaller quantities as summarized in Table 22.1.

THE USE OF MICRO-ORGANISMS TO RESHAPE MOLECULES

One process, already described (p. 682) which might also be included here is the application of acylase enzymes in conjunction with 6APA to produce the new penicillins. In this type of process, the organisms employed do not metabol-ize the substances presented to them but merely have the enzymic capacity to change certain groupings within the molecule. Steroids have frequently been treated in this way and there are a large number of organisms available to carry out a wide range of chemical changes.

Steroids are compounds of extreme importance in the economy of the animal body. Steroid hormones are secreted by the adrenal cortex and are responsible for the regulation of carbohydrate and mineral metabolism. Other well-known examples of steroid-secreting organs are the ovaries and testes that regulate the animal's sexual activities. Recently, steroid compounds have been administered for therapeutic reasons. Examples of such applications are the use of cortisone as an anti-inflammatory agent, the menstrual cycle regulators incorporated into the 'pill' for contraceptive use, and more recently their administration as post-operative immune reaction suppressors in organ transplant patients.

Steroids are complicated molecules with the basic structure as illustrated in Fig. 22.5.

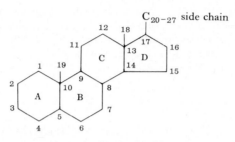

Fig. 22.5   The basic steroid structure.

To synthesize a steroid from suitable starting material takes about 32 steps and even to move an oxygen from C_{11} to C_{12} may take up to 12 steps. These lengthy chemical syntheses are not practicable for producing steroids on a commercial scale, and new methods were therefore sought. The first micro-biological steroid transformation was made in 1937 by Mamoli and Vercellone, and since that time a whole range of steroids have been changed in chemical structure by an equally wide range of micro-organisms. The advantage of the microbiological method is that highly specific changes in the steroid molecule can be effected in a single step. The known range of chemical changes includes hydroxylation, dehydrogenations, etc. The organisms concerned include many moulds, particularly species of Zygomycetes (p. 393) and Deuteromycetes (p. 395) and Actriomycetes and a few species of yeasts and unicellular bacteria.

Although the first microbial transformation of a steroid was made as long ago as 1937, one factor that discouraged such processes on a large industrial scale was the relative insolubility of these complex compounds. One of the factors that stimulated industrial-scale steroid transformations was the discovery that oxygen at the C_{11} position imparted important anti-inflammatory properties. This gave rise to a steady demand for such compounds as cortisone and stimu-lated new ways of manufacturing this and related compounds on a large scale. Thus, one of the early discoveries was the transformation of compound S (11 deoxycortisol) to cortisol and smaller amounts of cortisone by the mould *Cunninghamella blakesleeana*.

Such transformations as these are now commonplace, usually being performed on a 30,000 gallon scale. A suitable growth medium such as glucose, molasses, or corn steep liquor, with yeast extract added as a nitrogen source are provided for the transforming organism. The large-scale tanks are agitated and aerated, the organisms being allowed to reach maximum growth. Towards the end of the growth phase, the appropriate precursor is added in a non-toxic hydrophilic organic solvent such as ethanol, acetone, or propylene glycol. The chemical change can be followed chromatographically, and when complete the newly formed steroid is extracted with organic solvents. Although, owing to the insolubility of steroids, the total weight per batch of culture is small, losses are also small and a 95 per cent recovery is quite common.

There seems little doubt that the transformation of steroid compounds by micro-organisms is a process that will not be easily supplanted by chemical methods. The ease with which the substituent groupings of this complicated molecule are changed in a way that cannot be achieved by chemical means is a factor that appeals to manufacturers. In spite of the large range of transforming organisms known, little has been done to select high yielding mutants as has been successfully done in the antibiotics industry.

MIXED CULTURE FERMENTATIONS

There are a number of industrial fermentation processes that use mixed cultures. These include the production of fermented beverages (p. 659), vinegar-making, and ensilage of grass for cattle feed.

The manufacture of vinegar

This is prepared as the result of two microbiological processes, namely: alcoholic fermentation and acetification mediated respectively by yeasts and by bacteria of the genus *Acetobacter*. It is interesting to note that different countries have a preference for certain substrates. In Britain, malt wort is commonly used, whereas in France and other continental countries grape must is the usual starting material, whilst in the U.S.A. apple juice is the preferred substrate. Since alcoholic fermentation is discussed elsewhere (p. 660) it is the second phase that will be described in some detail.

All *Acetobacter* species are aerobic and usually found in the wild state growing as a pellicle at the air/liquid interface of alcoholic liquors. This is utilized in the old-fashioned French system called the Orleans process. In this, acetic acid bacteria are allowed to grow on the surface of wine in partly filled casks. When the wine has been slowly converted to vinegar, part is drawn off at the base of the cask, the original volume being re-established by replenishing with new wine. Although this is a slow process, it produces vinegar of the highest quality.

Perhaps the most widely used method for vinegar production is the so-called quick or trickling process. The apparatus consists of a large wooden tower packed with beechwood shavings, birch twigs or other non-compressible material. A sparge at the top of the tower distributes alcoholic liquor over the packing, which provides a support for the acetic acid bacteria. The liquor gradually acetifies as it trickles down the tower to a reservoir at the base from which it is recycled until the process is complete. Acetification being exothermic, the heat produced causes the tower to act as a chimney drawing cold air in at the base and discharging warm at the top. A series of vents around the base of the tower can be opened or closed to regulate air flow, while condensers at the exit trap all volatiles entrained in the warm air. This method, although slow, is simple in operation and requires the minimum of attention and mechanical maintenance.

The third method of vinegar manufacture is by cultivating acetic acid bacteria submerged in a suitable substrate. Constant and intense aeration is required to keep the organisms in a rapid phase of acetification and provision must be made for efficient dissipation of heat. This is usually effected by the presence of cooling coils in the acetifier that are supplied with coolant by way of a solenoid valve actuated by the temperature of the vinegar. The shape and size of vessels used for submerged acetification may vary, but the commonest is a tall narrow vat which will accommodate froth that accumulates on the surface of the liquor. These vessels may be of wood or stainless steel. The submerged system is not only used as a batch culture but also continuously when a rotating stirrer may also be incorporated to induce more thorough mixing and aeration. Such systems have the advantage of producing large quantities of vinegar in a relatively short time in a minimum of vat space. They do require, however, complex control mechanisms and suffer from the disadvantage that if the air supply is interrupted, even momentarily, the culture may take several days to recover.

Dilute acetic acid produced by synthetic means is sold as a substitute for vinegar. In Britain, at least, the condiment may not be called vinegar unless it has been produced by one of the microbiological processes here described. Furthermore, it is not recognized as vinegar unless it contains a minimum of 4 per cent of volatile acid.

Ensilage

The process of ensilage is an alternative to the drying of green crops for winter feeding of cattle and sheep. The resulting silage is stored in a succulent form. The process involves the controlled fermentation of green crops such as grass, maize, and occasionally sugar-beet tops or kale, and has the advantage of being largely independent of weather conditions. Processing can thus be done at the optimum time, the most nutritious product being achieved with younger, rather than mature crops. Success depends on adequate temperature control and exclusion of oxygen. The green crop is packed and compressed in containers (silos), in pits in the ground or in stacks above-ground enclosed in polythene. Molasses is sometimes added to promote fermentation and increase palatability. As the fermentation proceeds the temperature rises and the product becomes more acid. The acidity prevents the growth of many putrefactive organisms and, as the oxygen is used up, moulds and strictly aerobic bacteria are also inhibited. Only facultative and strict anaerobes continue to grow. At first, species of *Escherichia* and *Aerobacter* predominate, but as the pH falls these are replaced by lactobacilli and streptococci, which ferment the plant material to produce lactic, butyric, propionic, and acetic acids and esters giving flavours much relished by cattle. Active fermentation continues for three or four weeks, after which the temperature falls, and a fairly stable product results.

Sewage disposal

Perhaps the fermentation process of vital importance to the community as a whole, and one whose techniques are the least familiar, is the disposal of sewage.

All mixed culture processes have one feature in common, namely, the conditions to ensure that certain desirable organisms become dominant and, thus, perform the required chemical changes.

In highly industrialized countries such as Britain where large communities have developed, the disposal of industrial and domestic waste presents an enormous problem. The easiest means of sewage disposal, still practised by coastal and estuarine towns, is simply to run the raw sewage into the sea or nearby river. The larger inland communities have been forced, by lack of convenient dumping grounds, to develop sewage purification systems whose end-products are mainly pure water and some harmless solids. It is the purpose of this section to describe how this may be achieved.

Sewage consists mainly of water containing organic and inorganic dissolved and suspended substances along with many micro-organisms. A typical domestic sewage would have the composition shown in Table 22.2.

Table 22.2 Composition of a typical domestic sewage.

Constituent	% Organic carbon
Fatty acids	22·8
Carbohydrates	17·7
Proteins	10·0
Fatty acid esters	9·1
Soluble acids	6·8
Anionic surface active agents	4·5
Amino acids	1·6
Creatinine	1·1
Amino sugars	0·6
Amides	0·5

One of the important means of observing the quality of a sewage is by the measurement of its biochemical oxygen demand (BOD). This is defined as the weight of dissolved oxygen (usually in mg) required by a definite volume of liquid (usually 1 litre) during five days' incubation at 65°F. Nowadays, 20°C is recognized as the international standard temperature for incubation. The aim of purifying sewage is to reduce the BOD so that if effluent waters are run into a stream or river, the indigenous flora and fauna will not die from lack of oxygen.

When sewage is received at a processing plant, there is a preliminary screening to remove solid matter. Grit and stones are allowed to settle out from a slow-flowing stream of sewage. There is a regular removal of putrefactive sludge to prevent the onset of anaerobic digestion which will be discussed later. The remaining material which is removed after varying periods of sedimentation (usually 15 hours in Britain) consists of liquid with suspended flocs of organic matter. This may be treated in one of several ways.

Treatment of watery effluent

ACTIVATED SLUDGE TREATMENT. This process is used at large sewage works and consists of running the sewage into tanks where it is aerated either by compressed air or stirring. This vigorous agitation continues from 5 to 15 hours, and during this time flocculation of the organic matter takes place. The effluent is subsequently run into settling tanks, where the sediment is recycled back to provide an inoculum for the incoming sewage, whilst the water is sufficiently pure to be discharged into a river. When the sewage contains an excess of certain substances, e.g., carbohydrates in brewery effluent, it is necessary to add nitrogen to keep the balance correct.

The chemical and microbiological changes that take place in activated sludge are not fully understood. However, the dominant floc-forming bacterium is called *Zooglea ramigera*. Nitrifying bacteria are also present, and while these do not compete with the heterotrophs for nutrients, there is considerable competition for oxygen. Fungi do not become well established, and if *Geotrichum* species appear it usually indicates an upset in the balance of the process. Protozoa are present both as the free-swimming forms and the stalked species, the latter being dominant. Their presence is most important because they feed on bacteria, and effluent from this process lacking protozoa is often cloudy.

Some of the known chemical conversions are shown below:
1. Metabolism of urea.

$$CO(NH_2)_2 + H_2O \longrightarrow CO_2 + 2NH_3$$

2. Nitrification as carried out by *Nitrosomonas europeae*.

$$2NH_4^+ + 3O_2 \longrightarrow 2H^+ + 2H_2O + 2NO_2^-$$

$$2NO_2^- + O_2 \longrightarrow 2NO_3^-$$

3. Nitrate reduction.

$$2NO_3^- \longrightarrow 2NO_2^- \longrightarrow 2(HNO) \longrightarrow 2NH_2.OH \longrightarrow 2NH_3$$

$$
\begin{array}{c}
\downarrow \quad \nearrow \ N_2 \\
H_2N_2O_2 \\
\searrow \ N_2O
\end{array}
$$

The activated sludge process will remove 85–95 per cent of BOD and the same percentage of suspended materials. Thus, it enables the solids to be spread on the land without danger to health, and water to be returned to the rivers without causing pollution.

BIOLOGICAL FILTERS. These are a familiar sight often to be found near factories that purify their own effluent. They consist of a circular tank, above ground to a height of about 4 ft, packed with inert material such as coke that acts as a filter bed. Over the top of the bed, four sparge arms slowly rotate distributing effluent evenly over the whole surface. The micro-organisms become stratified within the filter bed, different types growing at a level where the oxygen concentration and composition of the effluent suits their particular requirements best. In the upper layers such heterotrophs as *Zooglea ramigera* are to be found and as the organic matter diminishes in the lower layers autotrophs such as *Nitrobacter* and *Nitrosomonas* grow. In addition, there are such organisms as *Fusarium* and *Geotrichium* growing on the surface, moulds being more numerous than in the activated sludge process. Protozoa abound, the motile species grow in the upper layers whilst the stalked kinds grow lower down the beds. A macrofauna of small animals graze upon the solid matter that gets caught in the interstices of the bed, thus preventing the plant from becoming blocked. Examples of the chemical changes that take place and some of the organisms responsible for them are shown below.

1. $NH_4^+ \xrightarrow{a} NO_2^- \xrightarrow{b} NO_3^-$ (a) *Nitrosomonas* sp., (b) *Nitrobacter* sp.
2. Phenolic substance $\longrightarrow CO_2 + H_2O$ *Vibrio* sp., *Pseudomonas* sp., *Actinomyces* sp.
3. $CNS^- \longrightarrow CO_2 + NH_3 + SO_4^{--}$
   $S_2O_3^{--} \longrightarrow SO_4^{--}$ } *Thiobacillus* sp.

Reduction of BOD is about the same as that of the activated sludge process, whereas the removal of suspended solids is slightly less, varying from 70–90 per cent.

Synthetic detergents have been added to sewage in recent years. These are surface active agents that cause considerable foaming and also reduce oxygen transfer by about 20 per cent in the activated sludge process. Their structure is that of an alkyl chain as illustrated below.

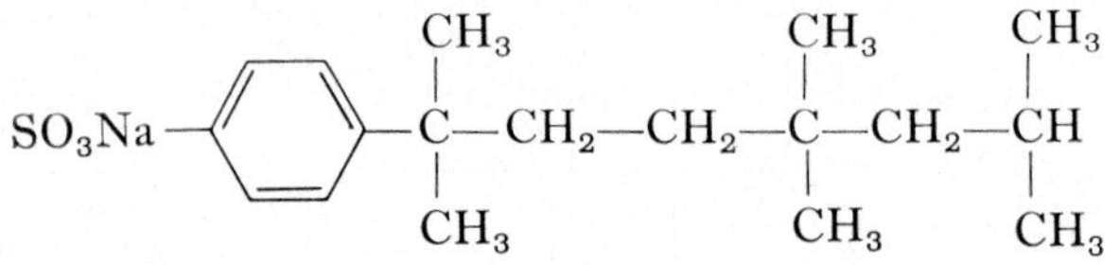

$$SO_3Na - \langle \bigcirc \rangle - \underset{\underset{CH_3}{|}}{\overset{\overset{CH_3}{|}}{C}} - CH_2 - CH_2 - \underset{\underset{CH_3}{|}}{\overset{\overset{CH_3}{|}}{C}} - CH_2 - \underset{\underset{CH_3}{|}}{\overset{\overset{CH_3}{|}}{CH}}$$

It has been shown that if the benzene ring contains an alternative substituent, biological attack takes place through this group. Furthermore, the straight-chain compounds are more readily attacked than branched chain ones. The tendency now is to produce detergents containing a higher proportion of straight-chain compounds, and some countries now have legislation prohibiting the manufacture of detergents resistant to biological degradation.

OXIDATION PONDS. Settled sewage is run into ponds and held for about 30 days. The action of bacteria produces carbon dioxide and ammonia. The presence of these substances encourages the growth of algae which releases oxygen during photosynthesis. These conditions permit aerobic bacteria to grow which decompose the organic matter further, leaving a minimum of residue.

LAND TREATMENT. Settled sewage is run on to arable land where accumulated organic matter is removed by microbiological oxidation. This system is not used today as a major method for dealing with effluent, but is usually applied to the cleaning of effluents from other processes such as activated sludge.

Treatment of solids

Sludge of this type is derived from the initial screening of raw sewage, from the settled deposits of activated sludge or as humus sludge from percolating filters. These are combined and placed in a 'digester'. This is the part of the sewage that may contain pathogens and it is, therefore, essential that this material is rendered innocuous before disposal. The digester, in reality, is nothing more than a large-scale septic tank whose contents are stirred and heated. The most common temperature range is 30–35°C which is applied for about one month. Thermophilic digestion using a temperature of 50°C is sometimes practised, but is less usual than the mesophilic digestion which is more easily controlled. During digestion, methane, with a calorific value of about 700–750 BThU/cu ft, is evolved and in a large sewage works this is used as a source of heat and power. Organisms producing methane belong to several genera as illustrated by the following list: *Methanobacterium formicum, Methanobacillus omelianskii, Methanosarcina barkerii*. They utilize a variety of substrates including alcohols, fatty acids, CO_2, and H_2 to produce methane and may be highly substrate specific. An example of a methane producing reaction by *Methanobacterium suboxydans* is illustrated below.

$$2CH_3CH_2CH_2COOH + 2H_2O + \overset{*}{C}O_2 \longrightarrow \overset{*}{C}H_4 + 4CH_3COOH$$

Butyric acid   Methane   Acetic acid

Experiments using labelled carbon (*) have shown that the CO_2 carbon becomes the methane carbon. Other changes taking place during digestion are the liquefaction of solids and the breakdown of carbohydrates, fats, and proteins. The digestion process can give rise to very obnoxious odours since indole, skatole, and compounds containing mercaptans are produced.

When sludge digestion is complete, the excess liquid is returned to the digester and the residual solid material is dried by vacuum filtration, by heat or in underdrained beds. The remaining dry solid can then be disposed of by incineration or applied to the land as manure.

FERMENTATION FOR THE PRODUCTION OF MICRO-ORGANISMS

Bakers' yeast

All the processes so far considered have been concerned with the production of an end-product from a substrate. The conversion of part of the substrate to biomass is incidental in all these processes, but once the optimum amount of end-product has been achieved the organisms are merely a waste product. In the preparation of bakers' yeast, the roles are reversed, for it is the aim of the producer to achieve the maximum quantity of micro-organisms from the substrate.

Pasteur in 1876 first showed that aerating a yeast culture would improve growth, and it is this principle that is used today for producing panary yeast. The fermentation is carried out as a batch culture in vats of 50,000–200,000 litres capacity. The yeast species used is *Saccharomyces cerevisiae* and this is grown in a medium consisting mainly of beet or cane molasses. This is often deficient in biotin or pantothenic acid, and it may be necessary to supplement the quantities naturally present. Other prefermentation treatments of the molasses involve its dilution, treatment with sulphuric acid and heating to remove sludge protein.

Inocula are built up initially from single cells until a sufficient quantity has been prepared to inoculate the vats used for commercial production. Various yeasts have been bred to fulfil the requirements of modern baking, and breeding programmes are in progress to produce yeasts with even more desirable characteristics

In the large-scale production of yeasts, various technical factors must be considered. The normal sequence of events in alcoholic fermentation is a rapid multiplication of yeast cells with a consequent depletion of oxygen. This halts cell multiplication and the yeasts switch to their anaerobic method of energy-gaining with the consequent production of ethanol and carbon dioxide. It is the aim of the yeast manufacturer to keep the cells in their aerobic multiplication phase, thus producing maximum biomass with minimum alcohol. About 12 per cent of oxygen is utilized in aeration so that about 40 times the absolute requirement must be supplied which is a total of 330,000 cu. ft. of air per ton of commercial yeast. The high rate of aeration presents problems, for example, the amount of energy produced per unit of sugar metabolized aerobically is about 1,384 Kcal compared with 117 Kcal for the same quantity consumed anaerobically. Some of this energy is lost as heat and it is necessary to maintain steady conditions and a working temperature of 30°C.

A further complication of a rich medium combined with a high rate of aeration is that the yeasts may grow so rapidly that they outstrip the available oxygen. In such a situation, oxygen starvation will occur and instead of the yeasts producing $CO_2 + H_2O$, as they do under aerobic conditions, they begin to produce CO_2 and ethanol, which is undesirable. To limit yeast growth and thus avoid oxygen starvation, the nutrients are controlled. At first, they are kept at a low level and as the cells increase exponentially they are added to keep pace with yeast growth. Thus, at any one time there is only a small excess of nutrients over the number of yeasts available to use them. Another variable requiring control is pH. Since nitrogen is often added in the form of $(NH_4)_2SO_4$, the removal of the ammonium ions tends to create acid conditions which are adjusted by periodical additions of alkali to keep the pH in the range 4·0–5·5.

At the end of yeast growth, when the nutrients are exhausted, aeration is continued for 30–60 minutes to 'ripen' the yeast. It has been found that this part of the process stabilizes the yeast and imparts better keeping qualities. After removal from the vats, the yeast is centrifuged and washed several times then chilled to 2–4°C. Water is removed by vacuum filter dehydrators and the yeast packaged. Packing material can be of importance in the preservation of bakers' yeast, and contamination can only be prevented by strict controls throughout the process (p. 659).

Mention has been made, in the early part of this chapter, of the diminishing role played by micro-organisms in industry. The trend seems likely to continue until there remain only those microbiological processes that can compete economically with the synthetic chemical industries or where the application of chemical methods is quite inappropriate, as in the disposal of sewage. It seems possible that there might be a resurgence of industrial microbiological methods in the distant future when our present reserves of fossil fuels have been exhausted. Micro-organisms can offer a means of converting plant and animal constituents into hydrocarbons should this type of compound still be required in large quantities when the naturally-occurring substances have disappeared.

BOOKS AND ARTICLES FOR REFERENCE AND FURTHER STUDY

CARR, J. G. (1968). *Biological Principles in Fermentation.* Heinemann Educational Books Ltd., London, 97 pp.

IIZUKA, H. and NAITO, A. (1967). *Microbial Transformation of Steroids and Alkaloids.* University of Tokyo Press and University Park Press, State College, Pennsylvania.

Process Biochemistry (1967). Volumes 1 et seq.

RAINBOW, C. and ROSE, A. H. (1963). *Biochemistry of Industrial Micro-organisms.* Academic Press, London and New York, 708 pp.

ROSE, A. H. (1961). *Industrial Microbiology.* Butterworths, London, 286 pp.

Index

Index

Principal references are in **bold** type, illustrations are in *italics*.